AF767170

Dogs, Past and Present: An Interdisciplinary Perspective

*To our fathers, Edoardo Fiore and Luciano Lugli who always respected their dogs and who taught us to love these splendid and extraordinary friends as well as all the animals.*

# Dogs, Past and Present

## An Interdisciplinary Perspective

Edited by

Ivana Fiore and Francesca Lugli

Archaeopress Archaeology

Archaeopress Publishing Ltd
Summertown Pavilion
18–24 Middle Way
Summertown
Oxford OX2 7LG

www.archaeopress.com

ISBN 978-1-80327-354-9
ISBN 978-1-80327-355-6 (e-Pdf)



Cover: Researcher Graziano Capitini with dogs Bankhar, Baatar, Baavgai, Arslan and Kurtan (camp 37/2011 of Narantsogt and Bolorman, N 48°03′158″; E 103°01′131″, 1736 metres above sea level, December 2012). Photo by F. Lugli.

The publication of this volume was supported by ISMEO - *Associazione Internazionale di Studi sul Mediterraneo e l'Oriente* through the MUR Project 'Storia, lingue e culture dei paesi asiatici e africani: ricerca scientifica, promozione e divulgazione.'

ISMEO

# Contents

## Section 3. *Dogs through Time: Role, Task and Position*



## Section 4. *Dogs: Archaeological and Archaeozoological Cases*

## Section 5. *Representation of Dogs in Different Cultures*

## Section 6. *Dogs: Myth and Symbolism*

# Acknowledgements

This volume originated from the conference 'Dogs, Past and Present: An Interdisciplinary Perspective' that was held at CNR (Consiglio Nazionale delle Ricerche) and at Sapienza University in Rome (14th-17th November 2018), promoted by the Italian Association for Ethnoarchaeology and organised by the editors. The conference was generously supported by CNR; Sapienza-Università di Roma; Museo delle Civiltà, Roma; ISMEO (Associazione Internazionale di Studi per il Mediterraneo e l'Oriente); AIAZ (Associazione Italiana di Archeozoologia); Is.I.P.U. (Istituto Italiano di Paleontologia Umana); SMA (Sistema Museale di Ateneo, Università di Ferrara); UNISS (Università degli Studi di Sassari); IIPP (Istituto Italiano di Preistoria e Protostoria); Eulabor Institute; Radmedica-Bologna; RadMed; Xilema; Lingue Merlis; Beta Analytic; ES srl Progetti, Sistemi and Agros, Luisa Migliorati, Alessandra Sperduti & Aki and Vienna Eleuteri & Sita.

The Conference had the patronage of Ministry of Foreign Affairs and International Cooperation – Italy MFA, Assessorato alla Cultura del Comune di Roma, Institute of Archaeology and Ethnography of SB RAS - Novosibirsk (Russian Federation), Universitat Pompeu Fabra - CaSEs (Spain), Università di Padova, ICOM-Italia, Ordine dei Medici Veterinari della Provincia di Roma, Land srl and Società Geografica Italiana.

We thank Umberto Albarella, Francesca Alhaique, Davor Antonucci, Marco Baldi, Stefano Biagetti, Alberto Cazzella, Catherine Filejski, Marco Giuman, Stefania Ghio, Tolga Kamil, Leonardo Salari, Raffaele Sardella, Assunta A. Stoppiello, Galina B. Sychenko, Ursula Thun Hohenstein, Alessandro Vanzetti, Massimo Vidale for their help and suggestions.

We thank Prof. Adriano V. Rossi president of ISMEO for endorsing our projects and ISMEO which generously supported the conference and made the open access publication of the volume possible.

A special thank to our husbands Gianfranco Calandra and Graziano Capitini for their help and support, Helena Lex who has meticulously proofread the volume. David Davidson for the opportunity to publish the volume with Archaeopresss and for his constant availability and Erin McGowan for her constant and patient help in the various steps for the publication of the volume.

We remember Jacopo De Grossi Mazzorin, Filippo Maria Gambari, Ettore Janulardo and Francesca Santini whose last projects we are pleased to present.

Last but not least we are grateful to all the dogs of the world.

# List of Authors

**Francesca Alhaique**

Degree in Natural Sciences and PhD in Prehistoric Archaeology, Sapienza University; MA in Anthropology, Washington University in S. Louis. She currently works at Museo delle Civiltà, teaches at Tuscia University and is Research Associate in the Department of Anthropology at Washington University in S. Louis (USA). For over 25 years she has been investigating faunal assemblages from Italian and foreign sites spanning from the Palaeolithic to the Early Modern period.

**Eugenia Andreeva**

Eugenia Andreeva studied ancient history at the Historical faculty of the Lomonosov Moscow State University (specialist degree; graduated with honours in 2011) and completed a postgraduate program at the Institute of the World History of the Russian Academy of Sciences (2015). She is currently a research associate at the Institute and is preparing a PhD thesis. She also teaches Latin (State Academic University for the Humanities) and Ancient Greek (Russian Presidential Academy of National Economy and Public Administration).

**Marco Apollonio**

Full Professor of Zoology at the Department of Veterinary Medicine of the University of Sassari, Italy. Born in Rome (Italy) in 1958, MS degree in Biological Sciences at the University of Milan (Italy) in 1981. School of Specialisation in wildlife management at the Italian Institute of Wildlife Management 1983–86. Main research interests: behavioural and animal ecology, wildlife management and conservation, ecological genetics and systematics of large mammals.

**Ilenia Azzena**

PhD student at the Departments of Biomedical Sciences and Veterinary Medicine of the University of Sassari, Italy. Born in Tempio Pausania (Italy) in 1993, graduated in Natural Sciences at the University of Sassari in 2015 and MS degree in Land and Environment Management at the University of Sassari in 2018. Main research interests: animal population genetics and molecular taxonomy.

**Saverio Bartolini Lucenti**

Postdoctoral researcher at the Earth Science Department and the Natural History Museum with a project on Virtual Palaeontology, science dissemination and valorisation of the palaeontological heritage. Member of the digital and virtual palaeontology laboratory of the Earth Science Department of University of Florence - Paleo [Fab] Lab. His main research topic focuses on the evolution of Neogene and Quaternary carnivorans, especially canids and mustelids.

**Manuel Bea**

Obtained his Doctorate in History (Prehistory) from the University of Zaragoza (2005), where he conducted a PhD focused on the study of Levantine rock art. He is currently senior lecturer at the University of Zaragoza (Spain), Researcher at the Group of First Settlers and Archaeological Heritage of the Ebro basin (University of Zaragoza), member of the Instituto Universitario de Investigación en Patrimonio y Humanidades (IPH), member of the National Scientific Rock Art Committee of ICOMOS and associated research at the Institut d'Arqueologia University of Barcelona. He has published more than 150 scientific publications and directed many archaeological projects.

**Simona Bealcovschi**

Simona Bealcovschi holds a PhD in Cultural Anthropology from the University of Montreal. She has worked as a researcher and visual anthropologist in a museum context. At the University of Montreal, she heads the Visual Anthropology Laboratory and teaches courses in visual culture. Research interests: visual anthropology, experimental ethnography, cultural and political ecology, pet-culture, Canada, Alaska.

**Federico Bernardini**

Federico Bernardini is an archaeologist with extensive experience in the application of scientific methods to archaeology to study and characterise ancient materials and investigate past landscapes. He currently holds a position as researcher at the Centro Fermi (Rome) and at the Multidisciplinary Laboratory of the Abdus Salam International Centre for Theoretical Physics (ICTP, Trieste), where he coordinates the ICTP-Elettra laboratory for cultural heritage and archaeology.

**Marco Bertolini**

Marco Bertolini is a zooarchaeologist. He graduated in Prehistoric Science and obtained a PhD in Human Science with a thesis on the management and exploitation of fauna during the Bronze Age in north-eastern Italy. He deals with experimental archaeology aimed at reconstructing the manufacturing of animal hard materials and their use. He has analysed the bone tool collections of several Bronze Age sites.

**Christophe Blanchard**

PhD graduate in sociology and anthropology, Christophe Blanchard has been Associate Professor in educational sciences at the University of Paris 13 - Paris Sorbonne Nord since 2014 and is a member of the EXPERICE laboratory. His personal background (he holds a dog-handler degree) and his university research have led him to focus very closely on issues relating to the status and place of animals, particularly dogs, in our society.

**Raquel Blázquez-Orta**

Degree in Geology and Master's in Palaeontology at Universidad Complutense de Madrid (UCM). Currently, she is a PhD student at UCM granted by the Atapuerca Foundation and her investigation focuses on the analysis of the craniomandibular morphological changes of the genus *Canis* through traditional and geometric (2D and 3D) morphometric techniques.

**Hervé Bocherens**

Hervé Bocherens is a palaeobiologist, professor for biogeology at the University of Tübingen and at the Senckenberg Centre for Human Evolution and Palaeoenvironment at Tübingen. His main areas of research include reconstructing the diet and habitat of ancient vertebrates in terrestrial ecosystems using isotopic tracking, especially in connection with human evolution.

**Evgeniy S. Bogdanov**

Bogdanov earned his PhD in History, (2004) on 'The image of a predator in the plastic art of the nomadic peoples of Central Asia (Scythian-Siberian art tradition)'. Bogdanov currently works as a senior fellow at the Institute for Archaeology and Ethnography SB RAS, Novosibirsk. The focus of his research is culture and art of the ancient nomads in Central Asia, Volga, and Ural regions in the era of the early Iron Age. Bogdanov has been a leader and participant in many expeditions and scientific projects in China, Mongolia, Kazakhstan, Altai Mountains, Caucasus, and the Caspian Sea region.

**Fabio Bona**

After graduating in Geology (2001 - UniMi) and completing a PhD in Earth Sciences (2005 - UniMi), both with paleontological-quaternary themes, he has been working as a freelance paleontologist and archaeozoologist. He carries out research in Quaternary paleontology with attention to the evolution and biochronology of mammals. He also deals with the study of Holocene faunas from anthropic contexts.

**Francesco Boschin**

Francesco Boschin is a zooarchaeologist at the University of Siena with extensive experience in the analysis of faunal remains from Palaeolithic and Mesolithic contexts from Italy. In the last years he focused his research on the study of Late Pleistocene hunter-gatherers from southern Italy and developed new applications of digital imaging techniques to taphonomy and zooarchaeology.

**Alessandra Bottari**

Alessandra Bottari trained in Numismatics obtaining the title of PhD in Archaeological and Historical Sciences at the University of Messina she specialised in Archaeological and Historical Sciences at the University of Naples Federico II. Her research focuses on Numismatics and in particular on the interpretation of monetary iconography.

**Juliane Bräuer**

Juliane Bräuer is the head of the Dog Studies Lab at the Max Planck Institute of Geoanthropology in Jena, Germany. She is a biologist working in comparative psychology with a special interest in investigating the cognitive skills that different species – particularly dogs – have evolved to survive in their ecological niche. Her research topics include communication, cooperation, perception, and domestication. Websites: https://www.shh.mpg.de/person/47501/2375; https://www.psychologytoday.com/intl/experts/juliane-br-uer-phd

**Graziano Capitini**

Architect. He is a member of the missions in the Russian Federation, Mongolia and Portugal promoted by the Italian Association on Ethnoarchaeology with the sponsorship of the Ministry of Foreign Affairs and International Cooperation – Italy MFA. The focus of his research is on the spatial organisation of the camps and their location, on the structure of Mongolian yurts; on the relationships between the nomads and their dogs in Mongolia and in Siberia and between the fishermen and their dogs in Portugal and in Italy.

**Daniela Castagna**

Daniela Castagna is a field archeologist who has worked in Northern Italy for over thirty years; she has participated in several joint operations with Soprintendenza Archeologia, particularly on prehistoric excavations. She is studying and publishing some neolithic contexts of the Mantuan territory.

**Marco Casu**

Associate Professor of Zoology at the Department of Veterinary Medicine of the University of Sassari, Italy. MS degree in Natural Sciences at the University of Sassari in 1997. PhD in Analysis and management of natural ecosystems at the University of Sassari in 2003. Main research interests: phylogeography, phylogeny and molecular taxonomy of marine invertebrates and fish.

**Eugenio Cerilli**

Eugenio Cerilli graduated in Geological Sciences from the University of Rome 'La Sapienza' with an experimental thesis in Vertebrate Paleontology. Later he specialised in vertebrate fossil conservation and archaeozoological analysis. He has participated in several archaeological excavations in Italy and abroad. The field of interest and paper topics are: paleontological and archaeozoological conservation, analyses of paleontological and archaeozoological remains and contexts, geoarchaeology and paleoenvironmental reconstructions, archaeological excavations, teaching, contributions for museums and temporary exhibits.

**Dmitry V. Cheremisin**

Dmitry Cheremisin has been Research fellow at the Department of the Bronze and Iron Ages in Siberia, Institute of Archaeology and Ethnography, Siberian Branch since 1990. From 2007–2017 he worked as a lecturer of Primitive Art at Novosibirsk State University. He has taken part in archaeological research in the Far East, Altai, Kazakhstan, and Mongolia. Since 1990 he has conducted independent research in the territory of Russian Altai. He is the author of over 200 publications in Russian, French and German. From 1990 - until the present - he has been a research fellow at the Institute of Archaeology and Ethnography, Novosibirsk.

**Piero Cossu**

Postdoctoral fellow at the Department of Veterinary Medicine of the University of Sassari, Italy. MS degree in Biological Science at the University of Sassari in 2001. PhD in Analysis and Management of natural ecosystems at the University of Sassari in 2005. Main research interests: phylogeography, seascape genetics, population genetics and phylogeny of marine organisms.

**Alessandro Crispino**

He has an undergraduate degree in Lettere Antiche and an M.A. in Classical Archaeology (with a focus on Greek and Roman Numismatics), and and he has a PhD in Archaeological and Roman Historical Research at the University of Bari. His research focuses on the relationship between archaeological contexts and ancient numismatics, the interpretation of coins from excavation, monetary circulation in antiquity, and ancient trade. He has authored papers on coin findings in particular from Egnazia.

**Franco D'Agostino**

Is associate professor of Assyriology at the Italian Institute of Oriental Studies, Sapienza University of Rome. He is co-director of the Iraqi-Italian Mission at Abu Tbeirah and director of the Iraqi-Italian Mission at Eridu.

**Ignacio De Gaspar Simón**

Anatomy and Embryology Associate Professor in the Faculty of Veterinary Medicine of Universidad Complutense de Madrid since 1990. Member of the Research Team of Pleistocene Sites of Atapuerca since 2007, focusing his interest on Holocene fauna and the process of domestication. He has also been a member of an international research team working in Middle Awash (Ethiopia) since 2019, studying primates of Kesem Kebena. He has been awarded several times for his teaching work.

**Jacopo De Grossi Mazzorin(†)**

Jacopo De Grossi Mazzorin was professor of Archaeozoology at the University of Salento (Lecce-Italy), in the Department of Cultural Heritage. His work has resulted in more than 240 scientific papers. He has been involved in numerous Italian and International archaeological excavations within the Mediterranean Area. He was specialised in the study of animal remains from archaeological sites and their cultural interpretation. His research also included the integration of different archaeological disciplines linked to the relationship between humans and animals (economy, rituals, etc.).

**Frédéric Devienne**

Sinologist and specialist in animal iconography, Frédéric Devienne is currently co-leading a research program on the subject of Dogs in ancient China at the East Asian Civilisations Research Centre (CRCAO) in Paris. He is the director of the East Asian Languages and Civilisations (LCAO) department's library at the University of Paris, where he also teaches.

**Gianpiero Di Maida**

Completed his PhD at the CAU Kiel (Germany) in 2018, discussing a Thesis on the Lateglacial rock and portable art record of Sicily, which has been awarded with the Johanna Mestorf Price 2019. Between 2018 and 2021 he has served as scientific manager of the DISAPALE and the Wendel collection projects at the Neanderthal Museum (Germany), focusing in particular on digital methods of recording applied to the archaeological record. Since 2023 he works as scientific coordinator of the 'Climate Change and Early Humans in the North' SPRUNG-Project at the Lower Saxony Cultural Heritage Office in Hannover. His papers on the Lateglacial art, the first peopling of Sicily, digital methods of recording and documentation, the typology and technology of palaeolithic lithic artefacts have been published on several peer-reviewed journals. (Quaternary Science Reviews, PLoS ONE, Antiquity, Journal of Island and Coastal Archaeology, Open Archaeology).

**Martina Di Matteo**

Has a Master's degree in Prehistoric Ethnography of Africa from Sapienza University of Rome, where she is currently attending the School of Specialisation in Archaeological Heritage. She has participated in excavations in pre- and protohistoric sites in central and Southern Italy as well as abroad, particularly in Africa (Ethiopia, Kenya, and Tunisia). Her main research interests concern the analysis of archaeozoological remains from Holocene contexts, mostly in Africa.

**Liubov Eliseeva**

Liubov Eliseeva studied ancient history at the Lomonosov Moscow State University (MA; graduated with honours in 2017); completed a postgraduate program at the Institute of the World History of the Russian Academy of Sciences (2020). She is currently a junior research associate at the Institute of the World History, a member of a research project team at the Lomonosov Moscow State University (2020–2022), and teaches Latin and Ancient Greek at the State Academic University for the Humanities.

**Marco Fatucci**

Marco Fatucci graduated in Archeology from the Faculty of Conservation of Cultural Heritage of the University of Tuscia (Viterbo, Italy) with a thesis on the medieval faunal remains from the Santa Severa Castle. He has participated in archaeological excavations in Italy and abroad. He collaborates with museums, dealing with research and studies in the prehistoric field. He has published several essays and articles on the archaeozoological analysis of faunal remains.

**Ivana Fiore**

Zooarchaeologist, who collaborates with the Bioarchaeology Service at the Museum of Civilisations, in Rome. She is currently enrolled in the Doctoral Program in Environmental and Evolutionary Biology (Sapienza University of Rome). Her main interests are Zooarchaelogy and Taphonomy. She has specialised in the study of faunal remains and bone artefacts found in archaeological sites. She has presented contributions at national and international conferences, authored many scientific papers, organised several conferences, and edited colloquium proceedings. She has taught Zooarchaeolgy at Sapienza University of Rome (Faculties of Natural Sciences and Humanities) and at the University of Cagliari (Scuola Specializzazione in Beni Archeologici).

**Paolo Francalacci**

Full Professor of Genetics at the Department of Life and Environmental Sciences at University of Cagliari, Italy. MS degree in Biological Sciences at the University of Pisa (Italy) in 1982 and PhD in Anthropological Sciences at the University of Florence (Italy) in 1986. Main research interests: human population genetics, mitochondrial DNA and Y chromosome molecular markers and ancient DNA.

**María Ángeles Galindo-Pellicena**

In 2007, Master's thesis 'Study of the micromammals from the Middle Pleistocene of the Covacha de Los Zarpazos, Atapuerca, Burgos', a four-year Pre-doc research grant from the Minister of Science and Competitiveness, part of the Atapuerca Project research team. In 2014, PhD on the study of macromammals from Holocene levels at El Portalón site (Atapuerca, Burgos), (Program of Geology and Geological Engineering at the Universidad Complutense de Madrid (UCM). She has also been a member of the excavation team of Atapuerca since 2003 and of the Pinilla del Valle team since 2006. Currently, she works at the Regional Archaeological Museum of Alcalá de Henares (Madrid).

**Nuria García**

Professor of Paleontology at the Faculty of Geology of Universidad Complutense de Madrid (UCM). Bachelor's degree in biology and PhD in Paleontology specialised in Mammal Paleobiology. Member of the Atapuerca Research Project since 1994. She leads a University research group named Quaternary Ecosystems, focused on evolutions of ecosystems in the Human Evolution conetxt. She collaborates with international teams which work in Europe and Africa (Ethiopia). In these projects, she is responsible for the paleobiological studies of order Carnivora and paleoecology of large mammals.

**Yuri N. Garkusha**

Garkusha Yuri Nikolaevich is a researcher at the Institute of Archaeology and Ethnography of the Russian Academy of Science SB RAS (Novosibirsk). He is a graduate of the Faculty of Humanities of Novosibirsk State University. He has taken part in expeditionary activities on the territory of the forest-steppe and south-taiga Ob-Irtysh, Northern Angara region, Northern and Middle Priobye, and the Altai Mountains He has participated in the excavation of burial complexes with permafrost in the Republic of Mongolia. Garkusha is the author and co-author of more than 70 scientific publications including 5 collective monographs. His scientific interests include archaeology and the paleometallic era of Western and Southern Siberia and dendrochronology.

**Mietje Germonpré**

Mietje Germonpré is a palaeontologist and archaeozoologist at the Royal Belgian Institute of Natural Sciences (OD Earth and History of Life) in Brussels. Her main areas of research include human-animal interactions from hunting to domestication during the Middle and Upper Palaeolithic in northern Eurasia.

**Claudio Giardino**

Claudio Giardino is professor of Prehistory at the University of Salento (Lecce, Italy). He took part in archaeological missions in Europe and Asia. His main focuses are European and Asian prehistory, archaeometallurgy, experimental archaeology and ethno-archaeology. He is the author of numerous works, monographs and papers; he is also active in archaeological dissemination.

**Marco Giuman**

Marco Giuman is Associate Professor in Classical Archaeology at the University of Cagliari (Italy), where he is director of the School of Specialisation in Archaeological Heritage. Author of nine monographs and many dozens of articles published in specialised journals, he mainly deals with rituality, icononography and iconology of the Ancient World.

**David Ian Howe**

David Ian Howe (B.A., Anthropology, University of Tennessee; M.A., Anthropology, University of Wyoming) serves as a Laboratory Manager for New South Associates, Inc. David's research focuses on lithic technology, dog domestication, and hunter-gatherer ecology. Prior to working for New South, David worked for universities and cultural resource management firms, including SWCA Environmental Consultants and the U.S. Army Corps of Engineers.

**Dawid Adam Iurino**

Dawid Adam Iurino graduated in Natural Sciences at the Sapienza University of Rome in 2010 and in 2014 discussed a PhD thesis in Virtual Paleontology. From 2014 he has dedicated his research to the application of tomographic techniques in the study of fossil vertebrates, especially focusing on Quaternary carnivorans, acquiring and developing technical and theoretical skills required to process and manage CT data. During his academic career, he has participated in numerous excavations with international teams in some of the most renowned archeo-paleontological sites in Italy (Pirro Nord, Grotta Romanelli, Collepardo) and Tanzania (Olduvai and Laetoli). He is the author of more than 40 scientific articles and actively collaborates with numerous national and international research institutions and museums.

**Maria Kudinova**

Candidate of Sciences (History) (2017). Researcher at the Institute of Archaeology and Ethnography, Siberian Branch of the Russian Academy of Sciences (2020 – present). Associate Professor at the Chair of Oriental Studies, Novosibirsk State University (2009 – present). Graduated from Novosibirsk State University in 2009. Postgraduate study at the Institute of Archaeology and Ethnography, Siberian Branch of the Russian Academy of Sciences (2009–2012) and the School of Archaeology and Museology, Peking University (2017–2023).

**Guy Lanoue**

Guy Lanoue, PhD, Social and Cultural Anthropology, University of Toronto; currently, professor and Head, Department of Anthropology, University of Montreal. Extensive field research in northern North America and Italy. Interests: political organisation, myth, art; Native North America, Italy, Mesopotamia, New Guinea.

**Tiziano Latini**

Ministry of Cultural Heritage and Activities and Tourism Archaeologist Officer (MiBACT) engaged with Sapienza - Rome University and institutional organisations in national and international cultural heritage research.

**Martina Lázničková-Galetová**

Martina Lázničková-Galetová works as a researcher in Prehistory at the Moravian museum in Brno (Czech republic). She is interested in the use, technology and economics of hard animal tissue of artefacts, ornaments and art objects, and the human-animal relationship during the Upper Palaeolithic in Europe.

**Jennifer A. Leonard**

Jennifer Leonard is an evolutionary biologist and expert in the study of DNA from ancient remains, including samples from museum specimens that are just a few decades old to archaeological remains, of bones from specimens that can be tens of thousands of years old. During her research career she has studied wolf and dog populations from around the world.

**Grégoire Leroy**

Grégoire Leroy holds a PhD in animal genetics. He has been assistant professor at INRAE/AgroParisTech Génétique Animale et Biologie Intégrative joint unit since 2008, and has been seconded to the Food and Agriculture Organisation of United Nations since 2015. His research activities focus on characterisation, sustainable management and conservation of animal genetic resources.

**Tom Lewis**

Tom gained his PhD in quantitative genetics at the Roslin Institute and Nottingham University before spending 6 years at the Animal Health Trust researching the genetics of complex inherited disease in pedigree dog breeds. In 2014 he joined the Kennel Club where he continues to use pedigree and screening data to research disease and the genetics of populations.

**Sophie Licari**

Sophie Licari has been a cynologist journalist for 23 years and has worked for over 13 years for the official journal of the Société Centrale Canine. She is a specialist of the canine breed differentiation history and the history of dog uses in human societies. She is also a consultant in strategic communication and a lecturer in several French universities.

**Alberto Lombo**

Alberto Lombo holds a PhD in History (Prehistory) from the University of Zaragoza (Spain) (2015). His PhD on the study of Palaeolithic art focused on caricatures and laughing and smiling representations both on rock and mobile art. He often collaborates with the Department of Prehistory of the University of Zaragoza and has published numerous scientific papers.

**Francesca Lugli**

Francesca Lugli is the president of the Italian Association for Ethnoarchaeology. Currently, she is leading ethnoarchaeological investigations in Portugal, Mongolia and the Russian Federation supported by the Ministry of Foreign Affairs and International Cooperation – Italy MFA, and ISMEO. Her research focuses on modern nomads, their campsites, their land use strategies, their intangible heritage and also on the relationships between humans and dogs in different cultural and geographical contexts.

**Lucia Maimone**

Lucia Maimone graduated in Nature and Environment Sciences at the University of Parma in 2017 with a thesis in Paleontology. She then continued his studies attending the Second Cycle Degree in Ecology and Nature Conservation graduating in 2020 with a research thesis in the field of Biodiversity and Ecosystem Functioning entitled 'Deviation from theoretical gas saturation values in lotic systems: methodological approach and possible interpretations'. During her studies she collaborated with the University of Ferrara for the development of new techniques for the calculation of gas concentration in samples of lotic water through mass spectrometry through Membrane Inlet Mass Spectrometer (MIMS).

**Olga V. Maltseva**

Olga Maltseva has a PhD in Historical Science (Ethnography, Ethnology and Anthropology, 2008). She is a senior researcher in the Institute of Archaeology and Ethnography of the Siberian Branch of the Russian Academic Science. Her research focuses on the problems of human-animal relationships in fishing and hunting communities.

**Marino Marini**

Marino Marini is an Italian medieval archaeologist and curator of the ceramic collections of National Museum of Bargello and of the National Museum of Palazzo Davanzati in Florence. His preferred study field is Tuscan Renaissance majolica (Florence, Montelupo, Bacchereto, Cafaggiolo, Siena area) but several studies have also been aimed at the production of other ceramic centres such as Faenza, Deruta and the ancient Duchy of Urbino. He organised the exhibitions 'Fabulae Pictae. Miti e storie nelle maioliche del Rinascimento' at the National Museum of Bargello and 'Passione e collezione. Maioliche toscane dal XIV al XVIII secolo' at Casa Buonarroti Museum in Florence. He is currently engaged in the preparation of the majolica catalog of the National Museum of Bargello.

**Beniamino Mecozzi**

Beniamino Mecozzi graduated with a Bachelor's degree in Natural Science in 2014 and a Master's degree in Science of the Sea and Natural Landscape in 2016 at the Sapienza University of Rome, with a thesis on Paleontology. He discussed a PhD thesis in vertebrate paleontology in 2020, with a focus on mammal faunas from the late Middle to Late Pleistocene. In particular, he has studied the fossil remains from one of the most important sites of Mediterranean Europe, Grotta Romanelli. During the last years, he has participated in new research with international teams at renowned archeological and/or paleontological sites, such as Cimitero di Atella, Collepardo, Grotta Romanelli, Grotta Santa Maria d'Agnano and Notarchirico. During his research activity, he has published many scientific and informative articles.

**Paolo Mereu**

Assistant professor in Biochemistry at the Department of Biomedical Sciences of the University of Sassari, Italy. MS degree in Biological Sciences at the University of Sassari in 2002. PhD in Biochemistry, Biology and Molecular Biotechnology at the University of Sassari in 2006. Main research interests: phylogeny of mammals and bird mitogenomes, molecular dating and taxonomy; epigenetic alterations, DNA methylation and gene expression in vitrified oocytes.

**Luisa Migliorati**

Associate Professor of Ancient Topography, and qualified as a Full Professor, now retired. Her teaching activity took place in the Department of Ancient World Studies at Roma Sapienza University. She still teaches Ancient Topography in the Master's course in Classical Archaeology in UNITELMA Sapienza. She is teaching member of the Specialisation course of Conservación del Patrimonio Arquitectónico (MCPA) at the Universidad Autónoma de Yucatán, Fac. Arquitectura. She is Vice-president of the International Council for Philosophy and Human Sciences, member of the Board of the World Philology Union.

**Claudia Minniti**

Claudia Minniti teaches Bioarcheology at the University of Salento (Lecce-Italy), in the Dep. of Cultural Heritage. She has been a staff member of the University of Sheffield (UK) as a Marie Curie Fellow. Her research interests include the study of animal remains from archaeological sites from Italy and other countries (Syria, Turkey, Iran, Iraq with integration of various disciplines (e.g. history, archaeology, ethnography, and zoology). Her work has resulted in more than 120 papers in national and international journals, conference proceedings and books.

**Giuseppe Minunno**

Giuseppe Minunno received his PhD in Ancient Near Eastern Studies from the University of Rome 'La Sapienza' in 2009, and his Specialisation degree in Archaeology from the University of Pisa in 2004. He is a member of the Editorial Scientific Board of the project 'An Encyclopaedic Dictionary of the Phoenician Civilisation'. His research interests include Phoenician culture and the religions of ancient Syria.

**Peter Mitchell**

Peter Mitchell read Archaeology and Anthropology at Cambridge and then completed his doctorate in Oxford (1987). After working in Cape Town and Wales, he returned there to take up his current post in 1995. He combines a long-term interest in the archaeology of southern African hunter-gatherers (including fieldwork in Lesotho) with broader themes in African archaeology as a whole and has more recently written extensively on aspects of the history of human relations with horses, donkeys, and dogs.

**Rachele Modesto**

Obtained a Master's degree in Archaeological Sciences (Prehistoric curriculum) at Sapienza University of Rome, then a Specialisation Diploma in Archaeological Heritage and a PhD in Prehistoric Archaeology with a dissertation focusing on Bronze Age cult/funerary structures. She has participated in many excavations especially in Southern Italy and in one case abroad (Malta); she currently works as a freelance archaeologist.

**Martin Mosser**

Completed his MA and PhD in Classical Archaeology at the University of Vienna. Already during his studies, he was employed at the Department of Urban Archaeology in Vienna, where he is still works today. He has directed numerous excavations in the city of Vienna and evaluated a lot of excavation records. His special field of expertise is the Roman Vienna, in particular the legionary fortress of Vindobona.

**Margherita Mussi**

Associate Professor, Univ. di Roma Sapienza, Director of the Italian Archaeological Mission to Melka Kunture and Balchit (Ethiopia). Fieldwork: Surveys and excavations since 1977 (Italy and Eastern Africa). Research interests: the Palaeolithic of the Horn of Africa the palaeolithic peopling of Italy (mainland and islands)- the earliest peopling of Europe- Palaeolithic art and burials.

**Miriam Napolitano**

Miriam Napolitano is a PhD candidate in Classical Archaeology at the University of Cagliari. Her research regards the Roman material culture, especially the Roman engraved gems and the pottery discovered in Sardinia. She has presented her research results at national and international conferences and spent a period of research at the Beazley Archive of Oxford University. She is a member of the editorial staff of 'Medea. Rivista di Studi interculturali' (Università di Cagliari)', http://ojs.unica.it/index.php/medea.

**Andrei V. Novikov**

Andrei V. Novikov, PhD, is a senior researcher at the Institute of Archaeology and Ethnography of the Russian Academy of Science SB RAS (Novosibirsk). He graduated in 1982 from the History Faculty of the Novosibirsk State Pedagogical Institute. Since 1982, he has been conducting expeditionary archaeological research in the Ob-Irtysh interfluve and Priobye. He has studied various archaeological sites in Priobye from the forest-steppe zone to the subarctic. His scientific interests include the archaeology of paleometals (mainly 1st - mid 2nd millenium AD) and the ethnography of the indigenous population of the north of Western Siberia. He teaches various disciplines at Novosibirsk State University. Novikov is the author and co-author of 190 publications in Russia and abroad (including 11 monographs, and 6 textbooks for students in archeology and ethnography).

**Silvia Nutini**

She graduated at the University of Pisa with a thesis in archaeozoology and recently obtained her master's degree in Museum Services Management at Palazzo Spinelli, Institute for Art and Restoration in Florence. She became a coworker of the Ministry of Cultural Heritage and Activities in 2011; worked for the National Museums of Lucca, specifically for the National Museum of Villa Guinigi; and is a coworker of the National Museum of Bargello in Florence; she is in a training project on the enhancement of the Medici Villas and Gardens. In 2018 she carried out an internship at the Florentine Civic Museums for the public qualitative and quantitative analysis for the development of strategies for promoting the different collections. She is the author of publications about archaeozoology and the iconographic study of ceramic decorations.

**Luca Pandolfi**

After getting a PhD in Vertebrate Palaeontology at the Roma Tre University in 2015, he had research experience in several institutions in Italy, the UK, and other countries around Europe. He worked as a postdoctoral researcher at the University of Florence from 2018 to 2022, and he presently holds a tenure-track position at the University of Basilicata. His research is focused on the study of large mammal fossils from Eurasia and Africa.

**Antonella Pansini**

She has a PhD and is an expert on the subject of the 'Survey and technical analysis of ancient monuments' at the Sapienza University of Rome. She has specialised in Ancient Topography at the 'Scuola di Specializzazione in Beni Archeologici' of the same university. She has experience as a freelance archaeologist, surveyor and collaborator at universities and public institutions, including international ones, in the field of archaeological excavations and research and enhancement projects. In 2019 she was a postdoctoral student at the Italian Archaeological School of Athens where she carried out various studies on Roman architecture in Greece.

**Davide Persico**

Davide Persico graduated in Natural Science at the University of Parma and discussed a PhD thesis on Micropaleontology from the Southern Ocean (Antarctica) to understand the paleoclimate and palaeoceanographic evolution. In 2006 and 2007 he was a member of the international program of Antarctic research 'ANDRILL' with which he participated in two scientific missions on ice. He is the author of 56 scientific papers in international and national journals. The life on the Po River, the naturalistic interests and the passion for paleobiology lead him to know and study the Quaternary paleofauna of the Po Valley publishing articles and monographs on the subject. He is currently Associate professor of Paleobiology and evolution of vertebrates and Naturalistic Museology at the University of Parma.

**Raffaella Poggiani Keller**

A specialist in pre-protohistory, she has worked for the Archaeological Superintendence of Lombardy since 1980 where she has also been involved with UNESCO sites and the design and management of museum networks and archaeological parks. She has taught seminars in Italian protohistory at the university of Pavia (2001–2006) and was the first Superintendent for archaeology in Abruzzo (2009) and then Lombardy (2009–2013). She is the author of over 300 publications including articles, monographs, and edited volumes.

**Ana Portillo Gómez**

Holds a PhD in Archaeology from the University of Córdoba (2016) and she has a degree in Art History from the University of Seville (2007). Professionally, her work as a Research Fellow in the Archaeology Area of the University of Córdoba and her experience as an archaeologist, working as a professional in this sector in different areas, stand out. Currently, she continues her research work within her research group 'Ancient cities of Andalusia' PAI HUM-882, to which she has belonged since 2009.

**Erich Pucher**

Obtained his PhD in Biology at the University of Vienna. After completing his studies, he worked as an archaeozoologist and director of the Archaeological-Zoological Collection at the Natural History Museum Vienna. During his career, he has analysed numerous faunal assemblages and he is author of many important archaeozoological publications, covering a wide spectrum of topics from the Neolithic period to Modern times.

**Laura Rodríguez**

Associate Professor in the Facultad de Cc Biológicas y Ambientales of Universidad de León. Degree in Biological Science at Universidad de Oviedo and PhD in Paleoanthropology at Universidad de Burgos. Part of the Atapuerca investigation team since 2006. Main investigation focus on the analysis of human fossil record from the anatomical and biomechanical point of view and using 3D methodologies such as CT scan, 3D scan and geometric morphometric analysis, in order to detect species and sexual differences.

**Licia Romano**

Completed an MA in Archaeology of the Ancient Near East at the Sapienza University of Rome (2007), followed by a PhD at the same university (2010). She now has a research contract at the Department of Oriental Studies, Sapienza University of Rome. After several years of field research at Ebla (Syria), she has been the field director and co-director of the Iraqi-Italian mission at Abu Tbeirah since 2011.

**Mikhail Sablin**

Mikhail Sablin is a Senior Scientist at the Zoological Institute of Russian Academy of Sciences in Saint-Petersburg (Theriology department). His main areas of research include zoology, palaeontology, systematic and zoogeography of mammals; problems of domestication and the history of the northern Eurasia fauna during the Quaternary.

**Konstantina Saliari**

Received her BA and MA degrees in archaeology at the University of Athens. She received her PhD on archaeozoology at the University of Vienna with distinction. Since 2019 she has been curator of the Archaeozoological Collection at the Natural History Museum Vienna. She has participated in numerous archaeological excavations and scientific conferences and she is involved in various interdisciplinary projects, courses and seminars.

**Daria Sanna**

Assistant professor of Genetics at the Department of Biomedical Sciences of the University of Sassari, Italy. MS degree in Biological Sciences at the University of Sassari in 2003. PhD in Environmental Biology at the University of Sassari in 2007. Main research interests: human and animal population genetics, phylogeography, phylogeny and molecular taxonomy.

**Francesca Santini** (†)

She graduated from the Sapienza University of Rome with a dissertation based on the animal remains from Pescorocchiano, an important Roman temple site in Central Italy. In 2013 she obtained a master's degree in Osteoarchaeology from the University of Sheffield with a dissertation focused on the analysis of dog remains outlined in this conference. In 2001 she started to collaborate with Soprintendenza Archeologica of Lazio as an archaeozoologist in several excavation and study projects. She has also worked with other institutions such as the British School at Rome, CNR and Soprintendenza of Rome. In 2017 she moved to England to work for the Museum of London Archaeology.

**Raffaele Sardella**

Raffaele Sardella graduated in Geological Sciences at the Sapienza University of Rome, with a thesis on Paleontology. He discussed his PhD thesis in 1994 on vertebrate paleontology entitled: 'Systematic and geographic distribution of machairodont from late Miocene to Pleistocene'. During his academic career, he participated in many research projects in Italy and abroad (Sweden, Spain, France, UK, and Germany). He is the scientific director of Grotta Romanelli (Lecce, southern Italy) excavations and the research project authorised by Soprintendenza – MiC. He is currently Full professor of paleobiology and evolution of terrestrial ecosystems at the Sapienza University of Rome.

**Massimo Scandura**

Associate professor of Zoology at the Department of Veterinary Medicine of the University of Sassari, Italy. MS degree in Biological Sciences at the University of Pisa (Italy) in 1996. PhD in Biology at the University of Bielefeld (Germany) in 2004. Main research interests: population genetics, phylogeography, sociality and ecology of wild mammals (in particular ungulates and wolves).

**Fabio Scarpa**

Postdoctoral fellow at the Department of Veterinary Medicine of the University of Sassari, Italy. MS degree in Land and Environment Management at the University of Sassari in 2010. PhD in Environmental Biology at the University of Sassari in 2014 and Second Level Masterorn in A in Bioinformatics at the Sapienza University of Rome (Italy) in 2015. Main research interests: phylogeny, phylodynamics and molecular taxonomy of aquatic animals.

**Tiziana Sgrulloni**

Archaeologist, Specialist in Archaeological Heritage and PhD in Ancient Topography (Sapienza University of Rome). She has been collaborating for several years with the chair of Ancient Urban Planning (Sapienza University of Rome) participating in the teaching, research and excavations of *Peltuinum* (Aq), in the field and dealing in particular with the study of ceramic materials. She is the author of several scientific publications.

**Alessandra Sperduti**

Bioanthropologist, head of the Bioarchaeology Service at the Museum of Civilisation in Rome. Adjunct Professor in Anthropology at the University of Naples 'L'Orientale'. She coordinates, promotes and conducts scientific research on human skeletal samples, cremated and inhumed, from archaeological contexts of various chronological horizons. She has published articles and book chapters on anthropological methodologies, funerary rituals, paleodemography, paleopathology, skeletal/dental indicators of occupational activities, paleogenetics, mobility, paleodiet, Science and Society.

**Maria V. Stanyukovich**

Maria V. Stanyukovich is a Chair of a research department of MAE (Kunstkamera), Russian Academy of Sciences. Her interests are in epic, ritual, shamanism, Philippine oral literature, anthropology, ethnolinguistics and ethnobotany, on which topics she has published extensively in Russian, English and Spanish. She conducted fieldwork in Central Asia, Kazakhstan, Dagestan, Cuba (1976–1993), Cambodia (2015, 2017) and the Philippines (1995–2020). She worked as a Visiting Professor in Japan and UK.

**Galina B. Sychenko**

Ethnomusicologist. Art Criticism Candidate (1998), Docent (2001). Since 1984 she has been working at Novosibirsk State Conservatory 'M. I. Glinka'; at the Institute of Philology, Siberian Branch of the Russian Academy of Sciences and other institutions. Now she is living in Rome and manages the Archivio Eurasia 'Romano Mastromattei'. Her main spheres of scientific interest are performing traditions (ritual, epic, song) of peoples of Siberia, Mongolia, Himalaya (field research) and Southeast Asia (comparative research). Since 1984 she has taken part and organised about 50 expeditions in Siberia, Mongolia and Nepal; and has participated in 30 research projects.

**Antonio Tagliacozzo**

Currently retired. He was Archaeologist Director at the Prehistoric Ethnographic Museum L. Pigorini (in Rome today the Museum of Civilisations). He directed the Museum's Archaeozoology Laboratory, developing new study methodologies and collaborating with the most prestigious international research bodies. He was President of the Italian Association of Archeozology. He has directed and participated in numerous excavation and research campaigns both in Italy and abroad. He has curated and participated in the creation of numerous exhibitions. He is the author of books and more than 200 articles in the main archaeological and scientific journals, both Italian nd foreign.

**Francesco Tanganelli**

Italian archaeologist, and a current PhD student in *History, culture and knowledge of Mediterranean Europe* at the University of Basilicata. He graduated from the University of Florence, with a degree in Greek Archaeology and a dissertation about dog typologies in Classical Attic tombstones. His interests range from Greek and Roman marble sculpture to ancient epigraphy and literature, with particular emphasis on animals in Antiquity.

**Ursula Thun Hohenstein**

Ursula Thun Hohenstein is Associate Professor in Methods for Archaeological Research at the University of Ferrara. She graduated in Natural Science and obtained a PhD in Anthropological Science analysing faunal assemblages exploited by Neanderthals. She is an archaeozoologist, specialised in taphonomy. She has been analysing fauna assemblages coming from important Pleistocene sites in Italy and he applied taphonomic study to fauna remains of protohistoric sites too, focusing on the manufacturing of bone tools from a technological point of view.

**Claudio Tuniz**

Claudio Tuniz is a Consultant at the Abdus Salam International Centre for Theoretical Physics in Trieste, and collaborator of the Enrico Fermi Centre in Rome. He has recently been promoting evolutionary perspectives to understand constraints and potentialities of our relationships with intelligent technologies.

**Anna Maria Tunzi**

Anna Maria Tunzi is an Archaeologist working for the Italian Ministry of Cultural Heritage in the Soprintendenza Archeologia, Belle Arti e Paesaggio per la città metropolitana di Bari. She has been the director of many archaeological excavations especially in prehistoric sites in Apulia, she has been curator of several exhibits and has been adjunct professor at the University of Foggia.

**Andrey V. Varenov**

Andrey V. Varenov has graduated from Novosibirsk State University in 1977. For 30 years he has worked as senior research associate at the Institute of Archaeology and Ethnography of the Siberian Branch of Russian Academy of Science. At the same time, since 1980 Andrey V. Varenov is the associate professor at the Department of Humanities of Novosibirsk State University. He lectures on Chinese, Japanese and Indian history.

**Blanca Vidal Orga**

Blanca Vidal Orga is a PhD researcher in the Dog Studies Lab at the Max Planck Institute of Geoanthropology in Jena, Germany. She is a psychologist with an additional interdisciplinary background in Neuroscience, Philosophy, and Evolutionary Science. Working in animal cognition -particularly dogs and non-human great apes -with a comparative approach, her research interests include communication, domestication, and a topic focus on interspecies cooperative behaviour.

**Carles Vilà**

Carles Vilà started his research career in 1986 studying the ecology of wolves. Later on, he introduced himself to the use of genetic tools to better understand the ecology, behaviour and evolution of natural population and to understand the domestication process, with a particular interest in dogs and wolves.

**Shi-Zhi Wang**

Shi-Zhi Wang graduated from AgroParisTech and Swedish University of Agricultural Sciences in 2018 with a PhD study in animal breeding and genetics. Currently, he is working on the genetic factors influencing hip and elbow dysplasia (HD, ED) in pedigree dogs in the Roslin Institute of the University of Edinburgh as a postdoctoral researcher.

**Clément Zanolli**

Clément Zanolli is a Paleoanthropologist at the PACEA Laboratory (UMR 5199 CNRS) of the University of Bordeaux in France and expert in the study of dental structural organisation, tooth-related biomolecular processes, and ontogenetic developmental/growth aspects. His research project aims to precise the hominid paleobiodiversity, the evolutionary trends and phylogenetic relationships, as well as to uncover some disregarded paleobiological aspects.

**Tiziana Zappatore**

She graduated from the 'School of Specialisation' in Archaeology of the University of Salento. She published studies concerning prehistoric archaeology, ethnoarchaeology and archaeometallurgy. She participated in archaeological excavations and surveys.

# Presentation

Adriano V. Rossi

President, ISMEO

a.rossi@ismeo.eu

When I brought my greetings to the participants of the opening session of the First International Conference 'Dogs, Past and Present: An Interdisciplinary Perspective' (15-18 November 2018, Roma, CNR-Sapienza), organised by the Italian Association of Ethnoarchaeology, with the patronage and support of many institutions, including the Italian National Council for Scientific Research, Sapienza University of Rome, Museo delle Civiltà, and ISMEO, I remembered the path traveled by Giuseppe Tucci's IsMEO (founded in 1933) to the new ISMEO re-founded in Rome in November 2012 as ISMEO - The International Association for Mediterranean and Oriental Studies.

Many of our members, having been members of the formerly dissolved Institute, have been involved during their scientific careers in studies and research relating to the different countries of the Asian world, especially those which had always been at the centre of Tucci's IsMEO, in all their linguistic, religious, geographical and cultural forms, i.e. Tibet, Nepal and Central Asia, including Mongolia and Siberia.

When Giuseppe Tucci and IsMEO in the 1950s marked the start of the Italian archaeological exploration of the Middle East and Central Asia, ethnoarchaeology as a scientific discipline (e.g. as formalised by David and Kramer in *Ethnoarchaeology in Action*, Cambridge 2001) had not yet been forthcoming: neither ethnographic data were gathered with specific archaeological goals in mind, nor vice versa.

Yet, as the young archaeologists who at the end of the 1960s and the beginning of the 1970s were involved with Maurizio Tosi and Shahr-e Sokhte excavations could confirm (particularly Lorenzo Costantini and Massimo Vidale, who sat in the scientific Committee of the Conference 'Dogs, Past and Present') ISMEO archaeological missions were on the forefront of this approach, and this remained the seal of IsMEO/ISMEO archaeological methodology in the next generations after that of Giuseppe Tucci's direct pupils.

This explains why the 5th Conference of the Italian Association of Ethnoarchaeology was held in Rome on the 13th-14th May 2010, in collaboration with IsIAO – the new incarnation of Tucci's IsMEO from which we are proud to descend as direct scientific heirs - on the subject: 'Ethnoarcheology: current research and field methods'; and on that occasion a specific session of the Conference, under the chair of Francesca Lugli and Maurizio Tosi, was dedicated just to the Ethnoarcheology of pastoralism.

The 2018 International Conference 'Dogs, Past and Present' (the Eighth Conference organised by the Association in its 25th-year) convened over 200 scholars from different fields of research (such as Genetics, Archaeology, Archaeozoology, Anthropology, Ethnoarchaeology, Folklore), ideally representing 20 countries and more than 60 Italian and foreign institutions, all of whom addressed(each one from its own perspective), the problems related to the role played by dogs in the past and in traditional societies of our days.

We think that the results of that Conference, now embodied in the present book, show that a concrete sharing and exchange between the different disciplines is indispensable in order to analyse and understand the human-dog relationship - which is at the heart of this book - in different cultural, chronological and geographical contexts. Bringing all these issues together makes this volume a unique work that appeals not only to academics, but to a wider audience as well.

It is in this spirit of international, interdisciplinary, and open cooperation that we are ready to work again, in future occasions, with the Italian Association of Ethnoarchaeology, and with all the scientific institutions involved in the organisation of the First International Conference 'Dogs, Past and Present'.

# Forewords

## Alberto Cazzella

'Former' Dipartimento di Scienze dell'Antichità, Roma, Università Sapienza
alberto.cazzella@gmail.com

Ivana Fiore and Francesca Lugli, organisers of the conference 'Dogs, Past and Present', in the first instance, and now editors of the homonymous volume, have dealt with both a very specific (dogs and their relationships with human beings) and wide theme, as it includes a wide variety of fields. Approaches of the various papers are manifold and suitably they are all interdisciplinary. Particularly (in my opinion), studies using the most recent bioarchaeological methods, on one hand, and research on symbolic aspects implying a specific role played by dogs (they are often confused in our thinking), on the other hand, are of greatest interest. I do not specifically discuss the various papers constituting the volume: there are too many and too varied. I limit myself to expressing my appreciation on the work carried out not only by the editors, but also by all of the authors. I am not an expert on dogs and their relationships with human beings through time and in various contexts, but as a prehistoric archaeologist particularly interested in an anthropological approach to research (a palaeoethnologist, using a customary term in Italy and France) I feel at home as regards many of the themes carried out. Obviously, also research which addresses post-prehistoric contexts are of interest to me from an ethnoarchaeological point of view, i.e. such as situations offering wider interpretive suggestions to archaeologists. In any case, beyond my personal perspective, layers of meaning may be manifold for all different scholars interested in the theme dealt with in this volume.

Continuing to express general opinions, I would like to highlight that on one side the volume cannot be considered a definitive study as many problems are still open (and this is a fine perspective for future research), and on the other side it is very wide, and going to become a benchmark for this theme in the next years. Its potential developments in various directions can be seen, for example, in the conference '*Canis* em Ambiente Aquatico', that took place in Lisbon the last May 5th, organised with the participation by one of the two editors of this volume. Although this is not my field, I would also like to suggest other specific items linked to the roles played by dogs: they are to be considered not as lacking aspects, but as some of many other possible developments of the general theme. 1) The effect of dogs on archaeological deposits (as regards not only remains of fauna, dogs can cause post-depositional disturbances to be aware of?); 2) dogs and archaeology of the senses (we know the particular sensitive skills of dogs, but how these animals are perceived by human beings in different cultural contexts by senses such as either smell or hearing); 3) the roles played by dogs in traditional medicine (for example, I think about the 'strange' belief of ancient Romans according to which Maltese dogs leant upon stomach can help digestion).

With my best wishes that this volume will give a strong stimulus to continue widening the research on dogs and their relationships with human beings through time and in various cultural contexts, I conclude this brief introduction.

## Simon JM Davis

Zooarqueologia, Laboratório de Arqueociências, DGPC, Lisbon
simonjmdavis@gmail.com

*That an association of wolf and man is perfectly conceivable, especially when pups are taken, is confirmed by the most interesting experiences of Mr and Mrs Crisler in northern Alaska ... These explorers ... adopted wolf pups which grew up in human company and revealed their social propensities. It is evident from Lois Crisler's report that hunters on the Mesolithic (and indeed Upper Palaeolithic) level would have found it easy to associate with wolves.* (Zeuner 1963: 84).

Where did dogs come from? How can dog remains be distinguished from those of their wild ancestor? When were dogs first domesticated? How have *our* ancestors treated them? Where and when was/iscynophagy practised? Why did it take so long for dogs to be present in tropical regions? How important have dogs been to people? These are some of the questions that articles in this book attempt to answer. They are written mainly by anthropologists, historians, zoologists, zooarchaeologists, and geneticists among others. There is a vast array of subjects treated here and so for dog lovers everywhere this book should serve as an important source of information.

For a long time, zoologists were unsure about the origin of our oldest friend. There were several contenders within the Canidae – a family that includes wolf, jackal, fox, coyote, as well as some extinct taxa. Charles Darwin (1885) for example was puzzled by the huge variety of dog breeds and suggested that the dog must be descended from several different species of canid, extinct and recent. Most zoologists today agree that the dog is descended from the wolf, a supposition confirmed by modern genetics as Carles Vilà and Jennifer A. Leonard write herein. One early study that indicated a closer link between dogs and wolves than between dogs and, say, coyotes, jackals, or foxes, was the electrophoretic work done by Vibeke Simonsen in the 1970s (1976: 7-18).

The dog is also considered to be man's first domesticated animal. By domesticated we mean that its evolution was, and is, largely the result of artificial selection by people with natural selection playing a somewhat less important role. Man's control of this animal has given rise to the vast array of breeds that range in size from the tiny Chihuahua to the Great Dane. Since the original domestication of the wolf probably happened many millennia ago, research concerning this event or series of events (since it may have happened more than once and in several locations) lies within the domain of zooarchaeology – the study of animal remains from archaeological sites. How then is it possible to distinguish archaeological remains of a dog from those of a wolf? This can be problematical as the teeth and bones of dogs and wolves are very similar. There are at least three indicators that help. The first is size. The wolf is generally larger than the dog, a difference that is readily apparent in the teeth and bones. The second is a cultural one. Occasionally archaeologists uncover a complete *Canis* skeleton, sometimes even buried alongside that of a human. The third has to do with how these animals eat. Both dogs and wolves, being carnivores, are known to swallow small bones of their prey such as phalanges, carpals, and tarsals, that are smaller than about 3 or 4 centimetres. Sometimes these are regurgitated and/or pass through the gut. Some partially survive the passage through the gut almost complete with characteristic signs of semi-digestion, a phenomenon described by Sebastian Payne and Pat Munson (1985: 31-39). Finds of all three indicators – reduced size, careful burial, and semi digested prey bones - in a site or several contemporary sites can be a strong indication that the *Canis* in question was a domesticated one, i.e., a dog rather than a wolf. In the 1970s François Valla uncovered a *Canis* puppy skeleton beneath the left hand of an elderly human skeleton, probably a woman, at the Natufian (a Mesolithic culture) site, Ein Mallaha in northern Israel. Small carnassial teeth of *Canis* similar in size to dogs but significantly smaller than wolves were also found at this and another contemporary Natufian site as were semi-digested bones of gazelles and caprines. Such semi-digested bones are unknown on earlier sites in the Levant. Thus, the combination of all three indicators suggested that 12,000 years ago *Canis* and people had an affectionate rather than a gastronomic relation (Davis and Valla 1978: 608-610). In this book Francesco Boschin and colleagues describe a new criterion for distinguishing wolf carnassial teeth from those of dogs using X-ray microtomography. They noticed that the lower carnassial teeth (the $M_1$) of dogs have a lower proportion of dentine than those of wolves. It will be interesting to apply their technique to the Natufian specimens as well as the so-called Palaeolithic dogs like the ones from Bonn-Oberkassel, Goyet cave, Eliseevichi, and others.

It seems that in many regions dogs played different roles. For example, according to Konstantina Saliari and colleagues, in Austria dogs were eaten but under Roman rule cynophagy became rare and Giuseppe Minunno writes that the Phoenicians sometimes consumed dog flesh but Darius, the Persian king, asked the Carthaginians to abstain –dogs had high status in Persia.

Francesca Lugli writes about dogs that serve a useful function among fishermen, they draw sleds in the Amur region, and as Olga Maltseva notes, wide paws are indispensable for dogs walking on snow in southern Siberia.

One little-known aspect of canine zoogeography is the dog's late appearance in the tropics - Peter Mitchell explains this as being due to the presence there of certain diseases like sleeping sickness.

While many if not most modern breeds of dogs probably originated two or three centuries ago, there is plenty of iconographic and osteological evidence described here in this book for different kinds of dog in antiquity. In terms of their overall body structure, many can be compared to modern breeds even though they may not necessarily be genetically related. The Iron Age and especially the Roman period, was characterised by a great increase in *Canis* variability and the Romans were among the first to breed lap dogs. Numerous examples of dog skeletons and burials in various periods and in many places are described, even in the Largo Amba Aradam on the 'C' line of the Rome underground! And many articles provide clear evidence that dogs have been our oldest friends for many millennia. Indeed, their great intelligence probably rendered them 'pre-adapted' to become part of human society: dogs can understand certain human cues like pointing to a cup with food hidden underneath and their sense of smell is one or two orders of magnitude greater than ours as Juliane Bräuer and Blanca Vidal Orga point out.

Congratulations to Fiore and Lugli for their efforts in amassing this array of canine articles!

**References**


Darwin, C.R. 1885. *The variation of animals and plants under domestication*. 2nd edn. London: John Murray.

Davis, S.J.M. and F. Valla 1978. Evidence for the domestication of the dog 12,000 years ago in the Natufian of Israel. *Nature* 276: 608-610.

Payne, S. and P.J. Munson 1985. Ruby and how many squirrels? The destruction of bones by dogs. In Fieller, N.R.J., D.D. Gilbertson and N.G.A. Ralph (eds), *Palaeobiological investigations; research design, methods and data analysis*: 31-39. Oxford: BAR International Series 266.

Simonsen, V. 1976. Electrophoretic studies on the blood proteins of domestic dogs and other canidae. *Hereditas* 82: 7-18.

Zeuner, F.E. 1963. *A history of domesticated animals*. London: Hutchinson.



## Dulam Sendenjav

Professor, National University of Mongolia, Ulaanbaatar
sosedula@gmail.com


It is my pleasure and privilege to present the volume 'Dogs, Past and Present: An Interdisciplinary Perspective'. From the very first pages, the interaction and synergy between the authors is apparent, who come from different disciplines and geographical areas. They all wish to reconstruct and to tell the history of the dog, which has been man's loyal companion since antiquity. Reading this book is a fascinating journey that retraces the ancient and long adventure of this animal. The authors begin by considering the dog's ancestors and the first steps of its life among human beings, and continue retracing the complex relationships that the dog has had with women and men throughout the ages. Dogs are present in many varied circumstances of humans' lives where they assume the most and often-antithetical positions in emotional, working and religious spheres. The book is certainly a crucial reference point for scholars, but it is also interesting for non-experts due to the breadth of topics, which cover both technical-scientific as well as historical-anthropological studies.

It was 17 years ago when Prof. G. Nandinbilig and I first met with Italian ethnoarchaeologists, Francesca Lugli and Graziano Capitini who became my closest friends. Their research topic was so intriguing and multidisciplinary as it required to study the Mongolian nomadic people's way of life in different geographical locations such as the Khangai mountainous region of northern, central, and western Mongolia, the Gobi semi-desert region in southern Mongolia, and the endless steppe in eastern Mongolia. Their research involved studying and tracking four seasonal camps of nomadic peoples across time and space. Their social anthropological research methods combined with archaeological ones allowed them to penetrate deeply into the local communities and the different historical eras of the nomadic peoples. This resulted in a fantastic study with tremendous results.

The academic conference of ethnoarchaeologists in Rome which we attended that was hosted by our friends, helped me understand this research discipline's latest developments. There is no doubt that Francesca Lugli and Graziano Capitini's studies have paved a new way in ethnoarchaeology. I want to highlight that their research of Mongolian nomadic people and pastoralism in the past, present, and future – or from a diachronic perspective - has revealed the vital qualities of the nomads, which are climate resilience and an adaptive capacity for change and uncertainty. Especially their study of the role of Mongolian dogs in a nomadic lifestyle, pastoralism, and the security of the camps was timely by filling the research gaps. This type of international research was new to Mongolia and had many challenges. I remember that we, were gathered together to request permission from the customs inspectors to send the samples for analysis abroad, collected from the dog dens of pastoralists living in the Mogod subdistrict of Bulgan, Mongolia. In this way, their work expanded, and served as the basis for more detailed studies, and further contributed to this edited book.

Ten years ago, three of us - Françoise Aubin, a respected Mongolia researcher, Isabelle Bianquis, a researcher of Mongolian culture, and I – co-authored the article 'Le chien et la bru, deux êtres liminaires en Mongolie (the dog and the daughter-in-law, two liminal beings in Mongolia)'[1] and it was a huge work for us from a social anthropological perspective. But Ivana Fiore and Francesca Lugli's edition of 'Dogs, Past and Present: An Interdisciplinary Perspective' is a complete book from various interdisciplinary perspectives including ethnoarchaeology and biology.

[1] Bianquis, I., F. Aubin and Dulam Sendenjav 2013. Le chien et la bru, deux êtres liminaires en Mongolie, in Buffetrille, K., J.L. Lambert, N. Luca, and A. de Sales (eds) *D'une anthropologie du chamanisme vers une anthropologie du croire. Hommage à œuvre de Roberte Hamayon*: 303–322. Ètudes Mongoles & Sibériennes, Centrasiatiques & Tibètaines, Centre d'Ètudes Mongoles et Sibériennes. Paris: Ecole Pratique des Hautes Etudes.

# Introduction

Ivana Fiore and Francesca Lugli

Dogs have currently many traditional and non traditional tasks and roles in human societies. Companions, guard dogs, shepherd dogs, hunting dogs, guide dogs for the blind and sledge dogs are among the most famous ones. Dogs also assist people with different physical disabilities or psychic disorders and Kea Grace published a list of more than one hundred examples on that topic subdivided into General Service Dog Tasks, Medical and Alert Service, Brace and Mobility Support Service, Virtual Assistance and Guide Service, Hearing Service, Psychiatric Service, and Other Services.[1]

Dog meat had an important role in many cultures worldwide and it is still currently consumed in various countries on all the continents, where it can also have a ritual significance, especially in Asia and Africa. According to the Humane Society International in Asia (especially in China, South Korea, Philippines, Thailand, Laos, Vietnam and Cambodia), perhaps more than 30 million dogs are killed every year for that purpose. China which is supposed to consume more than 10 million dogs per year is the biggest consumer of dog meat.[2]

In western countries, dog meat consumption is generally frowned upon and forbidden. Stefan Häne in 2012 inquired and wrote that in Switzerland perhaps 3% of the population, especially in rural areas, eat dog meat as jerky or traditional sausages.[3] The World Population Review also mentions Poland where dog fat is believed to have medicinal properties and the United Kingdom where the sale of dog meat is forbidden but it is allowed to eat it if the animal belongs to the killer/consumer and it is killed humanely. In the US it is forbidden, but the law does not include Native American rituals and traditions. In Canada and Australia, dog meat consumption is not explicitly illegal, but with stipulations that render it impossible.[4, 5]

Even if it is very difficult to know how many dogs there are worldwide, recent studies estimated that the global dog population was around 471 million in 2018[6] and around 900 million in 2020.[7] It means that the total dog population had nearly/almost doubled in two years. There were 7,631,091,040 in 2018 and 7,794,798,739 billion people in 2020 (that means around 1 dog per 8.66 persons).[8]

In the European Union, it seems there are around 89,821,000 dogs[9], 17.1 million in Russia[10], 89.7 million in the US[11], 5.9 million in Canada, over 5 million in Australia, and 52.2 million in China.[12] More than 75%-85%[13] of the worldwide dog population is free-ranging (wild, feral, stray, city, and village ones), which means that 200 million are stray dogs.[14]

Since the Victorian age, dogs have become more and more important as companions, especially in Western countries. The dog is certainly the species with the most macroscopic physical differences. Just think of toy dogs and Great Danes and how they differ in size and aspect. Nowadays, there are 340 dog breeds known throughout the world and the American Kennel Club recognises 199 breeds (https://www.akc.org/dog-breeds/). Many of these breeds are the result of selections that were carried out in the last two centuries.

---

1 All websites were viewed between December 2021 and May 2022. https://anythingpawsable.com/100-examples-service-dog-tasks/

2 https://worldpopulationreview.com/country-rankings/what-countries-eat-dogs

3 https://www.tagesanzeiger.ch/schweiz/standard/schweizer-sollen-keine-hunde-und-katzen-mehr-essen/story/19945914

4 https://worldpopulationreview.com/country-rankings/what-countries-eat-dogs

5 https://www.sbs.com.au/news/dateline/article/the-places-around-the-world-you-can-still-eat-dog-meat/1cudici96

6 https://www.statista.com/statistics/1044386/dog-and-cat-pet-population-worldwide/

7 https://pawsomeadvice.com/dog/how-many-dogs-are-in-the-world/woofdog.org/how-many-dogs-are-in-the-world/

8 https://statisticstimes.com/demographics/world-population.php#:~:text=The%20World%20population%20is%20projected,to%20more%20than%208%20billion

9 https://www.statista.com/statistics/515579/dog-population-europe/ The first five countries are Germany (10,700,000), United Kingdom (8,500,000), Poland (7,850,000), France (7,500,000) and Italy (8,300,000).

10 https://www.statista.com/statistics/515543/dog-population-europe-russia/#:~:text=Number%20of%20dogs%20in%20Russia%20 2010%2D2020&text=The%20gradually%20growing%20number%20 of,pet%20dogs%20in%20Russian%20households.

11 https://pawsomeadvice.com/dog/how-many-dogs-are-in-the-world/#:~:text=In%202020%2C%20the%20US%20had%20about%20 89.7%20million%20dogs.

12 https://www.statista.com/statistics/992408/china-number-of-dogs/

13 Canines that do not have a home and are not owned by people are referred to as free-range pooches. They are the most numerous group, estimated to make up between 75% and 85% of all dogs.

14 https://pawsomeadvice.com/dog/how-many-dogs-are-in-the-world/

Figure 1: Key words (a) and those of the titles of the article in this volume (b) in two cloud graphs (https://tagcrowd.com/).

A flourishing industry has exploded and grown in an exponential way related to the number of dogs and the increasing importance that they have in the lives of people. According to the American Pet Products Association (APPA), the Pet Industry reached $179.4 billion in 2020 with a Grow Rate (CAGR) of 5.2% from 2020 to 2027 and it is expected to reach $255.4 billion by 2027. The global dog food market was around $54.08 billion in 2020 and it is expected to grow up to $65.8 billion in 2026 (Expert Market Research).[15]

Over the last decades, countless studies have examined the current life of dogs and their new place in our societies as well as their crucial part in human life and history. Especially dog domestication and its success during prehistory is a fascinating theme that scholars of various disciplines are involved with.

In the last few years, data and hypotheses have progressively increased, sometimes controversially, in each field of investigation.

Thousands of texts are published every year and it is practically impossible to be up to date with the publications and there is not a real exchange between the various disciplines. The final effect is that scholars might often ignore what happens in the various fields of research that are not close to their own. Therefore, it is extremely complex to reach a complete perspective of dogs' importance and their history until today.

The volume is entirely dedicated to dogs and it is focused on the necessity of an 'interdisciplinary perspective' to reach a better comprehension of the phenomena. We hope that it will contribute to bridging the gap that is created by the lack of communication among the various disciplines. We also wish that it will be a platform for the exchange of practical and theoretical approaches to the problem for scholars from different fields of research.

It gives the chance to read articles that all together give a wide and diachronic viewpoint of the history of dogs since their domestication which occurred in the mists of time.

The dog is considered from the point of view of genetics, archaeology, archaeozoology, ethnoarchaeology, anthropology, ethnology, ethnography, history, linguistics, iconography, numismatics and art history. All the authors made an effort to have an interdisciplinary point of view so it was impossible to classify them according to strict and rigid subdivisions.

[15] https://www.expertmarketresearch.com/; https://woofdog.org/how-many-dogs-are-in-theworld/; How many Americans have pets? An investigation of fuzzy statistics - washingtonpost.com; Pet statistics–www.iii.org; Shelter Intake and Surrender - Pet Statistics - www.; aspca.org; Ending Pet Homelessness - www.humanesociety.org

In the opening, David Howe stresses the importance of a research journal on dogs but also with a 'strong public outreach component' as a topic that 'solicits higher engagement from students and the general public', in order to demonstrate that it is a field which warrants appropriate funding. Then, the articles, are divided into six sections (Dogs: Genetics, Microtomography and Morphometric Techniques; Wolf Versus Dog; Dogs through Time: Role, Task and Position; Dogs: Archaeological and Archaeozoological Cases; Representation of Dog in Different Cultures; Dogs: Myth and Symbolism) to accompany the reader across the long and intricate voyage that dogs and humans have covered together throughout the millenniums. It is important however to bear in mind that the sections are only generally indicative because each article considers various questions from an interdisciplinary and diachronic perspective.

In section 1 'Dogs: Genetics, Microtomography and Morphometric Techniques' the most widely used methodologies in the study of dog history - from the earliest days of its arrival in human communities to the present- are presented. Genetics and new non-invasive analysis technologies such as microtomography are crucial for the study of dog domestication and history. They enable the comparison of different findings and samples of different chronologies and origins. Every day precise and sophisticated analyses flow into general databases where they are easily accessible to the scientific community. So scientists progressively have new and often revolutionary data but, as Carles Vilà and Jennifer A. Leonard write, the huge amount of data often does not have a univocal answer and a constructive criticism of the results is indispensable.

The section is articulated between various paradigms that consider different methodologies and subject matters.

A brief history of dog breeds since antiquity is considered by Grégoire Leroy *et al.* who highlight how the great current variety of canine morphotypes was modelled for various functional and social roles. But breeders are more interested in beautiful rather than behavioural aspects. The same specimen is often used for many litters to maintain the purity of the breed and that is highly negative for the health and the character of dogs. The authors suggest how genomic and computer science discoveries should be applied to dog breeding. Daria Sanna *et al.* also work on the current dog population and they present a molecular survey on the dogs of the Mongolian and southern Siberian herders as well as the southern Siberian hunters. The ethnographic data shed light on the genetic ones and vice-versa and give a picture of the intentional and unintentional mechanisms that have occurred in central Asian pastoralism and hunting.

Other authors deal with innovative methodologies. For example, Francesco Boschin *et al.* provide a reliable and non-destructive method (such as thin sections of teeth) to distinguish between dogs and wolves that is based on the analysis of the dentine volume proportion. Analyses were carried out using X-ray microtomography and the lower dentine thickness of the dogs made it possible to distinguish them from the wolf.

Raquel Blázquez-Orta *et al.* are also interested in the wolf-dog distinction. They analyse their skulls and jaws by combining traditional and geometric morphometry.

In section 2 'Wolf versus Dog' various authors give great attention to the ancestor of the dog, its early stages of domestication and the spread of *Canis familiaris* throughout the different continents.

An extensive and detailed overview of the presence of wolves in Italy in the Pleistocene is proposed by Dawid Adam Iurino *et al.* who reconstruct a trend of the dispersal of large wolf forms from north to south.

One of the crucial issues in the study of domestication is the question of how this took place. The proposal of Mietje Germonpré *et al.* - analysing the various models of wolf domestication proposed in the literature - is that of the occasional capture of the fearsome Pleistocene predator and the consequent raising of its pups in captivity. Captive breeding could have produced a selection towards a more tame behaviour of the first dogs.

Also, Juliane Bräuer and Blanca Vidal Orga study the process of domestication but point out how the recent comparative psychology helps to understand that the dogs' selected skills during domestication were for communicating and cooperating with humans.

The work of Peter Mitchell is important in relation to the spread of *Canis familiaris* which suggests new arguments that had been given little consideration previously. Diseases may have limited the expansion of dogs into tropical environments.

Dogs have had a complex relationship with humans which means mutual changes and adaptations throughout time. The dog is present in many spheres of human life, including utility, affectivity and symbolism.

In section 3 'Dogs through Time: Role, Task and Position' the authors highlight how the dog has certainly enabled man to develop special economic activities exploiting various ecosystems as well as regulating human societal relations, from the very beginning. Guy Lanoue stresses that the relationship between the sexes is symbolically regulated by the dog. The role of pet dogs, how they have acquired a new symbolic position and how they now have a quasi-human status in Western society are examined by Simona Bealcovschi. Not only do humans influence and modify dogs, but dogs can be decisive in human societal changes.

The ethnographic perspective turns out to be crucial for obtaining a multifaceted and diachronic understanding of the phenomenon under consideration. The different degrees of the indispensability of the dog in the societies of shepherds and hunters in Mongolia and southern Siberia are highlighted by Francesca Lugli and Galina Sychenko. Lugli also focuses on the importance of the use of dogs in aquatic contexts and highlights how that topic has been neglected by scientific research.

But there is not only continuity in human-dogs relationships. Sometimes real and often dramatic fractures can also occur. Christophe Blanchard analyses French homeless people and their dogs and describes how they are often dramatically separated due to health issues. Another break in the fruitful relationship between man and dog is described in Lugli's work. Dogs were used on fishing vessels for defence and aid in fishing, but this practice has now fallen into disuse due to recent health regulations and new fishing technologies. Another caesura in the man-dog relationship is dealt with by Olga Maltseva, who tells the story of the Amur *laika* dogs that have become strays because they were no longer useful for fishing and hunting activities.

Section 4 'Dogs: Archaeological and Archaeozoological Cases', contains the most contributions, confirming the indispensability of collaboration between archaeologists and archaeozoologists for the understanding of the role of dogs in past societies. In particular, Andrei V. Novikov emphasises that not only is the archaeological context and faunal remains fundamental, but the integration and comparison of archaeological, ethnographic data and written sources is necessary to help improve and broaden the interpretation of ancient remains.

The dichotomy of the relationship between humans and dogs and its wide variability has been documented in many regions since prehistory. Dogs' remains allow researchers to reconstruct the history of dog from the Palaeolithic to modern times. Maria Kudinova highlights the relationship between past and present in how dogs or their representations are ritualised and how those identified as early as the Neolithic period in the spiritual culture of the peoples of China still survive today.

The cases analysed contextualise and provide valid support for the hypotheses put forward. They show us in concrete terms the role of the dog from both a utilitarian and a symbolic point of view and indicate which clues are useful or not for a correct interpretation. No element must be missed or underestimated, all data must be recorded, as noted in the work of Eugenio Cerilli and Marco Fatucci in order to have a plausible interpretative hypothesis.

The dog interprets its role alongside man as a guardian of property; a guardian and helper in the management of flocks; a companion and aid in hunting and fighting; a companion and guide in the afterlife; a sacrifice in special purification rites to female deities of the chthonic world; a ritual closure of sacred areas; and immolated as a guardian in foundation rites of structures or walls.

The association between man and dog in burials shows us the close emotional relationship between man and dog. But in emphasising these findings it is also necessary to emphasise the point of view of the dog, which was immolated at the death of its owner and therefore was a poor sacrificial victim. Archaeological data clearly demonstrates that already in Roman times there was a variety of breeds and sizes of dogs (Konstantina Saliari *et al.*), manipulated and selected by man for their aptitudes (dwarf companion dogs, slender and fast hounds, and powerful war dogs).

Humans' oldest and dearest friend becomes a source of food, as shown by the traces of slaughter found on the dogs studied by Francesca Alhaique, Marco Bertolini and Ursula Thun Hoenstein, in some cases perhaps out of necessity, but in others out of established tradition. Bogdanov's ethnographic data also confirm the consumption of dog meat, as the half-breed/hunting dogs that lived near the camps were killed and consumed when resources were scarce.

The hypothesis of stray 'street dogs,' who are probably driven off and hit, still retain wounds (Ivana Fiore *et al.*). This is evidence of the ambivalent relationship between man and dog, permeated with love, nevertheless most often exploited and battered.

Dog representation is crucial for a better understanding of the relationships between dogs and humans. The interpretation of figurative and written sources includes problems that range from the reality of the represented subject but also its symbolic value. In addition, the artist, craftsman or writer who created the representation is important as well as the cultural context and the collective imagination of the society to which the artefact is addressed.

Section 5 'Representation of the Dog in Different Cultures' shows the great variability of contexts which have produced depictions of dogs and also the many materials and techniques that were used for that purpose. The articles make clear the relevance of the data which can be obtained. In fact, it is possible to follow the history of dogs from the representation of its Palaeolithic ancestor (Gianpiero Di Maida *et al.*) to the Tuscan Renaissance ceramics (Silvia Nutini and Marino Marini) and the ancient and recent rock representations in Southern Siberia and Mongolia (Dmitry V. Cheremisin). It is possible to identify physical features (size, breed etc) but also the activities performed and the dogs' roles, especially in cultures without the use of writing as in some ethnographic or pre-protohistoric populations. In some cases, dogs are the favourite subject of artistic styles. Frédéric Devienne analyses how during the Chinese Han Dynasty there is an impressive variety of representations: engravings, paintings on stone and stone slabs for funerary structures such as tombs, sarcophagi, memorial shrines reliefs in bricks and perforated bricks, terracotta funerary figurines, decorations and ornaments of glazed and unglazed earthenware vessels and artefacts, bronzes, lacquers; paintings on silk; textiles, etc.

Numismatics is also a precious information source as it is easily possible to deduce from the research made by Alessandra Bottari and Alessandro Crispino who interpret the depictions of coins to determine the physical features of dogs as well their real and symbolic role in ancient societies.

Clearly, written sources are crucial. Liubov Eliseeva and Eugenia Andreeva consider the Greek epitaphs in a comparison with Latin inscriptions and literature and Greek funerary inscriptions and they demonstrate the abundance of data that can be deduced. The Antiquity is also analysed by Francesco Tanganelli who analyses the representation of dogs in Attic funerary monuments and takes into account depictions and written sources in the Greek and Roman world.

The study of myths, symbolism, legends and folklore gives important information about the complexity of the position of the dog in different cultures. They can be analysed from different sources such as archaeological remains, texts, iconography and the observation of current cultures. Some authors have focused their attention on myth and symbolism in a more comprehensive and detailed way, although it is a theme that recurs in most of the articles.

In section 6 'Dogs: Myths and Symbolism' the authors debate this topic according to different perspectives and provide a complex and articulate view. The three articles, respectively by Giuseppe Minunno, Ana Portillo-Gómez and Marco Giuman with Miriam Napolitano consider the everyday life of dogs and their symbolic value in the Phoenician and classical world and highlight the positive and negative roles and perceptions that they lived at the same time. The same dichotomy is evident in the article of Maria V. Stanyukovich who analyses the role of the dog in past and present everyday life in the Philippines but also folk beliefs, rituals and mythology. Ambivalence is also described and analysed by Galina B. Sychenko who studies the relationship between dogs, wolves and humans in the Turkic cultures of Siberia in a comparative perspective that considers texts of non-tale prose - myths, legends, everyday stories - of the Siberian Turk as well as ethnographic data.

We wished to see how the volume would appear to a reader reading the Index and doing a keyword search, to see what slant the volume takes after the contributions were delivered. We plotted the keywords and titles in two cloud graphs, selecting the words that were repeated at least twice. In the two graphs, the impact of the volume is evident:

- domestication, words wolf and domestication are often present
- breeds
- art and representations of the dog are recurrent themes, even in works that do not have this subject as their main focus,
- culture, human, burial, nomads are often present.

We close with a thought for all dogs, but especially for sad and abused dogs. To the sacrificed and slaughtered dogs described in the book, but above all to the ill-treated and malnourished dogs, to those whose tails and ears have been amputated, to those who have ended up in lager kennels, to those who are beaten or whose muzzle is clamped with scotch tape.

We hope that our volume will contribute to creating awareness and sensibility to these matters, besides giving a voice to dogs who cannot express their suffering.

## Opening

# Calling on a Favour from Human's Best Friend: Public Outreach in Science

David Ian Howe

Content Producer, Ethnocynology, 2004 Brewster Drive, Franklin, Tennessee, 37067, USA, david@davidianhowe.com

**Abstract**

Public outreach and communication is a difficult aspect facing all fields of science - especially anthropology. When the public has access to scientific information and new findings, the content is often misrepresented by the mediums in which it is provided. However, I have found that using dogs as a proxy for past human behaviour often solicits higher engagement from students and the general public. Using a social media blog, I have been able to maximise the effectiveness of this strategy and the success of the blog suggests there is significant interest in the subject. I propose that an interdisciplinary dog research journal be created with a strong public outreach component. In doing so, academic researchers and public readers alike can access this information, therefore demonstrating this is a field that warrants appropriate funding.

**Keywords**: dogs, social media, internet, Palaeolithic, public outreach.

### 1 Introduction

You have probably noticed that the internet is a place full of information - some of it accurate - some of it not so much. In these ever-changing times, the most pertinent challenge the scientific community faces is directing which of these information sets the public engages with. People outside our small academic circles don't engage with scientific content, perhaps because the information is difficult to access and not fun to read. This problem is made worse as misinformation is easier to produce, syndicate, and digest than primary literature, especially in digital spaces. While I personally love a good discriminant function analysis figure and reading thirty pages about strontium isotope levels as much as you, many members of the public don't. What I do know however, is that my creative and educational endeavours have taught me a critical lesson: stick a dog photo on it, and people will listen.

### 2 Participant observation

As an anthropologist who began their undergraduate studies in 2010, I know that I have first-hand ethnographic knowledge of what student and young adult culture is like in this peculiar, yet ever-evolving digital age. And one thing I observed as a subjective participant in this population, was that while students are certainly in classrooms and lecture halls to further their educations, it is *incredibly* easy for them to lose interest in subjects that are taught by uninspired lecturers presenting on obsolete technologies who then assign kilograms of reading assignments riddled with academic jargon.

As a student of anthropology at that time, I reluctantly agreed with my peers; with the exception of several gifted lecturers who continue to inspire my career, anthropologists as a whole need to work on their educational delivery. *But* there is some merit to the old ways, because I did learn something reading through all that dense theory - humans are social animals who constantly evolve their technologies (White 1943; Merrill 1972; Schiffer 2001, 2003; Kelly 2016). And just like we view humans conducting a Mode 4 stone tool workshop around a fire as an outdated institution, many young adults find sitting in classrooms and lecture halls to not only be outdated, but an ineffective way to learn. I noticed that this opinion held by many young adults today directly stems from the technology they have inherited - social media.

### 3 Using media to teach Anthropology

Social media is the technology through which people communicate in today's world. It is a biface-sized tool that fits in your pocket, that allows you to conduct business, summon food and transportation, and even date. It is also a revolutionary platform though which we can creatively spread information and the human experience. At face value, young adults today love seeing tidbits of peoples' lives on social media as they aspire

Figure 1. *The First Dog Burial*, illustrated by Ettore Mazza.

to live a similar lifestyle; but at a deeper, human level, they are able to observe different careers, cultures, and customs around the world through the eyes of someone their age with similar interests. I believe this global connection makes young adults today more empathetic than past generations (Vossen and Valkenburg 2016; Abdul-Mageed *et al.* 2017; James *et al.* 2017). Speaking of empathy - let's get back to dogs.

Direct engagement with a target audience through universal themes is an effective way of promoting science to people who are often unfamiliar with anthropology and its related fields. For example, when I present to the public, I begin with this image, titled *The First Dog Burial* (Figure 1).

I ask the audience to take a moment and think about the emotions the image invokes. When I ask them what they felt, the responses often include 'sadness', 'death', 'mourning', 'funeral', 'family', and 'sacrifice'. These are all excellent answers. I use the fact that they offered such obviously human emotions to explain to them that people in prehistory were intelligent, had sophisticated cultures, and were neither Spartan brutes nor 'cavemen'.

I tell them that I commissioned an artist, Ettore Mazza, to illustrate what an ancient dog funeral might look like. I ask them to look closely and notice that these Eastern European Upper-Paleolithic people are fully clothed, not shirtless. That there are men, women, and children. That there is a shaman adorning the dog with symbolic ochre, and that the people have adorned themselves with tattoos and jewellery. I explain to them that all of these *material* details in the image are supported by the archaeological and ethnographic record.[1] I ask them to look for the *postulated* emotion Mazza painted on their faces; the sadness they felt burying the dog. I ask them to look at the female consoling the child holding their leg while carrying an infant on her back. I point out the person behind them who cannot bear to look at the scene. Finally, I ask them to take note of the other clan members in the background respectably observing from afar as this family lays their loved one to rest. Then I tell them 'This is how people used to live. They weren't less than human - they were *us*'.

My goal here is to evoke empathy for prehistoric people in the audience. Once the audience members can *wear the shoes* of the men and women in the illustration, prehistory comes alive, so to speak. Now, the entire learning process has become more visceral, and therefore more memorable. I explain that, as we do today, people in the past likely had ambitions, anxieties, and certainly

[1] Shallow dog burials, red ochre, fine-tailored clothing, jewellery, and presumed body modifications inferred from carved figurines are all present in Upper- Paleolithic Europe (Morey 2006: 168; Gilligan 2010; Krutak and Deter-Wolf 2017: 245–247; Janssens *et al.* 2018; Velliky *et al.* 2019).

agencies affecting the course of human history. I do understand that this artwork and my subsequent lesson could be labelled as post-processual conjecture, or post-modern bias applied to the past.[2] My counter is that the lesson I am trying to teach is still logically delivered - these people mattered - and that's why *anthropology* matters. While it's unfortunate that this lesson comes at the expense of a dead dog, it beautifully illustrates another concrete lesson and scientific fact about the past: we've been burying dogs, for at least 14,000 years (Janssens *et al.* 2018).

## 4 Public outreach and Ethnocynology

Currently, there is no direct forum, publication, or avenue to specifically publish research and news on the science of dogs within anthropology. Because of this, I created a science communication (scicomm) blog to centralise, promote, and creatively display this research and engage with the public. I named it *Ethnocynology*, a neologism coined by Cummins (2002), which means the study of dogs within human cultural contexts (Figure 2).

The *Ethnocynology* blog presents the public with photography, videography, and paleoart, and is currently run on the Instagram social media application. Most posts feature my dog, Strider, exhibiting a behaviour alongside a caption with a 2200 -character write-up involving social cognition, archaeological information, ethnographic accounts, or perspectives from other fields discussing a cultural practice or myth related to the image (Figure 3).

While it is not a peer-reviewed publication, I am committed to providing citations and references when possible (Figure 4). The blog currently has tens of thousands of active followers who consistently comment, share, and engage. I point this out not to tout personal success, but because I believe it demonstrates the considerable potential that this niche field of research has. Another factor that I appreciate is that the general public can directly message my account with questions or to ask for clarification on a recent post. In the context of traditional academic dissemination, members of the public often do not have a means for asking these questions, or they at least do not know how to. While there are surely negative aspects of being so easily reachable, I think this is a critical step we often don't consider.

As *Ethnocynology* has expanded, I have been invited to collaborate with TedEd to write the script for *A Brief History of Dogs*, and was contacted by PBS Digital Studios to help oversee the script for their PBS Eons video *How Dogs (Eventually) Became Our Best Friends* (PBS 2020; TED 2019). These notable institutions curate educational content they deem to be scientific and entertaining. Most importantly, while not academic in the strictest sense, these outlets are able to reach wide and diverse audiences.

Earlier, I mentioned how young adults today love to follow the social media accounts of companies such as *TED*, *PBS*, and *Vox* for their concise, educational, and entertaining content. In addition, they also follow science influencers on social media by the millions. It's

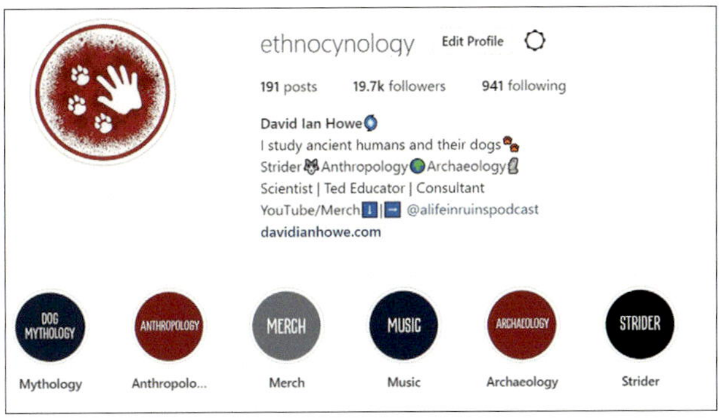


Figure 2. The home screen of the @Ethnocynology Instagram account.

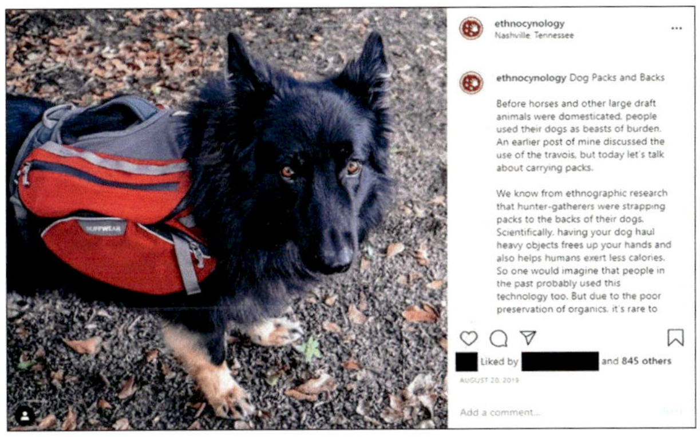


Figure 3. A post titled Dog Packs and Backs, discussing the use of packs and whether they cause spinal trauma (only some of the caption is shown) (@Ethnocynology Instagram account).

Figure 4. A post titled *Mesoamerican Mythology* showing the final two paragraphs with a call to introspection and list of references (@Ethnocynology Instagram account).

[2] Burials of any canid species could be a result of sacrifice or general consumption, they don't necessarily indicate emotional attachment (Geertz 1973; Schwartz 1998; Losey *et al.* 2013).

Figure 5. *Ancient Humans and Wolves* (2020) illustrated by Ettore Mazza. Used for a post discussing how humans, wolves, and domestic dogs may have interacted; the incorporation of myself and Strider helps the public better visualise and connect with a previously unfamiliar topic.

not just about high-quality productions, people like to see a personality behind the subject: Carl Sagan, Bill Nye, and Neil De Grasse Tyson excelled at this early on. Now YouTube personalities such as Vsauce, Veritasium, and the Vlog Brothers have all built award-winning science communication channels on their own. All of this was made possible by creating a lecture format that is easily digestible, unique, and most of all, *relatable*.

Much like we can conduct isotopic analyses on excavated dog remains as a proxy for understanding human diets in the past I use dogs as a lure to help people better understand and visualise the science of the human species as a whole (Figures 5 and 6) (Perri *et al.* 2019; Rogers *et al.* 2019; Burtt and Desantis 2020). And as we know from the conclusions of countless anthropological and psychological studies - *people love dogs* (Odendaal and Meintjes 2003; Morey 2010; Kaminski and Marshall-Pescini 2014).

Figure 6. *Of Mammoth and Man* (2020) illustrated by Ettore Mazza. Used for a post about Gravettian hunting of proboscideans. My dog and I are incorporated for the same purpose as Figure 5.

## 5 Conclusions

While *non-academic* avenues have proven beneficial to public engagement, I firmly believe that if we collaborate and centralise our research into an *academic* journal with a strong public outreach component, it would be much easier to locate current archeological, anthropological, and interdisciplinary research regarding dogs. This is a field the public wants to know more about, and as such, it is a field that deserves to be appropriately funded.

The next time someone asks you 'what do you do?' maybe don't explain the details of radiometric dating or isotopic sampling. Depending on your audience (I am sorry to tell you) you have already lost them. Instead, read the room, pull out your trusty social media biface, and do your version of this: show them a picture of the Natufian female and puppy burial from the Ain Mallaha site, near the Sea of Galilee, and explain the significance of the woman's hand being placed on the puppy at the time of burial (Davis and Valla 1978; Morey 2006: 165). Then tell them 'I study the ancient relationship between humans and domestic dogs. And I believe this is important because asking scientific questions about this fascinating relationship can teach us more about our modern society and our shared human past'.

If they're still not interested at this point, they're a cat person.

**Acknowledgements**

I'd like to thank my friend and colleague Ettore Mazza, he is truly a gifted artist. Most of all, I'd like to thank the organisers and the researchers of this volume. You all have made such wonderful contributions to science, and I would not be here without the diligent work and time you have put into this fascinating research.

**References**

Abdul-Mageed, M., A. Buffone, Hao Peng, S. Giorgi, J. Eichstaedt, and L. Ungar 2017. Recognizing pathogenic empathy in social media. International Conference on Web and Social Media.

Burtt, A., and L.R.G. Desantis 2020. Using dental microwear to understand the dietary behavior of domestic dogs in precontact North America. In B. Bethke and A. Burtt (eds) *Dogs: Archaeology Beyond Domestication*: 103–131. Gainesville: University Press of Florida.

Carrie, J., K. Davis, L. Charmaraman, S. Konrath, P. Slovak, E. Weinstein and L. Yarosh 2017. Digital life and youth well-being, social connectedness, empathy, and narcissism. *Pediatrics* 140 (Supplement 2): S71-S75.

Cummins, B. 2002. *First Nations, First Dogs: Canadian Aboriginal Ethnocynology.* Calgary: Detselig Enterprises.

Davis, S.J.M. and F.R. Valla 1978. Evidence for domestication of the dog 12,000 years ago in the Natufian of Israel. *Nature* 276 (5688): 608–610.

Geertz, C. 1973. *The Interpretation of Cultures.* New York: Basic Books.

Gilligan, I. 2010. The prehistoric development of clothing: archaeological implications of a thermal model. *Journal of Archaeological Method and Theory* 17(1): 15–80.

Kaminski, J. and S. Marshall-Pescini (eds) 2014. *The Social Dog: Behaviour and Cognition.* San Diego (CA): Academic Press.

Kelly, R.L. 2016 *The Fifth Beginning: What Six Million Years of Human History Can Tell Us about Our Future.* Berkeley: University of California Press.

Krutak, L. and A. Deter-Wolf 2017 *Ancient Ink: The Archaeology of Tattooing.* Seattle: University of Washington Press.

Losey, R.J., S. Garvie-Lok, J.A. Leonard, M.A. Katzenberg, M. Germonpré, T. Nomokonova, M.V. Sablin, O.I. Goriunova, N.E. Berdnikova, and N.A. Savel'ev 2013. Burying dogs in ancient Cis-Baikal, Siberia: temporal trends and relationships with human diet and subsistence practices. *PLOS One* 8(5): e63740

Luc, J., L. Giemsch, R. Schmitz, M. Street, S. Van Dongen, and P. Crombé 2018. A new look at an old dog: Bonn-Oberkassel reconsidered. *Journal of Archaeological Science* 92: 126–138.

Merrill, R.S. 1972. The role of technology in cultural evolution. *Social Biology* 19(3): 240–247.

Morey, D.F., 2006. Burying key evidence: the social bond between dogs and people. Journal of *Archaeological Science* 33: 158–175.

Morey, D.F.2010. *Dogs: Domestication and the Development of a Social Bond.* Cambridge: Cambridge University Press.

Odendaal, J.S.J. and R.A. Meintjes 2003. Neurophysiological correlates of affiliative behaviour between humans and dogs. *Veterinary Journal* 165(3): 296–301.

PBS, 2020, Eons, How Dogs (Eventually) Became Our Best Friends (viewed 15 May 2020).
<https://www.pbs.org/video/how-dogs-eventually-became-our-best-friends-hzxdyu/>.

Perri, A., C. Widga, D. Lawler, T. Martin, T. Loebel, K. Farnsworth, L. Kohn, and B. Buenger 2019. New evidence of the earliest domestic dogs in the Americas. *American Antiquity* 84(1): 68–87.

Rogers, B., K.J. Gron, J. Montgomery, P. Rowley-Conwy, G. Nowell, J. Peterkin, and D. Jacques 2019. Isotopic analysis of the Blick Mead dog: a proxy for the dietary reconstruction and mobility of Mesolithic British hunter-gatherers. *Journal of Archaeological Science: Reports* 24: 712–720.

Schiffer, M. 2001. The explanation of long-term technological change, in M. Schiffer (ed.) *Anthropological Perspectives on Technology*: 215–235. Albuquerque: University of New Mexico Press.

Schiffer, M. 2003. *Taking Charge: The Electric Automobile in America.* Washington D.C.: Smithsonian Scholarly Press.

Schwartz, M. 1998. *A History of Dogs in the Early Americas.* New Haven: Yale University Press.

TED, 2019, A brief history of dogs - David Ian Howe (viewed 15 May 2020).

<https://ed.ted.com/lessons/a-brief-history-of-dogs-david-ian-howe/>.

Velliky, E.C., A. Barbieri, M. Porr, N.J. Conard, and B.L. MacDonald 2019. A preliminary study on ochre sources in southwestern Germany and its potential for ochre provenance during the Upper Paleolithic. *Journal of Archaeological Science: Reports* 27: 101–977.

Vossen, H. and P. Valkenburg 2016. Do social media foster or curtail adolescents' empathy? A longitudinal study. *Computers in Human Behavior* 63: 118–124.

White, L.A. 1943. Energy and the evolution of culture. *American Anthropologist* 45(3): 335–356.

Detail from Leroy *et al.* fig. 1 (chapter 1.3): Dog diseases and health care of dogs. Livre de la Chasse, Gaston 'Phébus' comte de Foix, ms 616 folio 40, XVe siècle, Paris, BNF, Département des manuscrits.

# Section 1

## Dog Genetics, Microtomography and Morphometric Techniques

# 1.1 A Molecular View on the Domestication of Dogs

Carles Vilà and Jennifer A. Leonard

Conservation and Evolutionary Genetics Group, Estación Biológica de Doñana (EBD-CSIC),
Avd Americo Vespucio 26, 41092 Seville, Spain. carles.vila@ebd.csic.es; jleonard@ebd.csic.es
Corresponding author: Carles Vilà, carles.vila@ebd.csic.es

**Abstract**

The domestication of the dog was an important event in human history- it marked the first species to be domesticated. For this reason this process has generated a lot of interest, and a lot of questions: When? Where? Who? How? Why? These questions have been difficult to answer with high levels of confidence and accuracy, but the incorporation of genetic and genomic data can move us closer to some answers. Genetic data have firmly concluded that dogs were domesticated from the grey wolf and have also pushed back the date of domestication to several tens of thousands of years. Ongoing genetic and genomic research, including from ancient remains, continue to shed new light on these old questions. The study of the origin of dogs and the domestication process has attracted the attention of archaeologists and historians for a long time. The search and study of remains and documents has contributed to our understanding of how and when the domestication may have taken place and how dogs have spread around the world. However, since the mid-1990s the analysis of DNA sequences has become central in the study of animal domestication (Zeder *et al.* 2006). Genetic and genomic studies have multiplied, suggesting answers to many long-lasting questions. These studies seem to provide solid evidence and very robust data which allow the estimation of confidence intervals to historical events, facilitate hypothesis testing, and suggest exciting alternative hypotheses. However, these analyses also are based on assumptions and have limitations. In this article we will discuss some of the main results derived from the genetic analysis of dogs and wolves with regard to the domestication history and the origin of breeds. We will reflect on some lessons derived from molecular genetic analyses, we will consider some notions that seem very well supported by the genetic data, but we will also try to highlight aspects that deserve further attention. The goal of this article is not to summarise all existing publications. Instead, we will focus on aspects that are of general interest and that have attracted the attention of zooarchaeologists, evolutionary biologists, veterinarians, and geneticists.

**Keywords:** genetics, genomics, ancient DNA, dog domestication, grey wolf.

## 1 The genomics revolution

The comparison of DNA sequences has allowed the direct investigation of the evolutionary relationship between species and populations (based on DNA sequence similarity). Also, since mutations accumulate on DNA sequences with time, the magnitude of the differences between the DNA sequences in two lineages is related to the amount of time the two of them have been separated. Consequently, the study of DNA sequences can be used to understand the relationship between dogs and other species, between dog populations and breeds, or to evaluate the time since dogs and wolves separated. The analysis of DNA sequences has been possible since the last decades of the 20th century, and advances in molecular methods have allowed the extraction of minute amounts of DNA even from bone remains that are thousands of years old. This is called ancient DNA and has enormously facilitated the study of the evolutionary history of dogs. However, these analyses have increased even more in the last two decades, since we have entered a genomics revolution.

Advances in sequencing and DNA analysis technologies have resulted in a dramatic drop in the costs of DNA sequencing. The cost of sequencing a complete genome is more than a million times cheaper now than at the beginning of this century, when the first human genome sequence was published, and massive resources are available to understand the origin and evolution of dogs. This new generation of genetic and genomic approaches, together with ancient DNA technologies, have allowed studies on many aspects related to the domestication of dogs. Studies have addressed the ancestry of the dog, the place and time of domestication, how the domestication took place, the origin of dog breeds, the consequences of life with humans for dogs, genes that are important for the differentiation of dogs and wolves, etc.

## 2 The dog as a model organism

At the same time that genomic technologies have developed, the dog has advanced as a model organism in biomedicine. Different breeds of dogs have been bred in isolation, many since the mid-19th century. These different breeds are characterised by diverse traits.

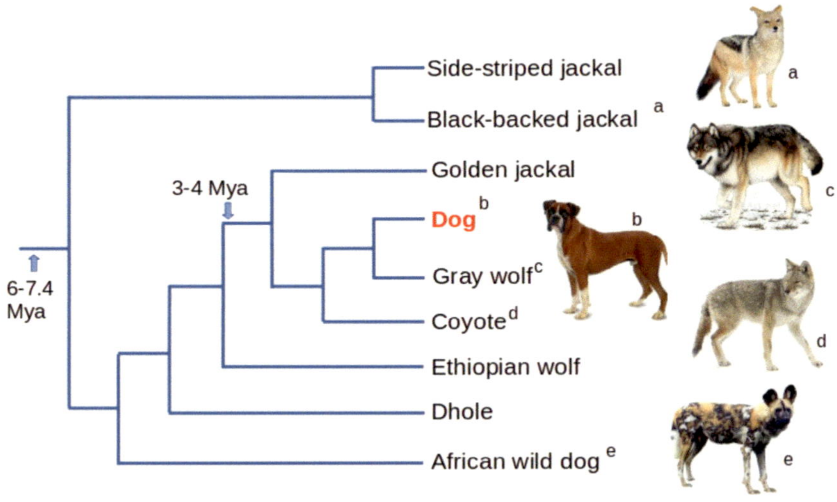


Figure 1. Evolutionary relationships between wolf-like canids based on genetic markers distributed across the genome (based on Lindblad-Toh *et al.* 2005). This tree shows the close proximity of dogs and grey wolves. Other analyses with different sets of genetic markers show the same close proximity, confirming that dogs derive from grey wolves.

Most apparently, they are characterised by different morphologies including size, hair length, texture and colour, head shape, leg length, etc. Many breeds are also characterised by high frequencies of genetic based conditions or diseases, such as deafness, hip dysplasia, or lymphoma. Some of these conditions develop in a way that is similar to their development in humans, making them good animal models to study these diseases and develop therapeutics.

Since purebred dogs represent genetically separate units and multiple breeds may share the same traits, breeds represent replicates in an experimental setup and the identification of the genetic basis for many of these disorders and morphological characteristics is easier than in humans (Vaysse *et al.* 2011; Plassais *et al.* 2019). For this reason, dogs were developed as a model organism even before the possibility of sequencing DNA. This positioned the dog to be one of the first animals to have their whole genome sequenced (Kirkness *et al.* 2003; Lindblad-Toh *et al.* 2005). The interest of the dog in biomedicine allowed the development of molecular genetic tools and this, together with the survival of grey wolf (*Canis lupus*) populations across much of their ancestral distribution (Mech and Boitani 2010), has allowed many evolutionary and anthropological studies on this species and on the domestication process, well before any other domestic animal or plant.

### 3 The wild ancestor species of dogs

Based on the archaeological record, and dogs' morphology and behaviour, many different theories about their origin were developed. It was very clear that they belonged to the genus *Canis* which includes several wild species such as the grey wolf, the coyote (*Canis latrans*), or multiple species of jackal. In fact, almost all of these species are known to occasionally hybridise in the wild and produce fertile offspring. An association with grey wolves was a common thread through the different theories on the origin of dogs, but some also called in hybridisation with other species. For Darwin, the diversity of dogs was an indication of a very diverse origin from multiple species: '*I do not believe, as we shall presently see, that all our dogs have descended from any one wild species*' (Darwin 1958). For some time, the Nobel laureate Konrad Lorenz also suggested that dogs derived from tamed grey wolves and golden jackals (*Canis aureus*). These alternative hypotheses were able to be rigorously tested with genetic data. C. Vilà and collaborators (Vilà *et al.* 1997) compared maternally inherited genetic material from many domestic dogs, grey wolves, golden jackals and other closely related species to show that all of the genetic material in dogs indicated that they were domesticated only from one species: the grey wolf. Subsequent studies involving almost complete genome sequences (Lindblad-Toh *et al.* 2005) have confirmed this view and, despite the occasional interbreeding with different wild species, all genetic evidence from dogs from all over the world supports the origin from grey wolves (Figure 1).

### 4 Time of domestication

Dogs were the first species to be domesticated, so the timing of this event is of great interest to understand not just the process of domestication, but also the social development and changes in our own ancestors. The ancestral species of the dog, the grey wolf, was widely distributed across Eurasia from before the arrival of modern humans in Eurasia offering ample opportunities for coexistence. A first approach to infer the time and place of dog domestication was the identification of the oldest dog remains in the archaeological record. However, this implies the ability to separate dog and wolf remains and this is only possible after those ancient dogs had been living separated from wolves for long enough to result in morphologically diagnosable remains. We have no idea how long this took. The first characters to change when dogs were domesticated were probably invisible in the archaeological record, such as behaviour, colour, tail and ear position. For other domestic animals used for food such as sheep, goats, cattle or horses, domestication could be inferred in the archaeological record by sudden increases in the number of remains or changes in the sex or age ratio (Zeder 2015). However, these approaches are not feasible with dogs because we do not know the reason for their domestication nor the selective forces that those ancient dogs probably experienced. Thus, osteological remains of ancient dogs and wolves could be very difficult to tease apart. This makes it very difficult to identify the time of domestication from the archaeological record alone. The estimate of the date of domestication based only on morphologically diagnosable traits is likely to result in an underestimation of the time since domestication. The earliest non-controversial evidence of domesticated dog is from the Pleistocene-Holocene boundary, but a number of earlier controversial remains have been described as domestic dogs (e.g. Germonpré *et al.* 2009; Druzhkova *et al.* 2013).

Genetic sequences accumulate mutations through the generations, and these genetic changes can be used to estimate dates of divergence between different sequences. This method was applied to sequences from grey wolves and domestic dogs in order to estimate the time of divergence between them, which may correlate with the date of divergence between them- the date of domestication. In the first study, this was applied to mitochondrial sequences, a part of the genome that is only inherited from the mother (Vilà *et al.* 1997). The study showed that the sequences from dogs around the world formed several clusters within the evolutionary tree of wolves (Figure 2). These different clusters indicate that different lineages of wolves participated in the origin of dogs. One of these clusters was very much larger than the others and included representatives from many different breeds. This was taken to represent the first lineage to be domesticated because this older time would give more time to accumulate mutations, and the authors tried to estimate the amount of time needed to generate as much diversity as observed among modern dogs from that group. The findings were sensational- the wolf and dog sequences had diverged long before the first non-controversial records of domestic dog in any archaeological site.

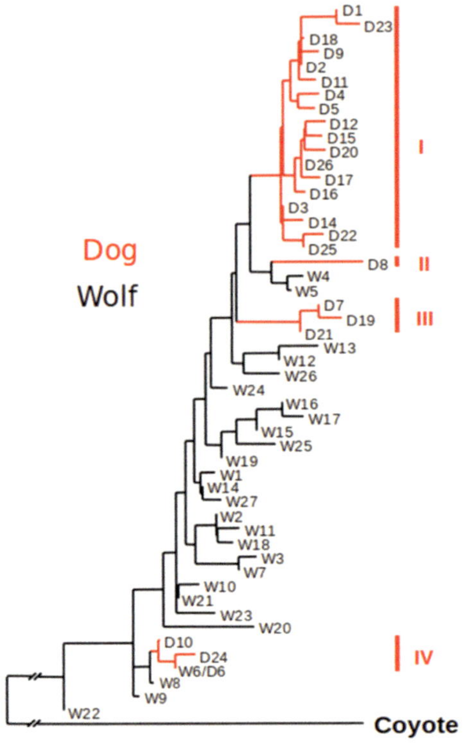


Figure 2. Phylogenetic tree of mitochondrial DNA sequences of dogs and grey wolves (based on Vilà *et al.* 1997). Dog sequences (marked in red) form four clusters within the tree of wolves indicating that multiple lineages (at least four) of wolves have participated in the origin of domestic dogs, probably representing several ancestral wolf populations.

This was exciting and spurred on much research to re-assess the faunal remains of many sites that had been dismissed as too old to be dogs. The date suggested by this study, however, should not be taken too literally for multiple reasons. One reason is that molecular clocks, the rate at which mutations accumulate along a DNA sequence, need to be calibrated for each dataset, ideally using fossil remains of known age and that could clearly characterise events within the evolutionary tree. In this case, the only reference point available was very distant and corresponded to the divergence between coyotes and wolves, more than a million years ago. Similarly, rates of mutation normally go with wide confidence intervals and in this study no confidence intervals were considered. The time of domestication should have been better represented by a very wide time interval.

Another important factor that needs to be considered is that the study was based on the assumption that the group of dog DNA sequences had one single origin, and this may not have been the case. Multiple similar sequences from wolves, from one or several populations, could have contributed to the origin of that diversity in dogs. Also, the study only took into account the diversity in living wolf populations, but it is clear that some populations of wolves have gone extinct since the time dogs were domesticated (Leonard *et al.* 2007). It is possible that one of these now-extinct populations that were not sampled had sequences that were more closely related to dogs than any of the living populations (Thalmann *et al.* 2013).

Considering that sequences from that group of dogs could have multiple origins, another group of researchers tried to estimate the date of divergence after dividing those sequences into smaller subsets of sequences (Savolainen *et al.* 2002). The dates obtained for the domestication event ranged between 15,000 and 40,000 years ago, but they did not provide robust justification on the selection of subsets of data. These two studies illustrate the importance of carefully considering the assumptions used in the analyses and show that considering only short mitochondrial DNA sequences from modern dogs was not likely to provide a definite answer to the date of domestication.

These studies were based on the study of just one short DNA sequence of maternal inheritance (a fragment of the mitochondrial DNA). However, one of the first lessons from the genomics revolution has been that different fragments of the genome are subject to very different selective forces and stochasticity, and the study of just one short DNA sequence is not likely to provide a full view of the evolutionary process. A more robust approach consists of integrating the information provided by very many sequences across the genome. Also, dogs and wolves are two very closely related lineages, from the evolutionary point of view. This implies that very few mutations are likely to have appeared since the separation of the two lineages, which makes the construction of robust evolutionary trees difficult. Instead, the relationship between these species is better characterised by differences in the frequency of different sequence variants (alleles) in the populations. These differences allow testing different evolutionary histories and even patterns of demographic change (Larson and Burger 2013).

One of the first attempts to estimate the date of domestication using whole genome data was part of the dog genome sequencing project (Lindblad-Toh *et al.* 2005). Modeling demographic declines associated with the origin of dogs and the formation of breeds the authors estimated that the domestication may have taken place about 27,000 years ago. However, this study was still constrained by some common problems. On one hand, the authors were using just purebred dogs to characterise modern dogs, without considering that most of the dogs worldwide were not purebred. On the other hand, the study did not have proper calibration points for the molecular clock. Fortunately, by including a whole genome of an ancient, radiocarbon dated wolf in the analysis a much more accurate date was possible (Skoglund *et al.* 2015). This supported that the time of domestication of about 27,000 years ago, substantially earlier than the date suggested by the archaeological record, but not as ancient as the first analyses of mitochondrial sequences suggested. Another study involving multiple complete mitochondrial genomes of ancient canids also provided dates between 18,800 and 32,000 years (Thalmann *et al.* 2013). Although the time frame is now in much better focus than it was just a few years ago, additional whole genome sequences from ancient, radiocarbon dated wolves and dogs from around their distribution are needed to really improve our knowledge about the time of domestication.

## 5 Place of domestication

Domestic dogs are very widely distributed, and before the modern time they were already found pretty much everywhere there were people. When Europeans arrived in the Americas in the late 1400s, they found people with dogs; when Europeans arrived in Australia little more than a century later, they also found people with dogs. The very wide distribution of domestic dogs has complicated the study of their place of origin. Once it was determined that they originated only from grey wolves, and not from any of their wild relatives distributed only in Africa or America, it was clear that the domestication must have taken place inside the distribution of the grey wolf, the wild ancestral species. Grey wolves, however, were very widely distributed at the time of domestication. They were distributed across nearly all of Europe, Asia, and North America (Mech and Boitani 2010). On the other hand, since the first analyses of mitochondrial DNA data (Figure 2) it was clear that because dogs grouped in several clusters within the evolutionary tree of wolves, multiple populations of grey wolves could have been involved in the origin of dogs.

Wolf populations in different geographic areas show some genetic differences (Leonard 2014). The sequences found in modern dog breeds, most of which originated in Europe, are clearly more closely related to sequences in Eurasian grey wolves than American grey wolves (Vilà *et al.* 1997; Koblmüller *et al.* 2016). Additionally, there are very divergent lineages of grey wolves in the Indian subcontinent that are clearly not ancestral to the domestic dog (Sharma *et al.* 2004). This, however, just narrows the area in which dogs could have been domesticated to Eurasia, except India.

Some authors have suggested that the centre of domestication could have been East Asia based on a supposedly higher diversity for a fragment of the mitochondrial genome in dogs from that area, suggesting that they had had more time to accumulate diversity (Savolainen *et al.* 2002). However, this result could be due to a sampling bias: most of the regions around the world were represented by purebred dogs, and these are characterised by very high inbreeding and limited diversity. Purebred dogs can not represent the overall diversity of dogs. A study involving village dogs (non-purebred dogs) found highest genome-wide diversity in Central Asia and suggested that this was the domestication origin (Shannon *et al.* 2015) while another study (focusing on purebred dogs) suggested a notable contribution of local wolf population to the origin of breeds from each region, with a remarkable contribution of wolves from the Near East (vonHoldt *et al.* 2010). However, all these studies were based on modern dogs and wolves. An analysis of mitochondrial DNA sequences of ancient canids suggested a European origin of domestic dogs (Thalman *et al.* 2013). It is very possible that the distribution and diversity of specific grey wolf populations has changed in the last tens of thousands of years, and that some populations went extinct, and so are not included in the analyses. Thus, the observation of discrepancies in the lineages observed in modern dogs and wolves has led to the suggestion that the domestication process could have been originated in a wolf population that has gone extinct (Thalmann *et al.* 2013; Freeman *et al.* 2014; Skoglund *et al.* 2015).

Each study is providing a different answer, and none of them seems to offer a final resolution. To untangle this story, extensive sampling of ancient populations of dogs and wolves, targeting the time frame of domestication, will be necessary. In any case, the evidence accumulates to suggest that multiple wolf populations in Eurasia have contributed to the diversity of dogs (Vilà *et al.* 2005).

## 6 American dogs

As mentioned above, Native Americans had domestic dogs when Europeans arrived in the New World. The presence of those dogs poses an interesting question: where did they come from? If the date of dog domestication was about 14,000 years ago, as often claimed in the literature, then it would be very unlikely that humans had dogs when they reached the Americas at about the same time or earlier. In that case, American dogs could be expected to derive from American wolves, which were distributed across North America. However, if the domestication of dogs was much earlier, as suggested by the genetic data, the dogs could have travelled with humans to the New World.

In order to test the alternative hypotheses that dogs were brought by humans or that dogs were independently domesticated in the Americas, ancient dogs, from before the arrival of Europeans to the Americas, were genetically analysed (Leonard *et al.* 2002). If the American dogs were more similar to other, modern, European dogs, then they came from the same domestication event. That would require that people already had domestic dogs when they reached the Americas, which in turn requires an early date of domestication. Alternatively, if the American dogs were more closely related to American wolves, that would imply that they were separately domesticated in North America. The results very clearly support the hypothesis that people brought already domesticated dogs with them when they colonised the Americas. The genetic sequences from the American dogs were much more closely related to other dogs than to American wolves. Further, many of the sequences from the ancient American dogs formed a group within the diversity of dogs, suggesting that they had been isolated for a long time (Leonard *et al.* 2002). A study with complete mitochondrial and nuclear genomes confirmed that all native American dog lineages before the arrival of Europeans shared a common origin that likely originated in Siberia and dispersed into the Americas alongside people (Leathlobhair *et al.* 2018).

Interestingly, those pre-contact American dogs diversified into lineages that are not present in purebred dogs (Leonard *et al.* 2002) and that may have disappeared also from modern mongrel dogs in the Americas (Castroviejo-Fisher *et al.* 2011; Figure 3). The mechanism that led to the disappearance of Native American dog lineages is not known but could be related to the arrival of diseases with European dogs. In any case, this implies that modern dog populations may not be representative of past populations and it is necessary to promote genomic studies based on historic and archaeological samples, which could allow for changes through time to be understood.

## 7 Origin of breeds

Dogs have fulfilled a variety of different functions in human society (Snyder and Leonard 2006). Different characters (morphologies or behaviours) in dogs may better match different functions. Many of these lineages, such as mastiff-like animals to protect livestock from predators or hunting dogs, have been maintained for millennia. However, the strict control

Figure 3. Machiguenga kids and dog from the Eastern slopes of the Peruvian Andes. Most current American dogs, even dogs from very isolated communities, do not carry mitochondrial DNA sequences similar to those found in ancient American dogs from before the contact with Europeans. This implies that those native American dog lineages have been replaced by the lineages that have arrived during the last centuries (Photo by C. Vilà).

of breeding and maintaining breed studbooks are more recent practices, starting during the 19th century. Modern dog breeds are characterised by being very uniform in morphology and behaviour as a result of intensive line breeding.

Purebred dogs represent highly inbred lineages resulting from the selection of a small founder population and promoting popular sires that are repeatedly mated with multiple females (Sundqvist *et al.* 2006). This is very different from the mating pattern in natural grey wolf or even feral dog populations and has resulted in a large number of diseases prevalent in many breeds (Ostrander and Kruglyak, 2000). However, purebred dogs do not represent the majority of dogs in the world and important levels of diversity persist in mongrels, stray dogs and village dogs (Boyko *et al.* 2009). This implies that these individuals should not be ignored when trying to understand the origin and evolution of dogs and can be important in maintaining healthy dog populations.

## 8 From wolf to dog

All these studies have helped to understand the evolutionary history of dogs and how intensive inbreeding can lead to the fixation of morphological and behavioural traits for the different breeds. However, one central question remains without answer: what makes dogs different from wolves? Both species can interbreed and produce fertile offspring. However, in areas where healthy wolf populations still survive, the two species remain separated. Hybridisation tends to occur only in areas where some individual wolves remain isolated or when the social structure of wolf packs is heavily disturbed by intensive hunting (Leonard *et al.* 2014).

Life with humans has transformed the genome of dogs. This has resulted in a relaxation in selective forces compared to wolves. Many individuals with anomalous morphologies or behaviours successfully reproduce (think for a moment of a chihuahua-like wolf or a bulldog-like wolf being the alpha male in a pack and leading the rest of the pack in a deer hunt). Also, the resources provided by humans have facilitated a very large population increase. Both processes have resulted in the accumulation of mutations in the genome of dogs, including deleterious mutations, which may have contributed to the exceptional diversity existing in dogs (Björnerfeldt *et al.* 2006; Cruz *et al.* 2008). Studies on the expression of genes in the brain of dogs suggest that a larger number of differences in the hypothalamus of dogs compared to that of wolves and coyotes (Saetre *et al.* 2004) could have resulted in a cascade of effects over the development of dogs leading to diverse morphologies.

Not a single gene or set of genes seems responsible for the divergence between dogs and wolves, but a higher ability to digest starch seems common place across many dog breeds and may have facilitated the expansion of dogs by making them able to exploit human resources (Axelsson *et al.* 2013), although other genes, including others related to brain function, may have also been important (Freedman *et al.* 2016).

## 9 Future prospects

While many studies are contributing to a better understanding of the evolutionary history of dogs, the analysis of the massive amounts of genetic data that are now available is not providing unique answers to these questions. The responses obtained are not always as robust as desired. Since just a few individuals are chosen for most of the analyses, it is necessary to be very aware of how samples are selected and what are the assumptions of the models used. Biases in the study design can result in apparently contradictory results. Robust inferences need to be based on study designs that take into account multiple facts. For example, it is more and more clear that purebred dogs do not represent the diversity of dogs; modern samples may not be representative of the dogs that existed in a given area in the past; dogs may not derive from a single domestication event, but multiple wolf populations may have been involved; admixture between dogs and wolves may have taken place frequently in the later millennia and in many areas around the world; current wolf populations are not representative of the diversity at the time of the domestication due to the massive population extinctions during the last two or three centuries, which may have led to the decline of more than 90% of the worldwide wolf population. This situation demands an explicit statement of the assumptions and limitations in each study as well as a critical interpretation of the results.

While generating large amounts of genetic data is becoming easier, it is apparent that the exclusive use of this data is not enough to understand the origin and evolution of dogs. We do not know what were the reasons that led our ancestors to domesticate wolves and include them in their societies. We do not understand either what roles they played in ancient human communities that led to their huge evolutionary success: dogs accompanied humans in their trans-continental migrations more than 10,000 years ago. While wild wolf populations have suffered dramatic declines worldwide, their descendants, the dogs add now to hundreds of millions of individuals. Archaeological and historical research can suggest hypotheses; molecular genetics can help to test them.

## Acknowledgements

We would like to thank Francesca Lugli and Daria Sanna for the invitation to participate in the '1st International Conference 'Dogs, Past and Present - an Interdisciplinary Perspective', an extraordinary conference. Also, CV would like to thank them and Graziano Capitini for their hospitality in Rome and their patience.

## References

Axelsson, E., A. Ratnakumar, M.L. Arendt, K. Maqbool, M.T. Webster, M. Perloski, O. Liberg, J.M. Arnemo, A. Hedhammar and K. Lindblad-Toh 2013. The genomic signature of dog domestication reveals adaptation to a starch-rich diet. *Nature* 495: 360–364.

Björnerfeldt, S., M.T. Webster and C. Vilà 2006. Relaxation of selective constraint on dog mitochondrial DNA following domestication. *Genome Research* 16: 990–994.

Boyko, A.R., R.H. Boyko, C.M. Boyko, H.G. Parker, M. Castelhano, L.Corey, J. D. Degenhardt, A.Auton, M. Hedimbi, R. Kityo, E.A. Ostrander, J. Schoenebeck, R. J. Todhunter, P. Jones, and C. D. Bustamante 2009. Complex population structure in African village dogs and its implications for inferring dog domestication history. *Proceedings of the National Academy of Sciences* 106: 13903–13908.

Castroviejo-Fisher, S., P. Skoglund, R. Valadez, C. Vilà and J.A. Leonard 2011. Vanishing native American dog lineages. *BMC Evolutionary Biology* 11: 73.

Cruz, F., C. Vilà and M.T. Webster 2008. The legacy of domestication: Accumulation of deleterious mutations in the dog genome. *Molecular Biology and Evolution* 25: 2331–2336.

Darwin, C. 1959. *On the Origin of Species. 1st Edition*. London: John Murray.

Druzhkova, A.S., O. Thalmann, V.A. Trifonov, J.A. Leonard, N.V. Vorobieva, N.D. Ovodov, A.S. Graphodatsky and R.K. Wayne 2013. Ancient DNA analysis affirms the canid from Altai as a primitive dog. *PloS ONE 8:* e57754.

Freedman, A.H., I. Gronau, R.M. Schweizer, *et al.* 2014. Genome sequencing highlights the dynamic early history of dogs. *PLoS Genetics* 10: e1004016.

Freedman, A.H., R.M. Schweizer, D. Ortega-Del Vecchyo, E. Han, B.W. Davis, I. Gronau, P.M. Silva, M.Galaverni, Z. Fan, P. Marx, B. Lorente-Galdos, O. Ramirez, F. Hormozdiari, C. Alkan, C. Vilà, K. Squire, E. Geffen, J. Kusak, A. R. Boyko, H. G. Parker, C. Lee, V. Tadigotla, A. Siepel, C.D. Bustamante, T. T. Harkins, S. F. Nelson, T. Marques-Bonet, E.A. Ostrander, R.K. Wayne, J. Novembre 2016. Demographically-based evaluation of genomic regions under selection in domestic dogs. *PLoS Genetics* 12: e1005851.

Germonpré, M., M.V. Sablin, R.E. Stevens, R.E. Hedges, M. Hofreiter, M. Stiller and V.R. Després 2009. Fossil dogs and wolves from Palaeolithic sites in Belgium, the Ukraine and Russia: osteometry, ancient DNA and stable isotopes. *Journal of Archaeological Science* 36: 473–490.

Kirkness, E.R., V. Bafna, A.L. Halpern, S. Levy, K. Remington, D.B. Rusch, A.L. Delcher, M. Pop, W. Wang, C.M. Fraser and J.C. Venter 2003. The dog genome: survey sequencing and comparative analysis. *Science* 301: 1898–1903.

Koblmüller, S., C. Vilà, B. Lorente-Galdos, M. Dabad, O. Ramirez, T. Marques-Bonet, R.K. Wayne and J.A. Leonard 2016. Whole mitochondrial genomes illuminate ancient intercontinental dispersals of grey wolves (*Canis lupus*). *Journal of Biogeography* 43: 1728–1738.

Larson, G. and J. Burger 2013. A population genetics view of animal domestication. *Trends in Genetics* 29: 197–205.

Leathlobhair, M.N., A.R. Perri, E.K. Irving-Pease, K.E Witt A. Linderholm, J. Haile, O. Lebrasseur C. Ameen, J. Blick, A.R Boyko, S. Brace, Y. Nunes Cortes, S.J . Guiry, A.J. Hansen, A. Hulme-Beaman, J. Johnson, A. Kitchen, A.K Kasparov, Young-Mi Kwon, P.A Nikolskiy, C. Peraza Lope, A. Manin, T. Martin, M. Meyer, K.N. Myers, Mark Omura, J.-M. Rouillard, E.Y Pavlova, P. Sciulli, M.-H.S. Sinding, A. Strakova, V.V. Ivanova, C. Widga, E. Willerslev, V.V. Pitulko, I. Barnes, M.T.P. Gilbert, K.M. Dobney, R.S. Malhi, E.P. Murchison, G. Larson and L.A.F. Frantz 2018. The evolutionary history of dogs in the Americas. *Science* 361: 81–85.

Leonard, J.A., R.K. Wayne, J. Wheeler, R. Valadez, S. Guillén and C. Vilà 2002. Ancient DNA evidence for Old World origin of New World dogs. *Science* 298: 1613–1616.

Leonard, J.A., C. Vilà, K. Fox-Dobbs, P.L. Koch, R.K. Wayne and B. Van Valkenburgh 2007. Megafaunal extinctions and the disappearance of a specialized wolf morph. *Current Biology* 17: 1146–1150.

Leonard, J.A. 2014. Ecology drives evolution in grey wolves. *Evolutionary Ecology Research* 16: 461–473.

Leonard, J.A., J. Echegaray, E. Randi and C. Vilà 2014. Impact of hybridization with domestic dogs on the conservation of wild canids, in M Gompper (ed.) *Free-Ranging Dogs and Wildlife Conservation,* Oxford University Press.

Lindblad-Toh, K., C. Wade, T. Mikkelsen, E.K. Karlsson, D.B. Jaffe, M. Kamal, M. Clamp, J.L. Chang, E.J. Kulbokas III, M.C. Zody, E. Mauceli, X. Xie, M. Breen, R.K. Wayne, E.A. Ostrander, C.P. Ponting, F. Galibert, D.R. Smith, P.J. deJong, E. Kirkness, P. Alvarez, T. Biagi, W. Brockman, J. Butler, C.-W. Chin, A. Cook, J. Cuff, M.J. Daly, D. DeCaprio, S. Gnerre, M. Grabherr, M. Kellis, M. Kleber, C. Bardeleben, L. Goodstadt, A. Heger, C. Hitte, L. Kim, K.-P. Koepfli, H.G. Parker, J.P. Pollinger, S.M.J. Searle, N.B. Sutter, R. Thomas, C. Webber, Broad Sequencing Platform members and E.S. Lander 2005. Genome sequence, comparative analysis and haplotype structure of the domestic dog. *Nature* 438: 803–819.

Mech, L.D. and L. Boitani 2010. *Wolves: behavior, ecology, and conservation.* University of Chicago Press.

Ostrander, E.A. and L. Kruglyak 2000. Unleashing the canine genome. *Genome Research* 10: 1271–1274.

Plassais, J., J. Kim, B.W. Davis, D.M. Karyadi, A.N. Hogan, A.C. Harris, B. Decker, H.G. Parker and E.A. Ostrander 2019. Whole genome sequencing of canids reveals genomic regions under selection and variants influencing morphology. *Nature Communications* 10: 1489.

Saetre, P., J. Lindberg, J.A. Leonard, K. Olsson, U. Petersson, H. Ellegren, T. Bergström, C. Vilà and E. Jazin 2004. From wild wolf to domestic dog: gene expression changes in the brain. *Molecular Brain Research* 126: 198–206.

Savolainen, P., Y. Zhang, J. Luo, J. Lundeberg and T. Leitner 2002. Genetic evidence for an East Asian origin of domestic dogs. *Science* 298: 1610–1613.

Shannon, L.M., R.H. Boyko, M. Castelhano, E. Corey, J.J. Hayward, C. McLean, M.E. White, M. Abi Said, B.A. Anita, N. Ikombe Bondjengo, J.Calero, A. Galov, M. Hedimbi, B. Imam, R. Khalap, D. Lally, A. Masta, K.C. Oliveira, L. Pérez, J. Randall, N. Minh Tam, F.J. Trujillo-Cornejo, C. Valeriano, N.B. Sutter, R.J. Todhunter, C.D. Bustamante and A.R. Boyko 2015. Genetic structure in village dogs reveals a Central Asian domestication origin. *Proceedings of the National Academy of Sciences of the United States of America* 112: 13639–13644.

Sharma, D.K., J.E. Maldonado, Y.V. Jhala and R.C. Fleischer 2004. Ancient wolf lineages in India. *Proceedings of the Royal Society B (Suppl.)* 271: S1-S4.

Skoglund, P., E. Ersmark, E. Palkopoulou and L. Dalén 2015. Ancient wolf genome reveals an early divergence of domestic dog ancestors and admixture into high-latitude breeds. *Current Biology* 25: 1515–1519.

Snyder, L.M. and J.A. Leonard 2006. Dog, in W.C. Sturtevant (ed.) *Handbook of North American Indians,* D. Stanford, B.D. Smith, D.H. Ubelaker and E.J.E. Szathmáry (eds) vol. 3 *Environment, Origins, and Population.* Smithsonian Institution.

Sundqvist, A.-K., S. Björnerfeldt, J.A. Leonard, F. Hailer, Å. Hedhammar, H. Ellegren and C. Vilà 2006. Unequal contribution of sexes in the origin of dog breeds. *Genetics* 172, 1121–1128.

Thalmann, O., B. Shapiro, P. Cui, V. J. Schuenemann, S.K. Sawyer, D.L. Greenfield, M.B. Germonpré, M.V. Sablin, F. López-Giráldez, X. Domingo-Roura, H. Napierala, H.-P. Uerpmann, D.M. Loponte, A.A. Acosta, L. Giemsch, R.W. Schmitz, B. Worthington, J.E. Buikstra, A. Druzhkova, A.S. Graphodatsky, N.D. Ovodov, N. Wahlberg, A.H. Freedman, R.M. Schweizer, K.-P. Koepfli, J.A. Leonard, M. Meyer, J. Krause, S. Pääbo, R.E. Green, R.K. Wayne 2013. Complete mitochondrial genomes of ancient canids suggest a European origin of domestic dogs. *Science* 342: 871–874.

Vaysse, A., A. Ratnakumar, T. Derrien, E. Axelsson, G. Rosengren Pielberg, S. Sigurdsson, T. Fall, E.H. Seppälä, M.S.T. Hansen, C.T. Lawley, E.K. Karlsson, LUPA Consortium; D. Bannasch, C. Vilà, H. Lohi, F. Galibert, M. Fredholm, J. Häggström, A. Hedhammar, C. André, K. Lindblad-Toh, C. Hitte, M.T. Webster 2011. Identification of genomic regions associated with phenotypic variation between dog breeds using selection mapping. *PLoS Genetics* 7: e1002316.

Vilà, C., P. Savolainen, J.E. Maldonado, I.R. Amorim, J.E. Rice, R.L. Honeycutt, K.A. Crandall, J. Lundeberg and R.K. Wayne 1997. Multiple and ancient origins of the domestic dog. *Science* 276: 1687–1689.

Vilà, C., J. Seddon and H. Ellegren 2005. Genes of domestic mammals augmented by backcrossing with wild ancestors. *Trends in Genetics* 21: 214–218.

vonHoldt, B.M., J.P. Pollinger, K.E. Lohmueller, E. Han, H.G. Parker, P. Quignon, J.D. Degenhardt, A.R. Boyko, D.A. Earl, A. Auton, A. Reynolds, K. Bryc, A. Brisbin, J.C. Knowles, D.S. Mosher, T.C. Spady, A. Elkahloun, E. Geffen, M. Pilot, W. Jedrzejewski, C. Greco, E. Randi, D. Bannasch, A. Wilton, J. Shearman, M. Musiani, M. Cargill, P.G. Jones, Zuwei Qian, Wei Huang, Zhao-Li Ding, Ya-ping Zhang, C.D. Bustamante, E.A. Ostrander, J. Novembre and R.K. Wayne 2010. Genome-wide SNP and haplotype analyses reveal a rich history underlying dog domestication. *Nature* 464: 898–902.

Zeder, M.A. 2015. Core questions in domestication research. *Proceedings of the National Academy of Sciences of the United States of America* 112: 3191–3198.

Zeder, M.A., D. Decker-Walters, D. Bradley and B.D. Smith (eds) 2006. *Documenting Domestication: New Genetic and Archaeological Paradigms*. California University Press.

# 1.2 Mitochondrial DNA Variation among Dogs of Mongolian, Tuvinian and Altaic Nomads

Daria Sanna[1], Ilenia Azzena[1,2], Piero Cossu[2], Fabio Scarpa[1], Massimo Scandura[2], Marco Apollonio[2], Francesca Lugli[3], Paolo Francalacci[4], Paolo Mereu[1], Marco Casu[2]

[1] Dipartimento di Scienze Biomediche, University of Sassari, Viale San Pietro 43/b, 07100 Sassari, Italy, darsanna@uniss.it, fscarpa@uniss.it, pmereu@uniss.it

[2] Dipartimento di Medicina Veterinaria, University of Sassari, Via Vienna 2, 07100 Sassari, Italy, iazzena@uniss.it, picossu@uniss.it, scandura@uniss.it, marcoapo@uniss.it, marcasu@uniss.it

[3] Associazione Italiana di Etnoarcheologia, Via Principe Umberto 41, 00185 Roma, Italy, luglifrance@gmail.com

[4] Dipartimento di Scienze della Vita e dell'Ambiente, University of Cagliari, Via Ing. Tommaso Fiorelli 1, 09126 Cagliari, Italy, paolo.francalacci@unica.it

Corresponding author: Daria Sanna· darsanna@uniss.it

**Abstract**

Dogs originated from the domestication of Eurasian grey wolves. From a genetic viewpoint, they can be grouped into two main clusters: the first is represented by several breeds obtained by artificial selection, whereas the second is of dogs that adapted to a human commensal lifestyle. Here we have provided a molecular survey aimed to infer on the genetic variability of dogs from nomadic camps in Mongolia, and the Republics of Tuva and Altai belonging to the Russian Federation. The results provided evidence of typical marks of expanding populations with multiple origins. Such a scenario could be the result of genetic exchanges among dogs from different camps, that were likely mediated by nomads.

**Keywords:** *Canis familiaris*, mtDNA, control region, genetic variability, Mongolia, Siberia.

## 1 Introduction

The dog (*Canis familiaris*) was the first domesticated species, which likely originated in the upper Paleolithic from the domestication of the Eurasian grey wolf (*Canis lupus*) (*i.a.* Shannon *et al.* 2015). From a genetic viewpoint, domestic dogs can be ranked into two main, highly divergent, groups: the first is represented by a large variety of pure breeds obtained by means of human-mediated artificial selection; the second encompasses large and strongly diversified groups of free-ranging animals adapted to a human commensal lifestyle (the so-called village dogs). Genetic data collected worldwide support a single geographical origin for domesticated dogs. In this context, the supposed first centre of domestication is located in Central Asia, as suggested by the highest levels of genetic variation that are generally reported in populations from this region.

Mitochondrial DNA (mtDNA) molecular markers were extensively used to infer on the phylogenetic relationships among canine populations distributed throughout the world. The mtDNA is a separate genome located inside cytoplasmatic organelles (the mitochondria) in all eukaryotic cells (Anderson *et al.* 1981). It is a small circular molecule, which is present in multiple copies per cell and is inherited maternally. Savolainen *et al.* (1997) described the occurrence of two highly informative, hypervariable regions (HVS-I and HVS-II) in the canine mtDNA. Pang *et al.* (2009) used these mtDNA regions to analyse 1,543 dogs spread across the Old World, evidencing the presence of six phylogenetic mitochondrial haplogroups (*i.e.* groups of similar sequences that share a common ancestor), named as clades A-F. Clades A, B and C occur at high frequencies in every canine population, suggesting the hypothesis of a possible common origin of these groups from a single domestication event. Conversely, the clades D, E and F showed a limited geographical dispersal and low frequencies of distribution.

The dogs of nomadic populations that live in areas near to the first centre of wolf domestication are generally poorly influenced by foreign gene flows and might show peculiar genetic traits that deserve to be investigated (Irion *et al.* 2005; Boyko *et al.* 2009; Pedersen *et al.* 2013; Shannon *et al.* 2015). In nomad camps, dogs are fundamental to protect livestock against wolves and predators. Therefore, nomads usually pick up dogs when they are puppies, preferring the bravest cubs with the strongest physical structures and peculiar morphological features, such as specific coat colours (Lugli 2016).

Current Mongolian and Siberian pastoralism can be considered the result of a multi-millenary process which started in prehistoric times. Current nomadism has had to face modernity and its market and social

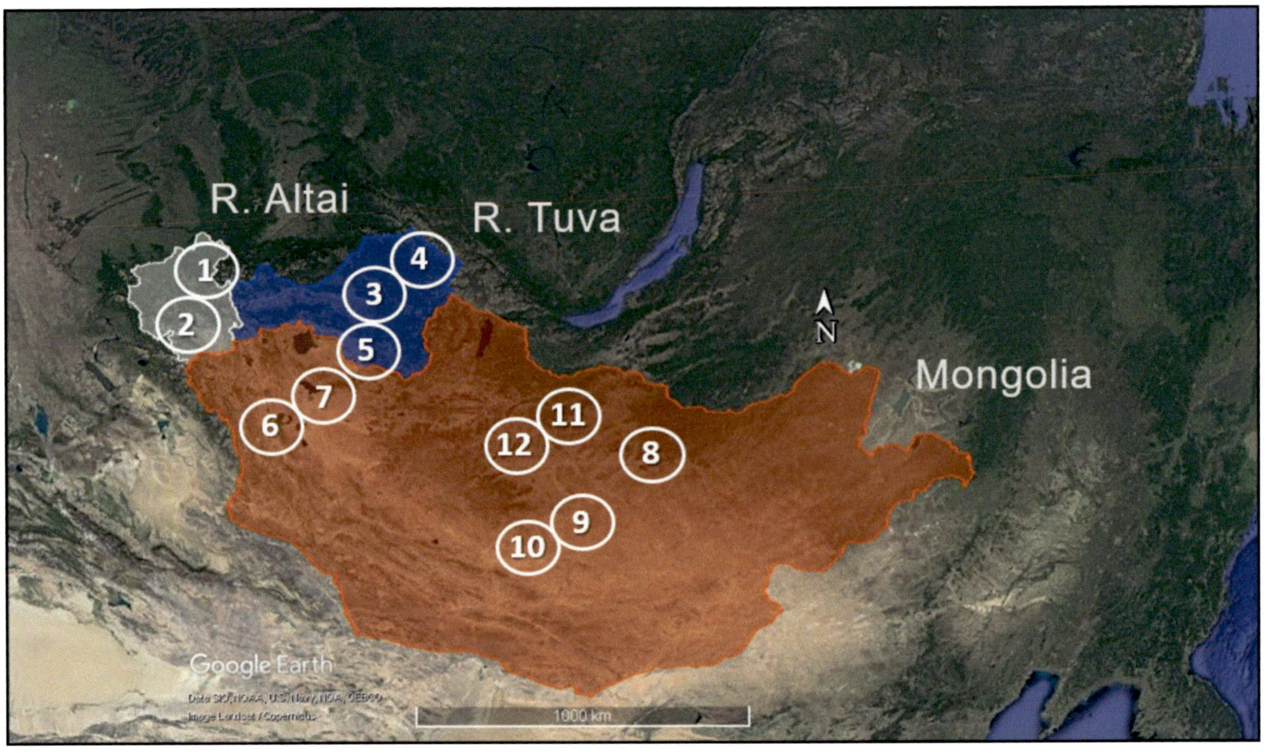


Figure 1. Sampling map showing the countries and the areas where dog hair samples were collected for DNA extraction. Arabic numbers inside white circles indicate the sites where sampling was performed. Republic of Altai:(1) Kurmach Baygol and (2) Kulada. Republic of Tuva: (3) Kyzyl, (4) Systig-Hem, Ador Kezhik, Toora Hem, (5) Erzin region. Mongolia: (6) Khovd aimag, (7) Zavkhan aimag, (8) Ulan Bator, (9) Delgerkhangai (Dundgovi aimag), (10) Övörkhangai aimag, (11) Mogod, Ulziit, Tal Bulag, Tsagaan Khust (Bulgan aimag), (12) Jargalant, Erdenet Mandal, UndurUlaan, BayyanUul (Arkhangai aimag).

models which push young people to abandon their traditional lifestyle.

In such a context, here we have provided a preliminary molecular survey, performed by means of the mitochondrial hypervariable region I (HVS-I) marker, aimed to infer for the first time on the genetic variability and the phylogeographic relationships among village dogs from nomadic camps of rural areas of Mongolia, and the Republics of Tuva and Altai (the Russian Federation).

## 2 Material and methods

In 2005 the Italian Association for Ethnoarchaeology (AIE) with the sponsorship of the Ministry of Foreign Affairs and International Cooperation – Italy (MFA) started the research project 'The camps of Mongolian nomads: an ethnoarchaeological perspective', which aimed to study the various crucial points of steppe pastoralism in different Mongolian ecosystems.[1]

This project also included research on dogs, which started in Mongolia in 2012. Consequently, the Italian Association for Ethnoarchaeology (AIE) with the sponsorship of Ministry of Foreign Affairs and International Cooperation – Italy MFA, started the mission 'Siberian nomads and their dogs' in 2013, which was conducted in the Republics of Tuva (2013, 2017), and Altai (2014, 2016) (the Russian Federation), and in the Kemerovo region (2015) in collaboration with the Novosibirsk State Conservatory, the Institute of Philology (SB RAS, Novosibirsk), the Institute for Humanities and Kyzyl College of Arts (RT), and the Institute of Altaistics (RA).[2]

The Mongolian and Siberian projects aimed to study and document the relationship between nomads, hunters and dogs in various socio-historical contexts and different ecosystems. Villages and camps of various regions were visited in order to document traditional situations. Hair samples from dogs owned by the families that were studied and interviewed were taken during the missions both in Mongolia and in Siberia.

[1] The project was conducted in collaboration with Tserenkhand (2002–2006) (National University of Ulaanbaatar; Academy of Sciences) and Dulam Sedenjav (after 2007) (National University of Ulaanbaatar). The missions were conducted by F. Lugli and G. Capitini and students and graduated of the National University. The research was conducted in various regions to register differences and variabilities.

[2] The project was conducted in Tuva (2013, 2017), Republic of Altai (2014, 2016), and Kemerovo region (2015) by F. Lugli and G. Sychenko (see Lugli and Sychenko in this volume) in collaboration with Novosibirsk State Conservatory, Institute of Philology (SB RAS, Novosibirsk), Institute for Humanities and Kyzyl College of Arts (RT), Institute of Altaistics (RA).

Table 1. Estimates of genetic diversity obtained for the mitochondrial HVS-I fragment of dog populations analysed here. n: sample size, S: number of polymorphic sites; H: number of haplotypes; h: haplotype diversity; and π: nucleotide diversity.

| Country | Sampling date | n | S | H | *h* | π |
|---|---|---|---|---|---|---|
| Mongolia | Total | 33 | 16 | 14 | 0.888 | 0.01328 |
| | Nov. 2013 | 14 | 15 | 9 | 0.923 | 0.01279 |
| | Oct. 2014 | 14 | 9 | 6 | 0.813 | 0.01137 |
| Republic of Tuva | Oct. 2013 | 23 | 18 | 12 | 0.917 | 0.01094 |
| Republic of Altai | Total | 23 | 11 | 11 | 0.806 | 0.01118 |
| | Oct. 2016 | 15 | 8 | 6 | 0.648 | 0.00832 |
| | Jul. 2014 | 8 | 10 | 7 | 0.964 | 0.01252 |
| **Total** | | 79 | 24 | 24 | 0.908 | 0.01276 |

Figure 2. Some of the individuals, from Mongolia, the Republic of Altai and the Republic of Tuva, whose hairs were collected to perform non-invasive DNA extractions in the present study (Photos by F. Lugli).

There was usually a preference to take the samples in traditional and isolated contexts. A few samples were also taken from a dog farm in Ulaanbaatar in Mongolia in order to analyse the Mongol Bankhar mastiffs.

The molecular analysis of a 348 base pairs-long HVS-I mitochondrial fragment was performed on 79 dogs from seven sites in Mongolia (33 individuals), and five sites from two states of the Russian Federation, being Republic of Tuva (23 individuals from two sites) and Republic of Altai (23 individuals from three sites) (Figure 1 and Table 1 for details). The individuals analysed included not only non-breed dogs, but also representatives of three canine breeds (Laika, Mongol

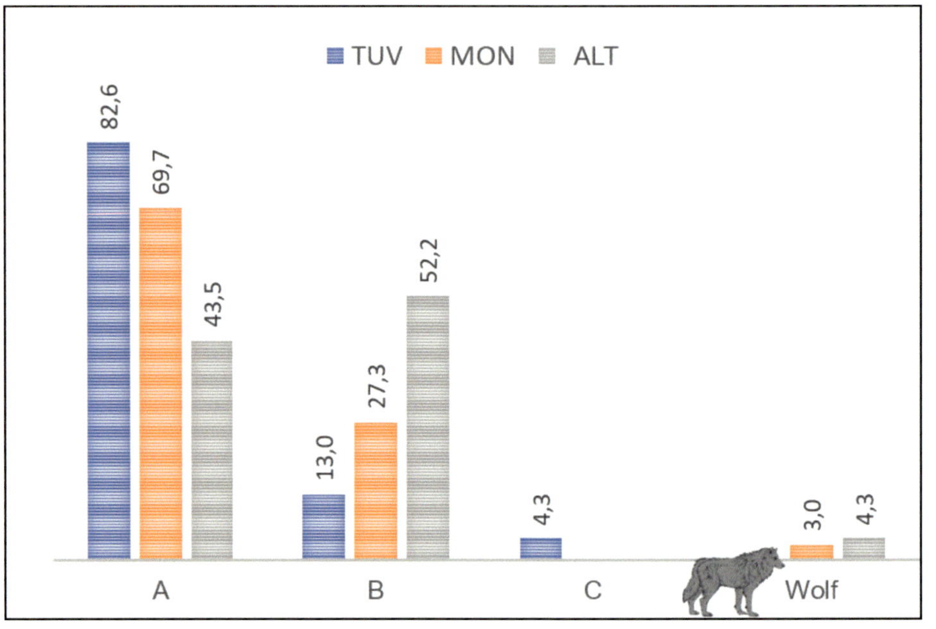


Figure 3. Distribution frequencies of the canine mitochondrial HVS-I haplogroups found for the dogs analysed here. MON: Mongolia; ALT: Republic of Altai; TUV: Republic of Tuva.

Bankhar, and Volkodav) (see Figure 2). Hair samples were collected from individuals with the help of their owners during 2013, 2014, and 2016. Genomic DNA was extracted from hairs by means of the InstaGene™ Matrix (Bio-Rad) according to the manufacturer's protocol. Sample quality and DNA concentration were determined via spectrophotometry using a ND-8000 (NanoDrop Technologies, Thermo Fisher Scientific Inc., Wilmington, DE). PCR was carried out using modified mtDNA HVS-I universal primers (Kocher *et al.* 1989) and sequencing was performed by an external service (Macrogen, The Netherlands). Newly generated sequences were aligned using the BioEdit 7.2.5. software package (Hall, 1999). The genetic variation was assessed estimating the number of polymorphic sites, the number of haplotypes (H), the haplotype diversity (*h*), and the nucleotide diversity ($\pi$) using the software package DnaSP 6.12 (Librado *et al.* Rozas, 2009). Genetic relationships among haplotypes were investigated by a Median-Joining network using the software package Network 10.2.0.0 (www.fluxus-engineering.com).

### 3 Results and discussion

High levels of genetic variability, resulting in a total of 24 haplotypes (*i.e.* nucleotide sequences corresponding to the same DNA fragment that differ among each other in some informative nucleotide sites), were found at each sampling location (see Table 1 for details). The lowest levels of genetic variation were found in the camps from the North of the Republic of Altai (Turochaksky district). In the present study, the two main worldwide distributed canine mitochondrial haplogroups (A, B) showed distributions of frequencies (Figure 3) that are

Figure 4. Images of the Laika-like individuals whose mitochondrial HVS-I sequence was also found in Siberian wolves. (a) a 3-year-old male from Mongolia; (b) a 1-year-old female from the Republic of Altai (Photos by F. Lugli).

consistent with those generally retrieved for almost all dog populations from the Old World (Savolainen *et al.* 2002 ; Ardaland *et al.* 2011; and references therein), with the haplogroup A showing the highest frequencies. Dogs from the Republic of Altai were the only exception as they showed the highest frequencies of distribution for the haplogroup B, in accordance with a similar trend already reported for Southwest Asian dogs (Ardalan *et al.* 2011). We hypothesise this latter finding may be a consequence of genetic drift mediated by human artificial selection. This evolutionary force may have acted on the Altaic population with repeated introductions of dogs carrying less common haplogroups just by chance. The haplogroup C was found only in dogs from the Republic of Tuva, and the genetic drift may be invoked in this case as well. As briefly outlined above, genetic drift is a stochastic evolutionary force whose strength depends on the population size. When a new population originates from a very small number of individuals (the so-called founders), genetic drift may trigger the loss of genetic variation changing the frequencies of haplotypes. As a consequence, the distribution of haplogroups in the new established populations may diverge from those reported for the original population.

It is noteworthy that a mitochondrial lineage, that is present in Siberian wolves, was found in two morphologically Laika-like individuals, one from Mongolia (a 3-year old male from the district of Bulgan) and one from the Republic of Altai (a 1-year-old female from the district of Kurmach Baygol) (Figure 4). Such a finding may be the result of past accidental domestications of wolf females or cubs in nomad camps. Indeed, although nomads usually prefer to select dogs directly from their canine families, they do not always follow this choice criterion and puppies can also be rescued from stray mothers or lost adults can be adopted (Lugli 2016).

However, it should be taken into account that the uncommon mitochondrial lineage found in these two individuals may also belong to the mitochondrial canine clade D, whose distribution is restricted to North Europe, Siberia, Southwest Asia and the Mediterranean Sea (Angleby and Savolainen 2005; Pang *et al.* 2009). Some sub-haplogroups of this clade are the products of a dog-wolf cross-breeding, rather than of independent domestication of wolves (Ardalan *et al.* 2011).

The network analysis (see Figure 5a and its legend for more details) evidenced that many sequences were exclusively found in single individuals, probably due to very recent multiple introductions of new dogs. The occurrence of star-like configurations in the plot suggests the lack of genetic divergence among areas, along with the occurrence of many founder effects. Indeed, here the network star-like configurations are represented by a common central haplotype, usually shared among individuals from many regions, that is surrounded by many lesser-frequent (and private to single individuals) haplotypes differing by a few mutations. The most common haplotypes in the network likely correspond to sequences belonging to the first dogs introduced in the nomads'camps, which had the opportunity to breed extensively. Such findings are consistent with the general trend of genetic homogeneity worldwide reported for dog populations (*i.a.* Pang *et al.* 2009).

A less frequent and highly divergent haplotype (Figure 5b) was found exclusively among dogs from a Mongol Bankhar breeding farm in Mongolia: we hypothesised the occurrence of a mtDNA matrilinear relationship among all individuals born in the farm, which likely descend from a group of related females.

The network analysis also evidenced that two Volkodav dogs from the Republic of Tuva show private-owned haplotypes (not shared with the other breeds). This finding suggests that the genetic divergence reported for these dogs is likely consistent with the different history and geographic origin of their breed.

## 4 Conclusions

The study of the genetic makeup of village dogs and central Asian local breeds represents an important step to depict the complex evolutionary history of these animals (Shannon *et al.* 2015).

In such a context, we have reported the first and preliminary molecular inference on dogs from the mountains of Mongolia and from the Russian Republics of Altai and Tuva. The results pointed out high levels of genetic divergence at each sampling site, and a lack of geographic differentiation among regions. Our findings reflect the typical marks of expanding populations with multiple origins. We hypothesise that such a scenario could be the result of repeated genetic exchanges among dogs from different nomads'camps, which were likely mediated by human activities. Indeed, Mongolian nomads usually move to villages to pick their dogs (Lugli 2016). Accordingly, the haplotype distribution frequencies and the founder effects evidenced by the network analysis, further account for the signature of artificial selection, which drastically skewed the genetic diversity within village dogs and local breeds such as the Mongol Bankhar mastiff. Within this framework, it should also be considered that the mitochondrial genetic variability reported for village dogs from nomadic camps may be sex-biased because of the maternal inheritance of the mitochondrial molecular marker here used. In the present study,

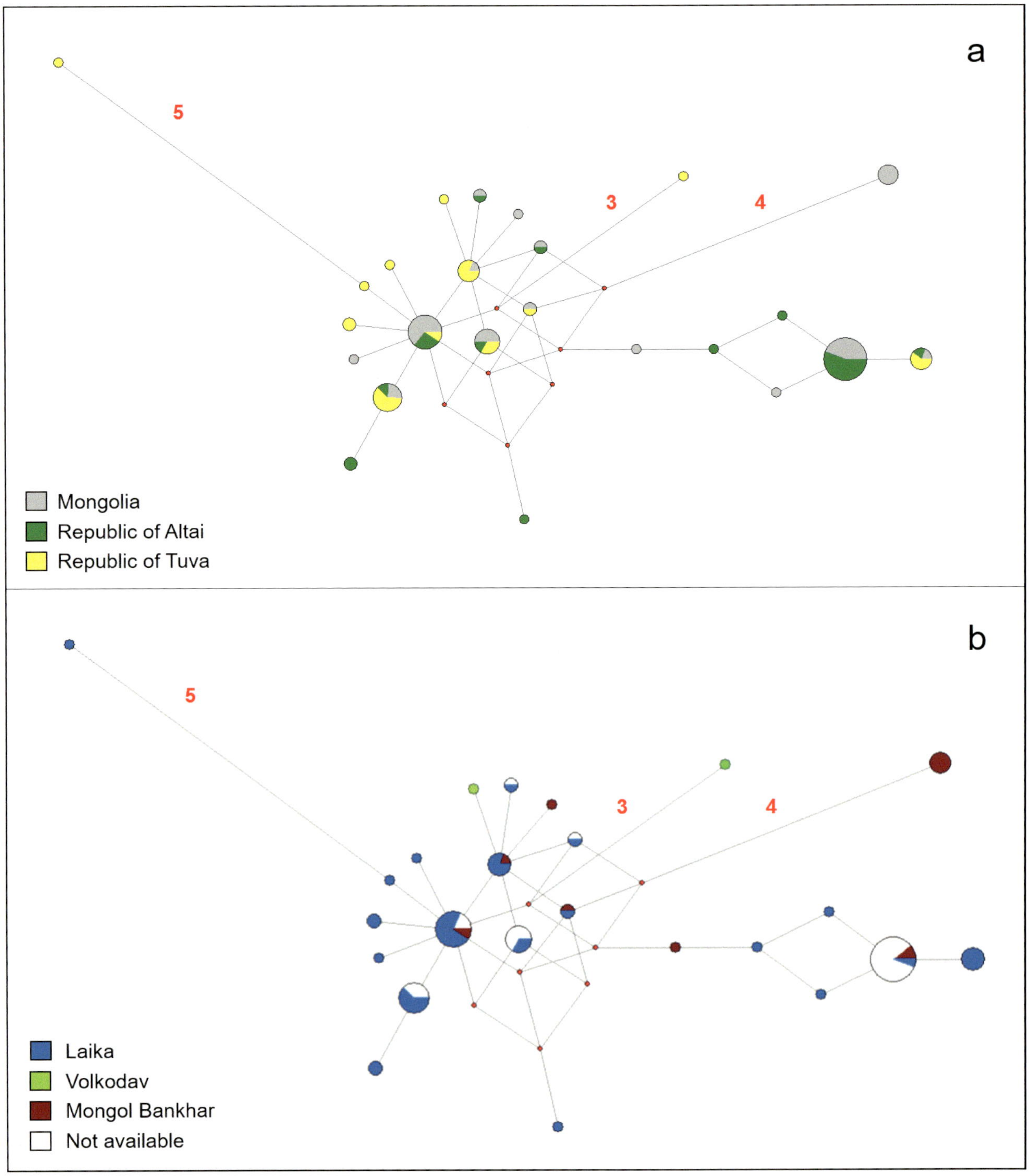


Figure 5. Median-Joining networks showing the phylogenetic relationships occurring among the mitochondrial haplotypes found in the present study. Haplotypes are represented by the circular spots on the graphic; the diameter of spots is proportional to the number of individuals that share the haplotype; the length of branches in the graphic are proportional to the number of nucleotide mutations occurring between the two haplotypes at the edges of the branch. The spots are coloured according to the characteristics of the individuals that share the haplotype. Little red spots in the graphic represent the median vectors that are crucial sequences likely existing in nature but not found among the specimens collected for the present study. In the present graphic all haplotypes diverge from each other for a single mutation except for the cases where Arabic red numbers on network branches indicate the occurrence of a higher number of point mutations. (a) the spots on the network are coloured according to the geographic origin of dogs; (b) the spots on the network are coloured according to the breed of dogs. Please note that it was not possible to identify the breed for all individuals.

the general trend of genetic homogeneity evidenced among areas may be the result of the sex-based choice criterion that nomads apply to pick their dogs. For example, 99% of Mongolian nomads' families decide to only have male dogs because they consider females too difficult to manage during their oestrus cycle (see Lugli 2016 for more details). As a consequence, the common ancestors of the dogs considered in the present study could be represented by a reduced number of females that were likely used as breeders. This choice might has decreased the effective population size (*i.e.* the number of mating individuals that contribute to the genetic pool of the next generation) and increased the genetic homogeneity evidenced by the mitochondrial DNA, which is matrilineally transmitted to the offspring.

The main mitochondrial haplogroups found in the present study among dog populations suggest a recent origin, common to other European canine populations.

In the future, the analysis of a larger number of individuals from further Mongolian, Altaic and Tuvinian sites, will shed further light on the evolutionary processes that might have shaped the genetic patterns of dog populations living in these Asiatic regions.

### References


Anderson, S., A.T. Bankler, B.G. Barrel, M.H.L. de Bruijn, A.R. Cuolson, J. Drouin, I.C. Eperon, D.P. Nierlich, B.A. Roe, F. Sanger, P.H. Schreier, A.J.H. Smith, R. Staden and I.G. Young 1981. Sequence and organization of the human mitochondrial genome *Nature* 290: 457–464.

Angleby, H. and P. Savolainen 2005. Forensic informativity of domestic dog mtDNA control region sequences. *Forensic Science International* 154(2–3): 99–110.

Ardalan, A., C.F. Kluetsch, A.B. Zhang, M. Erdogan, M. Uhlén, M. Houshmand, C. Tepeli, S.Z. Miraei Ashtiani and P. Savolainen 2011. Comprehensive study of mtDNA among Southwest Asian dogs contradicts independent domestication of wolf, but implies dog-wolf hybridization. *Ecology and Evolution* 1(3): 373–385.

Boyko, A.R., R.H. Boyko, C.M. Boyko, H.G. Parker, M. Castelhano, L. Corey, J.D. Degenhardt, A. Auton, M. Hedimbi, R. Kityo, E.A. Ostrander, J. Schoenebeck, R.J. Todhunter, P. Jones, and C.D. Bustamante 2009. Complex population structure in African village dogs and its implications for inferring dog domestication history. *Proceedings of the National Academy of Sciences* 106(33): 13903–13908.

Hall, T.A. 1999. BioEdit: a user-friendly biological sequence alignment editor and analysis program for Windows 95/98/NT. *Nucleic Acids Symposium Series* 41: 95–98.

Irion, D.N., A.L. Schaffer, S. Grant, A. N. Wilton and N.C. Pedersen, 2005. Genetic variation analysis of the Bali street dog using microsatellites. *BMC Genetics* 6: 6.

Kocher, T.D., W.K. Thomas, A. Meyer, S.V. Edwards, S. Pääbo, F. Villablanca and A.C. Wilson 1989. Dynamics of mitochondrial DNA evolution in animals: amplification and sequencing with conserved primers. *Proceedings of the National Academy of Sciences of the USA* 86: 6196 - 6200.

Librado, P. and J. Rozas 2009. DnaSP v5: a software for comprehensive analysis of DNA polymorphism data. *Bioinformatics* 25: 1451–1452.

Lugli, F. 2016. Mongolian nomads and their dogs, in S. Biagetti and F. Lugli (eds) *The Intangible Elements of Culture in Ethnoarchaeological Research*: 125–139. Springer: Cambridge.

Pang, J-F., C. Kluetsch, X-J. Zou, A-B., Zhang, L.Y. Luo, H. Angleby, A. Ardalan, A., C. Ekström, A. Sköllermo, J. Lundeberg, S. Matsumura, T. Leitner, Y.-P. Zhang and P. Savolainen 2009. mtDNA data indicate a single origin for dogs south of Yangtze River, less than 16,300 years ago, from numerous wolves. *Molecular Biology and Evolution* 26(12): 2849–2864.

Pedersen, N., H. Liu, G. Theilen and B. Sacks 2013. The effects of dog breed development on genetic diversity and the relative influences of performance and conformation breeding. *Journal of Animal Breeding and Genetics* 130(3): 236–248.

Savolainen, P., B. Rosen, A. Holmberg, T. Leitner, M. Uhlen and J. Lundeberg 1997. Sequence analysis of domestic dog mitochondrial DNA for forensic use. *Journal of Forensic Science* 42(4): 593–600.

Savolainen, P., Y.P. Zhang, J. Luo, J. Lundeberg and T. Leitner 2002. Genetic evidence for an East Asian origin of domestic dogs. *Science* 298(5598): 1610–1613.

Shannon. L.M., R.H. Boyko, M. Castelhano, E. Corey, J.J. Hayward, C. McLean, M.E. White, M. Abi Said, B.A. Anita, N.I. Bondjengo, J. Calero, A. Galov, M. Hedimbi, B. Imam, R. Khalap, D. Lally, A. Masta, K.C. Oliveira, L. Pérez, J. Randall, N.M. Tam, F.J. Trujillo-Cornejo, C. Valeriano, N.B. Sutter, R.J. Todhunter, C.D. Bustamante and A.R. Boyko 2015. Dog genetic diversity reveals Central Asian origin. *Proceedings of the National Academy of Sciences* 112.44: 13639–13644.

# 1.3 Ancient and Recent Changes in Breeding Practices for Dogs

Grégoire Leroy[1], Shi-Zhi Wang[2], Tom Lewis[3,4], Sophie Licari[5]

[1] GABI, INRAE, AgroParisTech, Université Paris-Saclay, 78350 Jouy-en-Josas, France.
gregoireleroyp@gmail.com

[2]State Key Laboratory of Genetic Resources and Evolution, Kunming Institute of Zoology, Chinese Academy of Sciences, Kunming, 650201, China. wangshizhi@mail.kiz.ac.cn

[3]The Kennel Club, Clarges Street, London, UK. Thomas.Lewis@thekennelclub.org.uk

[4]School of Veterinary Medicine and Science, The University of Nottingham, Sutton Bonington Campus, Leicestershire, UK.

[5]Journalist. Société Centrale Canine, 155 avenue Jean-Jaurès, 93300 Aubervilliers, France. sophielicari3@gmail.com

Corresponding author: Grégoire Leroy, gregoireleroyp@gmail.com

**Abstract**

Since its domestication, the dog species has been shaped by man to fulfil a diversity of functional as well as social roles. In this review, we investigate how socio-cultural and technical development have influenced its differentiation into a numerous number of more or less specialised morphotypes, and then, after the 19th century, into highly differentiated and homogenous breeds. Current dog breeding practices are characterised by features, which are more or less specific to the species and may have consequences on the health of dogs. Increasing concerns on welfare and health, combined with recent technological genomic and information breakthroughs are expected to affect the practices and governance of dog breeding, and drive the future evolution of dog breeds.

**Keywords:** dog, breeding practices, health, genomics.

## 1 Introduction

As a domestic species, the dog occupies a unique place regarding its relationship with man. Aside from being the first domesticated species, dogs largely share the same environment and food as humans, and is probably the mammal species that exhibits the largest diversity in terms of morphological features and uses (Sutter *et al.* 2008). This diversity is the product of a long and rich history of artificial selection across different geographical locations, from its early domestication to the creation of modern breeds in the Victorian era (Akey *et al.* 2010, vonHoldt *et al.* 2010). Dog breeding has long been a practice guided by the optimisation of functions devolved to the dog, applying for this purpose, opportunistically and empirically, a selective pressure on hereditary characteristics. The canine species evolved, from the ancestral morphotypes to a multiplicity of phenotypes, because the continuous search for performance in the accomplishment of these functions led to specialised forms. The cynophilists' work can be considered the culmination of the process, compartmentalising the species into many autonomous reproduction units, normalising for each the desired characteristics and the limits of the possible variations. According to the Fédération Cynologique Internationale (FCI), there are around 350 dog breeds currently recognised worldwide. Such diversity relates to various breeding objectives and practices, which may have also impacted the health of purebred dogs, especially regarding the dissemination of inherited disorders within breeds (Rooney and Sargan, 2010). In 2018, around 308 canine Mendelian disorders and traits were reported in the Online Mendelian Inheritance in Animals database (OMIA, 2018). In parallel with increasing concern among the public and stakeholders towards the welfare of domesticated species and companion animals in particular (Rooney and Sargan, 2009), recent technological developments, such as the genomic revolution, are likely to completely change the way humans select and pair breeding animals (Mellersh 2012; Stock and Reents 2013).

In this paper, we appraise the specifics and evolution of breeding practices in the domestic dog, based on a review of literature. We particularly focus on the most recent drivers of change and their consequences in terms of the challenges and prospects they present for the future of dog breeding.

## 2 Differentiation of specialised morphotypes from domestication to antiquity

The emergence of the dog is the result of the commensalism of small populations of wolves able to support the proximity of sedentary human settlements. Human communities integrated these canids into a later phase of voluntary domestication and then cynegetic use. The resulting primitive dog retained its conformation for thousands of years (medium size, mediolineal and mesencephalic format), such a homogeneity being due to a low selective pressure (Crockford 2006).

The process of differentiation began in the Ancient Near East (including Egypt), in the 5th-4th millennia BC, linked to the emergence of early civilisations. Two specialised morphotypes emerged from the optimisation of two functions: sight hunting and the protection off herds against predators. They were thus oriented in opposite directions: for the graioid (the sighthound), a thin body to gain speed, and for the molossoid, horizontal thickening to gain power. Some functional behavioural differences were also selected. Gradual variations provided the material for intentional selection. But with the canine populations not living in reproductive isolation, this selection was essentially post-zygotic, correcting the effects of unsatisfactory mating by the sorting of subjects to keep.

In the pre-Roman Antiquity, we have evidence of what is perhaps an organised breeding strategy, according to Herodotus (Hdt,. *I, 192*): four villages of Babylonia raised 'Indian' dogs for the King of Persia. For Herodotus, India extended to present-day Pakistan; and the term 'Indian dogs' referred to molossers. We can note that, according to the iconography, very powerful molossers existed in the Near and Middle Eastern Antiquity. Large-sized subjects suggest a sustained selective pressure.

It is important to avoid any anachronistic terminology. In his *History of the Animals* (*HA.*), Aristotle writes that the 'kinds' of dogs are numerous, using the term γένος (*genos*), which can be translated as 'breed' for convenience, but which does not cover the modern concept of breed. Aristotle also uses it to designate animal species as a whole. Like other authors in the period, it may be hypothesised that Aristotle uses this term with reference to the various intermediate variations presented by the canine species (Licari 2006: 9).

In Antiquity, pretty dogs seemed to be a social status marker, which suggests that selection is likely to produce, even occasionally, subjects of great value. According to Plutarch (Plut. *Alc, 9*), the Athenian general Alcibiades (5th century BC) possessed a dog of a remarkable size and beauty acquired at the exorbitant price of 70 minas. The fact that the social elites could have been attracted to the most differentiated dogs, could have resulted in the consolidation or even radicalisation of some specialised morphotypes. In imperial China, the aristocratic caste raised small pet dogs quite early. Some texts (*ErhYa*, 3rd century BC, *YiZhou Shu*, 4th century BC) mentioned '*short dogs*'. Some funerary statuettes from the Han period (256 BC -AD 220) represented small subjects with short snouts (Licari, 2012). Finally, we note the existence, mentioned by Roman authors and attested by archeology, of dwarf dogs among the Gauls. In this cynophagous society, they were not consumed, which suggests that they were valued.

According to the archaeological data, over the centuries and in the zone of Roman influence, a thrust was given to the differentiation of the species. The variability of the sizes, as implied by osteological markers of breed-type differentiation, increased strongly; dwarf dogs, and other subjects with large size or short snouts, became quite common. This evidence for the repeated occurrence of distinct morphotypes can be interpreted as the emergence of true dog breeding (Crockford 2000). The concern for selection is present among the cynegetic and agronomist authors, but it is based on a vast mixing from which proceeds this explosion of variability attested by archeology. They commonly recommended to cross varieties of different countries of the Roman world, to obtain the most effective auxiliaries. In the huge melting-pot of the empire, canine genes circulated. In addition, the authors provided advice for the choice of stud dogs that would not be disavowed nowadays, tending to show that if the selection is probably mostly post-zygotic, prezygotic selection was also be practiced. The idea of breed purity was present: Oppian (*Cyn.* I) advised not to cross the '*excellent breeds*' in order to preserve their qualities. Symmachus (Symm. *Ep. IV, 18*) mentioned the fact that some dog genealogies were established.

## 3 Breeding practices in the Middle Ages and Renaissance

As hunting was well anchored in the cultural norms of premedieval and medieval societies, the selection of associated varieties seems to have developed; the legal collections contain various mentions of hounds and falconry dogs (Licari 2010a, 2010b). The hound was of major importance in the cynegetic activity of the social elite, the number and abilities of packs constituting an aristocratic marker. The hound was subdivided into several formats, from the achondroplasic basset to the dog of great venery, and its head type radicalised towards a model with cutaneous laxity. The fact that the lips and ears drag to the ground on either side of the dog's nose as it lowers it to the ground would enhance

Figure 1. Dog diseases and health care of dogs. Livre de la Chasse, Gaston 'Phébus' comte de Foix, ms 616 folio 40, XVe siècle, Paris, BNF, Département des manuscrits.

its olfactory abilities, swirling the odorant molecules and guiding them towards the nostrils (Jensen 2007).

The breeding of the seigneurial packs tended to have much in common with modern practices. The feudal elite had the means to keep and house many dogs, and to dedicate staff to them. The description of kennels and daily care by the count of Foix Gaston 'Phoebus' (*Livre de la Chasse*, 1388) testifies to the comfort of their life conditions (Figure 1). The packs seemed to live in relative reproductive isolation, which accelerated the process of breed-type differentiation within them. Phoebus mentions the existence of an abortive medicine, dangerous and therefore probably rarely used, for the unwanted coverings, which suggests that the notion of prezygotic selection is not foreign to him. He insists on the fact that a good dog '*comes from a good father and a good mother*'. Phoebus describes the characteristics of several '*manners*' of dogs: the hound, the sighthound, the alan, the spaniel, the mastiff (guard of livestock and properties although sometimes used for hunting in a commoner context). As for the dwarf dogs, they reappeared in representations from the 13th century (dwarf spaniels, predecessors of bichons), with the female social elite. According to the philosopher Brunetto Latini (Le Livre du Trésor, 1260), the smallest being the most appreciated, they must be '*engendered from small parents*' (CLXXXVI).

Given the aristocracy's interest in dogs, the notion of 'purity' of bloodlines in canines resonated in the Middle Ages to that of the hereditary transmission of power in the dominant caste: the concept of noble 'blood' or 'race' to qualify it strengthened, at the same

Figure 2. Dog show, Crystal Palace, London, 1895. The Queen's London: a Pictorial and Descriptive Record of the Streets, Buildings, Parks and Scenery of the Great Metropolis, 1896.

time that it began to designate its dogs (Van der Lugdt and de Miramon 2008). The hunting treatises of the Renaissance then used the expression '*to make race*' in the sense of starting a lineage.

In the Renaissance, among the varieties of hounds for hunting deer, the most prestigious were designated according to their colour (black, grey and white, the other colours being considered as crossbreeds). Because colour was the most visible differentiation between dogs of similar morphotype, this served as the means of identification between 'breeds'. The aristocracy were not willing to 'make-do' with any colour, like peasant dogs, and selecting on colour led to the homogenisation of genetic heritages. The predilection of the seigneurial caste for hunting dogs and its practices to shape them, had a fertile posterity: the noble, selected dog, is the hunter. In *De Canibus Britannicis* (1570), the first cynological treatise, the doctor John Keys groups the varieties into three categories: the '*noble kind*' with hunting dogs, the '*ordinary kind*' for other utilities (shepherd, mastiff), and the '*mongrel*' kind with '*no notable shape*' and useless - except small services like turning the spits. This hierarchical vision of the species continued until the advent of cynophilia.

## 4 The cynophilist breed differentiation: a new paradigm

The naturalist Buffon (*Histoire Naturelle*, V, 1755), is a major source from which to apprehend the precynophilist state of the species. There were clearly more intermediate variations between utilitarian morphotypes than 'frank' models of them. Specimens embodying the radical versions of a morphotype well and truly existed, but without reproductive isolation allowing to continue their like-to-like breeding, they were not the norm. Buffon therefore anticipated the cynophilist breed differentiation: '*by keeping dogs of different breeds separately from each other*', one would prevent '*any alteration*'.

The acceleration given to differentiation comes from socio-historical conditions favourable to the increase of selective pressure and the decrease in contact between dog populations. In the second half of the 19th century, these conditions were optimally brought together with the industrial revolution, first in England where cynophilia emerged - with a chronological framework: first dog show in 1859, founding of the Kennel Club in 1873, and opening of the Stud Book in 1874 (Figure 2).

The cynophilist breed differentiation is based on a new paradigm: the breed is no longer a fuzzy notion, coming from a theoretical classification, but a concrete object answering a precise description and endowed with fixed characters. It emerges from a conjunction of factors that we retain, in a synthetic way: the standardisation of animal husbandry, the rise of the urban bourgeoisie, the expansion of the pet dog, the rise of nationalisms. The middle class developed new models of leisure as a means of definition and consolidation of its social position: concerning the dog, practices formerly divided between aristocratic hunting, agricultural shows, fights and comparative meetings of dogs of the lower classes, were mixed. Raising and exposing standardised dogs with genealogies was a form of social stratification in which the quality of the animal implicitly reflects that of its owner (Ritvo 1989).

The variability that already existed between morphotypes provided abundant material for this standardisation of the forms and the multiplication of newly recognised breeds. An aesthetic purpose was added to functional selection, and in some cases even replaced it, with some breeds becoming disconnected from the function of their predecessors: for example, the Yorkshire, a tiny silky-hairy terrier, two characteristics incompatible with an underground hunt for foxes or mustelids. The choice of characteristics to be set for each breed and achieved by reproductive isolation was partly arbitrary, given the continuum of secondary variations within precynophilist stocks. Laying the boundaries of breeds was therefore the subject of debate and even controversy. Every modern breed is thus the product of an intellectual construction, even when the morphotype which it prolongs has very old ancestry. But cynophilia has not only shaped old local varieties; recombinations were made to create new breeds from scratch. On the other hand, breeders made extensive use of mating between close relatives in this creation process, as underlined by Megnin (1883).

However, these modern practices are then not necessarily assumed for what they are. The concept of parentage to ancestral canine heritage remained an essential value for breeders, for whom the dogs of others without authenticity were therefore crossbreeds. In claiming a tradition, they did not realise that they were transporting at top speed the canine species into a new dimension. In sixty years, in place of a smaller number of 'varieties' carrying extensive 'within group' variation, there was now a large number of much more standardised breeds, separated by reproductive partitions - except misalliances or episodes of outcrossing.

Cynophilia is a decisive radicalisation of the species' differentiation process, a conceptual and methodological break leading to both an unprecedented phenotypic proliferation and a genetic bottleneck. Breedstocks were formed from a small number of founding subjects. Many secondary features and rural specimens have been left out at the creation of the breed or later after changes in the breeding goals and breed standards. It should be noted, however, that cynophilia is not a linear process; there are evolutions and involutions, accelerated modifications and conservatory practices.

## 5 Specific features of current dog breeding

Modern dog breeding practices are characterised by features, more or less specific to the species, and can be quite different to breeding livestock for example. First, as described above, the selection on aesthetics and morphology to conform to a recognised standard has taken a major importance as a selection objective. In a survey of French dog breeders undertaken in 2007, it was shown that the main breeding objective was related to morphology, ahead of behaviour, health and working ability, while selection on working capacities has remained predominant only for pointing dogs and scent hound breeds (Leroy *et al.* 2007). The main important stakeholders in dog breeding are individual breeders, whose activity can range from hobbyists producing occasional, or even only one, litter, to professionals who earn a significant income from selling puppies, and are generally organised in associations at national level (i.e. breed clubs and Kennel clubs). Dog breeders generally practice mass selection (selection on phenotypes) in contrast to livestock species, and quantitative genetic schemes have rarely been implemented until recent years (Wang *et al.* 2018). Traditionally, breeders chose selected breeding candidates based on information reported in pedigree documents, as well as on direct appraisal of the dogs and their offspring (Leroy *et al.* 2007). Canine breeding remains a craft, selection, the product of choices, trials and errors, applied to animals living in the close company of man and with whom they have strong emotional bonds.

One of the most specific features of dog breeding relates to the importance of pure-breeding as a paradigm. Indeed, in contrast to other species, such as cattle, horses, or cats, where pure-breeding and cross-breeding strategies constitute options which are privileged according to specific breeds or contexts (Danchin Burge *et al.* 2012; Pirault *et al.* 2013; Leroy *et al.* 2014), once a dog population is officially recognised as a breed, pure-breeding generally becomes the rule. In the pure-bred dog world, the rare official cross-breeding initiatives initiated by individuals or clubs are generally viewed with suspicion by many dog breeders (Farrell *et al.* 2015). At the same time, inbreeding, i.e. the intentional mating of related dogs, usually with the objective to increase the genetic contribution of an ancestor to a litter, or homogenise the puppies produced, has long remained a more or less common practice for dog breeders (Leroy *et al.* 2007). More importantly, another practice, which is common to most domesticated species under artificial selection, is the use of a limited number of sires to produce a large number of offspring, and in popular breeds the most popular male reproducers may produce more than 1000 progeny in their career (Calboli *et al.* 2008; Leroy and Baumung 2011).

A final specific feature of the dog, due to its proximity with humans, is the extent to which it is inured from environmental selective pressures. As a domesticated species, the dog is largely protected from the drivers of natural (or sexual) selection, since food and shelter is provided. Even beyond this, due to its status as 'primary' human companion animal (pet), meaning that - in western societies -the dog benefits from a high level of medical/veterinary care, management and intervention (e.g. caesarean sections, hip replacements, brachycephalic surgery, flea/tick treatment...). From a breeding perspective, this level of medical care enables individuals, which under other circumstances would have been incapable of surviving or reproducing, to produce offspring, and potentially transmit hereditary conditions to subsequent generations.

This feature, as well as those described above, are potentially detrimental to dog health. In particular, it has been underlined that the selection on specific morphological features can indirectly lead to various health and welfare problems. For instance, it has been advanced that the expression of Brachcephalic Obstructive Airway Syndrome was linked to different morphological traits, such as the ratio of muzzle length to skull length in animals, with differences according to breeds (Packer *et al.* 2015; Liu *et al.* 2017). In parallel, the overuse of a small number of reproducers in closed populations has contributed to the random dissemination of inherited disorders and potential inbreeding depression effect on fitness traits (Leroy and Baumung 2011). It has been estimated that out of 396 inherited disorders identified in dogs, 84 were associated, directly or indirectly, with breed specific morphological features (Asher *et al.* 2009), while 312 were not (Summer *et al.* 2010), with variable incidence and consequence for dog health and welfare according to disease and breeds.

## 6 Current drivers of change and consequences

Over the last years, dog breeding has been facing new challenges, which are likely to change the way dogs are selected. First, animal welfare has recently become a major issue for the public, especially when considering animals living in close proximity to humans. Given the issues outlined above, welfare associations and the media have strongly criticised dog breeding practices, citing their potential impact on health and welfare. Documentaries such as the BBC *Pedigree Dogs Exposed* have put a spotlight on the issue of dog welfare in relation to selection (Nicholas 2011). In parallel, the development of new technologies, such as Information and Communication Technologies (ICT) and genomics constitute complete game changers in the way dogs are selected. Globalisation of information exchange now allows breeders from all over the world to share information, and facilitates, at the same time, the gene flows across countries. In addition, because of its interest as a model for the study of human disease, the dog genome was first sequenced in 2005 (the third mammalian species, after humans and rats) (Shearin and Ostrander 2010). Consequently, hundreds of mutations have been identified, relating to morphological traits (coat colour and length, size...) or disorders, leading to the commercialisation of corresponding genomic tests. Private laboratories are currently offering owners and breeders panel tests allowing to genotype dogs for multiple traits, while in the same time providing information of the dog origins, at species and breed level.

The consequences have been heterogeneous across stakeholders and countries. Laying the blame at the pedigree dog system *per se*, some owners and breeders may have decided to not raise pedigree dogs anymore. As a consequence, following the example of the Labradoodle in the 70s, there have been an increase in the popularity of 'designer dog' F1 crossbreeds over the last years (e.g. Cockapoo, Jackapooor, Basador) (Caron-Lormier *et al.* 2016). Regarding purebred dogs, the Dutch government has recently implemented drastic rules for the breeding of brachycephalic breeds, with dogs with a snout length shorter than a third of the skull being banned for breeding (Limb 2019). Considering that, this ratio is on average under 20% for breeds such as the Pug or French Bulldog (Packer *et al.* 2015), such a decision actually puts the very presence of those breeds at stake in the Netherlands.

Direct stakeholders within dog breeding have also reacted in different ways to these challenges, with for instance some national kennel clubs imposing health requirements prior to breeding, or not allowing mating between close relatives. At breeder level, those practices seem to have become less frequent over the last years within breeds and countries (Wang *et al.* 2017), as illustrated by Figure 3, which shows the evolution of the proportion of dogs of inbred individuals accounting two generations of pedigree, for four breeds in France, Sweden and the United Kingdom over the period 1980–2014. Yet, if exaggerated morphological features and other inherited features now constitute clear common

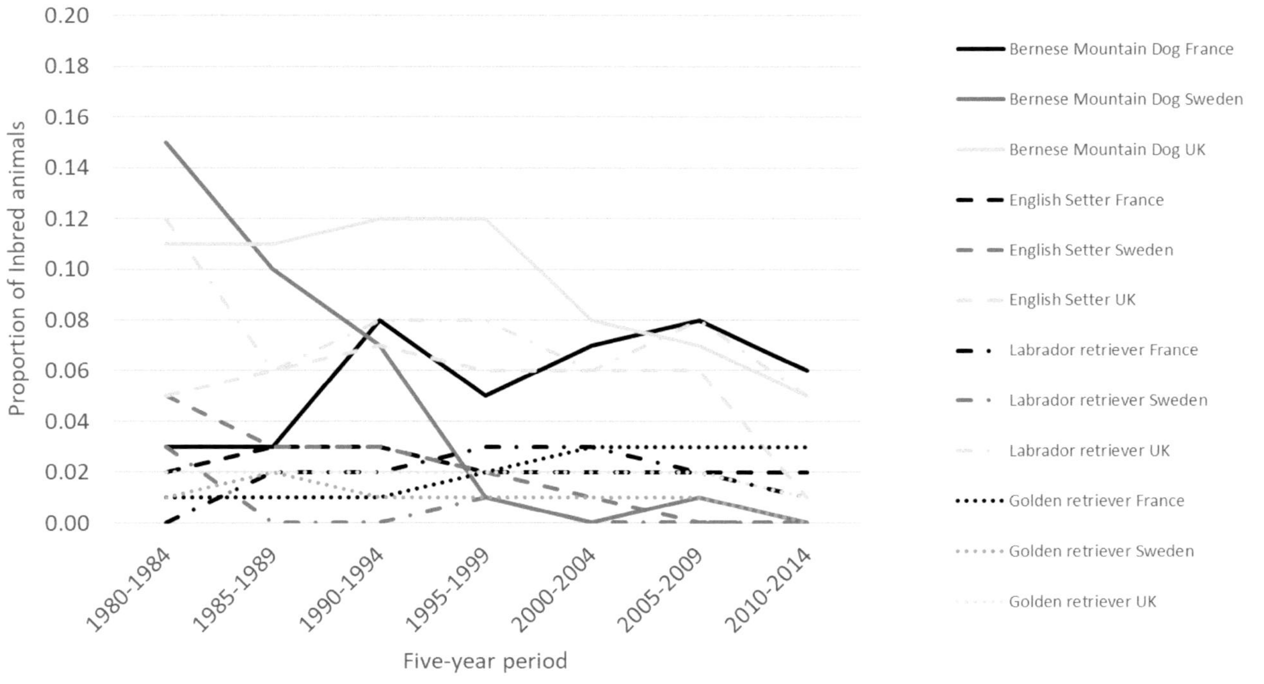


Figure 3. Evolution of proportion of inbred individuals (accounting two generations of pedigree) of the breeds studied considering different registration in France, Sweden and the UK.

concerns for national Kennel clubs, the tools, rules and strategies implemented largely differ across breeds and countries (Wang *et al.* 2018).

As previously stated, dog breeding remained largely empirical until recent years. However, with the increased availability of commercial genetic health tests, there has been an increasing incorporation of genetic testing in breeding policies at breed and national levels. In a survey conducted with 15 national Kennels Clubs in European and non-European countries, 40% reported that genetic tests results were made available for at least 50% of their breeds (Wang *et al.* 2018). For instance, in its online tool 'Mate Select', launched in 2011 (https://www.thekennelclub.org.uk/services/public/mateselect/), the UK Kennel Club publically provide results of clinical screening for over 16 disorders and results of DNA tests for over 90 disease-associated mutations (provision varying according to the relevance of the disorder to the breed). Yet, there is still a lack of standardisation and regulation to assess the quality and relevance of the growing number of DNA tests (O'Neil *et al.* 2017). Finally, in the last decade, quantitative genetic evaluation of complex traits has also been implemented by a small number of clubs and European Kennel clubs, mainly focusing on hip and elbow dysplasia, which are the most widespread complex inherited conditions in dogs.

With ICT development, access to information relative to a given dog has become available to anybody across the world through international databases. As a result, the internet is now a major source of information for breeders to find information on a dog, through websites and databases maintained by various stakeholders, from individual breeders to national or international organisations (breed and kennel clubs, laboratories, and universities). As an example, the Irish Wolfhound database (https://iwdb.org) is a platform which provides access to more than 150,000 individuals of the Irish Wolfhound breed from multiple countries, with various information for each dog, such as pictures, health status, ancestry and offspring, show results, age and cause of death. Paradoxically, the information, which before was scarce and difficult to obtain, has now become over whelming; the distinction between wanted and unwanted information and identification of priorities is now becoming more and more difficult for breeders and stakeholders.

## 7 Challenges and prospects

These different tendencies illustrate a shift from a selection focusing mostly on morphology, toward a growing importance of health as a selection objective, through the adoption, among other things, of new tools and technologies. These opportunities however come with some challenges. First, as national dog breed populations do not share the same genetic background, rarely more than a few of the hundreds of genetic tests that have been developed are of interest for a given breed in a given country. Out of the multiple information sources available on the internet, it is often difficult for dog breeders to identify the tools of real relevance for their own dogs. Therefore, there is a need for the development of holistic tools to (i) assess the complexity of the situation which most breeds are currently experiencing; (ii) define what are the priorities given the health or other more general challenges; (iii) propose strategies and adequate interventions that could be implemented by breed and kennel clubs to meet these respective challenges. It is also necessary to implement standardised international procedures and protocols for the use of genomic tools. One governance strategy for breed and kennel clubs could be to develop a standardised synthetic index, which integrate phenotypic, clinical and genomic results according to the priorities given to different concerns, following existing examples in livestock species.

In term of prospects, it is likely that the number of genomic tools (direct or risk tests, genomic estimated breeding value...) made available for breeders will continue to increase in the next years. Those tests may help to support breeders to select dogs on health, but also morphological or even behavioural or working ability traits. To that extent, it remains unknown to what degree genomics will help to decipher the heritability and genetic architecture behind behavioural traits (MacLean *et al.* 2019). However, the question of to what extent breeding will continue to be based on an empiric appraisal of dog qualities, or if dog breeders will mostly make use of the wide range of genomic tests to 'design' their dogs (purebred or not), remains. In addition, in relation to biotechnological advances, after the generation of the first gene edited dogs in 2015 (expressing muscle hypertrophy; Zou*et al.* 2015), several institutions have shown an interest in using such technologies to produce for instance working dogs with enhanced capabilities (Reardon 2016), even if the benefit and potential drawback of gene editing for dog breeding are still unclear.

The increased importance of biotechnologies in dog breeding are also likely to impact the governance of dog breeding. Indeed, through the different tests and tools commercially made available, and given the fact that the development of a specific test on a given trait may push breeders to select on this trait (even with incomplete penetrance mutation), it is likely that laboratories will take a growing interest in the orientation of breeding objectives. In parallel, the globalisation of exchanges and the subsequent availability of information will require the breed and national kennel clubs, traditionally undertaking the management of information and definition of breeding strategies, to adjust and adapt their roles. For individual dog breeders, these developments represent formidable opportunities for improving canine health, as well as presenting challenges to adapt their knowledge and practices to this changing framework.

This review illustrates, how, from the rise of the Roman Empire to the industrial revolution, socio-economic and technological changes have shaped the breeding practice and diversification of the dog species. In a similar manner, the most recent technological developments, as well as the increased interest towards animal welfare are likely to transform the way dogs are bred, with consequences on breeds' genetic structure and morphological diversity. To better understand those future evolutions, it is important to consider the whole social and organisational framework which will drive these changes. Sustainable management of dog breeds should try to avoid at least two opposite pitfalls: to ignore the new scientific tools to improve the selection, or to drift towards a hygienist vision which would try to create an impossible genetically perfect dog.

### Ancient sources

(Abbreviations after OCD https://oxfordre.com/classics/page/abbreviation-list/#s)

Hdt.-Herodotus, *Histories.*
*HA.*- Aristotle, *History of the animals.*
Plut. *Alc.*-Plutarch, Alcibiades.
*Cyn.*-Oppian, *Cynegetica.*
Symm. *Ep.*-Symmacus, *Epistulae.*
Gaston 'Phébus' comte de Foix, Le Livre de la Chasse, 1388.
Brunetto Latini, *Le Livre du Trésor,* 1260.
*Yi Zhou Shu, 4th century BC.* Quoted by V. W. Collier, Dogs of China and Japan in Nature and Art, 1921.
*ErhYa,* 3rd century BC. Quoted by V. W. Collier, Dogs of China and Japan in Nature and Art, 1921.

### References


Akey, J.M., A.L. Ruhe, D.T. Akey, A.K. Wong, C.F. Connelly, J. Madeoy, T.J. Nicholas and M. W. Neff 2010. Tracking footprints of artificial selection in the dog genome. *Proceedings of the National Academy of Sciences* 107(3): 1160–1165.

Asher, L., G. Diesel, G.F. Summers, P.D. McGreevy and L.M. Collins 2009. Inherited defects in pedigree dogs. Part 1: disorders related to breed standards. *The Veterinary Journal* 182(3): 402–411.

Calboli, F.C., J. Sampson, N. Fretwell and D.J. Balding 2008. Population structure and inbreeding from pedigree analysis of purebred dogs. *Genetics* 179(1): 593–601.

Caron-Lormier, G., G.C. England, M.Green and L. Asher 2016. Using the incidence and impact of health conditions in guide dogs to investigate healthy ageing in working dogs. *The Veterinary Journal* 207: 124–130.

Crockford, S.J. 2000. A commentary on dog evolution: regional variation, breed development and hybridization with wolves, in S.J. Crockford (ed.) *Dogs through time: an archaeological perspective*: 295–312. Oxford: Archaeopress.

Crockford S. 2006. *Rhythms of Life: Thyroid Hormone and the Origin of Species.* Victoria (Canada): Trafford Publishing.

Danchin-Burge, C., G. Leroy, M. Brochard, S. Moureaux and E. Verrier 2012. Evolution of the genetic variability of eight French dairy cattle breeds assessed by pedigree analysis. *Journal of Animal Breeding and Genetics* 129 (3): 206–217.

Farrell, L.L., J.J. Schoenebeck, P. Wiener, D.N. Clements and K.M. Summers 2015. The challenges of pedigree dog health: approaches to combating inherited disease. *Canine Genetics and Epidemiology* 2 (1):3.

Jensen P. 2007. *The behavourial biology of dogs.* Cabi Publishing.

Leroy, G., E. Verrier, C. Wisner-Bourgeois and X. Rognon 2007. Breeding goals and breeding practices of French dog breeders: results from a large survey. *Revue de Médecine Vétérinaire* 158(10): 496.

Leroy, G. and R. Baumung 2011. Mating practices and the dissemination of genetic disorders in domestic animals, based on the example of dog breeding. *Animal genetics* 42(1): 66–74.

Leroy, G., E. Vernet, M.B. Pautet and X. Rognon 2014. An insight into population structure and gene flow within pure-bred cats. *Journal of Animal Breeding and Genetics* 131(1): 53–60.

Licari S. 2006. Le chien chez les auteurs romains, *Cynophilie Française*, 132: 8–11.

Licari S. 2010a. L'émergence du morphotype braccoïde et son iconographie, 2e et 3e parties. *Cynophilie Française* 148: 18–22.

Licari S. 2010b. L'émergence du morphotype braccoïde et son iconographie, 2e et 3e parties. *Cynophilie Française* 150: 4–11.

Licari S. 1012. Emergence du morphotype du chien nain et son iconotraphie, 2e partie. *Cynophilie Française* 158: 26–29.

Limb M. 2019. Dutch crackdown on brachycephalic breeds. *Veterinary Records* 184 (23): 693.

Liu, N.C., G.I. Oechtering, V.J. Adams, L. Kalmar, D.R. Sargan and J.F. Ladlow 2017. Outcomes and prognostic factors of surgical treatments for brachycephalic obstructive airway syndrome in 3 breeds. *Veterinary Surgery* 46 (2): 271–280.

MacLean, E., N. Snyder-Mackler, B.M. vonHoldt and J. Serpell 2019. Highly heritable and functionally relevant breed differences in dog behavior. *BioRxiv*: 509315.

Megnin P. 1883. *Le chien, Histoire, Hygiène, Médecine, Vade-mecum de l'éleveur et de l'amateur de chiens.* Deyrolle, Paris.

Mellersh, C. 2012. DNA testing and domestic dogs. *Mammalian Genome* 23(1–2): 109–123.

Nicholas F.W. 2011. Response to the documentary pedigree dogs exposed: three reports and their recommendations. *The Veterinary Journal* 189(2): 126–128.

O'Neill, D. G., S.F. Keijser, A. Hedhammar, C. Kisko, G. Leroy, A. Llewellyn-Zaidi, S. Malm, P.N. Olson, R.M.A. Packer, J.F. Rousselot, I.J. Seath, J.W. Stull and B.N. Bonnett 2017. Moving from information and collaboration to action: report from the 3rd International Dog Health Workshop, Paris in April 2017. *Canine genetics and epidemiology* 4(1): 1.

Packer, R.M., A. Hendricks, M.S. Tivers and C.C. Burn 2015. Impact of facial conformation on canine health: brachycephalic obstructive airway syndrome. *PLoS One*0(10): e0137496.

Pirault, P., S. Danvy, E. Verrier and G. Leroy 2013. Genetic structure and gene flows within horses: a genealogical study at the French population scale. *PloS One* 8(4): e61544.

Reardon S. 2016. Welcome to the CRISPR zoo. *Nature News* 531(7593): 160.

Ritvo H. 1989. *The Animal Estate, The English and Other Creatures in Victorian England.* Harvard University Press.

Rooney, N.J. and D.R. Sargan 2009. *Pedigree dog breeding in the UK: a major welfare concern.* Hosham, UK: Royal Society for the Prevention of Cruelty to Animals.

Rooney, N.J. and D.R. Sargan 2010. Welfare concerns associated with pedigree dog breeding in the UK. *Animal Welfare* 19(S): 133–140.

Shearin, A.L. and E.A. Ostrander 2010. Leading the way: canine models of genomics and disease. *Disease models & mechanisms* 3(1–2): 27–34.

Stock, K.F. and R. Reents 2013. Genomic selection: status in different species and challenges for breeding. *Reproduction in Domestic Animals* 48: 2–10.

Summers, J.F., G. Diesel, L. Asher, P.D. McGreevy and L.M. Collins 2010. Inherited defects in pedigree dogs. Part 2: Disorders that are not related to breed standards. *The Veterinary Journal* 183(1): 39–45.

Sutter, N.B., D.S. Mosher, M.M. Gray and E.A. Ostrander 2008. Morphometrics within dog breeds are highly reproducible and dispute Rensch'srule. *Mammalian Genome* 19(10–12): 713–723.

Van der Lugt M. and C. de Miramon 2008. *L'hérédité entre Moyen-Âge et Époque moderne, Perspectives historiques.* Florence, Sismel, Micrologus Library.

Vonholdt, B., J.P. Pollinger, K.E. Lohmueller, E. Han, H.G. Parker, P. Quignon, J.D. Degenhardt, A.R. Boyko, D.A Earl, A.Auton, A. Reynolds, K. Bryc, A. Brisbin, J.C. Knowles, D.S Mosher, T.C. Spady, A.E. Eli Geffen, M. Pilot, W. Jedrzejewski, C. Greco, E. Randi, D. Bannasch, A. Wilton, J. Shearman, M. Musiani, M. Cargill, P.G . Jones, Zuwei Qian, W. Huang, Zhao-Li Ding, Ya-Ping Zhang, C.D. Bustamante, E.A Ostrander, J. Novembre and R.K. Wayne 2010. Genome-wide SNP and haplotype analyses reveal a rich history underlying dog domestication. *Nature* 464 (7290): 898.

Wang, S., G. Leroy, S. Malm, T. Lewis, E. Strandberg and W.F. Fikse 2017. Merging pedigree databases to describe and compare mating practices and gene flow between pedigree dogs in France, Sweden and the UK. *Journal of Animal Breeding and Genetics* 134(2): 152–161.

Wang, S., D. Laloe, F.M. Missant, S. Malm, T. Lewis, R. Verrier, E. Strandberg, B.N. Bonnett and G. Leroy 2018. Breeding policies and management of pedigree dogs in 15 national kennel clubs. *The Veterinary Journal* 234: 130–135.

Zou, Q., X. Wang, Y. Liu, Z. Ouyang, H. Long, S Wei, J. Xin, B. Zhao, S. Lai, J. Shen,, Q. Ni, H. Yang, H. Zhong, L. Li, M. Hu, Q. Zhang, Z. Zhou, J. He, Q. Yan, N. Fan, Y. Zhao, Z. Liu, L. Guo, J. Huang, G. Zhang, J. Ying, L. Lai and X. Gao 2015. Generation of gene-target dogs using CRISPR/Cas9 system. *Journal of molecular cell biology* 7(6): 580–583.

**Website**

OMIA, 2018, Online Mendelian Inheritance in Animals, viewed 15 September 2018,http://omia.org/home/.

# 1.4 Using X-ray Microtomography to Discriminate Between Dogs' and Wolves' Lower Carnassial Tooth

Francesco Boschin[1], Federico Bernardini[2,3], Clément Zanolli[4], Antonio Tagliacozzo[5], Claudio Tuniz[3]

[1] Università degli Studi di Siena, Dipartimento di Scienze Fisiche, della Terra e dell'Ambiente, U.R. Preistoria e Antropologia. Via Laterina 8, 53100, Siena, Italia, francesco.boschin@unisi.it

[2] Dipartimento di Studi Umanistici, Università Ca' Foscari, Venezia , Malcanton Marcorà , Dorsoduro 3484/D, Calle Contarini, 30123 Venezia, Italia. federico.bernardini@unive.it

[3] Multidisciplinary Laboratory, The 'Abdus Salam' International Centre for Theoretical Physics, Trieste, Italia. ctuniz@ictp.it

[4] Univ. Bordeaux, CNRS, MCC, PACEA, UMR 5199, Pessac, 33600, France, clement.zanolli@gmail.com

[5] Laboratorio di Bioarcheologia, Museo delle Civiltà, Piazza Guglielmo Marconi 14, 00144, Roma, Italia, antonio.tagliacozzo@cultura.gov.it

Corresponding author: Francesco Boschin, francesco.boschin@unisi.it

**Abstract**

Dogs and wolves exhibit similar dental features since they belong to the same species. Here we explore a new method to discriminate between wild and domestic forms, based on the analysis of the internal structure of the teeth. We analysed the lower first molar of 21 dogs and 17 wolves. X-ray microtomographic analyses were performed and tooth tissue proportions were assessed by the proportion of the dentine volume. As a result, dog molars show a lower percentage of dentine than those of wolves. This analysis offers promising applications in the study of dog domestication origins.

**Keywords**: dog domestication, wolf, lower carnassial tooth, X-ray microtomography, tooth tissue proportions.

## 1 Introduction

Dogs and wolves show similar skeletal features since they belong to the same species. Distinctions between the two groups can be related to different body proportions (e.g. length of limbs in relation to total body length) as well as to other specific features of peculiar skeletal elements. More diagnostic traits are located in the head region (skull and mandible) (Pluskowski 2006): for instance, dogs can show tooth crowding (even if this characteristic was also found in wolves and should be critically re-evaluated, e.g. Ameen *et al.* 2017), a lower orbital angle (e.g. Aaris-Sørensen 1977), smaller, compressed and crumpled tympanic bullae (e.g. Lawrence and Bossert 1967), a shortened facial part of the skull (and a consequently shortened mandible) (e.g. Clark 1996; Davis 1995), a steep frontal region (e.g. Lawrence and Bossert 1967) and a reduction of the relative length of both upper and lower carnassial teeth (e.g. Clark 1996; Davis 2003).

In spite of the above mentioned characteristics, the identification of prehistoric dogs is sometimes challenging due to the absence of clear diagnostic features or to the presence of a mosaic pattern of characteristics in the initial phases of domestication. In addition, diagnostic skeletal differences between dogs and wolves were often observed using modern individuals as a reference, but it has to be kept in mind that wolf populations lost both genetic and phenotypic variability through time. For instance, skulls of Late Pleistocene wolves from Beringia show a shortened rostrum (Leonard *et al.* 2007), and an overlap in carnassial tooth size was observed among wolves and dogs (e.g. Davis 2003). Difficulties in identifying the earliest domesticated populations is testified for instance by the scientific debate about the taxonomy of some *Canis* populations from a number of central and northern European Late Pleistocene archaeological sites (e.g. Crockford and Kuzmin 2012; Germonpré *et al.* 2009, 2012, 2015; Morey 2014). Since dogs were the first animals to be domesticated by humans, their history has great implications in the evolution of past human cultures and societies, and the identification of the first domestication centres and of the different domestication waves is of pivotal importance from an archaeological perspective (Frantz *et al.* 2016; Larson *et al.* 2012; Shannon 2015; Skoglund *et al.* 2015; Thalmann *et al.* 2013). For this reason, we propose here a new method to discriminate between wild and domesticated forms, based on the analysis of the internal structure of the lower carnassial teeth. In particular, this paper presents the study of the ratio between dentine volume and total volume in two selected parts of the tooth.

Table 1. Specimens considered in this work. Unisi: University of Siena; MNHT: Civic Museum of Natural History, Trieste; FA: Fisiocritici, Siena Academy of Science; PM: Bioarchaeology Lab. of the Museo delle Civiltà, Rome; L: length of the carnassial, measured at the cingulum (von den Driesch 1976).

| Taxonomy | ID | Sample location | Sample provenance/chronology | L | % of dentine | |
|---|---|---|---|---|---|---|
| | | | | | Slice 1 | Slice 2 |
| dog | 1 | Unisi | extant | 22.4 | 86.0 | 68.4 |
| dog | 2 | Unisi | extant | 22.2 | 85.6 | 69.7 |
| dog | 70 | Unisi | extant | 21.3 | 87.1 | 74.6 |
| dog | 196 | Unisi | extant | 25.3 | 87.3 | 73.0 |
| dog | 757 | MNHT | extant | 27.5 | 87.7 | 75.5 |
| dog | 1359 | Unisi | extant | 21.4 | 86.6 | 72.6 |
| dog | 95F | Unisi | extant | 24.0 | 87.0 | 77.2 |
| dog | chiostraccio | Unisi | extant | 20.5 | 88.6 | 74.5 |
| dog | M766 | MNHT | extant | 25.2 | 87.1 | 72.9 |
| dog | TS nonum | Unisi | extant | 19.9 | 86.4 | 71.3 |
| dog | TS3 | Unisi | extant | 22.0 | 85.6 | 74.9 |
| dog | TS6 | Unisi | extant | 22.7 | 87.9 | 71.9 |
| dog | TS7 | Unisi | extant | 23.5 | 88.7 | 73.9 |
| dog | TS8 | Unisi | extant | 25.6 | 87.1 | 72.9 |
| dog | TS9 | Unisi | extant | 23.8 | 87.4 | 74.4 |
| dog | TS10 | Unisi | extant | 23.0 | 88.6 | 76.3 |
| dog | TS11 | Unisi | extant | 25.0 | 87.8 | 75.7 |
| dog | TS13 | Unisi | extant | 18.5 | 88.7 | 74.2 |
| dog | SC1 | MNHT | Holocene - Slovenia | 19.8 | 86.6 | 71.8 |
| dog | SC3 | MNHT | Holocene - Slovenia | 21.4 | - | 74.9 |
| dog | Vpa6831 | MNHT | Holocene - Slovenia | 23.2 | 86.8 | 74.4 |
| wolf | 52 | Unisi | extant - Zoo | 30.0 | 89.1 | 77.8 |
| wolf | 353 | Unisi | extant - Central/Southern Italy | 27.3 | 91.6 | 79.2 |
| wolf | 357 | Unisi | extant - Central/Southern Italy | 30.4 | 90.2 | - |
| wolf | 358 | Unisi | extant - Central/Southern Italy | 28.0 | 90.2 | 79.1 |
| wolf | 359 | Unisi | extant - Central/Southern Italy | 28.6 | 91.0 | 84.5 |
| wolf | 375 | Unisi | extant - Central/Southern Italy | 26.4 | 90.7 | 78.5 |
| wolf | 376 | Unisi | extant - Central/Southern Italy | 29.5 | 90.1 | 79.7 |
| wolf | 377 | Unisi | extant - Central/Southern Italy | 27.0 | 90.4 | 81.1 |
| wolf | 378 | Unisi | extant - Central/Southern Italy | 27.5 | 89.0 | 78.9 |
| wolf | fis_139 | FA | extant - Central/Southern Italy | 26.5 | 88.8 | 78.9 |
| wolf | fis_135 | FA | extant - Central/Southern Italy | 26.5 | 89.9 | 80.2 |
| wolf | 551 | MNHT | extant - North-eastern Italy | 25.0 | 89.0 | 79.8 |
| wolf | 17775 | Unisi | Grotta Paglicci - MIS 2 (layer 12d) | 31.3 | 90.5 | - |
| wolf | R38 | Unisi | Grotta Paglicci - MIS 2 | 29.8 | 88.8 | 77.4 |
| wolf | P6265 | PM | Grotta Romanelli - MIS 2 | 30.4 | 88.8 | 77.6 |
| wolf | 877 | Unisi | Grotta Paglicci - Middle Palaeolithic | 26.4 | - | 80.7 |
| wolf | 3596_3 | PM | Grotta Romanelli - Terre Rosse | 24.7 | - | 82.1 |

## 2 Material and methods

In this exploratory study, the lower first molar of 18 recent dogs, three archaeological Holocene dogs from Slovenia, 11 extant Italian wolves, one extant zoo-wolf originating from a population of northern Europe and five Middle to Upper Palaeolithic wolves from Southern Italy were analysed (Table 1).

Almost all of present-day dogs were collected in the field; two specimens are from the zoological collection of the Civic Museum of Natural History of Trieste; the breed of all specimens is unknown. The three archaeological dog remains are stored at the Civic Museum of Natural History of Trieste and are from Holocene archaeological sites near Škocjan in South-western Slovenia; two are from old excavations and the context is unknown; one (Vpa6831) is from Grotta delle Ossa (Riedel 1977). Among present-day wild wolves, 8 are from Central-southern Italy and are part of the osteological collection of the Research Unit in Prehistory and Anthropology of the University of Siena; two are from Central Italy and are part of the zoological collection of the Siena Academy of Science (Accademia dei Fisiocritici), whilst one is from North-eastern Italy and is part of the zoological collection of the Civic Museum of Natural History of Trieste. Among archaeological wolves, all of them are from Apulia (Southern Italy) and in particular from two well-known sites: Grotta Paglicci and Grotta Romanelli.

The Paglicci site is located on the Gargano promontory (Foggia) and the remains studied in this paper come from three distinct excavated areas: one tooth (R38) is from the present-day cave and it was discovered in a Late Glacial context between the atrium and an inner room. Even if sediments from this area were reworked by looters, only Epigravettian remains were yielded (Arrighi *et al.* 2008; Ricci *et al.* 2016). One tooth, discovered in the main trench excavated in the cave's atrium, is from the Early Epigravettian layer 12d (Boschin 2019); this layer is dated between about 18–19 ky cal. BP (Boschin *et al.* 2018); The third tooth (877) is from a Middle Palaeolithic context (layer 1d) from the external rock shelter (Crezzini *et al.* 2016; Mezzena and Palma di Cesnola 1971). These remains are stored at the University of Siena.

Grotta Romanelli is located in Southern Apulia and is characterised by a stratigraphy composed of an upper part called 'Terre Brune', where Late Upper Palaeolithic evidence was detected (dated between about 13,800 and at least 8,600 cal. BP) (Calcagnile *et al.* 2019; Sardella *et al.* 2018; Tagliacozzo 2003), and a lower part called 'Terre Rosse', that lays under a stalagmite dated to 40,000+/-3,250 with the 230Th/238U method (Cassoli *et al.* 2003; Sardella *et al.* 2018). Among specimens analysed in this paper, two teeth are from the 'Terre Brune' and one is from the 'Terre Rosse'. All specimens are stored at the Bioarchaeology Lab of the Museo delle Civiltà in Rome. Wolves from the Upper Palaeolithic (MIS2) of Apulia (Epigravettian contexts at Grotta Paglicci and 'Terre Brune' at Romanelli) are generally characterised by a large size, whilst those from Middle Palaeolithic contexts (the external rock shelter at Grotta Paglicci and the 'Terre Rosse' at Grotta Romanelli) are characterised by a reduced size. This pattern was confirmed by matching together the evidence from other Apulian sites (Mecozzi and Lucenti 2018). It has to be highlighted that older (and smaller) Apulian wolves were previously considered as possibly belonging to *Canis mosbachensis*. Recently they were reassessed to belong to *Canis lupus* (Sardella *et al.* 2014). Given the fact that these small wolves overlap in size with dogs, their analysis is of great interest to understand if some differences in the lower carnassial tooth internal structural signature can be found between domesticated and wild individuals of similar body size.

The specimens were analysed by means of microfocus X-ray computed tomography using a system designed for the study of cultural heritage at the Multidisciplinary Lab of the 'Abdus Salam' International Centre for Theoretical Physics of Trieste (Italy) (Tuniz *et al.* 2013). The microCT acquisitions of the teeth were performed using a Hamamatsu L8121-03 sealed X-ray source with a focal spot size of 5μm. Sets of 1440 or 2400 projections, depending of the sample's characteristics, were recorded over a total scan angle of 360° using a Hamamatsu C7942SK-25 flat panel detector. The resulting microCT slices were reconstructed using the software DigiXCT (DIGISENS) in a 32-bit format. Once the 3D reconstruction of each specimen was completed, different tissues (enamel and dentine) were separated carrying out a semi-automatic threshold-based segmentation (e.g. Coleman and Colbert 2007).

Since the ratio between the volume of different tissues composing the tooth's structure can be affected by wear, a first effort was attempted to avoid this problem and to analyse all teeth in a homogeneous and reproducible way. For this reason, it was decided to analyse selected sub-volumes of each tooth, located in those portions not affected or less affected by use-wear. Starting from protocols already developed in virtual palaeoanthropology (e.g. Zanolli *et al.* 2018, 2019), we fitted a tooth cross-section to the cervix and we set it as a reference to extract two sub-volumes (hereafter volumes 1 and 2). Moving the reference cross-section through the tooth's crown, we selected four plans to extract the two volumes: a first plane (cross-section 1) was set tangent to the uppermost part of the cervix; a second plane (cross-section 2) was set at the point when the hypoconulid is closed; a third plane (cross-section 3) was set at the bottom of the fossa between the paraconid and the protoconid; finally, a fourth plane (cross-section 4) was set at the separation between the paraconid and the protoconid (Figure 1). Volume 1 is the part of the tooth comprised between cross-sections 1 and 2, whilst volume 2 is the part of the tooth comprised between cross-sections 3 and 4. A standardised ratio (expressed in percentage) between dentine volume and total volume was recorded for both volumes. Due to the presence of contact facets on the cingulum, or to the presence of worn surfaces on the paraconid and protoconid, the ratio wasn't calculated for both volumes in all samples (Table 1). In addition, the length (L) of each tooth (von den Driesch 1976) was measured with a calliper (Table 1).

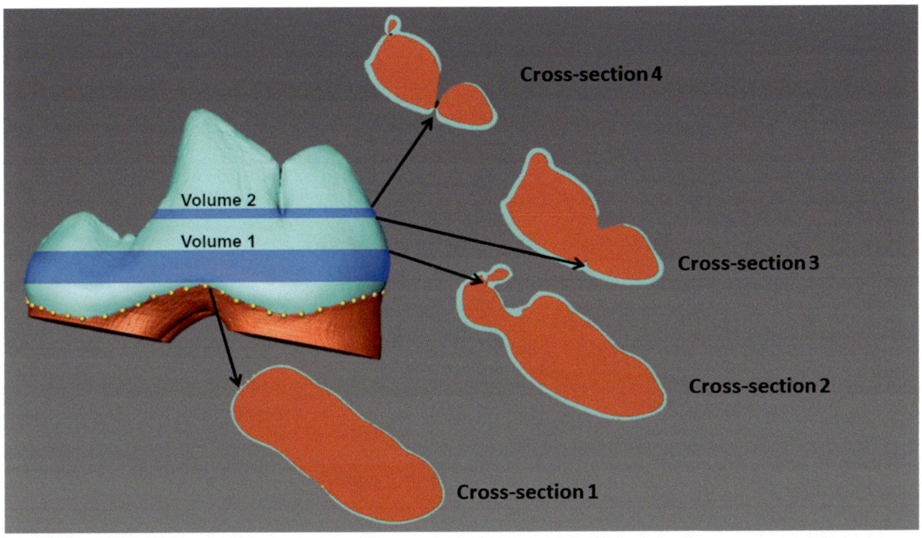


Figure 1. Cross-sections used to extract the two sub-volumes. Volume 1 located on the cingulum; volume 2 located on the cusps (by F. Boschin).

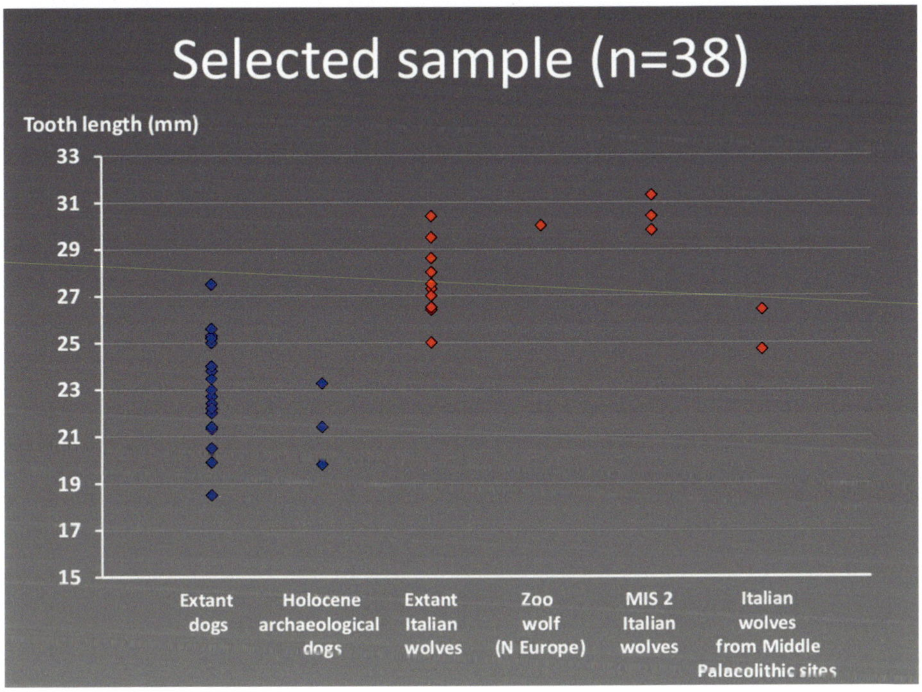


Figure 2. Length of teeth analysed in this paper. Image by one of the authors (by F. Boschin).

### 3 Results

Metric analysis reveals an overlap between wolves and dogs (Figure 2). In particular, larger dogs' teeth show a size that is comparable with that of both smaller extant Italian wolves and small-sized individuals from Middle Palaeolithic contexts. The teeth of the zoo-wolf originating from Northern Europe, as well as of the Italian wolves from the MIS 2 are larger than those of all considered dogs. If the analysis moves to the tooth internal structural signature the picture changes: the proportions of dentine are different, both in volume 1 and 2, between the wild and the domestic form. Considering volume 1, dogs show a lower proportion of dentine, thus indicating a thicker enamel. The values range between 85.6% and 88.7% in dogs and between 88.8% and 89.9% in wolves. As for volume 2, the distinction between the two groups is even clearer, as the ratio falls between 68.4% and 77.2% in dogs and between 77.4% and 84.5% in wolves (Figure 3). Also in this case wolves show a thinner enamel, as expressed by a higher proportion of dentine. Considering dogs and wolves as separate groups, the enamel thickness was tested in relation to the size of the teeth. Neither in domesticated, nor in wild individuals, was a clear correlation between the two parameters found. Linear correlation is very low in dogs both in volume 1 ($p=0.79$, $r2=0.003$) and in volume 2 ($p=0.28$, $r2=0.05$). In wolves the picture is similar: there is no correlation in volume 1 ($p=0.94$, $r2=0.0004$), and a not significant result was also found in volume 2 ($p=0.15$, $r2=0.14$). Even if statistics reject the hypothesis of a correlation between tooth size and enamel thickness, a negative trend can be observed in the volume 2 among wolves

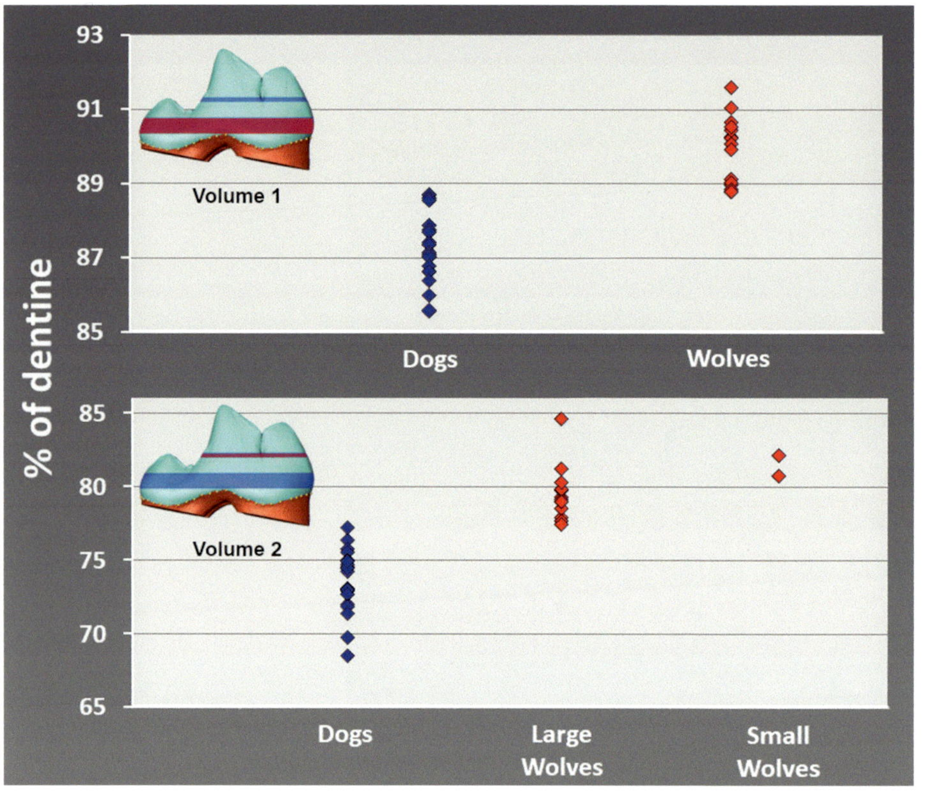


Figure 3. Proportion of dentine (% dentine volume/ total volume) in each volume (indicated in red on the tooth model). Image by one of the authors (by F. Boschin).

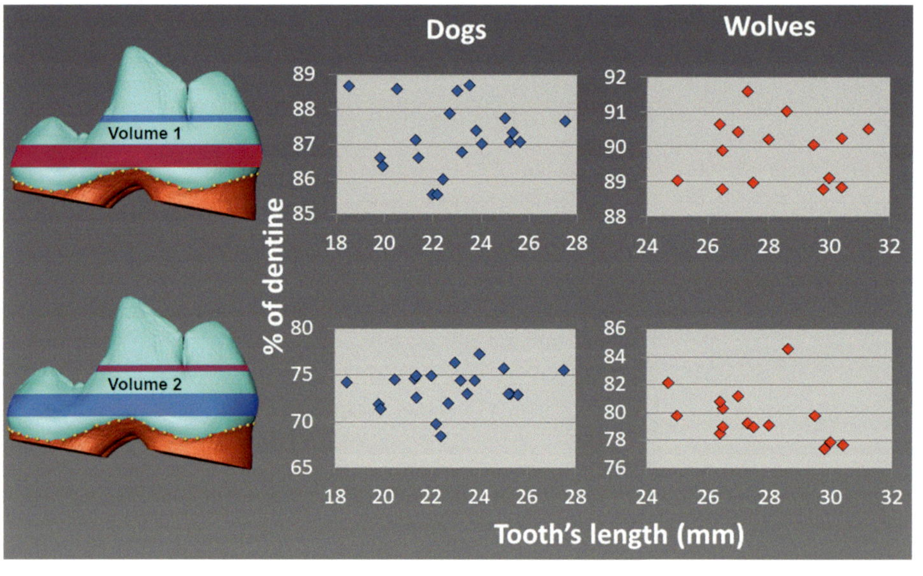


Figure 4. Correlation between tooth length (X-axis) and % of dentine (Y-axis) in dogs and wolves in volume 1 (top) and volume 2 (bottom). Image by one of the authors (by F. Boschin).

(Figure 4), with smaller individuals showing a thinner enamel. This is relevant from a taxonomic perspective, since wild individuals closer to dogs from a biometric point of view, can be better discriminated observing the proportion of dental tissues.

### 4 Discussion and conclusions

Our results demonstrate that the wolf domestication process did not only affect the relative size of the lower carnassial tooth but also its internal structure. An increase in enamel thickness is visible both in the area of the cingulum and in the main cusps. In particular, it seems that the difference is more pronounced in the latter region, where the minimum percentage of dentine volume reached by dogs is 68.4% and the maximum reached by wolves is 84.5%. In the cingulum, the range of variability is more compressed and varies from a minimum of 85.6% in dogs and a maximum of 89.9% in wolves. At the present stage of research it is difficult to assess whether the different tooth internal structure between dogs and wolves is related to a different masticatory behaviour or if it has been triggered by the shortened rostrum and mandible. The latter option could be argued due to the greater difference in enamel thickness observed between dogs and wolves in the 'more functional' area of the tooth (i.e. the cusps). Even if the breed of studied dogs is unknown, the great variability of the tooth's length (from 18.5 to 27.5 mm) could reflect a high canine diversity in the sample, and the absence of clear trends in the pattern of enamel thickness could reject a relation between the internal structural signal and masticatory mechanics. Also the structural difference detected in the cingulum, a region less involved in mastication, could suggest that changes in enamel thickness could be more related to a reorganisation of the tooth internal structure due to the reduction of the tooth's size triggered by domestication. Only further analysis of teeth belonging to dogs of known breeds could shed light on this issue. At the present stage of research, it can only be highlighted the valuable help given by microCT studies to the problem of the identification of domesticated individuals among faunal remains; indeed, regardless of whether the changes in enamel thickness from wolves to dogs are a matter of masticatory/feeding behaviour or not, the difference observed between the two groups is very clear, also when small wild individuals are analysed.


## Acknowledgements

The authors are grateful to dr. D. Arbulla (Civic Natural History Museum of Trieste), for the access to zoological collections and to the Slovenian archaeological specimens; to dr. F. Alhaique (Museo delle Civiltà in Rome) for the access to materials from Grotta Romanelli; to prof. G. Manganelli (University of Siena and Siena Academy of Science) for the access to zoological collections; to prof. Annamaria Ronchitelli and Paolo Boscato (University of Siena) for the access to archaeological materials from Grotta Paglicci and to the Soprintendenza Archeologia, Belle Arti e Paesaggio per le Provincie di Barletta-Andria-Trani e Foggia for supporting research at this site.


## References


Aaris-Sørensen, K. 1977. The subfossil wolf in Denmark. *Videnskabelige Meddelelser fra Dansk naturhistorisk Forening* 140: 129–146.

Ameen, C., A. Hulme-Beaman, A. Evin, M. Germonpré, K. Britton, T. Cucchi, G. Larson and K. Dobney 2017. A landmark-based approach for assessing the reliability of mandibular tooth crowding as a marker of dog domestication. *Journal of Archaeological Science* 85: 41–50.

Arrighi, S., V. Borgia, F. d'Errico and A. Ronchitelli 2008. I ciottoli decorati di Paglicci: raffigurazioni e utilizzo. *Rivista di Scienze Preistoriche* 58: 39–58.

Boschin, F. 2019. Exploitation of carnivores, lagomorphs and rodents at Grotta Paglicci during the Epigravettian: The dawn of a new subsistence strategy? *Journal of Archaeological Science: Reports* 26: 101871.

Boschin, F., P. Boscato, C., Berto, J. Crezzini and A. Ronchitelli 2018. The palaeoecological meaning of macromammal remains from archaeological sites exemplified by the case study of Grotta Paglicci (Upper Palaeolithic, southern Italy). *Quaternary Research* 90: 470–482.

Calcagnile, L., R. Sardella, I. Mazzini, F. Giustini, M. Brilli, M. D'Elia, E. Braione, J. Conti, B. Mecozzi, F. Bona, D.A. Iurino, G. Lembo, B. Muttillo and G. Quarta 2019. New radiocarbon dating results from the Upper Paleolithic-Mesolithic levels in Grotta Romanelli (Apulia, southern Italy). *Radiocarbon* 61: 1211–1220.

Cassoli, P.F., M. Gala and A. Tagliacozzo 2003. La caccia e l'utilizzo alimentare degli uccelli a grotta Romanelli durante le fasi finali del Pleistocene, in: P. Fabbri, E. Ingravallo and A. Mangia (eds) *Grotta Romanelli nel centenario della sua scoperta (1900-2000)*: 91–111, Lecce: Congedo Editore.

Clark, K.M. 1996. Neolithic dogs: A reappraisal based on evidence from the remains of a large canid deposited in a ritual feature. *International Journal of Osteoarchaeology* 6: 211–219.

Coleman, M.N. and M.W. Colbert 2007. CT thresholding protocols for taking measurements on three-dimensional models. *American Journal of Physical Anthropology* 133: 723–725.

Crezzini, J., P. Boscato, S. Ricci, A. Ronchitelli, V. Spagnolo and F. Boschin 2016. A spotted hyaena den in theMiddle Palaeolithic of Grotta Paglicci (Gargano promontory, Apulia, Southern Italy). *Archaeological and Anthropological Sciences* 8: 227–240.

Crockford, S.J. and Y.V. Kuzmin 2012. Comments on Germonpré *et al.* (2012) Journal of Archaeological Science 36, 2009 'Fossil dogs and wolves from Palaeolithic sites in Belgium, the Ukraine and Russia: osteometry, ancient DNA and stable isotopes', and Germonpré, Lázkicková-Galetová, and Sablin, Journal of Archaeological Science 39, 2012 'Palaeolithic dog skulls at the Gravettian Predmostí site, the Czech Republic'. *Journal of Archaeological Science* 39: 2797–2801.

Davis, S.J.M. 1995. *The Archaeology of Animals*. London: Batsford.

Davis, S.J.M. 2003. Faunal remains from Alcáçova de Santarém, Portugal. *Trabalhos do CIPA* 53. Lisboa: Instituto Português de Arqueologia.

Driesch von den, A. 1976. A guide to measurement of animal bones from archaeological sites. *Peabody Museum Bulletins* 1: 1–148.

Frantz, L.A.F., V.E. Mullin, M. Pionnier-Capitain, O. Lebrasseur, M. Ollivier, A. Perri, A. Linderholm, V. Mattiangeli, M.D. Teasdale, E.A. Dimopoulos, A. Tresset, M. Duffraisse, F. McCormik, L. Bartosiewicz, E. Gál, É.A. Nyerges, M.V. Sablin, S. Bréhard, M. Mashkour, A. Bălăşescu, B. Gillet, S. Hughes, O. Chassaing, C. Hitte, J-D. Vigne, K. Dobney, C. Hänni,

D.G. Bradley and G. Larson 2016. Genomic and archaeological evidence suggests a dual origin of domestic dogs. *Science* 352: 1228–1231.

Germonpré, M., M.V. Sablin, R.E. Stevens, R.E.M. Hedges, M. Hofreiter, M. Stiller and V.R. Després 2009. Fossil dogs and wolves from Palaeolithic sites in Belgium, the Ukraine and Russia: osteometry, ancient DNA and stable isotopes. *Journal of Archaeological Science* 36 (2): 47–490.

Germonpré, M., M. Lázničková-Galetová and M.V. Sablin 2012. Palaeolithic dog skulls at the Gravettian Předmostí site, the Czech Republic. *Journal of Archaeological Science* 39: 184–202.

Germonpré, M., M.V. Sablin, M. Lázničková-Galetová, V. Després, R.E. Stevens, M. Stiller and M. Hofreiter, M. 2015. Palaeolithic dogs and Pleistocene wolves revisited: a reply to Morey (2014). *Journal of Archaeological Science* 54: 210–216.

Larson, G., E.K. Karlsson, A. Perri, M.T. Webster, S.Y.W. Ho, J. Peters, P.W. Stahl, P.J. Piper, F. Lingaas, M. Fredholm, K.E. Comstock, J.F. Modiano, C. Schelling, A.I. Agoulnik, P.A. Leegwater, K. Dobney, J-D. Vigne, C. Vilà, L. Andersson and K. Lindblad-Toh 2012. Rethinking dog domestication by integrating genetics, archeology, and biogeography. *Proceedings of the National Academy of Science of the United States of America* 109: 8878–8883.

Lawrence, B. and W.H. Bossert 1967. Multiple character analysis of *Canis lupus, latrans*, and *familiaris*, with a discussion of the relationships of *Canis niger*. *American Zoologist* 7: 223–232.

Leonard, J.A., C. Vilà, K. Fox-Dobbs, P.L. Koch, R.K. Wayne and B. Van Valkenburgh 2007. Megafaunal Extinctions and the Disappearance of a Specialized Wolf Ecomorph. *Current Biology* 17: 1146–1150.

Mecozzi, B. and S.B. Lucenti 2018. The Late Pleistocene *Canis lupus* (Canidae, Mammalia) from Avetrana (Apulia, Italy): reappraisal and new insights on the European glacial wolves. *Italian Journal of Geosciences* 137: 138–150.

Mezzena, F. and A. Palma di Cesnola 1971. Industria acheulena 'in situ' nei depositi esterni della Grotta Paglicci (Rignano Garganico – Foggia). *Rivista di Scienze Preistoriche* 26: 3–30.

Morey, D.F. 2014. In search of Paleolithic dogs: a quest with mixed results. *Journal of Archaeological Science* 52: 300–307.

Pluskowski A. 2006. Where are the Wolves? Investigating the Scarcity of European Grey Wolf (*Canis lupus lupus*) Remains in Medieval Archaeological Contexts and its Implications. *International Journal of Osteoarchaeology* 16: 279–295.

Ricci, S., G. Capecchi, F. Boschin, S. Arrighi, A. Ronchitelli and S. Condemi 2016. Toothpick use among Epigravettian Humans from Grotta Paglicci (Italy). *International Journal of Osteoarchaeology* 26: 281–289.

Riedel, A. 1977. I resti animali della grotta delle Ossa (Škocjan). *Atti del Museo Civico di Storia Naturale di Trieste* XXX (2): 125–208.

Sardella, R., D. Bertè, A.D. Iurino, M. Cherin and A. Tagliacozzo 2014. The wolf from Grotta Romanelli (Apulia, Italy) and its implications in the evolutionary history of *Canis lupus* in the Late Pleistocene of Southern Italy. *Quaternary International* 328–329: 179–195.

Sardella, R., I. Mazzini, F. Giustini, B. Mecozzi, M. Brilli, D.A. Iurino, G. Lembo, B. Muttillo, M. Massussi, D. Sigari, S. Tucci and M. Voltaggio 2018. Grotta Romanelli (Southern Italy, Apulia): legacies and issues in excavating a key site for the Pleistocene of the Mediterranean. *Rivista Italiana di Paleontologia e Stratigrafia* 124: 247–264.

Shannon L.M., R.H. Boyko, M. Castelhano, E. Corey, J.J. Hayward, C. McLean, M.E. White, M. Abi Said, B.A. Anita, N.I. Bondjengo, J. Calero, A. Galov, M. Hedimbi, B. Imam, R. Khalap, D. Lally, A. Masta, K.C. Oliveira, L. Pérez, J. Randall, L.M. Tam, F.J. Trujillo-Cornejo, C. Valeriano, N.B. Sutter, R.J. Todhunter, C.D. Bustamante and A.R. Boyko 2015. Genetic structure in village dogs reveals a Central Asian domestication origin. *Proceedings of the National Academy of Science of the United States of America* 112: 13639–13644.

Skoglund, P., E. Ersmark, E. Palkopoulou and L. Dalén 2015. Ancient Wolf Genome Reveals an Early Divergence of Domestic Dog Ancestors and Admixture into High-Latitude Breeds. *Current Biology* 25: 1–5.

Tagliacozzo, A. 2003. Archeozoologia dei livelli dell'Epigravettiano finale di Grotta Romanelli (Castro, Lecce) strategie di caccia ed economia di sussistenza, in: P. Fabbri, E. Ingravallo and A. Mangia (eds) *Grotta Romanelli nel centenario della sua scoperta (1900-2000)*: 169–216, Lecce: Congedo Editore.

Thalmann, O., B. Shapiro, P. Cui, V.J. Schuenemann, S.K. Sawyer, D.L. Greenfield, M.B. Germonpré, M.V. Sablin, F. López-Giráldez, X. Domingo-Roura, H. Napierala, H-P. Uerpmann, D.M. Loponte, A.A. Acosta, L. Giemsch, R.W. Schmitz, B. Worthington, J.E. Buikstra, A. Druzhkova, A.S. Graphodatsky, N.D. Ovodov, N. Wahlberg, A.H. Freedman, R.M. Schweizer, K.-P. Koepfli, J.A. Leonard, M. Meyer, J. Krause, S. Pääbo, R.E. Green and R.K. Wayne 2013. *Science* 342: 871–974.

Tuniz, C., F. Bernardini, A. Cicuttin, M.L. Crespo, D. Dreossi, A. Gianoncelli, L. Mancini, A. Mendoza Cuevas, N. Sodini, G. Tromba, F. Zanini and C. Zanolli 2013. The ICTP-Elettra X-ray laboratory for cultural heritage and archaeology. *Nuclear Instruments and Methods in Physics Research Section*

*A: Accelerators, Spectrometers, Detectors and Associated Equipment* 711: 106–110.

Zanolli, C., L. Pan, J. Dumoncel, O. Kullmer, M. Kundrát, W. Liu, R. Macchiarelli, L. Mancini, F. Schrenk and C. Tuniz 2018. Inner tooth morphology of *Homo erectus* from Zhoukoudian. New evidence from an old collection housed at Uppsala University, Sweden. *Journal of Human Evolution* 116: 1–13.

Zanolli, C., O. Kullmer J. Kelley, A-M. Bacon, F. Demeter, J. Dumocel, L. Fiorenza, F.E. Grine, J-J. Hublin, A.T. Nguyen, T.M.H. Nguyen, L. Pan, B. Schillinger, F. Schrenk, M.M. Skinner, X. Ji and R. Macchiarelli 2019. Evidence for increased hominid diversity in the Early to Middle Pleistocene of Indonesia. *Nature Ecology & Evolution* 3: 755–764.

# 1.5 The Skull Shape of *Canis lupus*. A Study of Wolf and Dog Cranial Morphology

Raquel Blázquez-Orta[1], Laura Rodríguez[2], María Ángeles Galindo-Pellicena[3], Ignacio De Gaspar[1,4] and Nuria García[1,3]

[1]Dpto GEODESPAL. Grupo UCM-Ecosistemas Cuaternarios – Facultad de Ciencias Geológicas. Universidad Complutense de Madrid, C/ José Antonio Novais, 12, 28040 Madrid, Spain, rborta@ucm.es, igaspars@ucm.es, nugarcia@ucm.es

[2]Laboratorio de Evolución Humana, Dpto de Geografía e Historia, Universidad de Burgos, Plaza de Misael Bañuelos s/n, 09001 Burgos, Spain, lrgagosto@gmail.com

[3]Centro (UCM-ISCIII) de Evolución y Comportamiento Humanos. C/ Monforte de Lemos 5, pabellón 14. 28029 Madrid, Spain, mariangalindo79@gmail.com, nugarcia@ucm.es

[4]Sección Departamental de Anatomía y Embriología, Facultad de Veterinaria, Universidad Complutense de Madrid, Av. Puerta de Hierro s/n, 28040 Madrid, Spain, igaspars@ucm.es

Corresponding author: Nuria García, nugarcia@ucm.es

**Abstract**

The aim of this research is to analyse craniomandibular features in contemporary wolves and dogs in order to study evolutionary changes that are assumed to be related to domestication. We compared these modern canids with four fossils from different Upper Pleistocene (Grotta Romanelli, Terrasses de la Riera dels Canyars) and Holocene (Portalón) sites of the Mediterranean region. The specimens were analysed using both traditional and geometric (2D) morphometric techniques. Our results characterise wolves' greater mandible size (dental series), greater cranial width and length, and less elongated snout.

**Keywords**: domestication, *Canis lupus*, morphometry, skull, Pleistocene.

## 1 Introduction

The *Canis* domestication process is riddled with controversy (Müller 2002: 34–39) and the classification of the wolf-like canids remains uncertain (Boudadi-Maligne and Escarguel 2014: 80–81). Dogs/wolves have a deeper relationship with humans than any other mammal, and the origins of dogs are interesting to a very wide audience. Various morphological methods have illuminated the process of domestication and, recently, so have ancient DNA analyses of Pleistocene wolf-dogs (Germonpré *et al.* 2009: 473–74, 2015: 261–62). Lindblad-Toh *et al.* (2005) studied the structure of the haplotype and the phylogeny of the dog, to observe the relationship with the domestication of the dog. However, it is not so simple to genetically distinguish proto-dogs from contemporary wolf populations by genomic sequences (Druzhkova *et al.* 2013: 5; Thalmann *et al.* 2013: 873–74). After many years of intense DNA work, morphology is still the best witness of dog origins from wolves.

Research based on morphometric studies (Drake and Klingenberg 2010, 2017; Germonpré *et al.* 2009, 2012, 2015, 2017; Janssens *et al.* 2016; Morey 2014; Sardella *et al.* 2014) reveal morphological differences between dogs and wolves. Drake and Klingenberg (2010) analyse, in 106 breeds of domestic dogs, cranial diversity using geometric morphometry. The variation observed in the skulls, with respect to its length, forehead and neurocranium, is also found in the wolf, which these authors interpret as changes already occurring before domestication. In the archeo-paleontological record, complete skulls are not always recovered in good condition, therefore we also studied mandibles because they also show distinct morphology (Drake *et al.* 2017; Germonpré *et al.* 2015). Dogs are characterised by a shorter snout, a more pronounced forehead area and a wide palate (Germonpré *et al.* 2009: 481; Sablin and Khlopachev 2002: 796; Morey and Jeger 2015: 427) and, therefore, we propose these features diagnose domestic animals. The shortening of the snout in dogs implies a strong mandible and well-developed carnassial ($P^4$ and $M_1$) (Germonpré *et al.* 2009: 481). A skull of short length (condyle-basal length) is also identified as a typical feature of paleolithic dogs (Germonpré *et al.* 2012: 196). Another criterion that is used as evidence of domestication is the crowding of teeth due to the absence of diastema (or shortening of the snout) (Germonpré *et al.* 2015: 276). Similarly, metric data indicate that the mandibles of paleolithic dogs are shorter. In addition, they present a high frequency of very close premolars with cusps oriented towards the posterior mandibular region (the coronoid process) (Germonpré *et al.* 2015: 277).

Skulls and mandibles are complex biological forms with diet-related adaptations that respond quickly to selective pressures. One way to compare the morphologies of dogs, wolves and their kin is through different morphometric techniques: traditional and geometric. Combining both, it is possible to obtain a more complete morphometric analysis and also a comparison between techniques and their effectiveness in extracting morphological information in the study of the bone. Traditional morphometry basically describes simple changes in size and shape as mathematic regressions (allometries), which then focus on specific areas of variability. Multivariate data processes have enjoyed continuous improvement from the late 20th Century through to the present. Measured variables have been subjected to increasingly powerful factorial analysis, with current geometric morphometry (2D) that allows more complex observation of changes of shape, through building visual representations of morphological variation (Zelditch *et al.* 2004: 2). For this, anatomically homologous landmarks are located in 2D space for all analysed specimens. In geometric morphometry, the shape is defined as 'all the geometric information that remains when the location, scale and rotational effects are filtered from an object' (Kendall 1977: 428). By performing Multivariate techniques like Principal Components Analysis (PCA), variability can be represented numerically and graphically. PCA is a method to simplify descriptions of variation among individuals and make them easier to interpret (Zelditch *et al.* 2004: 156). As a result, PCA converts the original variables into a set of new variables, Principal Component (PC). PCs are linear combinations of the original variables and uncorrelated with each other. Wireframe graphs allow for better visualisation of changes in shape, connecting landmarks by straight lines. Two graphs are generated, one with the initial shape (or average form of the study samples) and another with the final shape (Klingenberg 2013: 19).

In this research, skulls and mandibles of modern Iberican wolves (*Canis lupus signatus*) and dogs (*Canis lupus familiaris*) are studied using the techniques described above. We analyse the morphological features most useful in differentiating wolf-like canids and characterise them in descriptive visual and numerical terms. In addition, in this work a preliminary analysis is carried out with a fossil sample of Paleolithic dogs and Pleistocene wolves from the Mediterranean region (Grotta Romanelli, Terrasses de la Riera dels Canyars and Portalón). This material is complete and well preserved and provides a base for future projects where the fossil record sample will be expanded.

Grotta Romanelli is a Late Pleistocene site located in the region of Apulia (Southern Italy) (Bertè 2013: 156; Sardella *et al.* 2014: 180). In this research we analysed the skull of a wolf from Level G (dated at 69 ka to 40 ka ±3250 years by Fornaca-Rinaldi and Radmilli, 1968) of '*terre rosse*', interpreted by Blanc (1928) as an eolian deposit. El Portalón de Cueva Mayor (Sierra de Atapuerca, Burgos, Spain), dated at 30 ka to 1000 BP years (Carretero *et al.* 2008: 74), is one of the most important archaeo-paleontological sites of the Meseta. The material analysed in this work corresponds to two mandibles of paleolithic dogs of the Bronze Age level. Broadly speaking, the Bronze Age layer is constituted by greyish silt-clay and sandy sediments, clasts, organic matter and coals (Carretero *et al.* 2008: 71; Pérez-Romero *et al.* 2016: 3–4). Terrasses de la Riera dels Canyars is a Pleistocene site located near Gavà (Barcelona) in an abandoned gravel pit dated to 39.6 ka cal BP (Daura *et al.* 2013: 26–27). The lithological sequence described in this site corresponds to coarse-grained fluvial deposits. We analysed a skull and a mandible of a Pleistocene wolf.

Our results suggest that wolves are characterised by a greater size of the mandible (dental series length), a greater cranial width-length and less elongation of the snout.

## 2 Material and methods

### *2.1 Material*

We studied 41 modern skulls of *Canis l. signatus* and *Canis l. familiaris*. Furthermore, we analysed fossil material from different Pleistocene and Holocene sites (Table 1). All this material is well preserved. In order to study *Canis lupus* from Grotta Romanelli (Level G) we used the high quality pictures figured in Sardella *et al.* (2014: 183, fig.

Table 1. Material (modern and fossil specimens) studied.

| **Modern material** | 41 specimens: 21 Iberian wolves and 20 dogs | | |
|---|---|---|---|
| | **Material** | **Site** | **Age** |
| **Fossil material** | 1 Pleistocene wolf (skull)<br>- P3580 | Grotta Romanelli (GR) | 69–40 ± 3250 ka[1] |
| | 1 Pleistocene wolf (skull and mandible)<br>- TC'07.M242401 (mandible)<br>- TC'07.N231644 (skull) | Terrasses de la Riera dels Canyars (TC) | 39.6 ka cal BP[2] |
| | 2 Paleolithic dogs (mandibles)<br>- CMIA8.51.1<br>- CMIA6.48.4 | El Portalón de Cueva Mayor (ATP) | Middle Bronze Age |

[1] Sardella *et al.* (2014: 183, fig. 3); [2] Daura *et al.* (2013: 26)

3). In this publication, the skull can be analysed in all the anatomical views (dorsal, ventral, lateral, frontal and caudal) and thus can be included in our comparative study. For the study of the Pleistocene wolf from Terrasses de la Riera dels Canyars we also analysed pictures figured in Daura *et al.* (2013: 40, fig. 11) and additional photos.

The modern material studied is stored at different museums and institutions: Museo Nacional de Ciencias Naturales (MNCN, Madrid), Laboratorio de Evolución Humana (LEH) from University of Burgos (UBU), Museo de Anatomía Comparada de Vertebrados (MACV, Complutense University of Madrid) and Facultad de Veterinaria of the Complutense University of Madrid (UCM). We used size as a criterion to select the skulls of dogs, thus controlling for allometry by selecting those of similar size to wolves.

### 2.2 Methods

#### 2.2.1 Traditional morphometric

For traditional analysis we followed Driesch (1976: 42–45, 60–61) using a digital caliper to the nearest 0,01 mm. We measured 25 variables on mandibles and 27 on skulls: we focused both on specific areas like the dental series (especially the carnassial and canine) and on generalised measures of the mandible (such as length, height of the body and height of the vertical ramus). We also measured several different cranio-facial (frontal bone, snout) and neurocranial chords. We used *STATISTICA* (version 12.5) to perform the factorial analyses of principal components. Finally, we studied the differences and variability of wolves and dogs with scatterplots.

#### 2.2.2 Geometric morphometric

For geometric (2D) morphometric analysis we took photographs of all the specimens. Skulls and mandibles were carefully placed on a base, always in the same position to obtain accurate results.

We selected several landmarks from different regions of skulls and mandibles (Tables 2–4 and Figure 1) and analysed them using tpsDig2 (version 2.3; Rohlf 2017). The landmarks chosen were previously-defined anatomic homologous points (Zelditch *et al.* 2004: 24) that allow a better analysis of the shape. Bookstein (1991: 63–65) describes three types of landmarks: Type I (classic craniometric points/discrete juxtapositions of tissue, for example, points of contact between bones), Type II (points of maximal curvature) and Type III (extreme points). In this study, most of the landmarks digitised are of Type I and II given their more conventional biological meaning.

We used *MorphoJ* (version 1.06d; Klingenberg 2011) to perform several analyses. We carried out a Procrustes superimposition to homogenise the data (Rohlf and Marcus 1993: 130). This consisted of translating, scaling and rotating landmark configurations to eliminate information unrelated to shape (Zelditch *et al.* 2004: 113). We thus excluded size differences, location and orientation differences, and then transformed the original landmarks in Procrustes space, generating new analagous points. Secondly, a PCA was applied to reduce dimensionality of the data and to simplify its representation. Finally, for a better display of the shape changes we used wireframe graphs. With these graphs we observed the differences and variability between both types of canids.

## 3 Results

### 3.1 Traditional morphometry

We performed a factorial analysis of the mandibles, analyzing 25 variables. We obtained two factors that represent 85.56% of the total variance. The first factor provides the highest proportion of the variability and explains 77.71% of the total variance, while the second factor explains 7.85%. The more significative variables in Factor 1 are those related to measurements that express the length and height of the mandibles, and dental series length. The more relevant variables from Factor 2 are the molar row ($M_1$-$M_3$) and the carnassial length ($M_1$). Figure 2A shows both factors with visible graphical discrimination between wolves and dogs.

For the skulls we performed a factorial analysis, where we analysed 27 variables. We obtained three factors that represent 72.42% of the total variance. Factor 1 (56.14%) and Factor 2 (10.18%) are graphically represented in Figure 2B. The most significative variables from Factor 1 are measurements related to dental series length (Prostion-$M^2$, $P^1$-$P^4$, etc.) and measurements of skull length. The most relevant variables from Factor 2 are measurements related to the frontal bone and muzzle breadth. Factor 2 does not discriminate between wolves and dogs, but it provides morphological details. Factor 3 only represents 6.10% of the total variance, not providing relevant information.

### 3.2 Geometric morphometry

A PCA for the mandibles was performed. We analysed the first four Principal Components (PC) because they show the maximum shape variation, with 75.06% of the total variance. However, only PC1 (37.98%) discriminates between wolves and dogs and is the one that better analyses the shape variation. By observing the landmarks in the wireframe of PC1, PC2, PC3 and PC4 (6.9%), we selected PC1 *vs.* PC4 (Figure 3A) to analyse mandibular differences. Given the variation of the shape observed in the PC1 wireframe, the final

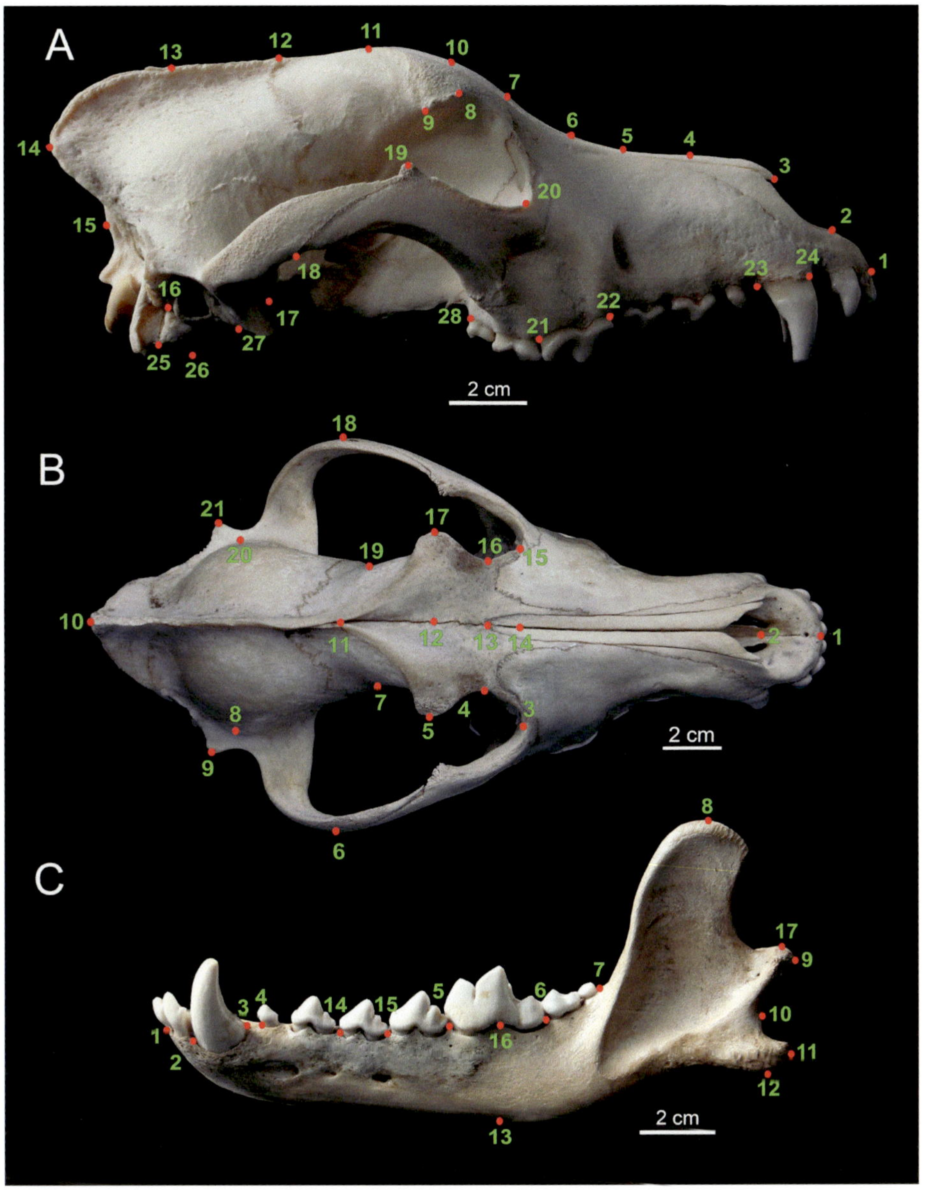


Figure 1. Position of the landmarks in the skull in lateral view (A), in dorsal view (B) and the mandible (C).

Table 2. Position and type of landmarks placed on mandibles. (Definitions according to Driesch 1976: 60–61).

| Landmark | Position | Type |
|---|---|---|
| 1 | Infradentale: the most prominent median point at the oral border of the alveoli of the incisors | I |
| 2–3 | Alveolus of the canine (oral-aboral points) | I |
| 4 | Oral point of the alveolus of the first premolar ($P_1$) | I |
| 5–6 | Alveolus of the carnassial ($M_1$) (oral-aboral points) | I |
| 7 | Aboral point of the third molar ($M_3$) | I |
| 8 | Coronion: the highest point of the coronoid process | I |
| 9 | Midpoint of the condylar process | II |
| 10 | Point between the condylar process and the angular process | II |
| 11 | Midpoint of the angular process | II |
| 12 | Basal point of the angular process | II |
| 13 | Midpoint at the base of $M_1$ | II |
| 14–15 | Alveolus of the third premolar ($P_3$) (oral-aboral points) | I |
| 16 | Basal border of body mandible (below $M_1$) | II |
| 17 | Higher point of the condylar process | II |

Table 3. Position and type of landmarks placed on skull (dorsal view). (Definitions according to Driesch 1976: 42–45).

| **Landmark** | **Position** | **Type** |
|---|---|---|
| 1 | Prosthion: the medial point of the line joining the most oral points of the premaxillae | I |
| 2 | Rhinion: the median point of the line joining the most oral points of the nasals | I |
| 3–15 | Lacrimal | I |
| 4–16 | Entorbitale: the naso-medial indentation of the orbit that corresponds with the most lateral point of the braincase | I |
| 5–17 | Ectorbitale: the most lateral point of the frontal bone on the occipital side of the orbit | I |
| 6–18 | Zygion: the most lateral point of the zygomatic arch | I |
| 7–19 | Frontostenion: point at the postorbital constriction | I |
| 8–20 | Euryon: the most lateral point of the braincase | I |
| 9–21 | Otion: the most lateral point of the mastoid region | I |
| 10 | Acrocranion: the most aboral point on the vertex of the skull in the median plane | I |
| 11 | Point between parietal and frontal bones | I |
| 12 | Frontal midpoint between Ectorbitale-Ectorbitale | I |
| 13 | Nasion: the median point of the naso-frontal suture | I |
| 14 | Point in the nasal bone between oral border of the orbits (median) | I |

Table 4. Position and type of landmarks placed on skull (lateral view). (Definitions according to Driesch 1976: 42–45).

| **Landmark** | **Position** | **Type** |
|---|---|---|
| 1 | Prosthion: the medial point of the line joining the most oral points of the premaxillae | I |
| 2 | Lowest point of the nasal opening (at the clivus), at the height of $I^3$ | III |
| 3 | Most anterior point of the nasal bone | III |
| 4–5 | Curvature of the nasal bone | II |
| 6 | Midpoint of the nasal bone at the level of the infraorbital | II |
| 7 | Nasion: the median point of the naso-frontal suture | I |
| 8 | Higher point of the orbit | I |
| 9 | Ectorbitale: the most lateral point of the frontal bone on the occipital side of the orbit | I |
| 10 | Midpoint- in the frontal bone- at the height of the ectorbitale | I |
| 11 | Most posterior point of the frontal (before the beginning of the sagital crest) | II |
| 12 | Coronal Suture (that separates the parietal bones and the frontal bone) | I |
| 13 | Origin of the external sagittal crest | II |
| 14 | Acrocranion: the most aboral point on the vertex of the skull in the median plane | I |
| 15 | Midpoint of the occipital bone | III |
| 16 | Point joining the postorbital process and the tympanic bullae | II |
| 17 | Point at the base of the skull (on basioccipital) | III |
| 18 | Temporal process of the zygomatic bone | I |
| 19 | Frontal process of the zygomatic bone | I |
| 20 | Oral point of the orbit | I |
| 21–22 | Alveolus of the carnassial ($P^4$) (aboral and oral point) | I |
| 23–24 | Alveolus of the canine (aboral-oral points) | I |
| 25 | Aboral point of the tympanic bullae | II |
| 26 | Midpoint of the bullae tympanic base | II |
| 27 | Oral point of tympanic bullae | II |
| 28 | Aboral point of the second molar ($M^2$) | I |

FIGURE 2A

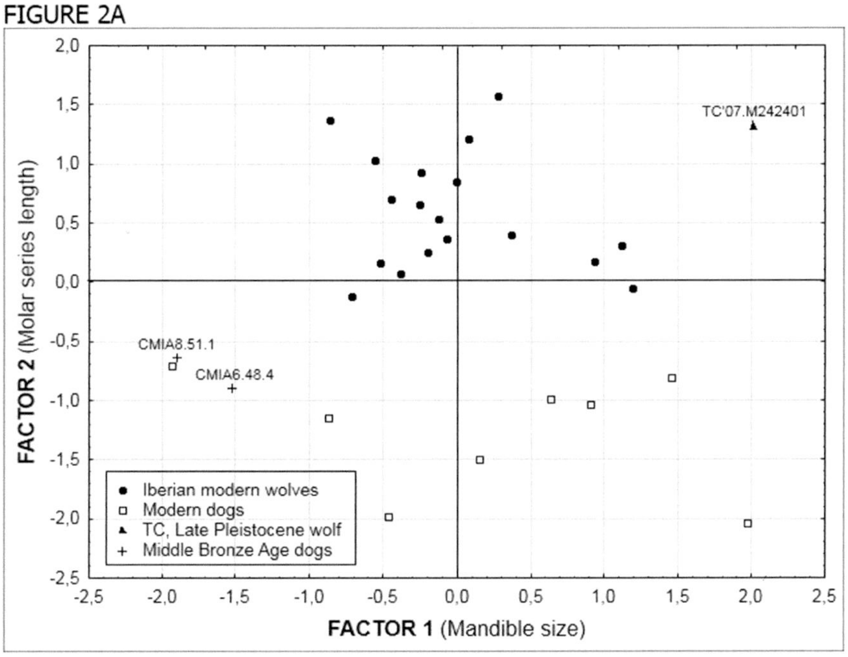


FIGURE 2B

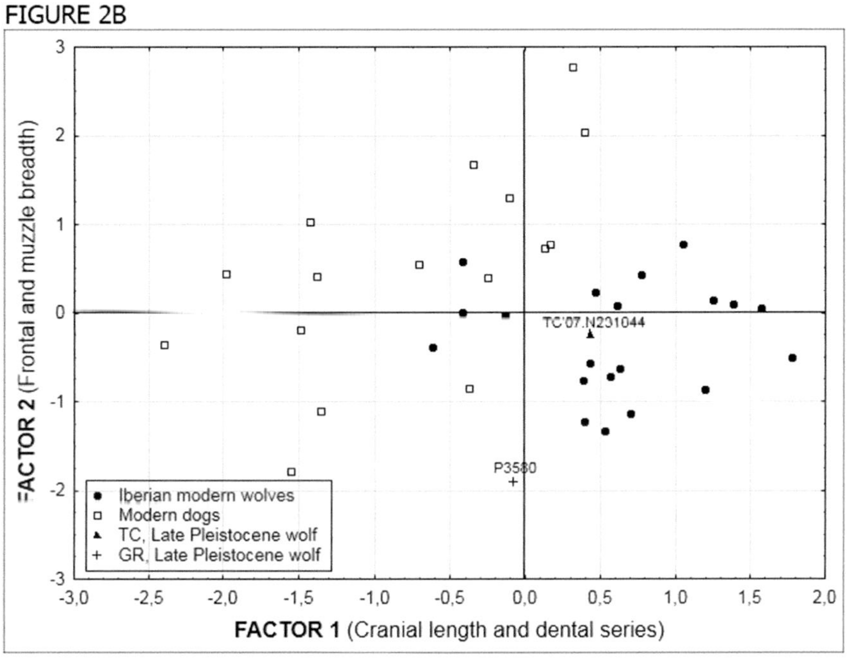


Figure 2. Scatterplots that analyse the mandible variables (2A): Factor 1 (size: length and height of the mandible and dental series length) *vs.* Factor 2 (molar series length) and the skull variables (2B): Factor 1 (cranial length and dental series) *vs.* Factor 2 (frontal and muzzle breadth).

shape reflects a mandible with a greater ascending ramus height, oriented towards the aboral, and a lesser molar series length. The shape of PC4 reflects a more elongated mandible with a lesser mandibular height (lesser ascending ramus height and lesser mandibular body height). In Figure 3A, PC1 shows a wider dispersion of dogs, presenting the more positive values of the PCA and PC4 shows some overlapping of dogs and wolves but also displays differences.

For the skulls we carried out a PCA, analyzing the variation of the shape in dorsal view and lateral view. From the dorsal view PCA, PC1 *vs.* PC3 (PC1: 32.90% and PC3: 13.6%) were selected (Figure 3B) because these are the ones that better express the shape differences. Given the variation of the shape observed in the PC1 wireframe, the final shape reflects a skull with a greater skull length and a less frontal bone breadth. The shape of PC3 reflects a skull with a lesser neurocranium (cranium) and postorbital constriction breadth. In Figure 3B, PC3 discriminates wolves and dogs. From the lateral view PCA, we selected PC1 *vs.* PC2 (Figure 3C), which explain 42.79% of the total variance (25.07% and 17.72% respectively). The shape of PC1 reflects the landmarks located in the snout that show greater variation, as a result of lesser facies height. The shape of PC2 reflects greater variability in the cranium. The PC1 (Figure 3C) suggests two groups

FIGURE 3A

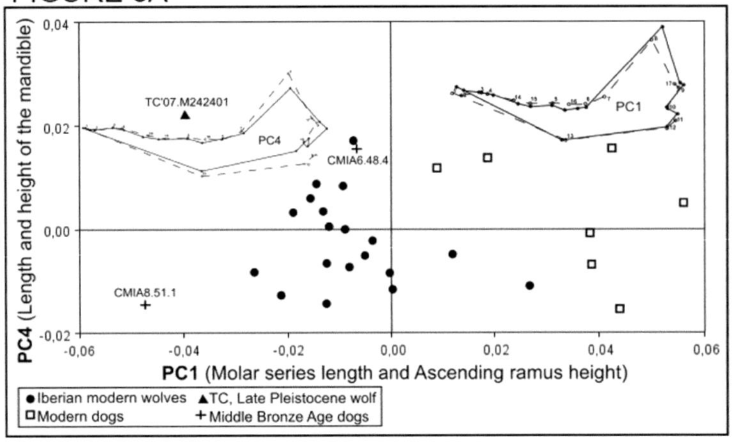


FIGURE 3B

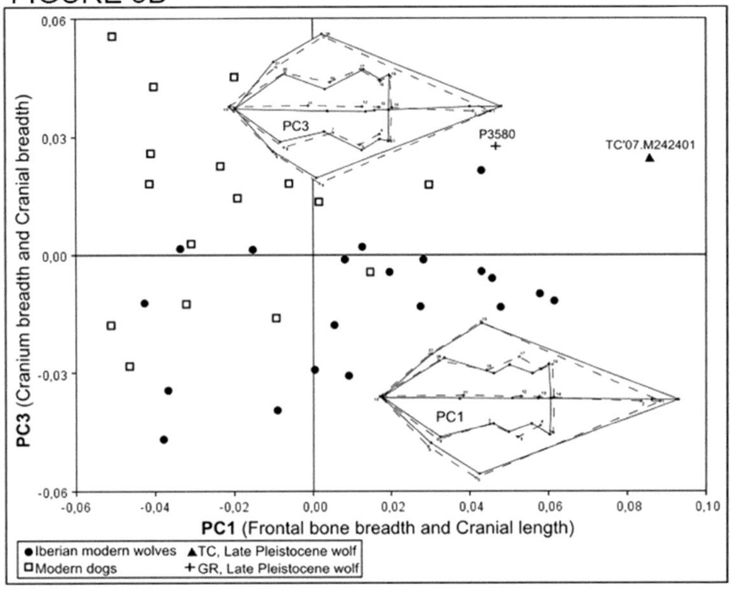


FIGURE 3C

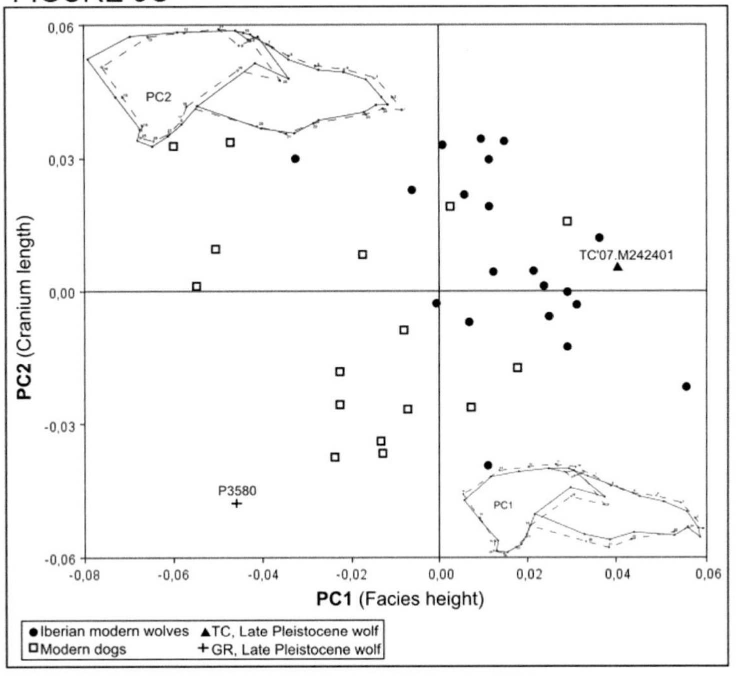


Figure 3. PCA that analyse the mandible landmarks (3A): PC1 (Molar series and ascending ramus height) *vs.* PC4 (Length and height of the mandible), the skull landmarks in dorsal view (3B): PC1 (Frontal bone breadth and Cranial length) *vs.* PC3 (Cranium breadth and Cranial breadth-at postorbital constriction) and the skull landmarks in lateral view (3C): PC1 (Facies height) *vs.* PC2 (Cranium length).

are distinguished. However, in PC2 they overlap.

## 4 Discussion

First we analysed the shape variations on the modern *Canis* according to the results obtained with the traditional (TM) and geometric morphometry (GM). Then we incorporated the study of fossil specimens to the morphology variation discussion.

### *4.1 Traditional morphometry*

Considering the results obtained in the factorial analysis of the mandibles (Figure 2A), Factor 1 is positively dominated by the general size of the mandible, that is, by length and height of the mandible and dental series length. Therefore, we interpret the positive values of Factor 1 as larger mandibles. In this case, there is no clear separation between dogs and wolves. From the most important variables in Factor 2, we can deduce that molar series length, especially by the size of the carnassial ($M_1$), is the one with the greatest burden on this factor. In other words, the most positive values (the modern wolves) in this axis are interpreted as individuals with larger molar series.

Analyzing the skulls' graph (Figure 2B), positive values in Factor 1 display a greater cranial length and dental series (snout and facial length). Most modern wolves fall into positive Factor 1 values, indicating that they have an elongated snout. In Factor 2, which expresses frontal and muzzle width, there is greater dispersion of dogs and wolves, but in general, it seems to indicate that dogs have greater frontal and muzzle

Table 5. Mandible features.

| Mandible Features | Iberian Modern Wolves | Modern Dogs | TC'07.M242401 (Wolf-TC) | CMIA8.51.1 (Dog-ATP) | CMIA6.48.4 (Dog-ATP) |
|---|---|---|---|---|---|
| Mandible size and dental series length | Greater | Lesser | Greater | Lesser | Lesser |
| Molar series length (carnassial size) | Greater | Lesser | Greater | Lesser | Lesser |
| Ascending Ramus height | Lesser | Greater | Lesser | Lesser | Medium |

Table 6. Skull features.

| Skull Features | Iberian Modern Wolves | Modern Dogs | P3580 (Wolf-GR) | TC'07.N231644 (Wolf-TC) |
|---|---|---|---|---|
| Snout elongation (cranial length) | Greater | Lesser | Medium | Greater |
| Dental series length | Greater | Lesser | Greater | Greater |
| Cranium breadth | Greater | Lesser | Greater | Greater |
| Facies height | Lesser | Greater | Greater | Lesser |
| Cranial breadth (at postorbital constriction) | Greater | Lesser | Greater | Greater |
| Frontal bone breadth | Lesser | Greater | Lesser | Lesser |
| Muzzle breadth | Lesser | Greater | Lesser | Lesser |

width. Even the largest wolves, do not show great frontal and muzzle width.

### *4.2 Geometric morphometry*

We analysed the shape variation of the mandibles with PC1 and PC4 (Figure 3A) and their respective wireframe. In PC1, the greatest variation appears in the molar series (especially by a larger size of the carnassial) and the coronoid process. Therefore, the positive values in PC1 are characterised by a greater ascending ramus height (coronoid process oriented towards aboral) and a shorter molar series length, in this case, it corresponds to the modern dogs. In the wireframe of PC4, a shape variation in the length and height of the mandible is observed. Dogs and wolves seem to overlap in this component, showing little differences between them.

We analysed the PCA of the skulls in the dorsal view (Figure 3B). The specimens located with positive values of PC1 are characterised by a greater cranial length (more snout elongation) and lesser frontal bone width. In general, most wolves falling into more positive values of PC1 display those characteristics. Whereas the dogs, located with negative values of PC1, display opposite features. Skulls with positive correlations in PC3, in general, are characterised by a lesser cranium breadth and a lesser cranial breadth (breadth in the postorbital constriction). Therefore, *a priori*, dogs would have a lesser cranium breadth and a lesser cranial breadth since they are placed in more positive values of this component. Whereas the wolves display the opposite traits.

Finally, we analysed the skulls in the side view (Figure 3C). Observing the wireframe, the positive values in PC1 are characterised by representing skulls of specimens with a lesser facies height, in this case, the wolves. With respect to PC2, skulls that have more positive values of this component are characterised by a greater cranium length; however, this component does not clearly discriminate between wolves and dogs.

### *4.3 Morphological differences between dogs and wolves. A comparison with fossil records from three ancient sites*

We have summarised in two tables the analyses of shape reflected through the different methods (TM and GM) regarding the most significant differences between modern wolves and dogs. The results obtained from the mandible analysis (Table 5) suggest that wolves are distinguished by a greater mandible size and a greater dental series length (especially the carnassial) and a lower ascending ramus height. Dog's skulls (Table 6) are distinguished by a lesser degree of snout elongation, a lower dental series length, and a lower width of the cranium. The muzzle and frontal bone of the skulls of the dogs studied are wider.

Finally, we carried out a preliminary study comparing modern canids with two Pleistocene wolves and two Paleolithic dogs. The Pleistocene wolf from Canyars (TC) is larger than the average modern wolves and yields undoubtful wolf craniomandibular features. The skull from Grotta Romanelli (P3580) is classified as a wolf by Sardella *et al.* (2014: 186), who performed a biometric study including several canid remains

(GR), and wolves from the Middle-Late Pleistocene of Apulia, revealing a wide overlap between *C. lupus* and *C. mosbachensis*. However, they concluded that the general morphology and proportions of the wolf from GR fall within the variability of *C. lupus*. Also, they argue that the small size of the GR specimen can be explained as southern wolves are smaller than northern wolves. Sardella *et al.* (2014: 190, 2018: 256) description of GR canids is: slender muzzle, short palatal area, tooth crowding, rounded neurocranium with a gently sloping forehead and low sagittal and nuchal crests, and reduced tympanic bullae. In our results, P3580 is characterised by a greater dental series length, greater cranium, and cranial breadth, in addition to a lesser frontal bone and muzzle breadth. All these features are present in modern wolves. This fossil has an average lengthening of the snout when compared to the rest of the wolf specimens. Furthermore, modern wolves present a smaller facies compared to this specimen.

The mandibles of paleolithic dogs are smaller and morphologically yield a lesser dental series size (especially a lesser carnassial size), than modern dogs. The paleolithic dog from Portalón (CMIA6.48.4) shows a slight 'crowding of the premolars', a feature that differentiates domestic dogs (Germonpré *et al.* 2015: 276). The chronology of this site (Middle Bronze Age) supports this evidence.

## 5 Conclusions

This work analyses the distinctive features observed in skulls and mandibles from wolves and dogs, by combining two morphometry methods (traditional and geometric) reinforcing the efficiency of the analyses. Both methods are useful for clarifying morphological differences between the two canids, however applying geometric morphometry, more complex changes can be observed. The results obtained from the mandibles analysed suggest that wolves are distinguished by a greater size of the mandible and a greater length of the dental series (specially the carnassial) and lower ascending ramus height. The dogs' skulls are distinguished by a lesser degree of snout elongation, lower dental series length and lower width of the cranium. Both snout and forehead of the dogs' sample studied, are wider.

The sample of modern wolves needs to be enlarged for a better knowledge of its variability. Although in this study we have selected the most complete fossil material available, fossils are usually incompletely preserved. The fossil collections we are studying (Portalón, Canyars, among other Pleistocene and Holocene sites) include an important number of fragmentary material that, nevertheless preserve important information on the undamaged regions. We will focus our research on the analysis of fossil remains of *Canis lupus* (especially dogs) which are damaged from excavations (sometimes consumed by humans) and this excludes them from studies, as they miss a complete set of variables to perform statistical analyses. We propose to focus the morphometric study on informative regions of mandible and maxilla (more likely to survive) that, as derived from the current study, have demonstrated to preserve valuable information to discriminate between dogs and wolves.

## Acknowledgments

The authors received funding from the Ministerio de Economía y Competitividad, Spain: PGC2018-093925-B-C33 (MCIU/AEI/FEDER,UE). We thank the Museum of Burgos (with special mention to M. Negro) for providing access to fossil materials recovered in digs between 1972 and 1983. We also thank Dra M. Sanz and Dr. J. Daura, responsible of the Terrasses del Canyars research project. We would like to thank Dr. W.H. Gilbert for his review and contribution to improve this paper.

## References

Bertè, D.F. 2013. L'evoluzione del genere *Canis* (Carnivora, Canidae, Caninae) in Italia dal *wolf-event* a oggi, implicazioni biocronologiche, paleoecologiche e paleoambientali. PhD thesis, Dipartimento di Scienze della Terra, Sapienza Università di Roma: 398.

Blanc, G.A. 1928. Grotta Romanelli. II. Dati ecologici e paletnologici. *Archivio per l'Archeologia e l'Etnologia* 58: 1–50.

Bookstein, F.L. 1991. *Morphometric Tools for Landmark Data: Geometry and Biology*. Cambridge: Cambridge University Press.

Boudadi-Maligne, M. and G. Escarguel 2014. A biometric re-evaluation of recent claims for Early Upper Palaeolithic wolf domestication in Eurasia. *Journal of Archaeological Science* 45: 80–89.

Carretero, J.M., A.I. Ortega, L. Juez, A. Pérez González, J.L. Arsuaga, R. Pérez Martínez and M.C. Ortega 2008. A Late Pleistocene-Early Holocene archaeological sequence of Portalón de Cueva Mayor (Sierra de Atapuerca, Burgos, Spain). *Munibe* 59: 67–80.

Daura, J., M. Sanz, N. García, E. Allué, M. Vaquero, E. Fierro, J.S. Carrión, J.M. López-García, H.A. Blain, A. Sánchez-Marco, C. Valls, R.M. Albert, J.J. Fornós, R. Julià, J.M. Fullola and J. Zilhão 2013. Terrasses de la Riera dels Canyars (Gavà, Barcelona): the landscape of Heinrich Stadial 4 north of the 'Ebro frontier' and implications for modern human dispersal into Iberia. *Quaternary Science Reviews* 60: 26–48. DOI: 10.1016/j.quascirev.2012.10.042

Drake, A.G. and C.P Klingenberg 2010. Large-scale diversification of skull shape in domestic dogs:

disparity and modularity. *The American Naturalist* 175 (3): 289–301.

Drake, A.G., M. Coquerelle, P.A. Kosintsev, O.P. Bachura, M. Sablin, A.V. Gusev, L.S. Fleming and R.J. Losey 2017. Three-Dimensional Geometric Morphometric Analysis of Fossil Canid Mandibles and Skulls. *Scientific Reports* 7 (9508): 1–8.

Driesch von den, A. 1976. *A guide to the measurement of animal bones from archaeological sites: as developed by the Institut für Palaeoanatomie, Domestikationsforschung und Geschichte der Tiermedizin of the University of Munich.* Peabody Museum Press, 1: 1–358.

Druzhkova, A.S., O. Thalmann, V.A. Trifonov, J.A. Leonard, N.V. Vorobieva, N.D. Ovodov, A.S. Graphodatsky and R.K. Wayne 2013. Ancient DNA Analysis Affirms the Canid from Altai as a Primitive Dog. *PLoS ONE* 8 (3): e57754.

Fornaca-Rinaldi, G. and A.M. Radmilli 1968. Datazione col método Th230/U238 di stalagmiti contenute in depositi mousteriani. *Atti della Società Toscana di Scienze Naturali* serie A 75: 639–646.

Germonpré, M., M.V. Sablin, R.E. Stevens, R.E.M. Hedges, M. Hofreiter, M. Stiller and V.R. Després, 2009. Fossil dogs and wolves from Palaeolithic sites in Belgium, the Ukraine and Russia: osteometry, ancient DNA and stable isotopes. *Journal of Archaeological Science* 36: 473–490. DOI: 10.1016/j.jas.2008.09.033

Germonpré, M., M. Lázničková-Galetová and M.V. Sablin 2012. Palaeolithic dog skulls at the Gravettian Předmostí site, the Czech Republic. *Journal of Archaeological Science* 39: 184–202. DOI: 10.1016/j.jas.2011.09.022

Germonpré, M., M. Lázničková-Galetová, R.J. Losey, J. Räikkönen and M.V. Sablin 2015. Large canids at the Gravettian Předmostí site, the Czech Republic: The mandible. *Quaternary International* 359–360: 261–279. DOI: 10.1016/j.quaint.2014.07.012

Germonpré, M., S. Fedorov, P. Danilov, P. Galeta, E-L. Jimenez, M. Sablin and R.J. Losey 2017. Palaeolithic and prehistoric dogs and Pleistocene wolves from Yakutia: Identification of isolated skulls. *Journal of Archaeological Science* 78: 1–19. DOI: 10.1016/j.jas.2016.11.008

Janssens, L., I. Spanoghe, R. Miller and S. Van Dongen 2016. Can orbital angle morphology distinguish dogs from wolves? *Zoomorphology* 135 (1): 149–158.

Kendall, D. 1977. The diffusion of shape. *Advances in Applied Probability* 9 (3): 428–430. doi:10.2307/1426091

Klingenberg, C.P. 2011. MorphoJ: an integrated software package for geometric morphometrics. *Molecular Ecology Resources* 11: 353–357.

Klingenberg, C.P. 2013. Visualizations in geometric morphometrics: how to read and how to make graphs showing shape changes. *Hystrix, the Italian Journal of Mammalogy* 24 (1): 15–24. DOI: 10.4404/hystrix-24.1-7691

Lindblad-Toh, K., C.M. Wade, T.S. Mikkelsen, E.K. Karlsson, D.B. Jaffe, M. Kamal, M. Clamp, J.L. Chang, E.J. Kulbokas III, M.C. Zody, E. Mauceli, X. Xie, M. Breen, R.K. Wayne, E.A. Ostrander, C.P. Ponting, F. Galibert, D.R. Smith, P.J. deJong, E. Kirkness, P. Alvarez, T. Biagi, W. Brockman, J. Butler, C-W. Chin, A. Cook, J. Cuff, M.J. Daly, D. DeCaprio, S. Gnerre, M. Grabherr, M. Kellis, M. Kleber, C. Bardeleben, L. Goodstadt, A. Heger, C. Hitte, L. Kim, K-P. Koepfli, H.G. Parker, J.P. Pollinger, S.M.J. Searle, N.B. Sutter, R. Thomas, C. Webber, Broad Institute Genome Sequencing Platform and E.S. Lander 2005. Genome sequence, comparative analysis and haplotype structure of the domestic dog. *Nature*, 438: 803–819. DOI: 10.1038/nature04338

Morey, D.F. 2014. In search of Paleolithic dogs: a quest with mixed results. *Journal of Archaeological Science* 52: 300–307.

Morey, D.F. and R. Jeger 2015. Paleolithic dogs: Why sustained domestication then? *Journal of Archaeological Science: Reports* 3: 420–428.

Müller, W. 2002. The domestication of the wolf - the inevitable first? in J.D. Vigne, J. Peters and D. Helmer (eds) The First Steps of Animal Domestication: New archaeozoological approaches: 34–40. Oxford: Oxbow.

Pérez-Romero, A., A. Alday, E. Iriarte, M. Francés-Negro, M.A. Galindo-Pellicena, A. Álvarez, L. Juez, J.L. Arsuaga and J.M. Carretero 2016. La cerámica de la Edad del Bronce en el yacimiento de El Portalón de Cueva Mayor (Sierra de Atapuerca, Burgos, España). *Munibe Antropologia-Arkeologia* 67: 1–22.

Rohlf, F.J. and L.F. Marcus 1993. A revolution in morphometrics. *Trends in Ecology & Evolution* 8 (4): 129–132.

Rohlf, F.J. 2017. tpsDig2. Versión 2.30. *Ecology & Evolution and Anthropology, Stony Brook University*. http://life.bio.sunysb.edu/morph/index.html

Sablin, M.V. and G.A. Khlopachev 2002. The earliest ice age dogs: Evidence from Eliseevichi 1. *Current Anthropology* 43 (5): 795–799.

Sardella, R., D. Bertè, D.A. Iurino, M. Cherin and A. Tagliacozzo 2014. The wolf from Grotta Romanelli (Apulia, Italy) and its implications in the evolutionary history of *Canis lupus* in the Late Pleistocene of Southern Italy. *Quaternary International* 328–329: 179–195. DOI: 10.1016/j.quaint.2013.11.016

Sardella, R., I. Mazzini, F. Giustini, B. Mecozzi, M. Brilli, D.A. Iurino, G. Lembo, B. Muttillo, M. Massussi, D. Sigari, S. Tucci and M. Voltaggio 2018. Grotta Romanelli (Southern Italy, Apulia): legacies and issues in excavating a key site for the Pleistocene of the Mediterranean. *Rivista Italiana di Paleontologia e Stratigrafia* 124 (2): 247–264.

Thalmann, O., B. Shapiro, P. Cui, V.J. Schuenemann, S.K. Sawyer, D.L. Greenfield, M.B. Germonpré, M.V. Sablin, F. López-Giráldez, X. Domingo-Roura, H. Napierala,

H-P. Uerpmann, D.M. Loponte, A.A. Acosta, L. Giemsch, R.W.Schimitz, B. Worthington, J.E. Buikstra, A. Druzhkova, A.S. Graphodatsky, N.D. Ovodov, N. Wahlberg,. A.H. Freedman, R.M. Schweizer, K.P. Koepfli, J.A. Leonard, M. Meyer, J. Krause, S. Pääbo, R.E. Green and R.K. Wayne 2013. Complete mitochondrial genomes of ancient canids suggest a European origin of domestic dogs. *Science* 342: 871–874.

Zelditch, M.L., D.L. Swiderski, H.D. Sheets and W.L. Fink, 2004. *Geometric Morphometrics for Biologists: A Primer*. San Diego: Elsevier Academic Press.

Detail from Germonpré et al. fig. 1 (chapter 2.2). Lateral view of the skull from the Pleistocene wolf from Trou des Nutons, Belgium (Royal Belgian Institute of Natural Sciences).

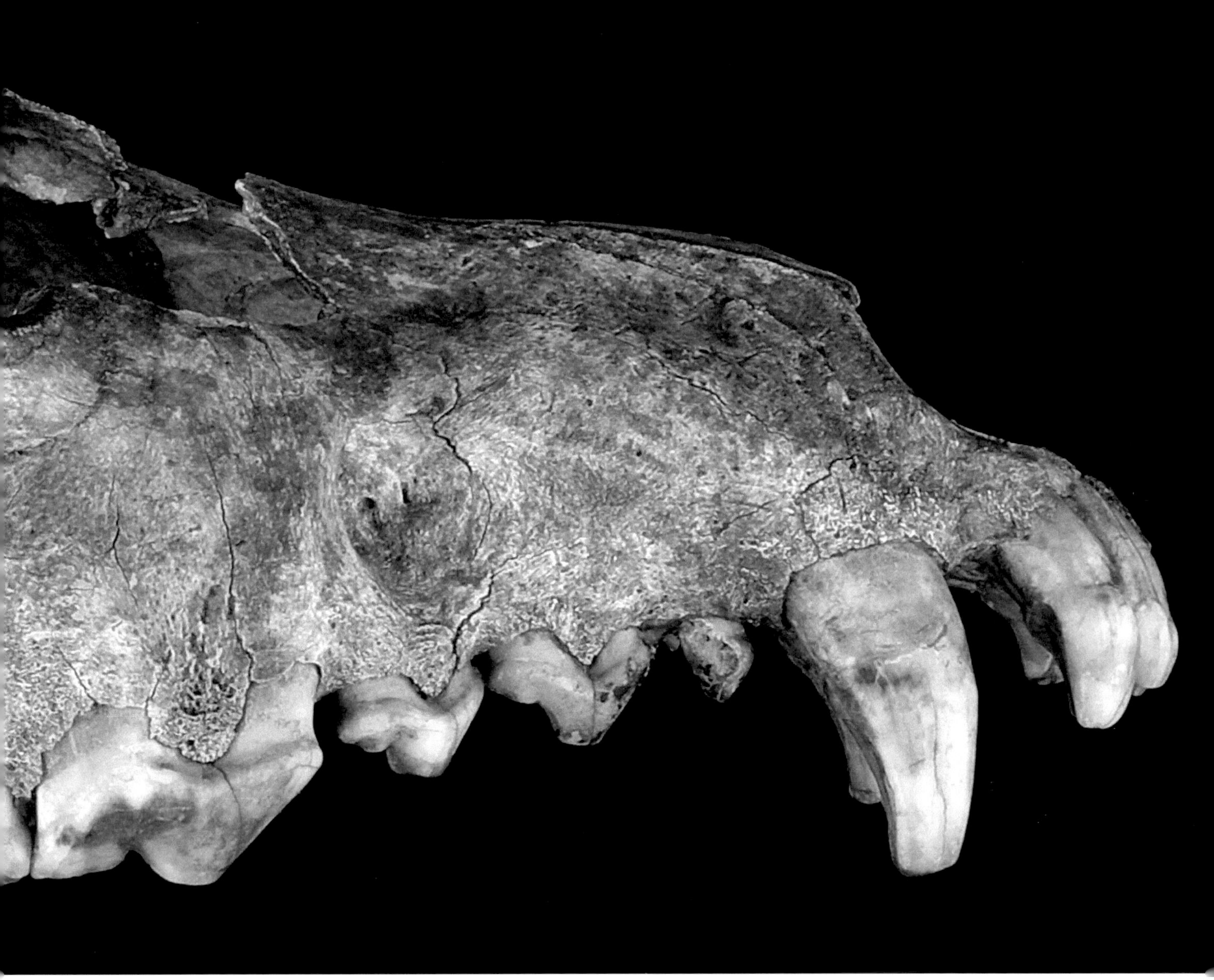

# Section 2
## Wolf Versus Dog

# 2.1 Size Variation of the Middle-Late Pleistocene Grey Wolf (*Canis lupus*) from the Italian Peninsula


Dawid Adam Iurino[1], Beniamino Mecozzi[1], Davide Persico[2], Lucia Maimone[2], Raffaele Sardella[1]

[1] Dipartimento di Scienze della Terra, PaleoFactory, Sapienza Università di Roma, Roma, Italy; dawid.iurino@uniroma1.it, beniamino.mecozzi@uniroma1.it, raffaele.sardella@uniroma1.it

[2] Dipartimento di Scienze Chimiche della Vita e della sostenibilità ambientale, Università di Parma, Parma, Italy; lucia.davide.persico@unipr.it, maimone@studenti.unipr.it

Corresponding author: Beniamino Mecozzi, beniamino.mecozzi@uniroma1.it



#### Abstract

Here we present a preliminary report on the biometric variability of the upper and lower carnassial of *Canis lupus* from several late Middle to Late Pleistocene sites of Apulia (southern Italy). Our results indicate a dimensional trend of wolves which allows two distinct chronological morphotypes to be identified, represented by medium-sized forms, reported until the MIS 4 followed by larger forms starting with the MIS 3.



**Keywords**: canids, carnivoran, Mediterranean, Apulia, biochronology.


## 1 Introduction

During the Late Pleistocene, the grey wolf (*Canis lupus*) was one of the most common species in the Eurasian carnivoran guild. In literature the earliest occurrence of *C. lupus* in Europe was reported in France from the late Middle Pleistocene site of Lunel Viel (MIS 11) (Brugal and Boudadi-Maligne, 2011). This attribution was widely accepted by specialists until the recent chronological revision of the levels 9–5 of this site, currently referred to MIS 9-7 (Uzunidis-Boutillier 2017, 2020) or MIS 9-8 (Brugal *et al.* 2019). According to this, the first occurrence of *C. lupus* in Europe would undergo a substantial change and the well-preserved skull from the Italian site La Polledrara di Cecanibbio (dated at 340–320 ka; MIS 9; Anzidei *et al.* 2012) could therefore represent one of the earliest and most reliable evidence of this taxon in Europe, but unfortunately a detailed description of this specimen is still lacking. Other remains referable to MIS 11 and ascribed to *C. lupus* consist of few isolated remains from Atapuerca TD10 (Spain, Cuenca-Bescos and Garcia 2007) and a lower $M_1$ from Castel di Guido (Italy, Sala and Barbi 1996; Petronio *et al.* 2019). The stratigraphy sequence of the Castel di Guido was dated between 327 ± 34 ka and 260 ± 37 with the US-ESR method (MIS 9, Michel *et al.* 2001, 2009), while other sequences of the area have been recently dated suggesting an older age for the deposit (412 ± 2 ka with $^{40}Ar/^{39}Ar$ , MIS 11, Marra *et al.* 2018). During this time span (MIS 11-9) the disappearance of *Canis mosbachensis* and the earliest dispersal of *C. lupus* occurred in Europe. The strong morphological affinity, the large biometric overlap and the intraspecific variability of the teeth documented in both species (Jiangzuo *et al.* 2018; Mecozzi *et al.* 2020a), make the attribution of the canid remains from MIS 12-8 still a problematic issue. The isolated teeth recovered from Atapuerca TD10 and Castel di Guido could represent the earliest occurrence of the species in Europe, but the scarcity of the material requires caution in taxonomic attribution. According to this, at present we prefer to consider the complete cranium from La Polledrara (Anzidei *et al.* 2012) the most reliable specimen to represent the first record of *C. lupus* in Europe.

Starting from the late Middle Pleistocene (early Aurelian), the grey wolf became a common element of the carnivoran guild in Europe (Sardella *et al.* 2014; Bertè and Pandolfi 2014; Mecozzi and Bartolini Lucenti 2018, Mecozzi *et al.* 2020a).

The taxonomic status of the late Middle to Late Pleistocene canids from the Italian Peninsula has long been debated, as the case of the well preserved material from the Level G of Grotta Romanelli site (Castro, Lecce), which was initially attributed to *Canis aureus* by Blanc (1920; 1928), successively referred to the small-sized *Canis mosbachensis* (Masini *et al.* 1991; Sala *et al.* 1992) and finally classified as *C. lupus* through morphological, biometric and CT-scan analysis (Sardella *et al.* 2014). The sample from Grotta

Figure 1. Fossil remains of *Canis lupus* from the Apulian Peninsula. (a) P3592, left hemimandible from Grotta Romanelli G; (b) P3590, left hemimandible from Grotta Romanelli G; (c) A8-25, right hemimandible from Avetrana 8; (d) AND17, right hemimandible from Avetrana 8; (e) CC394, right hemimandible from Cardamone; (f) CC10, right maxillary from Cardamone; (g) IN634, left hemimandible from Ingarano, (h) IN635, left hemimandible from Ingarano; (i) IN504, cranium from Ingarano; (j) MPND860, left hemimandible from Melpignano (Cava Bianco); (k) MPND859, right hemimandible from Melpignano (Cava Bianco); (l) MPND842, left maxillary from Melpignano (Cava Bianco); (m) M7449, cranium from Grotta della Jena; (n) M7450, mandible from Grotta della Jena; (o) ZINZ.B6 3-1-59, left hemimandible from Grotta Zinzulusa; (p) SSND201, left maxillary from San Sidero (SS3); (q) PU100373, left hemimandible from San Sidero 1. Scale bar: 3 cm.

Romanelli (Level G) is characterised by reduced size and slender cranial morphologies especially if compared with the Late Pleistocene wolves from Italy (Sardella *et al.* 2014).

Some authors suggest that an increasing of size during the last 300 ka can be recognised, (Bonifay 1971; Sansalone *et al.* 2015; Mecozzi and Bartolini Lucenti 2018), and some chronosubspecies were instituted on the basis of biometric analysis: *C. lupus lunellensis* (late Middle Pleistocene), *C. lupus santenaisiensis* (early Late Pleistocene) and *C. lupus maximus* (late Late Pleistocene) (Brugal and Boudadi-Maligne 2011; Boudadi-Maligne 2012).

In recent years, several biometric studies on fossil wolves from the Italian Peninsula have been performed, attesting the presence of two distinct chronological morphotypes: slender forms spanning from the late Middle to early Late Pleistocene, and robust ones from the second part of the Late Pleistocene (after MIS 3) (Sardella *et al.* 2014; Bertè and Pandolfi 2014; Sansalone *et al.* 2015; Mecozzi and Bartolini Lucenti 2018; Mecozzi *et al.* 2020a). Despite this, further studies are needed to clarify the time and mode of the dispersal of the robust *C. lupus* in the Italian Peninsula. The peculiar geographical position of the Italian Peninsula and the richness of the canid fossil record represents a great blend to depict the evolutionary history of this iconic predator in the framework of Mediterranean Europe.

Here, we present a large sample of *C. lupus* from late Middle to Late Pleistocene sites of Apulia (southern Italy) (Figure 1). Biometric comparison of the upper and lower carnassial provides preliminary remarks on the biochronology and body size variation of Italian wolves.

Table 1. Fossil material of *Canis lupus* from the late Middle to Late Pleistocene sites of the Italian Peninsula.

| Sites+A1:G33 | Abbreviations | Region | Age | MIS | References | Repository |
|---|---|---|---|---|---|---|
| La Polledrara di Cecanibbio | LPo | Latium | Middle Pleistocene | 9 | Anzidei *et al.* (1989) | |
| Melpignano - Cava Bianco | CBa | Apulia | | 9-8? | This work | PL |
| Melpignano - Cava Nuzzo | CNu | Apulia | | 9-8? | This work | PL |
| Melpignano - Mirigliano | Mir | Apulia | | 9-8? | This work | MPUN |
| San Sidero - SS3 | SS3 | Apulia | | 9-8? | This work | PL |
| Grotta del Poggio | GPo | Campania | | 6 | Sala (1979) | |
| Grotta dei Ladroni | GLa | Apulia | Late Pleistocene | 5 | This work | IsIPU |
| Monte Tignoso | MTi | Tuscany | | 5 | Del Campana (1909) | |
| Grotta Laceduzza | GLc | Apulia | | 4 | This work | MPCCSM |
| Grotta Romanelli - Level G | GRo | Apulia | | 4-3 | Sardella *et al.* (2014); this work | MUCIV |
| Grotta di Sant'Agostino | GSA | Campania | | 4-3 | Tozzi (1970) | |
| Grotta Tina | GTi | Campania | | 4-3 | Martini *et al.* (1974) | |
| Riparo Mochi | RMo | Ligury | | 3 | Arellano (2009) | |
| Buca della Iena | BdI | Tuscany | | 3 | Pitti and Tozzi (1971) | |
| Cava Spagnulo | CSp | Apulia | | 3 | Mecozzi *et al.* (2017) | |
| Caverna Pocala | CPo | Friuli-Venezia Giulia | | 3 | Bertè (2013) | |
| Grotta del Broion | GBr | Veneto | | 3 | Bertè (2013) | |
| Grotta del Principe | GPr | Ligury | | 3 | Arellano (2009) | |
| Grotta Mora Cavorso | GMC | Latium | | 3 | Salari *et al.* (2017) | |
| Ingarano | Ing | Apulia | | 3 | This work | PL |
| Grotta della Masseria del Monte | GMM | Apulia | | 3 | Anelli (1959) | |
| San Sidero 1 | SS1 | Apulia | | 3 | This work | MGP |
| San Sidero 2 | SS2 | Apulia | | 3 | This work | MGP |
| Sternatia | Ste | Apulia | | 3 | Rustioni *et al.* (1994) | |
| Avetrana 8 | Ave | Apulia | | 3 | Mecozzi and Bartolini Lucenti (2018); this work | MUST |
| Riparo Fumane | RFu | Veneto | | 3 | Cassoli and Tagliacozzo (1994) | |
| Grotta della Jena | GJe | Apulia | | 3-2 | This work | MPUN |
| Buco del Frate | BFr | Lombardy | | 2 | Bertè (2013) | |
| Cardamone | Car | Apulia | | 2 | This work | ITCGC |
| Covoli di Velo | CVe | Veneto | | 2 | Bertè (2013) | |
| Grotta Paglicci | GPa | Apulia | | 2 | Bertè (2013); Boscato (1994) | |
| Grotta Zinzulusa | GZi | Apulia | | 2 | This work | IsIPU |

## 2 Material and methods

The studied material was collected from localities in Apulia (Figure 1) and is housed in different Italian Institutions and Museums (Table 1): PaleoFactory Laboratory, Department of Earth Science, Sapienza, University of Rome (PL); Paleontological Museum of the University of Naples Federico II (MPUN); Italian Institute of Human Paleontology (IsIPU); Commercial Technical Institute 'Galilei-Costa' (ITCGC); Museum of Geology and Paleontology of Turin (MGP); Museum of Pre-Classical Civilisations of Southern Murgia (MPCCSM), Museum of Civilisation (former National Museum of Prehistory and Ethnography 'Luigi Pigorini') (MUCIV), University Museum of Earth Sciences, Sapienza University of Rome (former Museum of Paleontology, MPUR) (MUST).

Set up our comparison dataset by considering the available studies on *C. lupus* specimens from the late Middle to Late Pleistocene of the Italian Peninsula (Table 1).

The length of the upper ($P^4$) and lower ($M_1$) carnassial were taken to the nearest 0.1 mm with a digital caliper following Driesch (1976).

The average values of the teeth of the Apulian specimens were compared with literature data on late Middle to Late Pleistocene specimens from Italy standard univariate plot.

## 3 *Canis lupus* from several late Middle to Late Pleistocene sites of Apulia

### 3.1 The Apulian region

Apulia, which extends for more than 350 km from the north-west to the south-east, and with a minimum extension of 30 km along the south-west-north-east axis, occupies the southern part of the Italian Peninsula and, due to its conformity, can be considered as a peninsula within a peninsula. In this region many fossiliferous sites have been discovered since the end of the 1800s. Many of these, also include evidence of human occupation in this area with fossils attributed to *Homo neanderthalensis* and *Homo sapiens* (e.g., Fondo Cattiè, Corridi, 1989, Grotta Romanelli, Sardella *et al.* 2019), and a rich amount of artefacts attributable to the Middle and Upper Palaeolithic.

From a paleontological perspective, mammal faunas recovered from Apulian deposits are a reference point for the Italian large mammal biochronological scale, especially for the late Middle to Late Pleistocene (Grotta Romanelli, Ingarano, Melpignano, San Sidero; Petronio *et al.* 2007). Difficulties in reconstructing the evolution of mammal faunas in Apulia persist since most of the sites lacks clear chronological constrains. A new project to study the Pleistocene mammal faunas of Apulia started in 2016, with the aim to revise the fossil samples from several localities and their chronological context. For instance, many deposits have long been chronologically referred to the early Late Pleistocene (MIS 5, e.g., Melpignano, San Sidero, Grotta Romanelli; Sala *et al.* 1992; Bologna *et al.* 1994; Petronio *et al.* 2007), but the ongoing revision would suggest an older age (Meccozzi *et al.* 2021a).

### 3.2 Apulian sites

Here we discuss the age of *C. lupus*-bearing sites from the late Middle to the Late Pleistocene of Apulia.

The older remains were found in the karst fissures of Melpignano and San Sidero (fossiliferous area of Maglie, Lecce). The age of the deposits was generally attested at the early Late Pleistocene (MIS 5). A recent revision of the large mammal remains from these localities expanded their chronological attribution, with some faunal assemblages attributed to the late Middle Pleistocene (MIS 9-8, Mirigliano Collection, Cava Nuzzo, Cava Bianco of Mepignano and SS3 of San Sidero), whereas others were referred to the Late Pleistocene (MIS 3, Fissure 1 and Fissure 2 of San Sidero) (Mecozzi *et al.* 2021a; Table 1).

A preliminary faunal list from the lower deposit of Grotta Laceduzza was reported by Mecozzi *et al.* (2019). The study of mammal remains is still in progress (Table 1).

The canid sample from the level G of Grotta Romanelli was described by Sardella *et al.* (2014), including a nearly complete cranium. This cave is long considered a key site for the study of the past Mediterranean ecosystems thanks to its archaeological and palaeontological content, and the relative stratigraphical, geomorphological and radiometric data (Sardella *et al.* 2014, 2018, 2019). The ongoing revision of the stratigraphical succession and chronological framework of the deposit does not exclude a possible late Middle Pleistocene age for the lower complex (including the level G). Despite this, the canid sample from the level G is referred to MIS 4-3, following the historical framework.

Mammal remains from Ingarano have been studied in a number of works (Capasso Barbato *et al.* 1992; Petronio *et al.* 1996; Petronio and Sardella 1998; Bedetti and Pavia 2007; Iurino 2014; Iurino *et al.* 2015; Mecozzi *et al.* 2020b, 2021b). The level B was dated with the $^{329}$Th/$^{234}$U method providing an age of 40 ± 2 ka. According to Bedetti and Pavia (2007), the vertebrate fossil remains and lithic artefacts were accumulated during the MIS 3.

The canid sample from Avetrana analysed here, was collected from bed 8. The revision of the mammal fauna from this level suggests a chronological attribution to MIS 3 (Mecozzi and Bartolini Lucenti 2018; Salari *et al.* 2019).

Remains of *C. lupus* from Grotta della Jena were figured by Giuscardi (1875) (Figure 2). Fieldwork activities on this deposit were conducted by F. Anelli during the 1950s, who also reported a preliminary faunal list (Anelli 1956, 1959), but the material was not formally described and figured. The faunal composition suggests an attribution to the latest Pleistocene (end of MIS 3-2). Here, we present the material of Giuscardi's collection.

Grotta Zinzulusa is one of the most renowned caves of Apulia, annually visited by approximately 70,000 people (Sardella *et al.* 2019). The cave includes several infilling successions which yielded rich vertebrate fossil and lithic samples (Iannucci *et al.* 2020). The studied material come from the upper levels B5-3 of the Antro B succession, chronologically referred to MIS 2.

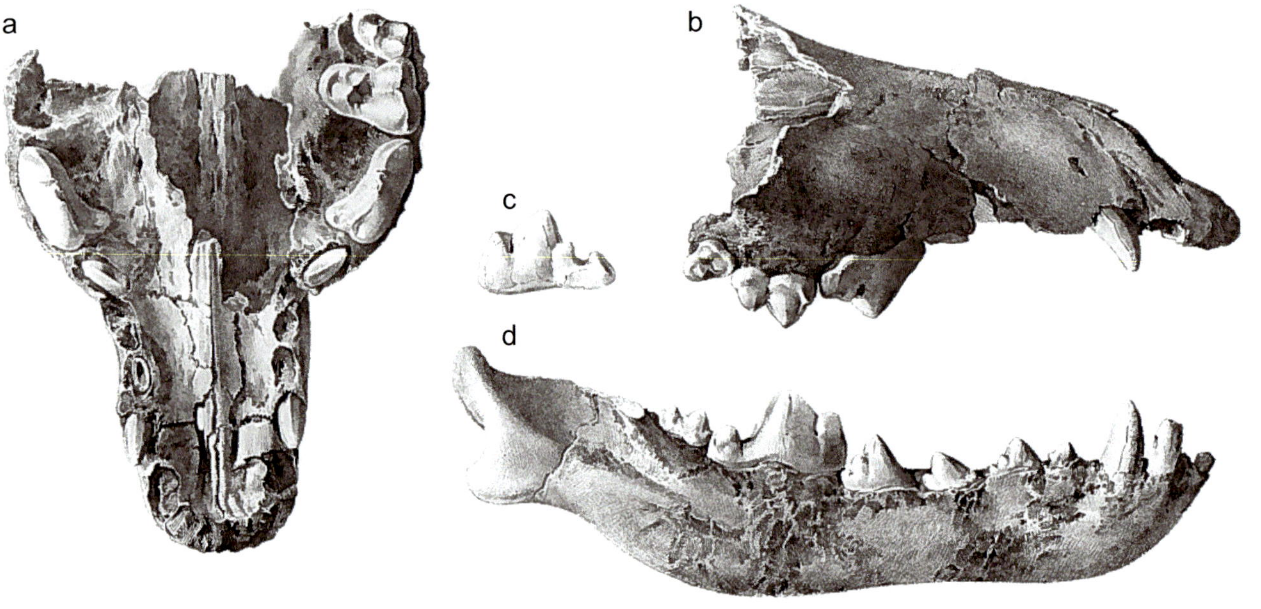


Figure 2. Historical representation of *Canis lupus* from Grotta della Jena. (a) rostrum in occlusal view; (b) rostrum in right lateral view; (c) $M_1$ in lingual view; (d) right hemimandible in right lateral view. After Giuscardi (1873).

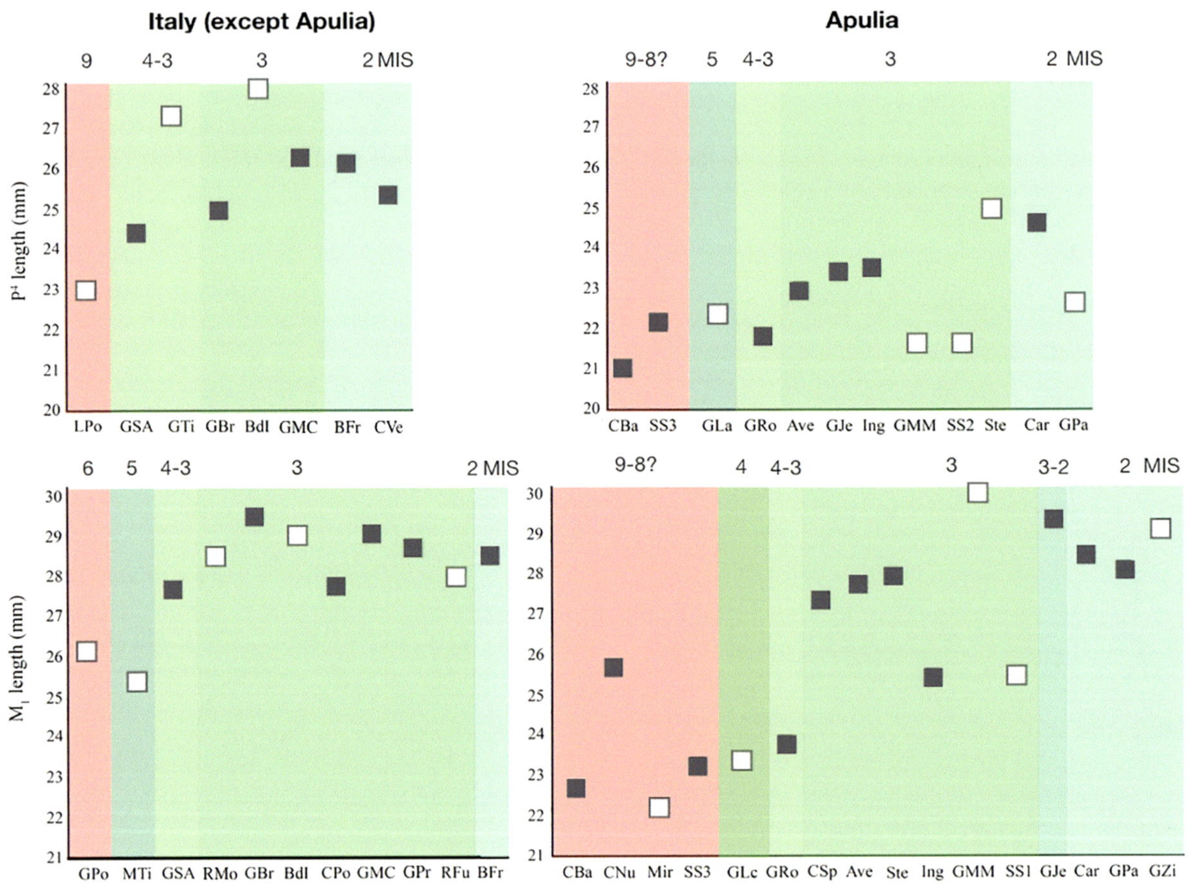


Figure 3. Variation in average length of the $P^4$ and $M_1$ of the Italian fossils of *Canis lupus* from the late Middle to Late Pleistocene sites. White squares indicate the presence of only one specimen per site, while grey squares indicate the presence of several specimens per site. For abbreviations see Tab. 1.

The faunal assemblage recovered from the karst infilling deposit of Cardamone was recently revised by Rustioni *et al.* (2003). The mammal fauna includes the woolly rhino (*Coelodonta antiquitatis*) and the woolly mammoth (*Mammuthus primigenius*), typical elements of the '*Mammuthus-Coelodonta* Faunal Complex'. Following this, the faunal assemblage was chronologically attributed to the climax of the Last Glacial Maximum (22–18 kyr).

### 4 Discussion and conclusions

According to our results, the relative increasing body size of the studied sample fits the general trend reported in literature (Bertè and Pandolfi, 2014; Sardella *et al.* 2014; Sansalone *et al.* 2015; Mecozzi and Bartolini Lucenti 2018), but with a different resolution between the upper and lower carnassial. The size variation of the $P^4$ during the time is less informative if compared to the $M_1$ in both the Italian and Apulian wolves (Figure 3). Focusing on the Italian specimens, those referred to MIS 9-5 would seem to be smaller compared to those from MIS 4-2, although the former are represented only by 1 $P^4$ and 2 $M_1$ respectively. More data for MIS 9-5 are required to better evaluate the time related variation of the size in Italian wolves.

Conversely, for the Apulian record the dimensional trend seems to be clearer, especially if the lower carnassial is compared to the sample from MIS (9-4), smaller than that from MIS 3-2 (Figure 2). Moreover, the specimens from Avetrana 8, Cava Spagnulo, Grotta della Jena and Grotta della Masseria del Monte show a larger size than those from Ingarano and San Sidero 1 (single $M_1$) indicating a considerable variation of the $M_1$ size during the MIS 3.

Excluding the sites of Ingarano and San Sidero 1, the length of the lower carnassial of the Apulian sample referable to MIS 3-2 overlaps with the Late Pleistocene (MIS 4-2) sample from Italian deposits (Figure 2).

However, among the other Italian sites, three are located in southern Italy, Grotta del Poggio (Sala 1979), Grotta di Sant'Agostino (Tozzi, 1970) and Grotta Tina (Martini *et al.* 1974). The first is referred to the late Middle Pleistocene (MIS 6), whereas the age of the other two is still unclear (probably MIS 4-3). The $P^4$ from Grotta Tina shows the largest size when compared to the whole Apulian record, whereas the dimension of the specimens from Grotta di Sant'Agostino are closer to those of Sternatia and Cardamone (Figure 2). For the $M_1$, the size of the specimen from Grotta del Poggio is smaller than those of MIS 4-2 from the Italian and Apulian sites. A few $M_1$s recognised from Grotta di Sant'Agostino are similar to those from Avetrana 8, Cava Spagnulo, Sternatia and Grotta Paglicci (Figure 2).

The biometric variation, especially of the $M_1$, can be interpreted as a consequence of a possible dispersal event from North to South of larger wolves which occurred during MIS 3 (Bertè and Pandolfi 2014; Sansalone *et al.* 2015; Mecozzi and Bartolini Lucenti 2018). Starting from MIS 2, the dimensions of the $M_1$ are very similar throughout the Italian Peninsula.

Unfortunately, Avetrana 8, Cava Spagnulo, Grotta della Jena and Grotta della Masseria del Monte have only been dated on a biochronological basis, preventing a detailed chronological attribution of this dispersal event.

In conclusion, the difference in the carnassial size, and in particular of the $M_1$, observed from the late Middle Pleistocene to early Late Pleistocene of the Italian record, can be interpreted as a dimensional trend of wolves, with medium-sized forms reported until the MIS 4 followed by larger ones starting with MIS 3. Compared to European fossil wolves, the size of the specimens from MIS 3-2 of the Italian Peninsula is closer to the large forms recognised in the second part of the Late Pleistocene (generally named *C. lupus maximus*, *sensu* Boudadi-Maligne 2012). The analysis of a wider sample and more craniodental characters is required to confirm such a trend.

### Acknowledgements

We wish to thank Ivana Fiore, Francesca Lugli and the Organising Committee of the Meeting 'Dogs, Past and Present - an Interdisciplinary Perspective', Rome, National Research Council (CNR) (Italy), 14th 17th November, 2018.

We are deeply indebted to the Soprintendenza Archeologia, Belle Arti e Paesaggio delle province di Brindisi, Lecce e Taranto for the permission of the research and field activities (2015–2017 and 2018–2020, resp. R. Sardella)

The authors thank: Donato Coppola (MPCCSM); Maria Carmela Del Re (MPUN); Addolorata Mazzotta and Carlo Viva (ITCGC), Francesca Alhaique (MUCIV), Marco Pavia (MPG), Linda Riti and Michele Macri (MUST) for granting access to the material and for their courtesy and helpfulness. We would like to thank the Istituto Italiano di Paleontologia Umana (IsIPU), and L. Bruni, for the access to the fossil collection from several sites of the Apulian Peninsula. This research was funded by Sapienza University of Rome 'Grandi Scavi' 2016-2017-2018-2019-2020 grants (RS).

## References


Anelli, F. 1956. *Un raro reperto di stambecco nella grotta della Jena presso Castellana (Bari).* 7 Congresso Nazionale di Speleologia, 105–107.

Anelli, F. 1959. Prime ricerche paleontologiche della grotta della Masseria del monte Conversano Murge in Bari. *Le grotte d'Italia*, 3, 1–34.

Anzidei, A.P., A. Angelelli, A. Arnoldus-Huyzendveld, L. Caloi and M.R. Palombo 1989. Le gisement pléistocene de la Polledrara di Cecanibbio (Rome, Italie). *L'Anthropologie* (Paris), 93(4), 749–781.

Anzidei, A.P., G.M. Bulgarelli, P. Catalano, E. Cerilli, R. Gallotti, C. Lemorini, S. Milli, M.R. Palombo, W. Pantano and E. Santucci 2012. Ongoing research at the late Middle Pleistocene site of La Polledrara di Cecanibbio (central Italy), with emphasis on human-elephant relationships. *Quaternary International*, 255, 171–187.

Arellano, A. 2009. Les grands mammifères des niveaux moustériens de l'Abri Mochi (Grotte de Grimaldi, Ventimille, Italie). Fouilles de 1949. *Bulletin du musée d'Anthropologie préhistorique de Monaco*, 49, 29–39.

Bedetti, C. and M. Pavia 2007. Reinterpretation of the late Pleistocene Ingarano cave deposit based on the fossil bird associations (Apulia, South-Eastern Italy). *Rivista Italiana di Paleontologia e Stratigrafia, 113, 487–507.*

Bertè, D.F. 2013. L'evoluzione del genere *Canis* (Carnivora, Canidae, Caninae) in Italia dal wolf-event a oggi: implicazioni biocronologiche, paleoecologiche e paleoambientali. 390 pp. PhD Dissertation, Sapienza, Università di Roma, Roma.

Bertè, D.F. and L. Pandolfi 2014. *Canis lupus* (Mammalia, Canidae) from the late pleistocene deposit of Avetrana (Taranto, Southern Italy). *Rivista Italiana di Paleontologia e Stratigrafia*, 120(3), 367–379.

Blanc, G.A. 1920. Grotta Romanelli I. Stratigrafia dei depositi e natura e origine di essi. *Archivio per l'Antropologia e la Etnologia* 50: 1–39.

Blanc, G.A. 1928. Grotta Romanelli. II. Dati ecologici e paletnologici. *Archivio per l'Archeologia e l'Etnologia* 58, 1–50.

Bonifay, M.F. 1971. Carnivores quaternaires du Sud-Est de la France. Mémoires du Museum national d'histoire naturelle, Série C, 21, 43–377.

Boscato, P. 1994. Grotta Paglicci: la fauna a grandi mammiferi degli strati 22–24 (Gravettiano antico-Aurignaziano). *Rivista di Scienze Preistoriche*, 46, 145–176.

Boschian, G., D. Caramella, D. Saccà and R. Barkai 2019. Are there marrow cavities in Pleistocene elephant limb bones, and was marrow available to early humans? New CT scan results from the site of Castel di Guido (Italy). *Quaternary Science Reviews*, 215, 86–97.

Boudadi-Maligne, M. 2012. Une nouvelle sous-espèce de loup (*Canis lupus maximus* nov. subsp.) dans le Pléistocène supérieur d'Europe occidentale. *Comptes Rendus Palevol*, 11(7), 475–484.

Brugal, J.P. and M. Boudadi-Maligne 2011. Quaternary small to large canids in Europe: taxonomic status and biochronological contribution. *Quaternary International*, 243, 171–182.

Brugal, J.P., A. Argant, M. Boudadi-Maligne, E. Crégut-Bonnoure, R. Croitor, P. Fernandez, J.B. Fourvel, P. Fosse, J.L. Guadelli, B. Babe, P. Magniez and A. Uzunidis 2019. Pleistocene herbivores and carnivores from France: An updated overview of the literature, sites and taxonomy. *Annales de Paléontologie*, 106(2), 102384.

Capasso Barbato, L., P.F. Cassoli, M.R. Minieri, C. Petronio, R. Sardella and M. Scarano, 1992. Note preliminari sulla fauna pleistocenica di Ingarano (Apricena, Foggia). *Bollettino Società Paleontologica Italiana*, 31, 325–334.

Cassoli, P.F. and A. Tagliacozzo 1994. Considerazioni paleontologiche, paleoecologiche e archeozoologiche sui macromammiferi e gli uccelli dei livelli del Pleistocene superiore del Riparo di Fumane (VR) scavi 1988e91. *Bollettino del Museo Civico di Storia Naturale di Verona*, 18, 349–445.

Cuenca-Bescós G. and García, N. 2007. Courier Forschunginstitut Senckenberg, 259, 99–110.

Del Campana, D. 1909. Vertebrati fossili di Monte Tignoso (Livorno). *Bollettino Società Geologica Italiana*, 28, 349–388.

Driesch von den, A. 1976. *A guide to the measurement of animal bones from archaeological sites.* Peabody Museum Bulletins 1: 1–148.

Gatta, M., M.F. Rolfo, C. Petronio, L. Salari and L. Silvestri 2016. Late Pleistocene skeleton of *Canis lupus* L., 1758 from Grotta Mora Cavorso (Jenne, Latium, central Italy). *Comptes Rendus Palevol*, 15, 941–949.

Giuscardi, C. 1873. Di una grotta con ossami nella provincia di Bari. *Atti della Regia Accademia delle Scienze Fisiche e Matematiche*, Napoli, 8, 1–13.

Iurino, D.A. 2014. Body size reduction and tooth agenesis in late pleistocene *Meles meles* (Carnivora, Mammalia) from Ingarano (southern Italy). *Rivista Italiana di Paleontologia e Stratigrafia*, 120, 109–118.

Iurino, D.A., A. Profico, M. Cherin, A. Veneziano, L. Costeur and R. Sardella 2015. A lynx natural brain endocast from Ingarano (southern Italy; late pleistocene): taphonomic, morphometric and phylogenetic approaches. *Hystrix*, 26, 110–117.

Jiangzuo, Q., J. Liu, J. Wagner, W. Dong and J. Chen 2018. Taxonomical revision of fossil *Canis* in Middle Pleistocene sites of Zhoukoudian, Beijing, China a review of fossil records of *Canis mosbachensis variabilis* in China. *Quaternary International.* 482, 93–108.

Martini, F., B. Sala, G. Bartolomei, M. Tonon and L. Cattani 1974. La Grotta Tina a Marina di Camerota (Salerno). *Bullettino di Paletnologia Italiana Roma.* 81, 27–79.

Masini, F., B. Sala, P. Ambrosetti, A. Azzaroli, G. Ficcarelli, T. Kotsakis, L. Rook and D. Torre 1991. *Mammalian faunas of selected Villafranchian and Galerian localities.* INQUA SEQS, Subcomission for European Quaternary Stratigraphy, Cromer Symposium, Norwich.

Mecozzi, B. and S. Bartolini Lucenti 2018. The Late Pleistocene *Canis lupus* (Canidae, Mammalia) from Avetrana (Apulia, Italy): reappraisal and new insights on the European glacial wolves. *Italian Journal of Geosciences*, 137, 138–150.

Mecozzi, B., D.A. Iurino, D.F. Berte and R. Sardella 2017. *Canis mosbachensis* (Canidae, Mammalia) from the Middle Pleistocene of Contrada Monticelli (Putignano, Apulia, southern Italy). *Bollettino della Società Paleontologica Italiana*, 56 (1), 71–78.

Mecozzi, B., A. Chakroun, H. Baills, D.A. Iurino, R. Coppola and D. Coppola 2018. Late Pleistocene mammal faunal assemblage from Cava Spagnulo (Grottaglie, Apulia, SE Italy). *Alpine and Mediterranean Quaternary*, 31, 5–19.

Mecozzi, B., D. Coppola, D.A. Iurino, R. Sardella and A.M. De Marinis 2019. The Late Pleistocene European badger *Meles meles* from Grotta Laceduzza (Brindisi, Apulia, Southern Italy): the analysis of the morphological and biometric variability. *Science of Nature* 106: 13.

Mecozzi, B., D.A. Iurino, A. Profico, C. Rosa and R. Sardella 2020a. The wolf from the Middle Pleistocene site of Ostiense (Rome): the last occurrence of *Canis mosbachensis* (Canidae, Mammalia) in Italy. *Historical Biology*, 1–12. DOI:10.1080/08912963.2020.1769090

Mecozzi, B., S. Bartolini Lucenti and D.A. Iurino 2020b. *Cuon alpinus* (Pallas, 1811) from the Late Pleistocene site of Ingarano (Foggia, southern Italy) and insights on the Eurasian Middle to Late Pleistocene record. *Alpine and Mediterranean Quaternary*, 33(1), 89–98.

Mecozzi, B., L. Belluci, F. Giustini, A. Iannucci, D.A. Iurino, I. Mazzini, F. Strani and R. Sardella 2021a. A reappraisal of the Pleistocene mammal assemblages from the karst infilling deposits of Maglie area (Lecce, Apulia, Southern Italy). *Rivista Italiana di Paleontologia e Stratigrafia*, in press.

Mecozzi, B., R. Sardella, A. Boscaini, M. Cherin, L. Costeur, J. Madurell-Malapeira, M. Pavia, A. Profico and D.A. Iurino 2021b. The tale of a short-tailed cat: New outstanding Late Pleistocene fossils of *Lynx pardinus* from southern Italy. *Quaternary Science Reviews*, 106840. DOI: 10.1016/j.quascirev.2021.106840

Michel, V., G. Boschian and P. Valensi 2009. Datation US-ESR de dents d'aurochs du site Paléolithique inférieur de Castel di Guido (Italie). ArcheoSciences. *Revue d'archéométrie*, 32, 51–58.

Petronio, C. and R. Sardella 1998. Remarks on the stratigraphy and biochronology of the late pleistocene deposit of Ingarano (Apulia, Southern Italy). *Rivista Italiana Paleontologia e Stratigrafia*, 104, 287–294.

Petronio, C., E. Billia, L. Capasso Barbato, G. Di Stefano, M. Mussi, S.J. Parry, R. Sardella and M. Voltaggio 1996. The Late Pleistocene fauna from Ingarano (Gargano, Italy): biochronological, palaecological, palaeoethnological and geochronological implications. *Bollettino della Società Paleontologica Italiana*, 34, 333–339.

Petronio, C., D.E. Di Canzio and L. Salari 2007. The Late Pleistocene and Holocene Mammals in Italy: new biochronological and paleoenvironmental data. *Palaeontographica Abteilung* A., 279, 147–157.

Pitti, C. and C. Tozzi 1971. La Grotta del Capriolo e la Buca della Iena presso Mommio (Camaiore, Lucca). *Rivista di Scienze Preistoriche,* 26: 213–258.

Rustioni, M., P. Mazza, L. Abbazzi, M. Delfino, L. Rook, S. Petrucci and F. Vianello 1994. The würmian fauna from Sternatia (Lecce, Apulia, Italy). *Bollettino della Società Paleontologica Italiana*, 33 (2), 279–288.

Rustioni, M., M.P. Ferretti, P. Mazza, M. Pavia and A. Varola 2003. The vertebrate fauna from Cardamone (Apulia, southern Italy): an example of Mediterranean mammoth fauna. *Deinsea*, 9 (1), 395–404.

Sala, B. 1979. Le faune pré-würmienne des grands mammifères de la Grotte du Poggio (Marina di Camerota, Salerno). *Atti della società toscana di Scienze Naturali*, 86, 77–99.

Sala, B., F. Masini, G. Ficcarelli, L. Rook and D. Torre 1992. Mammal dispersal events in the Middle and Late Pleistocene of Italy and Western Europe. *CFS Courier Forschungsinstitut Senckenberg*, 153, 59–68.

Salari, L., K.F. Achino, M. Gatta, C. Petronio, M.F. Rolfo, L. Silvestri and L. Pandolfi 2017. The wolf from Grotta Mora Cavorso (Simbruini mountains, Latium) within the evolution of *Canis lupus* L., 1758 in the Quaternary of Italy. *Palaeogeography, palaeoclimatology, palaeoecology*, 476, 90–105.

Salari, L., C. Petronio, T. Kotsakis, G. Di Stefano, F. Grossi, L. Maiorino, L. Pandolfi, M.F. Rolfo, F.D. Ruiu, G. Sansalone and A. Tagliacozzo 2019. Reassessing the faunal assemblages of the Late Pleistocene stratified karst filling from Avetrana (Apulia, southern Italy): the bed 8, palaeoenvironment and biochronology. *Alpine and Mediterranean Quaternary*, 32 (2), 1–55.

Sansalone, G., D.F. Bertè, L. Maiorino and Pandolfi, L. 2015. Evolutionary trends and stasis in carnassial teeth of European Pleistocene wolf *Canis lupus* (Mammalia, Canidae). *Quaternary Science Reviews*, 110, 36–48.

Sardella, R., D. Bertè, D.A. Iurino, M. Cherin and A. Tagliacozzo 2014. The wolf from Grotta Romanelli (Apulia, Italy) and its implications in the evolutionary history of *Canis lupus* in the Late Pleistocene of Southern Italy. *Quaternary International*, 328, 179–195.

Sardella, R., I. Mazzini, F. Giustini, B. Mecozzi, M. Brilli, D.A. Iurino, G. Lembo, B. Muttillo, M. Massussi, D. Sigari, S. Tucci and M. Voltaggio 2018. Grotta Romanelli (Southern Italy, Apulia): legacies and issues in excavating a key site for the Pleistocene of the Mediterranean. *Rivista Italiana di Paleontologia e Stratigrafia*, 124, 247–264.

Sardella, R., D.A. Iurino, B. Mecozzi, D. Sigari, F. Bona, L. Bellucci, M. Coltorti, J. Conti, G. Lembo, B. Muttillo and I. Mazzini 2019. Grotta Romanelli (Lecce, Southern Italy) Between Past and Future: New Studies and Perspectives for an Archaeo-geosite Symbol of the Palaeolithic in Europe. *Geoheritage*, 11(4), 1413–1432.

Tozzi, C. 1970. La Grotta di S. Agostino (Gaeta). *Rivista di Scienze Preistoriche,* 25, 3–87.

Uzunidis-Boutillier, A. 2017. Grands herbivores de la fin du Pléistocène moyen au début du Pléistocène supérieur dans le sud de la France. Implications anthropologiques pour la lignée néandertalienne (Unpublished PhD thesis). Aix-Marseille University.

Uzunidis-Boutillier, A. 2020. Large ungulates mobility and Neanderthal subsistence behaviours: A preliminary tooth microwear analysis. *Journal of Archaeological Science*: Reports, 29, 102084.

# 2.2 The Advantages of Owning a Palaeolithic Dog

Mietje Germonpré[1], Martina Lázničková-Galetová[2], Mikhail V. Sablin[3], Hervé Bocherens[4]

[1]Operational Direction 'Earth and History of Life', Royal Belgian Institute of Natural Sciences, Vautierstraat 29, 1000 Brussel, Belgium, mgermonpre@naturalsciences.be

[2]Moravian Museum, Zelný Trh 6, 65937 Brno, the Czech Republic, laznicko@yahoo.fr

[3]Zoological Institute RAS, Universitetskaya nab. 1, 199034 Saint-Petersburg, Russia, msablin@yandex.ru

[4]Department of Geosciences and Senckenberg Centre for Human Evolution and Palaeoenvironment (HEP), Universität Tübingen, Hölderlinstrasse 12, 72074 Tübingen, Germany, herve.bocherens@uni-tuebingen.de

Corresponding author: Mietje Germonpré, mgermonpre@naturalsciences.be

**Abstract**

Pleistocene wolves are the single ancestors of dogs. Several hypotheses have been proposed to explain the initial steps in the domestication process of the wolf. We favour a human-initiated model in which wolf pups were adopted by Upper Palaeolithic people. Captive wolf pups could then have been raised at Upper Palaeolithic camps for several motives and it is likely some pups, the most docile and less fearsome ones, could have survived until adulthood and reproduced, permitting a new selection on every next generation, ultimately leading to Palaeolithic dogs. We propose that the initial beginning of the wolf domestication process was linked with the cultural traditions of some Upper Palaeolithic societies. We review here the close relationships that existed between prehistoric humans and the first domestic canids. It can be expected that the presence of Palaeolithic dogs at camp sites and gathering localities conveyed some selective advantage to their owners. The Palaeolithic dogs could have been very useful as guards, by warning of the approach of predators or unfamiliar humans through vocalisations; this would have provided protection to the inhabitants of the camps and to the gatherers away from the settlements when accompanied by Palaeolithic dogs. Furthermore, Palaeolithic dogs could potentially have been suitable to increase the level of mobility of their people, helping with the transportation of firewood, lithics, gear, body parts of prey, etc. In addition, the anthropogenic manipulations of several Palaeolithic dog skulls, such as the perforation of the brain case or the insertion of objects in the mouth cavity, testify of the special social standing these canids held within some Upper Palaeolithic societies.

**Keywords:** Pleistocene, wolf domestication, Palaeolithic dog, Europe.

## 1 Introduction

Morphological and genetic analyses have shown that the Pleistocene wolf is the unique ancestor of the dog (Thalmann *et al.* 2013; Skoglund *et al.* 2015; Frantz *et al.* 2016; Germonpré *et al.* 2009, 2017a; Pilot *et al.* 2019). Most researchers accept that the beginnings of this domestication took place towards the end of the Pleistocene when human populations still lived in small groups as hunter-gatherers and when agriculture was not yet practiced (Larson *et al.* 2012; Thalmann *et al.* 2013; Morey and Jeger 2015; Freedman and Wayne 2017).

## 2 Evidence and methods

In this contribution, we use archaeological and zooarchaeological data gathered from Upper Palaeolithic sites in combination with ethnographic information on forager societies from the circumpolar north. First, we look at different morphotypes among large canids of the Pleistocene and their possible connection with the so-called domestication syndrome. Then we give a short review of the two main domestication models. Thereafter we consider some costs and advantages Palaeolithic dogs could have presented to their masters. We conclude that when enough food would have been available, the integration of Palaeolithic dogs in the daily life of Upper Palaeolithic hunter-gatherers could have brought many advantages to their owners.

## 3 Results

### *3.1 The domestication syndrome*

The modern dog is a domestic mammal that shows several of the features of the so-called domestication syndrome, such as a decreased skull length, a reduced snout length, a diminished body size, an increased variability in the number of vertebrae and an increased tameness (Wilkins *et al.* 2014; Sanchez-Villagra *et al.* 2016). Animal studies have shown that facial, integumentary and behavioural systems are integrated

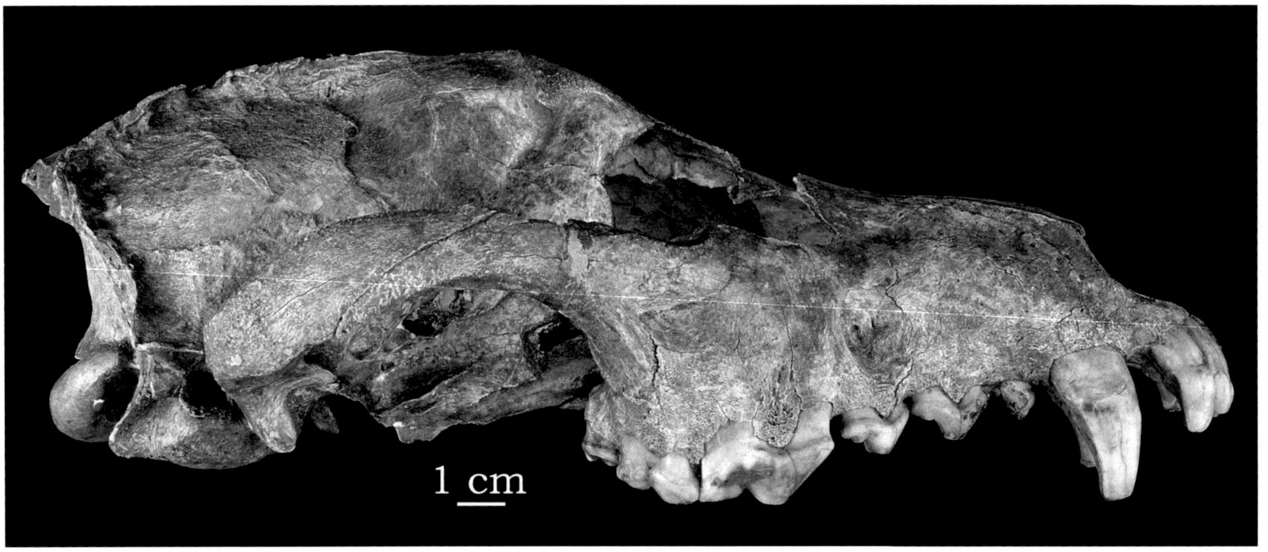


Figure 1. Lateral view of the skull from the Pleistocene wolf from Trou des Nutons, Belgium. Total Skull Length: 261 mm (Royal Belgian Institute of Natural Sciences).

by the neural crest, an embryonic population of cells. These cells possibly played an important role in the domestication process of the dog and other animals. The bones in the snout are derived from the neural crest, the bones of the back and the base of the skull originate from the mesoderm. The greatest shape and size changes in the skull of domestic animals compared to their wild forbears that ensued through domestication occurred in the snout, while the posterior part of the skull remained relatively stable (Lord *et al.* 2017 and references herein). Others have proposed that the domestication phenotype was brought about not by modifications of the neural crest but could have been driven by genes that tend to group together in clusters in the genome and that could be related to different phenotypic traits of domestication (Wright 2015).

At the Institute of Cytology and Genetics of the Russian Academy of Sciences in Novosibirsk, an interesting experiment has been conducted since 1959. The silver fox (*Vulpes vulpes*), like the dog a member of the Canidae, was selected for a reconstruction of the early beginnings of the wolf domestication process (Belyaev 1979; Trut 1999; Trut *et al.* 2012). This experiment started with a small founder population (c. 150 individuals) of farm foxes; only those animals were selected that reacted with a friendly response and reduced aggression towards humans, resulting in a strain of tame foxes (Nelson *et al.* 2016; Kukova pers. comm.). Already after the ninth generation, changes in the behaviour in a number of foxes could be noted: these foxes behaved more affectionately towards people than the individuals from the founder population. By selecting only for tame behaviour in the breeding pairs, changes in skull shape, colouration and behaviour appeared in their offspring (Trut 1999). Most interestingly, the skull and snout of several foxes from these later generations were shorter and wider compared to the skull and snout of the founders (Trut 1999), features that mirror characteristics of domestic dog skulls and that could be related to the domestication syndrome.

We have proposed, on the basis of multivariate analyses of the traditional morphometry of canid skulls and mandibles, that in several sites of the Upper Palaeolithic in Europe, two morphotypes of large fossil canids can be distinguished. A first morphotype is quite similar in size and shape to the recent wild wolves of northern Eurasia, although the muzzle of the fossil morphotype is on average longer and wider than the muzzle of recent Nordic wolves (Germonpré *et al.* 2017a). We named this morphotype 'the Pleistocene wolf' (Figure 1). The second morphotype has a unique morphology that falls outside the variability of the size and shape of the Pleistocene wolves and recent Nordic wolves (Germonpré *et al.* 2009, 2012, 2015a, b, 2017a) (Figure 2). This morphotype has, similar to domestic mammals, a smaller skull size and a shortened snout with a proportionally wide palate and brain case and a shorter lower jaw compared to the wild type (Galeta *et al.* 2021; Germonpré *et al.* 2009; 2012; 2015a; 2017a), features that resemble some characteristics of the domestication syndrome. Nevertheless, the skulls of the Palaeolithic dog morphotype differ from that of recent northern indigenous dogs (Galeta *et al.* 2021; Germonpré *et al.* 2017a). We used the term 'Palaeolithic dogs' to distinguish this type from modern dogs which are among other characterised by a smaller dentition.

Figure 2. Oblique view of the skull from the Palaeolithic dog from Předmostí (Czech Republic) with a bone fragment inserted between its teeth. Total Skull Length: 232 mm (Moravian Museum, Brno).

### 3.2 Two domestication models

Two main hypotheses have been proposed to explain the initial stages of the domestication process of the wolf (Germonpré *et al.* 2018). The human-initiated model of the wolf domestication must have begun with the arrival of young wolves in the camp sites (Germonpré *et al.* 2018). A parallel of a practise of the integration of young wild animals in the daily life of Palaeolithic people can be found in the ethnographic literature of northern people where many examples are recorded of the raising of wild animals at the settlements (e.g. Batchelor 1901; Drucker 1951; Prokof'yeva 1964; Sokolova 2000; Hamayon 2012). Wolf pups could have been captured and brought to the Upper Palaeolithic camps by the hunters for several reasons and probably some wolf pups, the most docile and less fearsome ones -the more aggressive ones being culled -could have survived to adulthood and reproduced, thus allowing selection in the behaviour of the canids from every next generation (Clutton-Brock 1995; Morey 2010; Germonpré *et al.* 2018). It is likely that the Upper Palaeolithic owners selected from the standing genetic variations in the human-directed social and communicative behaviour among their captive pups (cf. Hansen Wheat *et al.* 2020) This means that in the context of a domestication based on human adoption of wolf pups, the capture and rearing of young animals, in order to have a sufficient number of founders, must have been a rather widespread cultural tradition during the Upper Palaeolithic. Indeed, according to Porr and de Kara (2015), people during the early Upper Palaeolithic placed particular value on the personal interaction with animals. In the human-initiated model, the prehistoric hunter-gatherers must have checked a large number of individual Pleistocene wolves on their behaviour. In addition, they must have accepted that some docile wolves became sexually mature. Probably several communities repeated this effort over long periods of time and vast areas. As adduced by Lien (2015, p.165): '... domestication is not about one trajectory, it is about many'. The offspring of these friendly wolves passed the same type of selection process so that certain individuals could live, and the unwanted ones were killed, leading gradually to incipient dogs. Genetic data from extant dogs confirm that the founding stock of ancestors consisted of at least 500 wolves (Pang *et al.* 2009; Niskanen *et al.* 2013). Probably some of the early domesticated wolf lineages became extinct and could have been replaced by incoming dogs that accompanied migrant peoples during and after the Upper Palaeolithic (e.g. Frantz *et al.* 2016).

The second hypothesis of wolf domestication is the model of self-domestication. This hypothesis considers that some wolves adapted to the human niche by feeding on waste or food stored in Upper Palaeolithic camps. Wolves who were neither fearful nor aggressive evolved gradually, from generation to generation, to the first dogs (Zeder 2012; Larson and Burger 2013; Morey and Jeger 2015). However, it is very likely that the Palaeolithic hunter-gatherers had little surplus or waste that could have provided food for the scavenging wolves (Germonpré *et al.* 2018; Lupo 2019). In addition, they likely stored food in protected places to avoid unwanted visits from carnivores and commensal animals (Germonpré *et al.* 2018). Indigenous peoples in the North build raised supports and above-ground caches to keep food, so as not to attract predators (Fair 1997). The remains of prey animals are often deposited out of reach of the carnivores (Tanner 1979; Lavrillier 2011). Another essential criticism of this model of domestication is that wolves used to associating humans with food are dangerous: they can attack and kill people (Linnell *et al.* 2003), especially unattended children (Rajpurohit 1999; McNay and Mooney 2005; Behdarvand and Kaboli 2015). We believe that the Upper Palaeolithic people tried to avoid being visited by prowling predators and that therefore they stored food and waste in protected places. In addition, the hunters not only captured but probably even killed young carnivores in nearby dens (Germonpré *et al.* 2018).

### 3.3 The Palaeolithic dog

Coppinger and Coppinger (2016) note that the owning of Palaeolithic dogs has a cost and therefore should be balanced by advantages to their owners. This cost probably depended on the value and importance of the products and the roles that Palaeolithic dogs could deliver and fulfil when their owners needed them (cf. Sigaut 1980). Ethnographic data related to the keeping of dogs in the north underscore the important quantity and quality of food these animals require (Lupo 2019). In the Upper Palaeolithic camps, the Palaeolithic dogs could have been fed or could possibly have had access to remains of prey and offal; in this way they could have served as waste cleaners. Probably Palaeolithic dogs could have received a selection of food deemed undesirable by Palaeolithic men and women. Reconstructing the diet of several Palaeolithic dogs from the Gravettian Předmostí site, a mammoth site in the Czech Republic, revealed that Palaeolithic dogs mainly consumed meat of reindeer and musk ox. The absence of mammoth meat in their diet suggests that, unlike other carnivores, these Palaeolithic dogs did not have access to mammoth carcasses and were therefore probably tied up for at least part of the time (Bocherens *et al.* 2015). Also, in Late Glacial sites in Germany and Switzerland, it seems that the diet of Palaeolithic dogs was controlled and that they did not consume mammoth meat in large quantities (Baumann *et al.* 2020).

In various nomadic societies of Asia, the dog's main mission is to guard the camps and its inhabitants. Dogs warn of the approach of wolves, bears and strangers and their barking has a deterrent function (Shirokogoroff 1929; Lescureux 2007; Vaté 2013; Samar and Kim 2017; Klokov and Davydov 2018). Palaeolithic dogs would have been very useful as sentinels, warning through vocalisations of the approach of carnivores or unfamiliar humans and the presence of these dogs at the camp site or at berry gathering locations could have conveyed some selective advantage to their owners. A recent study of Zapata *et al.* (2016) which compared a genome wide association mapping of dog breed stereotypes for fear and traits related to fear and aggression - across several hundred dogs from diverse breeds with the genetic variation in extant wolf populations - revealed that reduced fear/aggression alleles are much more frequent in modern dog breeds than in wolves. Such high frequencies of these alleles are consistent with the selection of reduced fear and aggression variants among the captive wolf pups living at the Upper Palaeolithic camps during the domestication process. In addition, the reduced fear/aggression allele is often in perfect linkage disequilibrium with the allele for increased-body size. This could suggest that a selection of less fearful/aggressive individuals was of animals that were at the large end of the size range. The domestication of the least fearful/aggressive and largest wolves probably was suitable for the protection against apex predators (Zapata *et al.* 2016).

Dogs from the Far North were often used as pack animals. In this way, they carried two large bags on the left and right side of the back, filled with provisions (Nelson 1983; Balikci 1989). Estimates, based on ethnographic data from North America, for the weight of a filled backpack range from 15 to 20 kg (Speth *et al.* 2013; Loovers, personal communication, 2016). The dogs of the Siberian Khanty and Mansi carried water and firewood to the dwellings (Prokof'yeva *et al.* 1964). Furthermore, pack dogs have been shown to permit long hunting expeditions since hunters can stay out overnight thanks to the supplies carried by the dogs (Sharp and Sharp 2015). According to Maier *et al.* (2016), the Upper Palaeolithic hunter-gatherers living at higher latitudes, north of the treeline and in the continuous permafrost zone, had to adapt to a cold, dry climate in a treeless landscape and had to travel long distances to meet their daily needs. Palaeolithic dogs would have been very useful for the logistical and residential mobility of the hunter-gatherers and could have assisted their masters by helping them to transport firewood, lithics, body parts of hunted prey, etc. (Germonpré *et al.* 2017a; 2020).

Dogs play an important role in many hunter-gatherer societies as an aid to hunting. They can reduce search costs, increase prey encounter rates, keep dangerous animals away, track down injured prey and locate animal carcasses (Balikci 1989; Abe 2005; Grøn and Turov 2007; Vaté 2013; Perri 2016; Lupo 2017; Samar and Kim 2017). Perri (2016) and Lupo (2017) give detailed analyses concerning the use of Palaeolithic dogs for hunting. Nevertheless, whether Palaeolithic dogs were already hunting companions of Upper-Palaeolithic hunter-gatherers is difficult to answer (Morey 2010).

The oldest recognised dog burial is the one from the Magdalenian site of Bonn-Oberkassel (Germany) (Nobis 1986). This double human burial, dating from around 14,500 years ago, contains the remains of two dogs. One of these dogs suffered from a fatal infection. Several defects in the development of the enamel of its teeth suggest that this dog was seriously ill when he was between five and six months old. His owners must have taken good care of this puppy. Eventually, this dog died when it was about seven months old. According to Janssens *et al.* (2018), the care the animal received was based on compassion or empathy.

Body products from Palaeolithic dogs may have provided benefits, such as fur for clothing, meat and grease for food, long bones and teeth as a raw material for making tools. The Inuit use wolverine, wolf, fox and dog skins to make pants and as rims for the hoods and sleeves of parkas, since the long uneven hairs of these carnivores repel frost (Balikci 1989; Issenman 1997). In times of famine, the Inuit ate dogs (Laugrand and Oosten 2015). Mongolian nomads sometimes consumed dog meat as a medicine (Charlier 2015). People from north western North America made hooks from dog bones (Teit 1900). The interest of Upper Palaeolithic people in the body products of Palaeolithic dogs can be deduced from the bones of canids bearing marks of human manipulation. So far, human traces on the remains of Palaeolithic dogs have been observed with regard to the recovery of meat (Pionnier-Capitan *et al.* 2011) and the use of bones as a raw material (Germonpré *et al.* 2017b).

Among the peoples of the Far North, the bones, teeth and blood of dead dogs have a special meaning. In some Siberian peoples, women who desired children wore dog teeth as an amulet (Black 1973); dog canines were hung above the cradle as protection for the babies (Samar 2009). Also, in Siberia, some peoples ritually killed dogs as part of the bear festival (Samar and Kim 2017). The dogs themselves could receive specific treatment at death. Nomadic Mongolian breeders, before displacing their dead dogs, put a piece of fat, butter or milk in the mouth of the animal (Lugli 2016). In addition, the ethnographic literature of the circumpolar North is replete with beliefs that human souls need the souls of dogs to accompany them beyond (Kretschmar 1938; Schwarz 1997). At the Předmostí site, a number of canid remains were handled and modified upon death by the Gravettian inhabitants of the site (Germonpré *et al.* 2012, 2017b). The skulls of Palaeolithic dogs were manipulated: their braincases were perforated (Germonpré *et al.* 2012) in a manner similar to the perforations performed during ceremonies held for bears and wolves by the Ainu (Akino 1999; Walker 2005). At Předmostí, another Palaeolithic dog had a fragment of bone inserted between its teeth (Figure 2) (Germonpré *et al.*, 2012), recalling the food that Mongolian dogs received after death (Lugli 2016). In addition, at Předmostí, perforated canines of Palaeolithic dogs, used as a pendant, have been identified (Germonpré *et al.* 2012). In Eliseevichi, a Russian Epigravettian mammoth site dating from around 17,000 years ago, a skull of a Palaeolithic dog was found in a hearth near a concentration of mammoth skulls (Polikarpovich 1968). Its braincase is perforated on both sides, its jugal teeth have been extracted by damaging the alveoli. The location of the skull and the manipulations undergone by this animal suggest a ritual context (Sablin and Khlopachev 2002; Germonpré *et al.* 2009). The anthropogenic manipulation of the remains of Palaeolithic dogs from these Gravettian and Epigravettian sites testifies to the particular social status of these canids in certain societies of the Upper Palaeolithic.

## 4 Conclusions

The ancestor of the first domesticated animal in human history is a fearsome predator: the Pleistocene wolf. The first stages of this domestication can be placed in the Upper Palaeolithic. We think that a self-domestication of the wolf is unlikely because access to surplus food or waste in the Upper Palaeolithic camps would probably have been obstructed to large carnivores by prehistoric people in order to avoid predatory attacks. Several intertwined patterns imply the complex interaction between humans and this carnivore during this process. We propose that the start of this domestication was an unforeseen side effect of the capture and raising of pups by hunter-gatherers, a practice that was probably part of the cultural tradition of some Upper Palaeolithic societies and during which selection was exercised on the genetic standing-variation in human-directed social behaviour among the captive pups (cf. Zapata *et al.* 2016; Hansen Wheat *et al.* 2020). This selection on tame behaviour in every next generation led ultimately to the first dogs. Palaeolithic dogs began to transform the way of life of the Upper Palaeolithic people with whom they resided and contributed with many advantages to improve the daily lives of their owners.

## 5 Acknowledgments

This paper was financially supported by the Ministry of Culture of the Czech Republic by institutional financing of long-term conceptual development of the research institution (the Moravian Museum, MK000094862). Mikhail Sablin was supported by ZIN RAS (state assignment N°122031100282-2). M. Germonpré acknowledges support from the Brain.be 2.0 ICHIE project (BELSPO B2/191/P2/ICHIE).

## References

Abe, Y. 2005. *Hunting and Butchering Patterns of the Evenki in Northern Transbaikalia, Russia*. New York: Department of Anthropology, Stony Brook University.

Akino, S. 1999. Spirit-sending ceremonies, in W.W. Fitzhugh and C.O. Dubreuil (eds) *Ainu. Spirit of a Northern people:* 248–255. Washington DC: Arctic Studies Center, National Museum of Natural History, Smithsonian Institute.

Balikci, A. 1989. *The Netsilik Eskimo*. Prospect Heights: Waveland Press.

Batchelor, J. 1901. *The Ainu and their Folklore*. London: The Religious Tract Society.

Baumann, C., B.M. Starkovich, D.G. Drucker, S.C. Münzel, N.C. Conard, and H. Bocherens 2020. Dietary niche partitioning among Magdalenian canids in southwestern Germany and Switzerland. *Quaternary Science Reviews* 227: 106032.

Behdarvand, N. and M. Kaboli 2015. Characteristics of gray wolf attacks on humans in an altered landscape in the west of Iran. *Human Dimensions of Wildlife* 20: 112–122.

Belyaev, D.K. 1979. Destabilizing selection as a factor in domestication. *Journal of Heredity* 70: 301–308.

Black, L. 1973. The Nivkh (Gilyak) of Sakhalin and the lower Amur. *Arctic Anthropology* 10: 1–110.

Bocherens, H., D.G. Drucker, M. Germonpré, M. Lázničková-Galetová, Y.I. Naito, C. Wissing, J. Brůžek and M. Oliva 2015. Reconstruction of the Gravettian food-web at Předmostí I using multi-isotopic tracking ($^{13}C$, $^{15}N$, $^{34}S$) of bone collagen. *Quaternary International* 359–360: 211–228.

Charlier, B. 2015. *Faces of the wolf. Managing the human, non-human boundary in Mongolia*. Leiden: Brill.

Clutton-Brock, J. 1995. Origins of the dog: domestication and early history, in J. Serpell (ed.) *The domestic dog: its evolution, behaviour and interactions with people:* 7–20. Cambridge: Cambridge University Press.

Coppinger, R and L. Coppinger 2016. *What is a dog?* Chicago: University of Chicago Press.

Drucker, P. 1951. The Northern and Central Nootkan tribes. *Smithsonian Institution Bureau of American Ethnology Bulletin* 144.

Fair, S.W. 1997. Story, storage, and symbol: functional cache architecture, cache narratives, and roadside attractions. *Perspectives in Vernacular Architecture* 7: 167–182.

Frantz, L.A.F., V.E. Mullin, M. Pionnier-Capitan, O. Lebrasseur, M. Ollivier, A. Perri, A. Linderholm, V. Mattiangeli, M.D. Teasdale, E.A. Dimopoulos, A. Tresset, M. Duffraisse, F. McCormick, L.F., Bartosiewicz, E. Gál, E.A. Nyerges, M.V. Sablin, S. Bréhard, M. Mashkour, A. Bălăşescu, B. Gillet, S. Hughes, O. Chassaing, C. Hitte, J.-D. Vigne, K. Dobney, C. Hänni, D.G. Bradley and G. Larson 2016. Genomic and archaeological evidence suggest a dual origin of domestic dogs. *Science* 352: 1228–1231.

Freedman, A.H. and R.K. Wayne 2017. Deciphering the Origin of Dogs: From Fossils to Genomes. *Annual Review of Animal Biosciences* 5: 281–307.

Galeta, P., M. Lázničková-Galetová, M. Sablin and M. Germonpré 2021. Morphological Evidence for Early Dog Domestication in the European Pleistocene: The Randomization Approach. *The Anatomical Record* 304: 42–62.

Germonpré, M., S. Fedorov, M. Danilov, P. Galeta, E.L. Jimenez, M.V. Sablin, and R.J. Losey 2017a. Palaeolithic and prehistoric dogs and Pleistocene wolves from Yakutia: identification of isolated skulls. *Journal of Archaeological Science* 78: 1–19.

Germonpré, M. and R. Hämäläinen 2007. Fossil bear bones in the Belgian Upper Palaeolithic: the possibility of a proto-bear ceremonialism. *Arctic Anthropology* 44: 1–30.

Germonpré, M., M. Lázničková-Galetová, E.-L. Jimenez, R. Losey, M. Sablin, H. Bocherens and M. Van den Broeck 2017b. Consumption of canid meat at the Gravettian Předmostí site, the Czech Republic. *Fossil Imprint* 73: 360–382.

Germonpré, M., M. Lázničková-Galetová, R.J. Losey, J. Räikkönen, and M.V. Sablin 2015b. Large canids at the Gravettian Předmostí site, the Czech Republic: the mandible. *Quaternary International* 359–360: 261–279.

Germonpré, M., M. Lázničková-Galetová and M.V. Sablin 2012. Palaeolithic dog skulls at the Gravettian Předmostí site, the Czech Republic. *Journal of Archaeological Science* 39: 184–202.

Germonpré, M., M. Lázničková-Galetová, M. Sablin and H. Bocherens 2018. Self-domestication or human control? The Upper Palaeolithic domestication of the dog, in C. Stépanoff and J.-D. Vigne (eds) *Hybrid Communities, Biosocial Approaches to Domestication and Other Trans-species Relationships:* 39–64. London: Routledge.

Germonpré, M., M. Lázničková-Galetová, M.V. Sablin and H. Bocherens 2020. Could incipient dogs have enhanced differential access to resources among Upper Palaeolithic hunter-gatherers in Europe? in Luc Moreau (ed.) *Social inequality before farming? Multidisciplinary approaches to the study of social organisation in prehistoric and extant hunter-gatherer societies*. Cambridge: McDonald Institute Conversations [eBook].

Germonpré, M., R. Losey, M. Lázničková-Galetová, P. Galeta, M.V. Sablin, K. Latham and J. Räikkönen 2016. Spondylosis deformans in three large canids from the Gravettian Předmostí site: Comparison with other canid populations. *International Journal of Paleopathology* 15: 83–91.

Germonpré, M., M.V. Sablin, V. Després, M. Hofreiter, M. Lázničková-Galetová, R.E. Stevens and M. Stiller 2013. Palaeolithic dogs and the early domestication of the wolf: a reply to the comments of Crockford and Kuzmin. *Journal of Archaeological Science* 40: 786–792.

Germonpré, M., V. Sablin, M. Lázničková-Galetová, V. Després, R.E. Stevens, H. Stiller and M. Hofreiter 2015b. Palaeolithic dogs and Pleistocene wolves revisited: a reply to Morey (2014). *Journal of Archaeological Science* 54: 210–216.

Germonpré, M., M.V. Sablin, R.E. Stevens, R.E.M. Hedges, M. Hofreiter, M. Stiller and V.R. Després 2009. Fossil dogs and wolves from Palaeolithic sites in Belgium, the Ukraine and Russia: osteometry, ancient DNA and stable isotopes. *Journal of Archaeological Science* 36: 473–490.

Germonpré, M. and M.V. Sablin 2017. Chapter 2. Humans and mammals in the Upper Palaeolithic of Russia, in U. Albarella, H. Russ, K. Vickers and S. Viner-Daniels (eds) *Oxford Handbook of Zooarchaeology*: 25–38. Oxford: Oxford University press.

Grøn, O. and M.G. Turov 2007. Resource 'pooling' and resource management. An ethno-archaeological study of the Evenk hunter-gatherers, Katanga County, Siberia. *Acta Archaeologica Lundensia* 26: 67–72.

Hamayon, R.N. 2012. The three duties of good fortune: 'Luck' as a relational process among hunting peoples of the Siberian forest in Pre-Soviet Times. *Social Analysis* 56: 99–116.

Hansen Wheat, C. and H. Temrin 2020. Intrinsic ball retrieving in wolf puppies suggests standing ancestral variation forhuman-directed play behavior. *iScience*, 23: 100811.

Issenman, B. K.1997. *Sinews of Survival. The Living Legacy of Inuit Clothing*. Vancouver: University of British Columbia Press.

Janssens, L., L. Giemsch, R. Schmitz, M. Street, S. Van Dongen and P. Crombé 2018. A new look at an old dog: Bonn-Oberkassel reconsidered. *Journal of Archaeological Science* 92: 126–138.

Klokov, K. and V. Davydov 2018. Human-dog-reindeer communities in the Siberian Arctic and Subarctic, in: C. Stépanoff and J.-D. Vigne (eds) *Hybrid Communities, Biosocial Approaches to Domestication and Other Trans-species Relationships:* 261–274. London: Routledge.

Kretschmar, F. 1938. *Hundestammvater und Kerberos* I. Stuttgart: Strecker und Schröder Verlag.

Larson, G. and J. Burger 2013. A population genetics view of animal domestication. *Trends in Genetics* 29: 197–205.

Larson, G., E.K. Karlsson, A. Perri, M.T. Webster, S.Y.W. Ho, J. Peters, P.W. Stahl, P.J. Piper, F. Lingaas, M. Fredholm, K.E. Comstock, J.F. Modiano, C. Schelling, A.I. Agoulnik, P.A. Leegwater, K. Dobney, J.-D. Vigne, C. Vilá, L. Andersson and K. Lindblad-Toh 2012. Rethinking dog domestication by integrating genetics, archeology, and biogeography. *Proceedings of the National Academy of Sciences* 109: 8878–8883.

Lavrillier, A. 2011. The creation and persistence of cultural landscapes among the Siberian Evenkis: two conceptions of 'sacred' space, in P. Jordan (ed.) *Landscape and culture in Northern Eurasia*: 215–231. Walnut Creek: Left Coast Press.

Laugrand, F. and J. Oosten 2015. *Hunters, predators and prey. Inuit perceptions of animals*. New York: Berghahn Books.

Lescureux, N. 2007. Maintenir la réciprocité pour mieux coexister ? Ethnographie du récit kirghiz des relations dynamiques entre les hommes et les loups. Muséum national d'Histoire Naturelle, ED 227 Sciences de la Nature et de l'Homme, Thèse Pour obtenir le grade de Docteur du Muséum National d'Histoire Naturelle.

Lien, M. 2015. *Becoming salmon: aquaculture and the domestication of a fish*. Oakland: University of California Press.

Linnell, J.D.C., E.J. Solberg, S. Brainerd, O. Liberg, H. Sand, P. Wabakken, and I. Kojola 2003. Is the fear of wolves justified? A Fennoscandian perspective. *Acta Zoologica Lituanica* 13: 27–33.

Lord, K., R.A. Schneider and R. Coppinger 2017. Evolution of working dogs, in J.A. Serpell (ed.) *The domestic dog: its evolution, behavior, and interactions with people*, second edition: 42–66. Cambridge University Press, Cambridge.

Lugli, F. 2016. Mongolian nomads and their dogs, in: S. Biagetti and F. Lugli (eds) *The Intangible Elements of Culture in Ethnoarchaeological Research*: 125–39. Basel: Springer International Publishing.

Lupo, K.D. 2017. When and where do dogs improve hunting productivity? The empirical record and some implications for early Upper Palaeolithic prey acquisition. *Journal of Anthropological Archaeology* 47: 139–151.

Lupo, K.D. 2019. Hounds follow those who feed them: What can the ethnographic record of hunter-gatherers reveal about early human-canid partnerships? *Journal of Anthropological Archaeology* 55: 101081.

Maier, A., F. Lehmkuhl, P. Ludwig, M.P., Melles, I. Schmidt, Y. Shao, C. Zeeden and A. Zimmermann

2016. Demographic estimates of hunter-gatherers during the Last Glacial Maximum in Europe against the background of palaeoenvironmental data. *Quaternary International* 425: 49–61.

McNay, M.E. and P.W. Moone 2005. Attempted Predation of a Child by a Gray Wolf, *Canis lupus*, near Icy Bay, Alaska. *Canadian Field-Naturalist* 119: 197–201.

Morey, D. 2010. *Dogs. Domestication and the Development of a Social Bond*. Cambridge: Cambridge University Press.

Morey, D.F. and R. Jeger 2015. Palaeolithic dogs: why sustained domestication then? *Journal of Archaeological Science: Reports* 3: 420–428.

Nelson, R.K. 1983. *Make prayers to the raven. A Koyukon view of the Northern forest*. Chicago: Chicago University Press.

Niskanen, A.K., E. Hagström, H. Lohi, M. Ruokonen, R. Esparza-Salas, J. Aspi and P. Savolainen 2013. MHC variability supports dog domestication from a large number of wolves: high diversity in Asia. *Heredity* 110: 80–85.

Nobis, G. 1986. Die Wildsäugetiere in der Umwelt des Menschen von Oberkassel bei Bonn und das Domestikationsproblem von Wölfenim Jungpaläolithikum. *Bonner Jahrbuch* 186: 368–76.

Pang, J.-F., C. Kluetsch, X.-J., Zou, A.-B. Zhang, L.-Y., Luo, H. Angleby, A. Ardalan, C. Ekström, A. Sköllermo, J. Lundeberg, S. Matsumura, T. Leitner, Y.-P. Zhang, and P. Savolainen 2009. mtDNA data indicate a single origin for dogs south of Yangtze river, less than 16,300 years ago, from numerous wolves. *Molecular Biology and Evolution* 26: 2849–2864.

Perri, A.R. 2016. A wolf in dog's clothing: initial dog domestication and Pleistocene wolf variation. *Journal of Archaeological Science* 68: 1–4.

Pilot, M., A.E. Moura, I.M. Okhlopkov, N.V. Mamaev, A.N. Alagaili, O.B. Mohammed, E.G. Yavruyan, N.H. Manaseryan, V. Hayrapetyan, N. Kopaliani, E. Tsingarska, M. Krofel, P. Skoglund and W. Bogdanowicz 2019. Global phylogeographic and admixture patterns in grey wolves and genetic legacy of an ancient Siberian lineage. *Scientific Reports* 9: 17328.

Pionnier-Capitan, M., C. Bemilli, P. Bodu, G. Célérier, J.G. Ferrié, P. Fosse, M. Garcià and J.-D. Vigne 2011. New evidence for Upper Palaeolithic small domestic dogs in South-Western Europe. *Journal of Archaeological Science* 38: 2123–2140.

Polikarpovich, K.M. 1968. *Paleolit Verhnego Podneprov'ya*. Minsk: Nauka i Technika.

Porr, M. and M. de Kara 2015. Perceiving animals, perceiving humans. Animism and the Aurignacian mobiliary art of Southwest Germany, in S. Sázelová, M. Novák and A. Mizerová (eds) *Forgotten times and spaces: New perspectives in paleoanthropological, paleoethnological and archeological studies*: 293–302. First Edition. Brno: Institute of Archeology of the Czech Academy of Sciences, Masaryk University.

Prokof'yeva, E.D., N. Chernetsov and N.F. Prytkova 1964. The Khants and Mansi, in M.G. Levin and L.P. Potapov (eds) *The peoples of Siberia*: 511–46. Chicago: University of Chicago Press.

Rajpurohit, K.S. 1999. Child lifting: Wolves in Hazaribagh, India. *Ambio* 28: 162–166.

Sablin, M.V. and G.A. Khlopachev 2002. The earliest Ice Age dogs: evidence from Eliseevichi. *Current Anthropology* 43: 795–799.

Samar, A.P. 2009. The role of dogs in the Nanai cults. *Senri Ethnological studies* 72: 145–151.

Samar, A.P. and A.A. Kim 2017. On the question of traditional dog breeding among indigenous peoples of the Far East. *Anthropology and Archeology of Eurasia* 56: 32–51.

Sànchez-Villagra, M.R., M. Geiger and R.A. Schneider. 2016. The taming of the neural crest: a developmental perspective on the origins of morphological covariation in domesticated mammals. *Royal Society Open Science* 3.

Schwartz, M. 1997. *A history of dogs in the early Americas*. New Haven: Yale University Press.

Sharp, H.S. and K. Sharp. 2015. *Hunting caribou. Subsistence hunting along the northern edge of the boreal forest*. Lincoln: University of Nebraska Press.

Shirokogoroff, S.M. 1929. *Social organization of the Northern Tungus*. Shangai: Commercial Press.

Sigaut, F. 1980. Un tableau des produits animaux et deux hypothèses qui en découlent. *Production Pastorale et Société* 7: 20–36.

Skoglund, P., E. Ersmark, E. Palkopoulou and L. Dalén. 2015. Ancient wolf genome reveals an early divergence of domestic dog ancestors and admixture into high-latitude breeds. *Current Biology* 25: 1–5.

Sokolova, Z.P. 2000. The bear cult. *Archaeology, Ethnology, Anthropology of Eurasia* 2: 121–130.

Speth, J.D., K. Newlander, A.A. White, A.K. Lemke and L.E. Anderson. 2013. Early Paleoindian big-game hunting in North America: Provisioning or Politics? *Quaternary International* 285: 111–139.

Tanner, A. 1979. *Bringing Home Animals. Religious Ideology and Mode of Production of the Mistassini Cree hunters*. London: C. Hurst and Company.

Teit, J. 1900. The Thompson Indians of British Columbia. *Memoirs of the American Museum of Natural History, Anthropology, The Jesup North Pacific Expedition*: 163–392.

Thalmann, O., B. Shapiro, P. Cui, V.J. Schuenemann, D.K. Sawyer, D.L. Greenfield, M.B. Germonpré, M.V. Sablin, F. López-Giráldez, X. Domingo-Roura, H. Napierala, H.-P., Uerpmann, D.M. Loponte, A.A. Acosta, L. Giemsch, R.W. Schmitz, B. Worthington, J.E. Buikstra, A. Druzhkova, A.S. Graphodatsky, N.D. Ovodov, N. Wahlberg, A.H. Freedman, R.M. Schweizer, K.-P. Koepfli, J.A. Leonard, J.A., M. Meyer, J. Krause, S. Pääbo, R.E. Green and R.K. Wayne 2013. Complete mitochondrial genomes of ancient canids suggest a European origin of domestic dogs. *Science* 342: 871–874.

Trutt, L.N. 1999. Early canid domestication: the fox-farm experiment. *American Scientist* 87: 169–169.

Trut, L.N., I.N. Oskina, and A.V. Kharlamova 2012. Chapter 2, Experimental studies of early canid domestication, in E.A. Ostrander and A. Ruvinsky (eds) *The Genetics of the Dog:* 12–37. New York: CABI Publishing.

Vaté, V. 2013. Le chien chez les éleveurs de rennes chouktches, in C. Stépanoff, C. Ferret, G. Lacaze and J. Thorez (eds) *Nomadisme d'Asie centrale et septentrionale*: 206–207. Paris: Armand Colin.

Walker, B.L. 2005. *The Lost Wolves of Japan.* Seattle: University of Washington Press.

Wilkins, A.S., R.W. Wrangham and W. Tecumseh Fitch. 2014. The 'domestication syndrome' in mammals: a unified explanation based on neural crest cell behavior and genetics. *Genetics* 197: 795–808.

Wright, D. 2015. The genetic architecture of domestication in animals. *Bioinformatics and Biology Insights* 9 (S4): 11–20.

Zapata, I., J.A. Serpell and C.E. Alvarez. 2016. Genetic mapping of canine fear and aggression. *BMC Genomics* 17: 572.

Zeder, M.A. 2012. Pathways to Animal Domestication, in P. Gepts, T.R. Famula, R.L. Bettinger, S.B. Brush, A.B. Damania, P.E. McGuire and C.O. Qualset (eds) *Biodiversity in Agriculture: Domestication, Evolution, and Sustainability*: 227–259. Cambridge: Cambridge University Press.

# 2.3 Why Wolves Became Dogs: Interdisciplinary Questions on Domestication

Juliane Bräuer[1,2] and Blanca Vidal Orga[1,2]

[1]DogStudies Research Group, Max Planck Institute of Geoanthropology (MPI-GEA), Kahlaische Str. 10, 07745 Jena, Germany. braeuer@shh.mpg.de

[2]Department of General Psychology, Friedrich Schiller University Jena (FSU), Germany. vidalorga@shh.mpg.de

Corresponding author: Juliane Bräuer braeuer@shh.mpg.de

**Abstract**

During domestication, dogs have evolved human-like skills for functioning effectively in human societies. Findings of comparative psychology have shed light on the question of what skills dogs were selected for during domestication - namely skills that made them better able to communicate and cooperate with humans. However, if we want to understand the whole process of dog domestication, there are many compelling questions. Here we propose three lines of inquiry that should be considered- the starting point, process, and outcomes of domestication. To answer these questions, we need an interdisciplinary approach in which scientists from the fields of archaeology, genetics, anatomy, psychology, sociology, and anthropology work together.

**Keywords**: domestication, dogs, cognition, behaviour, interdisciplinary.

## 1 What we know: dogs in comparative psychology

Over the past 20 years, 'man's best friend' - the domestic dog (*Canis familiaris*) - has increasingly become a subject of scientific study for comparative psychologists who study animal cognition. It has been revealed that, compared to other social animals, dogs show special skills. In particular they show remarkable skills in three social domains:

- First, they seem to possess special communicative skills. They are able to produce communicative signals such as barking (Feddersen-Petersen 2000; Pongracz *et al.* 2005) and gaze alternation (Miklósi *et al.* 2000; Kaminski *et al.* 2011; Heberlein *et al.* 2017). Additionally, they are also able to comprehend more typically human forms of communication such as words (Kaminski *et al.* 2004; see also Merola *et al.* 2012), iconic signs (Kaminski *et al.* 2009b), and gestures. One often-used setup is the so-called object choice design, in which a food reward is hidden in one of two cups, out of the dog´s view. When a human provides a communicative cue, such as pointing or gazing at the correct cup, it was found that dogs are better than any other animal species tested at using these cues (see Hare and Tomasello 2005; Bräuer *et al.* 2006; Miklósi and Soproni 2006; Kaminski and Marshall-Pescini 2014 for reviews, see Figure 1). A dog's ability to use these human gestures probably evolved during domestication (Hare and Tomasello 2005; Bräuer *et al.* 2006; see also Price 1984). Indeed, free ranging dogs and shelter dogs with limited human contact are also able to use

Figure 1. Object choice paradigm. (A) A food reward is hidden in or under one of two cups, so that the dog does not see in which one. (B) Human provides a communicative cue, such as pointing at the correct cup. (C) Dogs approach the correct cup using communicative cues. (D) Human provides no cue or a non-communicative cue (without looking at the dog). (E) Dogs are at chance level in the absence of obvious communicative cues. (Image Credits: Blanca Vidal Orga).

these gestures to locate hidden food (Hare *et al.* 2010; but see Bhattacharjee *et al.* 2017 and Udell *et al.* 2008), whereas apes, our closest living relatives, perform much worse at this task (Bräuer *et al.* 2006). More importantly, dogs are much better at this task than hand-reared wolves; and even six-week-old puppies are already able to use human pointing gestures to locate hidden food (Hare *et al.* 2002; Riedel *et al.* 2008; Gácsi *et al.* 2009a; Gácsi *et al.* 2009b; but see also Lampe *et al.* 2017).

- A second social-cognitive ability that has been intensely investigated in dogs is perspective taking - the ability to assess what others can perceive. In particular, researchers have investigated whether dogs know what humans can see. Dogs are especially sensitive to human attention, i.e., they are able to distinguish between situations in which a human is attentive to them or not (Call *et al.* 2003; Bräuer *et al.* 2004; Gácsi *et al.* 2004; Virányi *et al.* 2004; Kaminski *et al.* 2017). They are also sensitive to the human perspective (Kaminski *et al.* 2009a; Kaminski *et al.* 2013). Finally, dogs are also capable of auditory perspective taking - the ability to assess what humans can hear (Bräuer *et al.* 2013a; see Kundey *et al.* 2010).
- The third social-cognitive domain in which dogs seem to show special skills is cooperation, defined as a behaviour that is beneficial to another individual or to both individuals involved in a task (Melis and Semmann 2010; Amici 2015). One kind of cooperation is prosocial behaviour, defined as a cooperative behaviour on the part of one individual (the actor) that benefits another individual (the recipient) and occurs voluntarily (see Bräuer 2015; Melis and Warneken 2016 for reviews). Dogs cooperate with humans towards various objectives including protection, hunting, herding, rescuing, searching, servicing, and guiding (Miklósi 2007; Serpell 2016). In all these scenarios, dogs behave beneficially towards humans. However, in most of these cases it is unclear whether dogs actually understand human intentions and are motivated to cooperate with them, or whether they have simply been trained to follow specific commands or react to particular situations in certain ways (Bräuer *et al.* 2013b; Bräuer 2015; Marshall-Pescini *et al.* 2016b). Recent studies have produced mixed evidence (see Bräuer 2015 and Marshall-Pescini *et al.* 2016b for reviews). In the prosocial choice task, subjects are given a choice between two reward combinations, one of which delivers a food item to the subject *and* the recipient (prosocial choice) and the other, which rewards only the subject (selfish choice). Quervel-Chaumette *et al.* (2015) found that dogs showed prosocial preferences towards conspecifics whereas in another version of the prosocial choice task, Dale *et al.* (2016) did not find evidence for prosocial preferences.

Given that dogs prefer humans to other dogs as social partners (Miklósi *et al.* 2003; Gácsi *et al.* 2005; Topál *et al.* 2005) and that the dog-human bond is comparable to the attachment between human infants and their mothers (Prato Previde and Valsecchi 2014), it is likely that dogs might cooperate better in such a task with a human partner than with another dog. However, Kaminski *et al.* (2011) did not find evidence that dogs *helpfully* inform a human about a hidden object (Kaminski *et al.* 2011). Dogs also do not seek help when their owner is simulating a heart attack (Macpherson and Roberts 2006) when they do not have special training. It is possible, however, that in these cases the dogs simply did not understand *how* to support the human partner (Bräuer 2015), thus Bräuer *et al.* (2013b) tried to make the human problem as obvious as possible for the dog. In their study, dogs opened a door when a human expressed that she wanted to enter a target room. Interestingly, the dogs continued to open the door for the human over multiple trials without receiving any reward, indicating a high motivation to support her (Bräuer *et al.* 2013b).

In sum, dogs outperform other social species in their ability to pay attention, to communicate and to cooperate with humans. It is therefore generally accepted that during the process of their domestication, dogs have evolved human-like skills that help them to function effectively in human societies (Hare and Tomasello 2005; Kaminski and Marshall-Pescini 2014). Thus, humans might have selected them to be particularly good cooperative and communicative partners (see Bräuer 2015 for a review).

## 2 What we do not know

In the previous section, we summarised how the findings of comparative psychology in the last 20 years have shed light on the question of what skills dogs were selected for during domestication - namely, skills that made them able to better communicate and cooperate with humans. However, if we want to understand the whole process of dog domestication, there are many open questions that, in order to be answered, require an interdisciplinary approach.

One general question is *when and where* the process started. Although some authors claim that dogs were domesticated more than 30–40,000 years ago (Vilá *et al.* 1997; Thalmann *et al.* 2013; Wang *et al.* 2013), others doubt that. The common understanding is that dog domestication started *at least* 14,000 years ago, as there is clear social and cultural evidence of domesticated dogs from this time (as illustrated by the Bonn-Oberkassel dog mandible found in a late Paleolithic grave in Germany; Janssens *et al.* 2018). Thus, a more precise date and place of the first domestication event remain unclear. Findings from genetics and archaeology can elucidate when and where precisely dogs were first domesticated (Wayne and Vilà 2001; Larson and Bradley 2014; Serpell 2016;), and how they spread all over the world (Kaminski and Marshall-Pescini 2014; see also Mitchell 2017; Ní Leathlobhair *et al.* 2018).

Another sub-question might be more difficult to answer: *why* were wolves domesticated in the first place? Clutton-Brock (1977: 1342) has explained domestication as the 'exploitation of one group of social animals by another more dominant group, which maintains complete mastery over its breeding, organization of territory, and food supply'. Selective breeding over the course of many generations enhances various behavioural and physical characteristics conducive to domestic harmony and utility (Price 1984; Serpell 2016). Other authors, such as Zeder (2012) see domestication from a more mutualistic approach, where both human and domesticate benefit from the relationship (Price 1984; Zeder 2012). If domestication leads to a symbiotic relationship, then we would expect there to be advantages for both species - humans and wolves/dogs.

There is no doubt what the advantage was for the wolves that not only gained a new food resource in human camps but were also protected from predators. However, what was the initial advantage for the human to domesticate the wolf? Hale (1969) has suggested behavioural characteristics that facilitate the domestication process, such as, for example, a hierarchical group structure, a critical period in development of species-bond, being omnivorous, and a short flight distance - meaning they do not run far from humans when they encounter them (Hale 1969; see also Diamond 1997). However, a number of these characteristics do not apply to wolves, such as being precocial and having limited agility. Moreover, wolves and early humans were competitors in hunting the same kinds of prey. So why did we domesticate a species that was a food competitor before we even settled down? A tentative answer to this question might rely on various hypotheses about how wolves/dogs were initially used at the beginning of the domestication process. Humans might have taken advantage of their attentiveness and their hunting abilities. Wolves/dogs probably cleared camps of garbage and vermin, and they also might have been used as a source of meat, of heat, and as a means of transportation (Zimen 1992; Paxton 2000; Miklósi 2007; Serpell 2016). Therefore, wolves might have had specific traits that allowed humans to make advantageous use of them, but whether this is the case or whether they were coincidentally in the right place at the right time for being domesticated remains an open question (Kaminski and Marshall-Pescini 2014).

There are also multiple theories on *how* domestication started. Perhaps it was initiated by humans, as hunters brought wolf pups into the camp, or perhaps wolves approached human camps to feed on discarded food scraps. The first scenario paints a picture where humans actively selected particularly friendly and approachable wolf puppies for companionship (Zimen 1992; Kaminski and Marshall-Pescini 2014). In the second scenario, in line with the so-called self-domestication hypothesis, wolves that were less aggressive or less fearful towards humans would have had the selection advantage to approach and live in close proximity to humans, and so had the opportunity to exploit new and reliable food sources (Hare *et al.* 2012; Kaminski and Marshall-Pescini 2014). Today, many researchers suggest that it was a combination of these two scenarios - that wolves lived in close proximity to humans, and that some of them were tamed by humans and later humans selected for animals showing less aggression and fear (Coppinger and Coppinger 2001; Miklósi 2007; Galibert *et al.* 2011; Kaminski and Marshall-Pescini 2014). Some authors have speculated that social structure and hunting behaviour were similar for early humans and dogs' ancestors, as wolves also hunt and breed cooperatively (Clutton-Brock 1977; Coppinger and Coppinger 2001; Mech and Boitani 2003). Cooperation skills probably already existed in dogs' ancestors, as recent studies with hand-reared wolves socialised with humans have confirmed that these wolves not only show high social attentiveness and tolerance but are also highly cooperative. Such characteristics may have provided a good basis for the evolution of dog-human cooperation (see Range and Virányi 2015 for a review).

In short, both the reason for domestication and the way in which it started remain highly speculative. But they both depend on the place and timing of the initial domestication. For example, it would be crucial to know whether dogs were domesticated long before or during the Neolithic (see also Ben-Dor *et al.* 2011; Larson and Bradley 2014). Thus, unless the place and timing are not agreed upon, it is nearly impossible to draw conclusions about the reasons and process of initial domestication.

Another crucial point to understanding dog domestication is to take into account non-western cultures. Nearly all of the animal cognition studies mentioned above tested dogs owned by people from 'WEIRD' societies, i.e. Western, Educated, Industrialised, Rich, and Democratic societies (Henrich *et al.* 2010). However, the majority of dogs in the world - about 75% - are not kept in the same way as they are in western countries (Gray and Young 2011; Kaminski and Marshall-Pescini 2014). Gray and Young (2011) explored typical dog-human dynamics in 60 different societies, using the electronic global ethnographic database eHRAF. They found that, globally, some dogs served

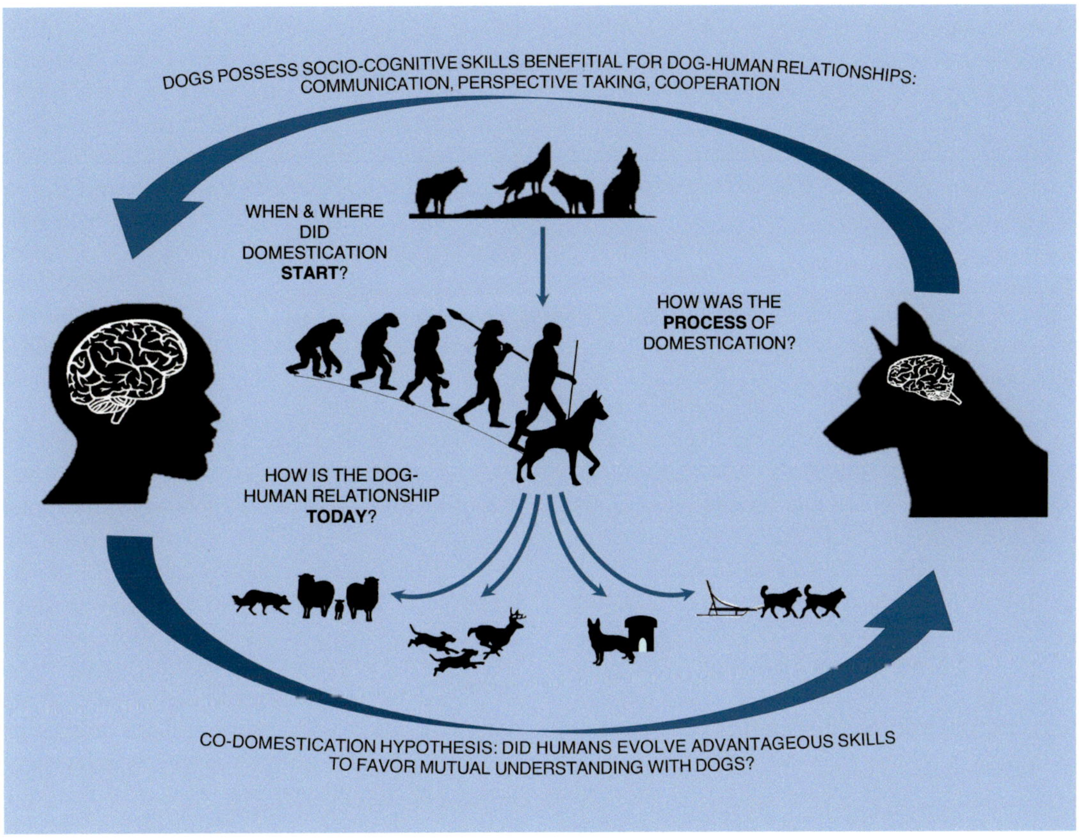


Figure 2. Figure represents the co-evolutionary relationship between dogs and humans. Several questions about dog domestication remain open and can be summarised under three main themes: the starting point, process, and outcomes of domestication. (Parts of this figure were created by Kathryn Kirby and are used with her permission).

practical functions such as aiding in hunting and pest removal, while some were simply kept for pleasure or companionship. Additionally, positive and negative interactions, feeding, and sleeping patterns varied substantially across societies (Gray and Young 2011).

Positive or negative attitudes toward dogs might also influence how dogs are understood. Amici *et al.* (2019) compared how groups of humans with different levels of experience with dogs rated dog emotions from pictures. They found that persons from cultures that are not generally dog-positive (for example, in this study, Muslim cultures in which dogs are often viewed as 'impure,' and are rarely integrated as part of the family) perform worse at accurately interpreting a dog's emotions from a photograph compared to persons who come from cultures that have a generally positive attitude towards dogs (Amici *et al.* 2019). The authors concluded that the ability to recognise dog emotions is mainly acquired through age and experience. However, happy and angry emotions were recognised well independently of participants' age and experience.

We know, therefore, that there are differences among human cultures in the ways dogs are kept, valued and communicated with. What we do not know is whether these differences act as different selective pressures on dog cognition and behaviour, and whether any resulting differences are likely to be heritable. Unfortunately, the absence of carefully planned cross-cultural studies means that we do not know whether dogs kept in non-western cultures perform similarly to 'western' dogs in cognitive tests.

## 3 New questions and interdisciplinary approaches

To answer the crucial questions about dog domestication, there has to be an interdisciplinary approach including archaeology, genetics, anatomy, psychology, and anthropology. Below we propose three lines of questions to be considered, see Figure 2.

### *3.1 The starting point of domestication*

Geneticists and archaeologists are continuing to investigate *when* and *where* wolf domestication took place (i.e. Ostrander *et al.* 2019). The particular question to be answered is whether it was only one or a very few domestication events in one place (i.e. Thalmann *et al.* 2013) or whether dogs were domesticated more than once in multiple places (Mech and Boitani 2003; Frantz *et al.* 2016). The latter scenario would better explain the huge variability in modern dogs that could potentially also involve their cognitive skills. This would raise the possibility that humans from particular cultural backgrounds may have selected their dogs for particular aspects of cognition.

A further related question is *why* wolves were domesticated at that time. More precisely: what skills made dogs valuable for humans so that they were domesticated? The question of what skills made dogs valuable for humans should be investigated. One possibility is their sense of smell. As dogs have an excellent sense of smell, which is 10,000 to 100,000 times better than that of humans (Dalziel *et al.* 2003; Walker *et al.* 2006; Green *et al.* 2012; Hall *et al.* 2015; Polgár *et al.* 2016; Bräuer and Belger 2018), it is feasible that early humans found this skill useful. Wolves probably perceived predators earlier with their nose, and humans might have noticed that. When the bond between the two species became closer, the sensitive nose of dogs may have been very helpful in cooperative hunting.

To answer this question about the benefit for early humans, it is crucial to investigate how dogs were initially used. Archaeological and anthropological findings about the earliest functions of dogs in various cultures can help to answer this question. Moreover, it is also important to directly quantify the benefit of dog keeping. For example, it was found that hunting dogs have a big impact on hunting success (Ruusila and Pesonen 2004). Further studies should also investigate whether there are societies with a long history of cooperating with dogs that have been more successful over evolutionary time periods than societies without dogs (i.e., Shipman 2017).

### *3.2 The domestication process*

One of the most crucial questions about the process of domestication is what skills dogs were selected for. The obvious approach to answer this question is a comparison with their closest living relatives, wolves. Regarding behaviour and cognition, there have been various studies in the last 20 years that directly compare the behaviour of dogs and wolves. In most of these studies, dogs and wolves were raised in an identical way to ensure that the study subjects were truly comparable (i.e. Lampe *et al.* 2017). Although dogs outperform wolves in their ability to use the human pointing gesture (see above), there are various tasks on which they underperform in comparison to their undomesticated relatives. Wolves outperform dogs in quantity discrimination (Range *et al.* 2014) and causal reasoning (Lampe *et al.* 2017). They are more persistent (Marshall-Pescini *et al.* 2017a; Rao *et al.* 2018; see also Miklósi *et al.* 2003) and more risk prone than dogs (Marshall-Pescini *et al.* 2016a). In their packs, wolves reconcile better than dogs (Cafazzo *et al.* 2018) and share monopolisable food more often (Dale *et al.* 2017). Regarding cooperation tasks, the results are mixed. On some tasks, wolves cooperate better with conspecifics than dogs (Marshall-Pescini *et al.* 2017b), while on other tasks they perform equally well (Bräuer *et al.* 2020). These findings illustrate that wolves have maintained skills that enable them to survive in the wild, and that they are adapted to their life in their cohesive social group.

To characterise changes during domestication, an interdisciplinary approach would be illuminating. For example, it was recently found by a group of anatomists and psychologists that dogs possess a muscle responsible for raising the inner eyebrow intensely that is not present in wolves (Kaminski *et al.* 2019). It is likely for two reasons that domestication transformed the facial muscle anatomy of dogs specifically for facial communication with humans. First, dogs only use this muscle to raise the inner eyebrow when a human is looking at them (Kaminski *et al.* 2017). Second, humans have a preference to adopt dogs that show the inner eyebrow raise (Waller *et al.* 2013). Why exactly humans prefer such dogs is still speculative. It might be related to paedomorphism, i.e., that dogs appear more like wolf puppies. These features were thought to have evolved as a byproduct of the domestication process, and arose accidentally when aggression was actively selected against (Hare *et al.* 2012), but it is also possible that dogs were selected for these features as human prefer paedomorphic characteristics (Kaminski *et al.* 2019). This combination of behavioural and anatomical studies seems to be very promising to shed light on the selection process during domestication.

Another less obvious approach to understand the domestication process is using linguistics. It is possible that linguistic patterns might provide clues regarding the history of dog-human-relationships in particular societies. For example, linguists often use patterns of colexification and cognacy to make inferences about the origins of particular terms and the extent to which they may have been borrowed from a neighbouring language vs. inherited from a linguistic ancestor. It might be interesting, for example, to examine patterns of colexification between lexemes for 'dog' and those for 'wolf' or other wild canids across as many of the world's ~7000 languages as possible. It could then be tested whether observed patterns of colexification reflect what is known from the historical and archaeological record about dog domestication and dog keeping across regions and language families. Similarly, it might be expected that the standard term for 'dog' in a language would reflect dogs' dominant function in societies that speak that particular language. This hypothesis could be tested by examining patterns of cognacy (shared word origin) among societies in which dogs share vs. differ in their dominant functions (e.g., hunting, herding, defence).

Finally, to fully understand the domestication process, one should not only investigate dog cognition and behaviour but also the human side of the equation and the possibility of co-domestication. According to the co-domestication hypothesis, not only have dogs evolved special skills to assess humans, but humans may have also evolved special advantageous skills that favour mutual understanding between the two species (Kaminski and Marshall-Pescini 2014; Amici *et al.* 2019).

### *3.3 The result of dog domestication*

The third and final important line of inquiry into the question of dog domestication explores its outcome - the contemporary dog-human relationship. One lingering question is the universality of dog-human interactions, i.e., the extent to which the recently recognised typical social-cognitive skills of dogs are universal or whether they are influenced by the society in which the dogs live. For example, what patterns are found in the differences in keeping, treating, and perceiving dogs across cultures? One hypothesis would be that the dog-human relationship is closer in societies where the dominant dog function requires intense cooperation between humans and dogs (see also Bräuer *et al.* 2020). It is also likely that there is a more positive attitude towards dogs when they have predominantly cooperative functions (i.e., shepherd or hunting dogs), resulting in more careful treatment of these dogs, than when they have predominantly non-cooperative functions (i.e., guard dogs). Moreover, in particular in western societies, sociological studies can describe the influence dogs have on societies today (i.e. Sanders 1999).

Another question is whether there are differences in dogs' cognitive skills and dog-human interactions depending on the cultural practices of the society they live in and dogs' function(s) in that society. Have human cultural differences acted as different selective pressures on dog cognition, and have they produced detectable differences in dog-human interaction? To investigate this, dogs from different cultural contexts should be tested in cognitive experiments about dog-human communication and cooperation (see above).

Finally, to better understand the contemporary dog-human relationship, it would also be important to find ways to quantify the benefit of dog keeping today. A first approach could be to attempt to quantify the benefit of dog ownership, for example, in western societies (i.e. Cutt *et al.* 2007; Knight and Edwards 2008). By developing new techniques for measuring the benefit to humans of dogs in different contexts, as well as calculating their costs, we can assess whether the dog-human relationship is, as commonly believed, mutualistic or, as some argue, parasitic (Archer 1997).

## 4 Concluding remarks

We have seen how in recent years, research in comparative psychology has revealed that dogs evolved human-like skills in order to live functionally in human societies. Particularly, dogs show remarkable skills in three social-cognitive domains: communication, perspective taking, and cooperation.

We know that humans might have selected dogs to be good cooperative and communicative partners, however, there are several aspects of this process that are still unknown: When, where, and how did the process of domestication start? Why were wolves a good candidate to be domesticated? How do the results of this process look today and how does dog keeping differ across cultures?

In conclusion, although the research about dog domestication has made great leaps forward in the last 20 years, several open questions remain. These questions can be summarised under three main themes - the starting point, process, and outcomes of domestication. To answer these questions, we need an interdisciplinary approach in which scientists from archaeology, genetics, anatomy, psychology, sociology, and anthropology work together.

## Acknowledgements

We thank Kathryn Kirby and William Taylor for helpful comments on earlier versions of the manuscript, and we are grateful to Anne Gibson for proof reading.

## References

Amici, F. 2015. The evolution and development of human cooperation. *Interaction Studies* 16 (3): 383–418.

Amici, F., J. Waterman, C.M. Kellermann, K. Karimullah, and K.J. Bräuer. 2019. The ability to recognize dog emotions depends on the cultural milieu in which we grow up. *Scientific Reports* 9: 16414.

Archer, J. 1997. Why do people love their pets? *Evolution and Human Behavior* 18: 237–259.

Ben-Dor, M., A. Gopher, I. Hershkovitz and R. Barkai. 2011. Man the Fat Hunter: The Demise of Homo erectus and the Emergence of a New Hominin Lineage in the Middle Pleistocene (ca. 400 kyr) Levant. *PLoS ONE* 6(12): e28689.

Bhattacharjee, D., N., ND., S. Gupta, S. Sau, R. Sarkar, A.Biswas, A. Banerjee, D. Babu, D. Mehta and A. Bhadra. 2017. Free-ranging dogs show age related plasticity in their ability to follow human pointing. *PLoS ONE* 12(7): e0180643.

Bräuer, J. 2015. I do not understand but I care: the prosocial dog. *Interaction Studies* 16(3): 341–360.

Bräuer, J. and J. Belger. 2018. A ball is not a Kong: Odor representation and search behavior in domestic dogs (*Canis familiaris*) of different education. *Journal of Comparative Psychology* 132: 189–199.

Bräuer, J., J. Call and M. Tomasello. 2004. Visual Perspective Taking in Dogs (*Canis familiaris*) in the Presence of Barriers. *Applied Animal Behaviour Science* 88: 299–317.

Bräuer, J., J. Kaminski, J. Riedel, J. Call and M. Tomasello. 2006. Making inferences about the location of hidden food: Social dog, causal ape. *Journal of Comparative Psychology* 120: 38–47.

Bräuer, J., M. Keckeisen, A. Pitsch, J. Kaminski, J. Call and M. Tomasello. 2013a. Domestic dogs conceal auditory but not visual information from others. *Animal Cognition* 16: 351–359.

Bräuer, J., K. Schönefeld and J. Call. 2013b. When do dogs help humans? *Applied Animal Behaviour Science* 148: 138–149.

Bräuer, J., K. Stenglein, and F. Amici. 2020. Dogs (*Canis familiaris*) and wolves (*Canis lupus*) coordinate with conspecifics in a social dilemma. *Journal of Comparative Psychology*, 134, 211–221.

Cafazzo, S., S. Marshall-Pescini, M. Lazzaroni, Z. Virányi and F. Range. 2018. The effect of domestication on post-conflict management: wolves reconcile while dogs avoid each other. *Royal Society Open Science* 5: 171553.

Call, J., J. Bräuer, J. Kaminski and M. Tomasello. 2003. Domestic Dogs (*Canis familiaris*) Are Sensitive to the Attentional State of Humans. *Journal of Comparative Psychology* 117: 257–263.

Clutton-Brock, J. 1977. Man-made dogs. *Science* 197: 1340–1342.

Coppinger, R. and L. Coppinger. 2001. *Dogs: A Startling New Understanding of Canine Origin, Behavior, and Evolution.*New York, NY: Scribner.

Cutt, H., B. Giles-Corti, M. Knuiman and V. Burke, V. 2007. Dog ownership, health and physical activity: A critical review of the literature. *Health and Place* 13: 261–272.

Dale, R., M. Quervel-Chaumette, L. Huber, F. Range and S. Marshall-Pescini. 2016. Task differences and prosociality: Investigating pet dogs' prosocial preferences in a token choice paradigm. *PLoS ONE* 11: e0167750.

Dale, R., F. Range, L. Stott, K. Kotrschal and S. Marshall-Pescini. 2017. The influence of social relationship on food tolerance in wolves and dogs. *Behavioral Ecology and Sociobiology* 71: 107.

Dalziel, D. J., B.M. Uthman, S.P. Mcgorray and R.L. Reep. 2003. Seizure-alert dogs: a review and preliminary study. *Seizure-European Journal of Epilepsy* 12: 115–120.

Diamond, J. 1997. *Guns, Germs, and Steel: The Fates of Human Societies.* Scranton, PA: W. W. Norton.

Feddersen-Petersen, D. U. 2000. Vocalization of European wolves (*Canis lupus lupus* L.) and various dog breeds (*Canis lupus f. fam.*). *Archiv Fuer Tierzucht.*43: 387–397.

Frantz, L.A.F., V.E. Mullin, M. Pionnier-Capitan, O. Lebrasseur, M. Ollivier, A. Perri, A. Linderholm, V. Mattiangeli, M.D. Teasdale, E.A Dimopoulos, A. Tresset, M. Duffraisse, F. McCormick, L. Bartosiewicz, E. Gál, É.A. Nyerges, M.V. Sablin, S. Bréhard, M. Mashkour, A. Bălăşescu, B. Gillet, S. Hughes, O. Chassaing, C. Hitte, J.-D.C., Vigne, K. Dobney, C. Hänni, D.G. Bradley and G. Larson. 2016. Genomic and archaeological evidence suggest a dual origin of domestic dogs. *Science* 352: 1228–1231.

Gácsi, M., B. Gyoöri, Z. Virányi, E. Kubinyi, F. Range, B. Belényi and Á. Miklósi. 2009a. Explaining Dog Wolf Differences in Utilizing Human Pointing Gestures: Selection for Synergistic Shifts in the Development of Some Social Skills. *PLoS ONE* 4: e6584.

Gácsi, M., B. Gyori, A. Miklósi, Z. Virányi, E. Kubinyi, J. Topál and V. Csányi. 2005. Species-Specific Differences and Similarities in the Behavior of Hand-Raised Dog and Wolf Pups in Social Situations with Humans. *Developmental Psychobiology* 47: 111–122.

Gácsi, M., E. Kara, B. Belényi and Á. Miklósi. 2009b. The effect of development and individual differences in pointing comprehension of dogs. *Animal Cognition* 12: 471–479

Gácsi, M., A. Miklósi, O.,Varga, J. Topál, and V. Csányi. 2004. Are readers of our face readers of our minds? Dogs (*Canis familiaris*) show situation-dependent recognition of human's attention. *Animal Cognition* 7: 144–153.

Galibert, F., P. Quignon, C. Hitte and C. André. 2011. Toward Understanding Dog Evolutionary and Domestication History. *Comptes Rendus Biologies* 334: 190–196.

Gray, P. and S.M. Young. 2011. Human-Pet Dynamics in Cross-Cultural Perspective. *Anthrozoos: A Multidisciplinary Journal of The Interactions of People and Animals* 24: 17–30.

Green, P.A., B. Van Valkenburgh, B. Pang, D. Bird and A. Curtis. 2012. Respiratory and olfactory turbinal size in canid and arctoid carnivorans. *Journal of Anatomy* 221: 609–621

Hale, E.B. 1969. Domestication and the evolution of behavior, in E. S. E.Hafez (ed.) *The behaviour of domestic animals* 2nd edition: 22–42. London: Bailliere, Tindall and Cassell.

Hall, N., K. Glenn, D. Smith and C. Wynne. 2015. Performance of Pugs, German Shepherds, and Greyhounds (*Canis lupus familiaris*) on an odor-discrimination task. *Journal of Comparative Psychology* 129: 237–246.

Hare, B., M. Brown, C. Williamson and M. Tomasello. 2002. The domestication of social cognition in dogs. *Science* 298: 1634–1636.

Hare, B., A. Rosati, J. Kaminski, J. Bräuer, J. Call and M. Tomasello. 2010. The Domestication Hypothesis for Dogs' Skills with Human Communication: A Response to Udell *et al.* (2008) and Wynne *et al.* (2008). *Animal Behaviour* 79: e1-e6.

Hare, B. and M. Tomasello. 2005. Human-like social skills in dogs? *Trends in Cognitive Sciences* 9: 439–444.

Hare, B., V. Wobber and R.W. Wrangham. 2012. The self-domestication hypothesis: evolution of bonobo psychology is due to selection against aggression. *Animal Behavior* 83: 573–585.

Heberlein, M.T., D.C. Turner, and M. Manser. 2017. Dogs' (*Canis familiaris*) attention to human perception: Influence of breed groups and life experiences. *Journal of Comparative Psychology* 131: 19–29.

Henrich, J., S.J. Haine and A. Norezavan. 2010. The weirdest people in the world? *Behavioural Brain Science* 33: 61–83.

Janssens, L., L. Giemsch, R. Schmitz, M. Street, S. Van Dongen and P. Crombé. 2018. A new look at an old dog: Bonn-Oberkassel reconsidered. *Journal of Archaeological Science* 92: 126–138.

Kaminski, J., J. Bräuer, J. Call and M. Tomasello. 2009a. Domestic dogs are sensitive to a human's perspective. *Behaviour* 146: 979–998.

Kaminski, J., J. Call, and J. Fischer. 2004. Word learning in a domestic dog: Evidence for 'Fast Mapping'. *Science* 304: 1682–1683.

Kaminski, J., J., Hynds, P. Morris, and B.M. Waller. 2017. Human attention affects facial expressions in domestic dogs. *Scientific Reports* 7: 12914.

Kaminski, J. and S. Marshall-Pescini. 2014. *The social dog: Behaviour and cognition*. San Diego, CA: Academic Press/Elsevier.

Kaminski, J., M., Neumann, J. Bräuer, J. Call, and M. Tomasello, 2011. Domestic Dogs Communicate to Request and Not to Inform. *Animal Behaviour* 82: 651–658.

Kaminski, J., A. Pitsch and M. Tomasello. 2013. Dogs steal in the dark. *Animal Cognition* 16: 385–394.

Kaminski, J., S.,Tempelmann, J. Call and M. Tomasello. 2009b. Domestic dogs comprehend human communication with iconic signs. *Developmental Science* 12: 831–837.

Kaminski, J., B.M. Waller, R. Diogo, A. Hartstone-Rose and A.M. Burrows. 2019. Evolution of facial muscle anatomy in dogs. *Proceedings of the National Academy of Sciences* 116(29): 14677–14681.

Knight, S. and V. Edwards. 2008. In the Company of Wolves:The Physical, Social, and Psychological Benefits of Dog Ownership. *Journal of Aging and Health* 20: 437–455.

Kundey, S.M.A., A. De Los Reyes, C. Taglang, R. Allen, S. Molina, E. Royer and R. German. 2010. Domesticated Dogs (*Canis familiaris*) React to What Others Can and Cannot Hear. *Applied Animal Behaviour Science* 126: 45–50.

Lampe, M., J. Bräuer, J. Kaminski and Z. Virányi. 2017. The effects of domestication and ontogeny on cognition in dogs and wolves. *Scientific Reports* 7: 11690.

Larson, G. and G. Bradley.2014. How Much Is That in Dog Years? The Advent of Canine Population Genomics. *PLOS Genetics* 10: 1004093.

Macpherson, K. and W. Roberts. A. 2006. Do Dogs (*Canis familiaris*) Seek Help in an Emergency? *Journal of Comparative Psychology* 120: 113–119.

Marshall-Pescini, S., I. Besserdich, C. Kratz and F. Range, F. 2016a. Exploring Differences in Dogs' and Wolves' Preference for Risk in a Foraging Task. *Frontiers in Psychology*, 7: 1241.

Marshall-Pescini, S., R. Dale, M. Quervel-Chaumette and F. Range. 2016b. Critical issues in experimental studies of prosociality in non-human species. *Animal Cognition* 19: 679–705.

Marshall-Pescini, S., A. Rao, Z. Virányi and F. Range. 2017a. The role of domestication and experience in 'looking back' towards humans in an unsolvable task. *Scientific Reports* 7: 46636.

Marshall-Pescini, S., J.F.L. Schwarz, I. Kostelnik, Z. Virányi and F. Range. 2017b. Importance of a species' socioecology: Wolves outperform dogs in a conspecific cooperation task. *Proceedings of the National Academy of Sciences* 114: 11793–11798.

Mech, L.D. and L. Boitani, L. 2003. *Wolves: Behavior, Ecology, and Conservation*.Chicago, IL: University of Chicago Press.

Melis, A.P. and D. Semmann. 2010. How is Human Cooperation Different? *Philosophical Transactions of the Royal Society of London B Biological Sciences* 365: 2663–2674.

Melis, A.P. and F. Warneken. 2016. The psychology of cooperation: insights from chimpanzees and children. *Evolutionary Anthropology* 25: 297–305.

Merola, I., E. Prato-Previde, and S. Marshall-Pescini. 2012. Dogs' Social Referencing towards Owners and Strangers. *PLoS ONE* 7(10): e47653.

Miklósi, A. 2007. *Dog Behaviour, Evolution, and Cognition.* Oxford: Oxford University Press.

Miklósi, A., E. Kubinyi, M. Gácsi, Z. Virányi, and V. Csányi. 2003. A simple reason for a big difference: wolves do not look back at humans but dogs do. *Current Biology* 13: 763–766.

Miklósi, A., Polgárdi, J. Topál and Csányi, V. 2000. Intentional behavior in dog-human communication: An experimental analysis of 'showing' behaviour in the dog. *Animal Cognition* 3: 159–166.

Miklósi, A. K. and Soproni. 2006. A comparative analysis of the animals' understanding of the human pointing gesture. *Animal Cognition* 9: 81–93.

Mitchell, P. 2017. Disease: A Hitherto Unexplored Constraint on the Spread of Dogs (*Canis lupus familiaris*) in Pre-Columbian South America. *Journal of World Prehistory* 30: 301–349.

Ní Leathlobhair, M., A.R. Perri, E.K. Irving-Pease, K.E. Witt, A. Linderholm, J. Haile, O. Lebrasseur, C. Ameen, J. Blick, A.R. Boyko, S. Brace, Y.N.Cortes, S.J. Crockford, A. Devault, E.A. Dimopoulos, M. Eldridge, J. Enk, S. Gopalakrishnan, K. Gori, V. Grimes, E. Guiry, A.J. Hansen, A. Hulme-Beaman, J. Johnson, A. Kitchen, A.K. Kasparov, Y.-M. Kwon, P.A. Nikolskiy, C.P. Lope, A. Manin, T. Martin, M. Meyer, K.N. Myers, M. Omura, J.-M. Rouillard, E.Y. Pavlova, P. Sciulli, M.-H.S., Sinding, A. Strakova, V.V. Ivanova, C. Widga, E. Willerslev, V.V. Pitulko, I. Barnes, M.T.P. Gilbert, K.M. Dobney, R.S. Malhi, E.P. Murchison, G. Larson and L.A.F. Frantz. 2018. The evolutionary history of dogs in the Americas. *Science* 361: 81–85.

Ostrander, E.A., G.-D. Wang, G. Larson, B. M. Vonholdt, B. W. Davis, V. Jagannathan, C. Hitte, R.K. Wayne, Y.-P. Zhang, and D.K. Consortium. 2019. Dog10K: an international sequencing effort to advance studies of canine domestication, phenotypes and health. *National Science Review* 6: 810–824.

Paxton, D.W. 2000. A Case for a Naturalistic Perspective. *Anthrozoös* 13: 5–8.

Polgár, Z., M. Kinnunen, D., Újváry, Á. Miklósi, and M. Gácsi. 2016. A Test of Canine Olfactory Capacity: Comparing Various Dog Breeds and Wolves in a Natural Detection Task. *PLoS ONE* 11: e0154087.

Pongracz, P., C. Molnar, A. Miklósi and V. Csányi. 2005. Human Listeners Are Able to Classify Dog (*Canis familiaris*) Barks Recorded in Different Situations. *Journal of Comparative Psychology* 119: 136–144.

Prato Previde, E. and P. Valsecchi. 2014. The immaterial cord: the dog-human attachment bond, in J. Kaminski and S. Marshall-Pescini (eds) *The Social Dog:* 165–189. San Diego, CA: Academic Press/Elsevier.

Price, E.O. 1984. Behavioral Aspects of Animal Domestication. *Quarterly Review of Biology* 59: 1–32.

Quervel-Chaumette, M., R. Dale, S. Marshall-Pescini and F. Range. 2015. Familiarity affects other-regarding preferences in pet dogs. *Scientific Reports* 5: 18102.

Range, F., J. Jenikejew, I. Schröder and Z. Virányi. 2014. Difference in quantity discrimination in dogs and wolves. *Frontiers in Psychology* 5: 1299.

Range, F.Z. and Virányi. 2015. Tracking the evolutionary origins of dog-human cooperation: The 'Canine Cooperation Hypothesis'. *Frontiers in Psychology* 5: 1582.

Rao, A., F. Range, K. Kadletz, K. Kotrschal and S. Marshall-Pescini. 2018. Food preferences of similarly raised and kept captive dogs and wolves. *PLoS ONE* 13: e0203165.

Riedel, J., K. Schumann, J. Kaminski, J. Call and M. Tomasello. 2008. The Early Ontogeny of Human-Dog Communication. *Animal Behaviour* 75: 1003–1014.

Ruusila, V. and M. Pesonen. 2004. Interspecific Cooperation in Human (*Homo sapiens*) hunting: the benefits of a barking dog (*Canis familiaris*). *Annual Zoological Fennici* 41: 545–549.

Sanders, C.R. 1999. *Understanding Dogs: living and working with canine companions.* Philadelphia, PA: Temple University Press.

Serpell, J. 2016. *The domestic dog: Its evolution, behaviour and interactions with people.* Cambridge: Cambridge University Press.

Shipman, P. 2017. *The Invaders: How Humans and Their Dogs Drove Neanderthals to Extinction.*Cambridge, MA: Harvard University Press.

Thalmann, O., B. Shapiro, P. Cui, V.J. Schuenemann, S.K. Sawyer, D.L. Greenfield, M.B. Germonpré, M.V. Sablin, S. López-Giráldez, X. Domingo-Roura, H. Napierala, H.-P. Uerpmann, D.M. Loponte, A.A. Acosta, L. Giemsch, R. W. Schmitz, B. Worthington, J.E. Buikstra, A. Druzhkova, A.S. Graphodatsky, N.D. Ovodov, N. Wahlberg, A.H. Freedman, R.M. Schweizer, K.-P. Koepfli, J.A. Leonard, M. Meyer, J. Krause, S. Pääbo, R.E. Green and R.K. Wayne 2013. Complete mitochondrial genomes of ancient canids suggest a european origin of domestic dogs. *Science* 342: 871–874.

Topál, J., M. Gácsi, A. Miklósi, Z. Virányi, E. Kubinyi, and V. Csányi. 2005. Attachment to humans: A comparative study on hand-reared wolves and differently socialized dog puppies. *Animal Behaviour* 70: 1367–1375.

Udell, M.A.R., N.R. Dorey and C.D.L. Wynne. 2008. Wolves Outperform Dogs in Following Human Social Cues. *Animal Behaviour* 76: 1767–1773.

Vilá, C., P. Savolainen, J.E. Maldonado, I.R. Amorim, J. Rice, R.L. Honeycutt, K.A. Crandall, J. Lundeberg and R. K. Wayne. 1997. Multiple and ancient origins of the domestic dog. *Science* 276: 1687–1689.

Virányi, Z., J.,Topál, M. Gácsi, A. Miklósi, and V. Csányi. 2004. Dogs respond appropriately to cues of humans' attentional focus. *Behavioural Processes* 66: 161–172.

Walker, D.B., J.C. Walker, P.J. Cavnar, J.L. Taylor, D.H. Pickel, S.B. Hall and J.C. Suarez. 2006. Naturalistic quantification of canine olfactory sensitivity. *Applied Animal Behaviour Science* 97: 241–254.

Waller, B.M., K. Peirce, C.C. Caeiro, L. Scheider, AM. Burrows, S. McCune and J. Kaminski. 2013. Paedomorphic Facial Expressions Give Dogs a Selective Advantage. *PLoS ONE* 8: e82686.

Wang, G.-D., W. Zhai, H.-C. Yang, R.-X. Fan, X. Cao, L. Zhong, L.Wang, F. Liu, H. Wu, L.-G. Cheng, A.D. Poyarkov, N.A. Poyarkov Jr, S.-S. Tang, W.-M. Zhao, Y. Gao, X.-M. Lv, D. M. Irwin, P. Savolainen, C.-I. Wu and Y.-P. Zhang. 2013. The genomics of selection in dogs and the parallel evolution between dogs and humans. *Nature Communications* 4: 1860.

Wayne, R.K. and C. Vilá, 2001. Phylogeny and origin of the domestic dog, in A. Ruvinsky and J. Sampson (eds) *The Genetics of the Dog*: 1–14. Wallingford: CAB International.

Zeder, M. A. 2012. Pathways to Animal Domestication, in A.B. Damania, C.O. Qualset, P.E. Mcguire, P. Gepts, R.L. Bettinger, S.B. Brush, and T.R. Famula (eds) *Biodiversity in Agriculture: Domestication, Evolution, and Sustainability*: 227–259. Cambridge: Cambridge University Press.

Zimen, E. 1992. *Der Hund.*Munich: Wilhelm Goldmann Verlag.

# 2.4 Vector-Borne Diseases as Possible Constraints on the Spread of Dogs into the Tropics and Beyond

Peter Mitchell

School of Archaeology, University of Oxford and Rock Art Research Institute, University of the Witwatersrand; St Hugh's College, Oxford, OX2 6LE, United Kingdom, peter.mitchell@arch.ox.ac.uk

**Abstract**

Dogs were probably domesticated toward the end of the Pleistocene in both East Asia and Europe, subsequently moving into central Eurasia, North America, and North Africa. However, their subsequent expansion into the tropics was much later and frequently quite staggered compared to most of the northern hemisphere. This paper asks if novel disease challenges of which dogs had no prior experience might help explain this pattern, constraining their ability to move beyond areas to which their wolf ancestors were native and into environments hosting vector-borne infections against which, initially at least, they had no resistance.

**Keywords:** dogs, expansion, Tropics, vector-borne diseases, constraints.

## 1 Introduction

There is general agreement that domestic dogs (*Canis lupus familiaris*) descend from the Eurasian grey wolf (*C. lupus*), with domestication taking place in both eastern and western Eurasia, followed by partial replacement of western Eurasian dogs by those of East Asian ancestry (Frantz *et al.* 2016). Hunter-gatherers were keeping dogs in northern China, Russia's Far East, Europe, and parts of the Middle East before 10,000 years ago and they occur in pastoralist contexts in Egypt's Western Desert *c.* 6000 cal. BC. Newly published dates also confirm their spread as far as the Midwestern United States by 10,000 years ago (Perri *et al.* 2019), necessitating a still earlier entry from Siberia (Ni Leathlobhair *et al.* 2018). However, this widespread, presumably fast-moving late Pleistocene/early Holocene dispersal across the Holarctic (Arctic plus Nearctic plus Palearctic) region did not continue as they moved into the tropics (Figure 1). Instead, from Central and South America to Sub-Saharan Africa to Southeast Asia and Australasia, dogs are unknown until 5000 years ago at best. Indeed, in some places (e.g. Africa south of the Equator) they appear significantly later still, while in others (e.g. Amazonia) they were absent until European contact (Larson *et al.* 2012; Mitchell 2015, 2017; Piper 2017; Stahl 2012: 108).

Multiple factors might explain this situation. Taphonomic issues are certainly important as it may be difficult unambiguously to assign fragmentary or poorly preserved remains to dogs rather than wild canids, something not helped by the enormous size variation that domestication produces, a tendency to rely upon teeth and cranial parts rather than post-cranial material when making identifications (Stahl 2012), and a still limited application of aDNA. Nor can we assume that dogs conveniently chose to die where archaeologists dig; many animals likely perished at 'work' (e.g. on hunts) or on the periphery of human campsites and not all societies ate them. Deliberately burying dogs - with or without accompanying human individuals - is also far from universal (Morey 2006). It is, however, surely an exercise in special pleading to assume that taphonomy and human choice about what to eat and whom to bury suffice to explain what seems to be a robust pattern on a global scale. This is all the more so bearing in mind that questions abound over the accuracy of identifications and dates for claims of pre-5000 BP dogs in the tropics. For example, a 9000–7000-year-old age for five dogs buried at Cueva del Tecolote in central Mexico depends upon stratigraphic extrapolation and claimed archaeological affinities, not direct dating (Pompa y Padilla and Serrano Carreto 2001). Likewise, though Cranbrook (2014) suggests that a few undiagnostic bones from Madai Cave, Sabah, and Ille Cave, Palawan, point to their presence in terminal Pleistocene/early Holocene times in Island Southeast Asia, these more likely represent locally extinct dhole (*Cuon alpinus*) populations that reached Borneo and the Philippines from the mainland at times of low sea-level (Piper 2017).

If taphonomic factors cannot explain the much slower expansion of dogs into the tropics and beyond perhaps we should look to cultural variables, including the functions dogs fulfilled in human societies and the ways in which they were viewed? As hunting aids, for instance, their value certainly varies from one environmental setting to another, depending in part upon the prey species in question (Koster 2009; Lupo 2017). Thus, while they may be effective against

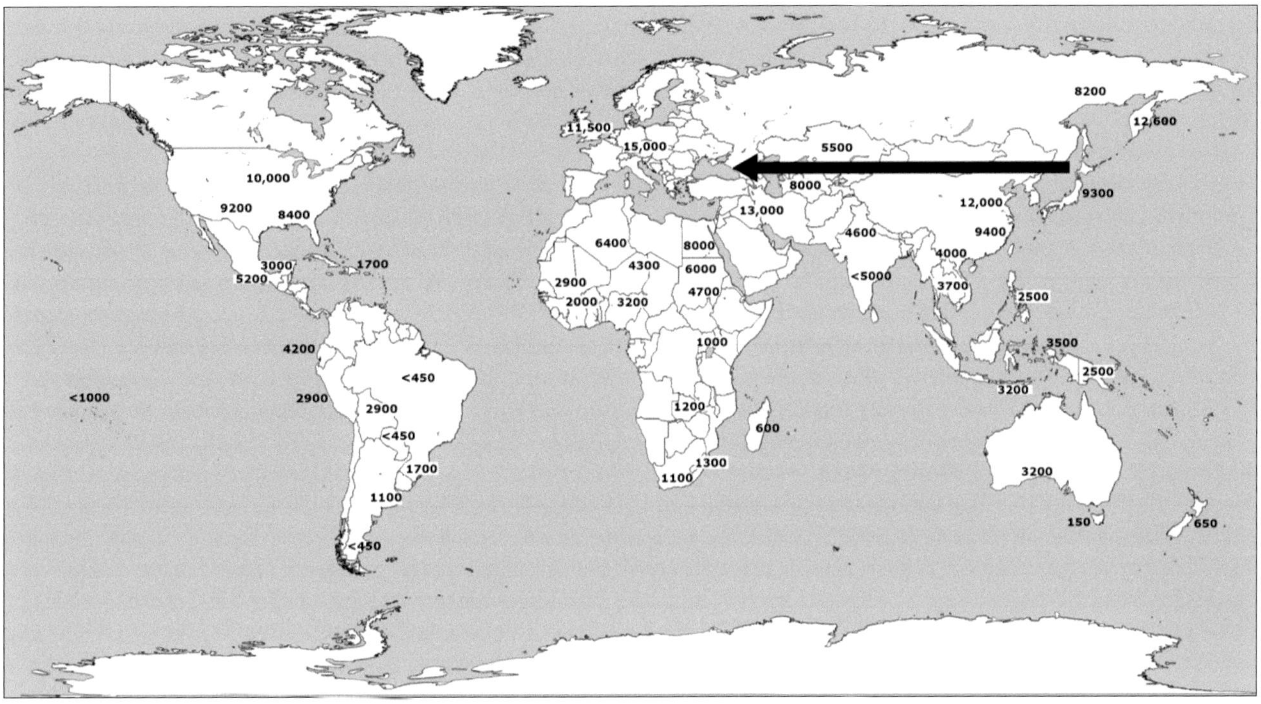


Figure 1. The global expansion of the dog (After Larson *et al.* 2012; Mitchell 2015, 2017; Frantz *et al.* 2016; Perri *et al.* 2019). Dates are all cal. BP.

terrestrial species, dogs may contribute little when hunting arboreal taxa (monkeys, birds) taken with projectile weapons like bows-and-arrows or blowpipes.

Even then, however, one might expect that they would have proven useful as watchdogs against predators or - perhaps - have been valued as companions, much like the many wild animals, canids included, that people keep as pets (e.g. Stahl 2013). Systems of value and symbolic associations must, however, also be considered. Vander Velden (2008), for instance, argues that their conceptual association with jaguars (i.e. predators) means that dogs cannot readily enter into many Native Amazonian societies, an understanding that perhaps deterred people from accepting them. On the other hand, culture can also change, as with the extraordinarily swift take-up of dogs by Indigenous Tasmanians following British settlement of the island at the start of the nineteenth century (Boyce 2006).

## 2 A role for disease?

The possibility that disease constrained the expansion of dogs into tropical environments has not yet been widely discussed in the archaeological and anthropological literature, although Koster (2009: 593) notes that they show 'high mortality' in such habitats in the New World. Accidents, attacks by wild animals, and malnutrition all contribute to this (Koster 2009: 591), but it is on vector-borne diseases that I want to focus here, arguing that across the tropics they once posed – and in many places still pose – serious impediments to the successful keeping of dogs. The high mortality rates reported from regions such as Panama (Pineda *et al.* 2011), Venezuela's *llanos* savannas (Yu 1997), and southeastern Bolivia (Fiorello *et al.* 2006) underline this well.

That biological constraints may have applied to the spread of dogs into environments quite different from those to which their ancestor, the grey wolf, was adapted should not occasion surprise (Figure 2). In Sub-Saharan Africa, for example, it is well established that while domestic livestock of ultimately Near Eastern origin - cattle (*Bos taurus*), sheep (*Ovis aries*), and goats (*Capra hircus*) - ultimately reached the Cape region of South Africa, their expansion was far from smooth. Instead, it was heavily influenced (and in some areas completely prevented) by the prevalence of previously unencountered pathogens easily transmitted by ticks or biting flies, pathogens that find their natural reservoirs in ungulates that, unlike livestock, are native to the region (Gifford-Gonzalez 2000, 2017). The expansion south of the Sahel of horses (*Equus caballus*), donkeys (*E. asinus*), and pigs (*Sus scrofa*) was even more stringently curtailed, with trypanosomiasis, African horse sickness, and African swine fever producing mortality rates that can range close to 100% (Mitchell 2018).

The possibility of encountering equivalent disease threats may have been even greater where other wild canid species were present since pathogens and vectors already adapted to those taxa may have found it relatively easy to transfer to the new arrivals and include dogs among their targets. Lacking any prior acquaintance, dogs will have found such pathogens much more dangerous than the native canids with which they had co-evolved. The susceptibility of modern wild canids to diseases transmitted to them in part by domestic dogs, such as rabies and canine distemper (Woodroffe *et al.* 2004), underlines how devastating these encounters could have been.

Direct evidence for the presence of infectious disease in prehistoric dogs is rare, notwithstanding fortuitous instances such as the identification of brown dog ticks (*Rhipicephalus sanguineus*) - an important vector discussed further below - on a mummified puppy of Roman date in Egypt's Kharga Oasis (Huchet *et al.* 2013) or of multiple parasites on mummified Chiribaya shepherd dogs in Peru's Atacama Desert (Richardson *et al.* 2012). Pending further such instances, the recovery of pathogen DNA from bone (Costa Junqueira *et al.* 2009) and dental calculus (Weyrich *et al.* 2015), or robust chronologies for the evolution of canine-specific pathogenic strains obtained via molecular phylogenies, we must turn to less direct lines of argument. That argument has three main planks:

1. the veterinary literature can help establish which diseases affect domestic dogs in the world's tropical regions today, where, how, and under what conditions they occur, and the impacts that they have;
2. where one species displays a more virulent form of the disease than another then it most likely received it more recently (Gifford-Gonzalez 2000: 109);
3. where breeds of dog, native to the tropics today, display tolerance to a particular disease, but dogs from elsewhere in the world (typically Europe or North America) do not, then it is likely that a) such resistance took some time to evolve; and b) the disease concerned once posed a much more serious threat to all dogs within the tropics themselves.

Drawing these strands of evidence together, I now explore the possibility that vector-borne infectious diseases native to tropical parts of the world constrained the expansion there of domestic dogs. In doing so, I summarise, revise, and develop points made in earlier papers with respect to Sub-Saharan Africa (Mitchell 2015, 2018) and South America (Mitchell 2017), while extending my argument to include the Indian sub-continent and Southeast Asia as well.

## 3 Sub-Saharan Africa

A cousin of the grey wolf, the African golden wolf (*Canis lupaster*), is native to northern Africa (Viranta *et al.* 2017), but genetic studies suggest only limited introgression after dogs were introduced there from the Middle East (Liu *et al.* 2018). That introduction is first evident in Middle Neolithic herder contexts in Egypt's Western Desert *c.* 6300–5600 cal. BC. Dogs may then have spread rapidly among - and as part of the expansion of - pastoralist societies up the Nile and across what is now the Sahara, reaching at least as far as Khartoum and southwestern Libya by 4500 cal. BC. However, they are not attested in the Sahel, the savanna region immediately south of the desert, before the mid-second millennium cal. BC. South of the River Niger and in West Africa's forest zone they remain undocumented before the Christian era (Mitchell 2015), although puzzlingly linguistic reconstructions imply that Bantu-speakers dispersing south from Cameroon kept them in the last millennium BC (Vansina 1990: 92). Even so, the archaeological records of eastern and southern Africa currently suggest a lag of several centuries between the arrival of Bantu-speaking Iron Age Farming Communities in these regions and their first evidence for dogs in the second half of the first millennium cal. AD (Mitchell 2015).

Sub-Saharan Africa hosts several infectious diseases that threaten canine survival. Trypanosomiasis (sleeping sickness) is among the most deadly. Infection with the protozoon *Trypanosoma brucei brucei* typically produces a rapid, acute infection that almost invariably results in death (Matete 2003; Nwoha and Anene 2011), while infection with *T. congolense* results in a more chronic condition that can also result in near-100% mortality (Ezeokonkwo *et al.* 2010). However, while dogs of European origin often die suddenly once exposed (Hörchner *et al.* 1985; Watier-Grillot *et al.* 2013), in several regions of Sub-Saharan Africa local dogs may remain asymptomatic and healthy after infection with *T. congolense* and, possibly, also *T. b. brucei* (Abenga *et al.* 2005; Keck *et al.* 2009; Lisulo *et al.* 2014). Evolution of trypanotolerance must nevertheless have taken some time, and its absence in European breeds underlines the severity of the threat posed by canine trypanosomiasis to dogs entering Sub-Saharan Africa for the first time.

*Trypanosoma* spp. are principally spread when their vector, the tsetse fly (*Glossina* spp.), bites and feeds on the blood of a mammalian host. Significantly from an archaeological standpoint, the flies need shady, bush environments in which to rest and reproduce, something that effectively restricts them to areas where annual rainfall exceeds 500–700 mm. The well-established southward movement of rainfall belts in the mid-Holocene both aridified the Sahara and opened up newly tsetse-free areas to pastoralist settlement to its south from around 4500 BP (Smith 1992). The lack of evidence for dogs south of 22°N before the second millennium BC fits this perfectly, as does the fact that their appearance then seems to track the orientation of the region's rainfall belts, from central Niger to southeastern Mauritania and eventually northeastern Nigeria (Mitchell 2015).

Trypanosomiasis is, however, just one disease among many. African horse sickness, which is caused by a virus of the genus *Orbivirus*, is also a major threat. While almost all known outbreaks result from dogs eating infected horsemeat, transmission by insect vectors is also likely (van Sittert *et al.* 2013). Left untreated, infection produces death rates of 20–78% (van Rensburg *et al.* 1981). As with *T. congolense*, indigenous 'Africanis' dogs in South Africa are much less likely to become infected than breeds of European origin (van Sittert *et al.* 2013). Antibodies against the disease are widespread in both jackals (*Canis* spp.) and painted wolves (*Lycaon pictus*) (Alexander *et al.* 1995), but its natural host is the zebra (*Equus* spp.), an animal that was almost certainly absent from North Africa and the Sahara when dogs first arrived there (MacDonald and MacDonald 2000; Faith 2014). In other words, whether from eating infected equid meat or, quite possibly, by direct infection via its midge vectors (*Culicoides bolitinos; C. imicola*), African horse sickness is also likely to have been a previously unknown and severe threat to domestic dogs entering Sub-Saharan Africa for the first time, with a degree of tolerance only gradually acquired.

Two further tick-borne diseases are also relevant elsewhere in the Old World tropics. Canine monocytic ehrlichiosis is caused by the bacterium *Ehrlichia canis*, infection with which proceeds through acute and then chronic stages leading to suppression of the immune system, multiple organ dysfunction, severe anaemia, and haemorrhaging (Davoust *et al.* 2014). While dogs do not always exhibit acute symptoms, haemorrhagic forms of the disease may be fatal and the condition produces mortality rates of 44–75% if animals are left untreated (Davoust *et al.* 2003). Once again, we find a pattern in which: a) native African dogs, though often showing high frequencies of seroprevalence, may be asymptomatic, while those of European origin typically become severely ill (Davoust *et al.* 2014); and b) while *E. canis* may be widely present in wild African canids, experimental studies produce at most limited symptoms, implying a much longer co-evolutionary history (van Heerden 1979).

*E. canis* is only transmitted effectively by the brown dog tick (*Rhipicephalus sanguineus*), which also infects black-backed jackals (*C. mesomelas*; Price and Karstad 1980). With a worldwide distribution, *R. sanguineus* divides into two clades, a temperate one present where mean annual temperature is <20°C and a tropical one where it is >20°C (Zemtsova *et al.* 2016). Under Holocene conditions the tropical clade's range therefore extends over the Sahara, the Sahel, West, Central, and eastern Africa (except for areas of high elevation), Arabia, most of the Indian sub-continent, Southeast Asia, low-lying parts of New Guinea, northern Australia, most of Central America, the circum-Caribbean lowlands of South America, and Brazil. Significantly, there is evidence that only this clade can transmit *E. canis* (Moraes-Filho *et al.* 2015). This fits neatly with the latter's relatively low pathogenicity in wild canids and in African dogs compared to breeds of European origin to suggest that when dogs first entered Sub-Saharan Africa they will have found in *E. canis* another serious threat, additional to canine trypanosomiasis and African horse sickness.

Yet another challenge probably came from canine babesiosis, a malaria-like disease caused by piroplasmid parasites of the genus *Babesia*. *B. rossi* is restricted to Sub-Saharan Africa as is its vector, the yellow dog tick (*Haemaphysalis elliptica, H. leachi*). Both species are now adapted specifically to feed on dogs, though they are also reported from all three Sub-Saharan wild canids: painted wolves, black-backed jackals, and side-striped jackals (*C. adustus*) (Penzhorn 2011). However, though they may act as carriers and reservoirs for babesiosis (van Heerden 1980), these taxa show no overt clinical signs when infected by *B. rossi*, exactly as one would expect if they share a long co-evolutionary history with it. Domestic dogs, on the other hand, experience a highly virulent infection in which mortality rates, especially for those of non-local origin or descent, can approach 100%, even with treatment, although indigenous West African breeds at least appear to have evolved a degree of resistance to it (Penzhorn 2011; Adamu *et al.* 2014). Given that evidence for painted wolves or jackals in the mid-Holocene Sahara or North Africa is scarce to non-existent (Mitchell 2015), it seems unlikely that dogs entering these regions would have been exposed to babesiosis before reaching areas south of the desert, where the modern situation suggests it will have posed another serious threat to their survival.

Summing up, dogs moving south of the Sahara likely encountered a series of highly pathogenic diseases against which they had no natural resistance: canine trypanosomiasis, African horse sickness, canine monocytic ehrlichiosis, and canine babesiosis. Others may also have existed, but in these four cases at least pathogen, vector, and host were - before the arrival of the dog - all restricted to Sub-Saharan environments. Dogs, a temperate species descended from Eurasian wolves, will not have had experience of them and - as with Euro-American breeds today - can be expected to have suffered severely as a result. We should expect that considerable time was required for such partial resistance to develop, leading us to predict a much slower and more staggered expansion of the dog south of the Sahara relative to the pattern seen in the northern hemisphere, precisely as the archaeological record reflects (Mitchell 2015).

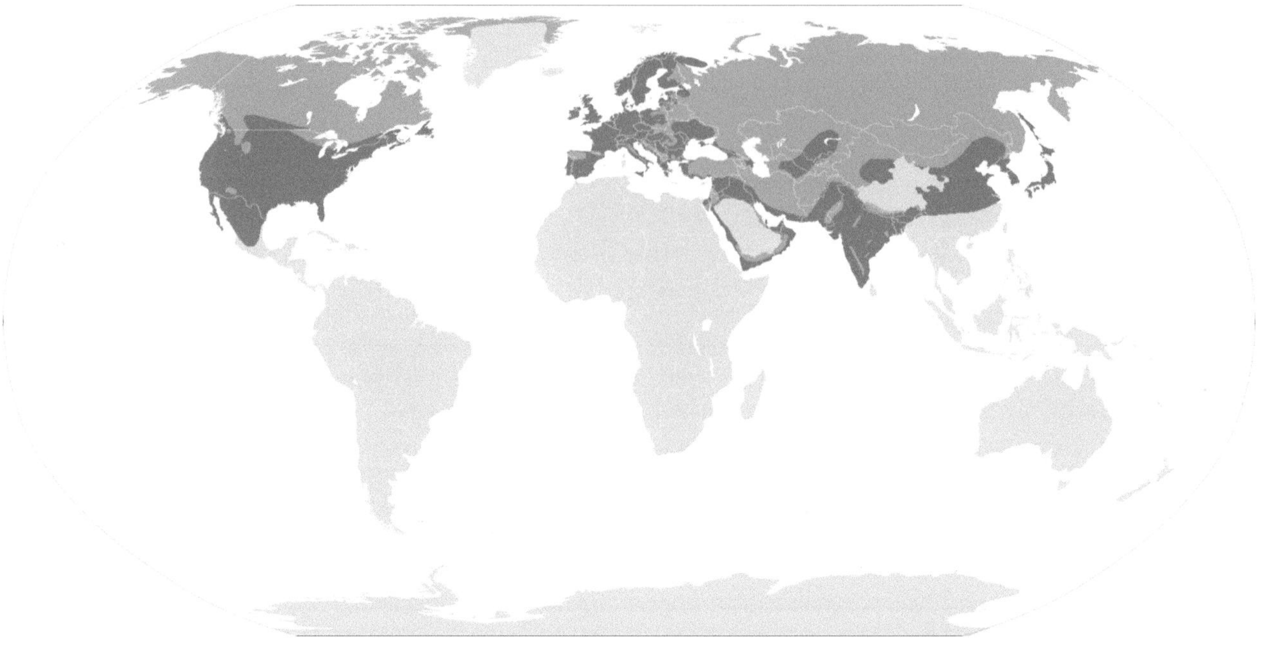

Figure 2. The global distribution of the grey wolf (*Canis lupus*) (After Morrison *et al.* 2007 and Wolf and Ripple 2017). Note that grey wolves in the Indian sub-continent (*C. lupus pallipes* (*C. indica*)), as well as Himalayan wolves (*C. himalayensis*), which are not shown here, are phylogenetically basal to and genetically distinct from all other wolves and not ancestral to the domestic dog. Dark grey – areas of former presence; mid-grey – areas of current presence; lightest grey – never present.

**4 South America**

In the New World, too, dogs spread much later into the tropics than into its northerly, temperate regions, and all claims for their presence in Central or South America before 3000 cal. BC are either plagued with chronological uncertainty or reflect confusion with indigenous wild canids (Stahl 2012; Mitchell 2017). This includes Quebrada de los Burros on Peru's southern coast where Rodríguez-Loredo (2012) attributes a deliberately buried, 9500-year-old canid skeleton to a domestic dog, though confusion with wild foxes (*Pseudalopex* spp.), which were also deliberately buried (Stahl 2012), cannot be excluded without detailed morphometric, ZooMS, or aDNA data. Instead, the earliest firm signal comes from Mexico's Tehuacán Valley *c.* 3825–2600 cal. BC with more widespread evidence in Central America and Mesoamerica after 1600 cal. BC (Rosenswig 2015) and finds from secure third-millennium cal. BC contexts in coastal Ecuador and Peru (Mitchell 2017). Expansion into the central Andes followed by 1500 cal. BC, northern Colombia by the first millennium cal. BC, and the grassland regions of southeastern Brazil, Uruguay, and north/central Argentina by the mid-first millennium AD, though dogs remained absent from Amazonia, the Gran Chaco, and central/southern Patagonia until after European arrival in the Americas (Mitchell 2017). Could disease have retarded their expansion into and beyond tropical America, as well as their distribution within it, just as in Sub-Saharan Africa?

In an earlier paper (Mitchell 2017), I identified a series of major disease challenges to domestic dogs in South America today, acknowledging that some (e.g. rabies, surra - another trypanosomal infection - canine babesiosis, and canine monocytic ehrlichiosis) are post-Columbian introductions and therefore irrelevant to the situation prevailing before 1492. Canine distemper, caused by a highly infectious virus of the same name and characterised by a fatality rate that some authorities place second only to rabies (Deem *et al.* 2000), was among those diseases that did seem likely to have been present in pre-Columbian times. However, Uhl *et al.* (2019) have now made a compelling case that it arose in Spanish America after European arrival via mutation of the human measles virus (HMV). While they emphasise just one (North American) dog population to argue that palaeopathological signs of canine distemper are not known before 1492, they combine a plausible scenario for its evolution with evidence of strong genetic similarities to HMV and rinderpest, both of which have exclusively Old World origins, noting that the genetics of an alternative potential ancestor, the DrMV virus found in common vampire bats (*Desmodus rotundus*; Nambulli *et al.* 2016: 102), are too poorly known for detailed comparison to be possible. Canine distemper, too, should now therefore join the list of post-Columbian pathogens.

However, other potential disease threats remain. Foremost among them is another trypanosomal infection, caused this time by *T. cruzi*, which in humans produces

Chagas disease, a condition that currently affects 16–18 million people in Central and South America (Eloy and Lucheis 2009). *T. cruzi* occurs in over 100 Neotropical mammals, including South America's two largest native canids, the crab-eating fox (*Cerdocyon thous*) and the maned wolf (*Chrysocyon brachyurus*), neither of which appears to suffer serious effects (Rocha *et al.* 2013). Transmission is largely via contamination with the faeces of at least 25 species of blood-eating triatomine insects, eating the vectors directly, or consuming infected meat; vampire bats can also spread the disease (Maywald *et al.* 1996; Herrera *et al.* 2011; Pineda *et al.* 2011). In its acute form infection with *T. cruzi* can produce sudden death because of cardiac arrhythmia, while over the longer term its chronic form leads to cardiac inefficiency and dysfunction (Eloy and Lucheis 2009). Mortality rates range from 25 to 68% (De Lana *et al.* 1992; Quijano-Hernández *et al.* 2012), with young animals at greatest risk (Kjos *et al.* 2008). Several different lineages of *T. cruzi* exist and may carry different health implications for dogs, just as for people (Duz *et al.* 2014), meaning that even if dogs evolved resistance to the dominant strain in one region they may have remained exposed to others elsewhere. The disease's pathogenicity and many hosts, which include wild canids, suggest that it may have posed a geographically widespread challenge to keeping dogs wherever it occurred.

A second disease, canine rangeliosis, resembles babesiosis in being caused by infection with a piroplasm, in this case *Rangelia vitalii*, which is transmitted by ticks, notably *Rhipicephalus sanguineus* (introduced to South America from Africa post-1492) and its natural vector, *Amblyomma aureolatum*, which is restricted to the eastern half of South America from Suriname to Argentina (Guglielmone *et al.* 2003). Initially recognised only in southern/southeastern Brazil (Franca *et al.* 2014), rangeliosis is now also known from Paraguay, northeastern Argentina, and Uruguay (Eiras *et al.* 2014; Inácio *et al.* 2019), and may be yet more widespread than this. Affecting young dogs much more often than adults, its symptoms, which include persistent bleeding from nose, ears, and mouth, are normally fatal in the absence of timely treatment (Franca *et al.* 2010). Wild canids, especially crab-eating foxes, are the one definite reservoir, but show no clinical, biochemical, or haematological alterations after infection, confirming that the disease is much less pathogenic in them than in dogs (de Souza *et al.* 2019). Given this, the strong congruence between its distribution and that of *A. aureolatum*, and the fact that *R. vitalii* is the only piroplasmid agent infecting dogs that is not reported outside the continent, it is highly likely that the latter 'coevolved with a native canid in South America' (Soares *et al.* 2014: 161) and that its high pathogenicity when it spills over into dogs helped exclude them from much of its range.

Leishmaniasis is also of interest and, like Chagas disease, of major zoonotic concern since *Leishmania* protozoa also infect people. While *L. infantum* (previously *L. chagasi*) is a post-Columbian arrival (Leblois *et al.* 2011), at least 14 other taxa are native to the New World tropics, all spread by sandflies (*Lutzomyia* spp.). Two merit attention with respect to infections of dogs as they can produce the more dangerous visceral form of the disease (Dantas-Torres 2009). One, *L. amazonensis*, occurs widely from Colombia to Argentina and is hosted by a wide diversity of mammals, including crab-eating foxes. The other, *Endrotrypanum* (previously *L.*) *colombiensis*, may be uniquely hosted by Hoffman's two-toed sloth (*Choloepus hoffmanni*) and is restricted to Colombia and Panama with probable extensions into the Brazilian and Peruvian Amazon. Both parasites were clearly well established in Neotropical America before dogs arrived. Moreover, since the cutaneous form of leishmaniasis produces considerable disfigurement in people - and did so in pre-Columbian times (e.g. Marsteller *et al.* 2011) - one wonders if human health concerns led some communities to reject keeping dogs, which are one of the parasite's principal hosts, just as some Indigenous Amazonian groups, like the Karitiana, today recognise them as transmitters of skin diseases (Vander Velden 2010: 137).

Neotropical America shares with Sub-Saharan Africa a much slower and later pattern of canine expansion than that found in the temperate regions to its north, with large areas still without dogs when European colonisation began in the sixteenth century. While canine distemper now seems to be a consequence of that colonisation - and thus not relevant to explaining the dog's spread through South America - other significant disease challenges remain. Canine trypanosomiasis caused by *T. cruzi*, canine rangeliosis, and visceral forms of canine leishmaniasis are all associated with high death rates among domestic dogs, but appear to produce few, if any, serious symptoms in the native mammals that host them, wild canids included. Arriving in the tropics, dogs - the descendants of temperate/Arctic-adapted Eurasian wolves with no prior experience of these infections - moved into areas beyond those inhabited by their grey wolf and coyote (*C. latrans*) cousins (Figure 2) and can be expected to have suffered considerable morbidity and mortality. Combined with potential losses from predators and perhaps being of less use for hunting key forest game species (cf. Koster 2009), this may have tilted the balance against keeping them in many areas, except where numbers could be maintained by imports from adjacent, relatively disease-free areas (as in Peru's Cordillera Oriental foothills; Schwartz 1997: 40).

It is thus notable that when dogs appeared in southeastern Brazil (Milheira *et al.* 2017), Uruguay

(López Mazz *et al.* 2018), and northern Argentina (Loponte and Acosta 2016) in the mid-first millennium AD they did so at a time of increased circulation of goods into the Pampas from the southeastern edge of the Andean world, were mostly deliberately buried - sometimes accompanying people - or had their teeth used as ornaments, and in some cases show what may be non-local isotope signals (e.g. Acosta *et al.* 2011). Dogs may, in other words, have been a rare, highly valued, and exotic item, consistent with the fact that - with one exception - they then went unobserved by sixteenth/seventeenth-century European visitors (Loponte and Acosta 2016) That this is the core area for canine rangeliosis is striking, and strengthens the case for thinking of vector-borne diseases as one constraint on when and where dogs were kept in the Americas.

### 5 India and Southeast Asia

I turn now to the third tropical region to which dogs were introduced: South and Southeast Asia. Dholes occur across both regions, though no longer in Island Southeast Asia, while the distribution of golden jackals (*C. aureus*) extends across India as far as Indochina. Grey wolves (*C. lupus pallipes* (*C. indica*)) are restricted to the whole Indian sub-continent (Figure 2). However, along with the Himalayan wolf (*C. himalayensis*), they are genetically distinct from - and phylogenetically basal to - all other wolves, being more closely related to jackals (Sharma *et al.* 2004; Aggarwal *et al.* 2007), and they did not contribute to the ancestry of dogs, which are archaeologically unknown in India/Pakistan before 5000 cal. BP (Frantz *et al.* 2016: Table S7). Nor, though documented in central China >8000 years ago (Jing 2008), are dogs attested in its far south or in Indochina, Thailand, and Burma before *c.*1800 cal. BC when rice and millet farmers moved into these regions from further north (Greig *et al.* 2016; Piper 2017). Expansion into the Philippines may have happened shortly thereafter, but associations older than 1000 cal. BC 'should be regarded with caution' (Piper 2017: 259). However, directly dated finds from Matja Kuru 2 in East Timor (Gonzalez *et al.* 2013) and Madura Cave in southern Australia (Balme *et al.* 2018) necessitate a very rapid movement through Island Southeast Asia and then across Australia itself in the centuries immediately before 1200–1000 cal. BC. Around the same time, mitochondrial DNA suggests a separate, Taiwanese-linked introduction to Near Oceania linked to the spread of the Lapita Complex, followed by another from southern Island Southeast Asia into the wider Pacific, Polynesia included (Greig *et al.* 2018).

Reviewing arthropod-transmitted diseases of dogs and other pets in Southeast Asia Irwin and Jefferies (2004: 27) note that the region remains 'relatively uncharted' and that detailed knowledge of its canine diseases is 'remarkably limited', reflecting poorly developed veterinary services and diagnostic infrastructure, lack of research, and attitudes to dogs that often differ from those found in the West. Much the same is true of the Indian sub-continent (Abd Rani *et al.* 2011), though both it and Southeast Asia (Island and Mainland divisions alike) favour the survival of a wide variety of ticks, flies, and other insect vectors able to transmit disease. Of these, *Rhipicephalus sanguineus* - here present in its tropical clade form (Zemstova *et al.* 2016) - is most prevalent and of greatest importance (Irwin and Jefferies 2004: 28). The diseases that it and other vectors spread are widely recognised as 'a significant cause of morbidity and mortality in dogs' in the region (Inpankeaw *et al.* 2016: 1), although for present purposes I focus on just two of them, both shared with Sub-Saharan Africa.

Canine monocytic ehrlichiosis is widespread in the Asian tropics, being endemic throughout the Indian sub-continent and Southeast Asia, including southern China and Taiwan, with a natural reservoir of infection in wild canids (Centre for Vector-Borne Diseases 2010). The parasite responsible, *Ehrlichia canis*, occurs widely in the region's dog populations, but without necessarily producing obvious clinical signs (Jirapattharasate *et al.* 2013; Konto *et al.* 2017; Piratae *et al.* 2015, 2019), suggesting that they have been exposed to it 'for an extended time' (Ahantarig *et al.* 2008: 1022). This is completely untrue, however, of dogs imported to the region from the outside, as the experience of British and American military and police dogs repeatedly showed in the 1960s and 1970s when many succumbed to an acute and frequently fatal haemorrhagic disease that was initially named tropical canine pancytopaenia but later recognised to be canine monocytic ehrlichiosis (Seamer and Snape 1972). Veterinary observations at the time, as well as earlier in India (Raghavachari and Reddy 1958), confirmed that 'the single most important factor influencing disease manifestation is the breed of dog infected...[and that] outbreaks such as those which were reported among military dogs in Southeast Asia occur when large numbers of these susceptible dogs are taken into an endemic area' (Huxsoll 1976: 54–55). In other words, Western military interventions in the region and the introduction of European breeds more generally during the twentieth century most likely replicated the consequences experienced when domestic dogs first encountered *E. canis* and its vector, the tropical clade of *R. sanguineus*, on their introduction to the Asian tropics 4000 or so years ago. Some considerable time was presumably required for local dog populations to then evolve the degree of resistance to ehrlichiosis that they, unlike dogs of Euro-American descent, now show.

Canine babesiosis is also endemic to South/Southeast Asia (Irwin and Jefferies 2004: 30), but unlike Sub-Saharan Africa the relevant parasite here is not *Babesia rossi*, but *B. gibsoni*, a smaller piroplasm transmitted principally by the tick *Haemaphysalis bispinosa* (Grove and Dennis 1972), which some sources (e.g. Inpankeaw *et al.* 2016) still confuse with the temperate taxon *H. longicornis*, although their distributions do not overlap; *H. longicornis* is native only to Japan, Korea, northeastern Russia, and China, while *H. bispinosa* occurs solely in the tropics, including the Indian sub-continent, Burma, Thailand, Peninsular Malaysia, and Indonesia (Fonseca *et al.* 2017). *R. sanguineus* is a secondary vector, with wild canids again a known host (CABI 2019). Once infected with *B. gibsoni*, dogs show symptoms that range from mild anaemia to widespread organ failure and death (Irwin 2009), but the disease is particularly severe among newly introduced animals, with native dogs showing a much less acute, more chronic infection that implies 'an apparent natural resistance' (Grove and Dennis 1972: 157). Once again, it would seem reasonable to suggest that before such resistance evolved the conjunction of *H. bispinosa* and *B. gibsoni* posed a not insignificant threat to the survival of dogs moving south of the Tropic of Cancer and entering tropical South/Southeast Asia for the first time. Co-infection with *Ehrlichia canis* will have intensified that threat as this results in a more severe anaemia than where dogs are infected with one agent alone (Niwetpathomwat *et al.* 2006; Inpankeaw *et al.* 2016). This is also true of other parasites, such as *Dirofilaria* spp., the pathogen responsible for canine heartworm, which also causes 'significant morbidity and mortality in dogs' in the region (Irwin and Jefferies 2004: 32).

## 6 Discussion

Whether at a personal or a community level, keeping dogs reflects a compromise between multiple considerations: the benefits they confer (companionship, protection, more effective hunting etc.) weighed against the costs incurred (food, commitment, the possible transmission of disease etc.) set within a matrix that is simultaneously symbolic (how they are perceived and valued), ecological (how easy is it for them to survive, thrive and reproduce), and social (the extent and ease of communication between different human groups). As this volume shows, different societies evaluate these considerations in different ways, but when viewed at a world scale such local variations are eclipsed by a robust, pan-tropical phenomenon, the relative speed with which dogs spread across temperate regions of the northern hemisphere and the much slower, later, and in places more staggered pace with which they expanded. Whether in Sub-Saharan Africa, South America, or South and Southeast Asia, this paper has argued for the importance of vector-borne diseases as one major factor in explaining this pattern. Today, canine trypanosomiasis, canine monocytic ehrlichiosis, canine babesiosis, African horse sickness, canine rangeliosis, and visceral canine leishmaniasis all constitute severe threats to dogs, especially those of non-local origin, and all find natural hosts and vectors among wild canids and other mammals native to the tropics. Having been domesticated from Eurasian grey wolves, which may themselves descend uniquely from a late Pleistocene Beringian source (Loog *et al.* 2018), when dogs moved beyond the Holarctic to enter the tropics (the Neotropical, Afrotropical, and Indo-Malaysian zoogeographical regions; Proches and Ramdhani 2012) they did so without prior experience of these diseases and likely suffered accordingly. Expansion would be expected to have been slow, or even curtailed or prevented all together, until a degree of resistance evolved, with knock-on effects for movement into sub-tropical parts of the southern hemisphere (Figure 3). The history south of the Sahara of domestic livestock taxa that were originally domesticated to its north (Gifford-Gonzalez 2000, 2017; Mitchell 2018) provides a ready parallel, as does the devastation that malaria (against which native Africans had partial immunity) wrought on Europeans settling on or visiting the West African coast in pre-modern times (Öberg and Rönnbäck 2016). Additionally, recent experience in the United States with *Hepatozoon americanum* confirms the devastating, frequently fatal, results that can ensue when pathogens cross from wild hosts (here possibly coyotes) into domestic dogs (Baneth and Vincent-Johnson 2016).

Infectious disease is certainly not the only factor to consider when explaining the spatiotemporal shape of canine expansion into the tropics. Tight associations between dogs and cultivators, or dogs and herders - and thus a link to the rate at which farming itself spread - were surely also at work, with Southeast Asia perhaps an excellent example of this (Greig *et al.* 2016; Piper 2017). But given the speed with which hunter-gatherers can successfully acquire dogs, even to the point of integrating them into their mythology (a few years in nineteenth-century Tasmania; Boyce 2006), and their ability to help secure food, warn of or protect against predators, and provide companionship, would tropical foragers have so consistently neglected them if the cost/benefit ratio of keeping them was not heavily skewed in a negative direction? And is it feasible that infectious disease, particularly diseases readily spread by tropical ticks, flies, and other vectors, did not contribute to that skewing, and on a global scale?

The argument I have advanced is open to evaluation in several ways. First, and most obviously, new finds may

Figure 3. The dog's global expansion (selected dates cal. BP only) compared to the world's zoogeographical regions (After Procheş and Ramdhani 2012).

change the chronological patterns I have rehearsed, perhaps even completely removing the case for a slower, later introduction into and through the tropics. For that to be convincing, however, robust identification of canid remains as those of domestic dogs rather than of locally endemic wild species plus secure dating, ideally on the bones themselves, are both needed. Neither criterion is yet universally present (Larson *et al.* 2012: Table 2; Stahl 2012; Piper 2017). Second, advances in veterinary research and more thorough exploration of the existing literature may be able to extend the evidentiary basis for arguing that local dog breeds have evolved a degree of tolerance to some tropical vector-borne diseases, with genetic studies potentially able to identify bottlenecks through which those dog populations have passed, bottlenecks that might have been caused by disease. The evidence of both ancient and modern DNA could, in principle, help date such events, and it is thus of interest that Liu *et al.* (2018) have recently reported a severe founder effect before population expansion among dogs in Africa. Third, in favourable circumstances pathogens found in ancient dog remains (coprolites, bones, dental calculus, soft tissues) may directly confirm infection, particularly as analytical techniques capable of recovering pathogen DNA in archaeological samples are more widely deployed. Fourth and finally, phylogenetic analyses of pathogenic organisms may help to determine their origin and antiquity, establishing whether they are indeed likely to have been present at the right time and in the right place to have constrained canine expansion; the recognition that Rift Valley Fever in Africa (which affects dogs as well as livestock) is, in fact, of late nineteenth-century origin (Bird *et al.* 2007) and thus irrelevant as a hindrance to the spread of prehistoric pastoralism (cf. Gifford-Gonzalez 2000) is a case in point.

Other questions also merit exploration. How far did human modification of the environment reduce risk to dogs, for example by using burning and tsetse-resistant breeds of goats to remove bush cover, thus limiting, or even eliminating, key insect vectors like *Glossina* spp.? How far was keeping dogs rejected because of perceived threats to human health (as I discussed for leishmaniasis above) or the health of livestock (*Ehrlichia ruminantium*, for example, the cause of heartwater, is widespread among dogs in Africa, making them a potential reservoir for this disease, which can produce mortality rates of up to 90%; Allsopp and Allsopp 2001). More detailed information on the susceptibility of wild canids to the diseases and the vectors I have discussed would also be welcome. And in the same vein, both ecologists and archaeologists might wish to explore the potential disruption that dogs caused when they were eventually successfully introduced to new ecologies (e.g. Balme *et al.* 2016). Hopefully, this paper will help spur research into all these topics, building a more holistic understanding of canine expansion into the tropics and beyond.

## References


Abd Rani, P.A.M., P.J. Irwin, G.T. Coleman, M. Gatne and, R.J. Traub 2011. A survey of canine-borne tick diseases in India. *Parasites & Vectors* 4: 141.

Abenga, J.N., K. David, C.O.G. Ezebuiro and F.A.G. Lawani 2005. Observations on the tolerance of young dogs (puppies) to infections with *Trypanosoma congolense*. *African Journal of Clinical and Experimental Microbiology* 6: 28–33.

Acosta, A.A., D. Loponte and C.G. Esponda 2011. Primer registro de perro doméstico prehispánico (*Canis familiaris*) entre los grupos cazadores recolectores del humedal de Paraná inferior (Argentina). *Antipoda: Revista de Antropología y Arqueología* 13: 175–199.

Adamu, N.B., M. Troskie, D.O. Oshadu, D.P. Malatji, B.L. Penzhorn and P.T. Matjila 2014. Occurrence of tick-transmitted pathogens in Jos, Plateau State, Nigeria. *Parasites & Vectors* 7: 119.

Aggarwal, R.K., T. Kivisild, J. Ramadevi, J. and L. Singh 2007. Mitochondrial DNA coding region sequences support the phylogenetic distinction of two Indian wolf species. *Journal of Zoological Systematics and Evolutionary Research* 45: 163–172.

Ahantarig, A., W. Trinachartvanit and J.R. Milne. 2008 Tick-borne pathogens and diseases of animals and humans in Thailand. *Southeast Asian Journal of Tropical Medicine and Public Health* 39: 1015–1032.

Alexander, K.A., P.W. Kat, J. House, C. House, C., S.J. O'Brien, M.K. Laurenson,, J.W. McNutt and B.L. Osburn 1995. African horse sickness and African carnivores. *Veterinary Microbiology* 47: 133–140.

Allsopp. M.T.E.P. and B.P. Allsopp 2001. Novel *Ehrlichia* genotype detected in dogs in South Africa. *Journal of Clinical Microbiology* 39: 4204–4207.

Balme, J., S. O'Connor and S. Fallon 2018. New dates on dingo bones from Madura Cave provide oldest firm evidence for arrival of the species in Australia. *Scientific Reports* 8: 9933.

Baneth, G. and N. Vincent-Johnson 2016. Hepatozoonosis, in M.J. Day (ed.) *Arthropod-Borne Infectious Diseases of the Dog and Cat:* 109–124. Boca Raton: CRC Press.

Bird, B.H., M.L. Khristova, P.E. Rollin, T.G. Ksiazek and S.T. Nicholl 2007. Complete genome analysis of 33 ecologically and biologically diverse Rift Valley Fever virus strains reveals widespread virus movement and low genetic diversity due to recent common ancestry. *Journal of Virology* 81: 2805–2816.

Boyce, J. 2006. Canine revolution: the social and environmental impact of the introduction of the dog to Tasmania. *Environmental History* 11: 109–129.

CABI. 2019. Babesiosis , viewed 30 September 2019, https://www.cabi.org/isc/datasheet/91723#tooverview

Centre for Vector-Borne Diseases. 2010. Canine erhlichiosis, viewed 24 September 2019, https://www.cvbd.org/static/documents/digest/CVBD_Easy-to-digest_no_7_ehrlichiosis.pdf

Costa Junqueira, M.A., C. Matheson, L. Iachetta, A. Llagostera and O. Appenzeller 2009. Ancient leishmaniasis in a highland desert of northern Chile. *PLoS ONE* 4(9): e6983.

Cranbrook, E.-O. 2014. Archaeology of the dog in Borneo: new evidence from the Everett Collection in the Natural History Museum, London, in K. Boyle, R. Rabett and C. Hunt (eds) *Living in the Landscape: Essays in Honour of Graeme Barker:* 171–182. Cambridge: McDonald Institute for Archaeological Research.

Dantas-Torres, F. 2009. Canine leishmaniasis in South America. *Parasites & Vectors* 2: S1.

Davoust, B., J.L. Marié, S. Mercier, M. Boni, A. Vandeweghe, D. Parzy and F. Beugnet 2003. Assay of fipronil efficacy to prevent canine monocytic ehrlichiosis in endemic areas. *Veterinary Parasitology* 112: 91–100.

Davoust, B., D. Parzy, J.-P., Demoncheaux, R. Tine, M. Diarra, J.-L. Marié, and O. Mediannikov 2014. Usefulness of a rapid immuno-migration test for the detection of canine monocytic ehrlichiosis in Africa. *Comparative Immunology, Microbiology and Infectious Diseases* 37: 31–37.

De Lana, M., E. Chiara and W.L. Tafuri1992. Experimental Chagas' disease in dogs. *Memórias do Instituto Oswaldo Cruz* 87: 57–91.

de Souza, V.K., B. Dall'Agnol, U.A Souza,. A. Webster, F.B. Peters, M.O. Favarini, F.D. Mazim, F.L. da Rocha, F.P. Tirelli, J.F. Soares, M.M. de Assis Jardim, T.C. Trigo and J. Recj 2019. Detection of *Rangelia vitalii* (Piroplasmida: Babesiidae) in asymptomatic free-ranging wild canids from the Pampa biome, Brazil. *Parasitology Research* 118: 1337–1342.

Deem, S.L., L.H. Spelman, R.A. Yates and R.J. Montali 2000. Canine distemper in terrestrial carnivores: a review. *Journal of Zoo and Wildlife Medicine* 31: 441–451.

Duz, A.L.C., P.M. de Abreu Vieira, B.M. Roatt, R.D.O. Aguiar-Soares, J.M. de Oliveira Cardoso, F.C.B. de Oliveira, L.E.S. Reis, W.L. Tafuri, V.M. Veloso, A.B. Reis and C.M. Carneiro 2014. The TcI and TcII *Trypanosoma cruzi* experimental infections induce distinct immune responses and cardiac fibrosis in dogs. *Memórias do Instituto Oswaldo Cruz* 109: 1005–1013.

Eiras, D.F., M.B. Craviotto, G. Baneth and G. Moré 2014. First report of *Rangelia vitalii* infection (canine rangeliosis) in Argentina. *Parasitology International* 63: 729–734.

Eloy, L.J. and S.B. Lucheis 2009. Canine trypanosomiasis: etiology of infection and implications for public health. *Journal of Venomous Animal Toxins including Tropical Diseases* 15: 589–611.

Ezeokonkwo, R.C., I.O. Ezeh, P.O. Obi, I.W. Onyenwe and W.E. Agu 2010. Comparative haematological study of single and mixed infections of mongrel dogs with *Trypanosoma congolense* and *Trypanosoma brucei brucei*. *Veterinary Parasitology* 173: 48–54.

Faith, J.T. 2014. Late Pleistocene and Holocene mammal extinctions on continental Africa. *Earth Science Reviews* 128: 105–121.

Fiorello, C.V., A.J. Noss and S.L. Deem 2006. Demography, hunting ecology, and pathogen exposure of domestic dogs in the Isoso of Bolivia. *Conservation Biology* 20: 762–771.

Fonseca. D.M., A. Egizi, and J. Occi 2017. Global health: the tick that binds us all. Review of the biology and ecology of *Haemaphysalis longicornis* Neumann, 1901, viewed 24 September 2019, https://fonseca-lab.com/research/globalhealth-the-tick-that-binds-us-all/ .

Franca, R.T., A.S. Silva, A.P. Loretti, C.M. Mazzanti and S.T.A. Lopes 2014. Canine rangeliosis due to *Rangelia vitalii*: from first report in Brazil in 1910 to current day - a review. *Ticks and Tick-Borne Diseases* 5: 466–474.

Franca, R.T., A.S. Silva, F.C. Paim, M.M. Costa, J.G. Soares, C.M. Mazzanti, and S.T.A. Lopes 2010. *Rangelia vitalii* in dogs in southern Brazil. *Comparative Clinical Pathology* 19: 383–387.

Frantz, L.A.F., V.E. Mullin, M. Pionnier-Capitan, O. Lebrasseur, M. Ollivier, A. Perri, A. Linderholm, V. Mattiangeli, M.D. Teasdale, E.A. Dimopoulos, A. Tresset, M. Duffraisse, F. McCormick, L. Bartosiewicz, E. Gál, E.A. Nyerges, M.V. Sablin, S. Bréhard, M. Mashkour, A. Bălăşeşcu, B. Gillet, S. Hughes, O. Chassaing, C. Hitte, J.-D.Vigne, K. Dobney, C. Hänni, D.G. Bradley and G. Larson 2016. Genomic and archaeological evidence suggests a dual origin of domestic dogs. *Science* 352: 1228–1230.

Gifford-Gonzalez, D. 2000. Animal disease challenges to the emergence of pastoralism in Sub-Saharan Africa. *African Archaeological Review* 17: 95–140.

Gifford-Gonzalez, D. 2017. 'Animal disease challenges' fifteen years later: the hypothesis in light of new data. *Quaternary International* 436: 283–293.

Gonzalez, A., G. Clark, S. O'Connor and L. Matisoo-Smith 2013. A 3000 year old dog burial in Timor-Leste. *Australian Archaeology* 76: 13–20.

Greig, K., A. Gosling, C.J. Collins, J. Boocock, K. McDonald, D.J. Addison, M.S. Allen, B. David, M. Gibbs, C.F.W. Higham, F. Liu, I.J. McNiven, S. O'Connor, C.H. Tsang, R. Walter and E. Matisoo-Smith 2018. Complex history of dog (*Canis familiaris*) origins and translocations in the Pacific revealed by ancient mitogenomes. *Scientific Reports* 8: 9130.

Greig, K., R. Walter and E. Matisoo-Smith 2016. Dogs and people in Southeast Asia and the Pacific, in M. Oxenham and H.R. Buckley (eds) *The Routledge Handbook of Bioarchaeology in Southeast Asia and the Pacific Islands*: 462–482. London: Routledge.

Grove, M.G. and G.L. Dennis 1972. *Babesia gibsoni*: field and laboratory studies of canine infections. *Experimental Parasitology* 31: 153–159.

Guglielmone, A.A., A. Estrada-Peña, A.J. Mangold, D.M Barros-Battesti, M.B. Labruna, J.R. Martins, J.M. Venzal, M. Arzua and J.E. Keirans 2003. *Amblyomma aureolatum* (Pallas, 1772) and *Amblyomma ovale* Koch, 1844 (Acari: Ixodidae): hosts, distribution and 16S rDNA sequences. *Veterinary Parasitology* 113: 273–288.

Herrera, H.M., F.L. Rocha, C.V. Lisboa, V. Rademaker, G.M. Mourao and A.M. Jansen 2011. Food web connections and the transmission cycles of *Trypanosoma cruzi* and *Trypanosoma evansi* (Kinetoplastida, Trypanosomatidae) in the Pantanal region, Brazil. *Transactions of the Royal Society of Tropical Medicine and Hygiene* 105: 380–387.

Hörchner, F., U. Zillmann, M. Metzner, A. Schönefeld and D. Mehlitz 1985. West African dogs as a model for research on trypanotolerance. *Tropical Medicine and Parasitology* 36: 257–258.

Huchet, J.B., C. Callou R. Lichtenberg and F. Dunand 2013. The dog mummy, the ticks and the louse fly: archaeological report of severe ectoparasitosis in Ancient Egypt. *International Journal of Paleopathology* 3: 165–175.

Huxsoll, D.L. 1976. Canine ehrlichiosis (tropical canine pancytopenia): a review. *Veterinary Parasitology* 2: 49–60.

Inácio, E.L., S. Pérez-Macchi, A. Alabi, P. Bittencourt and A. Müller 2019. Prevalence and molecular characterization of piroplasmids in domestic dogs from Paraguay. *Ticks and Tick-borne Diseases* 10: 321–327.

Inpankaew, T., S.F. Hii, W. Chimnoi and R.J. Traub 2016. Canine vector-borne pathogens in semi-domesticated dogs residing in northern Cambodia. *Parasites & Vectors* 9: 253.

Irwin, P.J. 2009. Canine babesiosis: from molecular taxonomy to control. *Parasites & Vectors* 2: S4.

Irwin, P.J. and R. Jefferies 2004. Arthropod-transmitted diseases of companion animals in Southeast Asia. *Trends in Parasitology* 20: 27–34.

Jing, Y. 2008. The origins and development of animal domestication in China. *Chinese Archaeology* 8: 1–7.

Jirapattharasate, C., J. Chatsiriwech, P. Suksai, T. Changbunjong, T. Rawangchue, W. Moonarmart and S. Sungpradit 2012. Identification of *Ehrlichia* spp. in canines in Thailand. *Southeast Asian Journal of Tropical Medicine and Public Health* 43: 964–968.

Keck, N., S. Herder, D. Kaba, P. Solano, J. Gomez, G. Cuny and B. Davoust 2009. Epidemiological study of canine trypanosomiasis in an urban area of Ivory Coast. *Parasite* 16: 305–308.

Kjos, S.A., K.F. Snowden, T.M. Craig, B. Lewis, N. Ronald and J.K. Olson 2008. Distribution and characterization of canine Chagas disease in Texas. *Veterinary Parasitology* 152: 249–256.

Konto, M., S.M. Tukur, M. Watanabe, P.A.M. Abd-Rani, R.S.K. Sharma, L.S. Fong and M. Watanabe 2017. Molecular and serological prevalence of *Anaplasma* and *Ehrlichia* sp. among stray dogs in East Malaysia. *Tropical Biomedicine* 34: 570–575.

Koster, J. 2009. Hunting dogs in the lowland Neotropics. *Journal of Anthropological Research* 65: 575–610.

Larson, G., E.K. Karlsson, A.R. Perri, M.T. Webster, S.Y.W. Ho, J. Peters, P.W. Stahl, P.J. Piper, F. Lingaas, M. Fredholm, K.E. Comstock, J.F. Modiano, C. Schelling, A.L. Agoulnik, P.A. Leegwater, K. Dobney, J.-D. Vigne, C. Vilà, L. Andersson and K. Lindblad-Toh 2012. Rethinking dog domestication by integrating genetics, archaeology and biogeography. *Proceedings of the National Academy of Sciences (USA)* 109: 8878–8883.

Leblois, R., K. Kuhls, O. François, G. Schönian and T. Wirth 2011. Guns, germs and dogs: on the origin of *Leishmania chagasi*. *Infection, Genetics and Evolution* 11: 1091–1095.

Lisulo, M., C. Sugimoto, K. Kajino, K. Hayashida, M. Mudenda, L. Moonga, J. Ndebe, S. Nzala and B. Namangala 2014. Determination of the prevalence of African trypanosomes in indigenous dogs in Mambwe district, eastern Zambia, by loop-mediated isothermal amplification. *Parasites & Vectors* 7: 19.

Liu, Y.-H., L Wang, T. Xu, X. Guo, Y. Li, T.-T. Yin, H.-C. Yang, Y. Hu, A.C. Adeola, O.J.Sanke, N.O. Otecko, M., Wang, Ma, O.S. Charles, O.S., M.-H.S. Sinding, S. Gopalakrishnan, J.A. Samaniego, A.J. Hansen, C. Fernandes, P. Gaubert, J. Budd, P.M. Dawuda, E.K. Rueness, L. Jiang, W. Zhai, M.T.P. Gilbert, M.-S. Peng, X. Qi, G.-D. Wang and Y.-P. Zhang 2018. Whole-genome sequencing of African dogs provides insights into adaptations against tropical parasites. *Molecular Biology & Evolution* 35: 287–298.

Loog, L., O. Thalmann, M.-S.S. Sinding, V.J. Schuenemann, A.R. Perri, M. Germonpré, H. Bocherens K.E. Witt, J.A. Samaniego Castruita, M.S. Velasco, I.K.C. Lundstrøm, N. Wales, G. Sonet, L.A.F., Frantz, H. Schroeder, J. Budd, E.-L. Jimenez, S. Fedorov, B. Gasparyan, B., A.W. Kandel, M. Láničková-Galetová, Napierala, Uerpmann, H.-P., Nikolskiy, P.A., Pavolva, E.Y., Pitulko, V.V., K.-H., Herzig, R.S. Malhi, E. Willerslev, A.J. Hansen, K. Dobney, M.T.P. Gilbert, J. Krause, G. Larson, A. Eriksson and A. Manica 2018. Modern wolves trace their origin to a late Pleistocene expansion from Beringia. *BioRxiv*, viewed 1 October 2019, https://doi.org/10.1101/370122

López Mazz, J.M., F. Moreno, R. Bracco and R. González 2018. Perros prehistóricos en el este de Uruguay: contextos e implicaciones culturales. *Latin American Antiquity* 29: 64–78.

Loponte, D. and A. Acosta 2016. Nuevos registros prehispánicos de *Canis familiaris* (Carnivora, Canidae) en la Cuenca del Paraná, Argentina. *Mastozoología Neotropical* 23: 431–454.

Lupo, K.D. 2017. When and where do dogs improve hunting productivity? The empirical record and some implications for early Upper Paleolithic prey acquisition. *Journal of Anthropological Archaeology* 47: 139–151.

MacDonald, K.C. and R.H. MacDonald 2000. The origins and development of domesticated animals in arid West Africa, in R.M. Blench and K.C. MacDonald (eds) *The Origins and Development of African Livestock: Archaeology, Origins, Linguistics and Ethnography*: 127–162. London: UCL Press.

Marsteller, S.J., C. Torres-Rouff and K.J. Knudson 2011. Pre-Columbian Andean sickness ideology and the social experience of leishmaniasis: a contextualised analysis of bioarchaeological and paleopathological data from San Pedro de Atacama, Chile. *International Journal of Paleopathology* 1: 24–34.

Matete, G.O. 2003. Occurrence, clinical manifestation and the epidemiological implications of naturally occurring canine trypanosomiasis in western Kenya. *Onderstepoort Journal of Veterinary Research* 70: 317–323.

Maywald, P.G., M.I. Machado, J.M. Costa-Cruz and M.R.F. Gonçalves-Pires1996. Leishmaniose teguementar, visceral e doença de Chagas caninas em municípios do Triângulo Mineiro e Alto Paranaiba, Minas Gerais, Brasil. *Cadernos Saúde Pública* 12: 321–328.

Milheira, R.G., D. Loponte, C.G. Esponda, A. Acosta, and P. Ulgium 2017. The first record of a pre-Columbian domestic dog (*Canis lupus familiaris*) in Brazil. *International Journal of Osteoarchaeology* 27: 488–494.

Mitchell, P.J. 2015. Did disease constrain the spread of domestic dogs (*Canis familiaris*) into Sub-Saharan Africa? *Azania: Archaeological Research in Africa* 50: 92–135.

Mitchell, P.J. 2017. Disease: a hitherto unexplored constraint on the spread of dogs (*Canis lupus familiaris*) in Pre-Columbian South America. *Journal of World Prehistory* 30: 301–349.

Mitchell, P.J. 2018. The constraining role of disease on the spread of domestic mammals in sub-Saharan Africa: a review. *Quaternary International* 471A: 95–110.

Moraes-Filho, J., F.S. Krawczak, F.B. Costa, J.F. Soares and M.B. Labruna 2015. Comparative evaluation of the vector competence of four South American populations of the *Rhipicephalus sanguineus* group for the bacterium *Ehrlichia canis*, the agent of canine monocytic ehrlichiosis. *PLoS ONE* 10 (0: e0139386.

Morey, D.F. 2006. Burying key evidence: the social bond between dogs and people. *Journal of Archaeological Science* 33: 158–175.

Morrison, J.C., W. Sechrest, E. Dinerstein, D.S. Wilcovem and J.F. Lamoreux 2007. Persistence of large mammal faunas as indicators of global human impacts. *Journal of Mammalogy* 88: 1363–1380.

Nambulli, S., C.R. Sharp, A.S. Acciardo, J.F. Drexler, and W.P. Duprex 2016. Mapping the evolutionary trajectories of morbilliviruses: what, where and whither. *Current Opinion in Virology* 16: 95–105.

Ni Leathlobhair, M., A.R. Perri, E. K. Irving-Pease, K.E. Witt, A. Linderholm, J. Haile, O. Lebrasseur, C. Ameen, J. Blick, A.R. Boyko, S. Brace, Y. Nunes Cortes, S.J. Crockford, A. Devault, E.A. Dimopoulos, M. Eldridge, J. Enk, S. Gopalakrishnan, K. Gori, V. Grimes, E. Guiry, A.J. Hansen, A. Hulme-Beaman, J. Johnson, A. Kitchen, A.K. Kasparov, Y.-M. Kwon, P.A. Nikolskiy, C. Peraza Lope, A. Manin, T. Martin, M. Meyer, K. Noack Myers, M. Omura, J.-M. Rouillard, E.Y. Pavlova, P. Sciulli, M.-H.S. Sinding, A. Strakova, V.V. Ivanova, C. Widga, E. Willerslev, V.V. Pitulko, I. Barnes, M.T.P. Gilbert, K.M. Dobney, S. Malhi, P. Murchison, G. Larson and L.A.F. Frantz 2018. The evolutionary history of dogs in the Americas. *Science* 361: 81–85.

Niwetpathomwat, A., S. Assarasakorn, S. Techangamsuwan, S. Suvarnavibhaja and M. Kaewthamasorn 2006. Canine dirofilariasis and concurrent tick-borne transmitted diseases in Bangkok, Thailand. *Comparative Clinical Pathology* 15: 249–253.

Nwoha, R.I.O. and B.N. Anene 2011. Clinical signs and pathological changes in dogs with single and conjunct experimental infections of *Trypanosoma brucei brucei* and *Ancyclostoma caninum*. *Journal of Veterinary Parasitology* 24: 91–102.

Öberg, S. and K. Rönnbäck 2016. Mortality among European settlers in pre-colonial West Africa: the 'white man's grave' revisited. *Göteborg Papers in Economic History* 20: 1–48.

Penzhorn, B.L. 2011. Why is southern African canine babesiosis so virulent? An evolutionary perspective. *Parasites & Vectors* 4: 51.

Perri, A.R., C. Widga, D. Lawler, T. Martin, T. Loebel, K. Farnsworth, L. Kohn and B. Buenger 2019. New evidence of the earliest domestic dogs in the Americas. *American Antiquity* 84: 68–87.

Pineda, V., A. Saldaña, I. Monfante, A. Santamaría, N.L. Gottdenker, M.J. Yabsley, G. Rapoport and J.E. Calzada 2011. Prevalence of trypanosome infections in dogs from Chagas disease endemic regions in Panama, Central America. *Veterinary Parasitology* 178: 360–363.

Piper, P.J. 2017. The origins and arrival of the earliest domestic animals in Mainland and Island Southeast Asia: a developing story of complexity, in P.J. Piper. H. Matsumara and D. Bulbeck (eds) *New Perspectives in Southeast Asian and Pacific Prehistory*: 251–274. Canberra: ANU Press.

Piratae, S., K. Pimpjong, K. Vaisusuk and W. Chatan 2015. Molecular detection of *Ehrlichia canis*, *Hepatazoon canis* and *Babesia canis vogeli* in stray dogs in Mahasarakham province, Thailand. *Annals of Parasitology* 61: 183–187.

Piratae, S., P. Senawong, P. Chalermchat, W. Harnarsa and B. Sae-chue 2019. Molecular evidence of *Ehrlichia canis* and *Anaplasma platys* and the association of infections with hematological responses in naturally infected dogs in Kalasin, Thailand. *Veterinary World* 12: 131–135.

Pompa y Padilla, J.A. and E. Serrano Carreto 2001. Los más antiguos Americanos. *Arqueología Mexicana* 9: 36–41.

Price, J.E. and L.H. Karstad. 1980. Free-living jackals (*Canis mesomelas*) – potential reservoir hosts for *Ehrlichia canis* in Kenya. *Journal of Wildlife Diseases* 16: 469–473.

Procheş, Ş. and S. Ramdhani 2012. The World's zoogeographical regions confirmed by cross-taxon analysis. *Bioscience* 62: 260–270.

Quijano-Hernández, I.A., A. Castro-Barcena, E. Aparicio-Burgos, M.A. Barbosa-Mireles, J.V. Cruz-Chan, J.C. Vásquez-Chagoyán, M.E. Bolio-González, M.E. and E. Dumontell 2012. Evaluation of clinical and immunopathological features of different infective doses of *Trypanosoma cruzi* in dogs during the acute phase. *The Scientific World Journal* 2012: 635169.

Raghavachari, K. and A.M.K. Reddy 1958. *Rickettsia canis* in Hyderabad. *Indian Veterinary Journal* 35: 63–68.

Richardson, D.J., S. Guillén, R. Beckett, W. Kyle, G. Conlogue and K. Harper-Beckett 2012. Archaeohelminthology of the Chiribaya shepherd, *Canis familiaris* (700–1476 A.D.) from Southern Peru. *Comparative Parasitology* 79: 133–137.

Rocha, F.L., A.L.R. Roche, R.C. Arrais, J.P. Santos, V. dos Santos Lima, S.C. das Chagas, S.C. Xavier, P. Cordeir-Estrela, P.S. D'Andrea and A.M. Jansen 2013. *Trypanosoma cruzi* TcI and TcII transmission among wild carnivores, small mammals and dogs in a conservation unit and surrounding areas, Brazil. *Parasitology* 140: 160–170.

Rodríguez-Laredo, C. 2012. La explotación de la fauna terrestre, in D. Lavallée and M. Julien (eds) *Prehistoria de la Costa Extremo-Sur del Perú*: 141–174. Lima: Institut Français d'Etudes Andines.

Rosenswig, R.M. 2015. A mosaic of adaptation: the archaeological record of Mesoamerica's Archaic period. *Journal of Archaeological Research* 23: 115–162.

Schwartz, M. 1997. *A History of Dogs in the Early Americas*. New Haven: Yale University Press.

Seamer, J. and T. Snape. 1972. *Ehrlichia canis* and tropical canine pancytopaenia. *Research in Veterinary Science* 13: 307–314.

Sharma, D.K., J.E. Maldonado, Y.V. Jhala and R.C. Fleischer 2004. Ancient wolf lineages in India. *Proceedings of the Royal Society B* 271: S1–S4.

Smith, A.B. 1992. *Pastoralism in Africa: Origins and Development Ecology*. Johannesburg: Witwatersrand University Press.

Soares, J.F., B. Dall'Agnol, F.B. Costa, F.S. Krawczak, A.T. Cornerlato, B.C.D. Rossato, C.M. Linck, E.K.O.

Sigahi, R.H.F. Teixeira, L. Sonne, M.K. Hagiwara, F. Gregori, M.I.B. Vieira, J.R. Martins, J. Reck and M.B. Labruna 2014. Natural infection of the wild canid, *Cerdocyonthous*, with the piroplasmid *Rangelia vitalii* in Brazil. *Veterinary Parasitology* 202: 156–163.

Stahl, P.W. 2012. Interactions between humans and endemic canids in Holocene South America. *Journal of Ethnobiology* 32: 108–127.

Stahl, P.W. 2013. Early dogs and endemic South American canids of the Spanish Main. *Journal of Anthropological Research* 69: 515–533.

Uhl, E.E., C. Kelderhouse, J. Buikstra, J.P. Blick, B. Bolon and R.J. Hogan 2019. New World origin of canine distemper: interdisciplinary insights. *International Journal of Paleopathology* 24: 266–278.

van Heerden, J. 1979. The transmission of canine ehrlichiosis to the wild dog *Lycaon pictus* (Temminck) and black-backed jackal *Canis mesomelas* Schreber. *Journal of the South African Veterinary Association* 50: 245–250.

van Heerden, J. 1980. The transmission of *Babesia canis* to the wild dog *Lycaon pictus* (Temminck) and black-backed jackal *Canis mesomelas* Schreber. *Journal of the South African Veterinary Association* 51: 119–120.

van Rensburg, I.B., J. De Clerk. H.B. Groenewald and W.S. Botha 1981. An outbreak of African horse sickness in dogs. *Journal of the South African Veterinary Association* 52: 323–325.

van Sittert, S.J., T.M. Drew, J.L. Kotze, T. Strydom, C.T. Weyerand A.J. Guthrie 2013. Occurrence of African horse sickness in a domestic dog without apparent ingestion of horse meat. *Journal of the South African Veterinary Association* 84: 948.

Vander Velden, F.F. 2008. Sobre a cães e índios. Domesticidade, classificação zoological e relação humano-animal entre os Karitiana na Amazônia Brasileira, in *IX Congreso Argentino de Antropología Social:* 1–20. Posada: Universidad Nacional de Misiones.

Vander Velden, F.F. 2010. Inquietas companhias: sobre os animais do criação entres os Karitiana. Ph.D. Dissertation, Universidade Estadual de Campinas.

Vansina, J. 1990. *Paths in the Rainforests: Towards a History of Political Tradition in Equatorial Africa.* London: James Currey.

Viranta, S., A. Atickem, L. Werdelin and N.C. Stenseth 2017. Rediscovering a forgotten canid species. *BMC Zoology* 2: 6.

Watier-Grillot, S., S. Herder, J.-L. Marié, G. Cuny and B. Davoust 2013. Chemoprophylaxis and treatment of African canine trypanosomiasis in French military working dogs: a retrospective study. *Veterinary Parasitology* 194: 1–8.

Weyrich, L.S., K. Dobney and A. Cooper 2015. Ancient DNA analysis of dental calculus. *Journal of Human Evolution* 79: 119–124.

Wolf, C. and W.J. Ripple 2017. Range contractions of the world's large carnivores. *Royal Society Open Science* 4: 170052.

Woodroffe, R., S. Cleaveland, O. Courtenay, M.K. Laurenson and M. Artois 2004. Infectious disease, in D.W. MacDonald and C. Sillero-Zubiri (eds) *The Biology and Conservation of Wild Canids*: 123–142. Oxford: Oxford University Press.

Yu, P. 1997. *Hungry Lightning: Notes of a Woman Anthropologist in Venezuela.* Albuquerque: University of New Mexico Press.

Zemtsova, G.E., D.A. Apanaskevich, W.K. Reeves, M. Hahn, A. Snellgrove, and M.L. Levin 2016. Phylogeography of *Rhipicephalus sanguineus* sensu lato and its relationships with climatic factors. *Experimental and Applied Acarology* 69: 191–203.

Detail from Lugli fig. 6 (chapter 3.3). Spillo bites the fin of the poor swordfish that was lifted aboard, Lipari Island-Messina, Sicily (Photo by Davide Dutto (https://blog.davidedutto.it)

# Section 3
## Dogs through Time: Role, Task and Position

# 3.1 Urban Nomads and their Dogs

Christophe Blanchard

Laboratory EXPERICE - Université PARIS 13 - Sorbonne Paris Nord (USPN),
99 avenue Jean-Baptiste Clément 93430 Villetaneuse, France. christophe.blanchard@univ-paris13.fr

**Abstract**

Many homeless people own a dog. This animal companionship has multiple influences in the daily lives of homeless people, and can help or complicate the numerous problems they encounter. My academic research aims to better understand and evaluate the issue and impact of this little-known social phenomenon.

**Keywords:** social exclusion, dog, human-animal relationships, vagrancy, poverty's treatment.

## 1 Introduction

In recent years, both in France and in Europe, there have been groups of young people, often homeless, accompanied by dogs (Figure 1). It seems that possessing a dog is a prerequisite for recognition as a member of the 'underground community'. It is an important means of identification that helps to shape a person's identity on the street. Dogs have a direct influence on the daily lives of their young owners. They allow them to survive more easily in a ruthless urban environment, whether as a practical help or an emotional outlet. In a multi-year anthropological survey of street youth, I have tried to show how the dog's presence helps young owners to ease their psychological pain and helps them to gain some recognition from the rest of society.

Figure 1. Dogs of homeless people on Boulevard Magenta, Paris. 2012 (Photo by Christophe Blanchard).

## 2 Methodology

PhD graduate in sociology and anthropology, my personal background (I hold a dog-handler degree) and my university research have led me to focus very closely on issues relating to the status and place of animals, particularly dogs, in our society (Blanchard 2014).

This made me decide to track down the path of homeless young people and their dogs (Figure 2). As a certified dog-trainer, it was relevant to me to explore the way these 'urban nomads' live.

Figure 2. Group of homeless young people with their dogs in front of the railway station Gare de l'Est, Paris. 2012 (Photo by Christophe Blanchard).

During this study, I have adopted an ethnosociological approach to life stories to better understand the 'subjective' reasons for the benefits of owning a dog. The approach chosen has been a qualitative method based on an individualised dialogue with the participants.

In this way, I have tried to evaluate the effects on objective (physiological) and subjective (self-report) indicators such as increased motivation, sociability,

Dogs, Past and Present (Archaeopress 2023): 98–100

cognitive capability and decreased stress in dog owners (Conein 1992: 87–104).

### 3 Companionship and unconditional love

The animal's 'support' is crucial in the construction of the identity of dog owners living on the street. In a non-judgmental way, the dog shows the homeless person unconditional love and acceptance. As such, it is therefore important to take a close look at the 'canine' figure. Moreover, street dogs have a specific personality, the importance of which is often overlooked by social science researchers.

If a precise taxonomy had to be developed, street dogs could be described as male, medium sized, robust and generally in good physical health despite a few fleas. Females are generally less numerous and less prised because their reproductive period is an additional constraint that is not always easy for the master to manage. Neither very big, very large, or very colourful, one is immediately struck by the great homogeneity of these animals, which form some kind of archetype.

For many of the homeless dog-handlers, the possession of a pet has a significant social effect that builds relationships with the rest of the population. As a reason for discussions with old ladies when leaving supermarkets or as empathic support to encourage contact, the presence of the dog among these audiences generally does not leave passers-by indifferent, especially in France where it remains people's second favourite pet. Better still, this intercession of the animal offers the owner the opportunity to enhance his self-esteem and make himself credible to the professionals he frequents (veterinarians, social actors). The dogs of homeless people are highly dependent on their owners for their safety, health and nutrition. Owning a pet is therefore a great responsibility.

Having a dog also helps to be taken seriously by other homeless people. Indeed, in a hostile environment like that of the street, the attachment to a group of peers is indispensable. The community assures the individual of having a protective bubble, both physical and psychological, against the dangers of the environment, but also against the risks of accentuation of the phenomena of desocialisation characteristic of certain individuals living on the margins of society.

However, contrary to established prejudices, the dog is not only a bait to get money or sympathy. Above all, the dog is a strong emotional support and a valuable aid to psychological survival.

For these homeless youngsters who have often experienced a chaotic life course, their dog is a kind of child substitute. T. did confirm: 'They are my kids; I pamper them, I care for them the way my parents should have taken care of me.'

As a 'child-dog', the animal represents the transitional object par excellence for homeless youngsters. Its vitality and autonomy make its presence important in the eyes of its owner. On the contrary, its privation (during a runaway or a seizure by the pound, for example) is particularly badly experienced. Its loss or disappearance feels like grief in its own right.

Friendly and non-aggressive, the dogs of the homeless are not at all like the dangerous animals that anxious people fear. Like other 'integrated' dogs, they have their legal documents, while their owners often have their identity papers. Furthermore, I have discovered that the dogs are part of an elaborate relational scheme. Dogs are for these young people the basis of a reinvented family where each puppy has an owner but also a godmother and godfather within the group.

### 4 The dog, an aggravating factor of exclusion

Even if the dog constitutes a functional aid that reassures and protects, as well as an emotional outlet allowing owners to live better in a context of proven exclusion, the presence of the dog undeniably remains a source of additional marginalisation for an already fragile population. The urban space and the network of institutional or social shelters remain largely unsuited to the needs of homeless young dog owners. Owning a dog on the street can be a real challenge. The majority of social structures do not allow pets and this situation makes it very difficult for homeless young people to find housing.

Systematically refusing to get rid, even episodically, of their pets, dog owners thus end up sacrificing their social or medical follow-up, reinforcing the implacable logic of marginalisation in which the street often locks them up.

For an often young and very precarious population, which does not necessarily benefit from social assistance, taking care of a dog is often expensive and has a direct impact on daily life.

Those who are wrongly called 'backpackers' thus remain little by little 'sedentary travelers', prisoners of the urban centres that they hardly ever leave. It should be remembered that the dog remains a cumbersome companion, which makes it difficult for these young people to travel by public transport or to take the train.

Thus, the urban itineraries of young homeless people accompanied by dogs are essentially structured around municipal or associative shelters, important attractive places that end up constituting, as sociologist Djemila

Zeneidi-Henry (2002) reminds us, a real 'geography of assistance', a coherent geographical system that maintains the identity of this marginalised population.

**5 Conclusions**

Claimed to be out of the norm, even deviant, homeless people with dogs are in reality the guilty conscience of a society that generally struggles to help some of its marginalised people. The dog is a crying symbol of this failure: socially adulated, generating billions of euros in profits for industry every year, it remains one of the main grains of sand that has been jamming the wheels of social support in France for more than twenty years. Most social structures refuse to accept the dogs of these precarious people, accentuating the infernal spiral of marginalisation from which they cannot escape. The 'dog problem' argument is an admission of powerlessness or incompetence, and in reality it carries within it the seeds of a vision of the world and of normative and coercive care. Any individual who does not integrate 'naturally' into the public policies of care and, consequently, into those advocated in social work schools, will be irremediably excluded from the social system. In this normative posture where the binomial man/dog is denied, it is ultimately up to the homeless to adapt, never to professionals and especially not to public opinion.

**References**

Blanchard, C. 2014. *Les maîtres expliqués à leurs chiens.* Paris: La Découverte.

Conein, B. 1992. Ethologie et Sociologie. Contribution de l'éthologie à la théorie de l'interaction sociale, *Revue Française de Sociologie* 23: 87–104.

Zeneidi-Henry, D. 2002. *Les SDF et la ville. Géographie du savoir-survivre.* Paris: Bréal.

# 3.2 'The Mayor is a Dog': The Coming of Age of Contemporary American Pet Culture

Simona Bealcovschi

Département d'anthropologie, Université de Montréal. C.P. 6128, succursale Centre-ville, Montréal, (Québec) H3C 3J7, Canada.
simona.elena.bealcovschi@umontreal.ca

**Abstract**

This article documents North American pet culture and explores the evolution of the ethical treatment of animals, which is deeply rooted in the emergence of a new American culture and identity in the latter half of the 18th century. Today, dogs as companion animals have acquired a new symbolic role and have been conferred quasi-kin and near-human status. Dogs have become the mirror of the imaginary because they embody an equation between what humans are, what they wish they could become and what they imagine as the nostalgic past. As political currents and ecological concerns become increasingly worrisome, the unwavering faithfulness of dogs is a solace.

**Keywords:** human/dog relationship, pet culture, North America, Anthropomorphism.

> *There are many people who when we ask them to join us say that they prefer to work for human beings. But are we not working for human beings? Are we not constantly striving to make men and women more humane and disposed to all kindly feelings and to teach children to become gentle and merciful? Is not everything which tends to elevate man in the mortal scale a benefit to him?*
>
> Caroline Earle White (1833–1916)

## 1 Introduction

This article[1] will develop and explore the hypothesis that contemporary pet culture is a consequence of an evolutionary progress in awareness and moral sensibility in America toward the protection and respect of sentience, and especially of dogs. I will survey how deeply rooted moral sensibilities in 18th and especially 19th century notions of progress ultimately led to political militancy. These actions led to legislation and to contemporary pet culture, with its blurring of lines between animals and humans. This has led to contemporary moral positions that are reshaping human-canine relationships. These new dimensions of attitudes to canines are expressed as daily domestic rituals of anthropomorphising our animal friends. Because culture is fuelled by symbols, by the dynamics of imaginary space and by consumption patterns, I will also explore how this new pet culture had created new social identities. What is the role of dogs in North American pet culture? This research is based on an ethnographic reading of popular media, as well as on the initial phases of a long-term anthropological field project in North America.

[1] I would like to thank my colleagues Luke Fleming and Guy Lanoue (Université de Montréal) for their valuable comments.

## 2 A quick theoretical overview of the question

Until recently, 'culture' was reserved exclusively for describing human societies, since culture was seen by most as an ideational construct of the human imaginary. Anthropology and sociology have paid more attention to how humans use animals and relatively little attention to how animals and humans live together. Here, I will draw a quick outline of how human-dog relationships have been theorised in social scientific frameworks which assume that their subject matter deals only with thinking subjects. Anthropology and sociology developed methodologies that mirrored the assumption that Nature and Culture formed a dichotomous pair, an assumption anchored in the Western tradition and which informs normative and often instrumental understandings of how humans should treat animals. Inspired by George Herbert Mead's theory of symbolic interactionism,[2] sociologists as well as anthropologists have avoided dismantling the frontier separating humans from other animal species (Alger and Alger 2003). No doubt, academic specialisation within established disciplinary boundaries contributed to discouraging the study of human and non-human relationships. In the 1970s, the seminal work of Jane Goodall and E.O. Wilson encouraged researchers such as Bruno Latour (1991), Philippe Descola (2005) and Tim Ingold (2000) to challenge the idea of an impenetrable frontier between Nature and Culture.

[2] Mead argues (1934) that people act toward others as they imagine others are acting toward them. In the ethos of the times, this capacity was seen as a uniquely human trait, whereas all dog owners know that this is exactly how dogs see their humans (Morris 2015).

As a result, a new sociological literature has emerged in the last several decades, where animals and pets are purported to transform human sensibilities to the point where they are seen as co-creators of the social world. For example, a series of ethnographic and sociological publications influenced by new interpretations of the old theory of symbolic interactionism explore animal shelters as social agents (Arluke and Sanders 1996; Alger and Alger 1999; Michalon 2013; Albert and Bulcroft 2015; Michalon, *et al.* 2016), or concentrate on how homeless people seek not only emotional comfort but try to create intimate worlds where people and animals are co-equal partners (Blanchard 2015).

A second category of thematic studies has analysed how dogs became a source of interactional identity resources that structure relationships within families (Tannen 2004). By documenting domestic rituals that anthropomorphise dogs, these studies show that people see their dogs not only as pets in the classic sense but as full fledged members of the family, as kin (Tannen 2004; Christen 2009; Charles 2014, Charles and Davies 2017, Haraway 2019).

Perhaps the views of a young historian (Brown 2016) will eventually prevail in the social sciences. Brown goes beyond a focus on the microdynamics of the individual and group to an analysis of the macro dimension in which animals are the co-creators of social spaces and cultures. As other authors have noted (eg. Michalon *et al.* 2016: 1), perhaps the time has come to examine 'anthropozoological relationships', in the sense of Guillo's approach, and in the sense of a 'sociology of anthropozoological relations', even if this implies an epistemological decentring, ever so slightly away from the human, and towards the animals with whom we share our lives and who 'participate ecologically, politically and interactionally' (Michalon *et al.* 2016: 20) in our daily lives.

Carter and Charles (2016: 1) critically question as well the limits of the discipline and its need for a reconceptualisation of society, considering, '(Sociology's) assumptions about human exceptionalism and its emergence in the context of industrialization and urbanization are key to understanding its lack of attention to animals and contribute to a limited conceptualization of society This can be remedied by viewing non-human animals as involuntarily embedded in social relationships, a move which involves a redefinition of the social and of what it means to be human; a revision of notions of agency, subjectivity and reflexivity; and a rejection of the speciesism and anthropocentrism on which sociology is based'. As a complement to this vision, Guillo (2009) has launched the study of 'anthropocanine' societies. In this view, dogs are not exterior to human society, since all social bonds are based on 'mutual adjustment' (Guillo 2009: 289). In this sense, it is interesting to note that in the majority of cultural analyses, animals are absent because they are not conceptual categories in the discipline's epistemological toolbox. Yet, creatures such as cows, horses, dogs and cats accompany humans in the creation of spaces where reified property and possessions become emotional and symbolic resources.

## 3 The birth of an ethic of emancipation: 'Animals Made Americans Human'

The growing sensitivity - despite being uneven in places and involving paradoxical erasures - to the fate and treatment of animals has been slowly developing over several centuries in North American culture. In his article 'Animals Made Americans Human: Sentient Creatures and the Creation of Early America's Moral Sensibility', historian B.L. Smith traces (2012) a well-documented portrait of the development of a moral and ethical emancipation in the United States. For example, in 18th century America it was unthinkable to denounce human cruelty and the torture of domestic animals, says Smith when citing a letter written by Thomas Jefferson (citing Ford 1904), one of the so-called Founding Fathers and President of the United States. Said Jefferson: 'I participate in all your hostility at dogs, and would readily join in any plan for exterminating the whole race. I consider them as the most afflicting of all the follies for which the men tax themselves'. Smith comments: 'It has long been assumed that actions and words such as Jefferson's represented in stark form, the negative attitudes that Americans held towards Animals around the turn of 19th century' (Smith 2012: 126).

Following on the heels of Jefferson, his contemporary Benjamin Rush (1745–1813), doctor, patriot, philosopher, writer, ardent Christian and a reformist politician, wrote about the cruelties visited upon animals by his fellow citizens who were, thought Rush, seemingly unaware that animals could feel pain just like humans. Erudite and cosmopolitan, Rush was aware of the progress made in Great Britain on the question of animal rights following public debates and discussions. In 1786, he presented a brief to the American Philosophical Society, in which he argued that cruelty to animals also destroyed a person's moral sensibility: 'If Americans' moral sensibilities were not developed and protected; it would ruin the young nation' (cited in Smith 2012: 127). He also denounced the lack of any legislation that protected animals, 'from outrage and oppression' (Rush 1789, Smith 2012). The fervour of the American revolutionary spirit led to a spate of utopian feelings, in which people like Rush sought to define the basis of an ideal society. These sentiments, however, often advocated a more humane world while including people who sought to create a republican national identity in anticipation of the second coming of Christ.

These were also turbulent times for British intellectuals that united philosophers such as Locke[3] with priests such as Humphrey Primatt[4] (Smith 2012). Cruelty to animals and children was regularly denounced, as much inspired by humanistic rationality as interpretations of biblical Genesis, in which God's exhortation to Adam was taken as an order to ensure that all creatures shared the earth in a peaceful manner.

The British moralising push certainly influenced Rush and his fellow Americans, who tended to see the abolition of cruelty towards animals as a Christian duty. In brief, the fight to eliminate the cruel treatment of children led to reforms to the American educational system that helped sensitise a new generation of young Americans towards the dehumanising effects of all forms of cruelty.

The British influence was especially strong in Philadelphia, not only because of its symbolic patriotic connotations but also because it was a port with a thriving commerce with the Old World. There was also a Quaker community in the northeast, whose members were particularly receptive to the British anti-cruelty crusade because of their opposition to all forms of violence. It is not surprising that the American Vegetarian Convention was founded in Philadelphia in 1850. Rush and others inspired various currents in the United States that generally coalesced into feelings of care and kindness towards animals. In the 19th century, this would come to be known as 'the culture of kindness' (Smith 2012: 136), which wove together many threads: abolitionism, conscientious objection to warfare, gender equality and respect of animals. These currents influenced another militant voice calling for ethical emancipation, Caroline Early White[5] (1833–1916), a Philadelphia activist and philanthropist who founded the first animal shelter in the United States. Born into wealth that allowed her to get an excellent education, a polyglot speaking six languages (including Latin), she directed her talents to a variety of liberal causes. An abolitionist and suffragette, in 1869 she created the WPSPCA (Women's Pennsylvania Society for the Prevention of Cruelty to Animals) as well as the American Anti-vivisection Society. This woman's group immediately focused on the problem of Philadelphia's numerous stray dogs. This cause was driven by the fact that at the time rabies was common, and there was no cure. In the 19th century, the prime vector of infection was dogs. Rather than seeking to kill sick dogs (who are asymptomatic in the early stages of the viral infection), the First American Animal Shelter[6] accepted all dogs and sought to have them adopted (infected animals usually died within a month, so survivors were rabies-free).

It is only in the second half of the 20th century that a fundamental change in human-animal relations began to take hold. The earlier movement, even with its laudable aims to modern ears, was based on largely religious values and beliefs that emphasised 'the culture of kindness'. Besides witnessing more and more examples of compassion for the plight of animals subjected to needless suffering, the 20th century also saw a revolutionary fervour and change as people sought to find legislative solutions to the problem of animal abuse. After two world wars irrevocably changed humanity's notions of violence, the 1960s saw a growing sensitivity to the effects of ignorance and indifference. In 1960, the WWF (World Wildlife Fund) was founded in Switzerland. The movement reached the United States in 1961, where the organisation adopted the iconic Giant Panda as its logo. In 1971, Greenpeace, another militant organisation, was born in Vancouver, Canada, but soon moved to Washington to promote its manifesto. The organisation was an alliance of Quakers, ex-hippies, journalists and American draft dodgers and deserters. One of its first targets (before it was actually called Greenpeace) was the series of underground nuclear tests conducted by the U.S. Department of Defence on a remote island in the Aleutian chain. While the US government voluntarily abandoned the test site in 1971, the organisation was launched and soon turned its attention to saving whales and to environmental issues in general.

It was in this effervescent atmosphere of militant protests that two books would appear and strongly influence animal ethics, Peter Singer's *Animal Liberation*[7] (1976) and Tom Regan's *The Rights of Animals* in 1983. The two major publications fueled this ethical movement. Regan is a major figure in moral philosophy

[3] Smith is referring to John Locke (1693) *Some Thoughts Concerning Education*.

[4] Smith refers to Humphrey Primatt the author of the tract, 'A Dissertation on the Duty of Mercy and Sin of Cruelty to Brute Animals' (1776).

[5] 'She was among the first to launch the Pennsylvania Society for the Prevention of Cruelty to Animals, and she created an offshoot of that organization to create a welcoming environment where women, too, could work for animal welfare. Later she became aware of the medical testing that was being done on animals, and she was first to establish the American Anti-Vivisection Society in the United States' (Kate Kelly, 2016 https://americacomesalive.com/2016/03/18/first-animal-shelter-u-s-due-caroline-earle-white/.

[6] At the third meeting of the women's branch of the PSPCA, the women passed a motion that'one of the objects of this Society shall be, to provide as soon as possible, a Refuge for lost and homeless dogs where they could be kept until homes could be found for them, or they be otherwise disposed of' (Kate Kelly, 2016 https://americacomesalive.com/2016/03/18/first-animal-shelter-u-s-due-caroline-earle-white/).

[7] In his book, *Animal Liberation*, Peter Singer states that the basic principle of equality does not require equal or identical *treatment*; it requires equal *consideration*.

and was influenced by Kantian thought. As such, he advocates animal 'rights' (and not mere empathy from humans) that are based on the fact that all species are 'subjects of a life'. All such subjects possess two fundamental rights: absolute respect (which is innate and can never be taken away), and the right to be free from suffering caused by pain or privation (for example, of food or of living space).

In March of 1980 PETA was launched in Rockville, Maryland. People for the Ethical Treatment of Animals is probably the most radical contemporary organisation fighting cruelty to animals. It has not hesitated to adopt new forms of combativeness to get its message across. Strongly influenced by Singer's writings, its founder Ingrid Newkirk adopted Singer's aphoristic declaration as its motto: 'Animals are not ours to eat, wear, experiment on, use for entertainment, or abuse in any other way'. It has become the largest worldwide organisation of its type, in its defence of ideals that are very similar to the English organisation Amnesty International (founded in 1961, with an American branch created in 1966). PETA focuses on morality and legislation in an effort to influence public opinion. It proposes to end animal abuse and seeks legal recognition of the rights of animals to life and freedom from exploitation. While earlier movements, laudable as their aims may have been, implicitly recognised a human-animal hierarchy, PETA and its adherents believe that all sentient life has the same fundamental right to life. Humans should not be merely empathic but are morally bound to recognise animals as other sentient species to be respected.

Recently, the movement has gained some traction. On November 25, 2019, the PACT law[8] was signed, which guaranteed animal rights by federal American law. This was the culmination of a nearly ten-year legal battle that began when President Obama signed into law in 2010 a statute that forbade the production of videos showing violence to animals.

## 4 Anthropomorphising dogs in pet culture

Americans are a nation of dog lovers. According to a pet owner's survey, there were approximately 89.7 million dogs owned in the United States in 2017. This is an increase of over 20 million since the beginning of the survey period in 2000, when around 68 million dogs were owned in the United States.[9] According to Canada's Pet Wellness Report-CVMA, there are 5.9 million dogs (and 7.9 million cats) in Canada.

Most dogs in North America are treated as members of the family. Dogs place at the pinnacle of the North American animal hierarchy in the near-human ways in which they are named and addressed - in the ways in which they are interpellated. They have a name, they are spoken to and spoken through (see Tannen 2004; Haraway 2010; Charles 2014, 2017). People dress them; they sometimes get better medical care than some humans do. Some have their own psychologists. There are doggie daycares and doggie parks where they go on play dates with their doggie friends. Recently, they have been allowed to accompany people on planes. Some dogs have become millionaires thanks to inheritance laws that allow humans to leave their fortunes to their pet. There are of course doggie hotels and professional dog walkers. Dog culture is supported by a huge consumer sector, which brings dogs even closer to the masters/consumers. It is not surprising that many dog owners humanise and anthropomorphise their pets.

Humanisation begins with giving the family dog a name. Of course, all pets tend to have names that denote some relationship of subservience to a human, but dogs are often though not always given peculiar names that could be human, but which normally do not denote a physical trait (some cats are Fluffy, but there are no dogs named Smelly or Drooly). Dogs are engaged in conversations, perhaps ritualised with baby

Figure 1. A Family Portrait – two cosplayers in Montreal (Photo by S. Bealcovschi 2018).

[8] The PACT Act specifically renders 'animal crushing' illegal on a federal level. Animal crushing is when a living bird, non-human mammal, amphibian, or reptile is purposely crushed, drowned, burned, suffocated, or impaled, or serious bodily harm is otherwise inflicted upon it. The law also makes it a nationwide felony to distribute or create videos of animal crushing.

[9] https://www.statista.com/statistics/198100/dogs-in-the-united-states-since-2000/.

words or cadences, but nonetheless very serious and earnest. Thus, dogs become a 'person' or an 'animal person' (Christen 2009), with a very human identity when they are registered at a vet's or on official travel documents. Dogs are uniquely part of the human social system (Christen 2009; Blanchard 2015) (see Figure 1).

Families with dogs see their pet as a younger brother or sister who is given the same care and attention as any human child. A dog's human 'siblings' often accept this situation, since another member reinforces the importance of their age-grade in the family hierarchy. Their birthdays are celebrated not only with special food, as a cat may receive, but also by giving the dog a human ceremony, with cake, singing and candles (blown out by a proxy, obviously). Not only do dogs have their own parks and walks, they are part of family vacations and road trips (again distinguishing them from house-attached cats). Dogs are like 'super siblings'.

Tannen (2004: 399) has described how dogs become important mediators in families, in which, 'speakers effect a frame shift to a humorous key, buffer criticism, deliver praise, teaches values to a child, resolve potential conflict with a spouse, and create a family identity that includes the dogs as family members'. This discursive practice puts words in the family dog's mouth and allows people to speak on behalf of a dog. The dog does not even have to be present in the conversation. Unbeknownst to the dog, its stock in the human family can be increased by this humanisation. This bonding of one human to another via a dog allows otherwise isolated people ('Daddy has been very, very bad. Mommy is angry at daddy') to define a family unit that otherwise stands on shaky emotional grounds. This discursive strategy, says Tannen, includes ventriloquism, baby talk and high-pitched registers, so people can 'speak *as* their pets'. These voices are sometimes more animated than normal speech: 'In other words, through realization of pitch, amplitude, intonational contours, voice quality, pronoun choice, and other linguistic markers of point of view, speakers verbally position themselves as their pets' (Tannen 2004: 403). This is similar to well-studied forms of indirect speech when people wish to avoid direct confrontation via dialogue. There may be a taboo against all forms of speaking, but sometimes people can get around the rules by using dogs as an intermediary.

Citing previously published research, Charles states (2014: 716) that 91% of American pet owners regard their pets as family members, and cites a national survey in Australia that found 'that 88% of the owners consider their pet as a member of the family', and that 'women are more likely than men to ascribe family membership to a dog'. As companion animals, dogs do not need to have a useful function in human society. The human-dog relationship is almost entirely based on emotional transfer and social interaction. This has become even more important as the postmodern condition changes the structure of American families. People are less connected to kin and are more individualist as a result, sometimes leading solitary lives in huge metropolises. Companion animals have become more important than ever to these people.

As family members, even if the 'family' consists of a human and his or her dog, dogs are formally introduced to others and enter wider social networks. They are in selfies, iconic family moments and anecdotes ritually recited at gatherings. Unlike human family members, dogs have access to all spaces in the house. They can enter bathrooms and other people's bedrooms that are normally off limits to humans. They are part of the decor, with their photos placed alongside human memorabilia. These, says Charles (2014: 718), are indicators that new hybrid family structures are emerging, 'post-humanist households (Smith 2003; Power 2008;) where humans are de-centred, and the species barrier has no meaning'.

### 5 'The Mayor is a Dog'; pet culture emerges in Hollywood ('Hollywoof')

The factory of dreams and of contemporary myths, Hollywood played a special role in creating new models of behaviour and thought for the 20th century American middle class, especially after the Second World War, when some dogs became major stars and major icons. Dogs came to national attention via Hollywood almost with the start of the movies. Hollywood has long had its canine stars, called 'actors', like humans. A few have received stars on the iconic Hollywood Boulevard: Strongheart, Rin Tin Tin and Lassie. All three dogs received their Walk of Fame star on the same day: February 8, 1960. This is as good a date as any to mark the humanisation of dogs.

There were other well-known dogs before, but these were heroes in the conventional sense: in 1925, during a hard Alaskan winter, Balto and Togo, two sled dogs who in five and a half days carried medicine for 674 miles (1,085Km) to the isolated community of Nome, are credited with having saved the lives of children from an incipient epidemic. These dogs, and hence all dogs, embodied force and exceptional courage. They are individual heroes who stood out from the crowd because of their exceptional qualities.

Balto and Togo were real dogs who embodied courage. Strongheart and Rin Tin Tin were actors who represented courage. Both were German Shepherd

Figure 2. This is a four-storey memorial for Tilly, a British bulldog who is no longer with us. Her master commissioned this mural by artist Kevin Ledo. A gift to dog lovers from Montreal (Photo by S. Bealcovschi 2019).

Dogs who came to America from Europe after World War I. Strongheart was the Americanised name of the German-born Etzel von Oeringen. Not only was he a silent movie star, he also was one of the first Hollywood celebrities to get an endorsement deal for Strongheart dog food. To add to his Hollywood mystique, he died at 12 years old because of an accident on a movie set. At the time, there were no laws protecting animals used in films.

Rin Tin Tin was another German dog who conquered Hollywood. Abandoned by the German Army in Lorraine, France, he was rescued by an American soldier who eventually had him buried in his 'native' France when he died in 1932, in the dog Cemetery in Asnières-sur-Seine outside of Paris. Despite owing his fame and fortune to Hollywood, Rin Tin Tin's owner Lee Duncan evidently had very strong emotional ties to the orphan he adopted.

Lassie, a Rough Collie, emerged as a film star in 1943 and was one of the first crossover actors to jump into the new medium of television, where she got her own show in 1954 that lasted 19 seasons. Coincidentally, all three of these early stars were associated with the military. Strongheart was trained in Berlin as a police dog, Rin Tin Tin was a war dog rescued by an American soldier, and Lassie was the companion of a young boy on a farm, and later was associated with the Forest Service. The story was based on an apocryphal story that a dog named Lassie had saved the life of a sailor believed to have drowned.

Nowadays, an unprecedented pet imagery is continually feeding social networks, the internet, advertisements (not only for dog products) and even as 'human interest' on television news and print journalism. The role of dogs in popular culture has increased as media have become omnipresent. Dog culture is no longer reserved to Hollywood mythmaking, as each owner can now use his cell phone camera to record and share his dog's antics. Some go well beyond a mere photo, as shown in Figure 2.

A Californian dog named Bosco, a Labrador and Rottweiler mix, has his own internet page because he was elected mayor in 1981 of the small town of Sunol. He was of course, an honorary major, but he kept the position until his death in 1994, even defeating two human political rivals for the title. Recently, a football player named Derrick Nnadi made the headlines when he became a Super Bowl champion, helping Kansas City to win against the San Francisco 49ers in January 2020. He celebrated his win by paying the adoption fees for more than 100 dogs at a shelter.

### 6 Conclusion

As part of the canonical representation of the good life, individual or familial, dogs occupy a special place in the imaginary of middle-class America, which means, given American social structure, that dogs play a special role in contemporary culture. Until the end of the 19th century, dogs may have been admired, but they were seen as utilitarian animals. As America became urbanised and industrialised, new sensibilities emerged that led to the foundation of various associations dedicated to preventing cruelty to animals and to the creation of the first shelters. Dogs are not only pets but they become the bedrock of pet culture. They become emotional companions and eventually, family members. They become surrogates for the protective intimacy of childhood. They can heal alienation and solitude. Semiotically, dogs are no longer true animals but become semi-human. As Donna Haraway notes (2010) when imagining what a biosocial history of humanity would look like, dogs and humans have evolved together.

At the same time, more and more advertisements depict lifestyles in which urban residents who live with dogs. In the 1950s, dogs came to define the new suburban American nuclear family. The presence of a dog as a family member allowed these families to project emotions onto their new lifestyles and new environs. This became a fundamental component of the post-war family. Dogs are thus intermediaries between close and far, between intimacy and the depersonalised outside world of consumerism. Not only is the dog in the home surrounded by toys like a baby, at least a quarter of dog owners I consulted cook for their pets. In effect, these dogs are isolated from the consumerism of the outside world, as they consume food prepared in the intimate space of the home.

In the contemporary consumer-oriented world, where individualism seems to triumph, where people are supposed to emulate popular idealised perfect perpetually healthy bodies and perfect lifestyles, companion dogs are even more important. More and more practices and attitudes will be attuned to humanising them by paradoxically treating them as special, non- animals, with their own pet foods, pet clothes, pet toys, etc. Even established dog shows such as the Westminster Kennel Club are enjoying a newfound popularity beyond the world of dog owners. Finally, the tactics of newer associations such as PETA have led to legislative protection for pets, at the moment when they seem to be transitioning (partially) into the world of humans, as witnessed by 'support animals' that are now considered regular passengers by some airlines. Nonetheless, to paraphrase Donna Haraway, is it part of our human nature to construct relations with others, including other species that are not based on domination, anthropocentrism and anthropomorphising, but rather on respect, affection and even love?

### References


Albert, A. and K. Bulcroft 2015. Pets and Urban Life. *Anthrozoös, A multidisciplinary journal of the interactions of people and animals* 28: 9–25. https://www.tandfonline.com/doi/abs/10.2752/089279388787058740

Alger J. and S. Alger 2003. Drawing the line between humans and animals: An examination of introductory sociology textbooks. *International Journal of Sociology and Social Policy* 23(3): 69–93.

Arluke, A. and C.R. Sanders 1996. *Regarding Animals.* Philadelphia: Temple University Press.

Bekoff, M. (ed.) 2010. *Encyclopedia of Animal Rights and Animal Welfare.* Santa Barbara: Greenwood Press.

Blanchard, C. 2015. Ce que les noms des chiens des sans-abris révèlent de leurs maîtres. *Anthropozoologica* 50(2): 99–107. https://doi.org/10.5252/az2015n2a3

Brown, F.L. 2016. Dogs and Cats: Loving Pets in Urban Homes, in F.L. Brown and P.S. Sutter (eds) *The City Is More Than Human: An Animal History of Seattle:* 149–188. Seattle: University of Washington Press.

Carter, B. and N. Charles 2016. The animal challenge to sociology. *European Journal of Social Theory* 21(1): 79–97. https://doi.org/10.1177/1368431016681305.

Charles, N. 2014. 'Animals Just Love You as You Are': Experiencing Kinship across the Species Barrier. *Sociology* 48 (4): 715–730.

Charles, N. and C.A. Davies 2017. My family and Other Animals: Pets as Kin. *Sociological Research Online* 13(5): 13–26. https://doi.org/10.5153/sro.1798

Christen, Y. 2009. *L'animal, est-il une personne?* Paris: Flammarion.

Descola, P. 2005. *Par- delà nature et culture.* Paris: Gallimard.

De Waal, F. 2016. *Sommes-nous trop 'bêtes' pour comprendre l'intelligence des animaux?* Paris: LLL Les liens qui libèrent.

Guillo, D. 2009. *Des chiens et des humains.* Paris: Le Pommier.

Guillo, D. 2016. Les recherches éthologiques récentes sur les phénomènes socio-culturels dans le monde animal: un regard renouvelé en profondeur. *L'Année Sociologique* 66(2): 351–384.https://www.cairn.info/revue-l-annee-sociologique-2016-2-page-351.htm

Haraway, D. 2019. *Manifeste des espèces compagnes: chiens, humains et autres partenaires.* Paris: Flammarion.

Ingold, T. 2000. *The Perception of the Environment: Essays on Livelihood, Dwelling and Skill.* London: Routledge.

Latour, B. 1993. *We have never been modern.* Cambridge: Harvard University Press.

Michalon, J., A. Doré and C. Mondémé 2016. Une sociologie *avec* les animaux: faut-il changer de sociologie pour étudier les relations humains/animaux? *SociologieS* Dossiers, Sociétés en mouvement, sociologie en changement, mis en ligne le 07 mars 2016. http://journals.openedition.org/sociologies/5329

Michalon, J. 2013. Fabriquer l'animal de compagnie. Ethnographie d'un refuge *S.P.A. Sociologie* 4(2): 163–181.

Morris, C.V. (ed.) 2015. *Mind, Self & Society. The definitive Edition.* Chicago and London: The University of Chicago Press.

Regan, T. 2012. *Les droits des animaux.* Paris: Hermann.

Smith, B.L. 2012. Animals Made Americans Human: Sentient Creatures and the Creation of Early America's Moral Sensibility. *Journal of Animal Ethics* 2(2): 126–140. https://www.jstor.org/stable/10.5406/janimalethics.2.2.0126.

Singer, P. 1975. *Animal Liberation: a new ethics for our treatment of animals.* New York: Avon Books.

Tannen, D. 2004. Talking the Dog: Framing Pets as Interactional Resources in Family Discourse. *Research on Language and Societal Interaction* 37(4): 399–420.

**Web sources**

https://americacomesalive.com/2016/03/18/first-animal-shelter-u-s-due-caroline-earle-white/ (viewed 1 November 2019).

https://awionline.org/content/preventing-animal-cruelty-and-torture-pact-act (viewed 25 November 2019).

https://en.wikipedia.org/wiki/Bosco_the_dog (viewed 25 November 2019).

https://www.ipnoze.com/joueur-football-paye-frais-adoption-chiens-refuge-derrick-nnadi/ (viewed 5 February 2020).

https://www.statista.com/statistics/198100/dogs-in-the-united-states-since-2000/ (viewed 30 November 2019).

# 3.3 Wolves, Dogs and Water – Dogs and Fishing Boats

Francesca Lugli

Associazione Italiana di Etnoarcheologia, Via Principe Umberto 41, 00185 Roma, Italy.
luglifrance@gmail.com

**Abstract**

It has not been evaluated in-depth yet if water may or may not have been a familiar element for dogs at the beginning of their 'arrival' among human beings. It is usually accepted that the first dogs probably helped humans with guarding and hunting (also for sea mammals (Rick *et al.* 2008: 1083), but dogs could have also helped men with fishing as Coppinger expressed in his book 'Fishing dogs' (2014). Dogs used to be present on fishing boats in many Mediterranean and non-Mediterranean regions until twenty years ago. The main tasks that dogs had were for companionship and to guard against thieves but they could also help to remove hawsers, point out shoals of fish and to help rescue a man or retrieve an object that had fallen into the sea. Even if dogs were certainly a traditional companion on Italian fishing boats, their presence and the relationship they had with fishermen is scarcely documented and has been completely ignored by studies on Italian fishing, fishermen and dogs.

**Keywords**: dogs, wolves, waterdogs, fishery, fishing boats.

## 1 Introduction

The role of dogs in aquatic contexts has not been evaluated in-depth. Dogs have helped with hunting in aquatic environments, including fishing since ancient times. The presence of dogs on fishing boats was a constant in many Mediterranean countries until the second half of the XX century. This important tradition has scarcely been recorded in Italy and must therefore be documented to ensure a record of this heritage is preserved for future generations.

### *1.1 Wolves and water*

Wolves are highly adaptable, exceptional predators who can live in various habitats all over the world. They mainly eat mammals of moderate to large size and to a lesser degree also animals of medium and small size (marmots, hares, badgers, foxes, polecats, squirrels, small rodents, insectivores and others). They can complete their diet with lillies, wild berries and fruits depending on their habitat and the season. Wolves also eat birds, reptiles, insects and even garbage when food is insufficient (Heptner and Naumov 1998: 213–221; Gable *et al.* 2017: 1; Homkes *et al.* 2020: 1). But wolves can eat also fish and it is well-known that grey wolves who live in British Columbia and Alaska catch and eat spawning salmon. Recent data of 'The Voyageurs Wolf Project: Understanding the Secret Lives of Wolves in the Northwoods' in Minnesota and the Yellowstone National Park documented that grey wolves are also very good swimmers; as an example, one monitored wolf swam 12 times across sections of Rainy Lake covering 2.6 miles over two days.[1] The project proved that wolves can also hunt freshwater fish as a seasonal food source and researchers made videos documenting it.[2]

### *1.2 Dogs and Humans - a significant dichotomy*

Dogs have been crucial for human life since ancient times and scholars have even proposed that dogs were the crucial point that allowed modern humans to drive Neanderthals to extinction (Shipman 2017).

Dogs have certainly been crucial for herding thanks to their capacity to protect and guide livestock and for hunting by helping hunters in their work. But they have also been fundamental for many rural sites for their multitasking behaviour. Sometimes, they have been indispensable for human survival. For example, Mongolian nomads who use dogs for guarding against wolves in their camps say that life without dogs is not possible in the steppe.[3] They are currently indispensable for carrying out many human tasks and making human life or jobs easier also in many western countries (Coppinger and Coppinger 2001).

Starting from the nineteenth century, dogs have progressively changed their status within the relationship that they have with humans in Western countries. The relationship dynamic between dogs and humans has changed towards a more and more emotional one. Dogs can be part of the family of their owner who often has an intense emotional involvement with them. In many countries, dogs are allowed to visit their human friends when they are hospitalised, they can enter restaurants,

[1] https://www.voyageurswolfproject.org/

[2] https://nywolf.org/the-wolves-of-voyageurs-national-park/

[3] It is possible to assume that ever since the bronze age the success of steppe pastoralism has been possible thanks to dogs (Lugli 2016).

shops and other important public sites. Their position in our societies is so important that a huge sector of economic production is completely dedicated to these 4 pawed friends with special food, medicines and accessories which are massively produced, sold and bought. They can have their own tombs and sometimes monuments are dedicated to special dogs.

Cruelty and joy have always characterised the long co-history that dogs have had with humans (Haraway 2003: 3). Nowadays, this dichotomy is stronger than in the past. If on one side dogs are special and precious friends for millions of people, on the other they are also considered as simply objects and tools without any right. So, the current life of dogs can be rich in love and consideration but also paved by thorns.

### *1.3 Dogs and water, waterdogs and fishing dogs*

It has not been evaluated in-depth yet if water may or may not have been a familiar element for dogs at the beginning of their 'arrival' among human beings. It is usually accepted that the first dogs probably helped humans with guarding and hunting (also for sea mammals (Rick *et al.* 2008: 1083), but dogs could have also helped men with fishing as Coppinger expressed in his book 'Fishing dogs' (2014). The fact that wolves can swim and fish could be considered a good prerequisite to think that the first dogs could also do those things.

But it is difficult to say when humans started using dogs in aquatic activities, dogs became undoubtedly crucial, sometimes indispensable, in exploitation and life in aquatic environments at some point of their lives with humans. For example, hunting and fishing in flooded areas could have been assisted by the use of dogs because they could swim to chase prey and to retrieve it when it was killed or wounded by the hunter or fisherman. It is also important to recall that dogs were introduced to many islands in prehistoric times e.g. California's Channel Islands (Rick *et al.* 2008: 1077–1087) and also that recent data show that Mesolithic dogs ate fish (Zhilin *et al.* 2014).

Various water dog breeds have been selected in different countries over the centuries. These can be considered the result of a focused selection for hunting activities in swamplands and marshlands where they are used by hunters. Tight, waterproof coats, usually medium-sized, good swimmers and good communicators are the current features that are usually requested of water dogs but also webbed feed are an important quality.

The American Kennel Club list the sixteen most famous breed swhich are excellent swimmers: American Water Spaniel, Barbet, Boykin Spaniel, Chesapeake Bay Retriever, Curly-Coated Retriever, English Setter, Flat-Coated Retriever, Irish Water Spaniel, Labrador Retriever, Lagotto Romagnolo, Newfoundland, Nova Scotia Duck Tolling Retriever, Otterhound, Portuguese Water Dog (Cão de Água Português), Spanish Water Dog (Perro de Agua Español) and Standard Poodle.[4]

But dogs of mixed and unknown breeds can also be splendid water dogs. For example, in the author's experience having rescued a puppy who was lost in the centre of Porto (Portugal) and returning to Italy with her in 2004, Francesca Lugli and her husband later discovered her natural abilities in water. Amalia was a small-sized mixed-breed black dog with short, waterproof hair and webbed feet. The Portuguese vet that was consulted told them that Amalia was a water dog even if it was impossible to recognise 'her ancestors' breeds. He supposed that somewhere and somehow she had a percentage of Cão de Água Português in her blood even if she did not look like it at all. Her origins remained a mystery but she certainly was an excellent swimmer and water dog (Figure 1).

Figure 1. Amalia was a dog of mixed and unknown breeds that Francesca Lugli and her husband found in Porto (Portugal) in 2004. She was a water dog with short, waterproof hair and webbed feet (Photo by F. Lugli).

### *1.4 Dogs on board*

The presence of dogs onboard ships and fishing boats is a particular aspect of the use of dogs in the marine world. It is rarely described and documented by the literature and scarcely documented by folkloristic, historic and scientific books but it is usually mentioned as they were an obvious presence. For example, in 'Robinson Crusoe'

[4] https://www.akc.org/expert-advice/lifestyle/breeds-that-are-born-to-swim/

(Defoe 1719), there were a dog and two cats on board who went ashore on the island with him and the dog was his 'pleasant and loving companion' for sixteen years. And when Robin found the Spanish ship which had smashed on the rocks, he saved a dog which was the only being alive onboard the wreck. But there is no information on the role that dogs played on the ships and the relationship that the crew had with them.

Even though information is scarce, the presence of dogs on board boats seems to have been known in many Mediterranean and non-Mediterranean regions since ancient times. For example, it seems that dogs could have been used to pass messages and to retrieve objects lost overboard by ancient mariners (Fogle 2000).[5] The body of a sailor and his dog who were found lying under the '*Nave B*' (Ship B) (1–15 AD) during the excavations in the harbour area of San Rossore (Pisa)[6] (Sorrentino *et al.* 2000) could be interpreted as the presence of the dog having been on board.

In recent times, dogs have been used in various kinds of maritime boats, such as cargo vessels, fishing boats and warships. Warships often have Mascot dogs which are considered important because they can be useful 'for their seemingly innate ability to build morale... and because dogs provide welcome relief from the monotony of being at sea for months on end'.[7] Dogs have been used as 'Soldier dogs' by various countries for a long time. During WWI and WWII they were used for various tasks such as detecting bombs, detecting hidden enemies, attacking, protecting, relaying messages and delivering first aid. Dogs are still used by various armies (Din 2013: 13).

Part of the official blog of the U.S. Naval Institute is dedicated to dogs on board with a long list of decorated dogs. The case of Sinbad is famous - a mixed-breed dog that served onboard USCG *Campbell* as the Coast Guard mascot during World War II - and was awarded six medals including the American Defense Service Medal and the World War II Victory medal. He has a granite monument at Barnegat Light (New Jersey) that was erected in his honour after his death.[8][9]

Nowadays dogs are often used in search and rescue maritime missions by many countries and in Italy, there is a famous training school for dogs - especially Newfoundlands, Labradors and Retrievers - which are used by the Coast Guard during its service.[10]

### *1.5 Dogs and Fishing boats*

Despite the scarcity of testimonies of dogs on fishing boats[11,12], those that exist document a very close relationship between dogs and fishermen[13], often of mutual respect and other times of great suffering for the dog (Figure 2).

Various waterdog breeds were used in different countries for that purpose, the most famous are the Cão de Água Português, the Perro de Agua Español, the Italian Lagotto Romagnolo, the Canadian Newfoundland, the Scottish Golden Retriever and the Canadian and English Labrador Retriever. These breeds are usually characterised by a tight waterproof coat which prevents thermal shock in freezing water, strong confidence with water and extraordinary ability in swimming. The Cão de Água Português is one of the most famous and it was well known in Portugal for its constant and specialised presence on fishing boats in the past. Raul Brandão in the book 'Os Pescadores' (1923: 274) described its tasks:

> '*Tripulavam-no vinte e cinco homens e dois cães, que ganhavam tanto como os homens. Era uma raça de bichos peludos, atentos um a cada bordo ao lado dos pescadores. Fugia o peixe ao alar da linha, saltava o cão ao mar e ia agarrá-lo ao meio da água, trazendo-o na boca para bordo*' (It was manned by twenty-five men and two dogs, who earned as much as the men. It was a breed of furry animals, watching one on each side beside the fishermen. The fish would run away from the wing of the line, the dog would jump overboard and would grab him in the middle of the water, bringing him on board in his mouth).

This breed became internationally famous when President Obama adopted a Cão de Agua named Bo occasionally called 'First Dog', in 2009.

But not all the dogs on board were used for helping fishermen in their activity. Sometimes the main task of dogs was essentially guarding. In this case, different breeds or mixed-breed dogs could be taken. Dogs were common on fishing boats in many Italian regions until a few years ago (as well as in other countries). Nowadays modern technologies have replaced dogs in their tasks and new health parameters often do not let fishermen have dogs on their boats.

[5] https://en.wikipedia.org/wiki/Water_dog
[6] https://www.archeostorie.it/museo-per-le-navi-di-pisa/
[7] The site of the U.S. Naval Institute provides many pictures of dogs on board (https://www.navalhistory.org/2018/04/27/sea-dogs)
[8] https://www.atlasobscura.com/articles/little-known-history-seafaring-pets-dogs-cats-chickens-war-exploration
[9] Sinbad was also the subject of the 1946 Universal Pictures short film 'Sinbad: Dog of the Seven Seas'.
[10] http://museumofmaritimepets.com/featuredbreed/newfoundlands.html

[11] For characteristics for EU fishing vessels see https://eur-lex.europa.eu/legal-content/EN/TXT/?uri=LEGISSUM%3A4304085&qid=1619555549420
[12] Fishery is also a lacustrine activity that was important for many Italian lakes. Nowadays lacustrine professional fishery is not intensely practised.
[13] Italian fishery has mostly been a male world. In recent times a few women are also actively involved in these activities. In the present article the masculine word 'fishermen' will be used for all the people involved on fishing boats.

Figure 2. A) Crew of the motor sailor Cigno that was confiscated by the Italian Navy (1941); B) Guido Bartoli, motorist of the motor sailor Cigno (1941); C) Crew of the trawler Don Abbondio with the dog Burgo (end of the '50); D) Crew of the motor sailor Marianna (1930) (Photos of the Photographic Archive of the Centro Culturale Polivalente (CCP) of Cattolica (Rimini, Emilia Romagna).

According to the fishermen that I was able to meet, at one time there was not a single fishing boat which did not have a dog. Usually, no special water dogs were used but the '*Lagotto Romagnolo*' - which is currently well known for being an exceptional truffle dog and a good water hunting dog - is said to be commonly preferred by the fishermen of the Central Adriatic coast.

This use was so common that even in recent times when local newspapers report the news that a dog was found and saved far from the shore, it is assumed that it has fallen from a fishing boat (eg. Dinoi 2017).[14]

The importance of dogs onboard has never been studied or documented. It is also curious that water dogs, fishing dogs and dogs onboard fishing boats have constantly been ignored by scientific studies on Italian fishing and also by the scholars of the disciplines which rely on dogs and their role in human history. Information on that topic is extremely scattered and fragmented. Coppinger[15] notes half seriously that there are a few anthropological studies about 'fishing dogs'...

[14] It is not unusual to find similar news stories from around the world. For example in 2016 the American Navy found and saved the German shepherd dog Luna on an island 80 miles off the coast of San Diego. She had fallen off a fishing boat in the Pacific Ocean five weeks earlier and was considered lost at sea. https://abcnews.go.com/US/navy-finds-missing-dog-fell-off-fishing-boat/story?id=37701943).

[15] Coppinger (2014) observed that when he started searching for news on 'fishing boats' there were no studies at all on this topic.

because anthropologists focus their studies on 'hunters and gatherers'and not on 'fishers and gatherers' and perhaps for that reason there are no studies on fishing dogs. On the contrary, it is astonishing how much has been written on hunting dogs (2014: 50–51).[16]

The sporadic testimonies show a very close relationship between dogs and fishermen, often of mutual respect and other times of great suffering for the dog. The use of dogs on fishing boats can now be investigated only by interviews with fishermen who have had or still have dogs on board, with people who know or knew the world of fishing boats, by very rare written sources and via the internet.

To document, study and analyse the presence and use of dogs on fishing boats is crucial to better understand the place of dogs in seafaring societies which depend on the fishing economy, to improve the knowledge of the History of fishing and, last but not least, to help fishermen who strongly want to have their dogs on board. The Italian Association for Ethnoarchaeology promoted the research which started in Italy in 2018 to document this tradition and its memory before it is completely forgotten.[17]

## 2 Evidence and methods. Dogs onboard fishing boats — tangibility and intangibility of a concrete presence

### *2.1 Abundance and lack of documentation*

Lately, the attention on dogs has progressively increased thanks to the important emotional role that dogs currently have in our societies. Consequently, many scientific reports have been written from various perspectives (archaeology, history, genetics, sociology, ethology, ecology, zoology, etc.) to improve the knowledge of this animal and also to analyse the dog-human dynamic, the origin and history of this coexistence and how it has changed over time. The scientific literature on dogs constantly grows and develops, it is so ample that it is difficult, perhaps impossible, to keep up-to-date with. But if dogs have been analysed and even micro analysed for what concerns their life, their history and the bond they have had with humans, the role of dogs in maritime and aquatic societies has inexplicably been almost completely neglected. There is a total lack of documentation on it and the few available data and information are scattered here and there. Therefore, there is not a solid base to start with for what concerns the study of dogs and fishing boats which is the topic of the present article.

If the presence of dogs on fishing boats was constant until the nineties, it is currently very rare because of new technologies and because it is often forbidden or opposed by recent health parameters. Therefore, fishermen who still use dogs on board are not very common and they are not always well -disposed to discuss this issue.

### *2.2 Research strategy*

The shortage of written information and the limited number of fishermen who still have dogs on board make finding reliable data extremely difficult. The research aimed to collect all the information in order to piece together the current and past life of dogs, their role, the dynamics, interaction and the mutual advantages of this cooperation and co-existence. So, a set of main problems were considered from a diachronic perspective:

- The extent of the phenomena;
- The role of dogs;
- The life of dogs;
- Regional differences and regularities of the use of dogs along the Italian coasts and islands;
- Differences and regularities depending on the various kind of fishing;
- Relationships between dogs and fishermen;
- Current national healthcare rules that allow or prohibit dogs on board.

Fishermen and their crew were contacted, consulted and when possible interviewed. But also various sources were considered and the 'investigation' moved along different courses of research. Different concepts were taken into account to obtain more data and above all a wide point of view of the problem.

Written sources and oral information were searched and used to obtain as many clues and data as possible. Ancient sources, literary works, scientific literature, administrative documents and the internet were taken into account. Ethnographic, archaeological and historical museums, photographic archives, libraries, ministries, harbourmasters and port captaincies[18],

[16] All over the world there are magazines dedicated to hunting dogs but not one to water and fishing dogs. Coppinger notes that his '...filed guide to the dogs lists 340 breeds of dog, and more than half of them (184, to be exact) are hunting breeds' (2014: 95).

[17] The present article is a preliminary report of the first step of the research which is currently being carried out in Italy and in Portugal (with the patronage and the sponsorship of the Ministry of Foreign Affairs and International Cooperation - Italy MFAI) and thanks to Konstantina Saliari, also in Greece. In the future, phylogenetic analyses performed on the main canine breeds with swimming skills will depict the evolutionary relationships among thes ekind of dogs from different geographic regions. Furthermore, molecular phylodynamic analyses will also be used to shed light on the human-mediated mechanisms of artificial selection which led from the domestication of the wolf to such highly specialised and selected dog breeds.

[18] All the harbourmasters and port captaincies of Italy were contacted.

harbour police, fishermen's associations and cooperatives, ship owners, fish markets and fishmongers, vets, canine associations, and hunters and hunters' associations (for what concerns hunting water dogs) were contacted.

Many Tirrenean and Ionic ports were visited or contacted: La Spezia, Lerici, Portovenere (Liguria), Carrara, Viareggio, Livorno (Tuscany), Fiumicino, Anzio (Lazio), Torre del Greco, Napoli, Salerno (Campania), Reggio Calabria, Roccella Ionica (Calabria), Bari, Taranto (Puglia) Siracusa, Catania, Mazzara del Vallo, Trapani, Palermo (Sicily), Cervia, Cesenatico, Pescara (Emilia Romagna), Ancona, San Benedetto del Tronto (Marche), Termoli (Molise).[19] A database based on the various sources and informers was created to register any clues, details and data we could find and a form was set up for 'No Available Information'. Hundreds of emails were sent and hundreds of calls were made to find any possible information.

### 2.3 Interviews - a crucial part of the research

As per Tim Ingold (MacDougall 2016), an effort was made not to consider the informers only as a standing reserve for collection. The interview had ten fixed sections that constituted the starting point for the 'conversation' with the informer.

1. Information on personal details, history of the informer, history of the family to contextualise the informer and to register if fishing has been a family activity for a long time, from generation to generation;
2. Information about the port, its history and its fishing activity to document traditional and non-traditional situations;
3. Information about the kind of practised fishing (Purse seine fishing, Trawl fishing, Gillnets fishing), daily or longer activity;
4. Area of activity;
5. Information about the fishing boats (dimensions, material, model, technology, how many people, how old it is, who the owner is, etc.);
6. Past and current dogs Presence/Absence on board;
7. Current rules and restrictions for what concerns dogs on aboard;
8. Dog's details (breed, age, where the dog comes from and when it arrived on board);
9. Dog's daily life; dog's station, dog's space and relationship with the seamen and their families;
10. Special stories and memories about dogs;

[19] Lacustrine fishery was also partially considered but the presence of dogs on board does not look to be part of a tradition. For example only one fisherman used to have his dog on board in the Bolsena lake (Viterbo, Lazio).

By the sections on dogs, the interview tried to have a description of the dynamics of the relationships between dogs and fishermen. So, it was considered if dogs have/had precise tasks, if they are/were important for the activity or emotional aspects, if they are/were only useful or if a complex and strong identity is conferred to them if they are nursed and if they are part of the family of their owner. All the sections were used to be expanded and to allow informers to feel free to speak.

## 3 Results

### 3.1 Health parameters

As was already mentioned above, recent health parameters seem to forbid fishermen to have dogs on board. But the reasons and the modalities are unclear. The Italian Ministry of Health was contacted by telephone and by email and after various messages and replies on the 16th July 2018 an email was received which explained that:

> 'However, the writer does not know any tradition about dogs on fishing boats.
>
> In any case, the fishing vessels are structures intended for food production.
>
> In food production facilities, EC Regulation 852/2004 imposes on food operators, including fishermen, the obligation to prevent the access of pets.
>
> In any case, previously, also the law 283/1962 did not contemplate the presence of pets in food facilities'.[20]

So, at the Italian Ministry of Health, the tradition of dogs and fishing boats is unknown and pets (not only dogs) are not allowed in food production. Fishermen are part of food production and work on boats that are food facilities and they cannot consequently have dogs with them. Although the 'writer' mentioned the law 283/1962, dogs were consistently present on fishing boats at least until twenty to thirty years ago, some decades after the year 1962.

Furthermore, herder dogs that live with sheep and goats can be said to be present in food production and live in food facilities. Italian herders serenely go on having dogs with their livestock. But it is common to have situations that contradict the laws. Therefore, the current situation in Italy is that: dogs are generally

[20] In Italian: 'Non risulta comunque allo scrivente nessuna tradizione inerente i cani a bordo dei pescherecci. In ogni caso i pescherecci sono strutture destinate alla produzione di alimenti. Nelle strutture produttive alimentari il Regolamento CE 852/2004 impone agli operatori alimentari, compresi i pescatori, l'obbligo di impedire l'accesso di animali domestici. In ogni caso, precedentemente, anche la legge 283/1962 non contemplava la presenza di animali domestici nelle strutture alimentari'

forbidden on board but there are also ports where fishermen are allowed to have dogs.

### *3.2 Fishermen and fisherwomen's interviews*

Because it is not always accepted that fishing boats have dogs on board, it is not always easy to speak about dogs. Fishermen who have dogs are often afraid to be reported, especially in Southern Italy. So, the first effort was often to persuade fishermen that the interviewer was not a spy from the Italian Ministry of Health. In one case, the wife of a fisherman of the port of Mazzara del Vallo (Sicily) who was certain that the interview was just a means of incriminating her husband, called Francesca Lugli and told her not to dare contact her husband anymore. She was vaguely threatening and sinister. This is just to describe the situation it was necessary to deal with sometimes. This being the case, the results of the interviews are globally debated and fishermen will mostly be nameless.

Various kind of nets were documented: *pesca con reti a circuizione (ciancioli)* (Purse seine fishing), *pesca con reti a strascico* (Trawl fishing), *pesca con reti da posta/(imbrocco, tramaglio e combinata* (Gillnets fishing), big and small professional fishing boats, daily fishing activity and up to three weeks on the high seas, with the immediate frozen process on board and without it, with between one and ten people working on board. The big boats usually sell their fish to stores, fishing cooperatives or big companies and small boats to small stores and/ or directly to buyers. In some cases, fishermen have a small shop in the port where they can sell their fish when they come back from fishing.

More than fifty fishermen who have, or have had dogs on board, or had information about past generations were contacted from various regions (Liguria, Tuscany, Latium, Campania, Calabria, Sicily, Molise, Marche, Emilia Romagna and Puglia). The author could not find any fisherman who currently have dogs on board in Adriatic central regions (Molise, Marche and Emilia Romagna). The informers were directly found and met in the ports or their contacts were passed by friends and colleagues or by private and public institutions. Thirty complete interviews were done, and in the other cases, it was possible to just have a short but often productive conversation.

Professional fishing is essentially a man's world and only three women were interviewed (from Lerici, Portovenere (Spezia - Liguria) and Marina di Carrara (Massa Carrara - Toscana)) (Figure 3). Almost all of the interviewees have a high school diploma. A few of them regret not having finished university. Only a few people lived and worked in a region that was not their native region. They usually remained in their hometown where the vessels were traditionally passed on from one generation to another.

Figure 3. Elena Ghersi and Wendy (Lerici 2020) (Photo by Franco D'Aiuto, husband of Elena).

Only four of them do not come from mariners and fishermen families, two fisherwomen[21], one of their husbands and a fisherman from Mazzara del Vallo (Trapani, Sicily). In this case, they started the activity when they were about twenty. More normally, they started learning the profession when they were twelve-thirteen years old. They went with their father or grandfather during the summer holidays. Sometimes it was considered an honour and fishing viewed as a magical world, sometimes only like a necessary apprenticeship for beautiful but hard work.

Almost all the informers stated that they love their work even if it is very hard and can be uncertain. They usually praised the contact with nature and the sense of freedom when they are offshore. They all only complained that the Italian government does not sufficiently protect the fishing industry. Also, their fathers and grandfathers loved their lives, even if it was harder than today.

[21] One fisherwoman from Lerici (Spezia, Liguria), one fisherwoman and her husband from Marina di Carrara (Massa Carrara, Tuscany), one fisherman from Mazzara del Vallo (Trapani, Sicily).

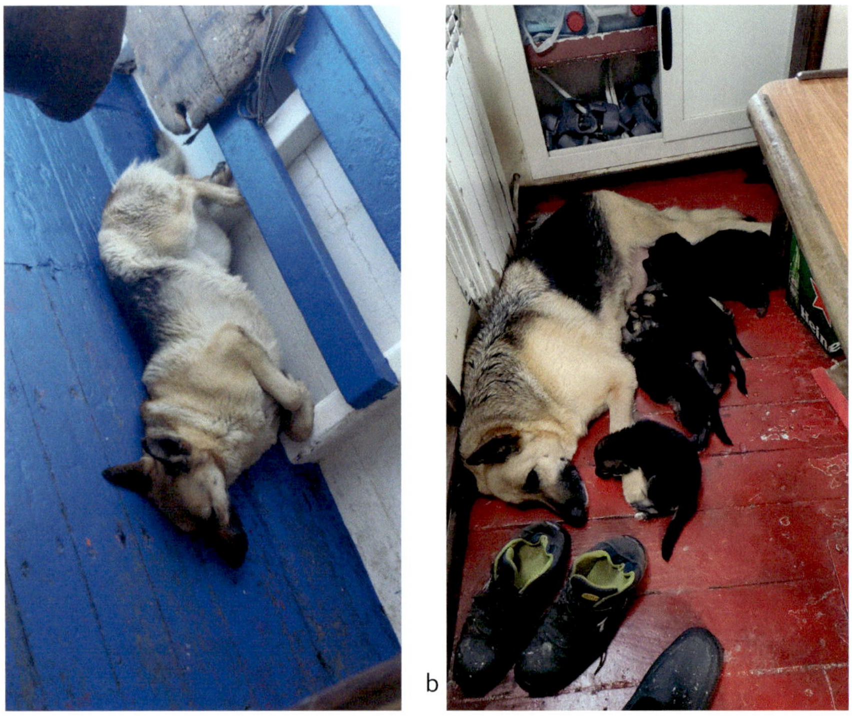


Figure 4. A) Max onboard (2020); B) Maia and her puppies onboard.
Fishing boat 'Lady Annamaria' of Amilcare Bronzino (2020)
(Photo by Amilcare Bronzino, Torre del Greco (Naples).

All the interviewees did not understand why boats are not allowed to have dogs on board. They said that dogs are not dangerous, do not touch the fish, do not carry parasites and they are constantly checked by their vets.

*3.2.1 Dogs' apprenticeship and tasks*

They all agreed upon the importance of getting the dog when it is a puppy (maximum 6 months old) because it can be hard to familiarise it with life at sea once it becomes an adult and it can also suffer seasickness. They can learn how to live on a fishing boat from an older dog or on their own.

Dogs have usually been part of their lives and their families' lives. So, in some cases it was also possible to have a few data about the dogs of their fathers and grandfathers but never of their great-grandfathers.

Many of them said that they love dogs and that vessel life without this extraordinary friend is not the same (Figure 4A-B). In the case of big vessels, many crews declared that the dogs' presence is pleasant, but it is not always easy to understand if they were sincere or they merely wanted to please their boss and captain. Dogs were also documented that 'belonged' to the captain and that were fed, looked after and loved by the rest of the crew. In one case, a mariner that was met at Roccella Ionica (Reggio Calabria) said that '*il cane di bordo è il mio migliore amico. Il cane è del comandante. Non vado d'accordo col comandante. Ma non me ne andrei mai senza portarmi via il cane*' (the dog is my best friend. The dog belongs to the captain. I do not get along with the captain. But I could never leave without taking the dog with me).

A few informers said the life of dogs could be full of pain and danger. Not all the crew members were nice to them and they often drowned offshore and also in the ports. The boats did not usually stop their work in order to save the dogs. Only one fisherman said that dogs are primarily found on big boats. All the others think that their presence does not depend on the dimensions of the vessels.

For past generations, dogs had a special task as essentially watchdogs. Nowadays they are usually watchdogs and companions or just companions (85%). Only in a few cases (15%), it was said that the dog is only for guarding like in the past and two fishermen added that the dog sleeps during the day and wakes up at dusk when the boat is in the port, so it can do its job.

But dogs can have additional and occasional tasks. They can bark to alert everybody that the boat is entering the port; or when they see big fish; or a school of fish; they can retrieve objects in the water and, in exceptional cases, save a man who falls in the water (but only water dogs or mixed-breed dogs with aquatic abilities).

A dog is also '*fortuna e speranza*' (good luck and hope). It is '*un dono di Dio*' (a gift from God). A fisherman from Torre Annunziata said that '*senza il cane non ho fortuna*' (without the dog I don't have good luck).

Dogs usually have a place on the deck where they can stay but they are often allowed to sleep in the cabin with the mariners. It seems that they have also been allowed to stay below deck in the past. Many fishermen who spend many days offshore said that the dog slept in the berth with them. When the boat returns ashore, dogs usually remain on board. Many fishermen said that they used to take the dog home with them before they got married. Fathers and grandfathers' dogs usually remained onboard.

*3.2.2 Dogs' breeds and provenance*

Many informers said that even though mixed-breed dogs were generally used by past generations, the *Lagotto Romagnolo* was frequently chosen for its aquatic capacities in the central northern Adriatic coasts (Emilia Romagna and Marche). Nowadays mixed-breed dogs are still the most numerous but various breeds are also documented: German Shepherds, Deutscher Boxers, Labradors, Pitbulls, Pomeranians and one Border Collie.

Only a few people think that a water dog is better, the majority of interviewees said that the breed is not important and there is no sex preference. What is important is that the dog is good, sweet, even-tempered and knows when to be aggressive and when to be quiet.

The dog is often a present, sometimes it is bought from a dog breeder and only in a few cases, it is taken from a rescue dog centre. For what concerns the past generations dogs were never bought, they were usually a gift or they were strays picked up on the docks.

*3.2.3 Dogs' names*

Many were called Argo (Odysseus' dog), Banco (Banquo), Birillo (Skittle), Bora (like the northern katabatic wind, because the dog was very lively), many called Lola, Maja, Pilù, Topazio (Topaz, for the hair colour), Zucchino (Courgette, because he was found abandoned in a case of courgettes). In most cases, dogs are not given traditional or non-traditional Italian names as it was in the past. So there are many foreign names like Jack, Jerry, Joseph, Ketty, more than one Lassie, many Max, many Rex (like the principal of the Austrian TV series 'Kommissar Rex'), Rocky, Rudy, and more than one Wendy.

*3.2.4 Dogs' food*

Dogs eat raw and cooked fish but also dry food. There is not a typical fish for dogs. They eat the damaged fish or what is abundant. But the fishermen said that dogs can be demanding and when possible they satisfy them. For example, some dogs love to eat crabs, slipper lobsters and prawns.

*3.2.5 Tales on dogs*

Sometimes the interviewees had special tales about their dogs, their courage and loyalty, their life on board and also on the importance that they can have for their masters. For example:

> '*Una notte del 1964 alcuni pescatori di posta che litigano sempre con i pescatori a strascico, salirono a bordo per uccidere il comandante che stava dormendo nella sua cuccetta mentre mio padre era al timone e l'equipaggio dormiva anche lui. Quando il cane si accorse che c'erano estranei a bordo immediatamente li attaccò e li mise in fuga*' (One night in 1964 a few gillnet fishermen (*pescatori di posta*) who always fight with bottom-trawling fishing people got on board to kill the captain who was sleeping in his berth while my father was in command (at the helm) (era al timone) and the crew were sleeping too. When the dog who was brave and strong realised that there were strangers onboard he immediately attacked them and they fled) (Torre del Greco, Naples).

> '*...in 1996, Cesare, un anziano pescatore con una gamba di legno, aveva un cane che sia chiamava Lassie che lui adorava e che lo riamava a sua volta. Un giorno il cane morì di vecchiaia. Il vecchio morì in mare una settimana dopo. Tutta la gente del porto ha pensato che si è suicidato perché la vita senza il suo migliore amico era insopportabile*' (...in 1966, Cesare, an elderly fisherman with a wooden leg, had a dog (Lassie) that he loved a lot who returned his love. One day the dog died of old age. The old man died at sea one week later. All the people of the port thought that he had committed suicide because life without his closest friend was unbearable) (Savona, Genoa).

> '*...una volta il mio cane ha attaccato un grongo che lo ha morso. Il pesce gli stava tagliando la lingua. Abbiamo dovuto usare un martello per staccarlo*' (...once my dog attacked a conger eel who bit it. The fish was hanging onto his tongue. We had to use a hammer to pry it off) (Mergellina, Naples).

> '*...il mio cane è buonissimo e dolcissimo. Ma la gente è cattiva e ha fatto un esposto alla polizia contro i cani a bordo. Ma il comandante del porto adora i cani. Ha difeso me e il mio cane e ha detto di non osare a provare a scacciare il mio cane o a fargli del male. Quando il*

*comandante del porto sale a bordo il mio cane gli fa un sacco di feste. Lo sa che è un suo amico e che lo difende'* (...my dog is very good and sweet. But people are nasty and complained against dogs being on board. But the harbour master loves dogs. He defended me and my dog and said that nobody should dare try to chase my dog away or hurt him. When the harbour master comes on board my dog makes a really big fuss of him. He knows that he is his friend and that he will defend him) (Torre Annunziata, Naples).

### 3.3 Written sources, photographic archives and online information

As it has already been mentioned, there are very few written sources and also the museums which are dedicated to fishing are very poor in information about dogs and fishing boats. The public and private institutions that the author contacted confirmed that this topic of research is absent in the history of the life of fishing boats.

Photographic archives sometimes have images that can provide useful information. Dorigo Vanzolini photographer of the Archivio fotografico del Centro Polivalente of Cattolica (CPT) did a valuable study. He went from house to house in order to find pictures and information about the fishing world of Cattolica and he managed to collect a huge amount of data. In the pictures of the CPC Archive the fishermen usually have their dog (or two dogs) in their arms with a tender attitude/behaviour which can be interpreted as a reflection of the close relationship that they had with them. The breeds are various, but often they are small and medium mixe dogs. Sometimes it is also possible to know the name of the dog as in the Figure 2.

#### *3.3.1 Websites*

On the internet, the presence of dogs is often taken for granted.

Dogs are usually mentioned as part of the life on board and considered as another member of the crew. For example, the news on the tragedy of the boat 'Francesco Padre' of the port of Molfetta (Bari, Puglia) always mentions the dog. On the night of the 3rd November 1994, the boat that was 20 miles from Budva (Montenegro) was damaged by an explosion that sank it. The captain, the crew (the machinist, two sailors and the head fisherman) and Leone, the 'loyal german shepherd dog' died (Figure 5). At that time, it was said that the boat was illegally carrying weapons and explosives. Only after twenty-five years could the families prove that the bow of the boat had been hit by numerous bullets and that it did not have explosives on board. It was the victim of an attack.[22]

In addition to the custom of dogs' being present on board, it is also possible to find some additional information. In the sections dedicated to pets on board of the website 'MTO nautica store' by Marco Scanu, the author writes that '... It has always been a common thing to see dogs on board fishing boats, large dogs sleeping on nets and eating raw fish. They were hardly newsworthy and no one asked the vet how to deal with them'.[23] And he adds that it was customary to take the dog when it was a puppy so that it was easy to teach it how to live onboard.[24]

In 2007 in his blog, David Dutto[25] wrote that he had met the dog *Spillo* (Pin) who lived on a boat used for swordfish fishing near Lipari island (Messina, Sicily). The dog is described as a very serious, busy dog. He was always actively sniffing the air, observing the sea and paying attention to all the smallest movements and sounds onboard.

Whenever swordfish were spotted and chased the dog would run up and down the deck. When they caught the fish he would stop where it would be lifted. He would bark without aggression to encourage and help his friends. When the big fish had been hoisted aboard Spillo would bite the fin and help to ensure its capture, then two more bites and he would return quietly to his place while the fishermen finished putting the fish in the hold. The fishing dog Spillo was respected by everyone on the boat, he had always been there and was part of the crew, eating with them, sleeping with them and fishing with them (Figure 6 A, B, C). The author concluded that Spillo reminded him of Jack, the dog of the tale '*Il cane che andava per mare*' by Stefano Malatesta (2000).[26]

Jack lived on Lipari island where he slept at Marina Corta. He had had a master but the wife did not want to have a puppy at home, so he was left under the bridge of Marina Corta which became his home. He knew the ferry timetable and he loved to move from one island to another even travelling as far as Naples and back. Everybody knew him and he was allowed to go on the ferries. He was not a '*randagio*' (stray dog) because he travelled for passion and not for necessity. Fishermen

22 https://www.open.online/2019/11/04/l-ustica-del-mare-25-anni-dalla-tragedia-del-francesco-padre-la-figlia-del-comandante-il-silenzio-li-ha-uccisi-una-seconda-volta/; https://www.youtube.com/watch?v=4eHi_3O0yuc

23 'È sempre stata cosa comune vedere i cani a bordo dei pescherecci, cani grandi e grossi che dormivano sulle reti e si cibavano di pesce crudo. Non facevano certo notizia e nessuno chiedeva al veterinario come fare'.

24 https://www.mtonauticastore.it/blog/animali_in_barca

25 https://blog.davidedutto.it

26 https://blog.davidedutto.it/?p=732

Figure 5. On the internet, the presence of dogs is usually mentioned as part of life on board and it is considered as a member of the crew. For example, the news on the tragedy of the boat 'Francesco Padre' of the port of Molfetta (Bari, Puglia) always mentions Leone, the 'loyal german shepherd dog' who died with the captain and the crew.
(Photo by Elvira Zaccagnino, Corriere del Mezzogiorno, 3 novembre 2019: 6)

Figure 6. A) Spillo on the bow of the Feluca observes the seawater; B) Spillo Participates in swordfish fishing; C) Spillo bites the fin of the poor swordfish that was lifted aboard (Lipari island, (Messina, Sicily) (Photo by Davide Dutto (https://blog.davidedutto.it).

fed him with fish. Once he disappeared for a long time and when he reappeared he was thin and had lost six-seven kilogrammes. The people called the vet who said that he was seriously ill but there was nothing to do to save him. One morning he dived from the dock, swam for fifty metres and then, looking back at the square, he let himself sink and died.

On the intenet, there are also various pictures of dogs on board and they confirm that there are no particular breeds that are currently used.[27]

*3.3.2 Zingarella e gli altri (Zingarella and the others)*

The short stories book 'Zingarella e gli altri' by Benedetta Trevisani (2004) is unique because it's the only text completely dedicated to the dogs of fishing boats, particularly off the coast of San Benedetto del Tronto (Ascoli Piceno, Marche) to Ravenna (Emilia Romagna) from the forties to the fifties. The author writes that all the fishing boats had one or more dogs on board. They were of different sizes and mostly mixed and unknown breeds. The dogs were different in their behaviours. They could be quiet and sweet but also impetuous and aggressive. It means that there was not a common ideal model dog. They lived in close contact with the crew who usually treated them well. But the dogs could also live a very unpleasant life with nasty masters.

The dogs were not allowed to go home with their masters, the boat was their home. But they could go ashore when the gangway was lowered. Sometimes the dogs drowned offshore but also in the ports.

*Tirompo* (I break you) was a black mixed shepherd dog who lived on the boat *Trionfale* in the sixties. He was energetic and spiteful. He fell in the water near Gibilterra where he presumably died. *Lola* was an unlucky dog who always was the victim of abuse by the crew at the end of the fifties. The captain could not bear that situation. So, he decided to abandon Lola in the port of Pescara but after twenty miles, he decided to go back to take *Lola*. He saw her on the dock in the same place where he had left her, she was waiting for the boat and started wagging her tail as she saw the boat coming back. *Zingarella* (little gipsy girl) was the dog of the fishing boat *Truentum* (the ancient name of Tronto). She was a shepherd dog and was so-called as she was born among the *zingari* (gipsies). The boat *Truentum* was from San Benedetto del Tronto (Ascoli Piceno, Marche) but it moved to the porto of Ravenna (Emilia Romagna) where it carried out trawling together with a second boat named *Pomello* (Pommel). On the 18th June1947, a mine got stranded in *Truentum*'s net and exploded. The crew of *Pomello* rescued the only two people of the boat who had survived and the remains of the dead mariners. Everybody was particularly sad for not finding Zingarella who was considered a member of the crew. After many days she reappeared at San Benedetto del Tronto. She was exhausted, had a broken paw and was looking for the port. Nobody knew how she survived and where she had come from.

**4 Conclusions**

The Italian Ministry of Health currently prohibits the access of pets to structures that are for food production (EC Regulation 852/2004 and the law 283/1962). Fishermen are considered to be part of food production. Consequently, they cannot have dogs with them. But the regulation is not clear and the current situation in Italy is that fishing boats cannot have dogs but in many ports they are welcome.

Also, new technologies have contributed to reducing the presence of dogs on board. In fact, boats are usually equipped with anti-theft devices and sophisticated systems to spot schools of fish or big fishes and to recover men and objects falling into the water which traditionally were the main tasks of dogs.

According to the informers, there wasn't one vessel which didn't have at least one dog on board until twenty-thirty years ago. At present, the tradition is still practiced in the Tyrrhenian ports whereas it is nearly absent in the Adriatic ones.

Dogs used to be indispensable. Nevertheless, the research on dogs and also on fishery and fishing boats never mentions them. Their role and the relationship they have or have had with fishermen has been constantly omitted. Therefore this research had to face a consistent lack of documentation. In order to find any possible data and clues many aspects were considered. Written sources, websites, photographic archives, museums, private and public institutions were taken into account and contacted. Interviews with fishermen or people involved in fishery were a crucial part of the research.

The data show that the tradition of the presence of dogs on fishing boats was well-established in Italy and that in the central Adriatic ports of Emilia Romagna and Marche it was stronger than in the other Italian regions.

Nowadays dogs are loved by fishermen who decide to have them despite the law in force. But they were often appreciated and loved by mariners and considered part of the crew also in the past. Their current main task is to be good companions. In a few cases, they are guards and companion dogs and they infrequently are only for guarding. Sometimes they can have additional tasks. Some dogs bark when they see schools of fish or big

[27] https://it.dreamstime.com/cani-sul-peschereccio-image128095933; https://it.dreamstime.com/fotografia-stockcane-sulla-barca-image89312608; https://www.pinterest.it/pin/807129564461901617/

fishes and others announce the entrance of the boat in the port.

Fishermen do not prefer any special breed and they have mixed-breed dogs as well as other breeds. In the past, mixed-breed dogs and various breeds could be used but the water dog *lagotto romagno* was often used for its aquatic capabilities by the fishermen of the Adriatic coast. The *lagotto* had presumably more specialised tasks than the other dogs, just as its Portuguese and Spanish cousins (the *Cão de agua* and the *Perro de agua*) had. It could help to recover the nets, the fish that had escaped and to haul the ropes.

Life on board is generally considered not 'natural' for dogs by the informers and by various sources. So, it has always been said that it is necessary to take the puppies before they are sixth months old in order to let them adapt to the fishing life.

The interviews, the current and the historical pictures, the tales and the websites reveal and confirm a very close dog/fisherman relationship. The emotional perspective seems to have had a preponderant position. Nowadays and in the past the fishing world has had its own identity which was almost independent of mainland life. In this parallel world, dogs and humans have certainly always shared all the difficulties as happy situations. Dogs were not only a crucial instrument for the vessels but in most cases, they were often considered as partners in life.

The use of dogs on fishing boats is also interesting because it can be considered a specific aspect of their presence in water and marine economies, a topic that has been largely ignored by scientific studies. The research on the presence of dogs on fishing boats allow us to bridge a gap in the documentation and at the same time to consider and analyse the importance of dogs in water environments.

## Acknowledgements

Many people helped me in my research. First of all, I want to remember Sebastiano Tusa who passed me many contacts, among them Don Ciccio of Mazzara del Vallo (the fisherman who discovered the 'Satiro danzante' statue) that I thank for his helpfulness and because he allowed me to speak with many people in Sicily. I thank Vienna Eleuteri for her important suggestions. I thank all the fishermen and fisherwomen who dedicated their time to my research, many times speaking by radio in the middle of the sea during their fishing expeditions. I thank Maria Luisa Stoppioni, Benedetta Trevisani, Dorigo Vanzolini and Davide Dutto for their exquisite helpfulness. I thank all the port authorities who helped me to contact fishermen who are both working and retired. Last but not least a special thanks to all the current and past dogs who have spent their lives with fishermen.

## References

Brandão, R. 1923. *Os pescadores.* Paris-Lisboa: Aillaud e Bertrand. Edicão de Victor Viçoso e Luis Manuel Gaspar. 2014. Lisboa: Relógio d'Água.

Coppinger, R. 2014. Fishing dogs. *A guide to the History, Talents, and Training of the Baildale, the Flounderhounder, the Angler Dog, and Sundry Other Breeds of Aquatic Dogs (canis piscatorius).* New York: Skyhorse Publishing.

Coppinger, R. and L. Coppinger 2001. *Dogs. A startling New Understanding of Canine Origin, Behaviour and Evolution.* New York: Scribner.

Defoe, D. 1719. *The Life and Strange Surprizing Adventures of Robinson Crusoe, of York, Mariner: who lived Eight and Twenty Years, all alone in an-inhabited Inland on the Coast of America, near the Mouth of the Great River of Oronoque.* Sidney: London: Printed by W. Taylorù.

Din, M. 2013. Pratically Emotional: An Anthropological Investigation of the relationship between Humans and Dogs. Unpublished PhD dissertation, Swarthmore College, Department of Sociology and Anthropology.

Dinoi, N. 2017. Stremato e solo in mare un cagnolino salvato a due miglia dalla costa. *IlMattino.it* 13 Ottobre 2017.

Fogle, B. 2000. *The New Encyclopedia of the Dog.* Verona: Dorling Kindersley Books.

Gable, T.D., S.K. Windels and J.G. Bruggink 2017. Estimating biomass of berries consumed by gray wolves. *Wildlife Society Bulletin* 41: 1–3.

Haraway, D. 2003. *The Companion Species Manifesto: Dogs, People, and Significant Otherness.* Chicago: Prickly Paradigm Press.

Heptner, V.G. and N.P. Naumov 1998. *Mammals of the Soviet Union. Sirenia and Carnivora* Vol.2, Part 1a: 164–270. New Delhi: Smithsonian Institution Libraries and The National Science Foundation Washington, D.C.

Homkes, A.T, T.D. Gable, S.K. Windels and J.K Bump 2020. Berry Important? Wolf Provisions Pups with Berries in Northern Minnesota. *Wildlife Society Bulletin* 44: 1–3.

Lugli, F. 2016. Mongolian Nomads and their Dogs, in S. Biagetti and F. Lugli (eds) *The Intangible Elements of Culture in Ethnoarchaeological Research*: 125–140. Springer.

Malatesta, S. 2000. *Il cane che andava per mare e altri eccentrici siciliani.* Vicenza: Neri Pozza Editore.

MacDougall, S. 2016. Enough about Ethnography: An Interview with Tim Ingold. Member Voices, *Fieldsights*, April 5. https://culanth.org/fieldsights/enough-about-ethnography-an-interview-with-tim-ingold.

Rick, T.C., P.L. Walker, L.M. Willis, A.C. Noah, J.M. Erlandson, R.L. Vellanoweth, T.J. Braje and D.J. Kennet 2008. Dogs, Humans and island ecosystems: the distribution, antiquity and ecology of domestic dogs (*Canis familiaris*) on California's Channel Islands, USA. *Holocene* 18,7: 1077–1087.

Shipman, P. 2015. *The Invaders. How Humans and Their Dogs Drove Neanderthalensis to Extinction.* Cambridge, MA: Harvard University Press.

Sorrentino, C., Z. Di Giuseppe and F. Manzi 2000. *Materiale osteologico animale,* in S. Bruni (ed.) *Le navi antiche di Pisa. A un anno dall'inizio delle ricerche*: 329–342. Firenze: Polistampa.

Trevisani, B. 2004. *Zingarella e gli altri.* San Benedetto del Tronto: Edizioni Circolo dei Sambenedettesi.

Zaccagnino, E. 2019. Francesco Padre, un'altra Ustica. Il peschereccio colpito 25 anni fa. Resta il mistero dei responsabili. *Corriere del Mezzogiorno,* 3 novembre 2019: 6.

Zhilin M., S.N. Savchenko, E.A. Nikulina, U. Schmölcke, S. Hartz and Thomas Terberger 2014. Eleven bone arrowheads and a dog coprolite - the Mesolithic site of Beregovaya 2, Urals region (Russia). *Quartär* 61: 165–187d.

### Web sources

http://museumofmaritimepets.com/home.html (viewed 31 May 2017).

https://abcnews.go.com/US/navy-finds-missing-dog-fell-off-fishing-boat/story?id=37701943 (viewed 16 March 2016).

https://www.akc.org/expert-advice/lifestyle/breeds-that-are-born-to-swim/ (viewed 31 May 2017).

https://www.archeostorie.it/museo-per-le-navi-di-pisa/ (viewed 19 August 2019).

https://www.atlasobscura.com/articles/little-known-history-seafaring-pets-dogs-cats-chickens-war-exploration (viewed 18 January 2019).

https://blog.davidedutto.it/?p=732 (viewed 18 october 2018)

https://it.dreamstime.com/cani-sul-peschereccio-image128095933 (viewed January 2019)

https://it.dreamstime.com/fotografia-stock-cane-sulla-barca-image89312608 (Viewed January 2019)

https://en.wikipedia.org/wiki/Water_dog (viewed on September 2018).

https://eur-lex.europa.eu (Viewed September 2018).

https://www.ilmattino.it/pelo_e_contropelo/stremato_e_solo_in_mare_un_cagnolino_salvato_a_due_miglia_dalla_costa-3301534.html (viewed 14 October 2017).

https://nywolf.org/the-wolves-of-voyageurs-national-park/ (viewed 6 December 2019).

htpps://www.mtonauticastore.it/blog/animali_in_barca (viewed 04 November 2019)

https://www.navalhistory.org/2018/04/27/sea-dogs (viewed 27 April 2018)

https://www.open.online/2019/11/04/l-ustica-del-mare-25-anni-dalla-tragedia-del-francesco-padre-la-figlia-del-comandante-il-silenzio-li-ha-uccisi-una-seconda-volta/ (viewed 4 November 2019)

https://www.pinterest.it/pin/807129564461901617/ (viewed January 2020)

https://www.youtube.com/watch?v=4eHi_3O0yuc (viewed 28 November 2019)

https://www.voyageurswolfproject.org/ (viewed 6 December 2019).

# 3.4 Dogs, Nomads and Hunters in Southern Siberia

Francesca Lugli[1] and Galina B. Sychenko[1]

[1]Associazione Italiana di Etnoarcheologia, Via Principe Umberto 41, 00185 Roma, Italy.
luglifrance@gmail.com, phurdok@gmail.com
Corresponding author: Francesca Lugli, luglifrance@gmail.com

**Abstract**

The article discusses some of the results of the research project supported by the Ministry of Foreign Affairs and International Cooperation – Italy MFA from 2013 to 2017, and focuses on the observation, documentation, and analysis of the relationships between man and dog in many pastoralist and nomadic cultures living in different ecosystems (steppe, *taiga*, mountains, mountain *tundra*, rivers and so on) in Southern Siberia. All of these societies are characterised by the presence of dogs which are always considered crucial and often indispensable. Different kinds of relationships between man and dog were observed and rich materials (photos, videos, audio recordings and drawings) were collected. On the basis of the complex methodology, proposed in Lugli (2016) and improved during the project, the authors analyse different aspects which characterise the existence of dogs and their relashionships with humans in different ecosystems as well as in economic and cultural contexts.

**Keywords:** dogs, southern Siberia, pastoralism, hunting, methodology.

## 1 Introduction

In 2013, the Italian Association for Ethnoarchaeology with sponsorship from Ministry of Foreign Affairs and International Cooperation – Italy MFA started the project 'Siberian nomads and their dogs'. From 2013 to 2017 a collaboration took place with Novosibirsk State Conservatory, Institute of Philology (SB RAS, Novosibirsk), Institute for Humanities and Kyzyl College of Arts (RT) and Institute of Altaistics (RA). The research project focused on the observation, documentation, and analysis of the relationships between man and dog in many pastoralist and nomadic cultures living in different ecosystems (steppe, *taiga*, mountains, mountain *tundra*, rivers and so on) in Southern Siberia.

Since 2013 five expeditions have been conducted by the authors with various participants: 2013 - Republic of Tyva (RT), Erzin district (A.Kh. Kan-ool, A.A. Khertek, A.S. Khertek); 2014 - Republic of Altai, RA (E.L. Tiron, A.V. Zolotukhina); 2015 - Kemerovo, K (L.N. Arbachakova, A.N. Arbachakov, K.A. Sagalaev); 2016 - RA (V.Ja. Sumachakova); 2017 - RT (A.Kh. Kan-ool, E.L. Tiron, Ch.T. Achity) (Figure 1).

During the missions, the authors managed to observe several different ethnic groups in southern Tuva (2013, Erzin Tuvas), central Altai (2014, Altai-Kizhi), southern Kemerovo (2015, Shors), northern Altai (2016, Chalkans) and north-eastern Tuva (2017, 'Taiga' Tuva-Tozhu, 'River' Tuva-Tozhu). All of them belong to the Turkic linguistic group, but from a cultural point of view, they represent different economic types.

Four main kinds of pastoralisms were documented: steppe nomadism, transhumant pastoralism, mixed pastoralism economy and reindeer breeding.

The Erzin region in southern Tuva is crossed by the Tesiin Gol river and is characterised by cold desert, dry steppes, desert steppes and meadow steppes (Golubyatnikov *et al.* 2020: 5). It is part of the ancient Uvsnuur lake basin that in 2003 was included as a world heritage site, representing one of the twelve protected areas that are the major biomes of eastern Eurasia.[1] The climate is dry continental, with cold winters.

It is a region of intense cultural exchange between Tuva and Mongolia and the Tore-Khol lake between the Russian and Mongolian borders is a crucial element of the local economy, such that Mongolian and Tuvinian herders are allowed to move along the border without particular restrictions.

Tuvinian steppe pastoralism went through important changes during the last century which modified its traditional social and economic organisation, especially concerning the property of livestock and the role of the households, but it maintained the main traditional crucial features of Central Asian steppe culture: the use of the traditional felt tent (*ger* in Mongolian, *yurta* in Russian taken from Turkic languages and *өг/ög* in Tuvinian), the sheep, goats, bovines, camels and horses breeding, the seasonal roaming, the use and reuse of the same *stoianka* (herders' camp) during the cold months.

[1] https://whc.unesco.org/en/list/769/

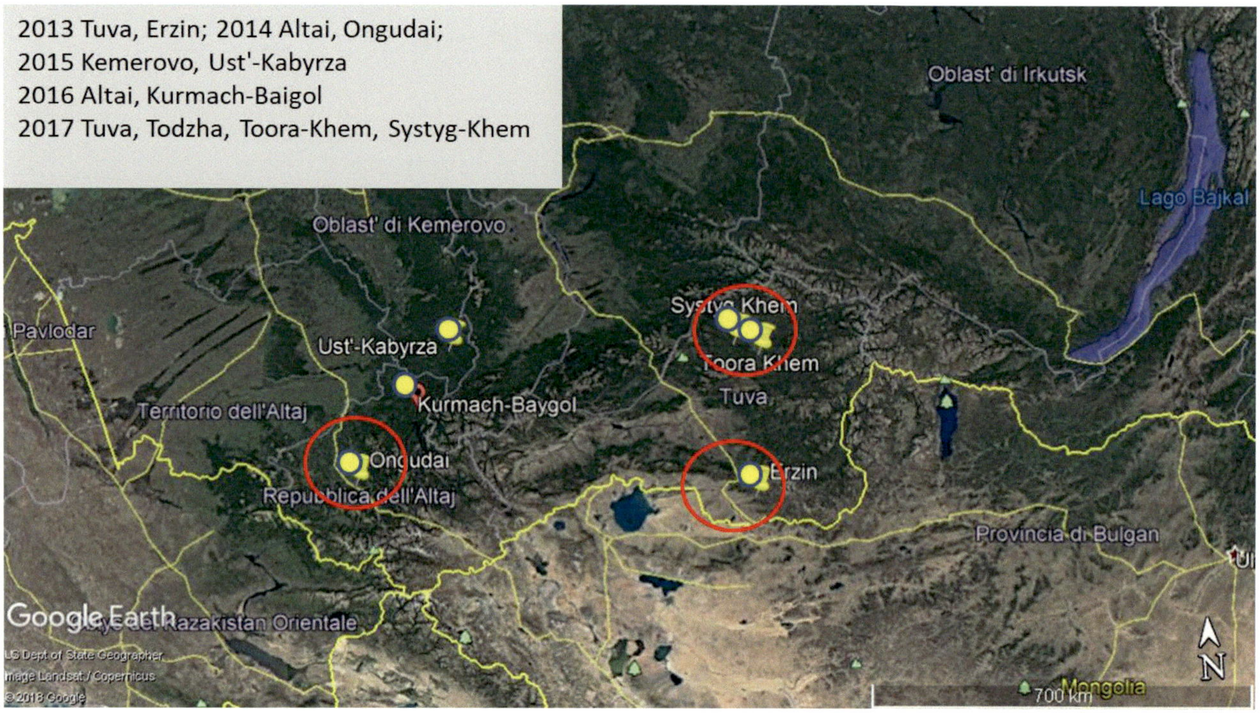


Figure 1. Southern Siberia: locations of the field research (from Google Earth).

Thanks to good pasture, water and salt availability that is abundant near the lake and in the mountains, nomads move only 3–5 times a year and the camps are usually around 18–25 km from each other.

Winter camps are not located close to the river or the lake, the traditional felt tent is often replaced by a wooden structure with two sloping roofs. The camps are not in the mountains where they used to be in the past, they are bigger than the traditional ones, with between two and six households living there, and may have covered and protected wooden structures for ovines and bovines.

Herders move to the lake in the warm season where they meet their Mongolian neighbours.

The complex mountain system of Altai is a crucial region for the study of dog domestication. In fact, the remains of a 33,000-year-old dog skull with evidence of domestication were found in the Razboinichya Cave in the western area of the Republic of Altai (Ovodov *et al.* 2011).

From a geographical perspective, these mountains are a natural interruption between the Siberian *taiga* and the steppe and arid lands of Mongolia and Kazakhstan. The Russian Altai mountains which can be more than 4000 metres high are extremely verdant with thousands of lakes and irregular steep rocks. The climate is severely continental with long cold winters and short warm summers.

In the Altai, herders mainly breed sheep and bovines, they spend the warm season in the mountains and the cold months in the valleys, the opposite model compared to the steppe model. In the Ongudai district (central Altai) the authors have observed the summer camps start from around 1200 metres above sea level and are essentially at an altitude between 1880 and 2100 metres. Pastoralism here has changed during the last century. In the past, families moved depending on the seasons. Nowadays, herders move with their livestock but have their own houses in the villages where their wives and children live all year long. Sometimes all the members of a family also move to the mountains. The livestock transfer usually takes no more than a couple of days from one camp to the following one.

The camps are usually of one family and there is only one housing structure. It is more and more uncommon to find the beautiful traditional hexagonal wooden *ail* that was covered with bark. The current lodges are usually quadrangular, sometimes pentagonal, and often covered with modern commercial materials.

The Todzha district (north-eastern Tuva) is situated in a wide basin surrounded by mountains, rich in dense forests, wide valleys and lakes. The area is characterised by a mixed economy that includes the coexistence and interaction of: reindeer breeders who lead a nomadic life; herders - essentially cattle and horse breeders - who usually have fixed camps for the various seasons; sedentary herders who live close to the villages and move the livestock to the pasture when necessary.

In the herders' camps, more than one family can live together. The current houses are usually wooden quadrangular structures.

Reindeer herders live nomadically in the northeastern *taiga* and mountain *tundra* where they continuously move in search of good pasture for the reindeers. Sporadically, they come to the villages or their vicinity to obtain necessities, and to exchange or sell reindeer skins, antlers or meat. They live in the traditional cone-shaped tents made of branches and bark and are currently covered by tarpaulin.

This socio-economic model is consistently supported and complemented by the activity of hunting, mainly for fur-bearing animals, which is carried out both by professional hunters, non-professional hunters and by shepherds who may be hunters as well and be away from their camps for months at a time, obviously leaving some family members behind to look after the livestock.

The coexistence in a single territory of different activities, characterised by varying degrees of mobility and the use of dogs, is one of the most interesting features of this region which, thanks to its isolation, presents remarkable lines of continuity with a more or less distant past. Furthermore, what makes the area even more interesting are the contacts that shepherds, especially reindeer herders, have had over time with the Tsaatan, the reindeer herders who inhabit northern Mongolia in the Khovsgol region, so much so that, in some cases, it has been reported to us that the dogs have been proven to be from the Tsaatan.

The village of Kurmach-Baygol in the Turochak district of the Republic of Altai is located in the north-east of the Republic of Altai, on the border with the Altai *krai*, Kemerovo *oblast'* and the Republic of Khakassia. It appears to be a typical settlement of the northern Altai, which differs significantly from its central and southern regions. This part of southern Siberia geographically includes the northern foothills of Altai, Kuznetzkii Alatau and the upper reaches of the River Abakan. It is characterised by a *taiga* landscape and snowy winters that contribute to the development of a special type of economy. Small-numbered Turkic-speaking groups - among them, the Chalkans and Shors we visited - have developed a rather complex type of economy consisting of hunting, fishing, gathering, cattle- and horse-breeding, farming, as well as a variety of ancient crafts (Sychenko 2021).

The Chalkans represent a typical model of the culture of the *taiga* zone. Hunting has been their basic economic activity since ancient times. Scientists assign hunting to have a paramount role in the traditional economy (Belgibaev 2001). Horse-breeding was also known amongst them even before colonisation by the Russian empire. The importance of the horse in traditional Chalkan culture is regarded as evidence of the nomadic past of their remote ancestors (Belgibaev 2001). Scientists believe that the Chalkans adopted cattle-breeding from the Russian settlers no earlier than the middle of the XIX century. It had become predominant among them in the XX century, due to the abundance of grass in the summer and feeding hay in the winter. Such husbandry is not subject to migrations; moreover, simple herding has no special needs: in many cases the cattle move freely in search of food, fairly close to settlements, with impassable *taiga* and rivers serving as a natural barrier (Belgibaev 2001).

Hunting in a *taiga* area includes several periods of different lengths. In the spring and summer periods, hunting is occasional and for some species it is forbidden. September is the month for hunting large ungulates (*maral*, deer). Most important is the one-month late autumn period of hunting with dogs for fur-bearing animals: squirrel, kolinsky, ermine, and sable in particular, as well as a longer period of winter hunting for different animals without dogs. The winter hunt for bears, which were fat and beautifully skinned, had a special and sacred significance. It was considered particularly prestigious among hunters. Our informants mentioned numerous small and medium-sized animals and birds, among which roe deer, fox, hare, grouse, hazel-grouse and others were considered most valuable.

In mountain areas of Altai, the hunting of wild goats, rams and other ungulates was important. In the steppes of the Altai and Tuva, cattle breeders hunted marmots, whose meat is considered a delicacy and has medicinal value.

Hunting wolves depends on a specific situation. In steppe and mountainous areas wolves are predators who prey on livestock, and are therefore to be exterminated. In addition their hides are used (although they are not considered very valuable), and various body parts have applications in folk medicine. In the *taiga* zone wolves are considerably fewer in number and do not cause much trouble. In general, the type of hunting is determined by the ecosystem, the type of economy and the presence of certain commercial animals.

Hunting is practised by professional hunters, amateurs and herders as well. Small animals are even hunted by children. During the Soviet era, many hunters were officially employed, such as Shor hunter S.I. Kurtegeshev from Kabyrza village, Kemerovo *oblast'*, or many informants from Systyg-Khem village in Todzha district. But nowadays, people hunt mainly for themselves, for meat, and also to sell their furs on the

| | Nomadism | Semi-nomadism | Transhumance | Settled way of life |
|---|---|---|---|---|
| **Pastoralists** | '*Taiga*' Tuva-Tozhu | Erzin Tuvas | Altai-Kizhi | Chalkans, Shors, 'River' Tuva-Tozhu |
| **Hunters** | '*Taiga*' Tuva-Tozhu | | Chalkans, Shors, 'River' Tuva-Tozhu | |

Table 1. Types of economic activities of observed ethnic groups.

free market. Therefore, many of our informants were professional but not official hunters. Hunting is strictly protected and regulated by governmental legislation.[2] It is typical for hunting communities to use dogs in certain types of hunting, especially for fur-bearing animals as well as bears. In rural settlements, 'other', 'regular', 'village' dogs perform guard duties, i.e. there is a specialisation in which hunting dogs constitute an elite.

In Table 1 we have attempted to present a typology of the hunter-gatherer communities we surveyed, depending on the predominant type of economic activity.[3] It should be stressed that economies are always complex and include many additional options to make better use of the resources of the areas they occupy (Table 1). All of these societies are characterised by the presence of dogs which are often considered indispensable. On the other hand, each observed case is characterised by a different kind of relationship between man and dog (Lugli and Sychenko 2018: 65–66).

## 2 Methods of the research

The research was based on the interdisciplinary approach previously elaborated in Mongolia (Lugli 2016) where the ethnoarchaeological mission 'The camps of Mongolian nomads - an ethnoarchaeological perspective' was conducted by G. Capitini and F. Lugli since 2005 (see Capitini and Lugli in the volume)[4] and where the study of the importance of dogs in Mongolian nomadism has become a crucial point of the research since 2012. In the missions, Ethnoarchaeology was used as a research strategy which analyses archaeological and historical problems from a diachronic and interdisciplinary perspective.

The presence of dogs must be analysed based on the economies, cultures and ecosystems that are taken into account. In Southern Siberia, different pastoralisms in the Tuva and Altai regions and different types of hunting in Tuva, Altai and Kemerovo were observed (Table 1). Consequently, the methodology previously elaborated in Mongolia was modified and enriched depending on the different contexts that were observed.

The authors believe that an interdisciplinary perspective is indispensable to address the importance of dogs in human history. Therefore, ethnoarchaeological, anthropological, ethnomusicological, folklore and ethnographic methodology were used.

Nomads' camps and also hunters' houses in their villages were georeferenced, observed, with a particular focus on dogs' spatial presence from an ethnoarchaeological perspective. The sound signals that the owners use to communicate with their dogs were recorded. The daily life of dogs was documented in herders' camps and hunters' villages, when possible.

The interview was a crucial part of the research and it aimed to investigate tangible and intangible aspects of the topic with special attention to differences and regularities in the various cultural, economic and geographical situations which were observed.

After preliminary geographical information, the first part of the interview was dedicated to the general information about the dog or dog's master and his/her family (names, surnames, patronymics, clan, birthplaces, ages, education etc.). Immediately after, economy, mobility, type of pastoralism or hunting, use and spatial organisation of the camp, the *yurta*/shelter/house were considered.

The second part of the interview was especially focused on dogs. The interviewed person was always free to talk about his/her dog but fixed questions were always posed:

> How many dogs do you have?
>
> For each dog: Name? Birthplace? Breed? Age? Sex? Colour? Size? How old was it when you took it?

[2] 'Legislation of the Russian Federation in the sphere of protection and use of the wildlife and its natural environment is based upon the provisions of the Constitution of the Russian Federation, the federal laws on environmental protection and it consists of this Federal law, laws and other normative legal acts of the Russian Federation adopted in relation thereto, and also by laws and other normative legal acts of the subjects of the Russian Federation on protection and use of the wildlife' (http://extwprlegs1.fao.org/docs/pdf/rus22375E.pdf).

[3] Table 1 was elaborated on the basis of models proposed in Fabietti and Remotti (2001).

[4] The mission was promoted by AIE with the sponsorship of the Ministry of Foreign Affairs and International Cooperation – Italy MFA.

Where and how did you take/buy the dog? Does a tradition or a special acquisition procedure exist?[5]

Are there special breeds in the area? Do you know if there were different breeds in the past?

Do you prefer female or male dogs? Why?

Do you prefer a particular colour? Why?

What are the choice criteria to select a puppy?

Do you have special traditions about puppies?

How do you choose the name?

What are the qualities of a good dog?

What are the dog's tasks (guarding, hunting, herd control, sledge dog, companion, other)?

Can you describe a dog's day and a day with your dog?

How is a dog trained?

What words do you teach the dog?

What signals do you teach the dog?

Why and how do you punish a dog?

Does the dog enter the tent/house/shelter?

Is the dog free or tied in the camp?

Where does the dog sleep?

What, where and when does the dog eat?

Is the dog free to mate whenever he/she wants?

What do you do with the puppies?

What do you do if a dog has health problems?

Can you tell me a special event that you experienced with your dog?

Do you know any local traditions about dogs?

Do you know any legends, epics or tales about dogs?

Do you have funeral rites and funeral formulas for a dog's death?

Is a dog a humans' best friend?

What is the relationship between dogs and wolves?

What about dog/wolf hybrids?

What do you think about dogs?

To conclude, is life possible without a dog? And, why yes, or why not?

Dogs were photographed with or without their owners, and when it was possible, dog's hair samples were taken to analyse their mitochondrial data (see the article by Daria Sanna *et al.* in the volume) to provide better knowledge and understanding of the dog's presence and its evolutionary process in Central Asia.

## 3 Materials and Results

The dogs' daily life, their tasks and the relationship they have with humans can considerably vary, depending on different economies, cultural contexts and ecosystems. Traditions are sometimes altered, modified or replaced above all by television and/or social media, depending on the internet connection availability. So, it is often possible to observe recent and unusual breeds and behaviour among herders and hunters.

### *3.1 Criteria of choice of puppies*

The choice of puppies is an important consideration for both herders and hunters. All respondents named several selection criteria - at least two or three. All the criteria they mentioned can be divided into different categories. The first category is the qualities of the puppy that will determine its success in future use. The second category is the puppy's gender, the third is the puppy's appearance and the fourth is the breed.

#### *3.1.1 Qualities of the puppy*

Our informants were almost unanimous in stating that the choice of dog was very important for its future use. Many said that dogs differ in their qualities according to their use. If a family's economic activity is limited to a single sphere - for example, traditional nomadic herding - then the choice of a puppy is conditioned by this. In other cases - such as the Kurmach-Baigol or Kabyrza hunters - hunting dogs are considered to be special, they do not serve as guard dogs in the village: there are other dogs for that purpose. Consequently, it can be assumed that the puppy's qualitative characteristics will be judged from these perspectives.

In fact, we found that on the one hand for both nomads and hunters of different types, the qualities of the puppy that are evaluated when choosing are roughly the same. These are courage, strength, intellect, and, in particular, having a low pain threshold. For example, a common selection method to establish a dog's pain threshold is to take the puppy by the tail (sometimes - withers) and lift it up. If the puppy remains silent, then this is considered a good sign.

On the other hand, there are also differences. Thus, Tuvan steppe nomads believe that a good dog should have a calm temperament, resting during the day and guarding at night. Therefore, when selecting a puppy, its character is assessed (Figure 2).

Vitality and willpower are very important for hunting dogs. They are tested in the following way: puppies, whose eyes have not yet opened, are carried a distance away from their mother, and whoever returns is considered to be the best. This puppy is seen to be smarter, faster and more tenacious. A similar method is to take the pups away from the kennel, and whoever crawls back first is considered better (Ongudai).

[5] E.g. for Mongolian nomads a dog cannot be bought, it cannot be a present, it must be simply taken (Lugli 2016).

Figure 2. Two puppies near the *yurta*. Erzin, 2013 (Photo by G. Sychenko).

A bark can be a valid sign to understand if a puppy will be a good hunter. If it barks at people it will be able to take on big wild animals like bears and boar, the opposite means that it will have a sweeter temperament and that it will hunt small animals (M.V. Irbilig Tuva 2017). Aleksei Dyganchi (Tuva 2017) said that he immediately understood that his dog was a hunter when he listened to it barking.

We haven't noticed any significant difference, except that strength and a well-balanced character are more important for nomads, but for hunters - speed and intelligence are the priority.

*3.1.2 Gender of the puppy*

When choosing a puppy, both shepherds and hunters look primarily at the mother. Already at the stage of selection, the qualities of the female are important. Some bitches that have produced good offspring have become famous.

Normally, neither nomads nor hunters neutered or spayed their dogs, so their gendered characteristics have relevance in use and preference. Thus, nomads and shepherds in the southern Tuva and central Altai tend to keep mainly male dogs near their *yurta* or shepherd's huts. The phenomenon is so consistent that of the twenty-five dogs we observed and sampled in the Erzin area, only five are female. In fact, owners argue that managing the female dogs' cycle is complicated because of the oestrus of the bitch and therefore the dog's removal from the camp for more or less prolonged periods, the arrival of male dogs fighting each other for her, ultimately, also pregnancy and the birth of puppies.

In the central Altai, this tendency manifests itself not only among shepherds, but also among hunters. We were told that the whole litter is never saved, only the best puppies are taken and females are eliminated first, although some are kept for good breeding and later become famous. 'Bitches are not liked here, they try to get rid of them. Only if she produces good puppies, which are taken away by hunters, they are valued, but often only the first litter is good, and then not. Unwanted puppies are killed or left in the village near the school, children pick them up' (A.A., Ongudai).

Quite the opposite is true for hunters. The prevailing opinion among the hunting communities (Kurmach-Baigol, Kabyrza, Todzha) was a preference for females as a hunting companion dog. In reality, both sexes are used, but it is universally believed that when hunting, bitches 'work better', are less tired and don't run away.[6] Also, females begin to hunt earlier than males. The male dogs lose interest in the work faster and can run away

[6] This opinion is shared by hunters from many different regions. For example, many Italian hunters believe that females are better for hunting because they are more serious, disciplined and sometimes also more intelligent (personal communication).

Figure 3. A hunter with his dogs. Todzha 2017 (Photo by F. Lugli).

for considerable distances. The opinion that bitches are more agile was expressed also in the Ongudai district.

In the communities we surveyed, if shepherds prefer male dogs, it is only because there is no shortage of good bitches in the surrounding villages and settlements who can produce healthy, hard-working offspring. The hunters themselves live in such settlements.

*3.1.3 Appearance of the puppy*

The puppy's external qualities include colour as well as individual physical characteristics. Regarding colour, most informants consider it unimportant, although some owners prefer dogs of a certain colour. In Erzin, a special emphasis is noted on black matted dogs with red spots and 'four eyes', i.e. light spots above the eyes. These traits are typical of Mongolian and Tuvinian sheepdogs which are rarely seen nowadays. If a puppy has even a purely outward resemblance to a traditional sheepdog, it is more highly valued.

It has been noted amongst Tozhu deer breeders and hunters that the most desirable colour for a dog is white, as this makes it less visible in winter (Figure 3).

It should be noted that northern Altai and Shor hunters do not use dogs during winter hunting, which involves deep snow cover and makes it difficult to move the dogs. Among the Tozhu people, a dog follows its master and reindeer all year-round, and in winter, which lasts several months, a light-coloured coat plays a crucial role. It is interesting to note in relation to this, one of the informants from Erzin had a white male dog, taken from the Tozhu of Kungurtug (eastern Tuva).

Even though one of the shepherds said that it is better if the dog does not look like a wolf in colour, most of the dogs we observed in the steppe area were of greyish and reddish colours.

Other external criteria are strong wide paws, a long bent tail, a scarlet tongue, 'antennae' whiskers, a bump on the skull, and transverse folds on the palate. Most of these criteria were listed by hunters.

Wide paws are indispensable for moving in the snow, so this criterion was particularly noted by Tozhu reindeer herders and hunters. The long bent tail has also been largely noted by hunters. It is used as a kind of 'rudder', allowing the hunting dog to manoeuvre better when chasing the beast. A scarlet tongue mentioned only once, demonstrates the healthiness of a dog, whereas long antenna whiskers are associated with a heightened sense of smell, which is essential for a hunting dog. 'A good *laika* doesn't need any training, it picks up the trail straight away. The main thing is a good sense of smell, then it will find it without a trace. Most of us hunt by trail, that's when the first snow falls' (P.D. Pustogachev, KB).

Figure 4. A Tuvan dog near the entrance of a *yurta*. Erzin 2013 (Photo by F. Lugli).

Figure 5. An old dog playing with children. Erzin 2013 (Photo by F. Lugli).

The presence of a bump on the skull is widely regarded as indicative of a dog's intelligence. 'When there are long whisker-like hairs near the ears - called 'antennae' - they are good at sniffing. When there is a bump on the head - 'cerebellum' - the dog has good coordination and is smart. The teeth are also checked - they should be healthy and the claws should be good' (Yu.P. Pustogachev, KB).

Another custom, which is quite common, seems strange at first glance. 'You open puppy's mouth, look - there should be dark stripes on the palate, the more the better' (S.I. Kurtegeshev, K). Many hunters insisted that the number of the folds of the palate should be odd and a minimum of 7 (9 or 11 is better). No explanation was given for this, the general motive being that this was the way older people did it. It can be assumed that this criterion, which was repeated very frequently, is related to the size of the dog's cranium and, therefore, higher intelligence, too.

*3.1.4 Breed of the puppy*

In all of the communities studied, there are notions of pre-existing local dog breeds. We particularly noted this in southern and north-eastern Tuva. In Erzin, local shepherd dogs that were very similar to the Mongolian *bankhar* were repeatedly mentioned. Some informants believe that Tuvinian dogs differed in their smaller size. One of the most characteristic features of Tuvinian sheepdogs was a strong build, long shaggy hair and a characteristic colour - black or black with red spots, including light dots above the eyes (Figures 4 and 5). In Tuva itself and Moscow, special breeders try to revive this breed[7], some of which can still be found in Tuva. According to our informants, such puppies are very rare and are of great value.

In Todzha, many informants reported a local *laika* breed, which differs from the modern *laika* in being smaller, with shorter but broader paws, thick fur and a distinctive light (white or light-grey) 'collar'.

Actually, local dog breeds in Southern Siberia have long been supplanted by the western-Siberian *laika*, well adapted to various conditions and quite versatile in their qualities.[8] It is currently the predominant canine breed (Figure 6). It is absolutely prevalent in all hunting communities as it perfectly suits the needs of hunters. They carefully select these dogs, and can travel especially to quite distant places to get them and will pay considerable sums of money for them. *Laikas* make up the 'elite' in hunting societies and do not mix with so-called 'village dogs', which are tied up on a lead outside the house and guard the dwelling.[9] Village guard dogs are most often mongrels, hybrids of different breeds.

Currently, some other breeds are also imported from the outside, such as the Caucasian shepherd dog, the East European shepherd dog, and the Siberian husky as well as decorative breeds for home keeping. Quite good qualities were noted in mestizos with hounds.

However, some dog breeds have not proved to be satisfactory. In the Erzin district we met a wolfhound called Masha, but she didn't do a good job as a guard. The Caucasian shepherd dog is considered overly aggressive, which is not a desirable characteristic of a sheepdog. In general, we haven't noticed any desire at all for breeds with pronounced fighting qualities neither among nomads, nor hunters.

[7] See for example http://mongun.ru/ (viewed 2 April 2021).

[8] Very informative article about this breed which gives precise ideas about its story and qualities can be found in Wikipedia.

[9] *Laikas*, even in the village, are generally not tethered.

Figure 6. *Laikas* of the village Kurmach-Baigol. 2016 (Photo by G. Sychenko).

### *3.2 Acquisition of the puppy*

It is common for both nomads and hunters to have several dogs of different ages to ensure continuity of their service. We are repeatedly told that it is better to take puppies from different parents, and by no means siblings. This is apparently motivated by concern for healthy offspring.

Nomadic pastoralists, who prefer male dogs, usually take their pups from relatives and acquaintances who live in nearby villages where they regularly come to stock up on food, clothing and fuel, and where their children often attend school. They usually just 'take' the puppies, sometimes for a symbolic fee. A puppy can also be 'given', but such a gift is not considered valuable. The custom of stealing a puppy is not practised by adults, but theft is a fairly popular method of getting a puppy by children. Naturally, this is not a criminal act, but a custom that has been going on since olden times. It is usually practised by children who 'steal' the chosen puppy and bring it to their parents or grandparents. There are very few cases when a stray dog associates itself with the *yurta*. In the cases we have been told about, this adaptation has been successful. Nowadays, puppies can be brought in from more distant places (for example from the capital of the region), of different breeds, but so far this trend is too exotic and does not always pay off.

Hunters, we have observed, are constantly on the lookout for opportunities to acquire a good hunting dog. They generally use the same methods, but their geographical acquisition is wider than that of nomads. Thus, they are willing to travel to another region, purchase a dog from kennels and even pay decent money for it. This may have to do with economics, as a good hunting season provides good financial support for the hunter's family.

### *3.3 Names of dogs*

The choice of a name for a dog is, on the one hand, very important; on the other hand, it often seems frivolous and even ironic. Dogs are given a wide variety of names - linguistically and semantically. We have grouped the names encountered into several tables (Tables 2–5) according to expeditions, distributing them according to the linguistic principle (in the native language or Russian) and providing a semantic translation. The colour highlights correspond to the different semantic categories analysed below.

The pastoral nomads of Erzin have Tuvinian and Russian names for their dogs with approximately equal frequency. The semantic categories in the two groups partially overlap. They are: a natural object or place (light blue cell); colour in general (green cells) or some body parts (light green cells); the name of a bird (blue cells) or an animal (dark pink cells). The name Laika is polysemous. This category is associated with a specific type of dog (peach cell). The right side of the table also highlights typical Russian nicknames whose meaning is unlikely to be fully understood by dog owners (lilac cells) (Table 2).

The colour highlights of Tables 2–5 correspond to the different semantic categories:
a natural object or place (light blue); a name of a bird (blue) or an animal (dark pink); a body part or some physical bodily characteristic (grey); some qualities of a dog (light yellow); colour in general (green) or some body parts (light green); specific type of dog (peach); Russian nicknames (lilac).

Table 2. Names of dogs by Tuvinian steppe pastoralists.

| Authentic Tuvinian names | | Loan names | |
|---|---|---|---|
| ***Name*** | ***Translation or significance*** | ***Name*** | ***Translation or significance*** |
| Taiga (2) | Forest / High mountain | Laika (3) | Popular dog-breed; derives from *лаять* 'to bark'; name of famous 'cosmic dog' |
| Ezir (2) | Eagle | Sharik (2) | Ball (dimin.) |
| Khartyga | Hawk / Falcon | Tuzik | Ace (dimin.) |
| Arzylan | Lion | Chaika | Seagull |
| As | Ermine | Belyi | White |
| Shangyr | Youthful (?) | Ryzhyi | Red-haired |
| Mel'der (2) | Bay | Rex | King |
| Kara | Black | Tarzan | Hero by E.R. Burroughs |
| Akkol | White arm / hand (paw) | Mishka | Masculine name (dimin. of Mikhail, or *Misha* 'Bear') |
| Moinakh | Neck (dimin.; usually white) | Masha | Feminine name (dimin. of Maria) |
| Tabak / Tavak | Plate, dish (in Tuv.), or Tobacco | | |

Table 3. Names of dogs by Altaian pastoralists.

| Authentic Altaian names | | Loan names | |
|---|---|---|---|
| ***Name*** | ***Translation or significance*** | ***Name*** | ***Translation or significance*** |
| Taigyl | Special kind of a grey-hound, *borzaia* (rus.); Enormous dog (myth.) | Mukhtar | Dog-hero of a Soviet movie and serial |
| Argut | Name of river in southern Altai | Marsik | Dimin. of Mars |
| Ak kol | White arm / hand (paw) | | |
| Moinakh (2) | Neck (dimin.; usually white) | | |
| Nai | Probably from Mong. 'friendship, friend' | | |

Table 4. Names of dogs by Tuva-Tozhu pastoralists and hunters.

| Authentic Tuvinian names | | Loan names | |
|---|---|---|---|
| ***Name*** | ***Translation or significance*** | ***Name*** | ***Translation or significance*** |
| Taiga (2) | Forest / High mountain | Tungus (2) | Ethnic name or particular kind of *laika* |
| Koigunak | Hare | Sharik | Ball (dimin.) |
| Küske | Mouse | Chernyi | Black |
| Kaldarak (3) | Diminutive of *kaldar* - variety of bay | Belka | Derives from Rus. *белая* 'white' or squirrel ; name of famous 'cosmic dog' |
| Saryg (4), Sagygbai | Yellow / light bay | Ukho | Ear |
| Shokarak (2) | Diminutive of *shokar* - spotted | Pal'ma | Palm-tree, or bladed weapon (?) |
| Kuuran | Light grey | Argo | Argo [ship] |
| Ak kol (4) | White arm / hand (paw) | Evro | Euro |
| Ak tösh | White breast | | |
| Moinakh (4) | Neck (dimin., usually white) | | |
| Chürek | Heart | | |
| Köstüg (5) | With four eyes; particular kind of a dog | | |
| Düktügür / Düktüg | Hairy | | |
| Taskarak | Bald | | |
| Cherlikpen | Wild | | |
| Borbak | Small ball (= Rus. *Sharik*) | | |
| N'azyi | From Mong. 'humble, small, nondescript' (?) | | |

Table 5. Names of dogs by Altaian hunters

| Authentic Altaian names | | Loan names | |
|---|---|---|---|
| *Name* | *Translation or significance* | *Name* | *Translation or significance* |
| Altai (2) | Name of the region, sacred area | Belyi | White |
| Taiga | Forest / High mountain | Lapka | Paw (dimin.) |
| | | Okhotnik | Hunter |
| | | Smelyi | Courageous |
| | | Dama (2) | Lady, Dame |
| | | Strelka | Arrow (dimin.); name of famous 'cosmic dog' |
| | | Pal'ma | Palm-tree, or bladed weapon (?) |
| | | Rex | King |
| | | Pirat | Pirate |
| | | Mukhtar | Dog-hero of a Soviet movie and serial |
| | | Tata | No exact meaning |
| | | Sh'ita | No exact meaning |

In pastoralists and hunters of the central Altai we find more national names of the already met categories (but not all of them). A noteworthy feature is the presence of mythological majestic names (Table 3). A great variety of names has been recorded for the Tuva-Tozhu people. Interestingly, they are dominated by national dog names and new categories appear: a body part or some physical bodily characteristic (grey cells) and some qualities of a dog (light yellow cell). Some exotic names (Argo, Evro) are present, too (Table 4).

The hunters of the northern Altai, on the other hand, have predominantly Russian names for their dogs. There are some fairly original female names (Dama, Tata, Sh'ita) (Table 5). Finally, we have recorded only two Russian dog's names (Naida and Sharik) among Shorian hunters.

### *3.4 Tasks*

Evidently, hunters and herders use dogs for different puposes and there are significant differences among the various pastoralist and hunting cultures that we could observe and that are essential to recognise.

The main task of dogs among the herders is usually to guard and alarm against predators and thieves. Therefore, dogs are essentially watchdogs and they are not requested to fight, their assignment is to bark in order to raise an alarm. But a few informers said that strong brave adult dogs can confront the wolves.

All the informers agree that their dogs are crucial because there are many wolves which can be extremely dangerous for humans and their animals. But in the southern Tuvinian steppe, dogs are always considered indispensable whereas in the eastern mountains they are deemed useful and crucial but not a hundred per cent indispensable.

#### *3.4.1 Tasks. Tuvinian Steppe pastoralism*

In the Tuvinian steppe – as in the mountains – dogs are never used as shepherd dogs. To guide the animals, dogs should grow up with them to feel part of the livestock. This tradition was never documented among our informers. Dogs usually remain in the camp but sometimes they go with the herders and/or with the animals. It is not a constant rule and it depends on the relationship they have with their master and on their own temperament.

Dogs are generally requested to be quiet and not aggressive with their family and many people agree that 'A good dog is quiet during the day and does its work during the night' and also 'It must frighten potential predators and bad-intentioned people and attack only when indispensable'. They are generally free and consequently very self-controlled. Their freedom lets them be part of the family and sweet companions of children (Figure 5).

But they are never allowed to enter the tent. Only in one case, a herder tied up Masha, a poor *volkodav* dog[10], with a very short rope. She was also subject to cruel treatment to make her mad and aggressive. Her ears had been cut. But this is not a local tradition. Masha

[10] The *volkodav* dog is a Central Asian breed that can guard livestock, be watchdogs and also fighter dogs.

Figure 7. Sharik 'smiling'. Erzin 2013 (Photo by F. Lugli).

was specifically bought to face wolves. The owner said that a female was chosen because males are too aggressive. He added that Masha did not do her duty well because she only barked at people who arrived on foot or riding horses, but she did not bark at cars.

Sometimes dogs can have different and facultative tasks which depend on their own 'personality' and 'talent'. For example, in southern Tuva, the dog Sharik was a good friend of horses and loved to stay with them also far away from the camp, without the herder. His master said that nobody had taught him to stay with the horses but he had grown up with them and for that reason, Sharik considered himself to be also part of the horses' family. He thought that it was a good feature and Sharik could also 'smile' if requested (Figure 7).

In the same family, Tusik was good at communicating with the herders and the other dogs. He used different barks depending on the situation and it was deemed a particular talent of Tusik. But this family loved its dogs in a very peculiar and specific way that was not common to the other families. It was the only family that neutered the dogs to make them quieter.

Many nomads told us that dogs often venture far from the camps to hunt little animals on their own because they have a special or peculiar temperament and love hunting.

The most important things that dogs are requested to carry out are: to be a quiet and good friend, especially for the children and to bark to warn that a danger is arriving.

*3.4.2 Tasks. Todzha and Altai mountains pastoralism*

In Todzha dogs also are requested to be good watchdogs and not aggressive but unlike in the steppe, they are almost always tied up during the day. Sometimes they can go with their masters but it is more frequent to find them chained in the *stoianka* than with the herders. During the night, they can be free to do their duty. Certainly, they can be good friends of their family but it seems not to be a crucial task of theirs as they do not live in close contact with their masters as dogs do in steppe pastoralism.

Herders sometimes use dogs to hunt in winter to take the fur and to improve their diet. But not all dogs are hunting dogs (see following chapter). Generally dogs do not have other tasks and even if they have particular talent and personality this is not considered important by their masters and they do not have the chance to choose what they like to do.

In the Ongudai region of Altai, dogs are 'multi-tasking' and live in very close contact with their masters whom they are always with. Dogs are universally recognised to be crucial and indispensable loyal friends. They are requested to guard against predators - especially wolves - to control and guide the animals even when the herder is not there and to hunt. The dogs are also members of the family and can be important for children too.

They do not have a particular specialisation and everyone of them can and must be able to do anything. Dogs bark when there are wolves but usually do not fight. Only when they are many dogs and the wolf is a solitary one.

Herders said that it is very difficult to manage the sheep without dogs and that they allow their master to be absent and not to stay with the livestock all day long. For example N.T. Topitonov said that 'In summer, early in the morning I send the sheep to the pasture and they come back in the hottest hours. Without dogs, I should be with the sheep all day long, but the dogs stay with them in my place'. 'Dogs are particularly indispensable when the herder and his animals move from one camp to another one depending on the season' (A.A., Ongudai).

A dog must be sweet and never aggressive, extremely obedient and understanding and they are never neutered. They are free when they are in the camps but they can be tied up when they are in the village. They are never admitted into the house.[11]

Reindeer breeders consider dogs to be indispensable but not to guard against wolves because they are not regarded as dangerous for the animals as they are in the steppe. A family usually has 3–4 dogs which are always nearby with the reindeers with whom they sleep on a fur that herders put to keep them warm. They are watchdogs and shepherd dogs as well. They warn when a predator such as a wolf or a bear arrives and they bark in a different way depending on the danger. The reindeer-breeders being at the same time good hunters also use a dog as a helper in hunting. The same attitude is typical for transhumance kind of hunting, both traditional and professional.

*3.4.3 Tasks. Hunting*

For hunters, dogs are useful but not always indispensable even if they are usually described as man's best friend. They are used when the snow is not too high and hunters usually go in the *taiga* with 2–3 dogs. Hunting dogs can have different specialisations (Tuva 2013).

A hunting dog must be brave and must not be afraid of fighting the sable (M.V. Irbilig). It must not bite and damage the prey and must not fidget when its master starts skinning the killed animal (Altai; Tuva). It must be quiet waiting to receive the meat from its master (Tuva 2013).

All the informers said that a good hunting dog must use various howls to communicate with its master. A specific task was documented in Ongudai where our informant Arman Anatolievic said that 'A good dog can find fish, spawning grounds, comes running with a special bark'. That is very important as it concerns the role that dogs have in aquatic contexts which has always been neglected by scientific studies (see Lugli in this volume).

***3.5 Training***

Both in Tuvinian steppe and mountains, dogs don't have special training and all the herders say that 'they know on their own what to do'. It is not necessary to teach them how to behave in the camp and they usually learn when they are puppies from adult dogs.

In the mountains of Altai, a puppy grows up with the animals and learns how to control and guide them when it is around six months old. 'When they are puppies it is like a game, it is important to teach them how to maintain the sheep all together without leaving anyone behind' and also 'A puppy must be trained not to chase sheep and not to be aggressive with them' (N.T. Topitonov). Many informers agree that 'If a puppy is intelligent it will quickly learn. If it is stupid it will never learn'.

All the herders say that dogs should not accept food from strangers, above all when they are absent. But dogs are not often trained for that and in our experience, it was possible to feed herders' dogs even without their master.

All the informers agree that training is indispensable for hunting because even if dogs' instinct is appropriate they need to learn to help their master, to follow and/or drive out the animals and last but not least, not to eat or damage the prey. Only a few hunters think that dogs do not need specific training (Altai 2016). But generally, hunters say that an 'apprenticeship' is necessary and that the best way to teach a puppy is having an adult dog as a model to imitate. There is not general agreement on the age to start training a puppy. Six months, at least 7/8 months old for some hunters and after the first year for others. If its parents are good hunters it will be a good hunter too and it will learn quickly (Tuva 2017).

It is also possible to understand if it will be courageous by shooting a sable at its paw to force it to fight. After

[11] During our visit in 2014 a particular law had been enacted whereby all dogs, except those accompanying shepherds in their work, had to be kept tied up. But, we were told that it was not part of the local tradition.

that the dog will be a splendid hunter or it will be afraid for all its life.

It is important to find the footprints of the animals (especially the sable) and to show them to the puppy (Tuva 2017). In Altai one of the hunters said that it is important to let a puppy lick the blood of a squirrel to teach it to hunt this animal.

Many hunters think that female dogs are better because males are lazy. But they indifferently use females and males and there is not a different training depending on the sex of the dog.

Hunters often say that it is important to respect the nature of the dog. Sometimes it learns rapidly, sometimes it needs a longer time. For example, one of the informers said that his last dog did not understand what it was requested to do for the first three years, but it eventually became an excellent dog.

Hunting dogs are never tied up (only when they are in the village if necessary) because they must know what they may and may not do.

### *3.6 Food*

All the herders say that dogs eat the leftovers of their masters (bones, meat, milk scraps and others) but sometimes a special food with an addition of flour can be prepared for them, both in Tuva and Altai. Industrial dry food is often used for puppies to let them grow up strong. Bones are considered crucial following weaning (Figure 8). All the bones are given even if sometimes they can wound the throat.

But there is not a general rule about dogs' food among the herders. For example, it is possible to document that 'dogs can eat everything' and also that 'it is important to give good food to the dog to let it be an affectionate friend who offers protection. Otherwise it is worthless to have dogs'. Or 'dogs are fed three times a day. It is better to feed them constantly. If the dogs are hungry, they walk away and run wild and even die' but also 'dogs must not eat in the evening. In the night they must be hungry to be aggressive'.

Herders often do not give water to their dogs even when they are tied up during the day. They are supposed to find it on their own, or they can drink the frost on the grass in the morning or the evening. For that reason, we always provided water for the dogs, and frequently we were able to quench the thirst of deprived dogs.

Reindeer herders prepare the food for their dogs in specific pots. They give them reindeer bones. Dogs must eat well because they work very hard. Hunters say that it is crucial that dogs eat properly, well and enough in order to let them be strong enough for hunting and not to induce them to eat the prey. Even if the hunters also say that dogs must know very well that they are not allowed to eat the killed or the wounded animal. In the *taiga*, the hunters give their leftovers and the remains of the killed animals which are usually cooked and mixed with wheat, flour and other ingredients. The hunters who are also fishers, when the hunting season is 'closed', also give fresh fish to their dogs. In Kurmach-Baygol the hunters explained that they prepare special dry fish which can be conserved for a long time for their dogs that they can also eat in winter.

### *3.7 Care and health*

Herders and hunters usually agree that dogs rarely get sick and that they know on their own which are the curative plants.

Dogs are currently vaccinated everywhere because it has been obligatory since the Soviet times, but there are different attitudes towards dog sickness.

Normally dogs are never gelded but in a few cases, it was possible to document castrated dogs in the Tuvinian steppe where B.A. Mandap, the owner of four dogs (Sharik, Tuzik, Kara and Tabak), said that it is necessary to make the dogs less aggressive.[12]

All the herders that we met said that traditionally dogs are not nursed at all when they get sick. As in Mongolia when a dog has health problems, it should not be cared for because that can bring bad luck to its family. In fact, the dog takes the disease in place of its owners (Lugli 2016) and if it has been cared for, somebody of the family will die. Some herders said that they were sorry for that tradition but that it is better to respect it. A few herders say that they think that it is a superstition and they care for their dogs and do call the veterinarian when it is necessary.

In the Altai mountains, herders say that dogs know how to take care of themselves. They usually do not nurse them but they do it if it is necessary and call the veterinarian if a dog has important health problems.

The tradition that prohibits dogs' care as in the Tuvianian and Mongolian steppe culture is not widely spread but it exists and a few people reported the same custom. If a dog is mortally injured it must be killed to not let it suffer, usually, it is suffocated.

The hunters from Tuva and Altai (2016) also say that dogs rarely get sick and that they know how to take care of themselves. Some hunters from various regions said that their dog had been bitten by a snake and that it

12 In Mongolian pastoralism dogs are not gelded.

Figure 8. Tuzik eats a bone. Erzin 2013 (Photo by F. Lugli).

was cured because it knows how to manage that. They nurse dogs when necessary. For example, in one case injections of penicillin were done and a dog was nursed when it had been attacked by a boar.

### *3.8 Funeral of a dog*

In all observed communities, it is believed that when death approaches, the old dog senses this and takes itself away from the people and disappears. Sometimes old dogs are killed with a shotgun or hung from a tree limb. However, when it comes to a 'good dog',[13] it is customary for both nomads and hunters to perform something resembling a burial ritual. Such rituals have first been found among the Mongols (Lugli 2016). In southern Tuva the ritual of burying the 'honourable' dog goes as follows. His tail is cut off, laid on his side, something like a pillow is put under his head and a piece of the sheep *kurd'uk* (rump) fat is put in his mouth. They are not buried in the ground but left in a secluded place on the surface of the ground. When saying goodbye, they say farewell words. Cutting off the tail, according to traditional beliefs, is connected to the Buddhist idea of reincarnation: the dog has a chance to be reborn as a human in the next life.

Describing the 'honourable death', another of the informants in central Altai believes that there is no single form of ritual, everyone has their own ritual. He personally saw a funeral performed by his grandmother's older sister. She took off the skin as it is a very powerful remedy for sciatica. The dog was put on a sledge and taken towards the cemetery, where a place was found. They threw stones in an old *kurgan*, put the dog on it, put a piece of bark on it and then threw stones on it again. And then a whole ritual was performed.

The hunters of the northern Altai and Shoria buried a hunting dog that died of old age or as a result of a fight with an animal (wolf, bear, wild boar) like a human being. They would take it to the forest or the *taiga*, find a secluded spot under a tree or bush, put it on its side and cover it with branches or bark to prevent predators from getting at it. They did not put anything in their mouths. When saying goodbye, the dog was called by name, his merits were listed, he was thanked, regretted and farewells were said.

In Todzha it was also said that not all dogs are buried, but only the best ones. It should be placed in a good place, laid on its side and covered with stones so that birds won't find it. It was customary for hunters to put bread or breadcrumbs in their mouths and for cattlemen a piece of fat. Here is an example of saying goodbye to a dog:

[13] That is, a dog playing a significant role in the owner's life, having special relationships, etc.

Эки ыдым чораан
Ам канчаар?
Чоруур чериӊге чоруп чор, - дээш.
Алгап-йөрээп чор мен, - дээш.
Дыштан! - дэпкаар.
My dog was good
What to do?
Go there where you need to go, - says.
I will glorify and remember you with good wishes, - says.
Have a rest! - have said.[14]

**4 Discussion. Relationships with dogs**

Everywhere traditional customs regulate dogs' lives and the relationships they have with their masters their whole life up until their death. But they can considerably vary depending on the different regions.

In the southern Tuvinian area of Erzin dogs usually live a free life in the camps with their masters who consider them crucial for their tasks against predators but also as good companions and loyal friends. Dogs are used to staying in the camps and they are not requested to move with the livestock. In summer they usually lay down in the shadow of the *yurta* and in winter they stay close to it to take advantage of the warmth coming from inside. Traditionally, adult dogs are not admitted inside the *yurta*, only puppies can sometimes enter, but in very few cases dogs were documented in the *yurta* for a short time.

Three situations are traditionally considered prohibited for the dogs in the camps. It is strictly forbidden to jump on the covering of the *yurta* and nomads kill the dogs who break this rule. Dogs can also be killed when howling too much. A bitch, as well as a whole litter could be killed if she had mated with a wolf. But except for these exceptional situations, a dog does not have particular restrictions and duties and above all, they are requested to be respectful and to obey when necessary. Nomads generally do not teach them many specific orders.

Dogs live in close contact with their masters and their family and often they spend a long time with children, especially when they are young. Nomads think that dogs have specific and exceptional features. They can perceive earthquakes but also inauspicious events and start howling to warn that something is happening. Sometimes they emit strange noises and that is because they can see or feel things that humans cannot detect. Dogs can be respected and considered close to a supernatural world (see Sychenko in this volume).

Dogs have a special place in the life of nomads and the complex traditions that articulate their life are a mirror of their importance. Dogs are usually respected. Nomads are the masters but dogs are not their slaves. They help their family to manage the dangers from possible predators (also humans) and nomads feed them in return. It is a mutual relationship. They have their place in their world, a different place that maybe will be better in the next life.

Dogs usually live their lives freely and are allowed to pursue their personal talents and nature which are esteemed by nomads as long as they are not counter-productive for the family or the livestock. On the whole it would appear that steppe nomads are usually well-balanced and live a good life.[15]

Sometimes they can be abused and suffer, depending on the families, but it is not a constant and that is a characteristic of the dichotomic relationships between dogs and humans almost all over the world.

In the region of Ongudai (Altai) dogs are also considered crucial by the herders but they are not only watchdogs like in the steppe, they also control and guide the livestock, they are good and assiduous companions for their masters and their families and sometimes they can also hunt.

The life of dogs is not so strictly regulated by the traditions as in the steppe, for example, there area variety of funeral rites as well as various beliefs about dogs. Thus, if it jumps on a hay barn it means that there will be good weather; if it lies down on its back scratching its head, it means that somebody will die soon; it is not good if a dog howls in a camp, etc. A dog should not jump on the roof of the *ail*, it is a bad thing and it could be killed for that but this is not as strictly forbidden as it is in the Tuvinian steppe.

Like in the steppe, they are not allowed to enter the house or the shelter, it is traditionally forbidden and they 'must learn that when they are puppies'. Dogs are always free, but not so free as to move too far from their masters or the camp. In 2014, a new law that imposed that all dogs had to be tied in the villages was respected, but not agreed upon by the herders.

The dogs of Ondugai have plenty of duties and are not allowed to live their particular inclinations if they are not being useful to their master and his family, even if there is no danger involved. Dogs must be the 'shadow' of their masters who usually claim that they love them not only for being useful but above all for being good and loyal friends, even if some families can be unkind

[14] Performed by R.T.-o. Dupchun (Todzha district, Een-sug). The text was transcribed and translated in Russian by A.Kh. Kan-ool.

[15] See also Capitini and Lugli in the volume for dogs in the steppe pastoralism.

to their dogs. Herders are the masters and dogs are their assistants who must obey like soldiers. So, the relationship between dogs and herders in this region looks more complex than in the steppe but at the same time it is simpler: the herder commands and the dog executes.

In the north-eastern Tuvinian mountains the semisedentary herders who live along the rivers certainly consider dogs to be crucial even if not a hundred per cent indispensable. They are guard dogs who are requested to bark if a danger moves close to the camp, they are never shepherd-dogs and do not have special training. Sometimes they can help their master in hunting. Here the dog's life is not traditionally regulated and there are no special customs and beliefs about it. During the day dogs are always tied up in the camp and sometimes can be free during the night, depending on the families. Dogs do not have special tasks or commands that they must learn. They are regarded as simple instruments to help the herders in their work. The dogs' talent and nature are not particularly considered and respected. Sometimes dogs play with children or have a place in the family's life but this is not the norm.

Todzha reindeer herders mostly combine reindeer husbandry with hunting. They move with the reindeer and their dogs accompany them throughout the year. For the winter season, the dog's special qualities, such as wide paws that allow movement in the snow, are important. According to informants with reindeer husbandry experience, wolves are not a source of great danger to reindeer. In this sense, we have not recorded anything similar to the nomads' claims about the crucial role of the dog for *taiga* reindeer husbandry, although a dog always warns of the presence of predators. The function of protection against strangers, thieves, etc. is also irrelevant in this area, due to their absence in the highland areas of the Sayan Mountains, where the population is extremely sparse.

The dog is primarily a companion, as well as a hunting tool, going hand in hand with reindeer husbandry. A dog in the *taiga* has a great deal of freedom and is not kept on a tether. It is treated as a member of the nomadic unit, which needs to be fed like any member of the family, including reindeer bones which are leftover after a human meal, in order for it to work well. The dog is trained, but the owners rely more on its innate qualities and intelligence. In contrast to the nomads of the steppe, there have been cases where reindeer breeders allow the dog to mate with a wolf. These offspring are treated ambivalently and the litter is not destroyed, though it is said that such pups are less docile. It seems that this model of interaction demonstrates a true symbiosis of man, dog and deer in the high mountain *taiga*.

Hunters in different parts of Southern Siberia show a very similar type of relationship with a dog. A dog as a companion is important to them, although they may be in the *taiga* without a dog. This is especially true during winter when there is deep snow. However, for hunters, the working qualities of a hunting dog are of great importance, chief among which are intelligence and tirelessness. A dog shelter made of twigs has been built for dogs near *taiga* huts, and some hunters may allow a dog to be inside their dwelling in case of severe frost. Hunters always have several dogs of different ages with them, because their training is also done with the help of a more experienced dog. A good hunting dog is highly valued, kept until old age and not killed. After death, which is often the result of a collision with a wild animal, the dog is buried with honour, calling it 'a faithful friend, a true companion'. Among hunters, stories of hunting incidents, including those involving a dog, are popular; in some cases, the dog has saved the life of its owner.

In the hunting community, it is not considered to be a bad sign if a dog jumps on the roof of a hunting hut, but the howling of a dog is considered to be a bad omen here as well. However, we have not come across any references to killing a dog for such reasons. The mating of a dog and a wolf is also not considered unacceptable.

In addition to hunting dogs, there are yard dogs in villages that guard modern houses by sitting on chains. Many informants said that these are different dogs, and that not every dog can hunt. A hunting dog is usually free in the countryside as well, and if it is, it is tied up so that it is not stolen.

It can be assumed that the hunting dog is seen as an equal partner of the hunter, who seeks not to dominate it, but to develop cooperation that often develops into a particularly close trusting relationship.

**5 Conclusions**

(1) The relationship between a human and a dog in Southern Siberia in its complexity had not yet been studied in the ethnographic literature. Therefore, an elaborated approach which could serve as a model of this kind of research is of great importance. Good results could also be achieved using a crosscultural perspective which allows to observe both the similiarities and diversity of different types of society (see authors' model). We hope that this kind of research can have relevance and importance from an historical perspective.

(2) As our research shows, a dog is an inseparable part of the life for both pastoralist and hunter societies. The fundamental difference between them is that for

pastoralism a dog is indispensable whereas for hunters it seems to be more crucial but non indispensable. Reindeer-breeders, who are at the same time hunters, have a particular attitude towards a dog which should be studied more.

(3) The 'Human - dog - wolf' relationship is an important indicator which distinguishes the observed cultures. It can be seen as an absolute enemy, like in southern Tuva and Mongolia, but is not so rigid in central Altai. In reindeer-breeders / hunting societies this relationship demonstrates the proximity between wolf and dog, which allows their hybridisation.

(4) An important aspect which in our opinion should be better studied is the gender one. As we noticed, the true relationships with a dog in all observed localities are almost exclusively the male's privilege. Females are very often rather indifferent, they may not even know the name of dogs, and they cannot answer many details about them. Children, in contrary, are closer to dogs, especially with puppies, with whom they very often play and spend their time.

### References


Belgibaev, E.A. 2001. *Traditsionnaya material'naya cul'tura chelkantsev basseyna reki Leb'ed' (vtoraya polovina XIX - XX veka).* Бельгибаев Е.А. *2001. Традиционная материальная культура челканцев бассейна реки Лебедь (вторая половина XIX - XX века).* Candidate of History diss. Omsk State University.

Fabietti, U. and F. Remotti (eds) 2001. *Dizionario di antropologia*. 5 edizione. Bologna: Zanichelli.

Golubyatnikov, L.L., I.N. Kurganova and V.O. Lopes de Gerenyu 2020. Estimation of C-CO 2 balance of natural steppe ecosystems: Khakassia and Tuva (Eastern Siberia, Russia) case studies. *IOP Conference Series Earth and Environmental Science* 606(1): 012013

Lugli, F. 2016. Mongolian Nomads and their Dogs, in S. Biagetti and F. Lugli (eds) *Intangible Elements of Ethnoarchaeological Research*: 125–140. Springer.

Lugli, F. and G. Sychenko 2018. Dogs, nomads and hunters in Southern Siberia, in I. Fiore and F. Lugli (eds), *Dogs, Past and Present: an Interdisciplinary Perspective*, Annali dell'Università di Ferrara. Sezione: Museologia Scientifica e Naturalistica 14: 65-66. Ferrara: Università degli Studi di Ferrara.

Ovodov, N.D., S.J. Crockford, Y.V. Kuzmin, T.F.G. Higham, G.W.L. Hodgins and J. van der Plicht 2011. A 33,000-Year-Old Incipient Dog from the Altai Mountains of Siberia: Evidence of the Earliest Domestication Disrupted by the Last Glacial Maximum. *PLOS One* 6(7): e22821

Sychenko, G.B. 2021. Migration of People and Melodies in the Taiga Area of Southern Siberia. *European Journal of Musicology* 20/1: 39–62.

# 3.5 The Dog – Human Interrelations in the Lower Amur Rural Regions (the Far East of Russia): Past and Present

Olga V. Maltseva

Institute of Archaeology and Ethnography, Siberian Branch, Russian Academy of Sciences.
Pr. Akademika Lavrentieva 17, 630090 Novosibirsk, Russia.

olymals@gmail.com

**Abstract**

The paper is dedicated to dogs of the Amur valley that became irreplaceable partners in hunting and transportation for the Amur peoples. The Amur dogs varied according to the character of their usage and location of the populations. In the 1930's as a result of aboriginal breeding, a cultural breed named Amur laika was formed. In the 19th century Amur hunters and fishermen suffered a breaking change in their lifestyle that had a dramatic effect on their relationship towards dogs. Since then the Amurlaika breed has become a stray dog in a human society.

**Keywords:** Amur laika, hunting, communications, stray dog, self-organisation.

## 1 History of the Amur dog research

The peoples of the Lower Amur district used the laika-dog as their indispensable helper in hunting and dog-drawn transportation. This type of dog comprises several breeds; the most famous among them are West Siberian, East Siberian, Karelian-Finnish, Yakut, Nenets, and Russian-European laika (Dmitrieva-Sulima 1911, Kon'kova 2009, Shirinsky-Shikhmatov 1895, Shirokii 2017). Since the late 19th century the outstanding handler and hunting expert A.A. Shirinsky-Shikhmatov (1895) included the Lower Amur district into the area of the Tungus laika habitat. However, many handlers considered that by this time cultured breeds prevailed and Tungus (or East Siberian) laika was included in their number (Geits 1968; Shirokii 2017). This breed was formed as a result of the aboriginal type breeding (Shiroky 2017: 40–45, 65–66). There were several tribal nests in the locations of the Nanai, Udege, Orochi, and Nivkh peoples' inhabiting the Amur valley region (Abramov 1940; Shirokii 2017: 85–90). Academic research of the Amur territory only started in the second half of the 19th century. Consequently, researchers started studying the Amur laika populations quite late. Only in the end of the 1930-s were the Amur laika dogs included into the classification of dogs. During this period K.G. Abramov created the Amur laika breed that was a result of interbreeding of a local sled dog (coastal) and hunting breeds (Abramov 1940). He founded a dogs' kennel that worked until the end of the 1940s, and since then the dog has been forgotten in handlers' circles. In 2003 Andrey Samar published a book 'Traditional dog breeding of the Nanai' that seemed to have resurrected the lost breed (2010).

## 2 A dog in the life of Amur fishermen and hunters

### *2.1 Siberian and Amur aboriginal dogs – differences in their types*

It can be inferred on the basis of all previous research that Amur dogs that participated in the selection were not of the same type according to their conformation and behaviour. Landscape and climatic conditions have made their own adjustments to the dogs' composition. In the districts of the Pacific coast and Arctic zone, sled handling where a dog was used for transportation and hunting was formed; hunting traditions were kept at the Amur tributaries, and the local population paid much attention to the development of hunting qualities among their dogs (Safonova and Santha 2016; Losey and Nomokonova 2018; Maltseva 2019). The Amur types such as 'sharp-tipped' dogs of Siberia have more similarities with their wild relatives. The conformation and behaviour of local canines were dominantly impacted by nature rather than by humans (Riabov 1939; Smirnov 1936). The process of evolution by natural selection resulted in the diversity of dog populations. Inside the Lower Amur district it is possible to outline two areas where dogs were tamed where they had different functions and consequently the approach of humans towards them was also different In due time, L.Yi. Shrenk pointed out that each Amur inhabitant used dogs differently depending on whether he was a hunter or a fisherman. The researcher emphasised that unlike the Siberian dogs the Amur dogs were not big and had a different colour (black, white, fawn, red), moreover, some of them looked like wolves (Shrenk 1899: 167–169, 176–179, 185–186).

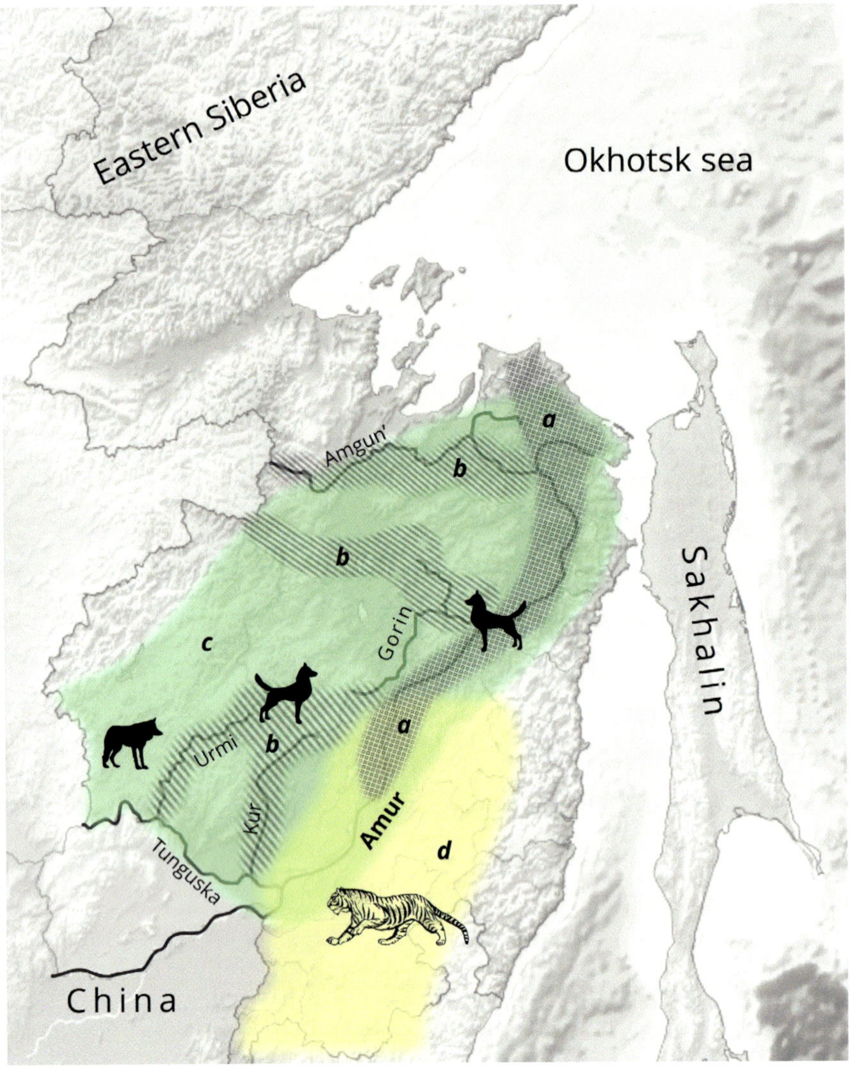


Figure 1. Dog, wolf and tiger ranges within the Amur territory: a, 'Ichthyophagus dog' native habit; b, 'dog-carnivore' (taiga meat-eater) distribution area; c, wolf species distribution area (from 1 to 5 entities per 1000 km$^2$ in 1940s) (Reimers and Bibikov1985); d, Amur tiger species distribution area (by O. Maltseva).

### *2.2 Specifics of the organisation of dogs' nutrition in the Amur River territory*

It should be emphasised that L.Yi. Shrenk did his research mainly at the Amur Estuary and northern part of the Tatar Strait, places which were inhabited by the Nivkhs (the Gilyaks), where sled handling and cynophagie (eating dog meat) were developed. Eating dogs even became a sacred tradition: this custom penetrated the rituals related to the worship of bears, the sea, the mountains and the taiga. These ceremonies opened the seasons of river and marine fisheries (Shrenk 1899: 124–126; Temina 2005: 66; Gasilin and Gorbunov 2018). The fact of using a dog as food was recorded among other peoples of the taiga and tundra zones, i.e. among the Ainu and the Eskimos. The organisation of the Amur dog's nutrition deserves special attention as a phenomenon of a fishing lifestyle (Shrenk 1899: 125–126, 144). During the salmon season, the inhabitants of the Amur villages harvested on a large scale migrating fish for winter. Salmon (chum) was gutted, plastered, i.e. divided into several layers, which later were dried on special hangers. Dried salmon carcasses were used for dog nutrition. Unlike the Siberian hunting dog's nutrition, the Amur dog's diet included river and sea food ingredients (salmon, saffron cod, seal and whale meat) that incited their separation from the family of ravenous (carnivorous) canids whose nutrition consisted of herbivores, rodents and small predators (Shrenk 1899: 116–117, 120–122, 149, 168–169; Abramov 1940: 20–24; Maltseva 2012). Another attitude to a dog was formed on the tributaries of the Amur River along which taiga hunters and reindeer herders lived. This district including the valleys of the Gorin, the Amgun and the Tunguska rivers, was a part of the so-called *Tungus wedge* (Patkanov 1906: 149–154) (Figure 1b). L.Yi. Shrenk outlined that in these territories a dog was considered as a working partner (Shrenk 1899: 146–147). Accordingly, it should not be excluded that the Siberian hunting groups migrated to the Amur river along its tributaries. Near the riverbed of the Amur, the distribution of 'meat-eating dogs' and the Amur 'ichthyophagous dogs' interbred.

Figure 2. The Amur laika. (After K.G. Abramov, album, no 14638-30, 1939. The Vladimir K. Arseniev Museum of the Far East History fund (Vladivostok).

### *2.3 Dogs vs. wolves at the Amur taiga territory: a question about communications*

The question about the wolf, jackal or wild dog species participation in the formation of the Amur dog stock has still not been solved. K. Abramov noticed that the Amur laika looked a bit like a wolf, which pointed to the admixture between domestic and wild animals: his cultural breed retained a fawn coloured coat, a strong croup, a broad forehead, little bat ears, a sword tail and occasionally yellow hawk prey eyes (Abramov 1940: 4–10) (Figure 2). A number of scientists argue that *Canis lupus* and *Canis familiaris* are irreconcilable antagonists in nature. A domestic dog, no matter what hunting skills it possesses, will be ware of a wolf (Kolosov 2009: 87). There is little evidence of wolves entering the territory of the Amur peoples. In the early 19th century, a wolf was a rare exception in the list of hunting trophies for the local population (Priamurie 1909: 268). We can see a few cases of the hunting this animal. Moreover, in the southern part of the Lower Amur area, a serious competitor of a wolf is a tiger that along with a bear stands on the top of the local food pyramid (Figure 1 c, d). Respectively, confrontation between the Amur dogs and their predatory relatives would not happen frequently. The rugged terrain covered with dense thicket and swamps that make flock hunting difficult could be a barrier for a wolf penetrating into the Amur taiga zone. Hunting experts notice that wolves were not observed in the taiga area thousands of years ago. The wolf's habitat was within tundra, forest-steppe, steppe, forest-tundra and semi-desert zones. These territories were favourable for ungulate hunting by pack predators. Only a female wolf or packs of young wolves could enter the forest, which consisted of the territories that other animals used for hunting. Wolf researchers also observed that wolves prefer living not far from the logged-off lands occupied by people. Similar places were usually chosen by inexperienced young beasts. The low number of wolves in human-populated areas led to the emergence of a large number of stray dogs. In this case there was an opportunity for crossbreeding (Cherenkov 2003: 9–22). Wolf-dog crossbreeding was observed in Siberia and the Russian Far East. Researchers highlighted that the hybrid species differed in their behaviour and conformation depending on their parental mix. As a rule, a female wolf raises its offspring from a dog alone and independently. A male wolf tries to keep a female dog with him and participates in the feeding of mixed offspring (Cherenkov 2003: 22; Samar 2010: 65–66). According to the observations of wolves in their own environment, it can be assumed that the Amur taiga territory is located in the periphery of the wolf range where the breed of 'northern dogs' was formed.

### *2.4 Communication between the Amur dog and taiga predators*

From this point of view, it is curious to analyse the relationship between the dog and other taiga predators and how this connection was regulated by humans. Another predator that the Amur dog often faced was a tiger. The Nanai treated it as a sacred animal, and there were cases when people raised the brood of a dead tigress. Initially, tiger cubs were kept along with dogs. The mature feline predators left people, and sometimes came back to them without harming the dogs that they had grown up with. According to the native people's belief, a tandem of a dog and a tiger provided human settlements with protection from attacks of other predators such as a bear which was the most dangerous predator of all (FMA: 2008, 2011).

Analyzing the specifics of keeping hunting dogs in the Lower Amur district, it is necessary to highlight that during their hunting training the dogs were not trained to attack tigers.

### *2.5 Specifics of the Amur hunting dogs*

In his work, A. Samar offers special names of dogs that are included into the hunting vocabulary of the Tungus-Manchurian peoples. Such terminology refers to several types of dogs. For instance, among the Evenks, a dog used for-hunting deer and elk was called *beiuman*; a hunting dog for wild boar was *torokiman*; a dog for hunting squirrels was *ulukimen* (while among the Negidal people it was called *beiuman*). The dog for sables was especially valued. The Udege called it *biaka* and the Negidal - *seiepman* (*elekhibechen*). The dogs that hunted predators were in a special category. The dogs that hunted bears were called *nakiman* (Evenk.) and *amakhaman* (Neg.); the dog-wolfhounds were named *n,onchakaman* (Evenk.). There was a special attitude to bear-hunters because bear hunting had a cult meaning.

During the process of hunting a bear the dog became equal to a human and was considered as a partner. A. Samar highlights that the Negidal term *amahaman* has ancient roots that come from the word *amaha* meaning 'an old man, a father' which is an allegorical designation of a bear (Samar 2010: 71–72). Such terminology indicates that within hunting specialisation the selection was purposeful when puppies with particular working qualities were selected for different types of hunting. During the process of raising the dogs for bear, sable and ungulate hunting much attention was paid to developing the puppies' skills to recognise the animals they hunted and, consequently, to react to them. Therefore, much attention was paid to the dogs' appearance. The Amur aboriginal hunting breed did not have a unique standard. There were several types of dogs - those that are large, with long body length; light and small for hunting fur animals, and stunted with short legs for hunting burrow animals (Samar 2010: 52–70). Cynologists point out that in different places in the Lower Amur territory original populations of dogs were formed. There were Orochen, Udege, Gorin laika, and Kur-Urmi laika has recently been added to the list of aboriginal breeds (Abramov 1940; Samar 2010: 55, 64; Shirokii 2017: 85–90, 121–127). A diverse composition of animals most likely determined the variation of aboriginal dogs. Each district had its own group of forest animals that the dogs had contact with and it affected the formation of their working qualities. For instance, in the South of the Amur region, in the foothills of Sikhote-Alin covered with broad-leaved forests, dogs were trained to catch Manchurian deer, in coniferous forests it was replaced by elk. However, where fur trading took place in the South East of Russia, dogs for sable hunting had priority. Searching for furs the hunters travelled long distances; they sometimes reached the coast of the Seaof Okhotsk or the Baikal region. At the stage of the development of trade connections, universal features of the hunting dogs of Siberia and the Far East emerged that can be explained by a mixture of litters from different geographical areas (Samar 2010: 65–68).

## 3 Factors that led to the population decline of the Amur laika

The end of the 19th - beginning of the 20th century became a turning point in the Amur dogs' lives. To follow the transformation it's important to consider some changes in the design of the Amur peoples' settlements. The local population changed to a settled way of life from the latter half of the 19th century when Russian settlers appeared on the banks of the Amur (Aliab'ev 1872). Before the 20th century small kin groups of hunters and fishermen lived in nomad camps in temporary buildings. They used to move to new places when nearby taiga or river resources depleted. This semi-sedentary lifestyle extended to the circle of domestic animals. They did not have strong connections to human dwellings. J. A. Lopatin in his work mentioned that in the Goldy (Nanai) villages a number of dogs exceeded the number of people. Most of the time they were out to pasture and they could move freely both inside and outside a village. Since the 20th century some fishing and hunting camps have been transformed into rural settlements the basis of which were peasant farmsteads with animals providing meat, and milk and doing arable work. Cattle, horse and pig breeding brought ~~a~~ serious problems for animal-antagonists, such as, for example, laika dogs (with wolf instincts) and herbivore animals coexisting in the borders of rural areas. Some Siberian examples proved that dogs and ungulates could coexist and even cooperate. Such collaborative activities of humans, dogs and reindeer appeared in northern Russia (Stepanoff *et al.* 2017; Klokov and Davydov 2019). The dogs of the Amur area could not improve their herding behaviour due to a lack of reindeer husbandry in most of the territory. Their hunting skills were a reason for conflict situations. There were cases when dogs mauled foals and calves taking them for prey, therefore the dogs were tethered (Lopatin 1922: 120–123). In contrast to wild ungulates, defensive behaviour is not developed among domestic animals and it is expressed in the 'escape distance' (the distance between a predator and a potential prey). The speed of movement developed within the populations of wild ungulates helps them avoid the attack of certain predators. Individuals have different speed limits inside the flock that depend on their gender, age, rank and type of predator (Bodridze 2016: 112–120).

The development of the agricultural sector has contributed to a change in the sled and hunting dog's status. If earlier it had been seen as a working tool, a hunting helper and a partner of a man the new economic practices weakened its importance in a human society. The Amur laika became a stray dog. In addition, by the 1960s the number of fish stocks in the Amur river significantly reduced as a result of overfishing that made it impossible for large dog packs to live on it. To minimise the number of dogs a regulated shooting of them was carried out (Samar 2010: 83–85).

## 4 New forms of communications between dogs, other domestic animals and humans

### *4.1 'Road dogs' of the Khabarovsk region*

Several wild islets have been preserved in the countryside of the Lower Amur district where we can still observe the relationship between man and dogs. Currently the East Russian region represents an area with a low population density and a looming picture of the wildlife expansion into the world that had previously been explored by the people and is represented by the cases when tigers and bears enter human settlements. Previously tame (domestic) animals that performed certain roles in households found themselves on the periphery of a wild world. As for dogs we can see more and more cases of their behaviour inherent to synanthropic species. Our monitoring in the Khabarovskyi region that has been conducted since 2003, represented flexible forms of animal behaviour that are connected with changes in their nutrition. It was observed that the dogs gather at the several bus stops along the Federal highway from Khabarovsk to Komsomolsk-on-Amur. They approached shuttle buses asking for food and then went away after their departure. The next picture could be observed in the village of Kondon in 2003. The local dogs had appeared at the bus stop before a bus arrived (a bus came at 10.25 on schedule). Their motive was clear because there was a small grocery store next to the bus stop and a small market where passengers could have a meal and share some food with the dogs at the same time. The next day at 10 a.m. the dogs reappeared near the store, but the bus did not come because it was broken. This is an example when animals use the infrastructure created by man in their daily activities. These situations showed that not only a place, but also a time, related to the human social sphere, were imprinted in the spatial cognitive map of dogs (Reznikova 2005: 75–84). In their mind bus stops and roads were associated with feeding places. In these groups, the relationships of dogs were built on subordination. Big dogs, as a rule, came first, and smaller dogs remained behind them and stayed a little on the side. Studies of the dogs' social behaviour show that a dog's ability to project their feeding spots onto the sites of human activities, and detect their relocation. These examples show that places like shops, markets, trash cans, centres of public nutrition attract dogs. They convey the information about locations of these objects with each other using various cues which indicates that they are social entities (Hare and Brown 2002; Mikolsi 2003; Hare and Tomasello 2005; Bräuer 2006; Kaminski and Nitzschner 2013).

#### *4.2 Monitoring the dogs in the countryside of Eastern Russia*

We observed a more complex configuration of the dog community in 2011, and 2014 in the Nanay village of Ulika-Natsional'noe with a population of 170 people. This village is located 75 km away from Khabarovsk on the tributary of the Tunguska River which flows into the Amur. The settlement is cut off from the world in the spring-autumn season and you can reach it only in the summer or in the winter by water and on ice. In the 1950-s the majority of the population of the Ulika village worked in the fishing collective farm 'Lenin's Way'. However, that collective farm had already become unprofitable for fishing by that time. Collective farmers started focusing on the development of subsidiary farming based on agriculture to get out of their predicament. In fact, a part of the Russian population in the neighbourhood village Ulika-Pavlovka controlled the agriculture operations among the Nanai in Ulika-Natsional'noe. In 1968 the village of Ulika-Pavlovka disappeared, and only the Nanai village Ulika-Natsional'noe remained at the tributary of the Tunguska river on the Kur river. The weakening of communications with other districts of the Khabarovsk region forced the residents of Ulika to return to subsistence-based practices. Hunting, fishing and gardening remain the only ways to survive. Nevertheless, recently traditional hunting has become complicated after new rules were introduced by the Ministry of Nature with large fines issued if they are violated, and these measures have made the hunting process non profitable. It has also led to a low demand for hunting dogs. After the closure of collective farms throughout the Amur region some locals tried to revive livestock and pig farming, but that attempt ended with the loss of control over animals. A Russian female migrant from Altai and a Korean businessman were the initiators of such processes. In both cases raising cows and pigs did not generate enough interest among the population: milk has not yet become a traditional product for locals with Nanai roots, and it has been difficult to sell pig meat due to the lack of a road network. Animals have begun to multiply rapidly without human intervention adapting to local environmental conditions as they were sent out to pasture. Currently they represent self-organised animal populations that get along with people. We can observe some analogues with the natural environment in the arrangement of their space. The increasing number of animals allows them to stray into packs where the relationship is built on the subordination principle. In the hierarchy of the animal community the top place is occupied by the dogs including two related groups where each of them has their own territory. Although human influence in regulating intraspecific and interspecific animal bonds is minimal, it is obvious that some segments of human activity have become significant for them in building hierarchical relationships. According to our observations, the building of the village administration located in the centre of the settlement was used as a marker that conventionally delimited the living spaces of the two rival groups of local dogs (Figure 3). It is a place where villagers gather for local events, share important news and get permission for fishing in autumn. The dogs started using the playground adjacent to this building for their interactions. Thus, in the evening representatives of two broods gathered there, these meetings often ended with a fight because of dogs trespassing into rival territory. Each member of

Figure 3. Dogs of st. Ulika-Natsional'noe (Photo by O. Maltseva, 2014).

Figure 4. Spatial arrangement of domestic animals inside Ulika-Natsional'noe territory (by O. Maltseva).

the dog packs had certain responsibilities according to its position in the established dog hierarchy. For example, only a certain dog was allowed to copulate with a bitch, while others with a lower rank protected it and repelled outsiders. Such demonstrative behaviour indicated the certain male dog's high status in the pack. The shouting of people, even notes of discontent in their voices were interpreted as a signal for the dogs to redirect their aggression to pigs. The following picture was observed once in the morning: the local dogs interpreted one man's anger at the broken roads as his command to attack the pigs. The lack of a leash and being allowed to roam free within the rural area makes them mobile and in their settled contacts with other herd animals (cows, pigs) they have a dominant position. The subordinate position of ungulates was expressed in the spatial-temporal displacement of their activities (Figure 4). T he pigs have developed a new tactic in their activities. The nighttime, when the dogs could not disturb them has become an active phase in their life. On the contrary, in the daytime they prefer hiding in bushes and leaving the village. To save their brood the cows have gone to abandoned, hard-to-reach areas where there is minimal risk to face domesticated predators.

Such examples demonstrate the establishment of networking in the domesticated animal community without human participation. A hierarchy has been structured in their self-organisation with the dogs at the top of it.

## 5 Conclusions

The history of the Amur laika shows one of the ways of how the interrelations between humans and dogs have evolved. Initially, there were several native dog populations in the Amur valley. They were formed within the hunting and fishing societies that affected the development and working qualities of local breeds. Thus, the differences between hunting and fishing economic activities and lifestyles determined the niches for dogs in human society. The dog performed a strictly practical role such as for sledding, for transporting or as a nutrition source among the Amur fishing communities that were typical for many peoples of the Pacific coast. The hunting groups of the Amur population treated their dogs as hunting partners. In this environment the dogs were selected according to the development of special qualities and external traits used for hunting particular animals. Such specialisation led not only to a variability of dogs within their populations, but also to the development of distinctive features between dogs from different territories depending on whether they were covered with broad-leaved or coniferous forests with a certain composition of animals. The specifics of Amur dogs nutrition in the territories near to and far from the Amur region also determined the division of aboriginal dogs into two large groups. Near the Amur channel where the population practiced salmon fishing the local dogs switched to fish as their food source and it was a reason to exclude them from the family of predatory canines. In the territories far from the river the dogs kept their preference to meat, and in this case the dogs were closer to their wild relatives (wolves, jackals) in contrast to the 'river' dogs.

Since the second half of the 19th century in the Amur valley, the migration activity of the population has been observed which has led to a new form of agriculture and to a change of the dog's niche in a human society. Since the 1930s attempts have been made to create a universal (cultural) breed of the Amur laika based on a mixture of dog litters of the river valley and taiga districts, but this work was interrupted as recently as in the 1960-s due to a lack of economic benefits of keeping that breed. In the countryside a dog with a hunting instinct could not coexist with domestic hoofed animals which it interpreted as prey. Thereby, the ceasing of hunting activity led to the loss of its role as a partner in human society and it took on a new status of a stray dog. Using the infrastructure created by humans, the dogs have been allowed to build their own communications that are similar to connections in the wild. In this model of communication we can observe the traits of a dog's behaviour that are close to synanthropic species which are tied to a human habitat, and at the same time, their self-organisation is typical for a wild animal community. It is built according to hierarchical relationships with domestic animals that are free of human control. In a new animal community the dogs have taken a dominant position and other animals (cows, pigs) have received a subordinate status.

## References


Abramov, K.G. 1940. *Hunting Laika of the Amur region. Promyslovaya layka Priamur'ya.* Khabarovsk: Dal'nevostochnoye knizhnoye izdatel'stvo. Khabarovsk: Far Eastern Publishing House. *Абрамов, К.Г.* 1940. *Промысловая лайка Приамурья. Хабаровск: Дальневосточное книжное издательство.*

Aliab'ev, G. 1872. *Distant Russia. Ussuri region.* Dalekaya Rossiya. Ussuriyskiy kray. Sankt-Peterburg: Tipografiya tovarishchestva 'Obshchestvennaya pol'za'. Алябьев, Г. 1872. Далекая Россия. Уссурийский край. Санкт-Петербург: Типография товарищества «Общественная польза».

Bodridze, Ia. 2016. *Wolf. Problems related to the reintroduction of large mammal predators. Volk. Problemy, svyazannyye s reintroduktsiyey krupnykh khishchnykh mlekopitayushchikh.* Moskva: Modern. Бодридзе, Я. 2016. Волк. Проблемы, связанные с реинтродукцией крупных хищных млекопитающих. Москва: Модерн.

Bräuer, J., J. Kaminski, J. Riedel and J. Call 2006. Making inferences about the location of hidden food: Social dog, causal ape. *Journal of Comparative Psychology* 120: 38–47.

Cherenkov, S.E. and A.D. Poiarkov 2003. *Wolf, jackal. Volk, shakal.* Moscow: ASTASTREL'. *Черенков, С. Е., А. Д. Поярков, 2003. Волк, шакал. Москва: АСТАСТРЕЛЬ.*

Dmitrieva-Sulima, M.G. 1911. *Laika and hunting with it. Layka i okhota s neyu.* Sankt-Petersburg. Дмитриева-Сулима, М.Г. 1911. *Лайка и охота с нею.* Санкт-*Петербург.*

FMA 2008, 2011. Field materials of the author 2008, 2011. Expeditions to the Nanai district of the Khabarovsk territory, informants M. Kimonko and S. Beldy.

Gasilin, V.V. and S.V. Gorbunov 2018. The Bear and Dog in Heathen Sanctuary Remains in the Mouth of the Agnevo River (Central Sakhalin). Medved' i sobaka v ostatkakh svyatilishcha v ust'ye reki Agnevo (tsentral'nyy Sakhalin). *Etnograficheskoye obozreniye* 3: 184–200. Гасилин, В.В. и С.В. Горбунов, 2018. Медведь и собака в остатках святилища в устье реки Агнево (центральный Сахалин). Этнографическое обозрение 3: 184–200.

Geits, A.V. 1968. *The East Siberian Laika. Vostochno-Sibirskaya layka.* Irkutsk. Гейц, А.В. 1968. Восточносибирская лайка. Иркутск.

Hare, B., M. Brown, C. Williamson and M. Tomasello 2002. The domestication of social cognition in dogs. *Science* 298: 1634–36.

Hare, B., and M. Tomasello 2005. Human-like social skills in dogs? *Trends in Cognitive Sciences* 9: 440–45.

Kaminski J. and M. Nitzschner 2013. Do dogs get the point? A review of dog-human communication ability. *Learning and Motivation* 44: 294–302.

Klokov, K. and V. Davydov 2019. Human-dog-reindeer communities in the Siberian Arctic and Subarctic, in Ch. Stepanoff and J.-D. Vigne (eds) *Hybrid Communities. Biosocial Approaches to Domestication and Other Trans-species Relationships:* 261–274. London: Routledge.

Kolosov, P.N. 2009. Domestication of a wolf or wild dog. *Science and education* 2: 86–88. *Колосов, П.Н.* 2009. Domesticating a wolf or wild dog. Domestikatsiya volka ili dikoy sobaki 2: 86–88. *Nauka i obrazovaniye Доместикация волка или дикой собаки. Наука и образование* 2: 86–88.

Kon'kova, E.Iu. 2006. *Laikas. Origin. Breeds. Using. Test rules.* Layki. Proiskhozhdeniye. Porody. Ispol'zovaniye. Pravilaispytaniy. Moskva: OOO 'Akvarium-Print'. *Конькова, Е. Ю.* 2006. *Лайки. Происхождение. Породы. Использование. Правила испытаний. Москва: ООО «Аквариум-Принт»*

Lopatin, LA. 1922. Amur, Ussuri and Sungarian Gol'ds, vol. 17. In *Notes of the society for the Amur region studying.* Vladivostok: (b.1). Gol'dy amurskiye, ussuriyskiye i sungariyskiye. Opyt etnograficheskogo issledovaniya, Zapiski obshchestva izucheniya Amurskogo kraya. Vladivostok: (b.i.). *Лопатин, И.А. 1922. Гольды амурские, уссурийские и сунгарийские. Опыт этнографического исследования, т. 17. Записки общества изучения Амурского края. Владивосток: (б.и.).*

Losey, R.J., T. Nomokonova, A.V. Gusev, O.P. Bachura, N.V. Fedorova, P.A. Kosintsev and M.V. Sablin 2018. Dogs were domesticated in the Arctic: Culling practices and dog sledding at Ust'-Polui. *Journal of Anthropological Archaeology* 51: 113–126.

Maltseva, O.V. 2012. Impact of salmon fishing on livelihood and social relations among the ulchi and the amur nanai (Second Half of the 19th - Early 20th Century). *Archaeology, Ethnology and Anthropology of Eurasia.* 40, 2: 124–132

Maltseva, O.V. 2019. 'Taiga' and 'River' Components in the Nanai Socio-Tribal Organization at Lake Bolon the Lower Amur. *Archaeology, Ethnology and Anthropology of Eurasia.* vol. 47, Is. 2: 131–139.

Miklosi, Á., E. Kubinyi, J. Topál, M. Gácsi, Z. Virányi and V. Cssáanyi 2003. A simple reason for a big difference: Wolves do not back at humans, but dogs do. *Current Biology* 13: 763–66.

Patkanov, S.K. 1906. *Experience of geography and statistics of Tunguska tribes of Siberia based on data from the 1897 Census and other sources. Ch. 2. Other Tunguska tribes.* St. Petersburg: Imperial Academy of Sciences. *Opyt geografii i statistiki tungusskikh plemen Sibiri na osnovanii dannykh perepisi naseleniya 1897 g. Idrugikh istochnikov. CH. 2. Prochiye tungusskiye plemena.* Sankt-Peterburg: Tip. Imperatorskoy AN. Патканов, С.К. 1906. Опыт географии и статистики тунгусских племен Сибири на основании данных переписи населения *1897* г. И других источников. Ч. *2.* Прочие тунгусские племена. Санкт-Петербург: Тип. Императорской АН.

Polner T.I. 1909. *Amur region. Facts. Figures. Observations.* Moscow: City printing house. *Priamur'ye. Fakty, tsifry, nablyudeniya.* Moskva: Gorodskaya Tipografiya 1909. *Приамурье. Факты, цифры, наблюдения. Москва: Городская типография.*

Reimers, N.F. and D.I. Bibikov 1985. Distribution area and its transformation. *Wolf*: 21–40 Moscow: Nauka. Areal i yego izmeneniye. *Volk:* 21–40 Moskva. *Реймерс, Н.Ф., Бибиков, Д.И.* 1985. *Ареал и его изменение. Волк*: 21–40 *Москва.*

Reznikova, Zh.I. 2005. *Intelligence and language of animals and humans.* Moscow: Akademkniga. *Intellekt i yazyk zhivotnykh i cheloveka.* Moskva: Akademkniga. *Резникова, Ж.И. 2005. Интеллект и язык животных и человека. Москва: Академкнига.*

Riabov, V.V. 1939. *The Evenki Laika and hunting with it.* Krasnoyarsk: Krasnoyarsk regional state publishing house. *Evenkiyskaya layka i okhota s ney.* Krasnoyarsk: Krasnoyarskoye krayevoye gosudarstvennoye izdatel'stvo. Рябов, В.В. 1939. Эвенкийская лайка и охота с ней. Красноярск: Красноярское краевое государственное издательство.

Samar, A.P. 2010. *Traditional dog breeding of the Nanai.* Vladivostok: Dalnauka. *Traditsionnoye sobakovodstvo nanaytsev.* Vladivostok: Dal'nauka. *Самар, А.П. 2010. Традиционное собаководство нанайцев. Владивосток: Дальнаука.*

Safonofa, T. and I. Santha 2016. Evenki Hunter Gathering Style and Cultural Contacts in K. Ikeya and R.K. Hitchcock (eds) *Hunter- Gatherers and their Neighbors in Asia, Africa, and South America*: 59–79. Senri Ethnological Studies 94.

Shrenk, L.I. 1899. *About foreigners of the Amur region*, vol. 2. St. Petersburg: Academy of Sciences Publishing House. *Ob inorodtsakh Amurskogo kraya,* t. 2. Sankt-Peterburg: Izd. AN. *Шренк, Л.И.* 1899. *Об инородцах Амурского края, т.* 2. *Санкт-Петербург: Изд. АН*

Shirinskii - Shikhmatov, A.A. 1895. *Album of Northern dogs (laikas).* Moscow: phototype Sherer, Nabgol'ts and Ko. *Al'bom severnykh sobak (layek).* Moskva: fototipiya Sherer, Nabgol'ts i Ko. *Ширинский-Шихматов, А.А.* 1895. *Альбом северных собак (лаек). Москва: фототипия Шерер, Набгольц и Ко.*

Shirokii, B.I. and O.B. Shirokii 2017. *Our Northern dogs. Introduction to the laika science.* Ekaterinburg: Publishing solutions. *Nashi severny yesobaki. Vvedeniye v laykovedeniye.* Yekaterinburg:

Izdatel'skiye resheniya. *Широкий, Б.И. иО.Б. Широкий.* 2017. *Наши северные собаки. Введение в лайковедение. Екатеринбург: Издательские решения.*

Smirnov, N.A. 1936. The basics of dog standardization and the standard of sled dogs compared to other laikas. *USSR laikas standards. Proceedings of the Arctic Institute,* vol. 56. Leningrad: Glavsevmorput': 101–177. Osnovy standartizatsii sobak i standart yezdovykh po sravneniyu s drugimi laykami. *Standartylayek SSSR.* Trudy Arkticheskogo instituta, t. 56. Leningrad: Glavsevmorput': 101–177. Смирнов, Н.А. 1936. Основы стандартизации собак и стандарт ездовых собак по сравнению с другими лайками. Стандарты лаек СССР. Труды Арктического института, т. 56. Ленинград: Главсевморпуть: 101–177.

Stepanoff, Ch., Ch. Marchina, C. Fossier and and N. Bureau 2017. Animal Autonomy and Intermittent Coexistences. North Asian Modes of Herding. *Current Anthropology* 58, 1: 57–81.

Temina, M. 2005. The bear's soul is on its way to its owner. *Wordbook of arts* 16: 63–66. Dusha medvedya na puti k svoyemu khozyainu 16: 63–66. *Slovesnitsa iskusstv* Темина, М. 2005. Душа медведя на пути к своему хозяину. Словесница искусств 16: 63–66.

# 3.6 The Mother of Dogs: Women, Power and Dogs in First Nations Societies in Northwest North America

Guy Lanoue

Département d'anthropologie, Université de Montréal,
Pavillon Lionel Groulx 3150 Jean-Brillant Montréal QC H3T 1N8, Canada.
guy.lanoue@umontreal.ca

**Abstract**

Dene dogs are treated with indifference, despite myths that affirm that humans are descended from dogs. Exceptionally, women often treat puppies well. These attitudes are related to Dene notions of personal power, which is considered a gift from animals. Dogs, however, are stripped of their symbolic power so they can play a vital role in establishing gender equality. All animals except dogs are considered superior to men and can be hunted only with their consent. Animal blood is therefore a sign of a sacrifice that ensures human survival. As a corollary, women's menstrual blood is therefore a sign of women's superior power, since it is a sign of reproduction and of survival. To establish social equilibrium between weak men and powerful women, men are therefore symbolically associated with powerful animals through the vision quest, while powerful women are associated with powerless dogs.

**Keywords**: Dene, animals, gender, agency, vision quest.

## 1 Introduction

Dogs have a long history within North American Native societies, especially the Athapascan-speaking Dene of Northwestern North America. The First Nations people of this region who dominated the forest and tundra generally hunted solitary animals such as moose, woodland caribou, and smaller species such as beaver and muskrat. They were contacted by Whites relatively late in the post-Columbian epoch. Until recently, Euro-Canadians had no real interest in these distant zones unsuited for agriculture. As a result, many Dene societies and traditions are relatively intact, and descriptions of their traditions and current lifeways are not dependent on archaeological reconstruction.

Dogs were and are extremely useful as pack animals and as part of dog teams pulling sleds, travois, or toboggans. In the late 1960s, some Dene began using snowmobiles in place of dogs but soon realised the limitations of being dependent on gasoline and costly machines that could sometimes break down in remote locations. In the 1980s, some people began returning to dogs as they compared the costs of owning and operating modern machinery to the relatively meagre returns from hunting and trapping. Economics, however, is not the only issue for Dene people.[1]

Dogs also play another role. Given that dogs are close to humans, they are often mediators between nature and culture. I shall explore the structure of the imaginary and the role of dogs among the Dene of northern British Columbia and the Northwest Territories. In particular, I will address two questions: why are dogs anomalous in the Dene bestiary - all animals have residual power from the time of creation, but dogs have none. Why are dogs more closely associated with women, especially since these societies largely depend on a (supposedly) male activity, hunting? I shall base my analysis on three sources: data from the early historical period (pre-1970), before northern Dene societies were swept up in a tide of massive change caused by Eurocanadian intrusion in the north; on nearly two years of field observation among the Hare (Sahtu) and Sekani (Tse'Khene and Kwadacha First Nation) in the late 1970s;[2] and on recent comparative analyses (Figure 1).

All animals, including dogs, are metaphors with which people construct an imagined community that counters the social fragility that emerges from Dene land-use patterns, where people are dispersed in small groups over territories measuring many thousands of

[1] While acknowledging that Dene 'living landscapes' may evolve to include Kevlar canoes and snowmobiles without compromising claims to 'authentic' Dene identity, some authors (eg., Andrews and Buggey 2008) overlook that modernisation is an economic choice with cultural overtones (see Jarvenpa 1977; Wilson 2014). Nelson *et al.* (2005) illustrate the difficulties surrounding modernisation in a northern Alberta Cree community. Savishinsky (1978) discusses how snowmobiles did not displace dogs in a Hare (Sahtu) community since the machines were unreliable; attitudes, however, are not all shaped by economics, as people still respected, 'good dog teams, their owners, and tough travelers in general' (Savishinsky 1978: 11). Hunting and trapping alone cannot finance modern technology. Michael Asch (1979) has shown that revenue from the modern sector finances 'traditional' activities among the Slavey (dene-Tha). See Loovers (2018), who documents how Gwich'in reject modern technology to reduce their dependency on the 'White' economy in preparation for a coming apocalypse.

[2] Here, I use conventional, anglicised names. I put the endonym in parentheses. 'Dene' is an ethnonym also used by many Athapascan speakers, meaning 'Human' or 'Man' (generic).

Figure 1. Dene people, northwest Canada. Sahtu Dene (Hare) mentioned in text are here classified as North Slavey (https://en.wikipedia.org/wiki/Athabaskan_languages).

square kilometres and see each other rarely. For most of the year, people lived in small, isolated, and dispersed hunting groups of 15 to 20 people. In the last 50 years or so, massive changes in Native societies caused by White intrusion has led to the loss of 'animal stories', meaning that a culture of individualism emerges, and animal metaphors become a less meaningful vehicle of creating community solidarity. Paradoxically, from a Eurocentric viewpoint, this weakening of social bonds has taken place as people have become more sedentary. Nonetheless, the underlying contradictions between individual and community persist. As one older man said, 'Today, the animals don't talk to us because we didn't respect them'[3] by allowing Whites to 'ruin the forest'. In fact, tensions may be worse today, with people spending unprecedented amounts of time in town rather than in the bush. Without their animal stories, they have fewer conceptual tools with which to confront previously unknown problems: new political hierarchies, ethnicity as a dangerous euphemism for Euro-Canadian racial discrimination and ethnonationalism, and increased tensions between men and women.

I believe that some tensions Indigenous people experience are exacerbated by their feelings of helplessness caused by the weakening of their relationships to tutelary animals. Belief in these creatures in the recent past led to practices that reinforced community solidarity, which was based on a desire for peace and equilibrium. Here, I will try to explain the anomalous role of dogs in the animal stories and why they were crucial for helping define a balance between men and women.

## 2 'Good dogs'

Unlike hunters of the Plains who were able to move their camp to be near the carcasses of the animals they killed in their brief but dramatic communal hunts, northern hunters generally target solitary animals: moose and

[3] When asked, he clarified that people had failed to stop the intrusion of Whites and deforestation, so the animals felt betrayed.

woodland caribou. They thus bring meat back to their hunting camps. People use dogs when they can (Sharp 1976: 27; Schwartz 1997: 52; Sharp and Sharp 2015: 116–7), either to move from the village to hunting camps or to bring meat back to camp or to the village. Unlike modern images of sleds pulled by a team of five to seven dogs, in the past, northern Dene people rarely used dogs this way (not to mention that there is no snow on the ground for five or six months of the year!) until they fully committed to the fur trade in the 19th century. It was more efficient to use these animals to carry canvas (earlier, caribou hide) packs, and for people, especially women, to pull toboggans (McCormack 2018) in winter. Archaeological evidence reviewed by Crellin (1994: 59–63) for Northern Athapaskan and Plateau peoples shows a 'scarcity' of dogs, which were 'lightly-built' and therefore not used for hauling.

While early explorers such as Daniel Williams Harmon sometimes used dogs in their explorations (Lamb 1957: 29), it seems Indians favoured travois or packs.[4] Fraser never mentions that Native dogs were used to 'carry loads', even though the Indian groups he encountered had them (Lamb 2007). Alexander Mackenzie's description of his first encounter with a small group of Sekani in 1792 is telling. This, as it turned out, was a typical hunting group: three men, three women, children, but no dogs (Mackenzie 1903: 90–1). Honigmann confirms (1946: 11) that dogs were not used for transport among the Slave (Dene-Tha), and that they only had small dogs for hunting (Sherwood 1958: 11). We can conclude there were few or no large dogs for transport. It was the White traders who had to find a way to move heavy trade goods long distances and adopted sleds for the purpose (Helm *et al.* 1981; Laurens Loovers 2018). Later, Canada's federal police (with various designations - today the RCMP) began more intensive patrolling of the north in the second half of the 19th century. Like the fur traders, they brought big dogs into the north since these were felt to be better suited to hauling sleds (Wishart 2018). These larger animals and sleds might have seemed appealing to Indigenous people as they too became immersed in the fur trade. They began travelling further afield from their home base and their seasonal camps, since they had to face three new problems. First, because the distribution of fur animals is very different from the distribution of meat animals, they sometimes had to travel much farther to trap than to hunt. Second, they had to bring the furs back to seasonal camps and also to the home base (community), or possibly to a distant trading post. Third, fur animals are not generally shared, unlike meat animals. To avoid conflicts, people trapped farther away from one another and from the trading post (Figure 2).

Figure 2. Sahtu (Hare) dog with pack, Mackenzie Valley north of the Arctic (Photo by G. Lanoue).

There is an economic trade-off between feeding five, six, or (rarely) seven large dogs for a team versus keeping only two or three packers and haulers. A team must be equipped, trained and fed. Pack dogs need no particular training. They carry multi-usage packs (that people can also sling over their shoulders, unlike a harness for a team), and smaller pack dogs can be left to scavenge at least part time, unlike sled dogs with their high caloric needs. Small size seems to have been the norm, even for hunting dogs. Simon Fraser mentions that groups he encountered in his 1806 voyage in northwest British Columbia, which he calls Atnah and Meadow (probably Upper Salish and southern Sekani, respectively) used dogs to run down deer (Lamb 2007: 147, 189), though not to attack it. The Dene may have favoured small dogs because they used them to signal the location of prey rather than attacking prey and hauling meat. Small dogs were less costly to feed and a small dog can bark just as effectively as a large one. Depending on size, a sled dog can eat 1.5 to 2.5 kilograms of food a day,[5] but a small dog would eat a fraction of that amount, not only because of its small size but because it has fewer caloric needs since it is only working sporadically. Since dogs were not primarily used to haul sleds, this means that there also fewer of them in an encampment. One or two would suffice a hunting group (Savishinsky 1974:

[4] Harmon's journal mentions two dogs hauling a sled loaded with 150 lbs (70 kg) of furs; elsewhere (Lamb 1957: 131) he mentions that two 'stout' dogs can pull 1000 lbs (450 kg) weight.

[5] Helm and Lurie (1961, cited in Crellin 1974: 62) calculate that a 5-dog team consume 3500 lbs (1885 kg) of food a year, or approximately 315 kg a year per dog.

Figure 3. Sahtu (Hare) dogs with sled, Mackenzie Valley north of the Arctic Circle (Photo by G. Lanoue).

165), rather than five to seven per sled, per family. This would mean less pressure on local resources.

Archaeological and ethnographic data for the Alaskan Koyukon Dene from the early contact epoch show that dogs were not brought into houses (today, bitches are allowed in houses in winter), but one might be kept in the tunnel entrance as a guardian (Clark and Clark 1974: 36). It is unlikely that Dene used dogs to hunt in the sense of helping them to bring down large game. For example, explorer Samuel Black mentions (1955) that the 'Theccanies' (Sekani) used dogs to hunt, but only to find prey and signal its presence. The dogs did not actually attack the animal but ran ahead of the hunter and signalled back when they found prey. Simon Fraser also mentions (Lamb 2007: 106) a dog sacrifice used to mark a funeral. Dogs were frequently eaten but only by his 'Canadian' employees, not by the Indians. On the western side of the Pacific divide, dogs were kept among the coastal tribes (belonging to the Coast Salish linguistic family, though Fraser does not call them by this name) as a source of fur with which to weave blankets (Lamb 2007: 121). Isham's 18th century account - earlier and perhaps more accurate as to pre contact attitudes vis-à-vis dogs - of life among the Cree of Hudson's Bay confirms (Rich 1949: 164) that hunting dogs were used to identify beaver houses. Tellingly, it is his only mention of dogs during his six-year sojourn (1743–9), which again suggests that dogs may not have been so important to survival as other contemporary writers assume.

As if to confirm this, recent genetic analysis has revealed that native dogs in the north are not derived from North American wolf populations but came from Siberia. Most indigenous genetic stock has disappeared and has been replaced by dogs brought over by Europeans (Leathlobhair 2018). At least one long-time observer (Sherwood 1958: 51–56) noted that Whites introduced the pack dogs used by First Nations people. Earlier accounts confirm this: Thorndike (1911, cited in Allen 1920: 445); trader Daniel Harmon in 1820 mentioned that Newfoundland dogs were brought in by traders to bolster local stock in Dene country. There may be exceptions, since northern Spitz-type dogs (Malamutes, Huskies) closely resemble large boned Spitz-type dogs found among Siberian groups such as the Chukchi. Some pre-contact DNA of Siberian origin has likely survived (Figure 3).

Northerners had also developed a small dog used to alert people of the presence of bears. It is now extinct among the Hare (Sahtu), where it was called the Hare Indian dog (Allen 1920: 491; Schwartz 1997: 33),[6] but in 1979 a regional variant (Tahltan bear dog) could still be found in northern British Columbia.[7] Another local breed was the Athabaskan, a long-haired sled dog found in the Mackenzie Valley (Walsh 1898: 122; this appears to be the same as the Hare Dog named by McCormack), though it too is considered extinct.[8] In other words, 'Indian' dogs have virtually disappeared because they were replaced with stronger, larger European dogs as the economy was reoriented to the demands of the fur trade in the 19th century. Heavier loads had to be carried greater distances. I suspect that today the dogs I saw belonging to the Hare of Fort Good

[6] This also refers to a local breed of hauler (McCormack 2018: 114–117, but Allen believes (1920: 491) that this is the 'Common Indian Dog', a 'large' breed that resembled 'black' or southern wolves (Ibid: 461); others also 'confused' the two (Ibid: 462–3).

[7] I am not claiming expertise, but the CKC (Canadian Kennel Club) and others affirm that it has been extinct since the 1930s or the 1950s. In the late 1970s, I saw at least one dog that resembled descriptions and pictures of this breed. The Sekani at the time had no means or motive to import dogs, so it must have been indigenous.

[8] Again, I saw examples that conformed to earlier descriptions. Although this is not proof of these dogs' local origin, people in the Mackenzie had no means of importing 'white' breeds.

Hope are Malamute-wolf crosses. There were several German Shepherds (or Shepherd-wolf hybrids) in the community that belonged to local Whites (there are about 30 Whites and 550 - 600 Aboriginal people in the community). Any puppies born to local ('Native') bitches that are suspected of being sired by these German Shepherds are immediately killed because, as one man put it, 'they are too intelligent'.[9] This suggests that Indigenous Dene are aware of differences between Native and European breeding stock.[10] I did not record any instances of older work dogs being put down among the Hare or the Sekani. Presumably, they live out their natural lifespans.

Dene people vary in how they treat dogs. Some like the Chipewyan (Denésoliné) treat dogs roughly. Dogs roam free and eat most of the village's waste (Sharp 1976). Chipewyan often neglect to feed their dogs, expecting them to scavenge. They are expected to dig shelters that partially protect them from black flies in early summer (Sharp 1976: 117). Sites for tethering dogs are chosen with this in mind, as well as the distance from a drinking source (in winter, dogs are not watered, as they eat snow). The fact that dogs eat anything and show no discrimination unlike humans is taken as proof of dogs' liminal status. They will sometimes be given offal from caribou because some parts (eg, the lungs, which are full of parasites) are considered unfit for human consumption (Sharp and Sharp 2015: 7). Among the Sekani and the Hare, however, fish are caught with nets and dried in the summer months as a source of dog food (cf. Walsh 1898: 122 for Alaskan dogs). Local stores used to sell 'broken rice', factory sweepings considered unfit for humans but given to dogs. Dogs are not allowed to run loose in the community. Among the Hare, at least, they will be shot. The northern Sekani may be more tolerant because they also let pack horses run free. Hare and northern Sekani pack dogs seem to have a healthy admixture of wolf genes and are aggressive. People believe that these dogs explicitly see people as food donors, so they can attack (especially small children) if they are hungry. Since the Chipewyan do not feed them, free-roaming dogs are not seen as aggressive because they do not associate food with humans. McCormack notes (2018) that many groups (eg, the Hare) treated puppies kindly, although this seems to be limited to women and very young children.

In brief, while dogs were important in Dene life, they were not primarily used for transport. They were used for their fur (by Pacific coastal groups only), as sentinels that warned of approaching threats, and especially as scouts to indicate the location of game. Since their bark was more important than their bite, most Indian dogs were small or medium sized. It is only with the arrival of Europeans and the development of the fur trade that large dogs were favoured for hauling, and these came from European breeding stock.

### 3 Ambivalent dogs

Dene use animals to create fields of shared meaning. Stories describing the lives of animals, especially of primordial animals that lived at the beginning of time, become metaphors for human agency and for a functioning community. Until recently, many northern hunters spent 80% of their time in small, isolated hunting camps with little interaction with others. These camps contained two hunters and their dependents, women and children. Sentiments of belonging to a larger community were therefore relatively weak, and communities were indeed fragile. The social landscape was threatened by individualism and egocentrism, both traits that encourage survival in the harsh environment but that obviously undermine sentiments of feeling an emotional attachment to the community. Dene seek to attenuate the effects of individualism by emphasising modesty, silence, humility, and verbal indirection. They also lessen feelings of individual accomplishment by attributing hunting success to power received by mythical animals. Even today, power is believed to reside with animals, who are the descendants of the powerful primordial creatures that dominated humans immediately after the earth was created. Men seek contact with animal power by going on vision quests (I will describe this below). It might seem that augmenting individual power will further weaken the sense of belonging to a community, but the reverse is true. Since people are forbidden to talk about their power received from animals, everyone in fact gossips about other people's powers. If a person becomes a better hunter (or, possibly, a healer), people will speculate, with admiration and with envy, about the identity of the 'animal doctor'. The benefits of enhanced individual abilities can only be validated by others, who gossip about other people's individual success. The community arises from the whispered gossip that fills the official silence surrounding contact with the world of animals. It is, in a sense, negative social space.

In this world of animal power, dogs are the exception. In an imagination populated with stories that relate how primordial animals shaped the world, there are relatively few dog stories that explain their origins. Dogs threaten the established worldview in which animals are considered superior: 1) many animals need no period of apprenticeship before adapting to

[9] Crellin (1974: 61) also suggests that Indigenous people did not breed their dogs with wolves or coyotes, since the resulting hybrids would have been much larger than the small dogs that were typical among Athapaskans.
[10] The standard northern Indian dog breed is also distinguished from European dogs by Greenland Inuit (Brown 1868: 347).

the environment; 2) they live in the same environment as humans, yet need no weapons to survive; 3) finally, humans need the collaboration of others to survive, yet most animals manage to survive alone (even wolves, who generally live in packs, can survive alone). In contrast, dogs are the living embodiment of everything that invalidates these views: 1) they are dependent on humans; 2) they understand human language; 3) dogs cannot live alone (or will not be allowed to live alone). If dogs are not human, neither are they normal animals. As if to confirm this, Clark and Clark report (1974: 36) that Koyukon affirm that dogs and children should not be raised together, since children will become quarrelsome like dogs. Even though Savishinsky calls (1974) dogs extensions of the self among the Hare, my own research experience revealed that anything used in food preparation and service touched by a dog should be thrown away. Among the Sekani, dogs are not allowed inside houses, though they are given a minimum of care.[11] This suggests there is some sort of barrier between the two categories, precisely because they are similar. In brief, dogs cannot confer primordial power on hunters who seek it by means of the vision quest, because they have no power to share. This is highly unusual, since the pantheon of 'power animals' are all effective hunters (or capable of hunting) - bears, wolves, eagles, spiders, coyotes, cougars, etc. Dogs are certainly capable of hunting,[12] and people tell stories of dogs attacking humans and sometimes even killing babies. Dogs should have primordial power, according to this logic, but they do not.

#### 4 'Human' Dogs

If dogs are not fully animal, how human are they? Today, some anthropologists treat all living creatures as sentient, or, rather, see no ontological differences between animals and 'post-human' humans (e.g., Charles 2014). Among the Quichua-speaking Runa in Ecuador's Upper Amazon, far from the Dene homeland, dogs also play a role in the local imaginary. Kohn describes (2007) an incident in which three dogs belonging to a family disappeared one morning. Their bodies were later found in the forest, where they had been dragged after being killed by a jaguar. The dogs had not barked while sleeping. People believe that when sleeping dogs bark, it is a sign they are dreaming. These dreams foretell the future. One type of bark indicates they are dreaming of chasing prey, which they will do the following day. In contrast, a more hi-pitched and whimpering bark (judging by how Kohn transliterated the sound - *cuai*) indicates a dream in which they are being chased and hunted by a jaguar. The family was puzzled that the dogs had not dreamed their own deaths. When dogs fail to dream, as in this case, it leads to a considerable crisis of faith in one's ability to deal with the future.

Kohn argues that, '... the biological world is constituted by the ways in which myriad beings - human and nonhuman - perceive and represent their surroundings. Significance, then, is not the exclusive province of humans' (Kohn 2007: 5). Granted that this story is a framing device, an entrance scenario meant to keep us reading and to hint at the argument that follows, Kohn's deeper analysis of how the Runa project agency onto animals and nature does not take account of what seems to me a fundamental feature: dogs are hunters, but sometimes they are prey. Nor does he mention the interesting fact that the story was related to him by a woman, a point to which I will return later.

I will not address Kohn's claim about the nature of signification, which eliminates the imaginary as a field of distinctly human action, nor will I pursue my criticism of the absence of community in his work, except perhaps as an agglomerations of selves coming into being (his vocabulary) that are framed by the repetition of key tropes and signifieds. I will, however, pursue the question of why women have a special relationship to dogs, at least in Dene society.

Unlike men, women cannot become powerful by acquiring animal power. They cannot receive power because they are believed to possess it from birth. Their menstrual blood is a tangible sign of human survival, just as animal blood is a sign of the animal's willingness to sacrifice itself to help humans survive. Because women in this sense are power, they cannot acquire it. Because their power is innate, however, and not acquired by the ritualised apprenticeship of the vision quest, they have no control over their power. This makes them dangerous. They are thus akin to both primordial animals (powerful) and dogs (powerless).

At the beginning of time, the earth was very different. Creation stories tell of a formless world inhabited by weak people and powerful animals. In fact, animals played the same role as people today: they spoke, sometimes they hunted humans and even married human women. This state of affairs continued until the Transformer intervened and gave the earth its actual form of rivers, forests and mountains. He also gave primordial animals their present-day biological form, usually smaller and weaker so they could no longer express their superior power. He did not fundamentally alter their essence: animals are still more powerful than people, as I mentioned above, but they can no longer speak, hunt humans or marry human females.

[11] In *Errances* (Desgent and Lanoue 2005), I describe an episode in which I was asked to help a dog who had a fishhook embedded in its jawbone. On the other hand, I also voluntarily intervened on several occasions to help dogs ignored by their owners.

[12] On one occasion, we had to hunt down a badly wounded dog who had escaped his leash because, said the chief hunter in our group, he would come back and hunt people when hungry, since he associated people with food.

In the north, Transformer is Beaver (Dzauya, which roughly means 'little beaver'); in the Southwest, the Basin and the Plateau, he is known as Coyote. In the east, Rabbit plays much the same role. Dene stories are silent about Transformer's origins. Many Dene are nominally Christian; they will freely state that God created Transformer and the earth. It is silly to call these affirmations inauthentic. Dene would not have blindly accepted Christian Biblical teachings if some important part did not already accord with their own ideas (cf. Mills 1986). The world as we know it is the result of Transformer's actions. They were not foreordained by the act of creation itself. Transformer is not all-powerful nor intelligent. His deeds are often the result of his greed, gross sexual appetite, and stupidity.

Beaver/Dzauya did not transform dogs because they had no innate power. Despite this, Dogs are nonetheless the protagonists of a very important Dene creation myth, Dog-Husband. This tale is widespread in the northwest, first noted by the missionary Emile Petitot in the 1860s (1886: 311). Morice mentions (1906: 264) a Dogrib (Tlicho, Tłı̨chǫ) belief that they are descendants of 'a big dog'. Sheppard (1983) has mentioned eleven variants, to which I added a twelfth collected by Diamond Jenness in 1927 but which was left unpublished until 2005 (Desgent and Lanoue 2005). The Dogrib variant concludes that Dog-Husband's offspring are the Dogrib. Boas cites (1891) three variants: Petitot's account of the Dog Husband tale among the Dogrib (Great Slave Lake); an unattributed version on Vancouver Island; and yet another among 'The Eskimo of Greenland and Hudson's Bay'.

Among the Beaver people (Dunne-za) of northeastern British Columbia, whose homeland is adjacent to the Sekani, the Dog-Husband myth noted by Ridington (1978: 68) is more allegorical than explicit. Ridington's narrator Charlie Yahey describes various aspects of how the local culture hero Saya (a sort of Transformer rather than Creator; he is similar to the Sekani culture hero Dzauya) created various aspects of the present-day world. In one story (idem: 64), the earth is an island that continually grows year by year. At some point, 'God' (Saya) made a dog whose task it was to explore the newly formed earth and report what he saw. Dog came back with a human bone in its mouth. God then banished the inedible animals below the earth (the narrator mentions it twice, which is a Dene narrative convention to emphasise a point). Because Saya 'did not like it' that dog came back with a human bone in his mouth, he 'throws dog away', creates a wolf from a dog (presumably, another), and sends it on the same exploratory mission.[13] Wolf, however, does not return. Saya claims to be unsure of the wolf's whereabouts, but the narrator states that he in fact knew but wanted wolf to be free and 'live with his teeth' (i.e., as a hunter). With wolf transformed into a hunter (his teeth function like knives), the world is now perfect. No mention is made of the whereabouts of the first, banished dog.

Dog is even more ambiguous in a second Beaver tale. The narrator seems to be speaking allegorically. If women leave their clothes lying about, dog will, 'smell them and piss on them' (Ridington 1978), a motif also mentioned by Goulet (1998: 99) for the Slavey (Dene-Tha). These are, 'no good' dogs 'made by the devil'. Then the woman will become, 'like drunk, like rabies' (Ridington 1978: 68). Ridington's editorial comment states that this is a version of the dog-husband myth, in which primordial dog returns to earth and sleeps with Dunne-za women, which, 'makes them *wechuge*' (Ridington 1978: 121); *wechuge* is the Beaver Cannibal Monster, akin to the better-known Algonquian Windigo (Lanoue and Ferrera 2004). A women's clothes 'sniffed' by a dog (otter or mink are also mentioned as possible offenders) will make the woman 'be rabies', i.e., transformed into a *wechuge*. The change is permanent; there is no cure. *Wechuge*/windigo typically tear off their clothes, are covered by hair, and are much stronger than normal humans. The narrator significantly ends the story on an apparent non-sequitur: there are two dogs that, 'go around to all the reserves', 'to make sure where all the people stay' (Ridington 1978: 68). Apparently, some primordial, powerful dogs are still about in the world.

Sharp (1976) describes the Dogrib exception: the first humans were the result of woman mating with a dog. She then gives birth to pups. When their coats are removed, they reveal their human forms. A supernatural man kills the dog-husband, and its dismembered body is transformed into the animals of the earth, birds, fish and terrestrial game. Eventually, the dog-man hybrid brothers marry their sister and give rise to the Dogrib people (Habgood 1970: 118). This is unique, since all other versions (including the Sekani) view mating with a dog, even inadvertently, as very negative and leading to dire consequences. Sheppard (1983: 96) mentions that in all variants except the Dogrib version, the events are precipitated by a woman's abusive treatment of dogs or other animals. The Dogrib present her as tricked by dog and suffering the consequences even if she is innocent. The same is true in the Sekani variant, where she seeks to identify the man who is creeping into her tent every night. It is even mentioned that her parents were strict with her - she was a 'good' (chaste) girl who was tricked. Nonetheless, her people abandon her and leave her to starve because of her relations with a dog.

[13] This motif evokes the biblical flood narrative, when Noah sends a raven to survey the earth after the flood. The bird, however, does not return. He then sends a dove, which does. Many First Nations' tales contain Christian motifs when these are consistent with indigenous notions of creation and destiny; see Lanoue (1993).

The Sekani Dog Husband myth is similar. Every night, an unknown man comes and visits a young woman in her tent. She never sees his face in the dark. One night, she throws a handful of ochre at his back as he is exiting the tent. In the morning, however, no man has a mark upon his back, but her father's dog's back is stained. Eventually, she gives birth to two male and one female pup. Everyone in the village leaves, horrified, extinguishing their fires and taking their food so the woman and her pups will starve. Primordial Crow, however, saved some fire in an old moccasin. The woman kills some rabbits and make snares from their sinew. The pups grow. One day, the woman notices that the pups have removed their dog skins and are playing as human children. She steals their furs and burns two of them, but the little girl saves her fur and remains a dog. The mother makes bows and arrows for her sons. They become good hunters. One day, they pursue six elk, who escape to the sky. The dog followed them, and so did the two boys and the mother. The mother and her sons were transformed into Orion's belt. The six elk become the Pleiades, and behind them is a single star, the dog.

At first glance, as others like Sharp have noted, the tale has several obvious themes: abandonment and togetherness; wild/nature versus culture/community; male versus female; incest (it is implied in the Dogrib version, since the ancestors are brothers and sisters); and, finally, reproduction. In fact, this seems to be the major theme, since the tale always begins with an inter-species illicit relationship that is unexpectedly fertile, and usually ends with a situation in which the protagonists are separated and no longer able to reproduce except, implicitly, in a 'normal' way - Dogrib create other Dogrib people.

Shepherd (1983: 90) states that the story has a four-part structure, like all Athabaskan tales (citing Scollon and Scollon 1981: 110–111): sexual maturation of a young woman; abandonment of the woman and her offspring; puppies become 'dead' to the 'dog world' when their skins are destroyed by their mother; finally, the children grow up but retain some dog traits. For Sheppard, dog is a negative symbol that explores the consequences of abnormal relationships. In this sense, she seems to be proposing a vision of myth akin to Lévi-Strauss' famous examination of the Gitksan (Gitxsan) story of Asdiwal, in which Lévi-Strauss claims (1973) that unusual or impossible geographic journeys are a metaphoric way of rejecting abnormal social configurations.[14] In other words, the adventures of Dog Husband's children, even if transformed into stars or people, form a 'just-so' story that is in essence a morality tale, albeit a complex one in which the unnamed women (either Dog Husband's wife or Asdiwal's mother, respectively), even as wrongly shunned victims, ultimately insure the survival of their people by creating the geographic and social arrangements that are necessary to life. Shepherd concludes (1983: 99) that, 'In performance, emotional and categorical ambiguities are resolved by the elaboration of culturally immediate relationships and images'. This is far from clear and begs the question of how ambiguities 'in performance' can be resolved by introducing an imaginary filled with even more ambiguity.

[14] A mother and daughter travel in the wrong direction, leading to the daughter's illicit rapport with an entity, who fathers Asdiwal; the women starve, and Asdiwal goes on to become a major culture hero only because of the unusual qualities of his conception, birth and upbringing.

## 5 Women and dogs

In these tales, there are two important protagonists, dogs and women. Kohn's Runa informants who are disturbed by dogs not dreaming their deaths are women. Closer to home, Adlam (1994/5) associates the Tahltan version of Dog Husband with menstrual taboos. The Sekani mother burned her sons' dog skins so they could not transform into dogs, but the female puppy saved her skin and presumably remained a dog. Significantly, the mother and her sons are transformed into stars of the constellation Orion, but the female puppy is singled out and becomes a particular, unique star. Finally, all versions of the Dog-Husband tale either mention or suggest that the human female protagonist is nubile and young. The Mother of Dogs is thus cast in the role of a proto-Eve. Clearly, women are not secondary characters in this drama.[15]

Significantly, women often train young dogs. McCormack's review of the evidence suggests that while sled dogs may be associated with men, pack dogs belong to women. In the past, these were more important than sled dogs. Among some groups, women train puppies by placing small packs on their backs (McCormack 2018: 112–113). It is women, after all, who sew packs from canvas (earlier, from caribou hide). Since women were responsible for hauling in earlier times, it is not surprising that they trained and cared for pack dogs when these were introduced in the northwest (Crellin 1994: 60). The popular association of dogsleds and men is likely a later development as trapping gained in importance at the expense of hunting. A typical fur trap 'line' (the trail, often circular, along which traps are set) is about thirty to sixty miles for the Sekani. A hunter will visit his traps every few days, maybe more often (trapped animals can chew their legs off and escape or be eaten by wolverines). Women and children stay in the hunting camp. A hunting expedition, however, is usually point to point, from the camp to a site determined to be a likely spot to find game. The hunter

[15] The Dene are not alone; Laugrand and Oosten report (2015: 163–5) that Inuit also associate dogs and women.

can usually return to his camp the same day. In brief, the fur economy lengthened distances from the home base, increasing the reliance on dogs as pack animals. Men may have spent more time trapping farther away, but this only made them dependent on 'feminine' dogs for transport.

It may also be significant that contemporary Dene have a mostly bilateral kinship nomenclature, but that proto-Athapascan was matrilineal, according to Dyen and Aberle (1974), though the connection to gendering of social roles is tenuous.

The key fact remains that Dene personhood is channelled through a series of myths and stories that establish power as an animal and not human quality. These myths consistently distinguish and support two different realities: men can acquire power, but women are born with power. As one hunter told me, men have power but women *are* power. While men use the vision quest to transform themselves into symbolic prey - they are alone, immobile, without weapons and food - to reproduce the primordial conditions when superior animals hunted humans, women are barred from seeking power (theoretically, at least till menopause, but I never observed nor heard of a woman engaging in a vision quest). If an animal approaches an immobile hunter, it does so because it is the animal's primordial power that is asserting itself. The animal's biological, post-Transformer nature would otherwise cause it to flee. When approached by an animal, the man as symbolic prey becomes imbued (or contaminated) by the animal's power. Contact can also occur through dreams (involuntary or directed) or, under unusual but not rare conditions, while hunting in the forest. Through contact, the person becomes contaminated by some of the animal's essential power and so becomes transformed into a human-hybrid that partly imitates the knowledge of the Animal Monsters of the mythical past. Men can only become aware of their powers through experience, especially by an increase in luck or specific abilities. Contact with the transcendental aspect of animals does not confer permanent abilities, and these animals are not tutelary spirits as such. The closest metaphor to my knowledge would be to say that contact initiates an on-going conversation between men and the transcendental aspect of animals, and that the 'messages' are only understood later, through evidentiary proof in a hunter's actions.

Moreover, women's monthly cycle is a sign that, like contemporary animals who sacrifice themselves and allow themselves to be killed,[16] they too shed blood to ensure human survival. Even though Sekani women are no longer isolated at first or subsequent menses, there are clear beliefs (at least among men) about the danger represented by female genitals and by menstrual blood in particular. However, many Athapaskan peoples have specific injunctions against female vision quests and combine these with various rules that aim at controlling contact with women's genitals or menstrual blood (for example, a menstruating woman cannot step over a prone man, nor can she step over his weapons).[17]

I am not the first to note that Dene women have a special tie to dogs. For example, Sharp (1976) proposes that the Chipewyan of northern Saskatchewan and the Northwest Territories see parallels between women and dogs: both are 'natural' and indiscriminate, whereas men are associated with 'culture' and wolves. Here, however, I argue the opposite: women and dogs are not similar but have contradictory and complementary traits.

## 6 Conclusions

Earlier, I noted what seems to be a contradiction: given the many uses to which dogs were (and are) put and the many resemblances between dogs and humans, dogs should be respected or at least viewed as a mirror of human society. They are, however, the only carnivorous animal in the Dene pantheon that is believed to have no primordial power. Furthermore, adult dogs are often badly treated and often ignored, though puppies are often well treated by women. In North America, dog sacrifices are well known among the Iroquois (White Dog Ceremony), though in the West I have only found one mention in the explorer literature: Simon Fraser mentions (Lamb 2007: 107) a dog sacrificed by the 'Rocky Mountain people' on the occasion of a funeral in which the unfortunate creature is casually described as hanging from a pole. His matter of fact description suggests he was familiar with this practice, which resembles dog sacrifices among the Koryak, Yukaghir and Gilyakin far-eastern (Kamchatka) Siberia,[18] who also sacrifice dogs to propitiate malevolent forces (Jochelson 1975: 91). Despite having no power and the ill treatment they often receive, dogs would not be sacrificed if they were totally without some symbolic value.

I have reviewed bits and pieces of the evidence to establish that: 1) Indian Dogs were probably of two types, a medium hauler and a small barker; 2) larger

[16] Sekani and Hare hunters tie the animal's feet into a bundle and suspend it from a branch, inviting the animal to come back to earth and reincarnate. The feet are the embodiment of mobility (a powerful symbol among nomadic hunters), so the ritual projects horizontal power onto the vertical axis of heaven and earth.

[17] There are occasional reports of isolating menstruating women (eg, Honigmann 1954: 124) among some nomadic Athapaskan-speaking peoples. Furthermore, among the Sekani, at least, the rule forbidding a woman from stepping over a man or his weapons is invoked at all times, not only during menstruation (see Legros 1999 for a Northern Tutchone analogy).

[18] One pre-DNA analysis suggests that all northern dogs are of one breed: '... I have seen dogs from Kamschatka [*sic*], Sitka, the western shores of Davis's Strait, and from Greenland which it was impossible to deny were of one species' (Brown 1868: 347).

sled dogs were probably introduced by Europeans; 3) the consensus is that few Indian dogs survived as the hunting economy was radically transformed by the fur trade, which favoured larger breeds able to haul more freight. I have also argued that their undoubted usefulness only underlines their puzzling semiotic ambivalence, since they are often treated badly. I argue that an explanation may lie in the fragility of Dene communities threatened by excessive individuality. The Dene solution to this problem is to attribute success and failure to non-human agency; in particular, to the world of animals. This however, leads to a logical problem: if animals are superior, as most Dene attest, a successful hunt can occur only because the animal sacrifices itself. Blood is the symbol of this and becomes a metonym for human survival. In the human world, menstrual blood is a metonym of human survival. This leads to many problems in drawing boundaries around men and women as symbols, logical problems that may manifest as ambiguous and tense behaviours, as I detailed elsewhere (Lanoue 2001).

To sum up, men are seen as naturally weak, and women as naturally powerful. Men, therefore, redress their weakness by developing special bonds with powerful animals. Women are forbidden to do so and were, until recently, subjected to special taboos and rules relating to their menstrual cycles. If women acquired animal power, their augmented power would lead to an imbalance that would threaten not only behavioural norms but also the entire mythical structure. Women are paired with 'powerless' dogs (they raise them; they play with them; they once mated with them) to contain symbolically their natural power. I think it is thus very likely that Dene stripped dogs of their power to allow them to play a role in countering women's excessive power. Despite their usefulness, I propose that in the imaginary of northern hunting societies dogs were attributed this ambiguous status to harmonise the semiotics of male and female inequalities within a framework where human agency and accomplishments do not undermine the fragile community. Weak men court powerful animals, while strong women become the mothers of powerless dogs.

## References


Adlam, R. 1994–5. The Dog Husband and the 'Dirty' Women: The Cultural Context of a Traditional Tahltan Narrative. *Igitur* VI (2)-VII (1): 39–57.

Allen, G.M. 1920. Dogs of the American Aborigines. *Harvard University Museum of Comparative Zoology Bulletin* 63: 431–517.

Andrews, T. and S. Buggey 2008. Authenticity in Aboriginal Cultural Landscapes. *APT Bulletin: The Journal of Preservation Technology* 39 (2/3): 63–71.

Asch, M. 1979. The Economics of Dene Self-Determination, in G.A. Smith and D.H. Turner (eds), *Challenging Anthropology*: 339–352. Toronto: McGraw Hill-Ryerson.

Black, S. 1955. A *journal of a voyage from Rocky Mountain portage in Peace River to the sources of Finlays Branch and north west ward in summer 1824*. Edited by E.E. Rich assisted by A.M. Johnson. London: Hudson's Bay Record Society.

Boas, F. 1891. Dissemination of Tales among the Natives of North America. *The Journal of American Folklore*. Vol. 4, No. 12 (Jan. - Mar., 1891): 13–20.

Brown, R. 1868. On the mammalian fauna of Greenland. *Proceedings Zoological Society of London* 28: 330–362.

Charles, N. 2014. Animals Just Love You as You Are: Experiencing Kinship across the Species Barrier. *Sociology* 48 (4): 715–730.

Clark, A.McF. and W.D. Clark 1974. Koyukon Athapaskan Houses as Seen through Oral Tradition and through Archaeology. *Arctic Anthropology* 11: 29–38.

Crellin, D.F. 1994. Is There a Dog in the House: The Cultural Significance of Prehistoric Domesticated dogs in the mid Fraser River Region of British Columbia. Unpublished M.A. thesis, Department of Archaeology, Simon Fraser University.

Desgent, J.-M. and G. Lanoue 2005. *Errances. Comment se pensent le Nous et le Moi dans l'espace mythique des nomades septentrionaux sekani*. Mercury Series, Cultural Studies Paper 78, Ottawa: Musée des Civilisations.

Dyen, I. and D.F. Aberle 1974. *Lexical Reconstruction: The Case of the Proto-Athapaskan Kinship System*. Cambridge: Cambridge University Press.

Goulet, J.-G. 1998. *Ways of Knowing: Experience, Knowledge, and Power among the Dene-tha*. Lincoln: University of Nebraska Press.

Habgood, T. (trans.) 1970. Indian Legends of North-Western Canada, by Emile Petitot. *Western Canadian Journal of Anthropology* 2 (1): 94–129.

Helm, J. and N. Lurie 1961. The Subsistence Economy of the Dog Rib Indians of Lac La Martre in the Mackenzie District of the Northwest Territories. Ottawa: Department of Northern Affairs and Natural Resources.

Helm, J., E.S. Rogers and J.G.E. Smith 1981. Intercultural Relations and Cultural Change in the Shield and Mackenzie Borderlands, in J. Helm and W.C. Sturtevant (eds), *Handbook of North American Indians, Volume 6 Subarctic*: 146–157. Washington DC: Smithsonian Institution.

Honigmann, J.J. 1946. *Ethnography and Acculturation of the Fort Nelson Slave*. Yale University Publications in Anthropology, no. 33. New Haven: Yale university Press.

Honigmann, J.J. 1954. *The Kaska Indians: An ethnographic reconstruction*. New Haven: Yale University Press (Publications in Anthropology 51).

Jarvenpa, R. 1977. Subarctic Indian Trappers and Band Society: The Economics of Male Mobility. *Human Ecology* 5 (3): 223–259.

Jochelson, W. 1975. *The Koryak, vol. 6 The Jesup North Pacific Expedition*. Memoir of the American Museum of Natural History. New York: AMS Press (1908).

Kohn, E. 2007. How Dogs Dream. *American Ethnologist* 34 (1): 3–24.

Lamb W.K. (ed.) 1957. *Harmon's Journal 1800 - 1819. Daniel Williams Harmon: A Partner in the North West Company*. Victoria : Touchwood Editions.

Lamb, W.K. (ed.) 2007. *The Letters and Journals of Simon Fraser, 1806 - 1808*. Dundurn Press (Macmillan, 1960).

Lanoue, G. 1993. Orpheus in the Netherworld in The Plateau of Western North America: The Voyage of Peni, in A. Masaracchia (ed.) *Orfeo e l'Orfismo*: 447–485. Rome: Gruppo Editoriale Internazionale.

Lanoue, G. 2001. Rites of Female Initiation and Seclusion: Women as metonyms among the Tsimshian, Tlingit, Nootka, Kwakiutl (West Coast) and Sekani (Athapaskan). *Storia, antropologia e scienze del linguaggio* XVI (3): 117–152.

Lanoue, G. and N. Ferrara 2004. Windigo 'Psychosis', Healing the Sick: Balancing the Self in Northern Canadian Hunting Societies. *Anthropologica* 46 (1): 69–83.

Laugrand, F. and J. Oosten 2015. Hunters, Predators and Prey. Inuit Perceptions of Animals. New York: Berghahn.

Laurens Loovers J.P and P. Jan 2018. 'Hard times are coming' Indeterminacy, prophecies, apocalypse and dogs, in Robert J. Losey, R.P. Wishart and J.P Laurens Loovers (eds) *Dogs in the North: Stories of Cooperation and co-domestication*: 191–211. London : Routledge.

Leathlobhair, M.N., A. Perri, E.K. Irving-Pease, K.E. Witt, A. Lindeholm, J. Haile, O. Lebrasseur, C. Ameen, J. Blick, A.R. Boyko, S. Brace, Y.Nunes Cortes, S.J. Crockford, A. Devault, E.A. Dimopoulos, M. Eldridge, J. Enk, S. Gopalakrishnan, K. Gori, V. Grimes, E. Guiry, A.J. Hansen, A. Hulme-Beaman, J. Johnson, A. Kitchen, A.K. Kasparov, Y. Mi-Kwon, P.A. Nikolovskiy, C. Peraza Lope, A. Manin, T. Martin, M. Meyer, K. Noack Myers, M. Omura, J.-M. Rouillard, E.Y. Pavlova, P. Sciulli, M.-H.S. Sinding, A. Strakova, V.V. Ivanova, C. Widga, E. Willerslev, V.V. Pitulko, I. Barnes, M.T.P. Gilbert, K. M. Dobney, R.S. Malhi, E.P. Murchison, G. Larson and L.A.F. Frantz 2018. The evolutionary history of dogs in the Americas. *Science* 361 (6397): 81–85. DOI: 10.1126/science.aao4776; 11-08-2018

Legros, D. 1999. Tommy McGinty's Story of Crow: A First Nations Elder Recounts the Creation of the World. Ottawa: Musée canadien des civilisations.

Lévi-Strauss, C. 1973. Le geste d'Asdiwal, *Anthropologie structurale deux:* 175–233. Paris: Plon. (1958).

Mackenzie, A. 1903. *Voyages from Montreal Through the Continent of North America: To the Frozen and Pacific Oceans in 1789 and 1793, with an account of the rise and state of the fur trade*. New York: A.S. Barnes.

McCormack, P. 2018. An Ethnohistory of dogs in the Mackenzie Basin (western Subarctic), in R.J. Losey, R.P Wishart and J.P Laurens Loovers (eds) *Dogs in the North: Stories of Cooperation and co-domestication*: 105–151. London: Routledge.

Mills, A. 1986. The Meaningful Universe: Intersecting Forces in Beaver Indian Cosmology. *Culture* VI (2): 81–91.

Morice, A.G. 1906. The Great Dene Race. *Anthropos* 1 (2): 229–277.

Nelson, M., D.C. Natcher and C.G. Hickey. 2005. Social and Economic Barriers to Subsistence Harvesting in a Northern Alberta Aboriginal Community. *Anthropologica* 47 (2): 289–301.

Petitot, É. 1886. *Traditions indiennes du Canada Nord-Ouest*. Paris: Maisonneuve.

Rich, E.E. (ed.) 1949. *James Isham's Observations on Hudson's Bay, 1743*. Toronto: The Champlain Society.

Ridington, R. 1978. *Swan People: A Study of the Dunne-za Prophet Dance*. Ottawa: National Museums of Canada (Canadian Ethnology Service paper no 38).

Savishinsky, J. 1974. The Trail of the Hare: Life and Stress in an Arctic Community. New York: Gordon & Breach.

Savishinsky, J. 1978. Trapping, Survival Strategies, and Environmental Involvement: A Case Study from the Canadian Sub-Arctic. *Human Ecology* 6 (1): 1–25.

Schwartz, M. 1979. *A History of Dogs in the Early Americas*. New Haven: Yale U Press.

Scollon, R. and S. Scollon 1981. Narrative Literacy and Face in Interethnic Communication. *Ablex, Advances in Discourse Processes* Vol. 8.

Sharp, H.S. 1976. Man: Wolf: Woman: Dog. *Arctic Anthropology*13 (1): 25–34.

Sharp, H. and K. Sharp 2015. *Hunting Caribou: Subsistence Hunting along the Northern Edge of the Boreal Forest*. Lincoln: University of Nebraska Press.

Sherwood, A. 1958. Some Remarks about the Athapascan Indians. *Anthropologica* 6: 51–56.

Thorndike, T.W. 1911. The Indian sled dogs of North America. *Recreation*: 74–77, 101–102.

Walsh, G. 1898. The Cult of the Dog. *The North American Review* 167 (500): 120–123.

Wilson, N. 2014. The Politics of Adaptation: Subsistence Livelihoods and Vulnerability to Climate Change in the Koyukon Athabascan Village of Ruby, Alaska. *Human Ecology* 42 (1): 87–101.

Wishart, R. 2018. The police and dogs during the early patrol years in the Western Canadian Subarctic, in R. J. Losey, R.P Wishart and J.P Laurens Loovers (eds) *Dogs in the North: Stories of Cooperation and co-domestication*; 152–171. London: Routledge.

# 3.7 Dogs through Time: An Ethno-Evolutionary Perspective

Tiziano Latini[1], Luca Pandolfi[2,3], Saverio Bartolini Lucenti[2]

[1] Sapienza University of Rome. Piazzale Aldo Moro 5, 00185, Rome, Italy.
tiziano.latini@uniroma1.it
Ministry of Culture (MiC). Via del Collegio Romano 27, 00186, Rome, Italy.
tiziano.latini@cultura.gov.it
[2] Earth Sciences Department, University of Florence. Via G. La Pira, 4, 50121, Florence, Italy. saverio.bartolinilucenti@unifi.it
[3] Department of Sciences, University of Basilicata, Campus di Macchia Romana. Viale dell'Ateneo Lucano, 10, 85100 Potenza, Italy. luca.pandolfi@unibas.it

Corresponding author: Tiziano Latini, tiziano.latini@cultura.gov.it

**Abstract**

The origin of dogs and *Canis lupus* domestication are still a matter of harsh debate among researchers. Osteological records of dog-like canids and genetic analyses suggests that wolf domestication could have taken place as early as 36,000 years BP. The domestication of dogs influenced human cultural behaviour. Throughout centuries and geographical contexts, dogs are well represented and related to human presence. The evolutionary relationship between dog and man has undergone several adaptations due to the different needs, of which Archaeologists and Ethnoarchaeologists must pay into consideration.

**Keywords:** Proto-dogs, mtDNA, domestication process, dog-man relationship, Ethnoarchaeology.

## 1 Introduction

The debate on the domestication of the wolf (*Canis lupus*) and, consequently, on the origin of the domestic dog is still controversial and involves experts in various disciplines, from archaeology to genetics, from ethology to palaeontology. The two major approaches that attempt to resolve this issue are archaeological research and the genetic reconstruction of the molecular steps from wolf to dogs.

One of the main discrepancies between the two lies in the timing of the origin of the dog. Classical estimations, based on archaeological evidence, suggest that the rise of the dog dates back to about 16–17,000 years BP (Morey 2014).

Early molecular clock approaches, instead, mark the genetic distinction between dogs and wolves much further back in time: Vilà *et al.* (1997) proposed an estimated time of divergence around 100,000 years BP, possibly even earlier.

More recent studies, that include some authors of this pioneering research, have lowered the estimation to the interval of 34–11,000 years ago (depending on the considered average mutation rate, Freedman *et al.* 2014). Thus, partially reconciling the gap between, morphometric and morphological studies on fossils and palaeogenetics. Nevertheless, the debate has not settled neither on the archaeological nor the genetic ground.

## 2 Past

### *2.1 Palaeolithic dogs or proto-dogs*

Several archaeological and molecular studies advocate that the origin of the dog is older than the Last Glacial Maximum (e.g., Germonpré *et al.* 2009). For instance, some canid remains, collected from European sites and referable to the first half of the Upper Palaeolithic (between 36,000 and 33,000 years ago), were attributed to the so-called Palaeolithic dogs (Germonpré *et al.* 2009) or proto-dogs (Figure 1).

Noteworthy, among these the canids (Germonpré *et al.* 2009; Germonpré *et al.* 2015), are those from the Goyet Cave (Belgium, 36,000–33,000 years BP, Figure 1), the Razboinichya Cave (Russia, 33,000 years BP, Figure 1) and the remains from Předmostí (Czech Republic, 27–26,000 years BP, Figure 1).

The peculiarity of the putative dog from the Razboinichya Cave (with a confirmed age of about 33,000 years BP) was also highlighted by molecular studies on mtDNA (Druzhkova *et al.* 2013): the analysis revealed a unique haplotype for this fossil, closer to dogs than to wolves. Alongside these studies, other researchers have questioned these attributions preferring more conservative interpretations (Boudadi-Maligne and Escarguel 2014) and others, such as Morey (2014), suggest that reliable evidence of *Canis lupus familiaris* fossils, and therefore for the timing of domestication, do not exceed 16–17,000 years BP.

Table 1. Main diagnostic morphological features between dogs and wolves.

| Section | Description | References |
|---|---|---|
| A | Proportionally short skull | Germonpré *et al.* 2009 |
| B | Proportionally short snout | Germonpré *et al.* 2009; Hare *et al.* 2012; Wilkins *et al.* 2014; Morey and Jeger 2015 |
| C | Wide palate | Germonpré *et al.* 2009 |
| D | Rising forehead area | Hare *et al.* 2012; Morey and Jeger 2015 |
| E | Teeth size reduction | Wilkins *et al.* 2014 |
| F | Reduction of the tympanic bullae | Evans and Lahunta 2013 |

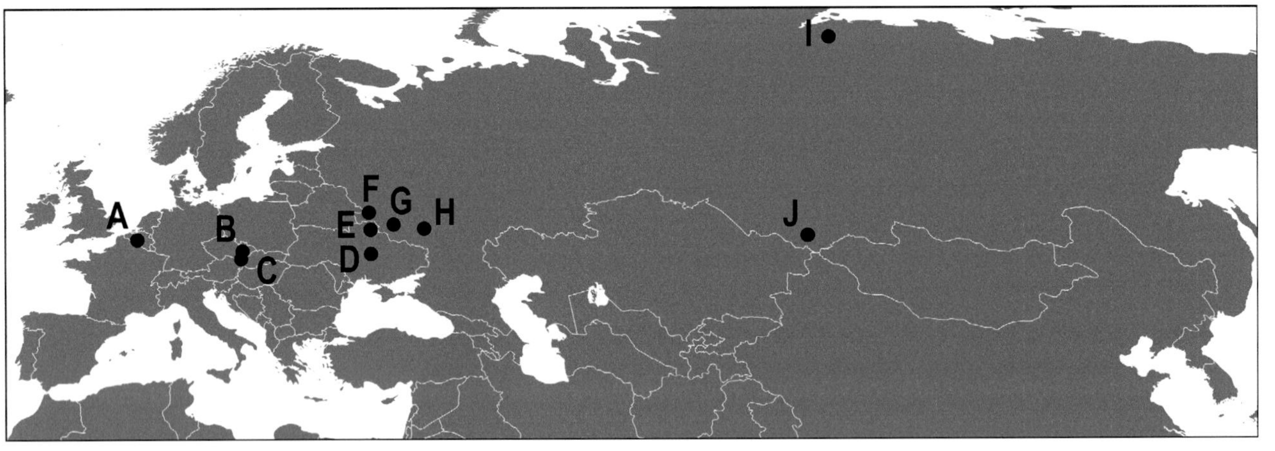


Figure 1. Selected fossil sites with proto-dog remains. A: Goyet Cave; B: Předmostí; C: Dolní Věstonice; D: Mezhirich; E, Mézin; F, Eliseevichi; G: Avdeevo; H: Kostenki 17; I: Anabar River (Yakutia); J: Razboinichya (After Germonpré *et al.* 2012).

The role played by proto-dogs in relation to human activities (in particular hunting) is a matter of debate, raising various hypotheses (Miklósi and Soproni 2006, Shipman 2015). In particular, Germonpré *et al.* (2017) suggested several roles for the proto-dogs, such as transporting materials, guarding camps and/or prey that have been hunted down by human beings, aid in hunting and defending large predators. The proto-dogs could represent an example of the failure of the first phase in the domestication of wolves (Ovodov *et al.* 2011; Thalmann *et al.* 2013); the latter hypothesis would explain the lack of proto-dog remains referable to the end of the Pleistocene (Boschin *et al.* 2020).

Part of the problem leading to such harsh debate is probably related to the techniques of investigation used in these methodologies, for instance the choice of the comparative specimens/samples used for biometric analyses. In addition, the possibility of recognition of valid diagnostic morphological features between dogs and wolves in fossil specimens (Table 1; Figure 2) may play an important role. Some of the most commonly used features to discriminate between dogs and wolves in Late Pleistocene sites are reported here.

Another unsettled debate is the geographic centre of the origin of dogs. Unlike the 'classical' theory of the origin of domestication in Europe (Thalmann *et al.* 2013), other scholars favour a Middle East (Holdt *et al.* 2010) or an East-Asian origin for the domestic dog (Wang *et al.* 2016). The former hypothesis finds support in the early fossil record of several European sites (before the LGM), whereas the latter deepens the roots in the discovery of a great genetic diversity in the sequenced genomes of East Asian dogs (in comparison to other ones). This intricate and confounded pattern of differences revealed by numerous studies may result from interbreeding between different dog populations and/or wild canids. Indeed, some scholars argue that domestication took place in several parts of Eurasia at different times in the last 30,000 years (Thalmann *et al.* 2013; Wang *et al.* 2016). Thalmann *et al.* (2013) suggested that the so-called Palaeolithic dogs may represent an example of the failure of the first phase in the domestication of wolves.

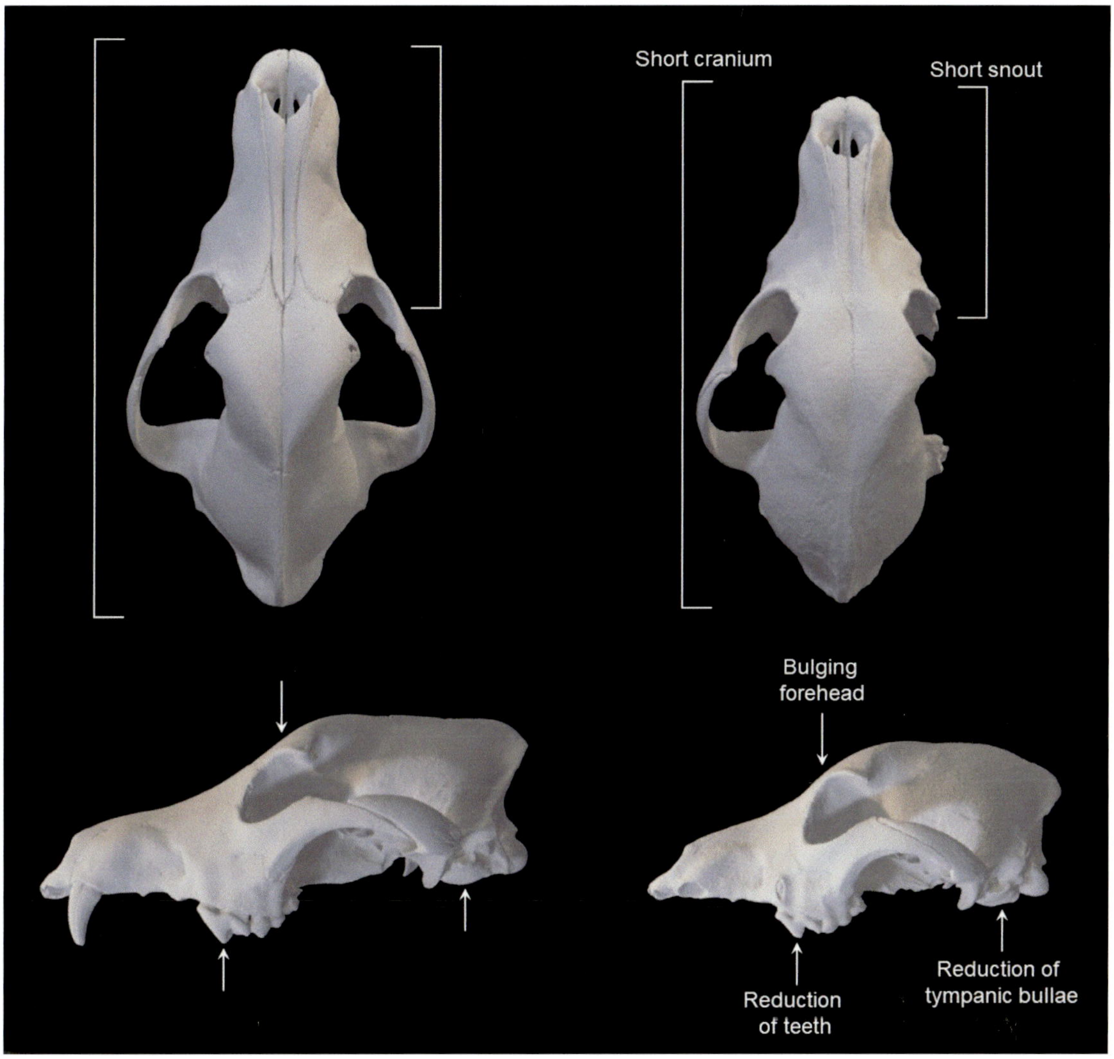


Figure 2. Relevant characteristics of a dog skull (right) compared to a wolf skull (left).

### *2.2 'Domestic' dogs*

The presence of small domestic dogs seems to be attested during the late Upper Paleolithic (i.e., 15–11,500 years BP).

An important example is the discovery of 49 small canid remains from three French sites (Pont d'Ambon, Le Closeau and Montespan), together with other records in the Iberian Peninsula (Pionnier-Capitan *et al.* 2011 and references therein) which show the presence in Western Europe of small dogs at least from the Middle Magdalenien to the end of the Epipalaeolithic.

Small-sized dogs appear to be contemporaneous with larger Russian Upper Palaeolithic dogs (between 13,000 and 17,000 years BP; Sablin and Khlopachev 2002), suggesting the presence of different centres of origins in Eurasia and the possible beginning of selective breeding.

Selective breeding seems to continue in the Early Holocene. Two different sizes of dogs have been recognised in the Early Holocene site of Zhokhov Island (around 9000 years ago) (Pitulko and Kasparov 2017): one with the shoulder height of about 60–70 cm, almost similar to the modern Alaskan malamute, whilst the other breed corresponds to the modern Siberian husky standard (Pitulko and Kasparov 2017).

Different forms of dogs were used by the ancient Zhokhov inhabitants for different purposes, including hunting and transportation (Pitulko and Kasparov 2017).

Dog burials have been recorded in several localities dating from the Mesolithic. The oldest example was found at Bonn-Oberkassel in Germany where a woman and a man were buried together with a dog between 12,650 and 11,280 cal. years BP (Grünberg 2013 and references therein). Dog burials are known from many sites in different countries: Italy, Denmark, France, Netherlands, Poland, Portugal, Spain, Serbia, Sweden, Russia, and others (Cencetti *et al.* 2006; Brea *et al.* 2010; Grünberg 2013; Daza Perea 2017; Hasler and Noret 2017; Losey *et al.* 2018; Albizuri *et al.* 2019 and others).

It should be noted that early dog burials are also known from the Natufian in Israel/Palestine, the Jomon culture in Japan (between 7500 and 7300 cal. years BP: Sato *et al.* 2015) from the Archaic complexes in North America (Morey 2006; Grünberg 2013), from early Neolithic in Siberia (Losey *et al.* 2011).

Dogs were buried at dwelling sites and in huts, but also at human burial grounds, interred alone or together with human remains.

Normally, in geographic areas where dogs were symbolised, they were not eaten but possible sacrifice rituals were found. Sacrificed dogs from the Bronze Age site of Krasnosamarskoe (1900–1700 cal. years BP) in the Russian steppes are explained as the remains of a coming-of-age ceremony in which boys were transformed into warriors by symbolically becoming dogs and wolves through the consumption of their flesh (Anthony and Brown 2017).

In some other areas, such as Thailand, domestic dogs were raised for meat and were consumed from about 3500 years BC up to modern times (Higham *et al.* 1980).

In the Bronze Age, it is possible to see a selection of features in domestic dogs (deduced for example by the strong variability observed on the limbs: De Grossi Mazzorin and Tagliacozzo 2000), with substantial morphological variability, but well-defined breeds seem to be distinguishable only in the late Roman period (dogs display different cranial forms and brachymelic dogs are present). The existence of different dog morphotypes during the Roman period has been confirmed by several studies (De Grossi Mazzorin and Tagliacozzo 2000; Meniel 2002; Baxter 2006, 2010; MacKinnon 2010; Colominas 2015; Pires *et al.* 2018). Indeed, during that time, small lap or toy dogs occurred for the first time (MacKinnon 2010; Pires *et al.* 2018 and references therein).

During the Roman period there is no unequivocal evidence of dogs being slaughtered, and no signs of butchery were found in a Roman context despite some sources reported the use of puppies in cooking (De Grossi Mazzorin and Tagliacozzo 1997, 2000).

In fact, the extent of morphological variation is most evident in Roman times and today perhaps because artificial selection was and is extensively exercised over dogs by humans for aesthetic/emotional reasons (e.g., dogs just for companionship such as lap dogs) and not only for functional/working purposes (dogs used for hunting or herding) (Pires *et al.* 2017).

## 3 Present

### *3.1 Enculturation*

Wolves could have become less and less wild in the past and humans could have employed proto-dogs to collaborate in their activity, especially in cooperative hunting (Lupo 2011: 4).

Some ethologists highlight the dog's specific ability to read and interpret human socio-communicative behaviour, may have evolved during domestication (Alleva *et al.* 2008). In fact, a non-social memory task has also been found, ruling out the possibility that dogs outperform wolves in all human-guided tasks (Hare *et al.* 2002). Seemingly, dog's socio-cognitive abilities emerged as a result of the selection of systems that mediate fear and aggression towards humans (Alleva *et al.* 2008).

The first proto-dogs would probably have approached man to consume his waste, as scavengers and not as wild predators:

> *They* [Dogs] *usually stay in close proximity to the village or camp and, surprisingly, often show reluctance to go on hunts.*
>
> (Lupo 2011: 7)

### *3.2 Domestication*

Little by little, man approached semi-wild dogs and utilised them for different purposes. The initial uses would in all likelihood have been connected to hunting:

> *On several occasions we observed hunters, dragging and even carrying the dogs into the forest to go hunting.*
>
> (Lupo 2011: 7)

A cross-cultural comparison between Lupo's case study and dog-man's contemporaneous behaviour reveals some differences about traditional societies and western urban non-traditional society. These various aspects could authorise to delineate some *buckshots* (Yellen 1977: 7) for a new kind of approach toward the relationship between man and his best friend.

Largely, familiarity between man and dog has been considered most probable during relatively recent times. In fact, during the last decades many countries have revised their laws concerning animal abuse[1], especially after the COVID-19 event. Modern hunter-gatherer societies do not seem to perceive dogs as domestic family members:

> *Bofi and Aka hunters laughed at the suggestion that dogs might be companions/friends or family members*
>
> (Lupo 2011: 6)

Dogs usually do not live in the internal boundary of the house:

> [...] *Dogs are generally roughly treated* [...] *kicked, hit or thrown out of huts.*
>
> (Lupo 2011: 6)

Dogs usually stayed outside of homes. Some traditional tribes consider the dog useful for cleaning the camps rubbish. The dog therefore becomes more of a scavenger than a hunter:

> *While all hunters acknowledged that dogs kept sites clean and worked as garbage disposals, they did not cite this as an important benefit to owning a dog.*
>
> (Lupo 2011: 6)

Dogs have been used to stave off annoying external nuisances in modern societies, and they guarantee protection at least to the family. Forest foragers in the Central African rainforest are able to guarantee protection on their own, without the help of the dog:

> *No hunter cited protection in the forest as a function fulfilled by dogs;*
>
> (Lupo 2011: 6)

Nobody feeds dogs:

> *Dogs are provisioned and what they eat depends on their hunting success.*
>
> [...]
>
> *Most people said that garbage was the food of the unsuccessful hunting dog.*
>
> (Lupo 2011: 7)

Industrial societies produce and forage products suitable for feeding dogs, and the sales of these articles are going up steeply.[2]

Nobody takes care of dogs and their health:

> *During the wet season when hunting returns are generally poor*
>
> [...]
>
> *All dogs are very lean and perpetually hungry.*
>
> (Lupo 2011: 7)

Nowadays, dogs are much more pampered than in the past. In Italy, adoptions have grown considerably in recent years. This occurrence is a symptom of greater care for the animal.[3]

### 3.3 Burial

The progressive spread of cremation practices with the conservation of ashes inside the house, including animals, such as dogs, denotes a highly emotional and symbolic value that goes beyond mere apotropaic superstitions regarding the proximity of death and its negative value. Now the dog is inside, it is inside the house, because it is part of it, without prejudice and without hesitation. The ashes of the dog (human remains as well) are often transformed into the most precious jewel there is: the diamond.[4] The death of a dog in traditional communities does not seem to have negative health or superstitious consequences. It is quite a neutral phenomenon. Dogs' burials, like human ones, are guaranteed and seem to have the same dignified level as man, without any implications of commonality or family relationships.

> [...] *dog burials will likely be found in close proximity to human burials and living spaces.*
>
> (Lupo 2011: 9)

Dog graves are close to living spaces and not within the site. The closeness to the human tombs does not seem to correspond to the affinity to the family unit or to any of its members:

> [...] *the proximity of burial location to the house and other family members*
>
> *does not reflect a similarly close position of the dog within the family.*
>
> (Lupo 2011: 9)

The dog seems to belong to nobody. This methodological perspective is an important factor for the interpretation of the archaeological

[1] No more dog meat in China: www.ansa.it/canale_terraegusto/notizie/cibo_e_salute/2020/04/09/-svolta-in-cina-vietato-mangiare-cani-e-gatti-_14799f8e-8b03-4c7d-a448-069e6829f42c.html (viewed 28 August 2020); Italian Law D.P.C.M. 11 March 2020 (GU n. 64, 11-3-2020: https://www.gazzettaufficiale.it/eli/id/2020/03/11/20A01605/sg (viewed 28 August 2020); Italian Law 20th July 2004, n. 189.

[2] Increase of dog and cat food sale from 1,792.70 Mio EUR (2013) to 1,971.40 Mio EUR (2016) in Italy: www.zoomark.it/media/zoomark/pressrelease/2017/Rapporto_Assalco-Zoomark_2017.pdf: 70 (viewed 30 January 2020).

[3] Zoomark Italy has recorded a gradual increase in the adoption of stray animals from 37.70% (2007) to 45.40% (2017): www.zoomark.it/media/zoomark/pressrelease/2017/Rapporto_Assalco-Zoomark_2017.pdf: 55 (viewed 30 January 2020).

[4] www.exequiapet.it (viewed 28 August 2020).

remains between man and dog. Archaeologists and Ethnoarchaeologists must be scrupulous about it.

## 4 Discussion

Traditional societies usually practised hunting for subsistence. Dog's brain/behaviour have changed over time (Horschler and MacLean 2019; Hecht *et al.* 2019) and the interaction between man and this animal is probably changing our attitude in the approach to all pets.[5]

The number of hunters has been decreasing for years now[6] whereas the ratio between hunters and dogs has increased from 7.84% to 15.01% units[7] over the past decade. If we focus our attention on the cited increase, and not on individuals, we can infer either that hunting has changed in a way that calls for more dogs, or that the dogs in excess could be additional dogs for affection; from dogs for hunting (Miklósi and Soproni 2006; Lupo 2011; Shipman 2015) we have probably come to consider dogs mainly for affection. In fact, in Italy for the first time during 2015, dogs have exceeded the number of children.[8]

## 5 Conclusions

Urban western society is changing. There is no longer an ethnocentric perspective. Dogs are taming/ educating their masters. While in the past man has given dogs the chance to be less animal, now dogs are giving man an occasion to become more human. A new ethno-evolutionary perspective has happened. For millennia dog has approached us, now we approach the dog for our well-being. Nowadays, dog is changing our habits. He (and not it) is changing our ethical vision, and therefore Ethnoarchaeologists must implement an emic view of the relationship between man and dog over the millennia.

Consequently, when we hypothesise our inferences, especially on *Naturvölker* and their relationship with dogs, we must be careful to invert our ethical vision, entering an emic perspective with our furry best friends.

## Acknowledgements

Thanks to the support granted by Linda Marie Amos.

[5] See Rome Mayor ordinance 30th December 2019, n. 244 about prohibition on the use of fireworks to preserve animals' safety as well: www.comune.roma.it/web-resources/cms/documents/ordinanza_botti_capodanno_2020.pdf

[6] See Hunting licences drop in Italy: www.istat.it.

[7] Dog-Hunter ratio in Italy: www.lav.it; www.lastampa.it - Regional registry.

[8] Dogs: 8.693.294; Children: 8.383.122 units (www.lav.it; www.istat.it)

## References

Albizuri, S., J. Nadal, P. Martín, J. F. Gibaja, A. Martín Cólliga, X. Esteve, X. Oms, M. Martí, R. Pou, D. López-Onaindia and M. Eulàlia Subirà 2019. Dogs in funerary contexts during the Middle Neolithic in the northeastern Iberian Peninsula (5th–early 4th millennium BCE). *Journal of Archaeological Science: Reports* 24: 198–207.

Alleva, E., F. Cirulli and N. Francia 2008. Rapporti uomini e cani, e cani e uomini nel terzo millennio: problemi emergenti. L'uso e l'abuso degli animali: spunti per un'azione didattica, in C. Bedetti, M.C. Barbaro and A.M. Rossi (eds). *Supplemento del Notiziario dell'Istituto Superiore di Sanità, Dispense per la scuola/2*: 1–11. Rome.

Anthony, D.W. and D.R. Brown 2017. The dogs of war: A Bronze Age initiation ritual in the Russian steppes. *Journal of Anthropological Archaeology* 48: 134–148.

Baxter, I.L. 2006. A dwarf hound skeleton from a Romano-British grave at York Road, Leicester, England, U.K., with a discussion of other Roman small dog types and speculation regarding their respective aetiologies, in M.S. Lynn and A.M. Elizabeth (eds) *Dogs and People in Social, Working, Economic or Symbolic Interaction* (Proceedings of the 9th ICAZ Conference): 12–23. Durham: Oxbow Books

Baxter, I.L. 2010. Small Roman dogs, viewed 11 February 2020. http://alexandriaarchive.org/bonecommons/archive/files/baxter_2010_small_roman_dogs_6dc7d64928.pdf.

Boschin, F., F. Bernardini, E. Pilli, S. Vai, C. Zanolli, A. Tagliacozzo, R. Fico, M. Fedi, J. Corny, D. Dreossi, M. Lari, A. Modi, C. Vergata, C. Tuniz, A. Moroni, P. Boscato, D. Caramelli and A. Ronchitelli Bernardini 2020. The first evidence for Late Pleistocene dogs in Italy. *Scientific Reports* 10, 13313. https://doi.org/10.1038/s41598-020-69940-w

Brea, M.B., P. Mazzieri and R. Micheli 2010. People, dogs and wild game: evidence of human-animal relations from Middle Neolithic burials and personal ornaments in northern Italy. *Documenta Praehistorica* XXXVII: 125–145.

Boudadi-Maligne, M. and G. Escarguel 2014. A biometric re-evaluation of recent claims for Early Upper Palaeolithic wolf domestication in Eurasia. *Journal of Archaeological Science* 45: 80–89.

Cencetti, S., P.P.A. Mazza, F. Chilieri and F. Cozzini 2006. Madonna del Piano (Sesto Fiorentino, Florence, central Italy) ox and dog: a case of intentional Iron Age inhumation. *Geobios* 39: 328–336.

Colominas, L. 2015. Morphometric variability of Roman dogs in Hispania Tarraconensis: the case study of the Vila de Madrid Necropolis. *International Journal of Osteoarchaeology* 26: 897–905.

Daza Perea, A., 2017. Preliminary Studies of Late Prehistoric Dog (*Canis lupus f. Familiaris* Linnaeus, 1758) Remains from the Iberian Peninsula: Osteometric and 2D Geometric Morphometric Approaches. *Papers from the Institute of Archaeology* 27(1): Art. 12, 1–21.

De Grossi Mazzorin, J., and A. Tagliacozzo 1997. Dog remains in Italy from the Neolithic to the Roman Period. *Anthropozoologica* 25–26: 429–440.

De Grossi Mazzorin, J., and A. Tagliacozzo 2000. Morphological and osteological changes in the dog from the Neolithic to the Roman period in Italy. Dogs Through Time: An Archaeological Perspective, J. Crockford (ed.). *British Archaeological Reports International Series* 889: 141–161. Oxford.

De Grossi Mazzorin, J., and C. Minniti 2006. Dog Sacrifice in the Ancient World: A Ritual Passage?, in L.M. Snyder and E.A. Moore (eds) *Dogs and People in Social, Working, Economic or Symbolic Interaction*: 62–66. Oxbow Books, Oxford.

Druzhkova, A.S., O. Thalmann, V.A. Trifonov, J.A. Leonard, N.V. Vorobieva, N.D. Ovodov, A.S. Graphodatsky and R.K. Wayne 2013. Ancient DNA Analysis Affirms the Canid from Altai as a Primitive Dog. *PLoS ONE* 8 (3): e57754.

Evans, H.E. and A. De Lahunta 2013. *Miller's anatomy of the dog-E-Book*. Elsevier Health Sciences.

Freedman, A.H., I. Gronau, R.M. Schweizer, D. Ortega-Del Vecchyo, E. Han, P.M. Silva, M. Galaverni, Z. Fan, P. Marx, B. Lorente-Galdos, H. Beale, O. Ramirez, F. Hormozdiari, C. Alkan, C. Vilá, K. Squire, E. Geffen, J. Kusak, A.R. Boyko, H.G. Parker, C. Lee, V. Tadigotla, A. Siepel, C.D. Bustamante, T.T. Harkins, S.F. Nelson, E.A. Ostrander, T. Marques-Bonet, R.W. Wayne and J. Novembre 2014. Genome sequencing highlights the dynamic early history of dogs. *PLoS genetics* 10(1): e1004016.

Germonpré, M., M.V. Sablin, R.E. Stevens, R.E. Hedges, M. Hofreiter, M. Stiller and V.R. Després 2009. Fossil dogs and wolves from Palaeolithic sites in Belgium, the Ukraine and Russia: osteometry, ancient DNA and stable isotopes. *Journal of Archaeological Science* 36 (2): 473–490.

Germonpré, M., M. Lázničková-Galetová and M.V. Sablin 2012. Palaeolithic dog skulls at the Gravettian Předmostí site, the Czech Republic. *Journal of Archaeological Science* 39 (1): 184–202.

Germonpré, M., M.V. Sablin, M. Lázničková-Galetová, V. Després, R.E. Stevens, M. Stiller and M. Hofreiter 2015. Palaeolithic dogs and Pleistocene wolves revisited: a reply to Morey (2014). *Journal of Archaeological Science* 54: 210–216.

Germonpré, M., S. Fedorov, P. Danilov, P. Galeta, E.L. Jimenez, M. Sablin and R.J. Losey 2017. Palaeolithic and prehistoric dogs and Pleistocene wolves from Yakutia: Identification of isolated skulls. *Journal of Archaeological Science* 78: 1–19.

Grünberg, J. M. 2013. Animals in Mesolithic Burials in Europe. *Anthropozoologica* 48 (2): 231–253.

Guagnin, M., A.R. Perri and M.D. Petraglia 2017. Pre-Neolithic evidence for dog-assisted hunting strategies in Arabia. *Journal of Anthropological Archaeology* 49: 225–236.

Hare, B., M. Brown, C. Williamson and M. Tomasello 2002. The domestication of social cognition in dogs. *Science* 298 (5598): 1634–1636.

Hare, B., V. Wobberand and R. Wrangham 2012. The self-domestication hypothesis: evolution of bonobo psychology is due to selection against aggression. *Animal Behaviour* 83(3): 573–585.

Hasler, A. and C. Noret 2017. Habitats et structures funéraires néolithiques sur le tracé du cadereau d'Alès à Nîmes (Gard): premiers résultats, in *6e Rencontres Méridionales de Préhistoire Récente, 2004, Coédition ADRAHP-PSO*. 2006: 171–190. Périgueux.

Hecht, E.E., J.B. Smaers, W.D. Dunn, M. Kent, T.M. Preuss and D.A. Gutman 2019. Significant neuroanatomical variation among domestic dog breeds. *Journal of Neuroscience* 39 (39): 7748–7758.

Higham, C.F.W., A. Kijngam and B.F.J. Manly 1980. An analysis of prehistoric canid remains from Thailand. *Journal of Archaeological Science* 7: 149–165.

Horschler, D.J. and E.L. MacLean 2019. Leveraging brain-body scaling relationships for comparative studies. *Animal cognition* 22 (6): 1197–1202.

Losey, R.J., V.I. Bazaliiskii, S. Garvie-Lok, M. Germonpré, J.A. Leonard, A.L. Allen and M.V. Sablin 2011. Canids as persons: Early Neolithic dog and wolf burials, Cis-Baikal, Siberia. *Journal of Anthropological Archaeology 30*(2): 174–189.

Losey, R.J., T. Nomokonova, A.V. Gusev, O.P. Bachura, N.V. Fedorova, P.A. Kosintsev, M.A. Katzenberg and M.V. Sablin 2018. Dogs were domesticated in the Arctic: Culling practices and dog sledding at Ust'-Polui. *Journal of Anthropological Archaeology 51*: 113–126.

Lupo, K.D. 2011. A dog is for hunting, in U. Albarella and A. Trentacoste (eds) *Ethnozooarchaeology*: 4–12. Oxford: Oxbow Press.

Lupo, K.D. 2017. When and where do dogs improve hunting productivity? The empirical record and some implications for early Upper Paleolithic prey acquisition. *Journal of Anthropological Archaeology* 47: 139–151.

MacKinnon, M. 2010. 'Sick as a dog': zooarchaeological evidence for pet dog health and welfare in the Roman world. *World Archaeology* 42(2): 290–309.

Meniel, P. 2002. Le chien en Gaule, in A. Curci and D. Vitali (eds) *Animali tra Uomini e dei. Archeozoologia del Mondo Preromano* (Studi e Scavi nuova serie 14): 45–52. Bologna: Ante quem.

Miklósi, Á. and K. Soproni 2006. A comparative analysis of animals' understanding of the human pointing gesture. *Animal Cognition* 9: 81–93.

Morey, D.F. 2006. Burying key evidence: the social bond between dogs and people. *Journal of Archaeological Science* 33(2): 158–175.

Morey, D.F. 2014. In search of Paleolithic dogs: a quest with mixed results. *Journal of Archaeological Science* 52: 300–307.

Morey, D.F. and R. Jeger 2015. Paleolithic dogs: Why sustained domestication then? *Journal of Archaeological Science: Reports* 3: 420–428.

Ovodov, N.D., S.J. Crockford, Y.V. Kuzmin, T.F. Higham, G.W. Hodginsand and J. van der Plicht 2011. A 33,000-year-old incipient dog from the Altai Mountains of Siberia: Evidence of the earliest domestication disrupted by the Last Glacial Maximum. *PLoS ONE* 6 (7): e22821.

Pionnier-Capitan, M., C. Bemilli, P. Bodu, G. Célérier, J.G. Ferrié, P. Fosse, M. Garcia and J.D. Vigne 2011. New evidence for Upper Palaeolithic small domestic dogs in South Western Europe. *Journal of Archaeological Science* 38: 2123–2140.

Pires A.E., I.R. Amorim, C. Borges, F. Simões, T. Teixeira, A. Quaresma, F. Petrucci-Fonseca and J. Matos 2017. New insights into the genetic composition and phylogenetic relationship of wolves and dogs in the Iberian Peninsula. *Ecology and evolution* 7(12): 4404–4418.

Pires, A.E., C. Detry, C. Fernandez-Rodriguez, S. Valenzuela-Lamas, A.M. Arruda, J. De Grossi Mazzorin, M. Ollivier, C. Hänni, F. Simões and C. Ginja 2018. Roman dogs from the Iberian Peninsula and the Maghreb–A glimpse into their morphology and genetics. *Quaternary International* 471: 132–146.

Pitulko, V.V. and A.K. Kasparov 2017. Archaeological dogs from the Early Holocene Zhokhov site in the Eastern Siberian Arctic. *Journal of Archaeological Science: Reports* 13: 491–515.

Sablin, M.V. and G.A. Khlopachev 2002. The earliest Ice Age dogs: Evidence from Eliseevichi 1. *Current Anthropology* 43: 795–799.

Sato, T., M. Hashimoto, Y. Abe and H. Ando 2015. Rediscovery of the oldest dog burial remains in Japan. *Anthropological Science* 123(2): 99–105.

Shipman, P. 2015. How do you kill 86 mammoths? Taphonomic investigations of mammoth megasites. *Quaternary International* 359360: 38–46.

Thalmann, O. and A.R. Perri 2018. Paleogenomic Inferences of Dog Domestication, in C. Lindqvist and O.P. Rajora (eds) *Paleogenomics: Genome-Scale Analysis of Ancient DNA*: 273–306. Springer: Cham.

Thalmann, O., B. Shapiro, P. Cui, V.J. Schuenemann, S.K. Sawyer, D.L. Greenfield, M. Germonpré, M.V. Sablin, F. López-Giráldez, X. Domingo-Roura, H. Napieral, H.P. Uerpmann, D.M. Lopont, A.A. Acost, L. Giemsch, R.W. Schmitz, B. Worthington, J.E. Buikstra, A. Druzhkova, A.S. Graphodatsky, N.D. Ovodov, N. Wahlberg, A.H. Freedman, R.M. Schweizer, K.P. Koepfli, J.A. Leonard, M. Meyer, J. Krause, S. Pääbo, R.E. Green and R.K. Wayne 2013. Complete Mitochondrial Genomes of Ancient Canids Suggest a European Origin of Domestic Dogs. *Science* 342 (6160): 871–874.

Vilà, C., P. Savolainen, J.E. Maldonado, I.R. Amorim, J.E. Rice, R.L. Honeycutt, K.A. Crandall, J. Lundeberg, and R.K. Wayne 1997. Multiple and ancient origins of the domestic dog. *Science* 276(5319): 1687–1689.

vonHoldt, B., J.P. Pollinger, K.E. Lohmueller, E. Han, H.G. Parker, P. Quignon, J.D. Degenhardt, A.R. Boyko, D.A. Earl, A. Auton, A. Reynolds, K. Bryc, A. Brisbin, J.C. Knowles, D.S. Mosher, T.C. Spady, A. Elkahloun, E. Geffen, M. Pilot, W. Jedrzejewski, C. Greco, E. Randi, D. Bannasch, A. Wilton, J. Shearman, M. Musiani, M. Cargill, P.G. Jones, Z. Qian, W. Huang, Z.L. Ding, Y.P. Zhang, C.D. Bustamante, E.A. Ostrander, J. Novembre and K. Wayne 2010. Genome-wide SNP and haplotype analyses reveal a rich history underlying dog domestication. *Nature*, 464 (7290): 898.

Wang, G.D., W. Zhai, H.C. Yang, R.X. Fan, X. Cao, L. Zhong, L. Wang, F. Liu, H. Wu, L.G. Cheng, A.D. Poyarkov, N.A. Poyarkov Jr, S.S. Tang, W.M. Zhao, Y. Gao, X.M. Lv, D.M. Irwin, P. Savolainen, C.-I. Wu and Y.P. Zhang 2013. The genomics of selection in dogs and the parallel evolution between dogs and humans. *Nature Communication* 4: 1860.

Wang, G.D., W. Zhai, H.C. Yang, L. Wang, L. Zhong, Y.H. Liu, R.X. Fan, T.T. Yin, C.L. Zhu, A.D. Poyarkov, D.M. Irwin, M.K. Hytönen, H. Lohi, C.I. Wu, P. Savolainen and Y.P. Zhang 2016. Out of southern East Asia: the natural history of domestic dogs across the world. *Cell research* 26(1): 21–33.

Wilkens, B. 2006. The Sacrifice of Dogs in Ancient Italy, in L.M. Snyder and E.A. Moore (eds) *Dogs and People in Social, Working, Economic or Symbolic Interaction*: 132–137. Oxford.

Yellen, J.E. 1977. *Archaeological approaches to the present*. New York: Academic Press.

# 3.8 Dogs and the Afterlife in Southern Italy between Ethnology and Archaeology

Claudio Giardino and Tiziana Zappatore

University of Salento, Via Dalmazio Birago, 64, 73100 Lecce, Italy.
claudio.giardino@unisalento.it, tiziana.zappatore.tz@gmail.com
Corresponding author: Claudio Giardino, claudio.giardino@unisalento.it

**Abstract**

It is well known that dogs are frequently related to the afterlife in many cultures. In some of the major Southern Italian regions (Campania, Basilicata, Calabria, Apulia) there are connections between dogs and the World of Death in modern popular traditions, but there are archaeological evidences too, that connect these traditions to a distant past. In these regions ethno-archaeological research has identified evidence of the possible survival of ancient cultural ideologies and modern folk beliefs.

**Keywords**: dogs, Ethnology, Archaeology, Mythology, South Italy.

## 1 Introduction

It is well known that dogs are frequently related to the afterlife in many ancient and modern cultures. The connection of dogs with the Underworld is, as a matter of fact, attested in many areas of Europe. Not only, as will be said later, in Greek-Roman mythology, but also in the Nordic one, there are several figures of dogs put on guard in places connected to the World of Death, such as Garmr, tied up in front of the Gnipahellir cave, inhabited by dead people; moreover, dogs accompany Odin to the afterlife (Chiesa Isnardi 2008: 570–571). These reasons have led the psychoanalytic schools of Jungian matrix to consider the dog as an archetype connected both with faithfulness and with death. As pointed out by the founder of Archetypal Psychology, James Hillman, the dog detects bones, burying them, digging them up, dealing in limbs that have been stripped of the flesh, just as its kin, the jackal, that prowls nightly among the tombs: the Egyptian god of the dead was in fact Anubis (Hillman 2008: 150–158).

Southern Italy is frequently a significant example of 'cultural conservatism' that has transmitted ancient beliefs and customs as social identities. Here some folk traditions have a substratum dating back to classical antiquity, and sometimes to pre-classical periods, despite Christianity and the succession and overlapping of foreign dominations since the Middle Ages (Lelli 2014: 13–17).

The dog is a subject of particular interest to explore this issue. In fact, contrasting aspects are at the basis of the iconic image of this animal. It is probably considered to be the result of the overlapping of different and sometimes antithetical conceptions. The most obvious aspect underlines the faithfulness of the dog to man; but Southern Italian lore still adds more obscure characters too, that tie it firmly in the World of Death.

In some of the major Southern Italian regions (Campania, Basilicata, Calabria, Apulia) there are connections between dogs and the World of Death in both archaeological evidences and modern popular traditions.

While for prehistory the evidences are obviously exclusively based on archaeological data, from the Classical Age the iconographic, literary and epigraphic sources become predominant. These sources became almost exclusive in the Middle Ages, also due to the absolute hegemony of Christian ideology.

The Dog-Death association appeared in Southern Italy as early as the Neolithic. This is confirmed by findings of dog bones in some graves, but the custom of burying dogs together with humans is also testified later, in the Copper and the Bronze Age, until the Roman period. The presence of a dog in a tomb necessarily implies the killing of the animal, maybe also with the value of a ritual sacrifice to the spirit of the dead, to accompany him to the afterlife.

Thus, according to some scholars, the presence of dogs seems to be extremely important in the man-death relationship, in some cultures of the Italian area as it appears in some crucial moments of human existence, in particular when man passes away and when he is transformed into a spirit.

Still in present times this connection can be noticed, such as in Southern Italian proverbs, according to which a crying dog announces death. In these regions (Campania, Basilicata, Calabria, Apulia, Sicily) ethno-archaeological research has identified evidence of the possible survival of ancient cultural ideologies and modern folk beliefs. (CG)

### *1.1 Methodological problems*

It is generally a difficult and risky exercise to combine the data of contemporary immaterial cultural heritage - typically the oral traditions such as proverbs and folk tales - with the information deriving from archeology and ancient literature.

Yet, in many European regions, some popular traditions have their roots in an ancient past that is sometimes millennia old.

The problem of relating sources of a different nature has long been debated and is still a source of discussion (Carandini 2000: 115–116). Frequently, approaches that combine historical, archaeological, ethnographic, and 'mythical' oral information are not taken into account, or they are discharged as methodologically unsafe (cf. Schmidt, Patterson 1995: 13–14).

In order to understand the reality of the past, it is however legitimate, and sometimes necessary, to use sources of information of a different nature to integrate one with the details of the other, providing insight into the world of social practice and reflecting on the actions and interactions of daily life. This however does not mean to 'flatten out' past societies by representing them in terms of unifying cultures, but rather to adopt a contextual and pluralistic approach (Funari *et al.* 1999; Paynter 2000: 13–17, 23–24).

Certainly, material, written, and ethnological indications constitute independent sets of evidence. However, only by connecting these data is it possible to shed light on otherwise obscure aspects of our past, such as the role of dogs in the perception and interpretation of the afterlife.

It is, therefore, necessary to have a global vision of the research topic, avoiding the dangers of being confined to the frontiers of hyper-specialisation, without losing sight of the universe of available information, albeit coming from different scientific sectors (Manacorda 2007).

## 2 Archaeological and ethnological indications

### *2.1 The archaeological evidence*

The Dog-Death association appears in Southern Italy as early as the Neolithic. Remains of dogs were recovered in some graves: for example at Cala Colombo near Torre a Mare (Bari, Apulia), in an artificial pit, parts of two human skeletons were placed, together with dog remains and grave furniture consisting of a bowl of the Late Neolithic Diana phase (Geniola 1977).

The custom of burying dogs together with humans is also testified in Copper Age graves too, like in other Italian regions like in the centre of Italy, in Ponte San Pietro (Viterbo) and in Casale del Dolce (Frosinone) or in Fontenoce (Recanati) (Miari 1993: 121; Silvestrini *et al.* 1992/93: 157–158, figs. 24–25; Fiore and Tagliacozzo 2000; Wilkens 2000a).

One of the most relevant examples is in a Gaudo Culture grave at Santa Maria delle Grazie near Mirabella Eclano (Avellino, Campania): here the remains of a dog were recovered next to the skeleton of a man in the so-called '*Tomba del Capo Tribù*' (Tribe chief's grave) (Onorato 1960: 29); the rich grave goods (vessels and weapons, both in copper and flint; a sceptre made of sandstone) indicate that he was a tribal leader. Bones of a dog were also found in other coeval graves in Campania, like in the grave 4 at Eboli (Bailo Modesti and Salerno 1998: 365) and in the grave 6517 at Pontecagnano (Bailo Modesti and Salerno 1998: tab. 39) (cf. Negroni Catacchio and Aspesi 2016: 630, fig. 6.2): a dog body was introduced inside the burial chamber together with its owner.

Sometimes in the Gaudo culture context only selected parts of dog skeletons were deposited in the tomb, like the skull in a grave near Tursi (Matera, Basilicata) (Cremonesi 1976: 111, 125, fig. 1c). On this site the skull of a dog was separated from the body of the dead and from his grave goods, and it was placed outside the perimeter of the grave, made by slabs of stone and placed on pebbles: it was a funeral offer. As for the man from Mirabella Eclano, also the dead from Tursi must have been a tribal leader.

Evidence of dogs used as a burial sacrifice were recovered in Bronze Age contexts too: an example is the remains of 6 dogs, among which 4 new-born puppies – they had been probably buried in a basket – that were recovered in a burial in Lavello (Potenza, Basilicata), tomb 743, a large tomb with several spaces with a long continuity of use, from the Middle Bronze Age to the Final Bronze Age (Cipolloni Sampò 1999: 161–165; Wilkens 2000b). Intriguing evidence comes from the Middle Bronze Age settlement of Roca Vecchia (Lecce, Apulia), where an old dog skeleton was found in a cult pit; in this case the sacrifice of dogs must have had a clear symbolic and cultural meaning (Wilkens 1995) (Figure 1).

The connection between dogs and the afterlife follows in Classical periods too: it is clear in mythology and in literature. Cerberus, the dog placed to guard the Underworld, was already present in Homer, even if he did not mention its name (Iliad 8, 368; Odyssey 11, 623); Hesiod mentioned its name and its genealogy for the first time, and he explained its function more precisely (Theogony, 315–318; 769).

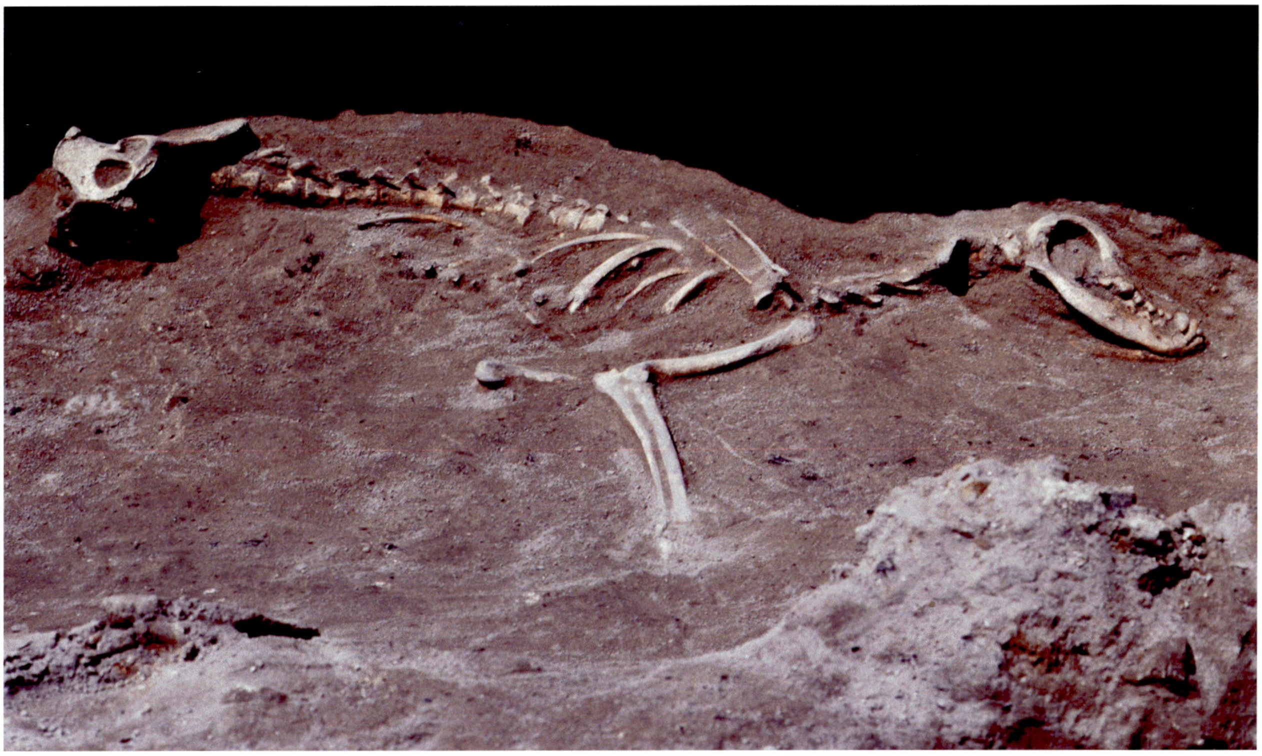

Figure 1. Roca Vecchia (Lecce, Italy), dog bones from a Bronze Age cult pit (Photo by Laboratorio di Scienze applicate all'Archeologia, University of Salento).

In Greek literary sources dog sacrifice is generally associated with ways of purification associated in passage rites, such as birth or the admission of boys into the warrior social group. Such rituals were generally associated with Chthonic gods (Mainoldi 1981: 51–59; Osanna 2001). According to some scholars, the dog represents the moments of transition of a spirit: the moment of its passing away, the entrance to the Underworld and its possible return in the shape of a spirit (Mainoldi 1981).

Sometimes such sacrifices were substituted with the votive deposition of clay figurines representing dogs. These were used to substitute the real animals with symbolic ones or testified the sacrifice that had happened previously. Archaeologically, these figurines were attested in various sanctuaries in Magna Graecia and Sicily, such as the one in Capua (late 4th-beginning 2nd century BC), in the Persephone sanctuary in Locri (late 4th-beginning 3rd century BC), in Morgantina (3rd century BC), in Agrigento (late 6th-middle 5th century BC). In Locri, in the area of Centocamere, other dog figurines were discovered, together with a *bothros* dedicated to Aphrodite and containing dog bones (De Venuto and Quercia 2006).

Etruscan paintings, such as the one of the Orcus (or Ogre) tomb in Tarquinia and the one of Golini tomb in Orvieto, depict Hades together with a dog's head. Hades himself, then, has as an attribute the Ἄϊδος κυνέη, the helm of Hades, made with dog skin, which confers invisibility (Pseudo-Apollodorus, Bibliotheca 1.6–7; Suidas s.v. Aidoskune, trans. Suda On Line).

In Etruscan-Latin culture the Lares (good spirits of deceased ancestors, from the Etruscan *lar*, father) were usually accompanied by dogs; in particular, *lares praestires*, guardians of the state, who wore dog fur coats and were depicted in association with dogs (Plutarch, Roman Questions 51–52). The same connection is found in Ovid, which connects dogs to crossroads (Fasti 6, 137–142) and in a passage from Tibullus (Elegies 1, 5, 56).

The goddess Hecate is particularly connected to dogs. She was the daughter of Zeus and Hera and the guardian goddess of dogs and of passage moments, including death. It was possible to meet her at crossroads, preceded by dogs howling. Dog sacrifices were dedicated to her and, according to ancient Greek beliefs, dogs announced Hecate's coming (Mainoldi 1981).

The dog bones discovered in a cave at Vaste, in Salento, together with the remains of domestic animals (De Grossi Mazzorin and Salinas 2010) are linked to Hera and Persephone worship, both goddesses with strong underworld characteristics.

Figure 2. Soleto (Lecce, Italy), Church of Santo Stefano, fresco, end of the 14th - beginning of the 15th century AD, (Photo by T. Zappatore).

Figure 3. Otranto (Lecce- Italy), Cathedral's mosaic, 12th century AD, (Photo by T. Zappatore).

Cerberus sometimes appears connected with Hecates, like in a 4th century BC Apulian red-figure volute krater from the Staatliche Antikensammlungen in Munich (Catalogue number: Munich 3297).

Just as in prehistory, and in Roman times, the use of burying dogs in tombs is not uncommon, such as in the necropolis of Fonte d'Amore near Sulmona (L'Aquila, Abruzzo) (4th-3rd century BC), where the presence of dogs is attested in at least three tombs (De Grossi Mazzorin 1995), or in the necropolis of Fidene (Rome, 2nd century AD) (De Grossi Mazzorin and Minniti 2000).

The Middle Ages Christian ideology, while actively working to erase previous pagan beliefs, retained some of its aspects, while adapting them to the new faith. The dog maintained, at least in part, the symbolic ambiguity that had distinguished it in the old civilisations, a symbol of faithfulness, but also of death. However, beyond the official doctrine, some ancient, deep-rooted beliefs were likely to be preserved in popular culture.

The custom of sacrificing dogs to accompany their owners in death was still attested in the Lombard period, in the necropolises of Povigliano (Verona), Testona (Turin) and Nocera Umbra (Perugia) (Giostra 2014: 268–269, fig. 13).

In the Middle Ages Cerberus maintained this position of 'guardian of Underword', but he was Christianised, becoming the guardian of Hell (see Dante, Divine Comedy, Inferno, Canto VI, 13–33). Cerberus is represented in this role in medieval mosaics, as in Otranto, and frescoes, as in the one from Soleto (Lecce), in the church of Santo Stefano, dating back to the end of the 14th– beginning of the 15th century (Figure 2).

Of particular interest are the floor mosaics of the Cathedral of S. Maria Assunta in Otranto, made in the mid-12th century by the monk Pantaleone: they are a precise representation of the medieval theological beliefs, full of symbolic meanings (Gianfreda 2008). The central figure consists of the *Arbor Vitae* (Tree of Life), from which the narration starts: here dogs are depicted several times in the act of biting the Tree (Figure 3). Their symbolic action against the Tree of Life attests to the survival, at a popular level, of beliefs that saw these animals as adversaries of life, as closely related to its opposite, death.

## 2.2 Ethnological traditions from Southern Italy

The connection between dogs and the afterlife is still present in the folklore of many Southern Italian regions. In some cases, the dog is also connected to hell and the world of devils. The nineteenth-century folklorist Giuseppe Pitrè in *Usi e costume credenze e pregiudizi del popolo siciliano*, reports an episode narrated in the *Historia Sicula* by the alleged Michele da Piazza, which probably occurred in 1341 in Messina:

> '*...in the form of dogs, many devils infested the cities, terrifying the citizens with frightening barks: a dog of which, all black, with a sword in his hand, entered the cathedral, the one where lost people also entered, broke and shattered silver vases, lamps, candlesticks and altars*' (Pitrè 1889: 105).

In Sicily, Apulia, Basilicata and Campania, spirits, especially of those who were murdered, chased wayfarers along the lonely roads in the shape of animals, in particular black dogs. They usually appeared at midday or midnight, threshold times that represent the transition from the morning to the afternoon or from one day to another. These beliefs were still attested at least until the middle of the last century. According to lore recorded in Campania, Calabria, Apulia, and Sicily, the ghosts of killed people took the appearance of dogs: if you met them in the street at night, you had to runaway to a crossroad, because they have to disappear in this place (Lelli 2012: 169). The symbolic meaning of crossroads was christianised and related to the sign of the Jesus' cross ('*crocivia*' in Salento dialect) (Rohlf 2007: 171, s.v. crocivia): in front of the holy cross bad spirits have to vanish (Colitti 2012: 37–42; Lelli 2016: 128–129).

In Salento (Apulia, province of Lecce), dog howling is an omen of death or a bad omen, as a proverb states '*cane ca chiance, morte ca vene*' ('when a dog cries, death comes') (De Donno 2005: 51). Here the relation to Hecate's myth is clear.

The ability of seeing the souls of the deceased was also conferred to dogs, as confirmed by late '800 direct evidences:

> '*Signor curato! Chi è morto stanotte? Menico. Ah ah... va bene! Già!... ecco!... il mio cane me ne ha dato il segno!... abbajava! Mugolava sta notte! Io- (sapete?) abito sulla strada, che mena alla casa di Menico... ah!... il mio Argante! Vedeva gli Spiriti, che andavano a visitare il povero Menico! Requiescal...*'('Mr. Curate! Who died tonight? Menico. Ha ha... all right! Right!... there it is!... my dog has shown it to me!... he barked! He whimpered tonight! I (you know) live on the road that leads to Menico's house... ha!... my Argante! He used to see the Spirits, that went to visit poor Menico! May he rest in peace!...') (De Simone 2006: 115).

A clear connection with the classical pre-Christian world is the custom, still attested in the 1900s, of offering bread to the Underworld dog, a food rich in symbolic and sacred values. In Calabria, in the area of Castrovillari (Cosenza), it was customary to put in the pocket of a deceased, together with a coin to pay the offering to Charon, also a piece of bread to satiate Cerberus (Lelli 2012: 179).

A recollection of that can be found also in Apulia, where people, in case of heavy thunderstorms, used to throw bread against the storm itself to calm it, as if it's a voracious wolf. Traces of this belief have been found in interviews recently carried out in the province of Lecce (Puglia). Mrs. Maria Carmina Zippo remembers seeing her neighbour when she was young, Mrs. Maria Villani, throwing pieces of bread during a storm, in order to calm it. The event occurred in April 1981, attesting to the preservation of the custom until fairly recent times.[1]

A possible reference to this can be also found in a common tongue-twister, based on breaking bread and throwing it against a dog's head: '*Sulla via de Culepazzu'n ce'n acapu de cane mazzu; spezzapane e mina pane a quiracapu de mazzu cane*' ('on the way to Collepasso – a town in province of Lecce – there is a thin dog's head; break bread and throw bread against that thin dog's head'). In some versions the Holy Bread from Maundy Thursday's Last Supper commemoration was thrown against the storm (Barletta 2006: 55).

Furthermore, the connection between dogs and the demonic forces had already been observed in the Middle Ages and it is still present in popular beliefs: for this reason dogs were driven away from ecclesiastical buildings, as affirmed by Mrs. Antonietta Rizzello from Montesano Salentino (Lecce).[2]

That the devil takes on the appearance of a dog is attested in the memoirs of Saint Pio from Pietrelcina (Pietrelcina, Benevento 1887 - San Giovanni Rotondo, Foggia 1968). The famous Catholic saint, of southern farm origins, wrote about an appearance of devil under the shape of a big dog: '*Non ottenendo nessuna risposta mi ritirai, ma con terrore dalla porta vidi entrare un grosso cane dalla cui bocca usciva tanto fumo. Caddi riverso sul letto e udii che diceva: "E' isso, è isso!". Mentre ero in quella positura vidi l'animalaccio spiccare un salto sul davanzale della finestra, da qui lanciarsi sul tetto di fronte, per poi sparire*' ('as I was not receiving an answer I moved away, but terrified I saw a big dog coming through the door and from its mouth a lot of smoke was coming out. I fell down and I heard him saying: "It's him, it's him!". While I was still in that position, I saw that terrible animal jumping on the window sill and from there he jumped on the roof in front of him, disappearing') (Tosatti 2003: 29).

Moreover, traditionally witches are connected with the Underworld deities. In Albano di Lucania (Potenza), the '*masciare*' (witches) used to rides dogs, also white ones; moreover, some witches used to turn into dogs (Bermani 2008: 154). The relationship between witches and dogs was also present in the Classical era: according to Lucano and Petronius, in fact, a witch can bark (*The Civil War 6, 688*; Satyricon, *63)*, while, for Apuleius, she turns into a dog before eating unburied corpses *(Metamorphosis 2, 22)*.

In Calabria, and in other areas of Southern Italy, dogs are associated with were wolves – demonic creatures– which preceded them and gave advance notice of their coming with loud barks. This belief is documented in the province of Catanzaro, in Satriano, where it was believed that the wolf was followed by ten or twenty dogs that barked and surrounded it (Bermani 2008: 292–293).

Werewolves were present both in Ancient Roman and Greek times (*Satyricon* 61–62) and are now widespread, with variations, in all of the southern regions (Zahler 2014: 25–28). (TZ)

### 3 Conclusions

The dog is a figure with extremely complex symbolic connotations, in which the relationship with death is linked to the characteristic of faithfulness. A connection that has been found since prehistoric times through multiple clues and which has survived in the popular traditions of many European regions until the Modern era. In the south of the Italian peninsula these beliefs have often been preserved, albeit mixed with the contributions of the Christian religion, in the form of stories, proverbs, superstitious and apotropaic gestures (Lelli 2014: 15). The examination of the articulated set of these evidences, which connect the dog with the afterlife and the supernatural world, allows us to frame more deeply the collective imagination relating to this animal in its diachronic and synchronic aspects. But also, through these beliefs, it is possible to explore the survival of rituals and traditions that are still well rooted in popular culture, but dating back to a past which, in particularly conservative areas such as the South of Italy, have been passed down to us through centuries and millennia. (CG)

### Acknowledgements

We would like to offer our special thanks to prof. Riccardo Guglielmino who provided the unpublished photograph of the dog from the excavations of Roca.

We thank the Diocese of Otranto that allowed the use of photos from Otranto Cathedral mosaics and Saint Stephan church at Soleto frescoes.

[1] Interview of Tiziana Zappatore with Mrs. Maria Carmina Zippo (Specchia -Lecce) March, 2018.
[2] Interview of Tiziana Zappatore with Mrs. Antonietta Rizzello (Montesano Salentino - Lecce) April, 2018.

## References


Bailo Modesti, G. and A. Salerno 1998. *Pontecagnano II.5. La necropoli eneolitica: l'età del Rame in Campania nei villaggi dei morti.* Napoli: Università degli Studi di Napoli l'Orientale.

Barletta, R. 2005. *Cicalata su diavoli, streghe, fate e orchi salentini. Credenze e pratiche popolari.* Lecce: Edizioni del Grifo.

Bermani, C. 2008. *Volare al Sabba. Una ricerca sulla stregoneria popolare.* Roma: Derive Approdi

Carandini, A. 2000. *Giornale di scavo: pensieri sparsi di un archeologo*, Torino: Einaudi.

Chiesa Isnardi, G. 2008. *I miti nordici.* Varese: Longanesi.

Cipolloni Sampò, M. 1999. Ipogeismo funerario e cultuale nella Daunia meridionale, in A. Gravina (ed.) *Atti 19° Convegno Nazionale sulla Preistoria - Protostoria - Storia della Daunia. San Severo 27–29 Novembre 1998. 2*: 155–188. San Severo: Archeoclub d'Italia.

Colitti, G. 2012. *Il tamburo del diavolo. Miti e culture del mondo dei pastori.* Roma: Donzelli Editore.

Cremonesi, G. 1976. Tomba della prima età dei metalli presso Tursi (Matera). *Rivista di Scienze Preistoriche* 31/1: 109–134.

De Donno, N. 2005. *Dizionario dei proverbi salentini*, Galatina: Congedo Editore.

De Grossi Mazzorin, J. 1995. Sepolture con cani nella necropoli pre- romana di Sulmona (AQ), in R. Peretto and O. De Cutris (eds) *Primo Convegno Nazionale di Archeozoologia. Padusa Quaderni 1*: 375–376.

De Grossi Mazzorin, J. and C. Minniti 2000. Le sepolture di cani della necropoli di età imperiale di Fidene- via Radicofani (Roma): alcune considerazioni sul loro seppellimento nell'antichità, in *Atti del 2° Convegno Nazionale di Archeozoologia. Asti, 14–17 novembre 1997*: 387–398. Forlì, Abaco.

De Grossi Mazzorin, J. and A. M. Solinas 2010. La fauna dei Bothroi di Vaste (Lecce) e sue implicazioni cultuali, in A. Tagliacozzo, I. Fiore, S. Marconi and U. Tecchiati (eds) *Atti del 5° Convegno Nazionale di Archeozoologia, Rovereto 10–12 novembre 2006*: 183–192. Rovereto (TN): Osiride.

De Simone, L. G. 2006. *La vita della Terra d'Otranto.* Lecce, Edizioni del Grifo.

De Venuto, G. and A. Quercia 2006. Le statuette fittili di cane in Italia meridionale in età preromana: la documentazione archeologica e il dato archeo-zoologico, in A. Tagliacozzo, I. Fiore, S. Marconi and U. Tecchiati (eds) *Atti del 5° Convegno Nazionale di Archeozoologia, Rovereto 10–12 novembre 2006*: 229–233. Rovereto (TN): Osiride.

Fiore, I. and A. Tagliacozzo 2000. Deposizioni di resti animali nelle tombe della necropoli di Casale del Dolce (Anagni, FR): l'esempio della Tomba 4, in *Atti del 2° Convegno Nazionale di Archeozoologia (Asti, 1997)*: 201–211. Forlì: ABACO Edizioni.

Funari, P.P.A., S. Jones, and M. Hall 1999. Introduction: archaeology in history, in P.P.A Funari, M. Hall and S. Jones (eds) *Historical Archaeology, Back from the edge*: 1–20, London: Routledge.

Geniola, A. 1977. Archeologia e cultura della comunità neolitica di Cala Colombo, in *La comunità neolitica di Cala Colombo presso Torre a Mare (Bari)*: 29–92. Bari: Società di storia patria per la Puglia.

Gianfreda, G. 2008. *Il Mosaico di Otranto. Biblioteca medievale in immagini*, Lecce: Edizioni del Grifo.

Giostra, C. 2014. La necropoli di Povegliano Veronese, loc. Ortaia, in E. Possenti (ed.), *Necropoli longobarde in Italia. Indirizzi della ricerca e nuovi dati, Atti del Convegno Internazionale (Trento 2011)*: 259–273. Trento: Provincia Autonoma di Trento.

Hillman, J. 2008. *Animal presences* (1st ed). Putnam, Conn: Spring Publications

Lelli, E. 2012. Folklorica IV. Briciole di folklore. *I quaderni del ramo d'oro on-line 5*: 166–175.

Lelli, E. 2014. *Folklore antico e moderno. Una proposta di ricerca sulla cultura popolare greca e romana.* Pisa: Fabrizio Serra Editore.

Lelli, E. 2016. *Sud Antico. Diario di una ricerca fra filologia ed etnologia.* Milano: Bompiani.

Mainoldi, C. 1981. Cani mitici e rituali tra il regno dei morti e il mondo dei viventi. *Quaderni Urbinati di Cultura Classica*, n.s. 8, 37: 7–41.

Manacorda, D. 2007. Fonti archeologiche e fonti scritte: vent'anni dopo Le vin de L'Italie romaine di André Tchernia, *Dimensioni e problemi della ricerca storica*, 2: 85–100.

Miari, M. 1993. La necropoli eneolitica di Ponte S. Pietro (Ischia di Castro, Viterbo), *RSP45*: 101–166.

Negroni Catacchio, N. and M. Aspesi 2016. Rinaldone e Gaudo: rituali a confronto, in N. Negroni Catacchio (ed.) *Preistoria e Protostoria in Etruria, Atti del 12° Incontro di studi. Ornarsi per comunicare con gli uomini e con gli Dei. Gli oggetti di ornamento come status symbol, amuleti, richiesta di protezione II. Valentano (VT)- Pitigliano (GR)- Manciano (GR), 12–14 settembre 2014*: 611–636. Milano: Centro studi di Preistoria e archeologia.

Onorato, O.G. 1960. *La ricerca archeologica in Irpinia, Avellino.* Avellino: Amministrazione provinciale.

Osanna, M. 2001. *Il sacrificio: cani per la dea*, in M. Osanna and M.L. Nava (eds) *Rituali per una dea lucana. Il santuario di Torre di Satriano.* Catalogo della Mostra, Potenza 2001: 107–109 Potenza: Cerbone Editrice.

Paynter, R. 2000. Historical and Anthropological Archaeology: Forging Alliances. *Journal of Archaeological Research*, 8/1: 1–37.

Pitrè, G. 1889. *Usi e costumi, credenze e pregiudizi del popolo siciliano 4.* Palermo: L. Pedone Lauriel.

Rohlf, G. 2007. *Vocabolario dei dialetti salentini (Terra d'Otranto), I*, Galatina: Congedo Editore.

Silvestrini, M., G. Cilla and G. Pignotti 1992/93. La necropoli eneolitica di Fontenoce (Recanati). *Picus. Studi e ricerche sulle Marche nell'antichità 12-13*: 127–185.

Schmidt, P.R. and T.C. Patterson 1995. Introduction: from constructing to making alternative histories, in P.R. Schmidt and T.C. Patterson (eds) *Making Alternative Histories: the practice of archaeology*: 1–24. Santa Fe (NM): School of American Research Press.

Tosatti, M. 2003. *Padre Pio e il diavolo. Gabriele Amorth racconta...*, Casale Monferrato, Edizioni Piemme.

Wilkens, B. 1995. Animali da contesti rituali nella preistoria dell'Italia centro- meridionale. In R. Peretto and O. De Cutris (eds) *Primo Convegno Nazionale di Archeozoologia. Padusa Quaderni 1*: 201–207.

Wilkens, B. 2000a. Il cane eneolitico di Fontenoce (Recanati-MC) in *Atti del 2° Convegno Nazionale di Archeozoologia. Asti, 14-17 novembre 1997*: 213–215. Forlì: Abaco.

Wilkens, B. 2000b. I resti faunistici della tomba 743 di Lavello Basilicata, in *L'ipogeismo nel Mediterraneo: origini, sviluppo, quadri culturali: atti del congresso internazionale Sassari-Oristano 23-24 maggio 1994, II*: 667–670. Sassari: Stampacolor industria grafica.

Zaher, N. 2014. La metamorfosi animalesca nella cultura italiana, in *Amaltea Trimestrale di cultura*, IX /1: 22–28.

# 3.9 Faithful unto Death. Burial, Legends and Heroism of the Dog from Antiquity to the Contemporary Age

Jacopo De Grossi Mazzorin (†)[1], Ivana Fiore[2,3], Claudia Minniti[1], Antonio Tagliacozzo[2]

[1]Università del Salento, via Dalmazio Birago 64, 73100, Lecce, Italy. claudia.minniti@unisalento.it.

[2]Bioarchaeology Section of Museo delle Civiltà, Piazza G. Marconi 14, I-00144, Roma, Italy. iva.fiore@gmail.com; antonio.tagliacozzo@cultura.gov.it

[3]PhD Programme in Environmental and Evolutionary Biology, Dip. di Biologia Ambientale, Sapienza University of Rome, P.le A. Moro, 5 Roma, Italy. iva.fiore@gmail.com

Corresponding author: Ivana Fiore, iva.fiore@gmail.com

**Abstract**

The dog has always played a special role in the relationship with humans. This work aims to assess the most significant findings of dogs associated with human burials in Antiquity from the Palaeolithic. Dog burials are well documented from the Neolithic and have continued until the contemporary age, providing evidence, along with the remains and sources in literature, frescoes and paintings, films and comics, the extent to which the link between this animal and humans is indissoluble. In modern times the cases of legendary dog loyalty remain very famous, like those of Achiko, Pal'ma, Bobby, and Lampo.

**Keywords:** Italy, Life/death, burial, sacrifice, legendary dog.

## 1 Introduction

The dog has always played a special role in the relationship with humans in terms of collaboration, trust, love, as well as friendship, even if humans can be ferocious and exploit the dog in various ways, sometimes with cruelty. The privileged relationship between human beings and dogs is archaeologically well attested in Antiquity from the Palaeolithic. Dog burials have continued until the contemporary age, demonstrating, along with the remains and sources in literature, frescoes and paintings, films and comics, the extent to which the link between this animal and humans is indissoluble. The aim of this work is mainly to investigate the most famous contemporary cases of legendary dog loyalty, like those of Achiko, Pal'ma, Bobby, and Lampo, just to name a few.

## 2 Archaeological findings

Some of the most famous dog remains were found at Bonn-Oberkassel more than a hundred years ago[1]. They are dated to the Upper Pleistocene and are 14,223± 58 years old. Recent re-examination has revealed the remains of two dogs. The best preserved dog was 7 months old at death and was buried with two humans. It was perniciously ill because of a distemper when it was 19 weeks old. This allowed researchers to determine how the represented dog was perceived and treated by Palaeolithic hunter-gatherers. In fact, a dog with a serious case of distemper dies in less than three weeks if it doesn't receive adequate care (Janssens *et al.* 2018). Dog burials are known from many sites in several European countries, from Italy to Sweden and from Portugal to Russia (Grünberg 2013; Hasler and Noret 2017; Losey *et al.* 2018; Albizuri *et al.* 2019).

The oldest emergence of such similar behaviour seems to be demonstrated in another famous case: that of the puppy skeleton of a dog or a wolf found close to the human skeleton of a woman at Ain Mallaha in northern Israel (Davis and Valla 1978). The skeleton was dated to around 12,000 years ago. Instead, at Hayonim Terrace, a man was found interred with two small dogs, about 13,000 years ago, similar to other Natufian burials found in the Near East (Tchernov and Valla 1997). The dogs were probably killed, testifying that even in these cases, humans had carried out cruel activities.

In Italy, numerous cases of dog burials have been documented, dating from the Neolithic to Late Antiquity. In this paper, we will mention the most

[1]This work aimed to assess all findings of dogs associated with human burials, through a detailed collection of data on dog remains, their position, completeness of the skeleton, description of the individual anatomical elements, number of individuals, age, sex, burning and butchery marks. We will try to highlight particularities and differences among the different contexts and periods. Several articles in this volume deal extensively with these issues, see Bona *et al.*, Latini *et al.*, Giardino and Zappatore, so we have preferred to cite these papers and make this article more publicly available in view of the emotive nature of the topic.

important cases and those we have studied. Famous Neolithic dog burials are documented in Valdaro (near Mantua) where a dog was found in a burial called 'the hunter', placed on his owner's feet, together with a set of arrowheads and blades (Castagna *et al.* 2014, Bona *et al.* this volume) and in Ripoli (Teramo), where it was buried together with a woman. In contrast, Chalcolithic dog burials are more frequent and located in central and southern Italy, such as the cases of Mirabella Eclano (Avellino), Tursi (Matera), Gaudo (Salerno), Casale del Dolce (Frosinone), Ponte S. Pietro (Viterbo), and Fontenoce (Ancona) (Onorato 1960; Cremonesi 1976; Miari 1993; Silvestrini *et al.* 1992/93; Bailo Modesti and Salerno 1998; Fiore and Tagliacozzo 2000; Wilkens 2000).

Two burials containing only dogs were found at Osteria del Curato-via Cinquefrondi (Rome). One of these was turned on its right side; the skeleton was still in anatomical connection with the limbs flexed and was devoid of the head. The other was turned on the left side and still in anatomical connection (Anzidei *et al.* 2007).

In the following centuries (Bronze and Iron age), dogs continued to be buried and among the numerous cases, we can mention the dog buried with a man at Cetona (Siena) and the adult dogs and puppies found in tomb 743 at Lavello (Potenza) or the puppy dog buried with a child at Pontecagnano-Colucci (Salerno) and the multiple burials of dogs and humans from Sant'Eufemia (Padua) (Guidi 1992; Cipolloni Sampò 1999; Facciolo *et al.* 2006; Negroni Catacchio and Aspesi 2016; Fiore 2016).

Among Etruscan communities, the dog was part of religion and myth to the point of being a status symbol, and for this reason, it is found in many princely tombs (Gambari and Tecchiati 2004). A significant discovery of dog burial is that of a dog buried together with an infant found in the necropolis of Amelia (Terni) and dated to the 4th-3rd century BC (Salari *et al.* 2014). The bronze bell-rattle lying in the grave together with the dog skeleton undoubtedly seals the emotional link between the animal and its owner, perhaps his playmate.

The dog takes on special significance in contexts where it is associated with newborns or aborted fetuses (Peltuinum-Aquila and Lugnano in Teverina), where this animal likely had magical or therapeutic value in addition to being a travelling companion; its sacrifice in other archaeological contexts (Kolonos Agoraios of Athens, Eretria, and Messenè) was considered as part of a purification ritual for the premature (Soren *et al.* 1995; Soren and Soren 1999; Fiore and Salvadei 2014; Migliorati *et al.* 2018; Liston *et al.* 2018; Sperduti *et al.* 2018).

The dog cemetery found at Ashkelon in Israel is particularly famous, where possibly thousands of dogs were interred from the 5th to the 3rd century BC (Edrey 2008). The majority of these dogs were puppies; they share many characters with the modern Canaan Dog, perhaps representing the ancestral population from which the modern breed descends. It is the largest known ancient dog cemetery known in the world. It is usually mentioned to refer to the reputed healing properties of dogs. Alternatively, it may have been the site of a facility for breeding dogs.

Dogs have continued to be buried in the historical period, both with the deceased and alone. In the case of dogs buried with humans, the meaning can reflect a sacrifice of the dog with the function of guardian to the burial of the master or even the extreme residence of the faithful companions of the deceased. In this case, the meaning can be read as an act of loving care to preserve the memory and affection of the dog, in the will to remember his loyalty to the owner (De Grossi Mazzorin and Minniti 2006).

Numerous dog burials, without any association with humans, were found in various necropolises of the Roman period located in the suburban area of Rome, such as in Fidene-via Radicofani and in Via Nomentana, at the junction with Via Palombarese) (De Grossi Mazzorin and Minniti 2000, 2001). Tombs dedicated only to dogs are also well documented in ancient Greece (Trantalidou 2006).

### 3 Ancient sources

Several writers mention the loyalty of dogs to their masters (Minniti 2022: 174–186). Plutarch (*Them.* 10, 9–10) reports the sacrifice of the dog of Xanthippe, father of Pericles who could not bear the abandonment of his master due to the invasion of the Persians and followed him by sea to Salamina, where he died from the effects of the journey[2].

Aelian tells of the affection of Polyarchos for his dogs (*VH* 8, 4) and tells of the funerals of the dogs of the philosopher Lakydès (*NA* 7, 41). Many qualities were recognised in the animal, defined in every age as the best friend of man. In this regard, Cicero (*nat. deor.* 2, 158) states: 'And let's not talk about dogs, their fidelity in guarding, and their affection for the master...' What does all this mean except that the dog was created to meet the needs of man?' (*Canum vero tam fida custodia*

[2] Ancient sources follow the Thesaurus Linguae Latinae (TLL - https://www.thesaurus.badw.de/en/tll-digital/index/a.html#a) the Liddell-Scott-Jones, A Greek-English Lexicon (LSL - http://stephanus.tlg.uci.edu/lsj/01-authors_and_works.html). Cicero: *De natura deorum*; Elian: *De natura animalium*; Lakydès: *De natura animalium*; Plato: *Republica*; Plutarch: *De amore prolis*; Plutarch: *vita di Temistocle*;Varro: *De re rustica*.

*tamque amans dominorum adulatio tantumque odium in externos et tam incredibilis ad investigandum sagacitas narium, tanta alacritas in venando quid significat aliud, nisi se ad hominum commoditates esse generatos).*

Elian (*NA* 7, 38), Quintus Smyrnaeus (16, 281), Plutarch (*moralia, De amore prolis* 2, 40), Lucretius (5, 862), and Varro (*res rusticae* 2, 9) all praise the memory of the dog, the only animal that recognises its name, as well as its loyalty and devotion to the master. Instead, the custom of honouring animals is clearly described in the life of the Roman Emperor Hadrian, who erected graves for his dogs and most loyal horses (HIST AVG., *Hadr.* 20). It is also well documented in some funeral inscriptions composed for companionship dogs.

Also in Greek literature, the courage and loyalty of dogs are praised. The episode of Ulysses when he had just arrived in Ithaca and was only recognised by his faithful dog Argo, has become proverbial. The specific ability of dogs to be simultaneously sweet and courageous was also remembered by Plato (*R.* 2, 374e–376c), who considered these animals to be a model for the guardians of his Ideal City. In Spartan society, hunting dogs were considered strictly private property, and anyone using them needed to receive the consent of their master.

## 4 Modern and contemporary times

The custom of dog burial has been practised in recent times with different motivations. From the end of the 15th century, it was also linked to the trend for breeding and the creation of dogs that were perfectly responsive in appearance and character to the needs of man. This is clearly expressed in portraiture but also in epitaphs that highlight the loyalty to the master, even when he has died; including the themes of defence and custody of the grave; and of voluntary death of the animal that has lost its master.

Among the most famous historical examples are the canine burials of Francesco Gonzaga and Isabella d'Este (1474–1539)[3] and those of Frederick the Great of Prussia (1712–1786)[4] (Santi 1999). A faithful dog is often represented on the funerary monuments of famous people, and since the 19th century, the custom of burying dogs in special cemeteries began to spread, such as in Hyde Park in London[5].

The loyalty of dogs is then largely testified by more recent events, including the cases of Bobbie, who was the puppy of Edinburgh, the Border collie Tip, and many others about which many books have been written[6]. Among these is the well-known story of Hachiko, the Akita dog who, following the unexpected death of his master, Prof. Hidesaburo Ueno, waited for him every day for ten years (from 1925 to 1935) at Tokyo's Shibuya railway station[7]. The story has had a huge resonance both in Japan and abroad, so much so that it has recently been revived in two successful films, one set in Japan and the other in the United States, both starring the same actor.

Always linked to the 'railway' is the famous case of Lampo, the travelling dog. On a sunny August day in 1953, Lampo arrived by a freight train at the station of Campiglia Marittima (Livorno). It was immediately adopted by the deputy station master, Elvio Barlettani, who called him Lampo because of his speed. Lampo had been riding on trains for many years, but he always returned to the station to spend the night. In the morning, he used to accompany Virna, Barlettani's daughter, to her school in Piombino, then return to Campiglia station. One day, by mistake, the dog caused a train to stop, and the railway compartment ordered the dog to be removed. It was brought to southern Italy and abandoned in the countryside, but after five months, the dog, sick and malnourished, reappeared in Campiglia. His fame grew suddenly; it got the attention of the national and international press and the dog was finally adopted officially by the Italian Railways. It will also be filmed by several television crews. Lampo died in 1961, killed by a manoeuvring train that was late. A statue was erected in his honour at the station of Campiglia Marittima (Barlettani 2014).

Another case is that of Pal'ma, a female German shepherd that was abandoned by its master in 1974 at the airport of Vnukov, even if it is not very famous in the West. At the time of boarding, the Iljušin 18 flight crew refused to accept the dog because it lacked a veterinary certificate. Its master left anyway, leaving the animal alone. From that day on, the dog stayed at the airport and ran to meet all the Il-18 in the hope of seeing its master again, who never returned. In 1976, a paper written by the reporter Jurij Michailovich Rost spread the story of Pal'ma throughout the Soviet Union[8]. Hundreds of touched readers wrote messages or sent rubles to feed Pal'ma. Fortunately, Vera Arseniivna Kotljarevskaja, an activist from the Society for the Protection of Animals and professor of biology at the Faculty of Education in Kiev, arrived at the airport and decided to stay a week on the spot to

[3] Viewed 10 May 2020, https://www.mantovaducale.beniculturali.it/it/news/519-i-cani-dei-gonzaga-da-rubino-a-tibris
[4] Viewed 10 May 2020, https://learnearnandreturn.wordpress.com/2011/05/25/old-fritz-and-his-dogs/
[5] Viewed 10 May 2020, https://funlondontours.com/the-victorian-pet-cemetery-of-hyde-park/

[6] Viewed 10 May 2021, https://www.farmersjournal.ie/greyfriars-bobby-the-most-faithful-dog-585174
[7] Viewed 30 December 2021, https://allthatsinteresting.com/hachiko-dog; https://it.wikipedia.org/wiki/Hachik%C5%8D
[8] https://en.wikipedia.org/wiki/A_Dog_Named_Palma

Figure 1. Bernardo Gengarelli and Tommy (Photo by D. Riparbelli).

gain the dog's trust. She was able to put a collar on it, sedate her and bring her to her home in Ukraine. Later, Pal'ma had three puppies and lived with Vera until the dog's death.

## 5 Story of Tommy

Finally, we wish to describe the unpublished story of Tommy (Figure 1). It was the dog of another railwayman, Bernardo Gengarelli, who worked at the station of Pescara (Abruzzi). As told to Jacopo De Grossi Mazzorin[9] by his nephew Davide Riparbelli, Bernardo, after life-changing events linked to the Second World War, managed to return to Pescara with his family at the end of hostilities. His newfound peace did not last long, because his wife Elvira passed away due to alymphoma. Bernardo was so overwhelmed with his loss; he moved house. Although the man was inconsolable, he could always count on his dog, Tommy's affection. Every day for five years, Bernardo went to the cemetery of San Silvestro on his Lambretta scooter, to kneel at Elvira's grave. On the 16th of May 1956, at the exit of the cemetery, he was struck by a car. Tommy did not resign himself to the death of his master. Every morning he walked to the cemetery to lie down on the graves of Bernardo and Elvira, and remained there till the evening. After a few months, however, he was also killed by a car on the way back.

[9] Jacopo De Grossi Mazzorin

## References

Albizuri, S., J. Nadal, P. Martín, J. F. Gibaja, A. Martín Cólliga, X. Esteve, X. Oms, M. Martí, R. Pou, D. López-Onaindia and M. Eulàlia Subirà 2019. Dogs in funerary contexts during the Middle Neolithic in the northeastern Iberian Peninsula (5th–early 4th millennium BCE). *Journal of Archaeological Science: Reports* 24: 198–207.

Anzidei, A.P., G. Carboni, M.A. Castagna, A. Celant, M. Cianca, R. Egidi, S. Favorito, R. Funicello, G. Giordano, M. Malvone and A. Tagliacozzo 2007. *L'abitato eneolitico di Osteria del Curato-via Cinquefrondi: nuovi dati sulle facies archeologiche di Laterza e Ortucchio nel territorio di Roma*, Atti della XL Riunione Scientifica dell'IIPP, Strategie di insediamento fra Lazio e Campania in età preistorica e protostorica, II, Roma, Napoli, Pompei, Firenze: 477–508.

Bailo Modesti, G. and A. Salerno 1998. *Pontecagnano II.5. La necropoli eneolitica: l'età del Rame in Campania nei villaggi dei morti.* Napoli: Università degli Studi di Napoli l'Orientale.

Barlettani, E. 2014. *Lampo il cane viaggiatore*. Edizione Illustrata. Piombino: La Bancarella.

Castagna D., V. Gazzoni, G.L.F. Berruti and M. De March 2014. Studio preliminare sulle sepolture neolitiche del territorio mantovano: i casi di Mantova, Bagnolo San Vito e San Giorgio, in M. Bernabò Brea, R. Maggi, A. Manfredini (eds), 5000–4300 a.C. *Il pieno sviluppo del Neolitico in Italia. Rivista di Studi Liguri* LXXVIILXXIX(2011–2013): 339–352.

Cipolloni Sampò, M. 1999. Ipogeismo funerario e cultuale nella Daunia meridionale, in A. Gravina (ed.) *Atti 19° Convegno Nazionale sulla Preistoria-Protostoria- Storia della Daunia 2*: 155–188. San Severo: Archeoclub d'Italia.

Cremonesi, G. 1976. Tomba della prima età dei metalli presso Tursi (Matera). *Rivista di Scienze Preistoriche* 31/1: 109–134.

Davis, S.J. and F.R. Valla, 1978. Evidence for domestication of the dog 12,000 years ago in the Natufian of Israel. *Nature*, 276 (5688): 608, 10.1038/276608a0.

De Grossi Mazzorin, J. and C. Minniti 2000. Le sepolture di cani della necropoli di età imperiale di Fidene-via Radicofani (Roma): alcune considerazioni sul loro seppellimento nell'antichità, in *Atti del 2° Convegno Nazionale di Archeozoologia*: 387–398. Forlì, Abaco.

De Grossi Mazzorin, J. and C. Minniti 2001. Caratterizzazione archeozoologica: le sepolture di cani, in P. di Manzano (a cura di), *Ad deverticulum. Scavi archeologici lungo la bretella Nomentana-GRA*: 81–93. Roma: Generalvie spa.

De Grossi Mazzorin, J. and C. Minniti 2006. Dog Sacrifice in the Ancient World: A Ritual Passage? in L.M. Snyder and E.A. Moore (eds), *Dogs and People in Social, Working, Economic or Symbolic Interaction*, Proceedings of the 9th Conference of the International Council for ArchaeoZoology: 62–66. Oxford: Oxbow Books.

De Grossi Mazzorin, J. and A. Tagliacozzo 2000. Dog remains in Italy from the Neolithic to the Roman period. *Anthropozoologica* XXVXXXVI, 1997: 429–440.

Edrey, M. 2008. The Dog Burials at Achaemenid Ashkelon Revisited. *Tel Aviv 35*: 267–282.

Facciolo, A., I. Fiore and A. Tagliacozzo 2006. Archeozoologia dei contesti rituali paleoveneti, in A. Curci and D. Vitali (eds) *Animali tra uomini e dei. Archeozoologia del mondo preromano* (Studi e scavi nuova serie 14): 53–76. Bologna: Ante quem.

Fiore, I. 2016. *Analisi archeozoologica dei resti ossei animali da alcune tombe della necropoli del Picentino, proprietà Colucci*, in S. De Natale, B. D'Agostino and B. Gastaldi (eds), *Pontecagnano. II 7. La necropoli del Picentino. Tombe della Prima Età del Ferro dalla proprietà* (Collection du Centre Jean Bérard): 139–143. Naples: Centre Jean Bérard.

Fiore, I. and L. Salvadei 2014. I resti ossei di cani e neonati rinvenuti nei pozzetti II e III del teatro romano di Peltuinum: analisi preliminari. *Rendiconti della Pontificia Accademia* 84, 2011–12: 387–402.

Fiore, I. and A. Tagliacozzo 2000. Deposizioni di resti animali nelle tombe della necropoli di Casale del Dolce (Anagni, FR): l'esempio della tomba 4, in *Atti del II Convegno Nazionale di Archeozoologia*: 201–211. Forlì, Abaco.

Gambari, F.M. and U. Tecchiati 2004. Il cane e il cavallo come indicatori di status nella preistoria e nella protostoria, in F. Marzatico, P. Gleirscher (eds), *Guerrieri, principi ed eroi fra il Danubio e il Po dalla Preistoria all'Alto Medioevo* (Catalogo della mostra): 231–241. Trento.

Grünberg, J.M. 2013. Animals in Mesolithic Burials in Europe. *Anthropozoologica* 48 (2): 231–253.

Guidi, A. 1992. Le età dei metalli in Italia centrale e Sardegna, in A. Guidi and M. Piperno, *Italia Preistorica*: 420–470. Roma-Bari: Laterza.

Hasler, A. and C. Noret 2017. Habitats et structures funéraires néolithiques sur le tracé du cadereau d'Alès à Nìmes (Gard): premiers résultats, in *6e Rencontres Méridionales de Préhistoire Récente*: 171–190. Pèrigueux: Coédition ADRAHP-PSO.

Janssens, L., L. Giemsch, R. Schmitz, M. Street, S. Van Dongen and P. Crombé 2018. A new look at an old dog: Bonn-Oberkassel reconsidered. *Journal of Archaeological Science*, 92: 126–138. https://doi.org/10.1016/j.jas.2018.01.004

Liston, M.A., S.I. Rotroff and L.M. Snyder 2018. *The Agora bone well* (Hesperia Supplement 50). American School of Classical Studies at Athens.

Losey, R.J., T. Nomokonova, A.V. Gusev, O.P. Bachura, N.V. Fedorova, P.A. Kosintsev, M.A. Katzenberg and M.V. Sablin 2018. Dogs were domesticated in the Arctic: Culling practices and dog sledding at Ust'-Polui. *Journal of Anthropological Archaeology* 51: 113–126.

Miari, M. 1993. La necropoli eneolitica di Ponte S. Pietro (Ischia di Castro, Viterbo), *Rivista di Scienze Preistoriche*, 45: 101–166.

Migliorati L., I. Fiore, A. Pansini, P.F. Rossi, T. Sgrulloni and A. Sperduti 2018. Sepolti nel teatro: il valore simbolico dei cani in sepolture comuni infantili. *Scienze dell'Antichità* 23 (3): 593–611.

Minniti, C. 2022. Pratiche funerarie e sacrifici, in J. De Grossi Mazzorin and C. Minniti (eds), *Gli animali a Roma. Tre millenni di interazione con l'uomo* (Futuro anteriore 5). Firenze: All'Insegna del Giglio.

Negroni Catacchio, N. and M. Aspesi 2016. Rinaldone e Gaudo: rituali a confronto, in N. Negroni Catacchio (ed.) *Preistoria e Protostoria in Etruria, Atti del 12° Incontro di studi. Ornarsi per comunicare con gli uomini e con gli Dei. Gli oggetti di ornamento come status symbol, amuleti, richiesta di protezione II*: 611–636. Milano: Centro studi di Preistoria e archeologia.

Onorato, O.G. 1960. *La ricerca archeologica in Irpinia, Avellino*. Avellino: Amministrazione provinciale.

Salari L., R. Sardella, E. Squazzini, A. Lisciarelli and T. Suoadoni 2006. *Il cane della necropoli di Amelia (Terni, Umbria)*, in A. Curci, D. Vitali (eds), *Animali tra uomini e dei. Archeozoologia del mondo preromano* (Studi e scavi nuova serie 14): 179–191. Bologna: Ante quem

Santi, F. 1999. Cadaveri e carogne. Per una storia del seppellimento animale, in Il cadavere, Antropologia e immaginario sociale, *Micrologus. Science Nature and Medieval Societies*, VII: 155–203.

Silvestrini, M., G. Cilla and G. Pignotti 1992/93. La necropoli eneolitica di Fontenoce (Recanati). *Picus. Studi e ricerche selle Marche nell'antichità 12- 13*: 127–185.

Soren, D., D. Fenton and W. Birkby 1995. The Later Roman Infant Cemetery near Lugnano in Teverina, Italy: some implications, *Journal of Palaeopathology* VII, 1: 13–42.

Soren, D. and N. Soren (ed.) 1999. *A Roman villa and a late Roman infant cemetery: excavation at Poggio Gramignano Lugnano in Teverina*. Roma: L'Erma di Bretschneider.

Sperduti, A., L. Migliorati, A. Pansini, T. Sgrulloni, P.F. Rossi, V. Vaccari and I. Fiore 2018. Differential burial treatment of newborn infants from Late Roman Age. Children and dogs depositions at Peltuinum, in V. Nizzo (ed.), *Antropologia e archeologia a confronto*: Archeologia.

Tchernov, E. and F.F. Valla 1997. Two new dogs, and other Natufian dogs, from the southern Levant. *Journal of Archaeological Science 24*(1): 65–95.

Trantalidou K. 2006. Companions from the oldest times: dogs in ancient Greek literature, iconography and osteological testimony, in Snyder L. M., E.A. Moore (eds), *Dogs and People in Social, Working, Economic or Symbolic Interaction*: 96–120. Oxford: Oxbow Books.

Wilkens, B. 2000. Il cane eneolitico di Fontenoce (Recanati- MC) in *Atti del 2° Convegno Nazionale di Archeozoologia*: 213–215. Forlì: Abaco.

### Web sources


https://www.mantovaducale.beniculturali.it/it/news/519-i-cani-dei-gonzaga-da-rubino-a-tibris (viewed 10 May 2020)

https://learnearnandreturn.wordpress.com/2011/05/25/old-fritz-and-his-dogs/ (viewed 10 May 2020)

https://funlondontours.com/the-victorian-pet-cemetery-of-hyde-park/ (viewed 10 May 2020)

https://www.farmersjournal.ie/greyfriars-bobby-the-most-faithful-dog-585174 (viewed 10 May 2021)

https://allthatsinteresting.com/hachiko-dog; https://it.wikipedia.org/wiki/Hachik%C5%8D (viewed 30 December 2021)

https://en.wikipedia.org/wiki/A_Dog_Named_Palma

https://lahteet.com/dog-waiting-for-the-owner-at-the-airport-for-two-years/ (viewed 30 December 2021)

Detail from Saliari *et al.* fig. 7 (chapter 4.4): Roman tile with footprints of middle to big-sized dogs from the Roman legionary brickyard in Vindobona (Steinergasse 17, 1170 Vienna) (Photo by Mario Mosser).

# Section 4
## Dogs: Archaeological and Archaeozoological Cases

# 4.1 *Ur-gir*[1] and the Other Dogs from Abu Tberah (Southern Iraq): Considerations on the Role of Dogs in Sumer during the 3rd Millennium BCE

Francesca Alhaique[1], Licia Romano[2], Franco D'Agostino[2]

[1]Archaeozoology Division, Bioarchaeology Service, Museum of Civilisations, Piazza G. Marconi 14, Rome, Italy, francesca.alhaique@cultura.gov.it

[2]Department of Oriental Studies, Sapienza University of Rome, P.le A. Moro, 5 Rome, Italy, licia.romano@uniroma1.it; franco.dagostino@uniroma1.it

Corresponding author: Francesca Alhaique, francesca.alhaique@cultura.gov.it

**Abstract**

The site of Abu Tbeirah is located about 15 Km NE of Ur (Southern Iraq) and is dated to the second half of the 3rd millennium. The large faunal assemblage includes mainly domestic animals associated with many fish and mollusk remains. Dog elements are in general rare a'nd were recovered in some cases associated with human burials, while in others they seem to represent isolated intentional interments of this animal; these findings will also be considered within the framework of the beliefs regarding the animal world in the Sumerian culture as we know it from literary sources and iconographic evidences.

**Keywords:** Sumerian period, burials, Southern Iraq, religion, rituals.

## 1 Introduction

The medium sized city of Abu Tbeirah, Iraq (30° 98′ 43.93″ E, 46° 26′ 97.35″ N) is situated about 15 Km NE of Ur (Nasiriya, DhiQar province, Southern Iraq) and covers a surface of about 42 ha. Since 2012, archaeological investigations have been carried out by an Iraqi-Italian archaeological mission in different areas of the site evidencing, so far, some buildings, several burials and a harbour dated to the second half of the third millennium BCE, between the end of the Early Dynastic and the beginning of the Akkadian period (D'Agostino and Romano 2018; D'Agostino *et al.* 2015; D'Agostino *et al.* in press); at this time the region was a marshy area near the ancient Gulf shoreline (Romano 2019; Milli and Forti 2019). One of the main goals of the interdisciplinary research project was to reconstruct human adaptations to this water-rich environment and to understand how people coped with changing climatic conditions.

In Area 1, located in the southeastern part of the site, a very large household (Building A, ca. 600 m$^2$) was discovered providing evidences of daily life activities as well as of burial practices, the latter because of the presence of sub-pavement graves both inside and outside the building. The structures of the household were then cut by several graves of a cemetery and garbage pits in the latest occupation phase of the area (Romano 2019). Other domestic structures belonging to the end of the third millennium BCE were uncovered in Area 2, in the northeastern part of the site, where a similar situation occurred: such structures were in fact cut by graves (one of them, Grave 100, was particularly rich) that were in turn severely disturbed by later activities, possibly belonging to a now eroded more recent phase (D'Agostino and Romano 2015).

## 2 Faunal background

A relatively large faunal assemblage was collected during all the field seasons in the different areas of the settlement. The archaeozoological and taphonomic analysis of the remains is still in progress, but has been completed and published for the material from the latest phases of Area 1 (Alhaique 2019); other faunal data relevant for this paper come from Area 2, especially from the burials.

In general, the study mainly identified domestic animals associated to many fish and mollusk remains which, as mentioned before, together with other archeological, geological and environmental evidences (Celant and Magri 2019; Jotheri 2019; Milli and Forti 2019; Romano 2019), show that the sea was much closer and therefore the environment in the surroundings of the site was very likely similar to the current situation in the Iraqi Marshes. The animal remains were collected in domestic contexts referred to the most recent phase (Phase 1) of

[1] The word for '(domestic) dog' in Sumerian is **ur**, often to be found with the specification **ĝi** (also read **ĝir**) meaning probably 'native', as **ur-ĝi** or **ur-ĝir**, translated with doubts as 'native dog, hound'. In the lexical texts we have information on many different dogs, for instance the ur-Elam, 'dog from Elam' (Persian dog) or ur-Marhaši (dog from Asia), etc.

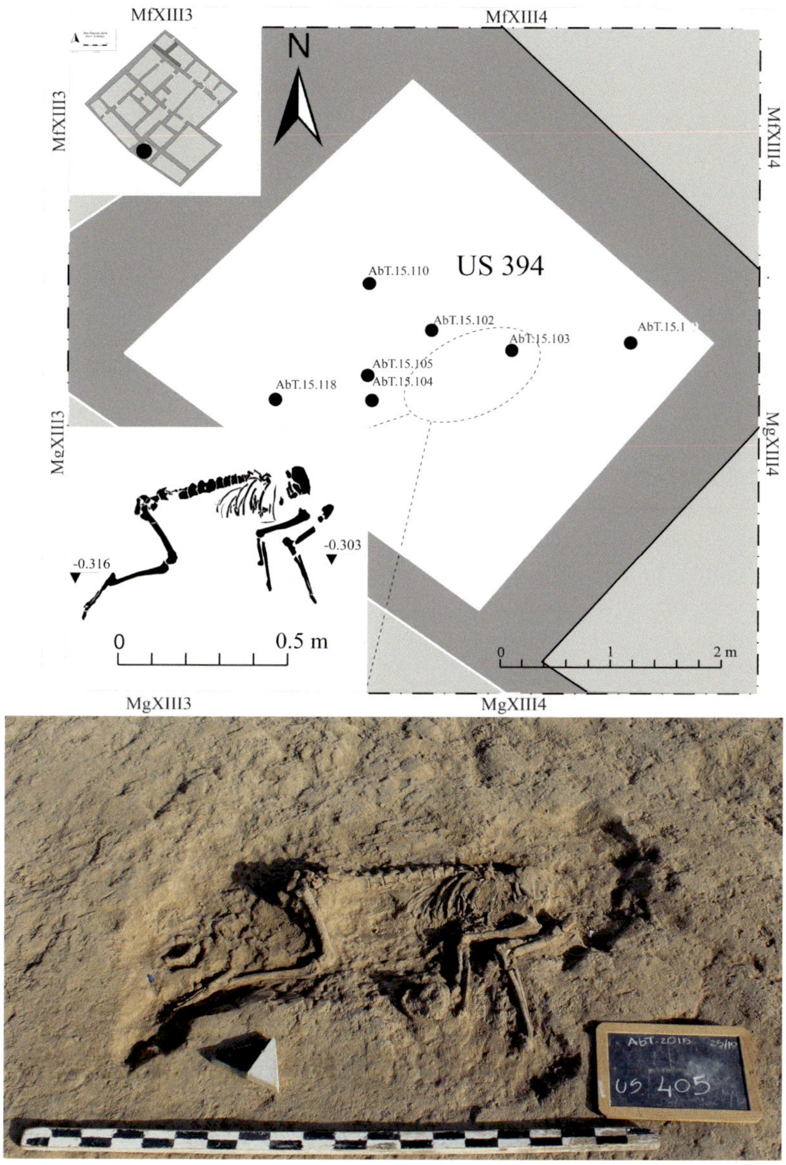


Figure 1. Dog Burial in Room 22, Building A, Area 1 (Images Iraqi-Italian mission in Sumer).

use of Building A as well as in association with human burials (both the sub-pavement graves of the Building and the Cemetery above it) as food offerings or leftovers of funerary banquets (Alhaique 2019; Alhaique *et al.* 2019; Alhaique, Tafuri *et al.* 2021). In general, the faunal composition does not show particular differences in the use of the main species in daily life and funerary rituals. Ovicaprines and pigs are the most common taxa in all contexts, but subtle differences between 'sacred' and 'profane' settings may be suggested by the age at death (Alhaique *et al.* 2021) with younger animals being more frequently associated to the deceased. In contrast, there is a general scarcity of cattle, both in ritual and domestic contexts; this could be explained by the ecological features of the territory surrounding the site, which were probably not appropriate for a large-scale cultivation for which such taxon could have been used. As far as the equids are concerned, both domestic donkeys and *Equus hemionus* or *Equus hemionus*-donkey hybrids have been discovered (Gabbianelli *et al.* 2015). With the possible exception of the skeletal elements collected from a pit below Graves 15 and 16, they have so far all been associated with human burials or were interred in their own grave. Wild mammals are extremely rare: wild boar was almost exclusively found in graves, while gazelle bones were only found in Building A. The occurrence of fox and tortoise only in Grave 100 may indicate some particular meaning for these species, although the specimens are too few to be confident with this interpretation. The constant presence in both domestic and funerary contexts, of marine and freshwater taxa (both fish and mollusks), suggests a strict relationship between humans and the surrounding environment with the exploitation of all the available resources. These aquatic species were used not only as food, but also as raw materials for making tools and objects or, in the case of mollusks, as containers (the so called 'cosmetic shells').

## 3 Discussion about dogs and conclusions

Dog remains are in general extremely rare in the faunal assemblage, although carnivore gnawing, probably produced by dogs, is documented in domestic contexts, suggesting the actual presence of these animals in everyday life of the people living at Abu Tbeirah. However, so far, all the specimens of this taxon were recovered either in association with human graves or their bodies were intentionally interred as isolated depositions.

In particular, during the 2015 field season, the almost complete skeleton of a single dog was found under the pavement of Room 22 of Building A in Area 1 (Figure 1). The animal was still in anatomical position with slightly flexed limbs and was laying down on its left side with a North-East/South-West orientation, facing North-East. Notwithstanding the general completeness of the skeleton, the head and neck vertebrae were completely missing.

This dog skeleton, although a well-defined pit was lacking, but given the absence of head and neck, very likely represents a ritual interment, possibly suggesting the sacrifice of the animal. This practice is widely attested in the ancient Near East (Ramos-Soldado 2016) and over all the Mediterranean region, and might be interpreted both as an offering and/or as protection for the building.

The archaeozooogical analyses evidenced that the individual was about 2 years old and had a withers height between 52 and 55 cm. It was not possible to unquestionably define the sex of the animal because on one end the absence of the *baculum* may suggest that it was a female, but this lack may just be the result of a loss of this small element during the excavations. No bone modifications were detected on the skeleton and the black colour of many of the elements is not related to burning, but to accidental manganese staining (E. Peverati, pers. comm.), as is the case of many other animal and human remains from the site.

The only other dog bones recovered so far at the site come from Area 2. At least one adult animal, represented by relatively few skeletal elements (Figure 2) and with

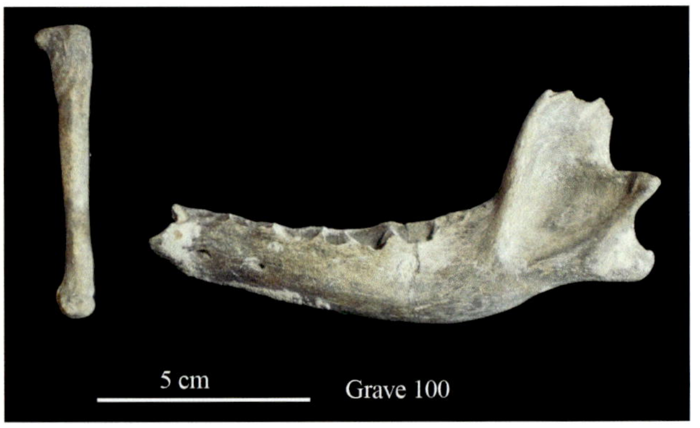


Figure 2. Selected dog elements from Grave 100, Area 2 (Photo by F. Alhaique).

a shoulder height of about 50 cm, was associated with Grave 100, a very rich human burial that was unfortunately heavily disturbed by later activities (D'Agostino *et al.* 2011). A second dog was a 5–6 months old puppy that was found in the fill of a pit (Figure 3); it may either represent an animal burial or have been associated with a disturbed human grave.

Textual sources attest to a wide range of attitudes towards dogs, based on their role in domestic contexts as well as on their healing properties connected to the cult of Gula (Nett 2021, Ramos-Soldado 2016, Tsouparopoulou 2020). Dogs are also present in Mesopotamian literature

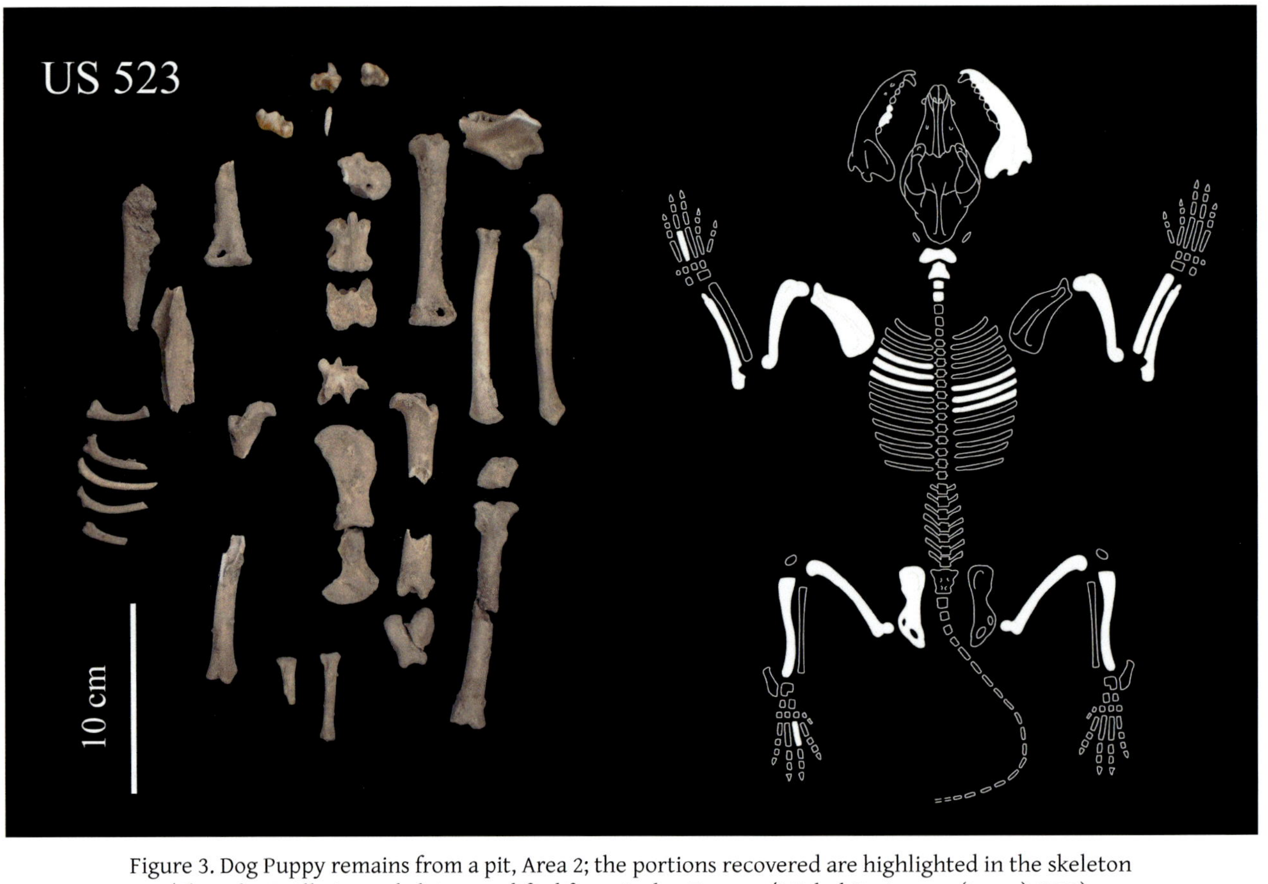


Figure 3. Dog Puppy remains from a pit, Area 2; the portions recovered are highlighted in the skeleton (Photo by F. Alhaique; skeleton modified from ArcheoZoo.org/Michel Coutureau (Inrap) 2013).

and are frequently mentioned in proverbs and fables, emphasising both their positive aspects (guarding, shepherding, hunting etc.) and negative ones (Gordon 1958, Wu 2001, Tsouparopoulou 2012, Tsouparopoulou and Recht 2021). Although the seated dog only clearly became a divine symbol in the Old Babylonian period, third millennium iconography also depicts dogs in a range of contexts. An Early Dynastic votive plaque from Nippur shows a dog in a typical domestic scene, under the chair of a banqueting character (Hansen 1963, Plate V); in contrast, the Sargon Stele, Louvre Sb1 (Nigro 1998, Figure 12), shows domestic dogs and vultures devouring and dismembering the bodies of the enemies (Tsouparopoulou and Recht 2021). In any case, besides the religious and cultural role of this species for the Sumerians, the data from Abu Tbeirah suggest a special care for this animal connected with the nature of the strict relationship, even in the afterlife, between humans and dogs.

### References


Alhaique, F. 2019. Faunal remains, in L. Romano and F. D'Agostino (eds) *Abu Tbeirah Excavations I. Area 1 Last Phase and Building A - Phase 1*, Rome: Sapienza University Press, 419–438.

Alhaique, F., M.A. Tafuri, L. Romano and F. D'Agostino 2021, *Cibo per i morti e cibo per i vivi, una prospettiva dalla Mesopotamia meridionale all'alba della storia. Atti L Riunione Scientifica dell'Istituto Italiano di Preistoria e Protostoria*. Studi di Preistoria e Protostoria, 6: 997–1004.

Alhaique, F., L. Romano and F. D'Agostino 2019. *Vita quotidiana e morte ad Abu Tbeirah (Iraq meridionale) nel III millennio a.C.: dati preliminari dalle faune dall'Area 1*, in M. Baldi, R. Dan, M. Delle Donne, G. Lucarini and G. Mutri (eds) *Archaeology of Food - New Data from International Missions in Africa and Asia*. Serie Orientale Roma, n.s. 17: 63–74.

Alhaique, F., L. Romano and F. D'Agostino 2021. *Between sacred and profane: human-animal relationships at Abu Tbeirah (Southern Iraq) in the 3rd millennium BCE*, in L. Recht and C. Tsouparopoulou (eds) *Fierce lions, angry mice and fat-tailed sheep: Animal encounters in the ancient Near East*. McDonald Institute Conversation Series: 63–75.

Celant, A. and D. Magri 2019. Palaeoenvironment, climate, and land use in Southern Mesopotamia/ Nasiriya Area, in L. Romano and F. D'Agostino (eds) *Abu Tbeirah Excavations I. Area 1 Last Phase and Building A - Phase 1*, Rome: Sapienza University Press, 39–48.

D'Agostino, F., M. Vidale, L. Romano and M. Angelozzi 2011. Abu Tbeirah. Preliminary Report of the First Campaign (January-March 2012), *Rivista degli Studi Orientali* 2011: 17–34.

D'Agostino, F. and L. Romano 2018. The Harbor of Abu Tbeirah and the Southern Mesopotamian Landscape in the 3rd Mill. BC: Preliminary Considerations, *Rivista degli Studi Orientali* 2018: 19–31.

D'Agostino, F., L. Romano, A. Kadhem 2015. *Abu Tbeirah, Nasiriyah (Southern Iraq). Preliminary Report on the 2013 Excavation Campaign*, in M.G. Biga *et al.* (eds), Homenaje a Mario Liverani, fundador de una ciencia nueva (II)/Omaggio a Mario Liverani, fondatore di una nuova scienza (II) (= ISIMU 13), Madrid 2011(2015): 209–221.

D'Agostino, F., L. Romano, A. Khadem Ghanim, F. Alhaique, A. Celant, G. Festa, V. Forte, C. Lemorini, L. Medeghini and M.A. Tafuri in press. Tell Abu Tbeirah. Preliminary Report of the First Four Campaigns, *Sumer.*

Gabbianelli, F., F. Alhaique, L. Romano, F. D'Agostino and A. Valentini 2015. mtDNA Analysis for the Characterization of Sumerian Equids, *Italian Journal of Animal Science* 14, 112.

Gordon, E.I. 1958. Sumerian Animal Proverbs and Fables: 'Collection Five' (Conclusion), *Journal of Cuneiform Studies* 12 (2): 43–75.

Hansen, D.P., 1963. New Votive Plaques from Nippur. *Journal of Near Eastern Studies* 22 (3), 145–166.

Jotheri, J. 2019. The environment and landscape archaeology of the Abu Tbeirah region, in L. Romano and F. D'Agostino (eds) *Abu Tbeirah Excavations I. Area 1 Last Phase and Building - Phase 1*, Rome: Sapienza University Press, 49–58.

Milli, S. and L. Forti 2019. Geology and Palaeoenvironment of Nasiriyah Area/Southern Mesopotamia, in L. Romano and F. D'Agostino (eds) *Abu Tbeirah Excavations I. Area 1 Last Phase and Building A - Phase 1*, Rome: Sapienza University Press, 19–38.

Nett, S. 2021. The dogs of the healing goddess Gula in the archaeological and textual record of ancient Mesopotamia, in L. Recht and C. Tsouparopoulou (eds) *Fierce lions, angry mice and fat-tailed sheep: Animal encounters in the ancient Near East*. McDonald Institute Conversation Series: 55–62.

Nigro, L. 1998. The Two Steles of Sargon: Iconology and Visual Propaganda at the Beginning of Royal Akkadian Relief, *Iraq* 60: 85–102.

Ramos-Soldado, J.L. 2016. *Structured Deposition of Animal Remains in the Fertile Crescent during the Bronze Age*, Archaeopress Publishing, Oxford.

Romano, L. 2019. Abu Tbeirah and Area 1 in the second half of the 3rd millennium, in L. Romano and F. D'Agostino (eds) *Abu Tbeirah Excavations I. Area 1 Last Phase and Building A - Phase 1*, Rome: Sapienza University Press, 59–91.

Tsouparopoulous, C. 2012. *The 'K-9 Corps' of the Third Dynasty of Ur: The Dog Handlers at Drehem and theArmy.* Zeitschrift für Assyriologie und vorderasiatische Archäologie,Volume 102 (1): 1–16.

Tsouparopoulou, C. 2020. The healing goddess, her dogs and physicians in the late third millennium

Mesopotamia. *Zeitschrift für Assyriologie und vorderasiatische Archäologie* 110 (1): 14–24.

Tsouparopoulou, C. and L. Recht 2021. Dogs and equids in war in 3rd millennium Mesopotamia, in L. Recht and C. Tsouparopoulou C. (eds) *Fierce lions, angry mice and fat-tailed sheep: Animal encounters in the ancient Near East*. McDonald Institute Conversation Series: 279–289.

Wu, Y. 2001. Rabies and Rabid Dogs in Sumerian and Akkadian Literature, *Journal of the American Oriental Society*, 1218 (1): 32–43.

# 4.2 Ritual Use of Dogs in the Neolithic Cultures of China

Maria Kudinova

School of Archaeology and Museology, Peking University, Yiheyuan Road 5, 100800 Beijing, China.
Institute of Archaeology and Ethnography, Siberian Branch of the Russian Academy of Sciences, Akademika Lavrentieva Avenue 17, 630090 Novosibirsk, Russia
maria-kudinova@yandex.ru

**Abstract**

The article summarises the data on osteological materials of domestic dogs unearthed from the archaeological sites in the territory of China, which reflect the ritual use of dogs in the Neolithic period. Judging by the context of these findings, it can be presumed that dogs were used in funerary rites (accompanying burials, sacrifices related to mortuary practices) and as building sacrifices, that were in turn based on the perception of a dog as a protector against evil forces. The images of dogs in Chinese Neolithic art also reflect the important role of this animal in the worldview of ancient people.

**Keywords**: China, Neolithic, funerary practices, building sacrifice, art.

## 1 Introduction

Domestication of dogs is one of the most important achievements in human history. According to paleogenetic data (Wu Zhuang 2014; Frantz *et al.* 2016; Wang *et al.* 2016), one of the dog domestication centres is located in Southeast Asia (including a part of China's territory south of Yangtze). But wherever the primary domestication of dogs occurred, dogs appeared in the territory of China in remote antiquity and played a specific role in the material and spiritual culture of peoples who inhabited this territory in different historical epochs. Various functions of dogs in Chinese culture (including ritual practices, customs, art etc., especially those of the Shang epoch) were the subject of inquiries made by Chinese and western researchers, including Ling Chunsheng (1957), Wang Lihua (1992), Li Xiangsheng (2006), Zhang Zhen (2006), Guo Zhiwei (2012, 2014), Gao Guangren and Shao Wangping (2013), Li Zhipeng (2011), Wu Zhuang (2014), Roderick Campbell (2015), etc. The origins of this tradition can be traced to earlier periods. Therefore, it seems highly important to make a study of the earliest periods of 'dog breeding' in China, when the economic and ritual practices of later epochs firstly emerged.

## 2 Materials

Current available archaeological materials contradict the paleogenetic data, which indicate the earlier taming of dogs in Southern China. The earliest osteological materials of domestic dogs from the territory of China were unearthed at the Nanzhuangtou site (Xushui County, Hebei province, in Northern China) and date back to the Early Neolithic period, *c.* 8000 BC (Wu Zhuang 2014: 66–67; Yuan Jing 2015: 89–90). Neither were dog bones found among the faunal remains in the Early Neolithic sites in Southern China. From the Middle Neolithic period (c. 7000–5500 BC) domestic dogs spread over the territory of nowadays China. There are only two archaeological cultures of this epoch, where dog bones were not found: the Shangshan culture (9000–6000 BC, Zhejiang Province) and the Pengtoushan culture (7500–5600 BC, Hubei and Hunan Provinces). During the Late and Final Neolithic period (*c.* 5500–2000 BC) regional differences finally resolved, the ratio of dog bones in the mammal remains is consistently equal to 5–10 % (Wu Zhuang *et al.* 2016: 157–158).

Osteological remains of domestic dogs were found at the archaeological sites of the Middle, Late and Final Neolithic periods (in total about 200 sites), which belong to the following cultures:

- in Central and Northwestern China: Peiligang (7000–4800 BC, Henan Province), Cishan (6000–4500 BC, southern part of Hebei Province), Dadiwan (6200–5000 BC, Gansu Province, northern part of Shaanxi Province), Majiayao (4000–2800 BC, Gansu Province, Qinghai Province, Ningxia Hui Autonomous Region), Machang (2400–2000 BC, Qinghai Province, Gansu Province), Yanghsao (*c.* 6000–2000 BC, the Upper and Middle Yellow River) and Longshan (c. 2900–1900 BC, the Middle Yellow River) cultures;
- in Northeastern China: Xinglongwa (6200–5300 BC, Inner Mongolia Autonomous Region, Liaoning Province, Hebei Province), Hongshan (4700–2800 BC, Inner Mongolia Autonomous Region, Liaoning Province, Hebei Province), Xinkailiu (4000–3000 BC, Heilongjiang Province), Houwa (4000–3000 BC, Liaoning Province), Fuhe (3500–3100 BC, Inner Mongolia Autonomous Region), Xiaoheyan (3500–2800 BC, Liaoning Province);

- in Eastern China: Houli (6500–5000 BC, Shandong Province), Qingliangang (5400–4400 BC, Shandong and Jiangsu Provinces), Dawenkou (4300–2200 BC, Shandong and Jiangsu Provinces);
- in Southeastern China: Kuahuqiao (6200–5000 BC, Zhejiang Province), Hemudu (5000–3400 BC, northeastern part of Zhejiang Province), Majiabang (5000–4000 BC, Zhejiang and Jiangsu Provinces), Songze (4400–3300 BC, Shanghai, Jiangsu Province, Zhenjiang Province), Longqiuzhuang (4300–3500 BC, Jiangsu Province), Dapenkeng (4300–2500 BC, Taiwan), Liangzhu (3200–2200 BC, Jiangsu and Zhejiang Provinces), Tanshishan (2500–2000 BC, Fujian Province);
- in Southern China: Daxi (4400–3300 BC, Sichuan, Hubei, Hunan Provinces), Xuejiagang (3500–2600 BC, southwestern part of Anhui Province, eastern part of Hubei Province, northern part of Jiangxi Province), Qujialing (3400–2600 BC, Hubei and Hunan Provinces, southern part of Shaanxi Province, southwestern part of Henan Province, northern part of Jiangxi Province), Shijiahe (2500–2000 BC, Henan, Hubei, Hunan Provinces);
- in Southwestern China: Dingshishan (6000–5000 BC, Guangxi Zhuang Autonomous Region), Zengpiyan cave site (7000–5000 BC, Guangxi Zhuang Autonomous Region) (Chen Wenhua 1992: 342–343; Li Xiangsheng 2006: 243–244; Ren Shinan, Wu Yaoli 2010: 127–134, 141–147, 278–331, 359–362, 381–383, 414–443, 451–476, 711–713; Wu Zhuang *et al.* 2016: 157–158).

Apparently, in some Neolithic cultures (Dawenkou, Liangzhu, Longshan cultures) a dog was an economically important animal and one of the most important food sources of protein (along with pigs). It has been proven by dog bones with cut marks unearthed from the sites (in some cases - particularly from waste pits). In a burial of Dawenkou culture (M47) at the Yedian site (Zoucheng County, Shandong Province) dog bones were found among pig and chicken bones and pottery fragments, that could have been the remains of sacrificial food (Zouxian Yedian 2005: 105). The routine and ritual consumption of dog meat by Asian peoples is also well-known from ethnographical and written sources. Concurrently, there are several sites (Jiahu, Longqiuzhuang), where the evidence of the ritual use of dogs (such as burials of dogs in graves and sacrificial pits) were discovered, but no traces of dog meat consumption (i.e. bones with cut marks) can be seen (Longqiuzhuang 1999: 492; Wuyang Jiahu 1992: 902), that can reflect the formation of a taboo of eating dog meat, which exists now among some peoples of China (Yao, Miao, She, Manchu). Probably, dogs were also used for hunting, guarding or as partners in children's games, but no evidence of these types of use in the Neolithic period have survived.

Over the course of millennia, dogs have not only become an important element of material culture, but have also been included in a system of spiritual culture of ancient people. In the Middle Neolithic (c. 7000 BC) a tradition of burying dogs in human's graves emerged. At least 65 graves with buried dogs (whole skeletons or parts of bodies - skulls, mandibles, limb bones) at 19 sites of different cultures are known to date (Table 1). Dog burials in sacrificial pits and ash-pits were found at more than 40 sites. Judging by the context of these findings, it can be presumed that dogs were used in funerary rites (as accompanying burials, sacrifices connected with mortuary practices) and as building sacrifices, that were in turn based on the perception of a dog as a protector against evil forces.

## 3 Results

The earliest evidence of dog sacrifices and burials were found at Jiahu site (Wuyang County, Henan Province) and belong to the Peiligang culture (7000–6600 BC). The site was discovered in 1961, the excavations were carried out in 1983–2001 and 445 burials were unearthed. In one tomb (M341) a small fragment of a dog's bone was found near the left leg of the tomb occupant (middle-aged male). Aside from the dog's bone, grave goods also included two bone flutes, a fragment of a tortoiseshell and some other items. The burial belongs to the Peiligang culture and dates back to the period of 7000–6600 BC. Besides that, dog burials at this site were also found in one ash pit (H139, in addition to a dog skeleton there was some pottery and a grindstone) and ten sacrificial pits (Figure 1). Six sacrificial pits were located at the burial ground or near it, and four - inside dwellings or near them, no artefacts were unearthed from these pits (Wuyang Jiahu 1992: 106, 130, 169–171; Guo Zhiwei 2012: 54).

The next stage of this tradition is represented by burials belonging to different cultures of the Early Yangshao period (*c.* 6000–4000 BC): the Xiawanggang site (Xiawanggang culture), the Jiangzhai site (Banpo culture), the Dadiwan site (Culture of the 2nd period of Dadiwan) and the Zhanmatun site (Qinwangzhai culture). Among them, the largest number of dog burials in human tombs was found at the Xiawanggang site (Xichuan County, Henan Province), moreover, the whole skeletons and not just fragments of bodies were placed into tombs, so the Xiawanggang site can be considered as the most representative site of this stage. Among 124 tombs of the Xiawanggang culture, dogs were buried in five tombs (one tomb M287) with two individuals and four tombs with one individual). All tombs are rectangular earthen pits, measuring 2.04–2.34 m (length), by 0.42–0.80 m (width), by 0.4–1 m (depth). All tomb occupants were male, middle-aged or elderly (Figure 2). In one tomb (M112) a tortoiseshell was found along with a dog skeleton (Xichuan Xiawanggang, 1989: 26–32).

Table 1. Dog remains in the Neolithic burials in the territory of China.

| Site | Culture and Dates, BC | Number of tombs | Number of individuals or bone fragments in a tomb |
|---|---|---|---|
| *Central and Western regions* | | | |
| Jiahu, Wuyang County, Henan | Peiligang, 7000–6600 | 1 | a fragment of a bone |
| Xiawanggang, Xichuan County, Henan | Xaiwanggang (Yangshao) | 5 | 4 tombs with 1 skeleton, 1 tomb with 2 skeletons |
| | Longshan | 1 | 1 |
| Dadiwan, Qin'an County, Gansu | 2nd period of Dadiwan (final period of early Yangshao), 4500–3900 | 1 | 1 |
| Jiangzhai, Lintong District, Shaaanxi | Banpo (Yanshao) | 1 | a fragment of mandible |
| Jiangjiaping, Huanghe County, Gansu | Machang | 1 | 1 |
| Daxincun, Fengxiang County, Shaanxi | Keshengzhuang (Longshan) | 1 | a fragment of mandible |
| Dachengshan, Tangshan City, Hebei | Longshan | 1 | 1 |
| *North-Eastern region* | | | |
| Diaosiguigou, Chifeng City, Inner Mongolia | Hongshan | 2 | fragments of humeri |
| *Eastern region* | | | |
| Liulin, Pizhou County, Jiangsu | Dawenkou | 8 | 1 |
| Dadunzi, Pizhou County, Jiangsu | Dawenkou | 9 | 5 tombs with 1 skeleton, 3 tombs with 2 skeletons, 1 tomb with 3 skeletons |
| Huating, Xinyi County, Jiangsu | Dawenkou | 8 | 1 |
| Yedian, Zoucheng County, Shandong | Dawenkou (middle or late period) | 1 | fragmented bones |
| *South-Eastern region* | | | |
| Nanhebang, Jiaxing City, Zhejiang | Songze | 1 | 1 |
| Longqiuzhuang, Gaoyou County, Jiangsu | Longqiuzhuang, 4300–3500 | 5 | 1 tomb with 1 mandible, 2 tombs with 1 skull in each, 2 tombs with 1 whole skeleton in each |
| Weidun, Changzhou City, Jiangsu | Majiabang | 1 | 1 |
| Saidun, Huangmei County, Hubei | Xuejiagang, 3500–2600 | 1 | 3 |
| Zhuangqiaofen, Pinghu County, Zhejiang | Liangzhu | 13 | 1 |
| Tinglin, Jinshan District, Shanghai | Liangzhu (late period), *c.* 1700 | 3 | 1 |
| Wujiachang, Qingpu District, Shanghai | Liangzhu (late period) | 1 | 6 |

In the Neolithic epoch the custom of dog sacrifice (in human graves as well, mostly whole skeletons) was mostly widespread in the Dawenkou culture (4300–2200 BC, the Liulin, Dadunzi, Huating and Yedian sites). Dog burials in human tombs of the Dawenkou culture are quite similar to Yangshao ones: rectangular earthen pits, one dog skeleton in a tomb, tortoiseshells were found in several tombs (Yin Huanzhang *et al.* 1962: 87; Yin Huanzhang *et al.* 1965: 15;) (Figure 3). Burials of dogs at the Longqiuzhuang site (the Longqiuzhuang culture) and the Weidun site (the Majiabang culture) are nearly synchronistic with the sites of the Dawenkou culture. There were five tombs with dog remains (mandibles or skulls) found at the Longqiuzhuang site (Gaoyou County, Jiangsu Province). Apart from the burial in tombs, dog skeletons at the Longqiuzhuang site were also found in two sacrificial pits inside dwellings F1 and F2 (Longqiuzhuang 1999: 18, 77) (Figure 4).

The late stage of the tradition in the Neolithic period was presented in the Liangzhu culture (3200–2200 BC) and the Longshan cultures (2900–1900 BC). Most of the sites, where dog burials of this period were discovered, are located in the Lower Yangtze region: the Guangfulin site (three tombs), the Zhuangqiaofen site (13 tombs), and the Wujiachang cemetery (one tomb). Among them the Wujiachang M207 tomb can be distinguished by its rich grave goods, which included ceramic vessels, jade artefacts (ritual implements *cong* and *bi*, an axe *yue* and numerous pendants and beads), ivory bracelets and sceptres *yazhang*, stone tools, etc. (Shanghai Fuquanshan 2011: 4). Besides that, quite a specific funerary rite is shown in a tomb at the Jiangjiaping site (Yongdeng County, Gansu Province). It was a rather big earthen pit tomb with three smaller pits at the bottom. The pit in the centre of the tomb (waist-pit, *yaokeng*) is bigger than the two others, four pigs, one dog, an elderly woman and a skull of

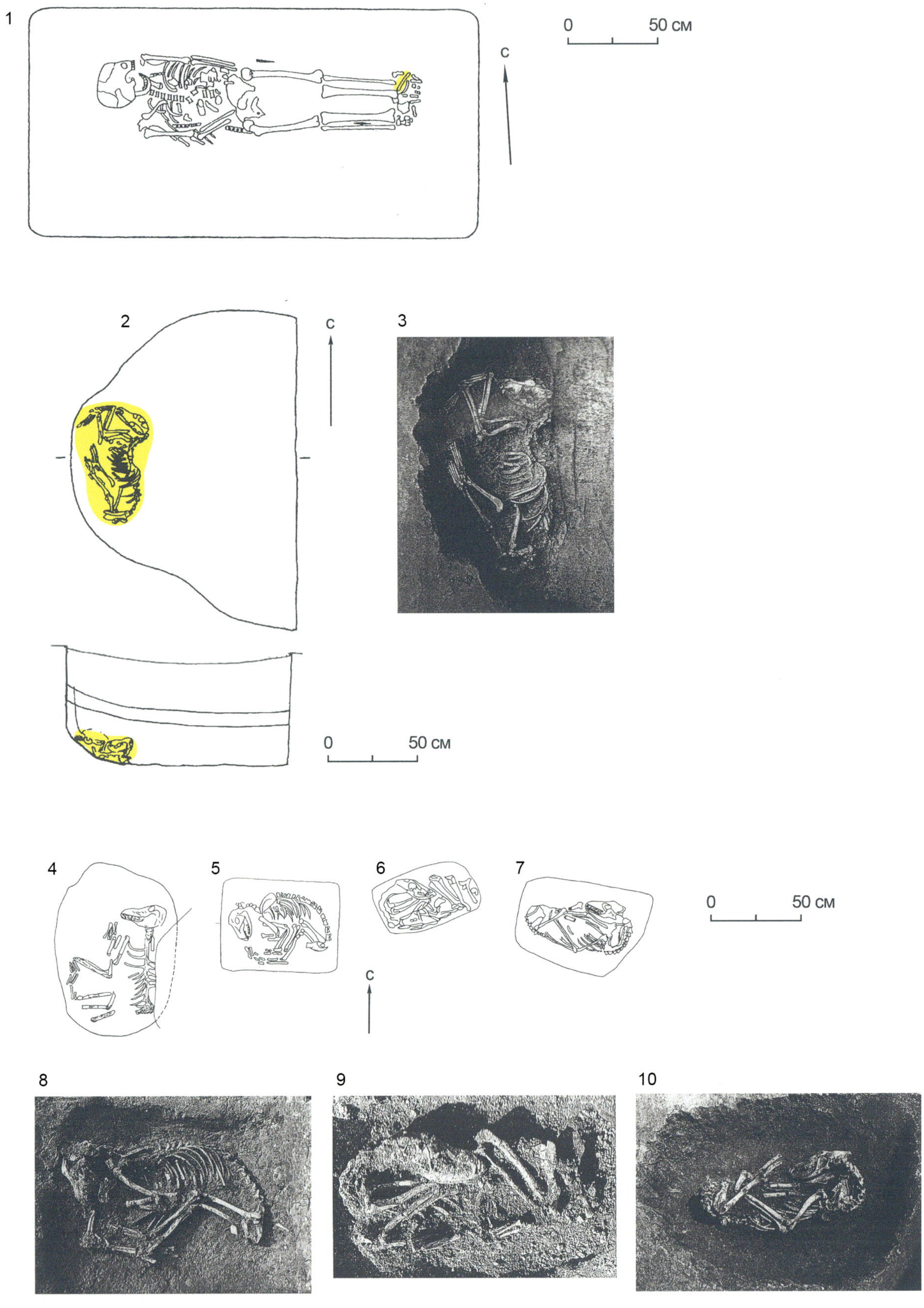


Figure 1. Dog remains at Jiahu site: 1-M341, 2-H139 (plan and section), 3-H139 (photo), 4-SK2, 5-SK8, 6-SK3, 7-SK9, 8-SK8 (photo), 9-SK3 (photo), 10-SK9 (photo), (After Wuyang Jiahu 1989: 105, 131, 169).

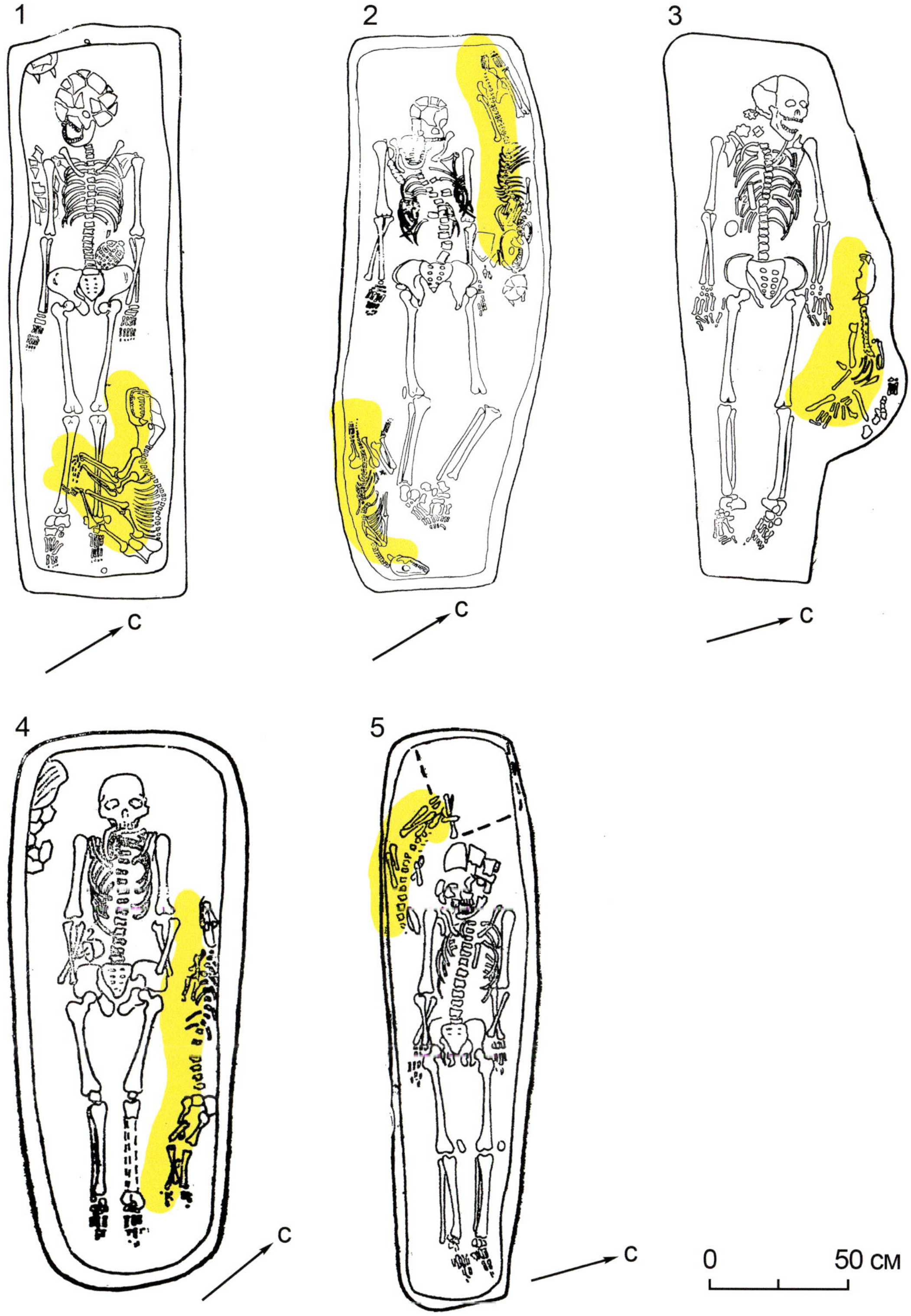


Figure 2. Dog burials at Xiawanggang site: 1 - M112; 2 - M285; 3 - M645; 4 - M224; 5 - M25 (After Xichuan Xiawanggang 1989: 25–27, 29, 30).

a ten-year-old child in a painted pottery vessel were placed there in layers. The tomb belongs to the Middle Machang culture (Guo Zhiwei 2012: 55–56; 2014: 58). A tradition of burying dogs in waist-pits became widespread in the Shang epoch, perhaps, that was one of its prefigurations.

In the matter of spatial coverage, the burials of dogs in the Neolithic epoch are mainly concentrated in the Huai River basin, the Upper and Middle Han River, the Northwestern region and the Lower Yangtze, but the boundaries of this area were not fixed. Dog burials of the Pre-Yangshao epoch were discovered only at the

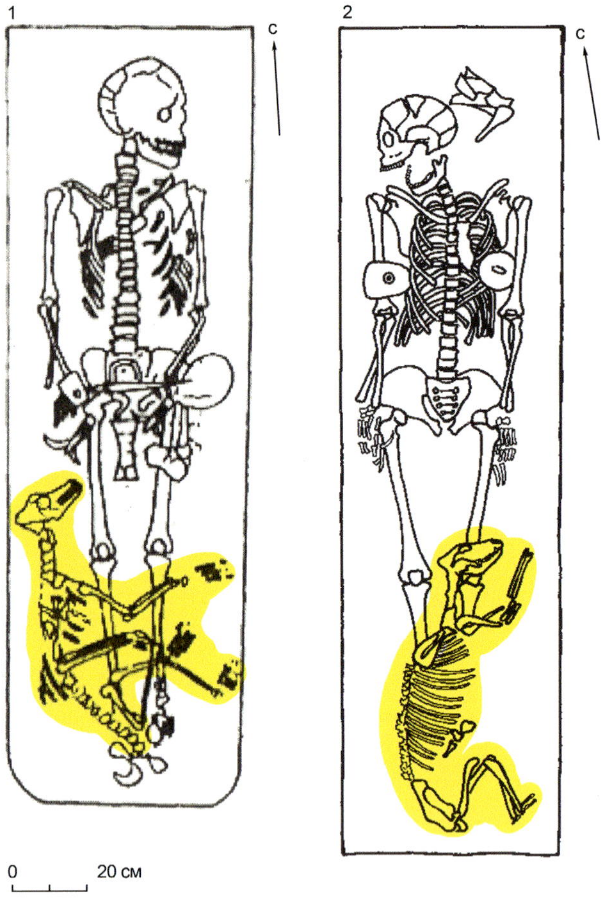


Figure 3. Dog burials at Liulin site: 1 - M25; 2 - M179 (After Yin Huanzhang and Zhang Zhengxiang 1962: 87; Yin Huanzhang *et al.* 1965: 17).

Jiahu site located in the Upper Huai River region. In the Early Yangshao period dog burials started to spread over the Han River basin (Xiawanggang) and Northwestern China (Dadiwan, Jiangzhai). In the time of the Dawenkou culture the area of the distribution of dog burials underwent a substantial modification: a significant number of burials spread over the territory of the Lower Huai River, where one of the centres of this tradition formed. At the same period dog burials emerged at several sites in the Middle and Lower Yangtze. During the middle and late periods of the Liangzhu culture and the Longshan epoch, dog burials were mainly spread over the Lower Yangtze (Tinglin, Zhuangqiaofen, Wujiachang), and only a small amount of burials were located in the Upper and Middle Han River region and in Northwestern China (Guo Zhiwei 2012: 55–56).

Burials of dogs in ash-pits and sacrificial pits were discovered at more than 40 sites of different cultures. A considerable part of these burials were located at the burial grounds or near them (at least eight sites are known so far), probably, they are the remains of funerary or obituary rites or the sacrifices offered to dead ancestors. There are also nine sites, where dog burials were found inside dwellings. These might have served a protective function. For instance, four dwellings were unearthed at the Longqiuzhuang site, and dog burials were discovered in two of them (F1 and F2). In dwelling F1 a dog was buried in a shallow oval pit (length 0.84 m, width 0.5 m) under a posthole D1, that indicates that the pit was dug at the beginning of the building process, and a dog most probably was used as a building sacrifice. The burial of a dog in dwelling F2 is quite similar to the previous one (Longqiuzhuang 1999: 18) (Figure 4, *1*, *2*). Other burials of dogs defy interpretation, but the context implies their ritual meaning. For example, the skeleton of a juvenile dog was found in ash-pit H23 of irregular form at the Pingzhai settlement (Gushi County, Henan Province). It laid in a round ash spot (diameter 0.64 m, thickness of ash layer 5 cm), with its skull oriented to the west. Under the ash spot a big irregular stone of approximately 10 cm in diameter was unearthed. Above the dog skeleton there was one small stone and one big stone with a carefully shaped stone arrowhead on it (Figure 5). The complex dates back to 2000–1700 BC and belongs to the Henan Longshan culture (Li Weiming *et al.* 2000: 333–334, 354–355).

A small amount of dog's images in Chinese Neolithic art is known up to the present moment. They include different types of depictions, such as: a painting on the surface of a Yangshao pottery vessel, a relief figure of a dog on the fragment of a pottery vessel (Hemudu culture), a dog-shaped pottery vessel (Dawenkou culture) and numerous small-sized clay sculptures of dogs found at the sites of the Shijiahe culture. The painting on the late Yangshao (3500–2900 BC) vessel found at the Dadiwan site (Qing'an County, Gansu Province) depicts the scenes of fighting between two dogs. In the centre of one of the compositions there is a fish (which could have been the reason for the fight). The fish, whose image

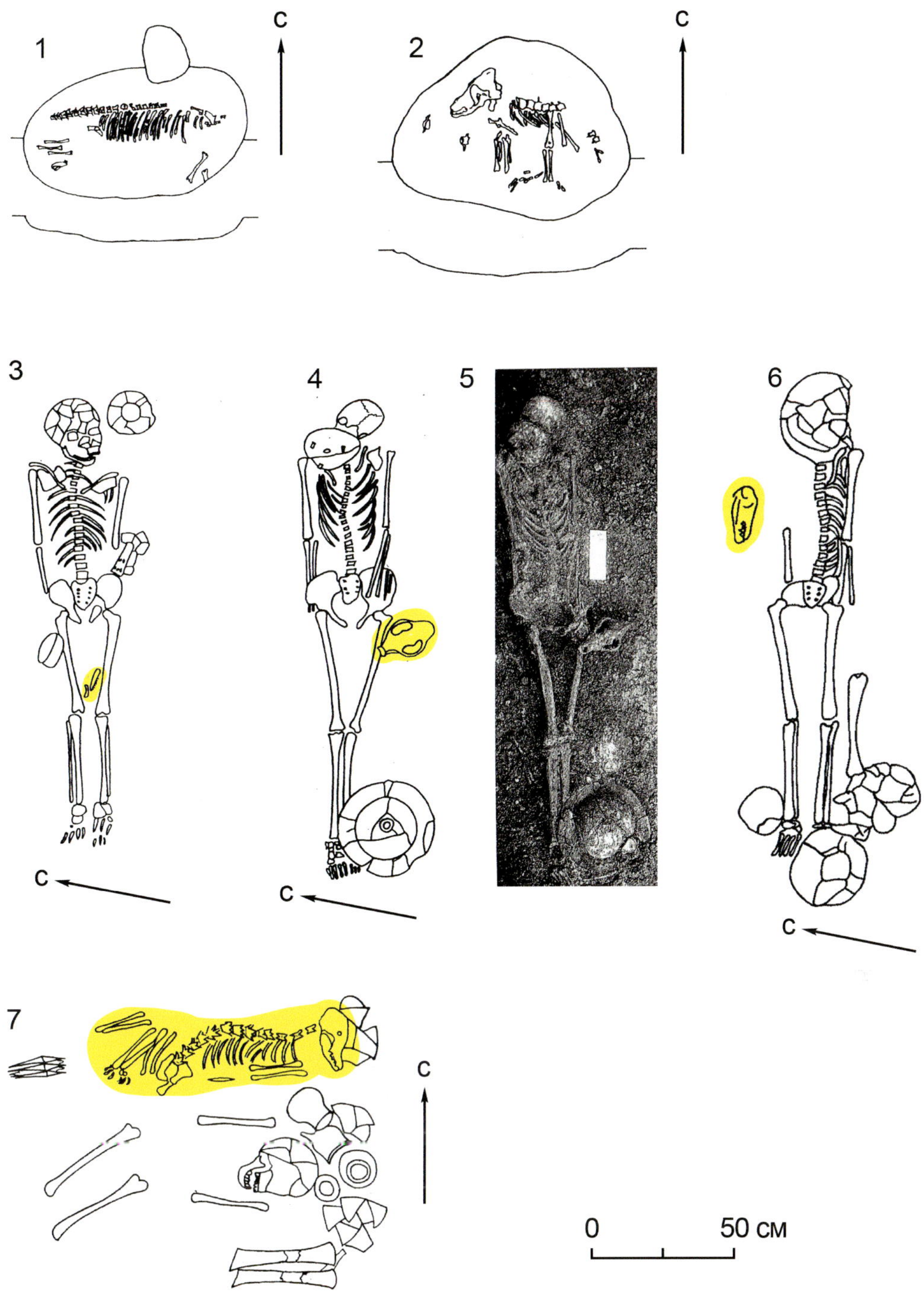


Figure 4. Dog remains at Longqiuzhuang site: 1 - F1; 2 - F2; 3 - M143; 4, 5 - M400; 6 - M198; 7 - M57 (After Longqiuzhuang 1999: 21–22, 54, 110; Guo Zhiwei 2012: 57).

is rather common for Yangshao painted pottery, could mark the whole scene as taking place in the underworld. Another possible interpretation is that the dogs stand as the mediators on the border of two worlds. Both explanations correspond to the archaeological materials of the Yangshao culture. Small clay figurines of dogs found at the Dengjiawan site (Tianmen County, Hubei Province) are numerous and multifarious (Figure 6). Dogs are depicted in various attitudes: standing, sitting, lying, dogs with bones in chaps, adult dogs with cubs, etc. There are also some figurines of humans stroking dogs. Analogies to dogs' figurines from Dengjiawan can be found in contemporary folk crafts from China (e.g. painted clay figurines of dogs and other animals *nigou* made in Huaiyang County, Henan Province) which prove the stability of ritual practices in traditional culture.

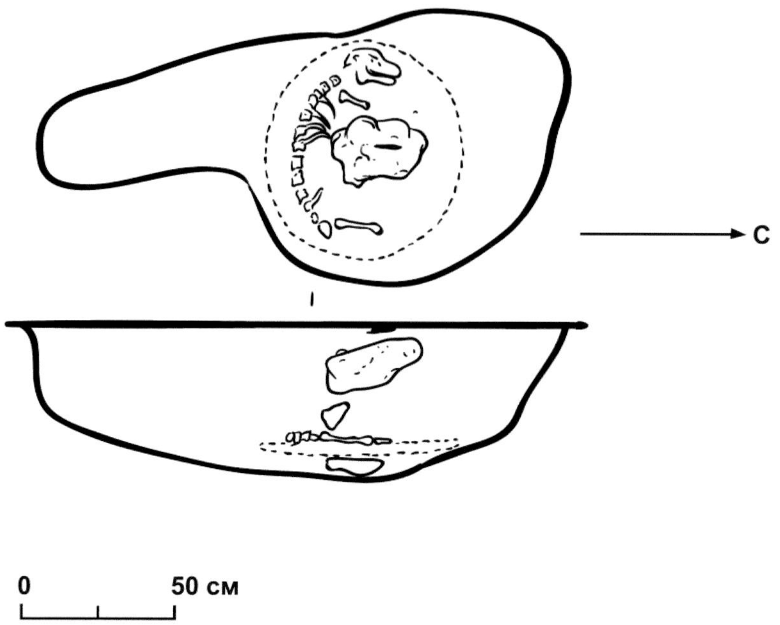


Figure 5. Ash-pit H23 at Pingzhai site (After Li Weiming *et al.* 2000: 334).

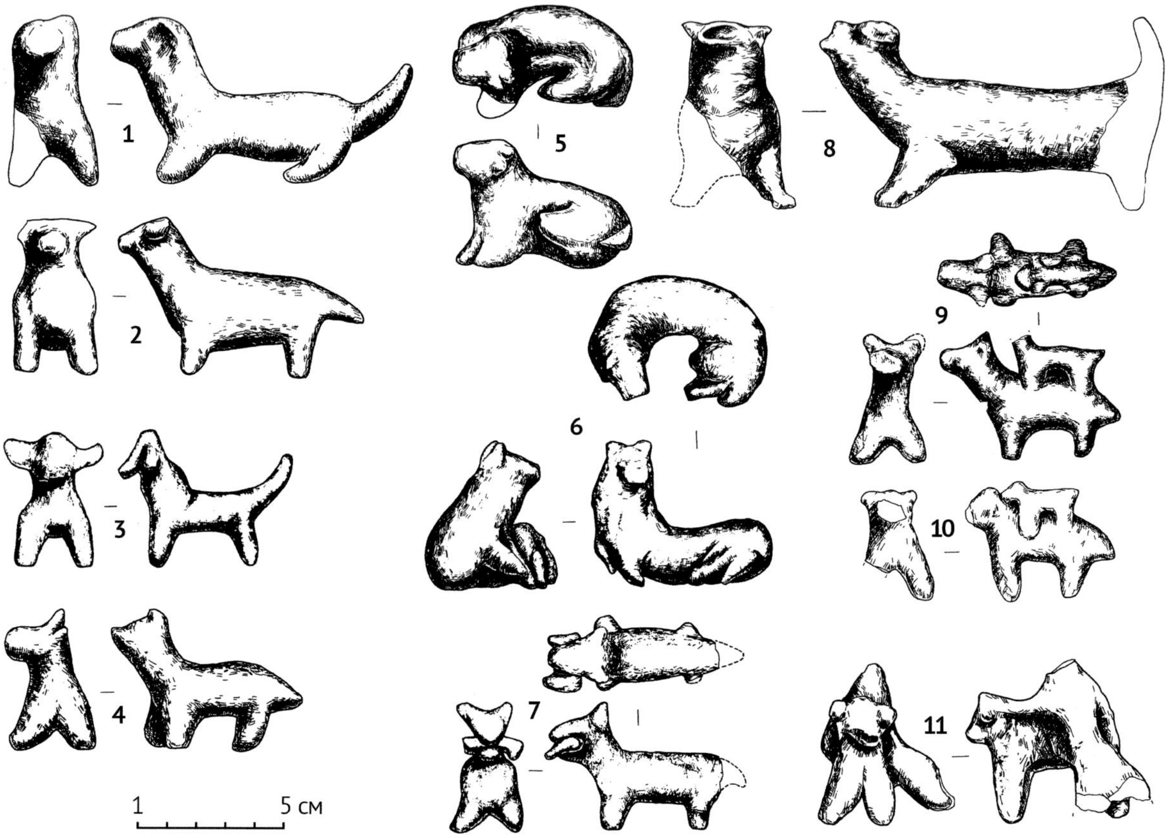


Figure 6. Clay figurines unearthed at Dengjiawan site (After Dengjiawan 2003: 188–194).

## 4 Discussion

Despite a small number of Neolithic dog burials known up to now, Chinese scholars have undertaken numerous attempts to interpret their meaning. Zhang Zhen (2006: 7) presumes that the burials of dogs at the Jiahu site were due to the use of dogs as guards: they were sacrificed to protect the settlement and its inhabitants. As for the burials of dogs in human graves, the animal could act as the owner's companion and hunting assistants in the underworld. Gao Guangren and Shao Wangping (2013: 387–388) argue that the burials of dogs in the Neolithic and Bronze Age (in the Shang-Yin culture) are of a similar nature: a dog was a tomb occupant's companion, it followed and guarded its owner after death. In their opinion, the Shang tradition of dogs accompanying burials was borrowed from the eastern part of China. In the Dawenkou culture there were three main sacrificial animals used in funerary rites and buried in human tombs: pigs (whole bodies, heads or mandibles), tortoises (or tortoiseshells) and dogs. According to Gao Guangren's point of view (2000: 136), the meanings of these three animals' burials were radically different: pig skulls and mandibles were the symbols of wealth and high social status; the tortoise was an object of worship, a divine animal; and dogs were probably considered as animate animals (animals who have souls) which could continue to serve their owners in the afterlife. Nevertheless, the scholars do not exclude the possibility of the use of dogs (dog meat) as a sacrificial meal served either for a tomb occupant or for the spirits of the underworld. Wang Lihua (1992: 270) assumes, that the tradition of dog burials in the Neolithic and later epochs was caused by the role of dogs in everyday life as guards and defenders. Guo Zhiwei (2012: 60–61) presumes, that the most probable function of accompanying dog burials was the protection of the owner's soul after death, but it is also possible that a dog could be a companion of its owner in the underworld or serve as sacrificial food.

It is now important to mark out some potential directions for future research of dog burials and reconstruction of rituals. Firstly, a study of co-burials of dogs and tortoises seems quite promising. This approach was firstly applied by Shao Wangping and Gao Guangren, who studied co-burials of dogs and tortoises in human tombs at the Liulin (three tombs) and Dadunzi (three tombs) sites of the Dawenkou culture (Pizhou County, Jiangsu Province). They discovered, that these tombs differed from others by having richer grave goods, which included among other items ritual ceramic vessels *zun*. Due to this, the researchers presumed, that these six tombs could belong to tribal elders or priests. It is remarkable that a tomb with the co-burial of a dog and a tortoise very similar to the tombs from Liulin and Dadunzi, was also discovered in the Early Yangshao culture layer at the Xiawanggang site, which could be the result of cross-cultural contacts. It is also noteworthy that stewed dog meat with soft-shelled turtle (*yuanzhi gourou*) is one of the most famous dishes of Xuzhou traditional cuisine (the region where Liulin and Dadunzi sites are located). It was first mentioned in Sima Qian's 'The Records of Grand Historian' ('*Shiji*') in the beginning of the 1st century BC: according to the legend it was firstly cooked by Fan Kuai, a close friend of the founder of Han dynasty Liu Bang, in the 3rd century BC, but in fact it could have even earlier origins. In traditional Chinese culture a dog and a tortoise represent male (*yang*) and female (*yin*) elements respectively. The combination of these two animals in one tomb could have represented the binary opposition (a kind of prototype of *yin-yang* dual model) as one of the basic elements of the universe model.

Secondly, a study of dog burials and sacrifices through the prism of the paleo-ethnological approach can also be productive. As M. Yu. Ul'yanov and D. V. Deopik (2012) point out, the members of the Dawenkou and Liangzhu cultures, where the most numerous dog burials were found, might belong to certain Austric ethnic groups, the ancestors of Hmong-Mien, Vietic, Tai and other peoples. The image of a dog is quite significant in the mythology and traditional culture of these peoples. The study of ethnic affiliation of the above-mentioned archaeological cultures may help to specify the range of possible ethnographic and folklore analogies that can be used for the interpretation and reconstruction of the ritual of burying dogs in human tombs.

## 5 Conclusions

So, it seems possible to conclude, that the image of a dog was an integral part of the culture of inhabitants of the territory of China in the Neolithic epoch. All the main types of ritual use of dogs, such as accompanying burials, sacrifices related to funerary and obituary rites and building sacrifices, were formed in the Middle Neolithic. Different depictions of dogs (paintings on the surface of a ceramic vessel, small figurines, etc.) still remain nowadays. The formation of a specific perception of a dog can be seen in Peiligang, Yangshao, Longqiuzhuang, Dawenkou, Liangzhu and Longshan cultures. The modes of ritual use of dogs or their depictions which formed during the Neolithic period have taken an important place in the spiritual culture of peoples of China over millennia, some of their survivals (relicts) can be seen even nowadays.

## References

Campbell, R. 2015. Animal, Human, God: Pathways of Shang Animality and Divinity, in B.S. Arbuckle and S.A. McCarty (eds) *Animals and Inequality in Ancient World*: 251–274. Boulder: University Press of Colorado.

Deopik, D.V. and M.Yu. Ul'yanov 2012. Istoricheskiye protsessy v drevnei Vostochnoi Azii v 3 — pervoi polovine 2 tys. do n.e.: skladyvanie 'dvuedinogo' Regiona. *State and Society in China* 42, 3: 7–32. Moscow: IVRAN (Деопик Д. В. и М.Ю. Ульянов 2012. Исторические процессы в древней Восточной Азии в 3 — первой половине 2 тыс. до н.э.: складывание «двуединого» Региона. *Общество и государство в Китае*, 3: 7–32. Москва: ИВ РАН).

Chen Wenhua 1992. Zhongguo nongye kaogu ziliao suoyin (10) gou. *Agricultural Archaeology* 3: 342–346, 341 (陈文华。1992。中国农业考古资料索引（十）狗。农业考古 3: 342–346, 341).

*Dengjiawan: Tianmen Shijiahe kaogu baogao zhi er* 2003. Beijing: Cultural Relics Publishing (邓家湾：天门石家河考古报告之二。2003。北京：文物出版社).

Frantz, L.A.F., V.E. Mullin, M. Pionnier-Capitan, O. Lebrasseur, M. Ollivier, A. Perri, A. Linderholm, V. Mattiangeli, M.D. Teasdale, E.A. Dimopoulos, A. Tresset, M. Duffraisse, F. McCormick, L. Bartosiewicz, E. Gál, E.A. Nyerges, M.V. Sablin, S. Bréhard, M. Mashkour, A. Balasescu, B. Gillet, S. Hughes, O. Chassaing, C. Hitte, J.-D. Vigne, K. Dobney, C. Hänni, D.G. Bradley and G. Larson 2016. Genomic and archaeological evidence suggests a dual origin of domestic dog. *Science* 352, 6290: 1228–1231.

Gao Guangren 2000. *Haidai qu xian-Qin kaogu lunji*. Beijing: China Science Publishing (高广仁。2000。海岱区先秦考古论集。北京：科学出版社).

Gao Guangren and Shao Wangping 2013. Zhongguo shiqian shidaide gui ling yu quan sheng, in ShaoWangping (ed.) *Shixue, kaoguxue wenxuan*: 375–391. Jinan: Shandong University Press (高广仁、邵望平。2013。中国史前时代的龟灵与犬牲。邵望平史学、考古学文选。济南：山东大学出版社).

Guo Zhiwei 2012. Shilun shiqian shiqi muzang xun quan xisu. *Cultural relics* 8: 54–62 (郭志委。2012。试论史前时期墓葬殉犬习俗。文物 8: 54–62).

Guo Zhiwei 2014. Shiqian shiqi yaokeng zangsu shixi. *Archaeology* 6: 56–63 (郭志委。2014。史前时期要坑葬俗试析。考古 6: 56–63).

Li Weiming, Zuo Chao and Niu Yumei 2000. Henan Gushi Pingzhai gucheng yizhi fajue baogao. *Acta Archaeologica Sinica* 3: 331–358 (李维明、左超、牛玉梅。2000。河南固始平寨古城遗址发掘报告。考古学报 3: 331–358).

Li Xiangsheng 2006. Jiantan xianmin yu jiachu goude guanxi. *Agricultural Archaeology* 1: 243–245 (李想生。2006。浅谈先民与家畜狗的关系。农业考古 1: 243–245).

Li Zhipeng 2011. Shang wenhua muzang zhong suizangde gou sheng yanjiu er ti. *Cultural Relics in Southern China*: 100–104 (李志鹏。2011。商文化墓葬中随葬的狗牲研究二题。南方文物: 100–104).

Ling Chunsheng 1957. Gudai Zhongguo ji Taipingyang qude can ji. *Bulletin of the Institute of Ethnology Academia Sinica*: 1–40 (凌纯声。1957。畲民图腾文化的研究。中央研究院历史语言研究所集刊 1–40).

*Longqiuzhuang: Jiang-Huai dong bu xinshiqi shidai yizhi fajue baogao* 1999. Beijing: China Science Publishing (龙虬庄：江淮东部新石器时代遗址发掘报告。1999。北京：科学出版社).

Ren Shinan and Wu Yaoli (eds) 2010. *Zhongguo kaoguxue: Xinshiqi shidai juan*. Beijing: China Social Sciences Press (中国考古学：新石器时代卷。任式楠、吴耀利主编。2010。北京：中国社会科学出版社).

Shanghai Fuquanshan yizhi Wujiachang mudi fajue. 2011. *Cultural Relics News* 21 October 2011: 4 (上海福泉山遗址吴家场墓地发掘。2011。中国文物报，2011年10月21日：4).

Wang, G., W. Zhai, H. Yang, L. Wang, L. Zhong, Y. Liu, R. Fan, T. Yin, C. Zhu, A.D. Poyarkov, D.M. Irwin, M.K. Hytönen, H. Lohi, C. Wu, S. Savolainen and Y. Zhang 2016. Out of southern East Asia: the natural history of domestic dogs across the world. *Cell Research* 26: 21–33.

Wang Lihua 1992. Zaoqi Zhongguo shehuide quan wenhua. *Agricultural Archaeology* 3: 265–270, 274 (王利华。1992。早期中国社会的犬文化。农业考古 3: 265–270, 274).

Wu Zhuang 2014. Xian-Qin shiqi jiaquan yanjiude xianzhuang yu zhanwang. *Cultural Relics in Southern China* 1: 65–73 (武庄。2014。先秦时期家犬研究的现状与展望。南方文物 1: 65–73).

Wu Zhuang, Yuan Jing, Zhao Xin and Chen Xianglong 2016. Zhongguo xinshiqi shidai zhi xian-Qin shiqi yizhi chutu jiaquande dongwukaoguxue yanjiu. *Cultural Relics in Southern China* 3: 155–161 (武庄、袁靖、赵欣、陈相龙。2016。中国新石器时代至先秦时期遗址出土家犬的动物考古学研究。南方文物3: 155–161).

*Wuyang Jiahu* 1999. 1. Beijing: China Science Publishing (舞阳贾湖（上卷）。1999。北京：科学出版社).

*Xichuan Xiawanggang*. 1989. Beijing: Cultural Relics Publishing (淅川下王岗。1989。北京：文物出版社).

Yin Huanzhang, Yuan Ying and Ji Zhongqing 1965. Jiangsu Pixian Liulin xinshiqi shidai yizhi dier ci fajue. *Acta Archaeologica Sinica* 2: 9–47 (尹焕章、袁颖、纪仲庆。1965。江苏邳县刘琳新石器时代遗址第二次发掘。考古学报 2: 9–47).

Yin Huanzhang and Zhang Zhengxiang 1962. Jiangsu Pixian Liulin xinshiqi shidai yizhi diyi ci fajue. *Acta Archaeologica Sinica* 1: 81–102 (尹焕章、张正祥、纪仲庆。1962。江苏邳县四户大墩子遗址探掘报告。考古学报1: 81–102).

Yuan Jing 2015. *Zhongguo dongwukaoguxue*. Beijing: Cultural Relics Publishing (袁靖。2015。中国动物考古学。北京：文物出版社).

*Zouxian Yedian* 1985. Beijing: Cultural Relics Publishing (邹县野店。1985。北京：文物出版社).

# 4.3 Neolithic Dogs in the Central Po Valley - A Review of Published Data and New Evidence

Fabio Bona[1,2,5], Daniela Castagna[3], Raffaella Poggiani Keller[4]

[1]Freelance. palaeontologist.fabgeo@libero.it
[2]Museo Civico dei Fossili di Besano, via Prestini 5, Besano (VA, Italia)
[3]Archaeologist. danielacastagna2017@gmail.com
[4]Soprintendente emerito per i beni culturali della Lombardia. rpoggianikeller@libero.it
[5]Associazione Culturale Amici di Castellaro (Castellaro Lagusello, MN). info@amicidicastellaro.it
Corresponding author: Fabio Bona palaeontologist.fabgeo@libero.it

**Abstract**

Four different ways in which dogs were used/exploited in the Neolithic were analysed. Two different modes are observed during the SMP: 1. A dog buried alone in a grave near polylobate structures, which could have a ritual value; 2. A dog as a companion in the grave of a human being, directly related to the funerary sphere. The other two forms of human-dog interaction come from the Recent to Late Neolithic period where dogs have been found in midden deposits: 3. the first case shows probable human care during the life of the animal; 4. the second case, with heavy traces of traumatic injuries, represents, on the contrary, a case of violent death, perhaps even slaughtering for meat. The study of the morpho-dimensional characteristics of these Neolithic Italian dogs allowed us to determine that all the dogs were very small or small in size and really very similar to each other regardless of their site of origin.

**Keywords**: Neolithic, dogs, Central Po Valley, burials, inhabited areas.

## 1 Introduction

The aim of the present work is to analyse the morphological and morphometric features of selected dog remains found in the central Po Valley as far north as the Pre-Alpine valleys, including the territories of Brescia, Mantova, Parma and Reggio Emilia (Figure 1). The archaeological contexts are dated between the Middle Neolithic (Square Mouthed Pottery culture - SMP - Vaso a Bocca Quadrata (VBQ) in Italian) and the Recent to Late Neolithic.

The first stepwas to study the bibliography of published, but never anatomically analysed, dog remains, bringing together as much data as possible on osteological elements of dogs from different Italian Neolithic contexts (De Grossi Mazzorin and Tagliacozzo 1997; De Grossi Mazzorin 2008).

The second step involved comparing morphometric data to identify the extent of homogeneity or heterogeneity of dog populations.

The third step, proceeding directly from the second, was to analyse the complex and composite picture of human-dog relationships- for which we currently only have limited available data - and to provide our own contribution to that understanding.

Finally, drawing together materials and knowledge from the various contexts, we consider the impact of selective breeding on the physical characteristics of the dogs and address questions concerning the selection of particular forms of dog to work in symbiosis with humans or to be employed as elements of ritual and burial practices.

## 2 Material and methods

The studied materials include four complete skeletons, one skull and some scattered remains of *Canis familiaris* from four different Neolithic north Italian sites (Figure 1):

- 1 complete skeleton from Valdaro (Mantova);
- 1 complete skeleton from Pontetaro (Parma) (Bernabò Brea *et al.* 2010a; 2010b);
- 1 complete skeleton from via Guidorossi - Parma (Bernabò Brea *et al.* 2010a; 2010b);
- 1 partial skeleton, 1 skull and some assorted bones from different individuals from Tosina di Monzambano (Mantova) (Bona 2014).

Initially, more sites known in the literature for the presence of buried dogs had been included in this study: Chiozza di Scandiano (RE) 1941-T2 (Bernabò Brea and Mazzieri 2014); Collecchio Cà Lunga T5

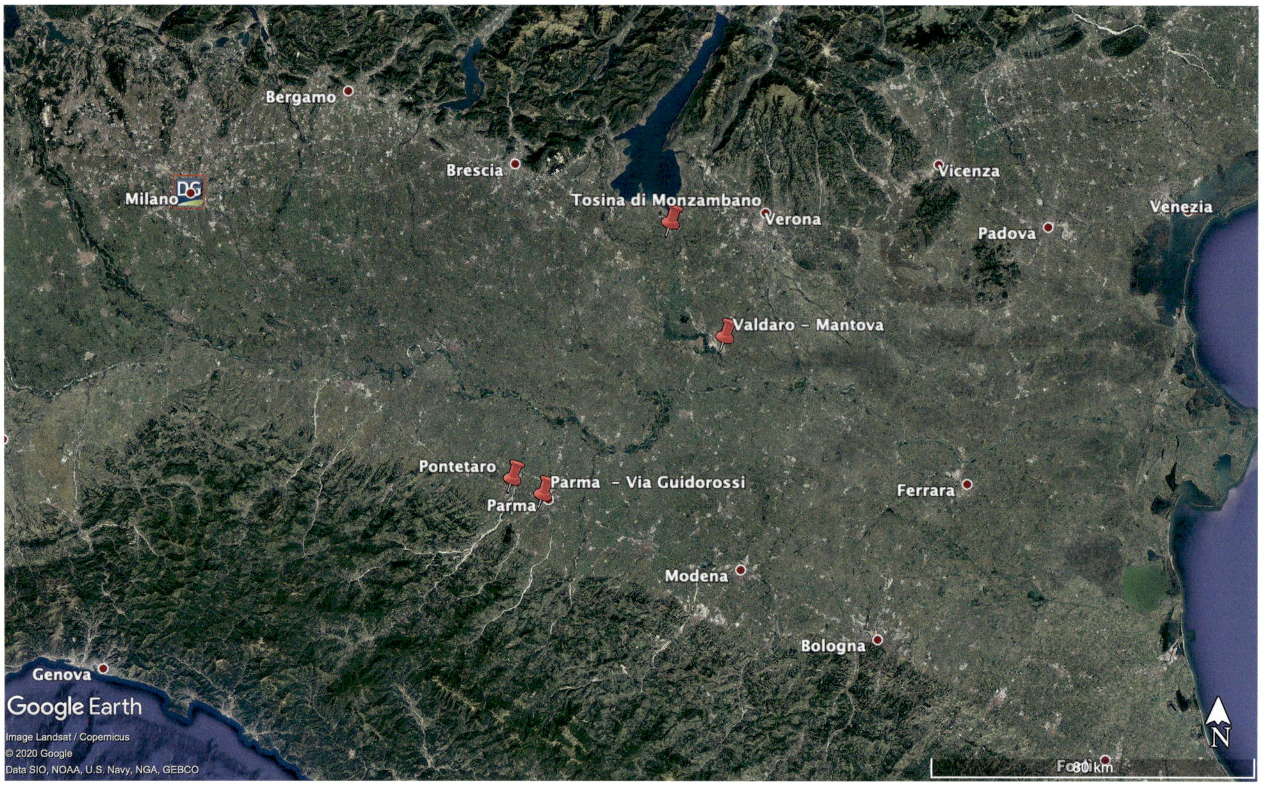


Figure 1. Geographical position of the four sites described in the text.

(Bernabò Brea and Mazzieri 2014); Bagnolo San Vito (Castagna 2014). However, after a zooarchaeological consideration of the remains, all these sites have been excluded from the study. At Chiozza T2 only a few disconnected, badly preserved bone fragments of an unknown small animal are present; at Collechio Cà Lunga T5 the 'scattered bones' in Bernabò *et al.* (2010a) turn out to consist only of an upper right third incisor; at Bagnolo T5 the identification of species was clearly wrong, some human phalanges having been identified as canine.

The minimum number identified (MNI) of the scattered dog remains was calculated following White (1953), Bökönyi (1970), Klein and Cruz-Uribe (1984) and De Grossi Mazzorin (2008 with references). Measurements were taken following von den Driesch (1976). To estimate the animals' dimensions we used parameters proposed by Koudelka (1884) and Harcourt (1974).

## 3 Chronology

The remains of dogs that are the subject of the present study are dated to a rather wide chronological interval. The chronological interval of the sites stretches, in fact, from the beginning of the Middle Neolithic (SMP 1) to the Recent Neolithic (Chassey-Lagozza culture) or Late Neolithic (*sensu* Maffi 2014). In Table 1 we present the 14C ages of the archaeological sites considered in this work.

## 4 Archaeological context

The materials examined in the present work come from various contexts:

- intentional burial in association with human inhumation (Valdaro, MN)
- in association with large polylobate pits and burials (Pontetaro, PR)
- single burials near funerary areas (via Guidorossi, PR)
- settlements - rare remains in anatomical connection and abundant disarticulated bones, sometimes with cut marks, found in midden deposits with other food remains characterised by the presence of butchery marks (Tosina di Monzambano, MN).

## 5 Results and discussion

### *5.1 Valdaro T5*

The Valdaro Neolithic site is located on a fluvial terrace and consists of several pits and some SMP1 burials.

Table 1. Age and dating of the sites involved in the work.

| Locality | Cultural phase | | $^{14}$C date | References |
|---|---|---|---|---|
| **Valdaro (MN)** | **SMP1** | | ***Tb 5: 5700 ± 50 BP*** | |
| | | | (4690–4400 BC 94.5% prob.) | |
| **Pontetaro (PR)** | **SMP2** | | | Bernabò Brea *et al.* 2010a |
| **Parma via Guidorossi (PR)** | SMP2 | | ***Tb 28: 5488 ± 45 BP*** | Bernabò Brea *et al.* 2010b |
| | | | (4449–4257 BC 94.5% prob.) | |
| | | | ***Tb 29: 5520 ± 45 BP*** | |
| | | | (4457–4267 BC 94.5% prob.) | |
| **Tosina da Monzambano (MN)** | Chassey-Lagozza | Sect. A | **US 110 base : 5120 BP** | Castiglioni and Rottili 2014 |
| | | | (3994–3794 BC 98% prob.) | |
| | | | **US 127: 5019 BP** | |
| | | | (3947–3706 BC 100% prob.) | |
| | | Sect. C | US 164: 4968 ± 45 BP | |
| | | | (3810–3650 BC 84% prob.) | |
| | | | **US 172: 5049 BP** | |
| | | | (3953–3774 BC 100% prob.) | |

The T5 burial represents the oldest example in northern Italy of a burial where a human being was buried with a dog (Castagna 2016). The human, a 40–50-year-old male, was arranged in a flexed position in an east-west direction with the head to the east. The skull is facing south. The upper limbs were slightly leaning forward with the forearms flexed to bring the hands in front of the face. The vertebral column and the rib cage were poorly preserved and very incomplete. The lower limbs are flexed, with the femurs forming an angle of ninety degrees relative to the spine. The accompanying items consist of two arrowheads (a geometric and a backed point) deposited near the face, whose arrow shafts, which are not preserved, were probably held in the hands. The geometric arrowhead is a rather wide spread type in SMP1 burials in the Mantua area and seems to persist until the transition to SMP2 (Castagna 2014).

At the feet of the human being lay the dog's skeleton. The dog's body was oriented north-south with the skull facing east. The vertebral column and the upper right limb of the dog are leaning on the feet of the man. According to the position of the upper and lower limbs, it is possible to infer that the animal was placed in a prone position (Figure 2).

The skeletal development implies that the dog was a mature adult: the vertebral epiphyses are fused, the third lower molar has erupted and the eruption/wear stage of the teeth is consistent with full growth.

A study conducted on C13 and N15 stable isotopes implies that the diet of the man and dog was similar. Specifically C13 indicates the exploitation of C3 plants (mainly wheat and barley) and N15 shows the exploitation of proteins (probably meat from sheep and pigs): the data imply that the dog lived in the same place and ate the same food as the human (Cavazzuti 2016).

The Valdaro burial resembles the famous one found in the Middle Neolithic funerary area near the village of Ripoli (Abruzzo, Central Italy), pertinent to the homonymous Culture (Grifoni Cremonesi 2014). In a pit containing five other individuals a woman was buried in a flexed position with very flexed legs and with a dog at her feet. The similarity between the case of Ripoli and that of Valdaro seems however to consist solely of the presence of both a dog and a human being, the rite of inhumation and the deposition method. In Valdaro we probably have a male individual (Cavazzuti 2016) with an adult dog placed on his feet (with the vertebral columns forming an angle of ninety degrees), while in Ripoli a woman was buried with the dog, with parallel vertebral columns and with the heads in opposite directions.

### *5.2 Pontetaro (PR)*

The Pontetaro dog was buried in a polylobate pit, a large negative structure shaped like a figure eight filled by a layer particularly rich in pottery

Figure 2. Valdaro (MN) Tb.5. In the red window a magnification of the dog (Photo © SAP, ATS SABAP Lodi, Cremona, Mantova).

fragments, faunal remains and pebbles. The bottom of the structure was marked by some pits or small depressions, one of which contained the dog skeleton. The study of materials dates this polylobate shaped like a figure eight structure to SMP2 (Mazzieri 2014a). Such big polylobate or articulated negative structures are typical of the SMP Po plain sites. Although their function is not well understood, they are the result of repeated human actions overtime. Frequently, these structures seem related to burials and votive deposits consisting of half vessels, millstones or pebbles (Bernabò Brea and Mazzieri 2009).

In Figure 3 A-C we show an hypothesised reconstruction of the position of this very strange dog burial. Starting from observation and a micro-excavation of the recovered skeleton it seems that the dog was thrown into the pit without any care. In fact, the body was bent over on itself with its hind legs covering its head. The dog is positioned on its right side and seems to be laid in such a way that immediately after the first cervical vertebrae the neck begins to twist the trunk. This causes the chest to rotate 180 degrees, so the limbs are in an anomalous position with respect to the rib cage. The vertebral column is preserved for a certain number of dorsal vertebrae then disappears (complete excavation of the find would be necessary to understand if they are preserved) and at the height of the skull a piece of pelvis reappears with the right femur still articulated, which points towards the skull. At the intersection of the distal part of the right femur and the proximal part of the right humerus the proximal portion of the left tibia begins and points towards the dog's muzzle. The left femur is present and lies directly below the right femur. The right tibia and both paws are absent.

The development of the skeleton suggests an adult age for the dog: vertebral epiphyses are fused, the third lower molar has erupted, and the wear of the teeth is consistent with adulthood.

### *5.3 Parma Via Guidorossi*

In a very significant Neolithic settlement located in the SW part of the city of Parma, characterised by the discovery of the biggest SMP necropolis in northern Italy with 56 burials dated to SMP2 (Bernabò Brea and Mazzieri 2009; Bernabò Brea *et al.* 2010b; Mazzieri 2014b), a well preserved dog burial has been found. The dog was buried in a 'normal' or sleeping position near a polylobate pit. According to the study of the position of the skeleton and the micro-excavation made on the preserved burial it is possible to understand that

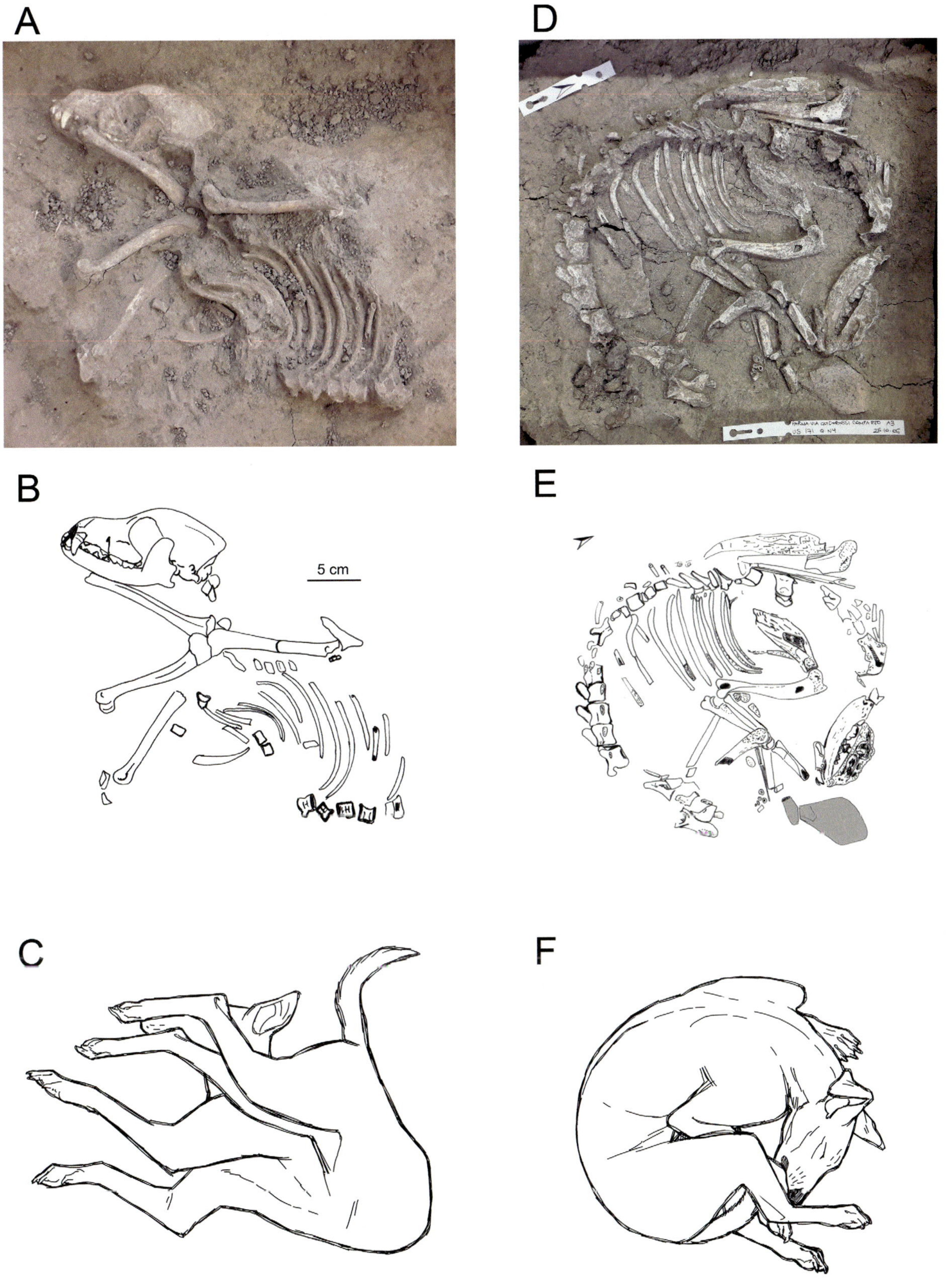


Figure 3. Pontetaro (PR) specimen. A, skeleton during excavation. B, skeleton drawing. C, specimen reconstruction hypothesis. Parma via Guidorossi specimen. D, skeleton during excavation. E, skeleton drawing. F, specimen reconstruction hypothesis (Photo and drawing of the buried dog © F. Bona; reconstruction drawing by Flavia Strani - PaleoFactory, Dipartimento di Scienze della Terra, Sapienza Università di Roma).

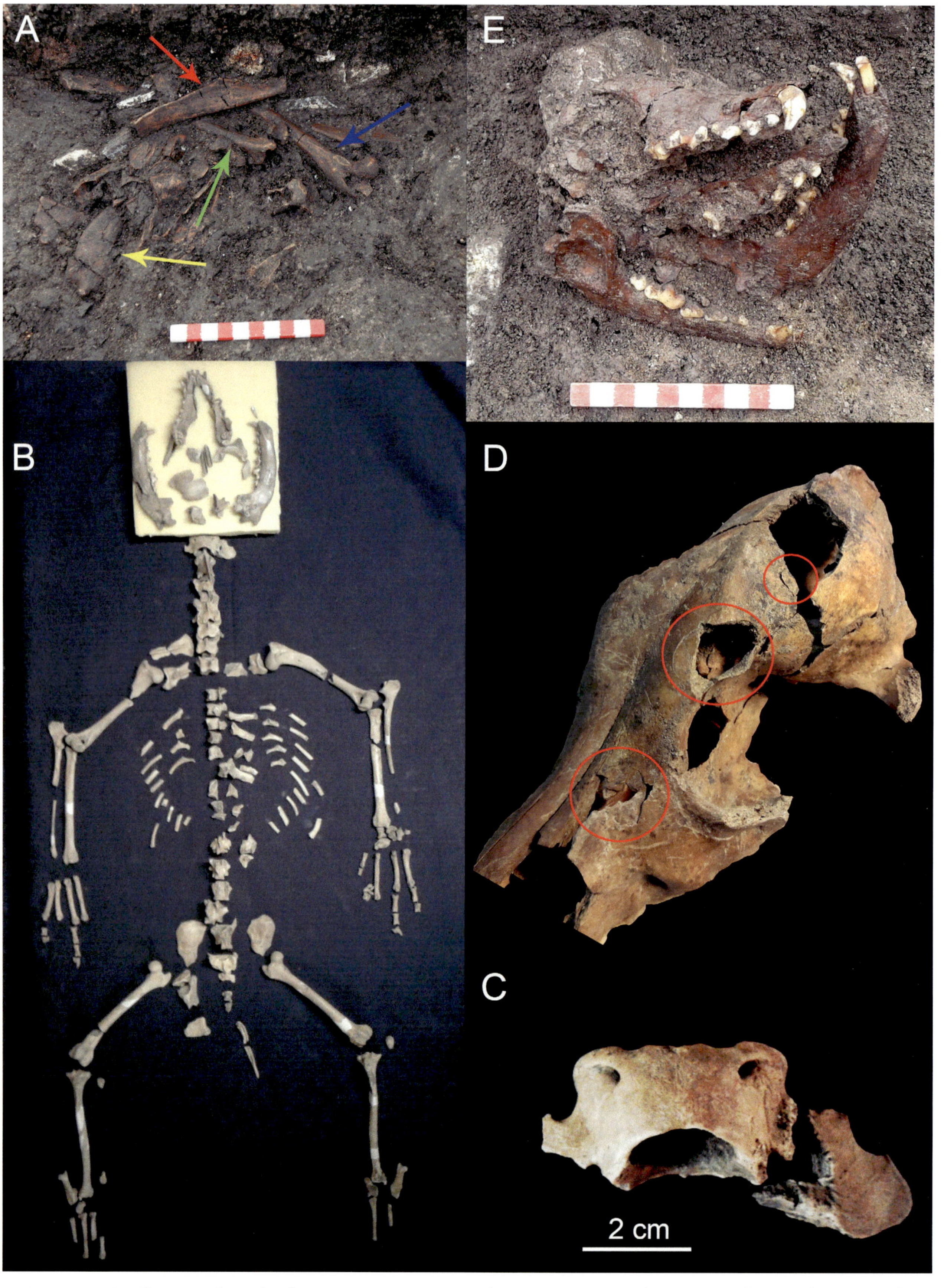


Figure 4. Tosina di Monzambano (MN). TOS16-2642. A, situation during excavation, red arrow *Sus* cubitus, yellow arrow the dog's right jaw, blue arrow dog's right femur, green arrow dog's right humerus. B, skeleton after cleaning and anatomical restoration. C, atlas with evidence of with evidence of the healed fracture. TOS18-6348. D, skull with in red circle underlined the 'green' fractures testifying the violent perimortem action on the dog; E, skull during excavation (Photos by F. Bona).

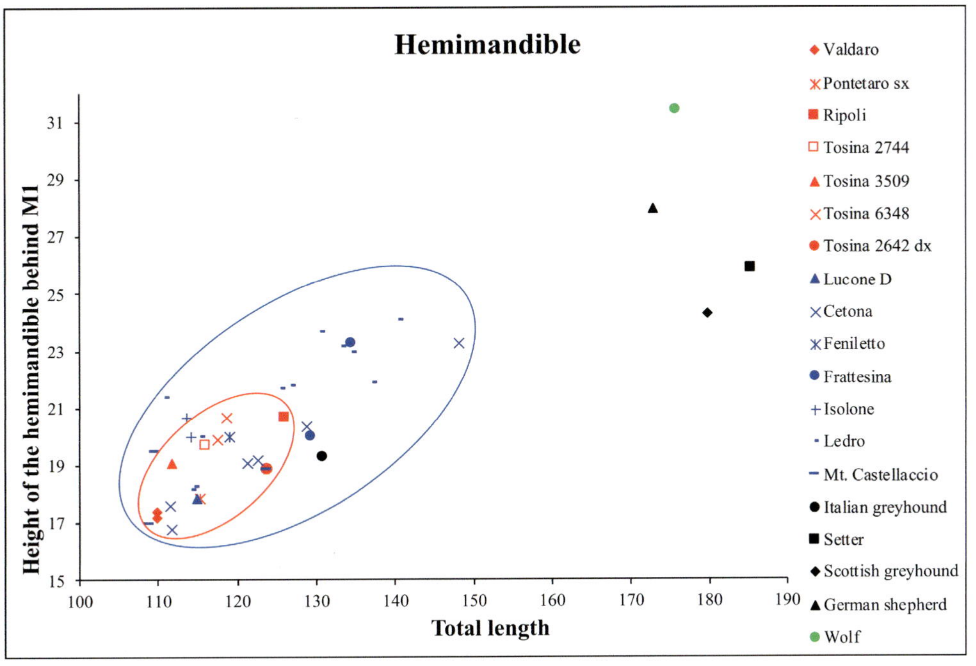


Figure 5. Comparison between hemimandible length and height of Neolithic dogs (Red), Bronze age dogs from selected sites (Blue), extant dog breeds (Black) and Wolf (Green).

the dog was buried curled up with the left forelimb positioned slightly outside. The spine is curved. The right forelimb is positioned above the body and the paw of the same limb appears to be placed between the two hind legs. The left forelimb is positioned 'under' the body. The whole skeleton, including the skull, is resting slightly to the left side, while the ribcage is positioned more ventrally (Figure 3D-F).

Skeletal development suggests an adult/old age for the dog: vertebral epiphyses are fused, the third lower molar has erupted, the carnassial teeth are heavily worn, and the general wear of the teeth is consistent with adulthood/old age.

### *5.4 Tosina*

Tosina di Monzambano is a site located in the morainic hills of southern Lake Garda (Figure 1). The excavations of the site are still in progress and a preliminary study shows a predominantly Recent to Late Neolithic presence (Lo Vetro 2014; Poggiani Keller 2014) (Table 1).

At the Tosina site several dog remains have been found in the garbage deposits. Here we describe two cases that are important for the completeness of the findings. The first specimen, TOS16-2642, consists of the scattered remains of a quite complete skeleton with pathologies (healed fractures at the atlas and partially at the right zygomatic arch). Could these traces of the onset of healing be linked to some form of human care? (Figure 4 A-C). Skeletal development suggests an adult age for the dog: the vertebral epiphyses are fused, the third lower molar has erupted, and the wear of the teeth is consistent with adulthood. The second specimen, TOS18-6348, consists of a complete skull with traces of traumatic injuries. Clearly, the injuries were the cause of its death (Figure 4D-E). It is possible to estimate an adult age for the dog based on the eruption of the second upper molar and general tooth wear.

### *5.5 Comparison of the dogs*

Having presented the contexts and forms of burial, we will now consider the morphometric features of these dogs in comparison to those of other Italian specimens, from the Neolithic to the Bronze age, that are found in the literature (Riedel 1995; De Grossi Mazzorin and Tagliacozzo 1997).

As one can see in Figure 5, the Neolithic dogs show greater morphological and metric homogeneity of the lower jaw than do dogs of the early Bronze age and other selected living breeds. So, although we have few data, it is possible to suggest a low morphometric variability for Neolithic dogs. We note that the dogs studied in this paper fall within the group of other Neolithic dogs. Overall, the Neolithic dogs are very small when compared to extant breeds or to wolves and are less differentiated than Bronze age dogs.

The same situation that we see in lower jaw dimensions is also visible in Table 2, where data of estimated withers height of the dogs are presented. It is possible to see that the dogs from Valdaro, Parma via Guidorossi and Pontetaro - the earlier dogs - are some

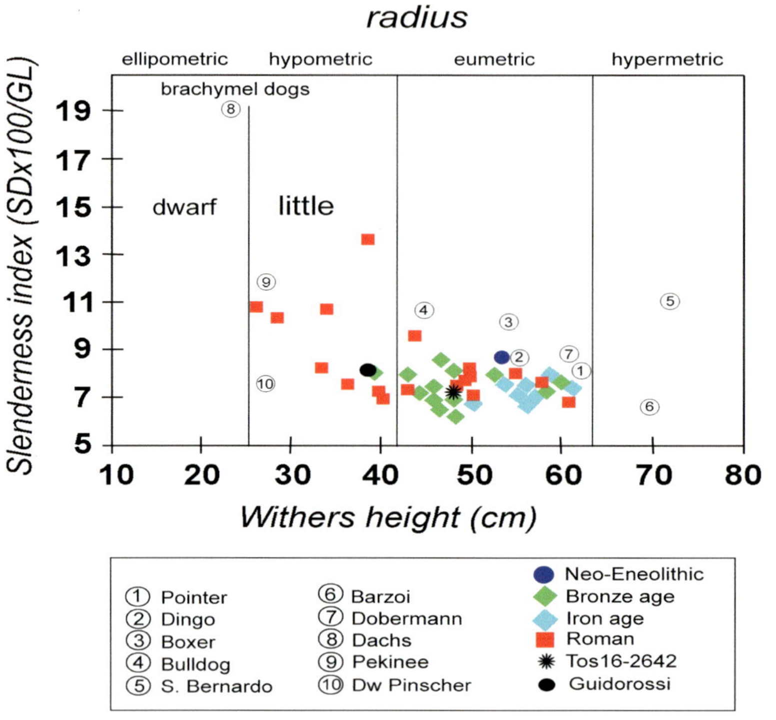


Figure 6. Comparison between different dog morphology according to radius dimensions (After De Grossi Mazzorin 2008).

Table 2. Dog height at the withers of Neolithic dogs and of Lucone A (Ancient Bronze age - Bona 2019) according to Koudelka (1885) (data courtesy of J. De Grossi Mazzorin).

| Koudelka (1885) | n | min | max | average |
|---|---|---|---|---|
| Valdaro | 1 | | | 400° |
| Via Guidorossi | 1 | 383 | 389 | 386 |
| Pontetaro | 1 | 404 | 444 | 424 |
| Ripoli° | 1 | 477 | 509 | 488 |
| Tosina 2642 | 1 | 473 | 480 | 476 |
| Tosina 1431 | 1 | | | 440 |
| Tosina 3496 | 1 | | | 423 |
| Tosina 4098 | 1 | | | 439 |
| Tosina | 5 | | | 444 |
| Neolitico | 2 | 426 | 528 | 477 |
| Lucone D (BA) | 13 | 410 | 532 | 448 |

what smaller than the other Neolithic examples. The dogs of Tosina are a little bigger but still fall within the Neolithic range.

In Figure 6 we see that the Parma via Guidorossi dog was very small and is smaller than the specimen from Tosina TOS16-2642, the tallest dog studied from the Late Neolithic.

## 6 Conclusions

In this work four different modes of exploitation/use of dogs in the Neolithic were analysed.

During the SMP period it was shown that some dogs were deliberately buried in two different ways: 1. The dog alone in a grave, unnaturally positioned (Pontetaro) or laid with care and attention (Parma via Guidorossi) near polylobate structures, that could have a ritual value; 2. The dog as a companion in the grave of a human being (Valdaro), directly related to the funerary sphere.

While not included in the analyses made for this contribution, it is important to recall that dog remains have been used as ornaments (teeth) in human burials from Emilia (Bernabò Brea *et al.* 2010a).

All these data suggest multiple nuances in the man-dog relationship, on which it is necessary that further reflection takes place. However, we can see that, to some extent, the dog had already become a life partner of man within villages, as evidenced by the diet of the specimen from Valdaro.

For the Recent Neolithic period, only the specimens found in Monzambano were examined. Some others had been literally thrown in with the organic refuse of the village and had probably been exploited for meat and skin.

In the case of the dogs found in the midden (Tosina) we clearly have two different forms of human-dog interaction: 3. the first dog suggests probable human care during the life of the animal; 4. the second case, with heavy traces of traumatic injuries, represents, on the contrary, a case of violent death maybe even slaughter for meat/skin.

All the dogs studied are adults. The study of the morpho-dimensional characteristics of these Neolithic dogs in comparison with data from the literature allowed us to state that all the Neolithic Italian dogs were very small or small in size and really similar to each other regardless of the site of origin. We can highlight how the selection process starting from the wolf was now particularly advanced and probably aimed at obtaining a light morphology of sheepdogs.

## Acknowledgments

The study of the dogs from Parma via Guidorossi, Pontetaro and Tb. 5 of Collecchio Cà Lunga was carried out with the authorisation of Complesso Monumentale della Pilotta (Parma) (Prot. n. 2449 and n. 2805).

The authors are grateful to Dr M. Bernabò Brea, Dr P. Mazzieri (SABAP Marche), Dr G.M. Facchinetti (SABAP CO, LC, MB, PV, SO and VA), Dr C. Longhi (SABAP Brescia e Bergamo), Prof. D. Lo Vetro (University of Florence), to Prof. E. Crosato (Cultural Association 'Amici di Castellaro') for everything that he has done for the study of the Tosina di Monzambano site. Thanks also go to Dr N. Cappellozza (SAP Società Archeologica), Dr A. Crosato, Dr R. Tremolada and all the students and graduate students who participated in the various excavation campaigns at Tosina. Special thanks to the Arieti family, owners of the field where the site of Tosina is located.

This work and all the excavation field activities at Tosina were carried out thanks to funding from the Lombardy Region, the Municipality of Monzambano (MN) and private sponsors.

## References

Bernabò Brea, M. and P. Mazzieri 2009. Oggetti e contesti rituali nella cultura VBQ dell'Emilia occidentale. *Padusa* XLV: 7–41.

Bernabò Brea, M. P. Mazzieri, and R. Micheli 2010a. People, dogs and wild game: evidence of human-animal relations from Middle Neolithic burials and personal ornaments in northern Italy. *Documenta Praehistorica* 27: 125–145.

Bernabò Brea, M. M. Maffi, P. Mazzieri and L. Salvadei 2010b. Testimonianze funerarie della gente dei Vasi a Bocca Quadrata in Emilia occidentale. Archeologia e antropologia. *Rivista di Scienze Preistoriche* 60: 63–126.

Bernabò Brea, M. and P. Mazzieri 2014. Osservazioni sulla sfera rituale del mondo VBQ in base ai dati forniti dagli insediamenti dell'Emilia occidentale. *Rivista di Studi Liguri* LXXVII-LXXIX 2011–2013: 315–321.

Bona, F. 2014. La fauna del sito di Tosina, in R. Poggiani Keller (ed.) *Contadini, allevatori e artigiani a Tosina di Monzambano tra V e IV millennio a.C. Una comunità neolitica nei circuiti padani e veneti*: 137–148. Brescia: Grafiche Tagliani Stampa e Comunicazione per Acherdo Edizioni.

Bona, F. 2019. The Ancient Bronze age pile-dwelling of Lucone lake (site D): preliminary archaeozoological data, in M. Baioni C. Mangani and M.G. Ruggiero (eds) Le palafitte: Ricerca, Conservazione, Valorizzazione: 185–193. Quingentole (MN): SAP editore.

Bökönyi, S. 1970. A new method for the determination of the number of individuals in animal bones material. *American Journal of Archaeology* 74: 291–292.

Castagna, D. 2014. Studio preliminare sulle sepolture neolitiche del territorio mantovano: i casi di Mantova, Bagnolo San Vito e San Giorgio. *Rivista di Studi Liguri* LXXVII-LXXIX 2011–2013: 339–352.

Castagna, D. 2016. Insieme per l'ultimo viaggio, in C. Longhi and D. Castagna (eds) *Protagonisti di 6.000 anni fa, cronache dal neolitico mantovano*: 6–7. Quingentole (MN): SAP editore.

Castiglioni, E. and M. Rottoli 2014. Le datazioni 14C dall'abitato di Tosina, in Poggiani Keller, R. (ed.) *Contadini, allevatori e artigiani a Tosina di Monzambano tra V e IV millennio a.C. Una comunità neolitica nei circuiti padani e veneti*: 165–166. Brescia: Grafiche Tagliani Stampa e Comunicazione per Acherdo Edizioni.

Cavazzuti, C. 2016. I risultati delle analisi paleoantropologiche, in C. Longhi and D. Castagna (eds) *Protagonisti di 6.000 anni fa, cronache dal neolitico mantovano*: 8–9. Quingentole (MN): SAP editore.

De Grossi Mazzorin, J. 2008. *Archeozoologia, lo studio dei resti animali in archeologia*. Bari: Laterza.

Driesh von Den, A. 1976. A guide to the measurement of animal bones from archaeological sites. *Peabody Museum Bulletin* 1: 1–148.

Grifoni Cremonesi, R. 2014. Aspetti ideologici e funerari nella cultura di Ripoli e nell'Italia centro meridionale. *Rivista di Studi Liguri* LXXVII-LXXIX 2011–2013: 265–274.

Harcourt R.A. 1974. The dog in Prehistoric and early historic Britain. *Journal of Archaeological Science* 1: 151–175.

Klein, R. and K. Cruz-Uribe 1984. *The Analysis of Animal Bones from Archaeological Sites. Chicago*: The University Chicago press.

Koudelka, F. 1884. Das Verhältnis der ossa longa zur Skeletthohebei den Saugetieren. Verhandlung des Naturforschung *Vereines in Brunn* 24: 127–153.

Lo Vetro D. 2014. Le industrie litiche di Tosina: un contributo alla definizione dell'identità culturale della Lagozza, in R. Poggiani Keller (ed.) *Contadini, allevatori e artigiani a Tosina di Monzambano tra V e IV millennio a.C.. Una comunità neolitica nei circuiti padani e veneti*: 67–108. Brescia: Grafiche Tagliani Stampa e Comunicazione per Acherdo Edizioni.

Maffi, M. 2014. Componenti culturali nei siti neolitici emiliani tra Neolitico recente e finale. Unpublished PhD dissertation, University of Trento, University of Lyon 2.

Mazzieri, P. 2014a. Il sito VBQ di Pontetaro (PR). *Rivista di Studi Liguri* LXXVII-LXXIX 2011–2013: 477–484.

Mazzieri, P. 2014b. Il sito VBQ di stile 'maendro-spiralico' di via Guidorossi a Parma. *Rivista di Studi Liguri* LXXVII-LXXIX 2011–2013: 447–456.

Mazzieri, P. and R. Micheli 2014. Tradizioni funerarie e ornamenti personali. Alcune osservazioni dalla sfera VBQ emiliana alla luce delle ultime scoperte. *Rivista di Studi Liguri* LXXVII-LXXIX 2011–2013: 323–330.

Poggiani Keller R. 2014. Tosina di Monzambano: prime considerazioni su cronologia e aspetti culturali, in Poggiani Keller R. (ed.) *Contadini, allevatori e artigiani a Tosina di Monzambano tra V e IV millennio a.C.. Una comunità neolitica nei circuiti padani e veneti*: 167–174. Brescia: Grafiche Tagliani Stampa e Comunicazione per Acherdo Edizioni.

Riedel, A. 1995. Le inumazioni di animali della necropoli Longobarda di Povegliano (Vr). *Annali del Museo Civico di Rovereto* 11: 53–98.

# 4.4 Evolution and Utilisation of Dogs in Austria: The Archaeozoological Record from the Neolithic to the Roman Period

Konstantina Saliari[1], Erich Pucher[1], Martin Mosser[2]

[1]Naturhistorisches Museum Wien, 1. Zoologische Abteilung, Archäozoologische Sammlung, Burgring 7, 1010 Wien, Österreich, konstantina.saliari@nhm-wien.ac.at, erich.pucher@nhm-wien.ac.at

[2]Museen der Stadt Wien - Stadtarchäologie, Obere Augartenstraße 26–28, 1020 Wien, Österreich, martin.mosser@stadtarchaeologie.at

Corresponding author Konstantina Saliari, konstantina.saliari@nhm-wien.ac.at

**Abstract**

This study investigates dogs in present-day Austria from the Neolithic to the Roman period. Dog remains in these periods were generally infrequent, rarely exceeding 5%. Cut and chop marks suggest the exploitation of dogs for meat and skin throughout prehistory. During the Roman period, indications of cynophagy became significantly rarer, but exploitation of dog skin continued. The morphometric examinations show that the size of dogs generally increased from the Neolithic period to the Iron Age. In the Roman period, urban sites display a comparably higher variation in morphology, whereas rural sites are mostly characterised by larger dogs and less variation. Finally, concerning the Roman period further information about the various morphotypes and uses of dogs derive from the archaeological record, also mentioned in the present study, including myths, epigraphic evidence and figural representations.

**Keywords:** *Canis lupus familiaris*, dog utilisation, butchery marks, morphometry, dog trackways.

## 1 Introduction

According to the consumer survey of *Statistik Austria* (the Austrian statistical office), 1,405,000 Austrian households owned one or more pets (37%), whereby 45% owned cats and 18% dogs. Many pet owners of the western world consider their animals as friends, companions and sometimes even family members (Blouin 2012; Meehan *et al.* 2017; Smolkovic *et al.* 2012; Petersen 2011; Tipper 2011). The increased presence of pets in western households and the typical emotional motives behind their keeping nowadays suggest important socio-politic and economic changes compared to the past. This has strongly influenced the new sociological role and treatment of animals, including dogs (Tague 2015; Tipper 2011).

In this work, we investigate the relationship between humans and dogs based on archaeozoological material dating from the Neolithic to the Roman period in present-day Austria. The study focuses on reconstructing the role and function of dogs in past human societies, their abundance in archaeological assemblages, as well as the appearance and spread of different morphotypes. For information on the sites mentioned in the text, see Figure 1 and Table 1.

## 2 Dog remains from the Neolithic to the Roman period in Austria

### 2.1 Occurrence and abundance of dog remains

The documentation of the presence of different species and their specific abundance is one of the first steps of archaeozoological analysis. This quantification usually already yields information about the economic organisation of a site. Nonetheless, archaeological and biological remains are also influenced by various other factors, including taphonomy, cultural background, archaeological context, excavation techniques, laboratory treatment and statistical methods (Davis 1987: 22–23; O' Connor 2000: 19–35; Reitz and Wing 2008: 192–193). All these parameters should be taken into consideration when analysing and interpreting species representation.

In Austria, based on the number of identified specimens (NISP) the frequency of dog bones among the domesticated species in the Neolithic period is less than 5–6%. Dog bones have already been found in Early Neolithic sites, but their abundance is very low. So far, only two Linear Pottery culture sites exhibit dog remains: in Poigen only one bone (Wolff 1977) and in Brunn 1 two bones (Pucher 2019a). Taphonomic factors are probably responsible for the underrepresentation of dogs in these early assemblages because archaeological and archaeozoological material from Early Neolithic sites is usually found in a very poor state of preservation (Lenneis pers. comm. 2017). Although Middle and Late Neolithic sites contain more dog bones (e.g. Pucher and Engl 1997; Pucher 2004a; Schmitzberger 2009a), they are still found at low numbers, hardly exceeding 3% (Kunst 2006a; Schmitzberger 2001). Indirect evidence such as gnawing marks provide additional indications of their presence (Schmitzberger 2001, 2009b: 50). Sites of the Baden culture show similar results (Kunst 2005; Pucher 2006a; Schmitzberger 2009c; Saliari and Pucher 2017).

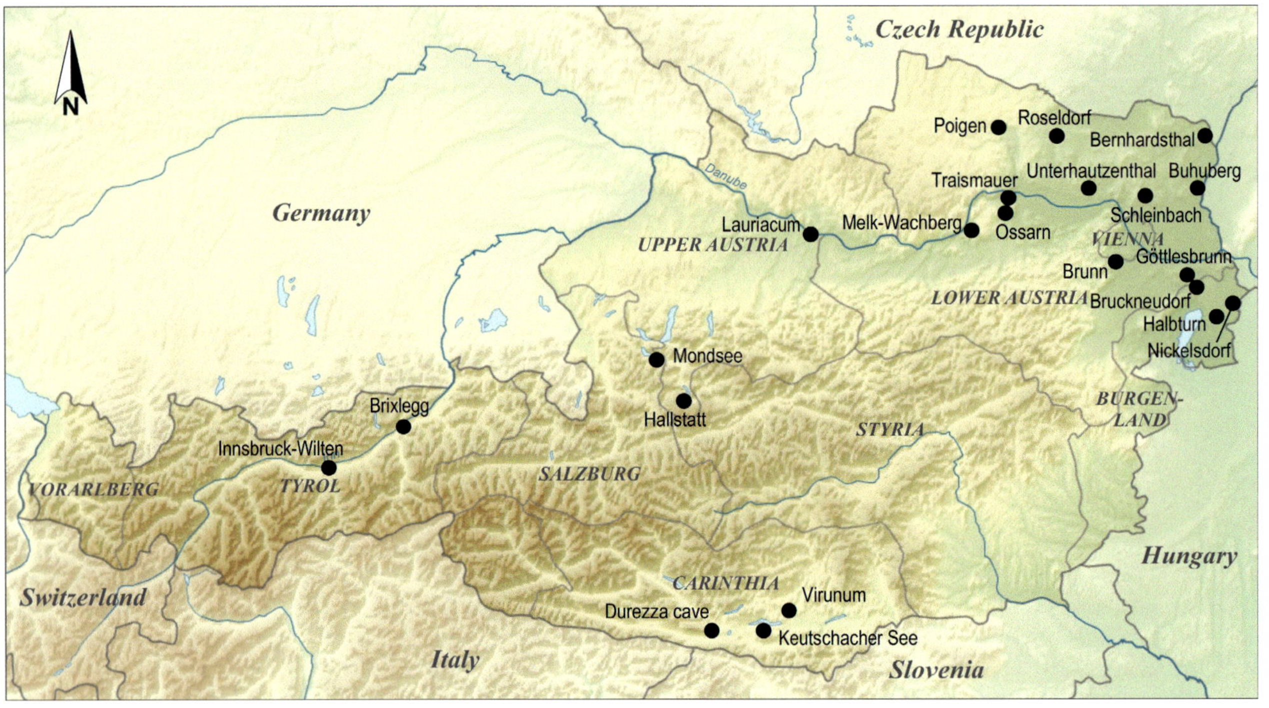


Figure 1. Relief map showing the sites with investigated dog remains, mentioned in the text. Map source: https://upload.wikimedia.org/wikipedia/commons/a/a7/Austria_relief_location_map.jpg. Map: E. Pucher.

During the Bronze Age, the frequency of dog bones usually ranges between 0.3% and 5–7% (Pucher 1987, 2019b; Riedel 1998; Schmitzberger 2008). Most Bronze Age sites examined in this study date mainly to the Early Bronze Age, and thus the occurrence of dogs in the Middle and Late Bronze Age still remains relatively unclear. An unusually high number of dog bones was recorded from Buhuberg, where they reached 7.1% among the domesticated species and 5.6% when also including the wild fauna (Pucher 1996a).

In Iron Age faunal assemblages, dog bones are mostly represented below 5% (Czeika 2006; Grill 2009: 21–25; Pucher 2004b; Saliari *et al.* 2016; Schmitzberger 2010a; Tecchiati 2012). Concerning Bronze and Iron Age mining sites, which clearly constitute exceptional archaeological circumstances, the frequency of dog bones is < 1% throughout (Abd el Karem 2009; Amschler 1939; Pucher 1999, 2014; Riedel 2003; Saliari *et al.* 2016; Saliari *et al.* 2020; Schmitzberger 2012).

During the Roman period, some sites show a high number of dog remains. At the settlement of Bruckneudorf, dog bones amount to 8.1% (Pucher 2018), and in the Roman city Virunum they reach almost 14% (Galik 2004). A dog bone value of 14% was reported from the villa rustica in Nickelsdorf (Riedel 2004). It should be noted however, that Pucher (2018) questioned these high values for Nickelsdorf because bones from dislocated skeletons were counted as isolated finds and therefore dog remains probably had a lower value around 7–8%.

Other Roman assemblages suggest that isolated dog bones usually have frequencies below 4–5%. Interestingly, the Roman period has yielded a considerable number of (partly preserved) skeletons (Galik 2004; Kunst 2006b; Pucher 2018; Schmitzberger 2007, 2010b).

### *2.2 Dog utilisation*

The archaeozoological record shows that dog meat formed part of the local diet even back in the Neolithic period, highlighting an important aspect of their role in society (Pucher 1997, 2004a; Schmitzberger 2009c). Butchery marks and fractures (Figure 2a-b) testifying to cynophagy are documented throughout prehistoric times within the borders of modern Austria (e.g. Galik *et al.* 2019: 60–69; Pucher 1987, 2004a, 2006a; Pucher *et al.* 2015; Riedel 1998; Schmitzberger 2012). In some Bronze Age and Iron Age sites (Pucher 2006b), consumption of dog meat seems to have been a regular practice. Additional evidence for the consumption of dog meat is provided by age profiles. Faunal assemblages from the Neolithic to the Iron Age suggest that dogs were slaughtered very often at the optimum age for meat consumption - mainly as young adults - similarly to pigs, which were raised for meat (Pucher 1987, 1999, 2004a, 2006b; Schmitzberger 2009a, 2009c).

Table 1. Archaeological sites with investigated dog remains mentioned in the text according to period and states of Austria (by Konstantina Saliari).

| Site | Period | State |
|---|---|---|
| Brunn 1 | Early Neolithic (Linear Pottery Culture) | Lower Austria |
| Poigen | Early Neolithic (Linear Pottery Culture) | Lower Austria |
| Ossarn | Late Neolithic (Corded Ware Culture) | Lower Austria |
| Keutschacher See | Late Neolithic | Carinthia |
| Mondsee | Late Neolithic (Mondsee Culture 3800–3200 BC) | Upper Austria |
| Wachberg bei Melk | Late Neolithic | Lower Austria |
| Brixlegg | Early Bronze Age | Tyrol |
| Schleinbach A | Early Bronze Age | Lower Austria |
| Unterhautzenthal | Early Bronze Age | Lower Austria |
| Buhuberg | Early/Middle Bronze Age | Lower Austria |
| Hallstatt | Late Bronze Age | Upper Austria |
| Göttlesbrunn | Iron Age (Hallstatt period and La Tène C) | Lower Austria |
| Durezza cave | Iron Age (Late Hallstatt-Early La Tène period) | Carinthia |
| Roseldorf | Iron Age (La Tène period, 4th - 2nd century BC) | Lower Austria |
| Bruckneudorf | 50–150 AD | Burgenland |
| Traismauer/ Augustiana | 1st-4th century AD | Lower Austria |
| Bernhardsthal | 2nd-3rd century AD | Lower Austria |
| Lauriacum | 2nd-3rd century AD | Upper Austria |
| Nickelsdorf | 2nd-3rd century AD | Burgenland |
| Innsbruck-Wilten | 2nd-4th century AD | Tyrol |
| Virunum | 2nd-4th century AD | Carinthia |
| Halbturn | 2nd-5th century AD | Burgenland |

The selection of body parts of dogs with high-quality meat and the continuous consumption suggest that cynophagy was neither a characteristic of social troubles nor an indicator of food shortage or hardship. As dietary practices constitute a key part of human identity (Hadjikoumis 2016; Twiss 2007), cynophagy should be viewed as a matter of cultural choice. Nevertheless, the generally low frequency of dog bones from the assemblages (based both on NISP and weight analysis) suggests their minor significance as a meat source in daily life. Furthermore, cynophagy might have had a ritual dimension. For example, cut marks indicating butchering were documented on dog bones from the Durezza cave (Carinthia), which were interpreted as cultic and/or sacrificial offerings (Galik 2000, 2002).

The variable butchery practices illustrate a variety of personal and/or local preferences and tastes. For instance, at the Late Neolithic site of Wachberg near Melk and at the Iron Age sites Göttlesbrunn and Roseldorf (Figure 2b), chop marks suggest smashing of the cranium for brain removal (Abd el Karem 2014; Pucher 1997, 2004b). At other sites, including the Early Bronze Age site Schleinbach A, the cranial region shows no butchery marks, even if other postcranial elements were chopped (Pucher 1996b).

From a technical point of view, certain observations allow insight into butchering techniques. Dog bones very often show marks similar to those found on species such as cattle, sheep/goats and pigs, which were a more staple part of the diet (e.g. Pucher 2006a; Schmitzberger 2009c). Nonetheless, the recorded marks are occasionally described as severe and brutal, suggesting very intensive smashing of bones and especially of skulls. Up to now, the data do not enable a geographical or chronological pattern regarding this behaviour to be distinguished. These marks and also those on at least several skulls are clearly related to butchery. This practice was exercised from the Neolithic to the Iron Age (e.g. Pucher 1996a, 1997, 1998, 1999, 2004b, 2006a). One simple but logical explanation is related to the robustness of dog crania. Because the cranial region is very compact and robust, probably more effort and power were needed to smash the skull. Even if this interpretation might explain the severe marks on the skull, however, it does not explain the intensive marks recorded more rarely on postcranial elements. Other interesting remarks on butchery practices were made for the material in Göttlesbrunn (Pucher 2004b); there, the upper jaws were chopped directly above the tooth sockets (alveoli).

In some assemblages with indications of cynophagy, there are dog bones that bear no marks, suggesting that some individuals were treated differently (Boschin and Riedel 2009). Without having archaeological evidence on ritual practices, it seems that certain dogs, which were disposed of separately (as cadavers?) and/or which exhibited undamaged bones, simply played a special role in that human society.

During the Roman period, the consumption of dog meat was significantly reduced. In several Roman assemblages, dog bones exhibit either no, or very rare butchery

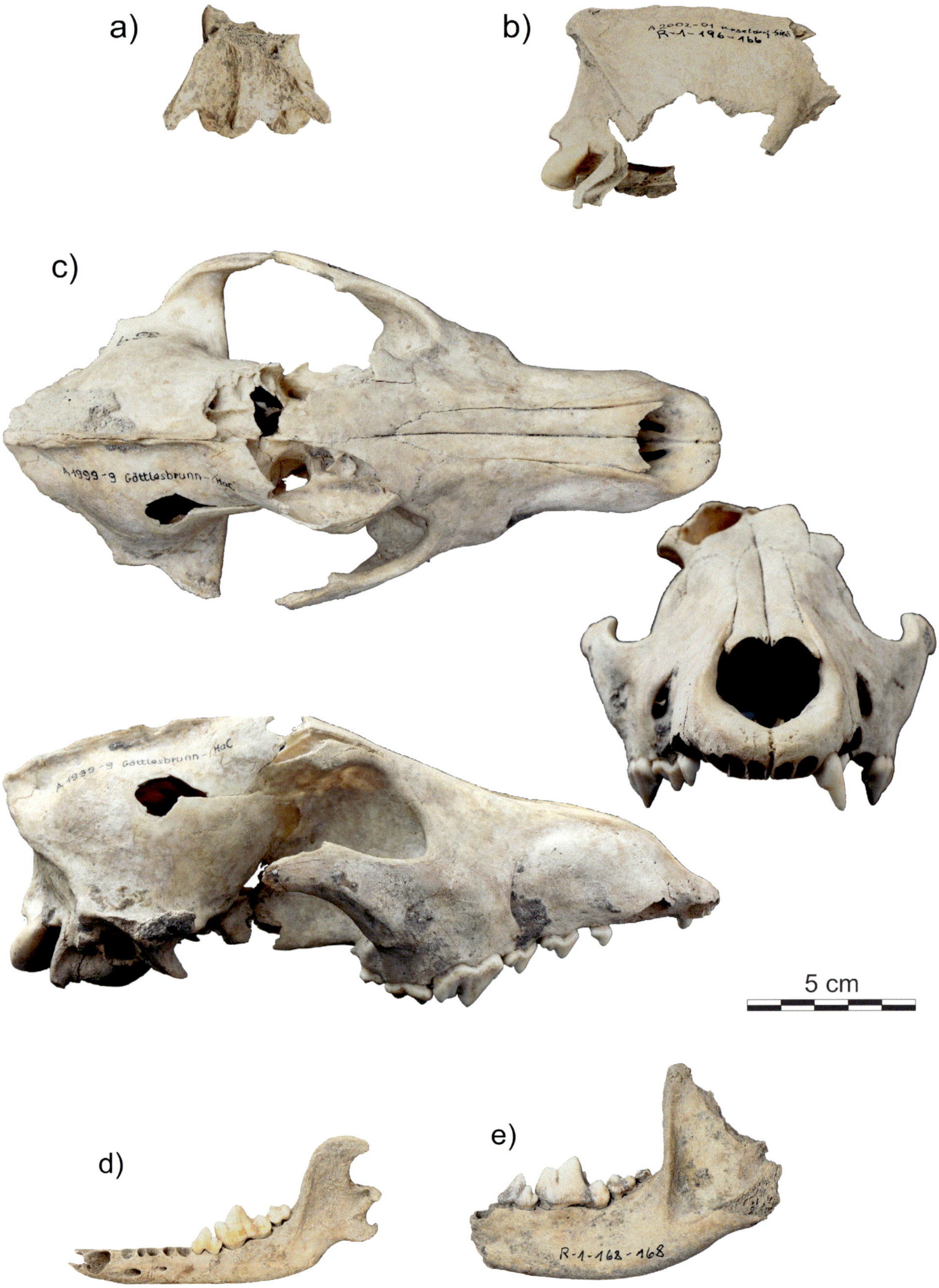


Figure 2. Iron Age dogs: (a) butchery mark on a dog cervical vertebra and (b) fractures on a dog neurocranium indicating brain removal from Roseldorf, La Tène, 4th-2nd century BC; (c) frontal, oral and lateral views of dog skull from Göttlesbrunn, Iron Age, Hallstatt period; (d-e) extremes of dog mandible size from Roseldorf, La Tène period, 4th-2nd century BC (Photos and figure by E. Draganits).

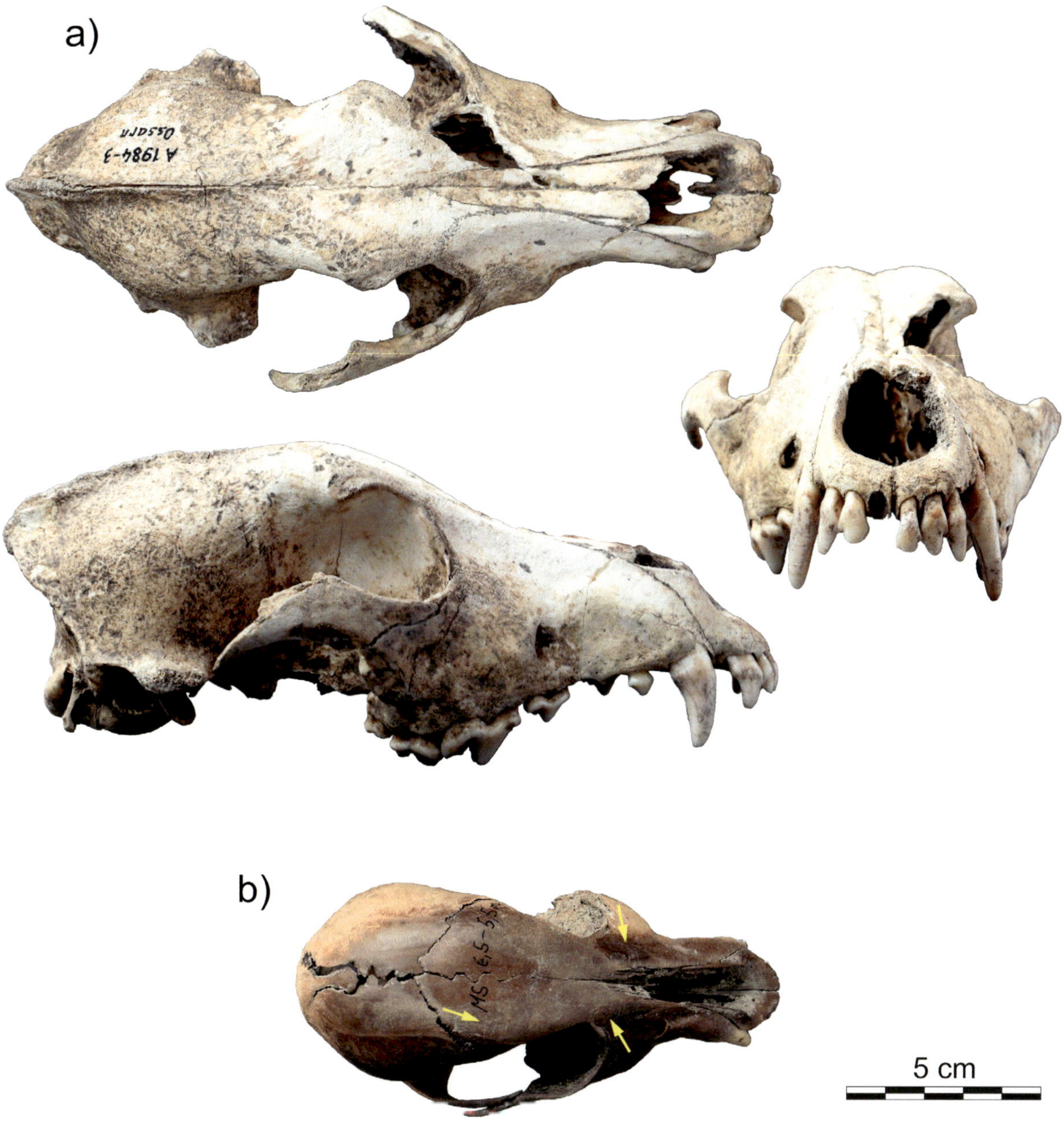


Figure 3. Neolithic dogs: (a) frontal, oral and lateral views of an adult dog skull from Ossarn (Late Neolithic) and (b) frontal aspect of a juvenile dog skull from Mondsee (Late Neolithic) both belonging to the 'palustris type', widespread during the Neolithic period. Cut marks on the snout and frontal bone at the skull of the juvenile individual from Mondsee (yellow arrows) indicate skin removal (Photos and figure by E. Draganits).

marks (Ambros 2006; Czeika 2003; Pucher 2018; Riedel 1993, 1996; Saliari and Pucher 2015; Schmitzberger 2007). In cases where cynophagy was documented, it is interpreted as an indicator of local dietary habits. Material from the Roman city Virunum suggests that occasionally even dog brain was consumed (Galik 2004). This visible change in diet may reflect the Roman presence and influence. Due to the very few butchery marks from this period, no specific pattern concerning the techniques employed can be established.

Although it is difficult to estimate the value of dog skin due to lack of organic remains, bones can provide information about skin exploitation (e.g. Boschin and Riedel 2009). Cut marks on the snout and frontal bone at a skull of a juvenile individual from the Late Neolithic site Mondsee (Figure 3b) were interpreted as an indication of skin removal (Pucher and Engl 1997: 37). The only dog bone that has been documented at the Bronze Age salt mine in Hallstatt was a talus. Because tali belong to the bones that are poor in meat, it was suggested that it probably arrived attached to fur (Pucher *et al.* 2013: 50–51).

During the Roman period, dog skin was apparently frequently exploited. The impressive number of 32 (partly preserved) dog skeletons was recorded at the settlement of Bruckneudorf (Pucher 2018). Although it is very challenging to interpret the finding, skin exploitation may have played an important role. At the villa rustica in Nickelsdorf, cut marks on dog bones suggest that skin of young dogs was favoured (Riedel 2004).

Dog bones from Bronze Age sites related to mining activities raised several questions concerning the role of dogs as working animals. As discussed below, large-sized dogs (around 70 cm) constitute an exception in the Bronze Age. The analyses of dog bones and their gnawing marks, however, do suggest the presence of some exceptionally large individuals in Bronze Age mining sites (Boschin and Riedel 2011; Pucher 2019b; Riedel 2003; Saliari *et al.* 2020). Even if these animals might have been used for guarding, other uses cannot be excluded. Later written sources offer some examples. Agricola (1556) mentions that dogs carried pigskin bags in mines. Despite the potential connection between Bronze Age mines and exceptionally big dogs, more faunal material and future research are necessary to understand the function of dogs in Alpine mining contexts (Saliari *et al.* 2020).

The use of dogs for hunting and guarding in prehistoric times has often been suggested (e.g. Schmitzberger 2009b: 50), but no osteological proof is available. During the Roman period, middle- to large-sized dogs - especially those found in rural sites - might have been used for guarding, protecting the livestock or as herding dogs (Riedel 2004). The very small dogs mainly favoured in Roman urban areas are interpreted to have been pets.

Dogs documented in ritual activities from cultural and funeral contexts are generally scarce throughout Austrian prehistory, yet with a few noteworthy exceptions (e.g. Bauer and Ruttkay 1974; Leskovar 1996; Galik 2002; Böhm 2010; Abd el Karem 2014). Iron Age burials almost never contain dog remains. Nevertheless, the few cases with dog bones are challenging to understand because grave material might be mixed with waste from domestic structures (Abd el Karem 2013, 2014; Saliari *et al.* 2016; Saliari 2021). The significant underrepresentation of dogs in burial contexts is somewhat surprising, considering the (chthonic) symbolism of dogs in Celtic mythology (Green 1992: 111–113).

The study of Roman graves suggests that dogs did not belong to the prevalent species that were deposited. Nevertheless, it often remains difficult or even impossible to understand and interpret the findings. In the case of the Roman cemetery of Halbturn, two skeletons of dwarf-sized dogs might represent a pet burial. However, due to lack of exact excavation data for this context, the interpretation of the findings is uncertain (Kunst and Doneus 2013).

### *2.3 Morphometric analysis of dog remains*

Neolithic dog bones suggest a polymorphic population with distinct size differences (Figure 3a-b). They were small to middle-sized individuals, with a height at withers between 35 and 50 cm (Pucher 2004a; Schmitzberger 2009b: 49). In some cases, dog bones were so small that it was difficult to distinguish them from foxes (e.g. Pucher 1997). The small (33–45 cm height at withers) and gracile dog bones found in many Neolithic assemblages present morphological similarities to the modern Spitz (Torfspitz). Slightly larger animals (45–55 cm height at withers) were formerly described as *Canis familiaris intermedius.*

The dogs from the pile dwelling sites of Mondsee were small-sized with gracile and slender bones, very similar to dog remains of this period elsewhere (Figure 3b). This is in strong contrast with other species from these sites such as cattle and small ruminants: they differ significantly from contemporaneous finds, for example from the Danube region (Pucher and Engl 1997: 37–38). Extremely small-sized dogs have been found in Keutschacher See. A mandible piece, deriving from an adult individual, was almost marten-sized. These bones represent the smallest dogs ever found in Austria (Pucher 2003: 269).

Although Bronze Age dogs are similar to Neolithic dogs morphologically, the former individuals with height at withers around 50 cm are significantly more common in the archaeozoological assemblages (Figure 4a-b). These dogs were morphologically similar to modern scent hounds. Dog bones from sites related to mining activities in western Austria constitute a distinct exception (Figure 4c), comprising considerably larger animals of about 70 cm height at withers (Boschin and Riedel 2011; Pucher 2019b; Riedel 2003; Saliari *et al.* 2020). These big individuals exhibit morphological similarities to modern Swiss mountain dogs and Rottweilers.

During the Iron Age, the average size of dogs increased and middle-sized individuals of about 60 cm height at withers dominated (e.g. Grill 2009: 229–234; Pucher 1999). Although the cranial region suggests wolf-

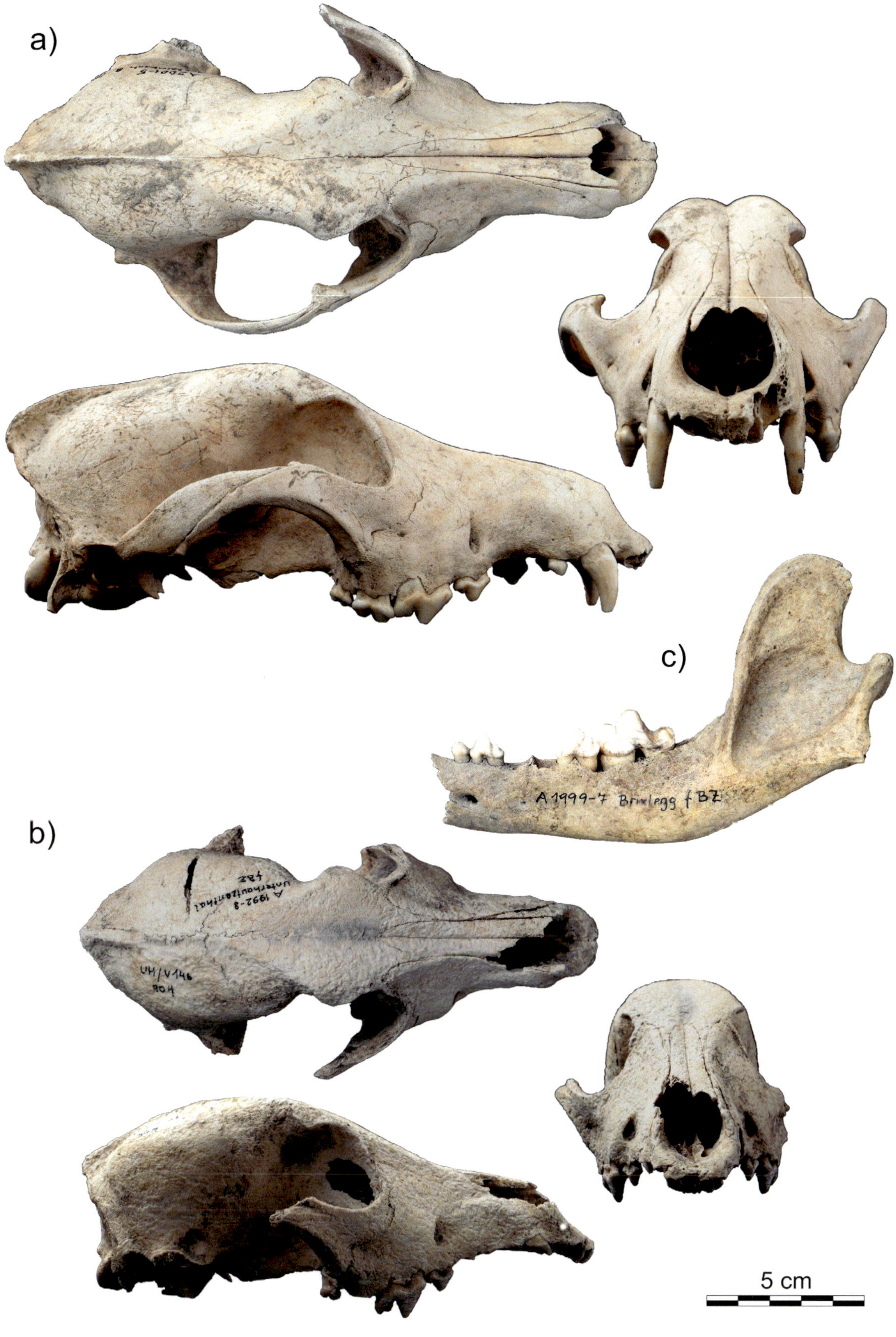


Figure 4. Bronze Age dogs: (a) frontal, oral and lateral views of dog skulls from Schleinbach and (b) Unterhautzenthal dating to the Early Bronze Age; (c) a mandible fragment of a considerably large-sized dog (around 70 cm) from Brixlegg (Early Bronze Age) (Photos and figure by E. Draganits).

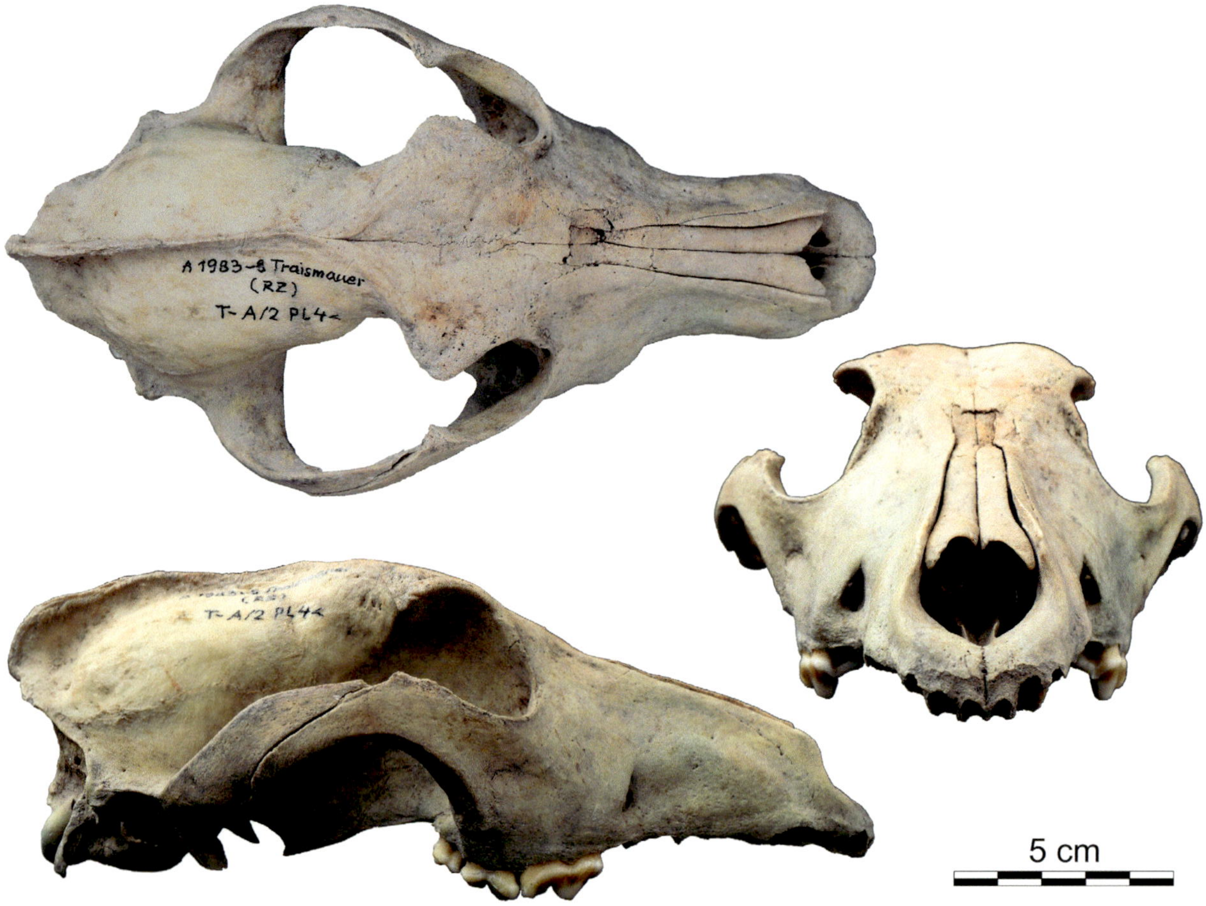


Figure 5. Roman dogs: frontal, oral and lateral views of dog skull from Traismauer/Augustiana (1st-4th century AD) showing a middle-sized individual similar to German shepherd dog (Photos and figure by E. Draganits).

like similarities, morphological variations among the different individuals have been documented (Figure 2c, e). These middle-sized animals were morphologically similar to German shepherd dogs. Although small-sized dogs became rare during the Iron Age (Czeika 2006; Schmitzberger 2010a), a few individuals of lapdog size (Figure 2d) were documented at the settlement and the Great Sanctuary in Roseldorf (Abd el Karem 2013; Bruckner-Höbling 2009). These animals probably served as a status symbol in the context of a larger settlement.

Roman period dog populations show a great size variation and many morphotypes. Especially sites with urban character display considerable size variations. At the civilian settlement Traismauer/Augustiana (Riedel 1993), sizes range from 46 to 61 cm, averaging 52.1 cm (Figure 5). The same site also yielded two brachymel dogs (33 cm). A similar average height at withers was recorded at Bruckneudorf (54.2 cm; Pucher 2018) and Lauriacum (Ambros 2006; Müller 1967). The bigger individuals bear morphological similarities to German shepherd dogs. The two skeletons of dwarf-sized dogs from the Roman cemetery of Halbturn have only 25 cm withers height. At this cemetery, the average height at withers, without the dwarf-sized dogs, was calculated at 58.5 cm (Kunst and Doneus 2013). Another brachymel dwarf-sized dog with 20 cm height at withers was found at Innsbruck-Wilten, where other isolated bone fragments also indicate the existence of large-sized animals (Pucher 2006c).

Rural sites generally exhibit more uniform sizes, indicating the presence of middle- to large-sized dogs. In comparison to urban sites, average individual size at rural sites was slightly bigger. Material from the villa rustica in Nickelsdorf (Riedel 2004) and the rural Germanic settlement in Bernhardsthal exhibited an average height at withers of 62 cm and 61.5 cm, respectively (Riedel 1996). At the former site, the size ranges between 55 and 67 cm, but smaller-sized animals between 36 and 42 cm were also found. Although ca. 60 cm-sized individuals are morphologically close to German shepherd dogs, some exhibited significant morphological variations.

### *2.4 Austrian dogs in a European context*

The evolution and utilisation of dogs from the Neolithic to the Roman period from present-day Austria show numerous connections and similarities with other European sites. Here the main focus is on the Roman period.

Abundant literature is available about the Roman impact and transformations connected with the socio-political and economic changes. Concerning dogs, two of the most important changes concern their function and morphometry. A major change is related to the consumption of dog meat. Cynophagy in Europe before the Roman expansion has been documented from several ritual and non-ritual contexts (e.g. Anthony and Brown 2017; Bălăşescu and Morintz 2018; Chrószcz *et al.* 2015; Ekroth 2007; Hadjikoumis 2016; Horard-Herbin *et al.* 2014; Škvor Jernejčič and Toškan 2018; Trantalidou 2006). Nevertheless, numerous studies show that the intensity, frequency and significance of dog meat consumption as well as the (economic, socio-political, ideological) motives behind this practice can vary significantly. Similarly to these finds from Austria, cynophagy almost ceased in several other regions during the Roman period (Argant 2017; De Grossi Mazzorin and Tagliacozzo 1997).

Concerning the morphometric evolution, the above-outlined high variability in size (20–23 cm to 65–68 cm height at withers) and shape of Roman dogs from Austrian faunal assemblages is also well documented from various sites in Italy and the Roman provinces (Argant 2017; Bartosiewicz 2000; Baxter 2006; Colominas 2016; Crabtree 2013; De Grossi Mazzorin and Tagliacozzo 2000; Harcourt 1974; Lepetz 1996; Peters 1998: 184–186; Trixl 2019).

## 3 'Cave canem' or searching for dog tracks in the Roman provinces

With only a few exceptions, for example dogs in Early Iron Age Situla art (e.g. Trebsche 2018: 211–244), archaeozoological research in Austria largely lacks figural, written or epigraphic evidence on their function, role and value compared with the subsequent Greek and Roman antiquity. This section presents these sources from Roman Austria. They extend our current knowledge, which is based solely on dog bones.

### *3.1 Roman dogs in mythological contexts*

The famous hunter Actaeon, son of Aristaeus and Autonoe, one day happened to see Artemis naked, while she was bathing. The goddess turned him into a deer, and he was killed by his own hunting dogs, who did not recognise him. The death of Actaeon caused such distress to his dogs that they only calmed down when the Centaur Chiron made a statue of their master (Grant and Hazel 2000: 33; Preston 1997: 86). This story from Greek mythology is one of many examples of dogs being mentioned and represented in ancient literature and art. A number of reliefs on grave monuments and votive stones showing the story of Actaeon suggest that the people in Roman Austria were familiar with this Greek myth. A grave(?) relief close to the auxiliary fort of Arelape (Pöchlarn, Lower Austria), today fixed into the western wall of the parish church of Pöchlarn, displays the dogs of the hunted Actaeon (lupa.at/388). The same motif is present in a relief in the Museum of Salzburg, which was used as spolia fixed into the graveyard wall of St. Martin in Lungau (Salzburg) (lupa.at/3637).

In classical mythology, dogs were companions to many deities. Artemis, Asclepius, Hecate, Hygieia, the guardian dieties Lares (Preston 1997: 85–87), Silvanus (Kremer 2012: 375–378) and Endymion are depicted with dogs. Additionally, a dog participates in the Mithraic tauroctony (the act of bull killing), in which the god Mithras, accompanied by a snake, a raven and a dog, stabs the bull with a dagger. This scene is depicted in many reliefs in Carnuntum, showing a dog together with a snake greedily licking the wound of the suffering bull (Kremer 2012: 103–108, tab. 52–56). In these mythological examples dogs were used as hunting companions and guardians, two functions that probably did not differ from their role in Prehistory.

### *3.2 Roman dogs in hunting scenes*

One of the most popular motifs found on terra sigillata in the entire area of the Roman provinces are reliefs of dogs participating in hunting activities. These scenes show dogs of different sizes and morphotypes (Oswald 1964: Nr. 1914-2039, Plate LXXVII-LXXIX, LXXXII). Furthermore, more than 80 stone monuments have been documented from the Roman provinces of Noricum and Pannonia Superior within the borders of modern-day Austria, showing scenes with hunting and shepherd dogs, or scenes with dogs in a mythological context (Figure 6a; Walde 2005: 181–185; lupa.at). Distinguishing between both spheres is not always simple.

In hunting scenes, their prey are usually hares, wild boar and red deer. The number and distribution of gravestones with hunting scenes along the Danubian Limes, around the Leitha Mountains, in the wooded areas of southern and eastern Styria, Salzburg, as well as in present-day Carinthia is remarkable. In contrast, in urban centres such as Carnuntum, Vindobona, Virunum, Teurnia, Ovilavis, Aelium Cetium or Lauriacum, where a high number of monuments are concentrated, hunting scenes with dogs are rare or even absent. One explanation for this difference might

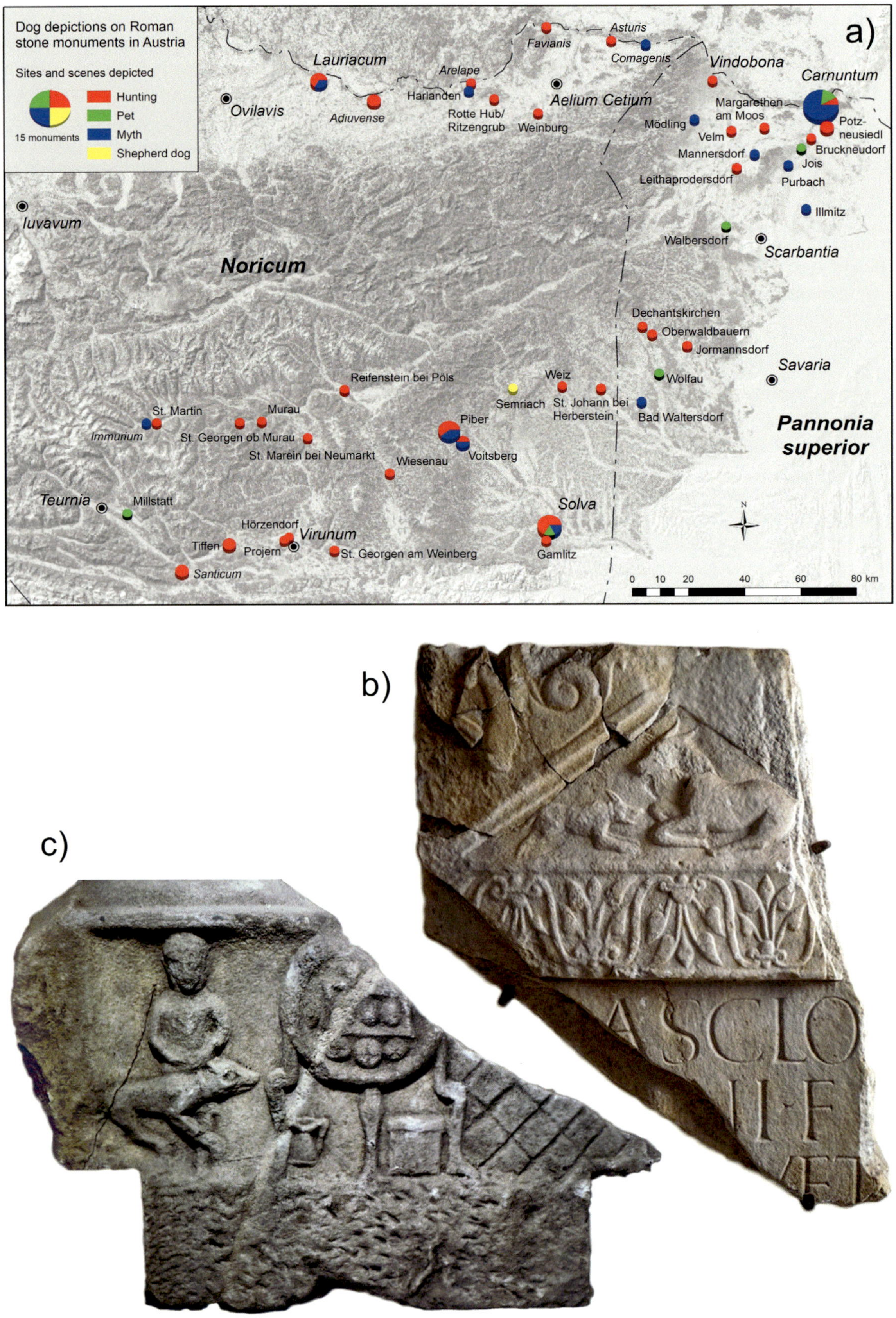


Figure 6. (a) Distribution map of dog representations on Roman stone monuments in Austria (map: M. Mosser; map source: Terrain Hillshade Tile Cache of Austria, published by geoland.at); (b) grave stele of Masculus with a dog barking at a lying goat in the pediment from Semriach, Styria (lupa.at/1359, photo: O a). Harl); (c) grave stele with commemorative meal for the dead and a servant with a dog from Wolfau, Burgenland (lupa.at/452; photo by O. Harl, © Landesmuseum Burgenland, Eisenstadt, BLM Inv.Nr. A-29360/ KG: Wolfau).

be that hunting was viewed as an activity of prestige exercised by privileged owners in rural areas (Ebnöther and Monnier 2002: 169). Possibly, these people tried to immortalise such noble hunting activities on stone monuments, close to their *villae rusticae*. Surprisingly, such reliefs with dogs are absent from western Austria (Roman province of Raetia) - at least in the '*ubi-erat-lupa*' database (lupa.at) - which might be attributed to a different workshop tradition.

### 3.3 Roman herding and guard dogs

In rural areas, dogs were also used as working animals (*canes pecuarii*) by shepherds for herding and to protect the flocks (Weeber 2005: 111, 249). A matching scene was recorded in the pediment of the grave stele of Masculus in Semriach (Styria), which depicts a lying goat and a barking dog (Figure 6b; lupa.at/1359). The significance of dogs as guardians of the house is shown by various sources, including the famous Pompeii mosaic with a chained dog and the inscription '*cave canem*' (Guzzo and d´Ambrosio 2002: 75; Weeber 2003: 172–173), and by a written record about a wall painting with a guard dog (Petronius, Satyricon 29, 1).

### 3.4 Roman dogs as 'best friends'

Roman grave reliefs from present-day Austria also indicate the role of dogs as companion animals. In Flavia Solva (Styria), there is a scene with a lying dog on a leash (lupa.at/5751), and on an ash chest from a grave from Millstatt am See, (Carinthia) a servant is depicted holding a sitting dog on a leash (lupa.at/3636). A similar scene was recorded on a grave relief showing a commemorative meal for the dead (Latin: *refrigerium*) from Wolfau (Burgenland), where a servant keeps a dog away from the table (Figure 6c; lupa.at/452). Representations of dogs on grave monuments of children have usually been interpreted as their pets. Some well-known examples are Marcellina, the five-year-old daughter of a soldier from the 15th legion in Carnuntum (lupa.at/121), who is sitting on a chair with a small dog next to her, and the grave stele of a boy from Jois (Burgenland) with a small dog wagging its tail at his feet (lupa.at/2252).

### 3.5 Roman dogs for entertainment

One of the most common characteristics of many Romans was their love for public games, spectacles and competitions such as chariot races, gladiator fights and animal hunts (*venationes*). The latter was especially favoured in the amphitheatres of the Roman provinces. Famous dedications are found on the altars of *venatores* at the amphitheatre of Virunum (Zollfeld, Carinthia), the capital city of Noricum. Experienced hunters (*venatores*) fought against bears, lions, bulls, wild boars and sometimes even against each other (Dolenz 2004: 303–306). Cut and chop marks on long bones of dogs from the amphitheatre of Virunum indicate that dog meat was probably used as fodder for the wild animals intended for the arena (Galik 2004: 436–437). However, it cannot be excluded that some dogs participated in such games (Köhne and Ewigleben 2000: 78; fig. 70). The murders of Christians initiated by Emperor Nero in 64 AD were an extreme example of the use of dogs; people were torn apart by dogs after they were covered with furs of wild animals (Tacitus, Annales XV, 44). Fighting dogs were already used by the Celts in battles and dogs of similar use were apparently imported to the Empire from Britain even before the Roman occupation (Strabon, Geographika IV, 5). From the military camp in Vindobona, 40 dog bones were recovered, indicating that Roman legions of the Danubian Limes commonly used dogs. Pathologies on dog bones suggest the use of dogs as draught - and pack animals (Czeika 2010: 922).

### 3.6 Footprints of Roman dogs

In addition to the above-mentioned written sources, figural representations as well as abundant archaeological and archaeozoological evidence such as gnawing marks and even dog faeces (Czeika 2010: 922; Galik 2004: 431) show that dogs of the Roman period literally left traces of their presence. It is not uncommon to find tracks of military sandal-boots (*caligae*) and animal tracks on Roman bricks. They were produced when the bricks were laid out for drying and the clay was still soft. Several examples of such dog footprints derive from the Roman legionary brickyard in Vindobona in today's 17th district of Vienna (Figure 7; Mosser *et al.* 2012: 104–108 fig. 19). Do these tracks belong to stray dogs? Did these dogs protect the brick production? What is the size range of these tracks? All these questions, together with the archaeological and archaeozoological finds from present-day Austria, clearly suggest the various functions that dogs served in the Roman culture as well as their significance for Roman society.

## 4 Conclusions

Generally, dogs from the Neolithic to the Roman period in present-day Austria are found in low numbers compared with other species such as cattle, small ruminants and pigs, which had great economic importance.

From the Neolithic period onwards, a wide spectrum of morphotypes can be observed. Already during the Bronze Age, dogs of around 50 cm height at withers become more common and, during the Iron Age, dogs of ca. 60 cm are the most frequent type. So far, no distinct dog breeds can be distinguished, and the relatively large-sized Iron Age dogs do not seem to represent a separate breed. The occurrence of wolf-sized dogs during the

Figure 7. Roman bricks (mainly roof tiles, *tegulae*) with footprints of middle to large-sized dogs from the Roman legionary brickyard in Vindobona (Steinergasse 17, 1170 Vienna) (Photos by M. Mosser).

Bronze Age from sites related to mining activities, and the appearance of lapdog-sized individuals in the Iron Age site Roseldorf, are notable exceptions. It remains unclear whether the latter reflect trade and exchange, because osteological evidence from Roseldorf suggests that some cattle were imported from the South.

In the Roman period, two enigmatic observations still require explanation: (i) in some cases isolated dog bones were found in high numbers and (ii) the relatively abundant (partly) preserved dog skeletons compared to other periods. Dogs probably had various functions in daily life. With the arrival of the Romans, for example, the higher diversity of sizes and shapes, especially in urban sites, points to a greater interest in dog pets. In the Roman period, beyond the osteological remains, different morphotypes and uses of dogs are also supported and highlighted by additional sources including myths, epigraphic evidence and figural representations.

## Acknowledgements

The authors would like to thank Erich Draganits for discussion and support in producing the figures as well as Michael Stachowitsch for improving the style of the text.

## References

### *Ancient sources*

Agricola, G. 1556. *De re metallica libri XII*. Schiffner, C. (transl. and ed.) 1928. Zwölf Bücher Vom Berg- und Hüttenwesen. Buch VI: Von den Werkzeugen, Geräten und Maschinen. Berlin: VDI-Verlag G.M.B.H.

Petronius. *Satyricon 29, 1*. Walsh P.G. (transl. and ed.) 2009. Petronius: The Satyricon. Oxford World's Classics. Oxford: Oxford University Press.

Strabon. *Geographika IV, 5*. Radt, S. (transl. and ed.) 2002. Strabons Geographika. Prolegomena. Buch I–IV: Text und Übersetzung,mit Übersetzung und Kommentar. Göttingen: Vandenhoeck & Ruprecht.

Tacitus. *Annales XV, 44*. Yardley, J.C. (transl. and ed.) 2008. Tacitus: The Annals. Oxford World's Classics. Oxford: Oxford University Press.

### *Modern sources*

Abd el Karem, M. 2009. Die spätlatènezeitlichen Tierknochenfunde des Simonbauerfeldes auf dem Dürrnberg, Salzburg. *Annalen des Naturhistorischen Museums in Wien* Serie A 110: 133–154.

Abd el Karem, M. 2013. *Keltische Festmähler und italische Rinder. Die tierischen Überreste aus dem 'Großen Heiligtum' der latènezeitlichen Siedlung Roseldorf. Archäologische Forschungen in Niederösterreich* 13. St. Pölten: Selbstverlag des NÖ Instituts für Landeskunde.

Abd el Karem, M. 2014. Die Gräbergruppe Hexenwandfeld, in G. Tiefengraber and K. Wiltschke-Schrotta (eds) *Der Dürrnberg bei Hallein*: 236–242. Rahden/Westfalen: Verlag Leidorf.

Ambros, C. 2006. Tierknochenfunde aus den römischen Schichten der Grabungen des Bundesdenkmalamtes in Lauriacum/ Enns, OÖ. 1972–1976 und 1997, in G. Winkler (ed.) *Schausammlung 'Römerzeit' im Museum Lauriacum Enns. Forschungen in Lauriacum* 12, 1/2006, Sonderband I/1: 103–119. Enns: Museum Lauriacum.

Amschler, J.W. 1939. Die Haustierreste von der Kelchalpe bei Kitzbühel, Tirol. *Mitteilungen der Prähistorischen Kommission der Akademie der Wissenschaften* 3/1–3: 96–121. Vienna: Verlag Hölder-Pichler-Tempsky.

Anthony, D.W. and D.R. Brown 2017. The dogs of war: A Bronze Age initiation ritual in the Russian steppes. *Journal of Anthropological Archaeology* 48: 134–148.

Argant, T. 2017. The Impact of Romanisation on Hippophagy and Cynophagy: A Long-Term Perspective from Lyon, France. *Journal of Historical Archaeology and Anthropological Sciences* 2/2. 00050. DOI: 10.15406/jhaas.2017.02.00050.

Bălăşescu, A. and A. Morintz 2018. A case of cynophagy at Radovanu-*Gorgana a doua* settlement, Călăraşi county (2nd-1st centuries BC). *Materiale și Cercetări Arheologice* (serie nouă) XIV: 133–148.

Bartosiewicz, L. 2000. Metric variability in Roman period dogs in Pannonia Province and the Barbaricum (Hungary), in S.J. Crockford (ed.) *Dogs through Time: an Archaeological Perspective. Proceedings of the 1st ICAZ Symposium on the history of the domestic dog*. (British Archaeological Reports International Series 889): 181–189. Oxford: Archaeopress.

Bauer, K. and E. Ruttkay 1974. Ein Hundeopfer der Lengyel-Kultur von Bernhardsthal, NÖ. *Annalen des Naturhistorischen Museums in Wien* 78: 13 27.

Baxter, I.L. 2006. A dwarf Hound Skeleton from a Romano-British Grave at York Road, Leicester, England, U.K., with a discussion of other Roman small dog types and speculation regarding their respective aetiologies, in L.M. Snyder and E.A. Moore (eds) *Dogs and People in Social, Working, Economic or Symbolic Interaction. Proceedings of the 9th Conference of the International Council of Archaeozoology*. Durham, August 2002: 12–23. Oxford: Oxbow Books.

Blouin, D.D. 2012. Understanding Relations between People and their Pets. *Sociology Compass* 6/11: 856–869.

Boschin, F. and A. Riedel 2009. Archäozoologische Untersuchungen an zwei Fundstätten der Aujnetitz-Kultur Niederösterreichs: Die Ziegelwerke von Stillfried und Schleinbach (Grabungen 1916–1939). *Annalen des Naturhistorischen Museums Wien* Serie A 110: 183–219.

Boschin, F. and A. Riedel 2011. Ein spätbronzezeitlicher Tierknochenfundkomplex aus der Kupferbergbausiedlung Brixlegg-Mariahilfbergl (Tirol). *Annalen des Naturhistorischen Museums Wien* Serie A 113: 561–618.

Böhm, H. 2010. Bestimmung und Erstbewertung ausgewählter Tierknochenfunde der Grabung 'Passauer-Hof' (Fundstelle 4) - Walterskirchen, NÖ. Unpublished results, University of Vienna.

Bruckner-Höbling, T. 2009. Bisherige Ergebnisse der Untersuchungen am Tierknochenmaterial aus der keltischen Seidlung Roseldorf-Sandberg in Niederösterreich, in V. Holzer (ed.) *Roseldorf Interdisziplinäre Forschungen zur größten keltischen Zentralsiedlung Österreichs*: 151–255. Wien: Österreichische Elektrizitätswirtschafts-Aktiengesellschaft.

Chrószcz, A., M. Janeczek, Z. Bielichová, T. Gralak and V. Onar 2015. Cynophagia in the Púchov (Celtic) Culture Settlement at Liptovská Mara, Northern Slovakia. *International Journal of Osteoarchaeology* 25: 528–538.

Colominas, L. 2016. Morphometric Variability of Roman Dogs in Hispania Tarraconensis: The Case Study of the Vila de Madrid Necropolis. *International Journal of Osteoarchaeology*. DOI: 10.1002/oa.2507.

Crabtree, P.J. 2013. A Note on the Role of Dogs in Anglo-Saxon Society: Evidence from East Anglia. *International Journal of Osteoarchaeology*. DOI: 10.1002/oa.2358.

Czeika, S. 2003. Tierreste aus dem Bereich der römischen Werkstätten am Michaelerplatz, Wien 1. *Fundort Wien - Berichte zur Archäologie* 6: 58–76. Wien: Phoibos Verlag.

Czeika, S. 2006. Hallstattzeitliche Tierreste der Ausgrabung Oberlaa, in C. Ranseder (ed.) *Eine Siedlung der Hallstattkultur in Wien 10, Oberlaa. Monographien der Stadtarchäologie Wien* 2: 349–363. Wien: Phoibos Verlag.

Czeika, S. 2010. Tierreste als Hinweise auf die Fleischversorgung, in *Die römischen Kasernen im Legionslager Vindobona. Die Ausgrabungen am Judenplatz in Wien in den Jahren 1995–1998. Monografien der Stadtarchäologie Wien* 5: 914–952. Wien: Phoibos Verlag.

Davis, S. J. 1987. *The Archaeology of Animals*. London: B.T. Batsford.

De Grossi Mazzorin, J. and A. Tagliacozzo 1997. Dog remains in Italy from the Neolithic to the Roman period. *Anthropozoologica* 25/26: 429–440.

De Grossi Mazzorin, J. and A. Tagliacozzo 2000. Morphological and osteological changes in the dog from the Neolithic to the Roman period in Italy, in S.J. Crockford (ed.) *Dogs through Time: an Archaeological Perspective. Proceedings of the 1st ICAZ Symposium on the history of the domestic dog.* (British Archaeological Reports International Series 889): 141–161. Oxford: Archaeopress.

Dolenz, H. 2004. Die Inschriften aus dem Amphitheater von Virunum, in R. Jernej and Ch. Gugl (eds) *Virunum. Das römische Amphiteater: die Grabungen 1998–2001*: 269–322. Klagenfurt/ Celovec: Wieser Verlag.

Ebnöther, C. and J. Monnier 2002. Ländliche Besiedlung und Landwirtschaft, in L. Flutsch, U. Niffeler and F. Rossi (eds) *Die Schweiz vom Paläolithikum bis zum frühen Mittelalter - Römische Zeit. SPM V*. Basel: Verlag Schweizerische Gesellschaft für Ur- und Frühgeschichte.

Ekroth, G. 2007. Meat in ancient Greece: sacrificial, sacred or secular? *Food & History* 5/1: 249–272.

Galik, A. 2000. Dog remains from the Late Hallstatt period of the chimney cave Durezza, near Villach (Carinthia, Austria), in S.J. Crockford (ed.) *Dogs through Time: an Archaeological Perspective. Proceedings of the 1st ICAZ Symposium on the history of the domestic dog.* (British Archaeological Reports International Series 889): 129–137. Oxford: Archaeopress.

Galik, A. 2002. The late Hallstatt and early La Tène animal bones assemblage from the vertical Durezza cave near Villach, in Carinthia. Unpublished Dissertation, University of Vienna.

Galik, A. 2004. Archäozoologische und kulturhistorische Aspekte der Tierknochenvergesellschaftungen aus dem Amphiteater von Virunum, in R. Jernej and Ch. Gugl (eds) *Virunum. Das römische Amphiteater: die Grabungen 1998–2001*: 395–482. Klagenfurt/ Celovec: Wieser Verlag.

Galik, A., S. Emra and M. Pacher 2019. Die tierischen Überreste aus der frühbronzezeitlichen Siedlung bei Drasenhofen, in K. Fiebig and A. Csaplaros (eds) *Trassenarchäologie 3*: 60–69. Pöttelsdorf: Eigenverlag.

Grant, M. and J. Hazel 2000. *Lexikon der antiken Mythen und Gestalten*. München: Deutscher Taschenbuch Verlag.

Green, M. 1992. *Animals in Celtic Life and Myth*. London and New York: Routledge.

Grill, Ch. 2009. Die menschlichen und tierischen Überreste aus dem spätlaténezeitlichen Heiligtum auf dem Frauenberg bei Leibnitz (Steiermark). Unpublished Dissertation, University of Vienna.

Guzzo, G.P. and A. d´Ambrosio 2002. *Pompeii. Führer durch die Ausgrabungen*. Napoli: Electa.

Hadjikoumis, A. 2016. Every Dog has Its Day: Cynophagy, Identity and Emerging Complexity in Early Bronze Age Attica, Greece, in N. Marom, R. Yeshurun, L. Weissbrod, and G. Bar-Oz (eds) *Bones and Identity. Zooarchaeological Approaches to Reconstructing Social and Cultural Landscapes in Southwest Asia*: 225–245. Oxbow and Philadelphia: Oxbow Books.

Harcourt, R.A. 1974. The Dog in Prehistoric and Early Historic Britain. *Journal of Archaeological Science* 1: 151–175.

Horard-Herbin, M.-P., A. Tesset and J.-D. Vigne 2014. Domestication and uses of the dog in Western Europe from the Paleolithic to the Iron Age. *Animal Frontiers* 4/3: 23–31.

Köhne, E. and C. Ewigleben 2000. *Caesaren und Gladiatoren. Die Macht der Unterhaltung im antiken Rom*. Mainz: von Zabern.

Kremer, G. 2012. *Götterdarstellungen, Kult- und Weihedenkmäler aus Carnuntum. Mit Beiträgen von Ch. Gugl, Ch. Uhlir und M. Unterwurzacher, CSIR Österreich, Carnuntum Suppl. 1.* Wien: Verlag der Österreichischen Akademie der Wissenschaften.

Kunst, G. K. 2005. Die Tierreste aus der Siedlung der Badener Kultur in Stoitzendorf, in O. Schmitsberger (ed.) *Eine Siedlung der klassischen Badener Kultur in Stoitzendorf im Weinviertel. Fundberichte aus Österreich* 43 (2004): 183–186.

Kunst, G.K. 2006a. Tierreste aus der endneolithischen Grubenhütte von Furth bei Göttweig, in A. Krenn-Leeb, K. Grömer and P. Stadler (eds) *Ein Lächeln für die Jungsteinzeit, Ausgewählte Beiträge zum Neolithikum Ostösterreichs, Festschrift für Elisabeth Ruttkay. Archäologie Österreichs*, 17/2: 153–163.

Kunst, G.K. 2006b. Tierreste aus ausgewählten Befunden der Grabungen 1997–1999 im Vicus Ost von Mautern a. d. Donau, in S. Groh and H. Sedlmayer (eds) *Forschungen im Vicus Ost von Mautern-Favianis. Die Grabungen der Jahre 1997–1999. Der Römische Limes in Österreich* 44: 637–708. Wien: Verlag der Österreichischen Akademie der Wissenschaften.

Kunst, G.K. and N. Doneus 2013. Roman graves and rural rubbish. Animal remains from the Roman cemetery of Halbturn, Austria. *Anthropozoologica* 48/2: 391–408.

Lepetz, S. 1996. L'animal dans la Société gallo-romaine de la France du nord. *Revue Archéologique de Picardie*. Numéro spécial 12.

Leskovar, J. 1996. Frühkeltische Siedlung und ein Frauengrab mit Hund und Schwein. *Linzer Archäologische Forschungen*, Sonderheft XVII, Linz.

Meehan, M., B. Massavelli and N. Pachana 2017. Using Attachment Theory and Social Support Theory to Examine and Measure Pets as Sources of Social Support and Attachment Figures. *Anthrozoös* 30/2: 273–289. DOI: 10.1080/08927936.2017.1311050.

Mosser, M., K. Adler-Wölfl, S. Czeika, I. Gaisbauer, S. Radbauer, H. Sedlmayer and S. Sakl-Oberthaler 2012. Befunde im Legionslager Vindobona. Teil VII: Der Abwasserkanal der via praetoria - Wien 1, Wipplingerstraße 6 (Altes Rathaus). *Fundort Wien 15*: 74–118.

Müller, R. 1967. Die Tierknochen aus den spätrömischen Siedlungsschichten von Lauriacum II, Wild- und Haustierknochen ohne die Rinder. Dissertation, Universität München.

O'Connor, T. 2000. *The archaeology of animal bones*. Texas A&M University Anthropology Series 4. College Station: A&M University Press.

Oswald, F. 1964. *Index of Figure-Types on Terra Sigillata ('Samian Ware'). Issued as a supplement to the Annals of Archaeology and Anthropology*. London: The Gregg Press Limited.

Peters, J. 1998. *Römische Tierhaltung und Tierzucht. Eine Synthese aus archäozoologischer Untersuchung und schriftlich-bildlicher Überlieferung. Passauer Universitätsschriften zur Archäologie* 5. Rahden/Westfalen: Verlag Marie Leidorf.

Petersen, A.T. 2011. From Fur Baby to Chick Magnet: A Sociological View of Dogs and Their People. Undergraduate Theses and Professional Papers, Paper 3, University of Montana.

Preston, P. 1997. *Metzler-Lexikon antiker Bildmotive*. Stuttgart-Weimar: Verlag J. B. Metzler.

Pucher, E. 1987. Tierknochen aus der Bronzezeit des Buhuberges (Niederösterreich). *Wissenschaftliche Mitteilungen aus dem Niederösterreichischen Landesmuseum* 4: 11–35. St. Pölten: Amt der NÖ Landesregierung.

Pucher, E. 1996a. Bemerkungen zur Auswertbarkeit kleiner Fundbestände anhand weiterer bronzezeitlicher Tierknochenfunde vom Buhuberg (Niederösterreich). *Forschungen in Stillfried* 9/10 (1990–1992): 101–148.

Pucher, E. 1996b. Die Tierknochenfunde aus der Schleinbacher Ziegelei, Bezirk Mistelbach, Niederösterreich (Grabung 1981 bis 1986). *Annalen des Naturhistorischen Museums Wien* Serie A 97: 21–54.

Pucher, E. 1997. Die Tierknochen aus der spätneolithischen Höhensiedlung auf dem Wachberg bei Melk an der Donau, in H. Schwammenhöfer, and E. Pucher (eds) *Die spätneolithische Siedlung am Wachberg bei Melk*: 41–6. Melk: Kultur- u. Museumsverein Melk.

Pucher, E. 1998. Der Knochenabfall einer späthallstatt-/latènezeitlichen Siedlung bei Inzersdorf ob der Traisen (Niederösterreich), in P.C. Ramsl (ed.) *Inzersdorf-Walpersdorf. Studien zur späthallstatt-/latènezeitlichen Besiedlung im Traisental, Niederösterreich. Fundberichte aus Österreich*, Materialhefte A6: 56–67.

Pucher, E. 1999. Archäozoologische Untersuchungen am Tierknochenmaterial der keltischen Gewerbesiedlung im Ramsautal auf dem Dürrnberg (Salzburg). Mit Beiträgen von Thomas Stöllner und Karin Wiltschke-Schtotta. *Dürrnberg-Forschungen* 2. Abteilung Naturwissenschaft. Rahden/Westfalen: Marie Leidorf.

Pucher, E. 2003. Einige Bemerkungen zu den bisher übergebenen Knochenaufsammlungen aus dem Keutschacher See in Kärnten, in B. Samonig (ed.) *Studien zur Pfahlbauforschung in Österreich, Materialien II - Die Pfahlbaustation des Keutschacher Sees. Mitteilungen der Prähistorischen Kommission* 51: 263–282. Wien: Verlag der Österreichischen Akademie der Wissenschaften.

Pucher E. 2004a. Der mittelneolithische Tierknochenkomplex von Melk-Winden (Niederösterreich). *Annalen des Naturhistorischen Museums Wien* Serie A 105: 363–403.

Pucher, E. 2004b. Hallstattzeitliche Tierknochen aus Göttlesbrunn, p.B. Bruck an der Leitha, Niederösterreich, in M. Griebl (ed.) *Die Siedlung der Hallstattkultur von Göttlesbrunn, Niederösterreich.*

*Mitteilungen der Prähistorischen Kommission der Österreichischen* 54: 309–328. Wien: Verlag der Österreichischen Akademie der Wissenschaften.

Pucher E. 2006a. Eine neuer Tierknochenfundkomplex aus einer Siedlung der Badener Kultur in Ossarn bei Herzogenburg in Niederösterreich. *Archäologie Österreichs* 17/2: 104–116.

Pucher, E. 2006b. Die Tierknochen aus einem keltischen Bauernhof in Göttlesbrunn (Niederösterreich). *Annalen des Naturhistorischen Museums Wien* Serie A 107: 197–220.

Pucher, E. 2006c. Das Tierknochenmaterial der Ausgrabung beim Bauareal des ARZ (Anton-Melzer-Straße 11), in Innsbruck-Wilten, in A. Picker, A. Höck, and E. Pucher (eds) *Die Rettungsgrabung des Tiroler Landesmuseums Ferdinandeum am Areal des Allgemeinen Rechenzentrums in Innsbruck-Wilten*: 163–202. Innsbruck: Veröffentlichungen des Tiroler Landesmuseums Ferdinandeum.

Pucher, E. 2014. Neue Aspekte zur Versorgungslogistik Hallstatts: Tierknochenfundkomplexe aus Pichl, Steiermark. *Fundberichte aus Österreich* 52/ 2013: 65–93.

Pucher, E. 2018. Der Tierknochenfundkomplex eines germanischen Dorfs im römischen Machtbereich: Bruckneudorf. *Fundberichte aus Österreich* 55/ 2016: Sonderdruck.

Pucher, E. 2019a. Remarks on the Animal Bone Finds from the Early Neolithic Settlement at Brunn am Gebirge, Wolfholz, in P. Stadler, and N. Kotova (eds) *Early Neolithic Settlement Brunn am Gebirge, Wolfholz, Site 2 in Lower Austria and the Origin of the Western Linear Pottery Culture (LPC). BUFM* 88 a, Volume 1/ part a: 517–531. Langenweißbach: Beier & Beran.

Pucher, E. 2019b. Die Tierknochen der mittelbronzezeitlichen Fundstelle Saalfelden-Katzentauern im Salzburger Pinzgau. *Annalen des Naturhistorischen Museums Wien* Serie A 121: 35–81.

Pucher, E. and K. Engl 1997. *Studien zur Pfahlbauforschung in Österreich. Materialien I - Die Pfahlbauten des Mondsees. Tierknochenfunde. Mitteilungen der Prähistorischen Kommission der Österreichischen Akademie der Wissenschaften. Das Altertum* 51: 1–150. Wien: Verlag der Österreichischen Akademie der Wissenschaften.

Pucher, E., F.E. Bart, R. Seemann and F. Brandstätter 2013. *Bronzezeitliche Fleischverarbeitung im Salzbergtal bei Hallstatt. Mitteilungen der Prähistorischen Kommission der Österreichischen Akademie der Wissenschaften* 80. Wien: Verlag der Österreichischen Akademie der Wissenschaften.

Pucher, E., K. Saliari and P.C. Ramsl 2015: Römische Haustiere eines Latènezeitlichen Hausherrn in Vindobona (Wien)?, in S. Flohr (ed.) *Beiträge zur Archäozoologie und Prähistorischen Anthropologie* X: 71–78. Langenweißbach: Beier & Beran.

Reitz, E.J. and E.S. Wing 2008. *Zooarchaeology*. Cambridge Manuals in Archaeology. Cambridge: Cambridge University Press.

Riedel, A. 1993. Die Tierknochenfunde des römerzeitlichen Lagervicus von Traismauer/ Augustiana in Niederösterreich. *Annalen des Naturhistorischen Museums Wien* Serie A 95: 179–294.

Riedel, A. 1996. Die Tierknochenfunde einer germanischen Siedlung an der Thaya bei Bernhardsthal im nordöstlichen Niederösterreich. *Annalen des Naturhistorischen Museums Wien* Serie A 97: 55–144.

Riedel, A. 1998. Archäozoologische Untersuchungen an den Knochenfunden aus der Věteřov-Kultur von Böheimkirchen (Niederösterreich). *Annalen des Naturhistorischen Museums Wien* Serie A 99: 341–374.

Riedel, A. 2003. Die frühbronzezeitliche Fauna von Brixlegg in Tirol. *Atti dell'Accademia Roveretana degli Agiati* 253. Serie VIII: III, B: 197–281. Rovereto: L'Accademia.

Riedel, A. 2004. Tierknochen aus der römischen Villa rustica von Nickelsdorf in Burgenland (Österreich). *Annalen des Naturhistorischen Museums Wien* Serie A 106: 449–539.

Saliari, K. 2021. Archäozoologische Analyse der Tierknochen aus dem Gräberareal Steigerhaushügel auf dem Dürrnberg, in R.B. Franke and K. Wiltschke-Schrotta (eds) *Der Dürrnberg bei Hallein - Die Gräbergruppe am Steigerhaushügel*: 160–166. Rahden/ Westfalen: Verlag Marie Leidorf.

Saliari, K. and E. Pucher 2015. Animal bones from the Roman assemblages in Weiden am See (Burgenland). Unpublished report, 1. Zoological Department, Archaeozoology, Natural History Museum Vienna.

Saliari, K. and E. Pucher 2017. Tierknochen erzählen, in N. Franz, J. Schwarzäugl and A. Tögel (eds) *Steinsichel und Bronzedolch. Urgeschichte in Weiden am See*: 48–51. Wien: Verlag Berger.

Saliari, K., E. Pucher and M. Kucera 2016. Archaeozoological investigations of the La Tène A-C1 salt-mining complex and the surrounding graves of Putzenkopf Nord (Bad Dürrnberg, Austria). *Annalen des Naturhistorischen Museums Wien* Serie A 118: 245–288.

Saliari, K., E. Pucher, M. Staudt and G. Goldenberg 2020. Continuities and changes of animal exploitation across the Bronze Age - Iron Age boundary at mining sites in the Eastern Alps. *Archaeofauna*: 77–106.

Schmitzberger, M. 2001. Die Tierknochen aus der mittelneolithischen Kreisgrabenanlage Ölkam (Oberösterreich). *Jahrbuch des Oberösterreichischen Musealvereines* 146/I: 43–86 + Ergänzungsheft.

Schmitzberger, M. 2007. Tierknochenfunde aus Potzneusiedl, in F. Sauer (ed.) *Die Archäologischen Grabungen auf der Trasse der A6. Fundstellen Potzneusiedl Wangheim*: 80–89 Wien: Bundesdenkmalamt und ASFiNAG.

Schmitzberger, M. 2008. Die Tierknochen, in P. Trebsche (ed.) *Die Höhensiedlung 'Burgwiese' in Ansfelden (Oberösterreich). Ergebnisse der Ausgrabungen von 1999 bis 2002. Linzer Archäologische Forschungen* 38/2: 284–306.

Schmitzberger, M. 2009a. Archäozoologische Untersuchungen an den Tierknochen aus den Rettungsgrabungen des Niederösterreichischen Landesmuseums in Michelstetten 1994–1999. *Annalen des Naturhistorischen Museums Wien* Serie A 110: 221–312.

Schmitzberger, M. 2009b. Haus- und Jagdtiere im Neolithikum des österreichischen Donauraumes. Dissertation, Universität Wien.

Schmitzberger, M. 2009c. Tierknochenfunde aus der Badener Kultur von Potzneusiedl, Burgenland. *Fundberichte aus Österreich* 47: 167–184.

Schmitzberger, M. 2010a. Die hallstatt- und latènezeitlichen Tierknochenfunde aus den Grabungen des Niederösterreichischen Landesmuseums 1994–1999 in Michelstetten., in E. Lauermann (ed.) *Die Latènezeitliche Siedlung von Michelstetten: Die Ausgrabungen des Niederösterreichischen Museums für Urgeschichte in den Jahren 1994-1999. Archäologische Forschungen in Niederösterreich* 7: 148–167. St. Pölten: Selbstverlag des NÖ Instituts für Landeskunde.

Schmitzberger, M. 2010b. Die Tierknochenfunde der Grabungskampagne 2008 auf der Keplerwiese in Linz. *Linzer Archäologische Forschungen*, Sonderheft 44: 75–94.

Schmitzberger, M. 2012. Die Tierknochen vom Ramsaukopf, Putzenkopf und Putzenfeld - neue Funde vom keltischen Dürrnberg bei Hallein. *Annalen des Naturhistorischen Museums Wien* Serie A 114: 79–138.

Škvor Jernejčič, B. and B. Toškan 2018. Ritual use of dogs and wolves in the Late Bronze and Early Iron Age in the South-Eastern Alpine region. New evidence from the archaeo(zoo)logical perspective, in S. Denis Vialou, L. Costamagno, C. Gourichon, and O.D. Dupont (eds) *Animal symbolisé, animal exploité: du Paléolithique à la Protohistoire*: 236–263. Paris: Éditions du Comité des travaux historiques et scientifiques. DOI: 10.4000/books.cths.4278.

Smolkovic, I., M. Fajfar and V. Mlinaric 2012. Attachment to Pets and Interpersonal Relationships. *Journal of European Psychology Students* 3: 15–23.

Tague, I. 2015. *Animal Companions: Pets and Social Change in Eighteenth-Century Britain.* University Park: Penn State University Press.

Tecchiati, U. 2012. Die Tierknochen aus der bronze- und eisenzeitlichen Siedlung auf dem Kiabichl bei Faggen (Tirol, Österreich). *Annalen des Naturhistorischen Museums Wien* Serie A 114: 21–78.

Tipper, B. 2011. Pets and Personal Life, in V. May (ed.) *Sociology of Personal Life*: 85–97. Basingstoke: Palgrave Macmillan.

Trantalidou, K. 2006. Companions from the oldest Times: Dogs in Ancient Greek Literature, Iconography and Osteological Testimony., in L.M. Snyder and E.A. Moore (eds) *Dogs and People in Social, Working, Economic or Symbolic Interaction. Proceedings of the 9th Conference of the International Council of Archaeozoology.* Durham, August 2002: 96–120. Oxford: Oxbow Books.

Trebsche, P. 2018. Die hallstattzeitlichen Jagddarstellungen der Kalenderberggruppe - zu einem Altfund von Rauheneck bei Baden (Niederösterreich), in K. Saliari, P. Trebsche, U. Tecchiati and A. Kroh (eds) *Von Keltenponys, Bergschecken und zahmen Hirschen: Festschrift für Erich Pucher. Annalen des Naturhistorischen Museums in Wien* Serie A 120: 211–244. Wien: Naturhistorisches Museum Wien Verlag.

Trixl, S. 2019. *Zwischen Wandel und Beständigkeit. Die Entwicklung der späteisenzeitlich-frührömischen Viehwirtschaft im Alpenraum und dem nördlichen Alpenvorland. Documenta Archaeobiologiae. Veröffentlichungen der Staatssammlung für Anthropologie und Paläoanatomie München* 14. Rahden/Westfalen: Verlag Marie Leidorf.

Twiss, K.C. 2007. We are what we eat. *The archaeology of food and identity*: 1–15.

Walde, E. 2005. *Im herrlichen Glanze Roms. Die Bilderwelt der Römersteine in Österreich.* Innsbruck: Institut für Klassische und Provinzialrömische Archäologie, Universität Innsbruck.

Weeber, K.-W. 2003. *Alltag im alten Rom. Das Leben in der Stadt.* Düsseldorf: Patmos Verlag.

Weeber, K.-W. 2005. *Alltag im alten Rom. Das Landleben.* Düsseldorf: Patmos Verlag.

Wolff, P. 1977. Die Tierreste aus den bandkeramischen Siedlungen Poigen und Frauenhofen, Ger. Bez. Horn, NÖ, in E. Lenneis (ed.) *Siedlungsfunde aus Poigen und Frauenhofen bei Horn. Prähistorische Forschungen* 8: 99–102. Wien: Verlag Berger.

### *Web sources*

Statistik Austria, Konsumerhebung 2014/15. Themenblatt Haustiere, viewed 24 September 2019. https://www.statistik.at/wcm/idc/idcplg?IdcService=GET_PDF_FILE&RevisionSelectionMethod-LatestReleased&dDocName=116074

Harl, F. and Harl O., Bilddatenbank zu antiken Steindenkmälern, viewed 2 January 2020. http://lupa.at

# 4.5 A Dog's Head in a House Pit at the Early Iron Age Site of Verucchio. Butchery Waste or Ritual Sacrifice?

Marco Bertolini and Ursula Thun Hohenstein

Università di Ferrara, Dipartimento di Studi Umanistici,
Laboratorio di Archeozoologia e Tafonomia, Corso Ercole I d'Este 32 - 44121 Ferrara
E-mail: marco.bertolini@unife.it; ursula.thun@unife.it
Corresponding author: Ursula Thun Hohenstein, ursula.thun@unife.it

**Abstract**

The settlement of Verucchio stands on a cliff in the Apennines, at 330 m a.s.l, characterised by an irregular plateau (Pian del Monte) surrounded by four hills not far from the Marecchia river. This area was occupied during the transition between the Late Bronze and the beginning of the Iron Age. During the 9th century BC, the protohistoric village became a central place for the nearby villages till the 7th century BC. After this period the village seems to have been scarcely populated. The site was again inhabited from the end of the 5th century, as testified by several buildings, including the famous House 4. The house is a rectangular-shaped building (20 x 18.5 m) and it is oriented along a NNE/SSW axis, divided into three rooms aligned on the eastern side (from north rooms A, B and C). During the recent archaeological excavations, carried out by the University of Pavia between 2012 and 2017, an oval ditch was recovered inside room C of the House 4 below the sub-foundations. The ditch develops along a NNE-SSW axis and contained fragments of Etrusco-Padano pottery (4th century BC), and abundant faunal remains including an upside-down dog skull. Excavations were carried out by the University of Pavia in 2011 and allowed to investigate three chronological phases of the inhabited area (D IX-VIII cent. BC, C VII-V cent. BC and B IV-III cent. BC). The archaeozoological analysis, still underway, involved a total amount of about 2700 remains. An interesting aspect of phase B is the presence of a young dog skull with deciduous dentition deposited in the house with other skeletal elements of domestic animals which could have several symbolic significances.

**Keywords**: dog, cut-marks, archaeozoology, Early Iron Age, Romagna, Northern Italy.

## 1 Introduction

The settlement of Verucchio rises on an Apennine cliff from which a ford of the river Marecchia to the north-west and a good part of the Adriatic coast to the east could be controlled, not far from Mount Titano that stands a few km south-east in a dominant position (Figure 1).

While for decades research was devoted to the famous necropolis of the Early Iron Age located at the steep sides of the plateau, for the settlement of Verucchio only synthetic reports of late-nineteenth century and preliminary news of some excavations conducted in the Sixties and Seventies of the last century were available.

The University of Pavia, in collaboration with the Soprintendenza Archeologica dell'Emilia-Romagna, launched the Verucchio-Pian del Monte Project in 2011 with the aim of updating the framework of the peuplement of the plateau between the first and second Iron Age (Harari *et al.* 2017). First a non-invasive field geomagnetic prospecting was carried out, which, despite the disturbances due to the intense urbanisation of the area, have produced a preliminary mapping full of anomalous signals, at least in part attributable to the buried anthropic structures. The first field excavations (2012) were carried out for surveys north of Via Nanni, in the public gardens (Saggio Beta), where it was possible to document few evidences related to the Iron Age, mostly in secondary deposition, directly in contact with geological levels. South of Via Nanni, within the fenced area (Saggio Alpha), a section of a foundation wall, only partially excavated in the late 1970s, was brought to light (Figure 2).

The excavations were however concentrated inside a masonry building of the Late Classical age, already excavated, and restored by the Soprintendenza, with test pits that first concerned the three eastern rooms and extended in 2015 to three other rooms. Almost everywhere it has been possible to verify that the past excavations had completely removed the uppermost layers, related to the occupation of the masonry house, and preserved the lowermost levels, referable to the Early Iron Age. The oldest occupations lie directly on the geological substrate of the hill. It is a complex series of canals and ditches, necessary to ensure the static and drainage of the slope. A series of at least four shallow parallel grooves, altogether more than 11 m long and about 10 cm broad and north-south oriented, appear to be far from immediate interpretation. The

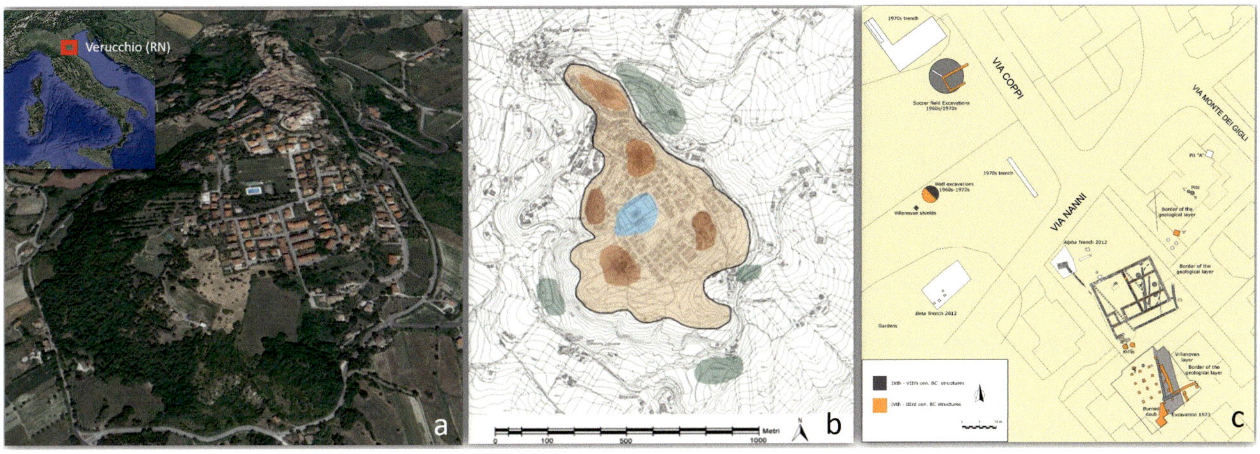


Figure 1. Location of the site of Verucchio (a) and the position of the archaeological area on the Plateu (Pian del Monte, b). Maps of the most important archaeological evidence (c) (images are modified from Harari *et al.* 2017).

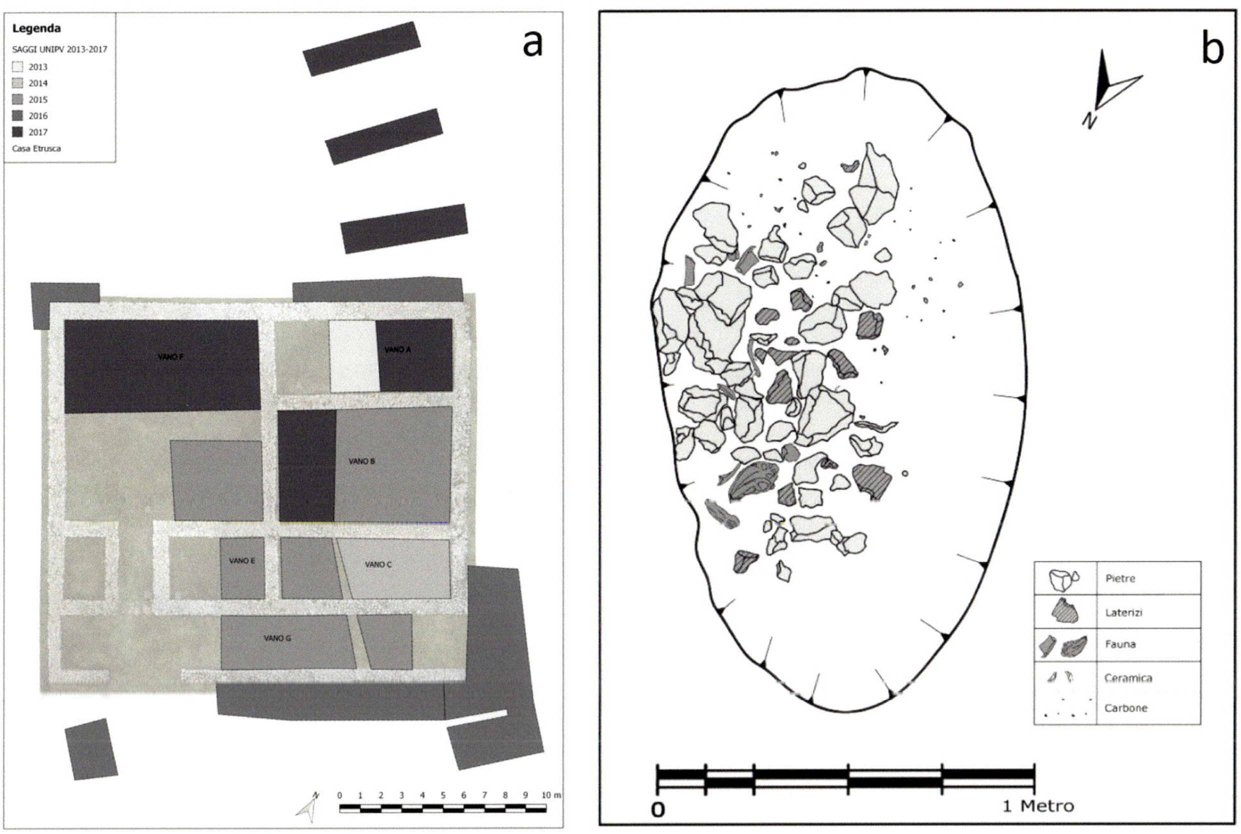


Figure 2. Map of the House 4 of Verucchio (a). Draw of the ditch (SU 1647) in the room C (b) (images are modified from Harari *et al.* 2017).

sections of these grooves are U-shaped, and the fillings are almost completely sterile.

On the left there are two large sub-excavated structures with the same north-south orientation. The two channels have been obtained inside a larger intervention, with wide V walls and stepped profile, up to 7–8 m wide and over 1 m deep for a length of about twenty metres. This evidence seems to document an impressive system of works which was useful for the management of soil and water and was conceived in the framework of a clear delimitation of the settlement area.

This large structure built in the 9th century BC, was abandoned between the end of the 9th century and the beginning of the 8th century with a series of overflows rich in materials of possible domestic origin, such as charcoals, wooden, fauna and pottery remains. After the obliteration of the ditch, the presence of a small

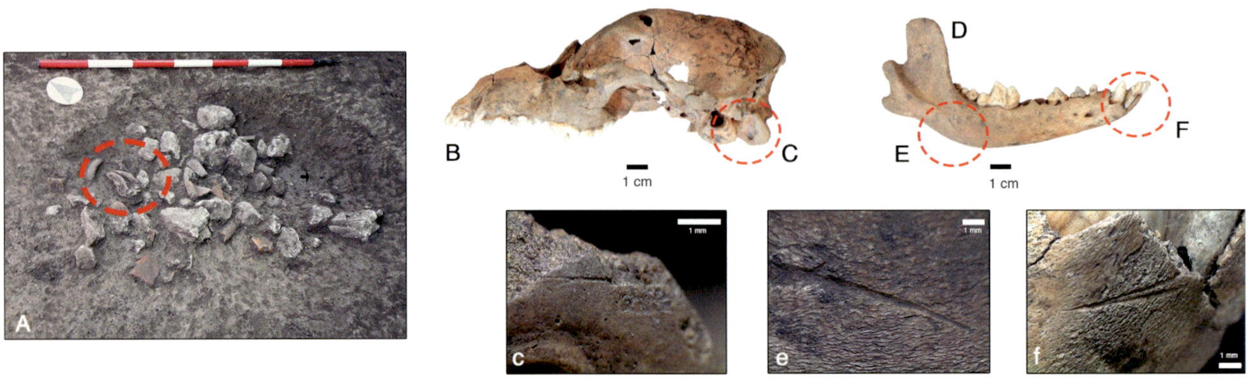


Figure 3. Pictures of the pit in the room C of the Late Iron age building of Verucchio (A) (courtesy of Zamboni and Rondini). The dog skull was in upside-down position. Left view of the skull (B). Disarticulation cut-marks are located on the occipital condyle (C-c stereomicroscope view). Right mandibole (D) with cut-mark correspond to skinning activity (E-F, e-f stereomicroscope image) (Photos by U. Thun Hohenstein).

oval structure (8x4m) always oriented north-south was recorded. The absence of layers of occupation related to this phase, probably removed during the previous excavations, prevent other considerations. Some pottery fragments testify the occupation of the area during the late 8th and early 7th century BC (Zamboni 2018).

After an apparent hiatus of over two centuries, from the middle of the 7th to the end of the 5th century BC, the settlement at Pian del Monte shows a reoccupation only in the late-classic age, from the end of the 5th century BC. For which it concerns this recent phase, there is little information reported by the excavations of the seventies, when portions of masonry buildings were recovered, including the famous 'House of Verucchio 4'.

Precisely in this building, from a stratigraphic point of view, some postholes have been identified inside the Villanovan levels.

The only closed context attributable to this late phase is a large oval-shaped pit, recognised within room C, which was filled by a discharge of stones and bricks, with abundant domestic material, including faunal remains and pottery that can be dated to the manufacturing of the Padana Etruria between the late 5th and the 4th century BC.

## 2 Methods

Faunal remains were identified and quantified (NISp, MNI), using osteological manuals (Pales and Lambert 1971; Schmid 1972; Barone 1976) and the reference collections of the Laboratories of Zooarchaeology and Taphonomy and of Large Mammals and Birds at the Department of Humanistic Studies of the University of Ferrara. Goats and sheep were distinguished according to the criteria elaborated by Zeder and Pilar (2010) for teeth and by Boessneck (1969) and Zeder and Lapham (2010) regarding the post-cranial skeleton. The discrimination between wild boar and pig was made according to the dimensions of the anatomical elements and the osteological reference collections. The minimum number of individuals (MNI) was estimated for each *taxon* combining age classes and laterality, and the data obtained from the teeth analyses. Age at death was estimated on the observation of tooth eruption and wear stage following Grant (1982) and Silver (1969) and on the epiphyseal fusion of long bones according to the methodologies proposed by Barone (1976) and Silver (1969). The taphonomic analysis was performed using a Leica S6D stereomicroscope (0.63x-4.0x magnification) equipped with an EC3 digital camera.

## 3 Archaeozoological analyses

The archaeozoological analysis, carried out on the faunal remains of Verucchio, has so far related to about 1548 osteological remains. The largest and most significant assemblage comes from the oldest phase of the settlement (D, 9th-8th century BC) with over 75% of the total remains recovered. The faunal assemblages from the phases C (7th-5th century BC) and B (4th-3rd century BC) are numerically poorly represented although the latter one has some peculiarities. For this reason, in this paper, we present the results of faunal remains coming from phases D and B (Table 1).

### *3.1 The faunal assemblage from the 9th-8th century BC*

From phase D, dated between the 9th-8th century BC, 1411 faunal remains were analysed: 42% has been identified taxonomically while the remaining 58% is mainly composed of unidentified fragments and

| *Taxon* | Phase D (9th-8th century BC) | | | | Phase B (4th-3rd century BC) | | | |
|---|---|---|---|---|---|---|---|---|
| | NISp | % | MNI | % | NISp | % | MNI | % |
| *Ursus arctos* | 2 | 0.47 | 1 | 3.9 | | | | |
| *Sus scrofa* | 3 | 0.71 | 1 | 3.9 | | | | |
| *Cervus elaphus* | 5 | 1.18 | 1 | 3.9 | 2 | 3.57 | 1 | 10 |
| **Total wild *taxa*** | **10** | | **3** | | **2** | | **1** | |
| *Canis familiaris* | 14 | 3.30 | 2 | 7.8 | 1 | 1.79 | 1 | 10 |
| *Equus caballus* | 1 | 0.24 | 1 | 3.9 | | | | |
| *Sus domesticus* | 142 | 33.49 | 9 | 34.3 | 17 | 30.36 | 3 | 30 |
| *Bos primigenius* | 101 | 23.82 | 4 | 15.6 | 23 | 41.07 | 1 | 10 |
| *Ovis* vel *Capra* | 107 | 25.24 | | | 10 | 17.85 | 2 | 20 |
| *Capra hircus* | 31 | 7.31 | 6 | 22.8 | 2 | 3.57 | 1 | 10 |
| *Ovis aries* | 18 | 4.25 | 2 | 7.8 | 1 | 1.79 | 1 | 10 |
| **Total domestic *taxa*** | **414** | **100** | **24** | **100** | **54** | **100** | **9** | **100** |
| Aves | 3 | | | | | | | |
| Pisces | 1 | | | | | | | |
| *Emys orbicularis* | 1 | | | | | | | |
| **Total other *taxa*** | **5** | | | | | | | |
| Cardiidae | 2 | | | | | | | |
| **Total NISp** | **431** | | | | **56** | | | |
| Carnivora | 10 | | | | | | | |
| Ungulata large sized | 70 | | | | 10 | | | |
| Ungulata m-l sized | 54 | | | | 11 | | | |
| Ungulata medium sized | 68 | | | | 12 | | | |
| Unidentified | 778 | | | | 48 | | | |
| **Total NUSp** | **980** | | | | **81** | | | |

Table 1. Composition of the fauna assemblage.

remains have only been determined anatomically and classified by size. The identified specimens consist mainly of domestic animals, followed by a smaller number of wild mammals. There are also other faunas such as marine mollusks, a vertebra of fish (probably marine), tortoise and birds. Among the wild ungulates red deer and wild boar were identified. Red deer is represented mostly by antler fragments and elements of the post-cranial skeleton. This allows us to assume the on-site transport of at least one carcass or part of it. Among the wild taxa the presence of two remains of brown bear is interesting, since this species is not very frequent in the fauna records of this period. Among the domestic animals, sheep and goats predominate with a clear prevalence of goats compared to sheep. Pigs follow with a slightly lower percentage while cattle are less represented. The horse is attested by a single tooth fragment, while the dog is present with few remains attributable to at least two individuals. The rate between the main domestic taxa changes if analysed on the basis of the Minimum Number of Individuals, for which pigs are more frequent than sheep or goats. Concerning the anatomical representation, cattle, sheep or goats, and pigs are in toto represented with almost all the elements of the axial skeleton while the cranial skeleton consists mainly of mandibles and teeth. Age classes, estimated from the analysis of the eruption stage and dental wear and the fusion degree of the epiphysis, allow us to surmise that the slaughter or killing of livestock took place. Pigs, as in the Bronze Age, were mainly used for meat. It is quite clear from the killing of young and sub-adults before reaching the third year of age. Cattle were exploited for both meat and labour power and by-products. While goats and sheep appear to have been mainly exploited for secondary products given the higher incidence of adult individuals.

### *3.2 The faunal assemblage from the 9th-8th century BC*

The fauna coming from phase B, dated between the 4th and the 3rd century BC, is interesting too. In fact, 137 fragments have been analysed, 41% of which have been identified at taxonomic level and almost all belong to domestic fauna.

It is difficult with such a small assemblage to obtain useful data concerning the management of animal resources. Cattle are however slightly preponderant compared to those of pigs and sheep/goat. Estimating the MNI, pigs and cattle are attested with the same number of individuals. This data must be considered preliminary because of the scarcity of the assemblage and other taphonomic factors, both edaphic and anthropic. Concerning the skeletal representation, sheep/goat and pigs are represented by all the skeleton districts. Pigs show some gaps especially in the forelimbs. The estimated age classes do not allow for any interpretation. An interesting aspect of this phase is the presence of a complete dog skull recovered inside a pit of the building. Four other bone fragments were found associated with the dog's skull: a sheep's scapula, an unidentified fragment of skull, a pig's mandible and a bovine horn core. It was a very young individual, less than one year old, with part deciduous dentition. An interesting detail is certainly the fact that the skull and mandibles were still found in anatomical connection documenting that the skull was thrown inside the pit still with the soft tissues attached.

### *3.3 Taphonomy*

The faunal assemblage presents a good state of preservation of the bone surfaces with a high degree of fragmentation. Taphonomic analysis has shown that bone surfaces are predominantly affected by manganese oxides and modified by root-etching. The low percentage of other modifications such as weathering, exfoliation, erosion is likely to indicate that osteological remains have been rapidly buried. This is confirmed by the scarce traces left by carnivores that affect about 3% of the fauna sample of phase D. Despite the good conservation of the surfaces, the number of anthropogenic traces is quite small, just over 4% of the remains, which were mostly unidentified fragments. These are mostly linear striae located near the joints and therefore traceable to disarticulation for the exploitation of the carcass in order to obtain smaller portions suitable for cooking, although there is very few evidence of heat exposure. Only on a few remains of wild mammals, butchery traces were found. It's interesting that the bear also has evidence related to its slaughter, reinforcing the hypothesis that it was hunted and slaughtered. On the remains of red deer traces were found which mainly related to the manufacturing of animal hard material to produce handles or composite elements, despite there being a fragment of tibia with traces attributable to defleshing.

### *3.4 The dog skull from 'House of Verucchio 4'*

The dog remains are in a good state of preservation and were found in anatomical connection, lying above several sheep/goat and pig bones and near a cattle horn core facing east (Figure 3A). The skull belongs to a puppy as many bones of the skull are not fused. The fourth lower premolar is still visible in the crypt and not yet erupted, while the upper one is halfway out. In the left jaw, moreover, the dp2 is flanked by the second premolar.

According to Hasebe (1952), the skull's maximum length (175 mm) allows classification of the animal as a small-medium or medium-sized dog, such as on the basis of the jaw's length (a medium-sized animal of 135 mm long). Considering the teeth eruption stage of the dog (Silver 1969), its age at death was about six months old.

In the upper part of the occipital condyles there are two cut-marks, which based on their direction and size are referred to a single slaughtering gesture (Figure 3C-c). On the internal margins of the occipital there are some small fractures produced by a forced disarticulation aimed at separating the skull from the column, or when the animal was killed. On the right jaw, just at the base of the canine, there is a cut-mark with a double exit point made by a metal blade. The mark shows the V-shaped section without secondary striations at the overlapping point of the two traces, suggesting a repeated action (Figure 3F-f).

The animal was intentionally slaughtered, skinned without removing the flesh. Moreover, the upside-down position of the skull seems not to be random since it is unusual and doesn't compare with other archeological contexts. The absence of other skeletal parts suggests a different treatment of those body portions. All these data allow us to suppose that the pit and its contents could have had a ritual purpose.

## 4 Conclusions and discussions

The new preliminary results, obtained from the study of the fauna of Verucchio Pian del Monte, give an important contribution to the reconstruction of the management of animal resources during the Early Iron Age in Emilia-Romagna. Preliminary data suggest that the economy of Verucchio was mainly focused on sheep or goat and pig farming.

A direct comparison with other sites was not possible because in the territory the published archeozoological data refer almost exclusively to settlements dated from the 6th century BC. Between the 8th and 7th century, breeding seems to concentrate on the exploitation of the sheep or goats and pigs while hunting plays a secondary role in the economy, perhaps mostly to defend cultivation. The presence of the bear is interesting, because it is a species not particularly frequent in the faunal assemblages of the period. At the moment it is not possible to assess the role of fishing in the urban economy. However, it is reasonable that this activity was practiced as a complement to the food resources. The ratio of the main domestic mammals indicates an economy dedicated to sheep or goat farming but with a clear tendency to increase the population, justified also by the number of pigs (NISp and MNI). The faunal assemblage from the 4th and 3rd century does not give new data about the economy, which, based on the edited (Farello 1997; 2006), shows an increase in the frequency of pigs, with respect to other domestic mammals, as happens in many other areas.

In the Ancient World the use of dogs in ritual practices could have played many symbolic roles, which can be synthesised in two general categories (Bodson 1980; Zaganiaris 1975; De Grossi Mazzorin and Minniti 2006; De Grossi Mazzorin 2008): the first one connects the animal sacrifice to Chthonic gods related to procreation, growth, and purification and the second one is linked to the role of a dog in everyday human life as a companion and guardian.

We know most of the information about Etruscan religion and ritual practices thanks to artistic representations, archaeological data and Roman texts.

In Etruscan culture, dogs are associated with Calu's cult, the god of the netherworld. Two paintings discovered in the tomb of Golini (Orvieto) and Orco (Tarquinia), represent the god of the dead bearing a dog-head hat (Kunée). Etruscan religion, like many aspects of culture, is enclosed in mystery and speculation, but we know some features thanks to the *Eugubinae Tabulae* (2nd century BC).

The remains of a dog, showing butchery marks, can be also referred to the agrarian rituals in which the processing of the animal and the use of some of its parts for the ceremony are often mentioned. In the Roman world, the sacrifice of a puppy near the city gates suggested the symbolic function of a dog-keeper (De Grossi Mazzorin and Minniti 2006; De Grossi Mazzorin 2008), but usually the entire skeleton of the animal was buried.

In this archaeological contest the dog's head deposited in the house with other skeletal elements of domestic animals could indicate several symbolic significances. Thus, the idea that it could be interpreted as an inaugural meal (Harari *et al.* 2017), related to the dog-keeper symbolism, may be the most plausible hypothesis.

### Acknowledgment

The authors are grateful to Prof. Maurizio Harari for the study of the fauna assemblages and Dr Lorenzo Zamboni and Dr Paolo Rondini for the archaeological documentation.

### References

Barone, R. 1976. *Anatomie comparée des mammifères domestiques*. Paris: Vigot Freres Editeurs.

Bodson, L. 1980. Place et fonction du chien dans le monde antique, *Ethnozootechnie* 25: 13–21.

Boessneck, J. 1969. Osteological differences between sheep (*Ovis aries*, Linné) and goat (*Capra hircus*, Linné), in D. Brothwell and E.S. Higgs (eds), *Science In Archaeology*: 331–358. Thames And Hudson, London.

De Grossi Mazzorin, J. 2008. L'uso dei cani nel mondo antico nei riti di fondazione, purificazione e passaggio, in: F. D'Andria, J. De Grossi Mazzorin and G. Fiorentino (eds) *Uomini, piante e animali nella dimensione del sacro*, BACT, 6: 71–81.

De Grossi Mazzorin, J. and C. Minniti 2006. Dog sacrifice in the Ancient World: a ritual passage?, in *Dogs and People in Social, Working, Economic or Symbolic Interaction*, 9th ICAZ Conference, Durham 2002: 62–66. Oxbowbook.

Driesch von den, A. 1976. A guide to the measurement of animal bones from archaeological sites. *Peabody Museum Bulletins* 1: 1–148.

Farello, P. 1997. Reperti faunistici da Verucchio - Scavo 1963, in *Archeologia dell'Emilia-Romagna* I (1): 41–52.

Farello, P. 2006. I reperti faunistici, in G. Bottazzi and P. Bigi (eds), *I primi insediamenti sul Monte Titano. Scavi e Ricerche (1997 2004)*: 185 190. All'Insegna del Giglio.

Grant, A. 1982. The use of tooth wear as a guide to the age of domestic animals, in B. Wilson, C. Grigson and S. Payne (eds), *Ageing and sexing animal bones from archaeological sites* (British Archaeological Reports British Series 109): 91–108. Oxford: Archaeopress.

Harari, M., P. Rondini and L. Zamboni 2017. L'abitato di Verucchio. Spazio insediativo e azioni cerimoniali, in E. Govi (ed.), *La città etrusca e il sacro. Santuari e istituzioni politiche*, Atti del Convegno: 25–50. BUP.

Hasebe, K. 1952. Dog skeletons, in Bunkazai Hogo Iinkai (ed.) *Board of the Protection of Cultural Properties*: 145–150 Yoshigo Kaizuka (in Japanese).

Menache, S. 1997. Dogs: God's worst enemies? *Society & Animals* 5(1): 23–44.

Pales, L. and C. Lambert 1971. *Atlas Ostéologique pour servir à l'identification des Mammifères du Quaternaire*. Paris: Editions du centre national de la recherche scientifique.

Schmid, E. 1972. *Atlas of animal bones for Prehistorians, Archaeologists, and Quaternary Geologists*. Amsterdam, London, New York: Elsevier Publishing Company.

Silver, I.A. 1969. *The ageing of domestic animals*, in D. Brothwell and E.S. Higgs (eds), *Science in Archaeology*: 283–302. Thames And Hudson, London.

Zaganiaris, M.J. 1975. Sacrifices de chiens dans l'Antiquité classique, *Platon* 27: 322–329.

Zamboni, L. 2018. L'abitato di Verucchio nella prima Età del Ferro, *Studi Romagnoli* LXVIII: 381–400.

Zeder, M.A. and H.A. Lapham 2010. Assessing the reliability of criteria used to identify postcranial bones in sheep, *Ovis*, and goats, *Capra*. *Journal of Archaeological Science* 37(11): 2887–2905.

Zeder, M.A. and S.E. Pilaar 2010. Assessing the reliability of criteria used to identify mandibles and mandibular teeth in sheep, *Ovis*, and goats, *Capra*. *Journal of Archaeological Science* 37(2): 225–242.

# 4.6 The Dogs from the Cult Layers of the *Ipogeo del Guardiano* (Trinitapoli, Barletta-Andria-Trani, Italy)

Martina Di Matteo[1], Anna Maria Tunzi[2], Rachele Modesto[3], Francesca Alhaique[4]

[1]Department of Classics, Sapienza University of Rome, Piazza A. Moro, 5, Rome, Italy, martina.dimatteo@uniroma1.it

[2]Soprintendenza Archeologia, Belle Arti e Paesaggio per la città metropolitana di Bari, Via Pier l'Eremita, 25/B, Bari, Italy, annamaria.tunzi@cultura.gov.it

[3]Department of Classics, Sapienza University of Rome, Piazza A. Moro, 5, Rome, Italy, rachele.modesto@gmail.com

[4] Archaeozoology Division, Bioarchaeology Service, Museum of Civilisations, Piazza G. Marconi 14, Rome, Italy, francesca.alhaique@cultura.gov.it

Corresponding author: Martina Di Matteo, martina.dimatteo@uniroma1.it

**Abstract**

The faunal assemblage from the 'Ipogeo del Guardiano' is characterised by the marked prevalence of domestic animals over wild ones. Among the domestic *taxa*, the dog is the most frequent followed by ovicaprines, cattle and pigs. Wild taxa are represented by cervids, foxes, hares, fish, tortoises, birds, and mollusks. All dog elements and human remains come from phase 2, suggesting an important role of the dog within funerary rituals. The interdisciplinary study of different aspects of the material culture will help to outline the dynamics of the rituals performed inside the hypogeum and to clarify their evolution through time.

**Keywords**: *Canis familiaris*, Bronze Age, hypogeum, funerary rituals, Southern Italy.

## 1 Introduction

In 2016 and 2017 two excavation campaigns were carried out in the *Parco Archeologico degli Ipogei* at Trinitapoli (Madonna di Loreto locality, Barletta-Andria-Trani province, Apulia), by the former Soprintendenza per i Beni Archeologici della Puglia in collaboration with Sapienza University of Rome (within the doctoral project of RM), in order to investigate the Bronze Age underground structure called *Ipogeo del Guardiano* (Figure 1).

The phenomenon of underground cult activities is well attested in Southern Italy during the 17th – 16th cent. BCE (e.g., Cipolloni 1986, 1998a; Recchia 1993, 1999a, 1999b; Tunzi Sisto 1990, 1998a, 1999; Tunzi Sisto, Langella 1995; Recchia, and Tunzi Sisto 2003).

In some cases, the hypogeal structures changed their function over time, shifting from places where ritual activities were carried out to collective tombs, generally around the 15th cent. BCE (Tunzi Sisto 1998b, 2001; Vanzetti 1999; Peroni *et al.* 2003).

## 2 The underground structure

The *Ipogeo del Guardiano* (Tunzi *et al.* 2017, 2018; Di Matteo 2018; Modesto 2019; Modesto *et al.* 2020) may be defined as a 'hypogeum with articulated plan', consisting of an open sloping entrance corridor (*Corridoio* 1, ca. 3 x 0.80 m) followed by an underground one (*Corridoio* 2, ca. 4.5 x 0.80 m) that finally opens into a Chamber (ca. 6 x 2.20 m; Figure 2), where ritual practices presumably took place. The corridors and the chamber are not aligned along an axis, suggesting a clear intentionality of the constructors, since there are no obstacles or traces that could suggest an adaptation to the available space or the inability to follow a straight line. Probably the construction choices are directly linked to the ritual that was practiced inside the hypogeum making the journey more complicated: moving from the entrance towards the central room, a person would go further and further away from the natural light passing through dark, narrow spaces.

Considering the dynamics related to the construction of the structure, it appears that the Chamber at first resulted in the filling of a Neolithic ditch intercepted in its southern portion. In fact it should be remembered that within the *Parco Archeologico degli Ipogei* at Trinitapoli there are not only evidences dated to the second millennium BCE (period of the hypogea), but also more recent ones (pits for planting, wells etc.) and more ancient structures. Among the earlier ones, there is a ditch referable to one of the many Neolithic villages characteristic of this area, which is dated approximately to the sixth millennium BCE (Tunzi Sisto 1999b). In the earliest phases of use, the terminal part of the Chamber of the *Ipogeo del Guardiano*, was dug into the filling of the aforementioned Neolithic ditch, but

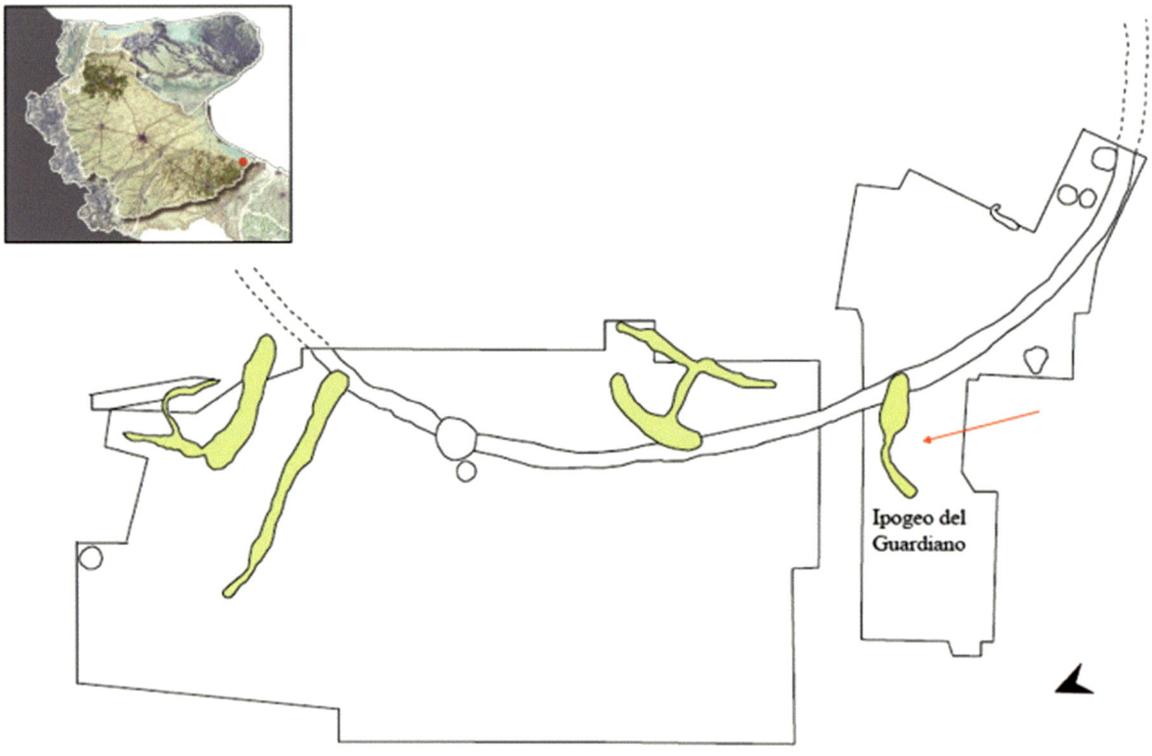


Figure 1. Map of the *Parco Archeologico degli Ipogei* at Trinitapoli (After Tunzi *et al.* 2017).

later, probably because of the instability of the walls, the area of the Chamber was reduced, moving back the limits to the area under the *crusta*.

On the basis of the ceramic assemblages, the *Ipogeo del Guardiano* may be placed within the context of the *Proto-apennine* culture (18th - 15th cent. BCE). In contrast to the other underground structures in the area, this hypogeum did not change its function through time, maintaining its ritual use until the last phases of utilisation.

Three Phases have been recognised during the excavations:

> ***Phase 1,*** the earliest one, is characterised by activities related to the use of fire, testified by the presence of a combustion structure; another peculiar occurrence in this phase is the high frequency of red deer antlers.
>
> ***Phase 2,*** is characterised again by activities related to the use of fire, but in this case combustion structures are absent. This phase yielded many human remains, especially skull portions along with scattered remains of the post-cranial skeleton, never in anatomical connection. Most of the faunal remains were collected from this phase.
>
> ***Phase 3,*** represents the final 'filling' of the structure; this may have occurred either naturally or as a result of voluntary action, perhaps connected to the 'closure' ritual that formally determined the end of the utilisation of the hypogeum.

### 3 The faunal remains

The faunal assemblage collected during the excavations at the *Ipogeo del Guardiano* includes a total of 1009 remains and 906 of them belong to the three phases (Di Matteo *et al.* 2018, Tunzi *et al.* 2018). The distribution of the faunal remains in the different phases is not homogeneous and most of the specimens were recovered in Phase 2. The preliminary archaeozoological analysis evidenced a high degree of fragmentation. Considering only the sample from the three phases, 52.1% of the specimens have been taxonomically identified to species or at least class level.

In general, domestic mammals show a marked prevalence over wild ones. Among the domestic *taxa*, the dog is the most frequent, although its remains are referable to relatively few individuals; this species is followed by ovicaprines, cattle and pigs. All the dog specimens were recovered in phase 2 in association with the human skeletal elements. As far as the wild *taxa* are concerned, it is worth mentioning that in phase

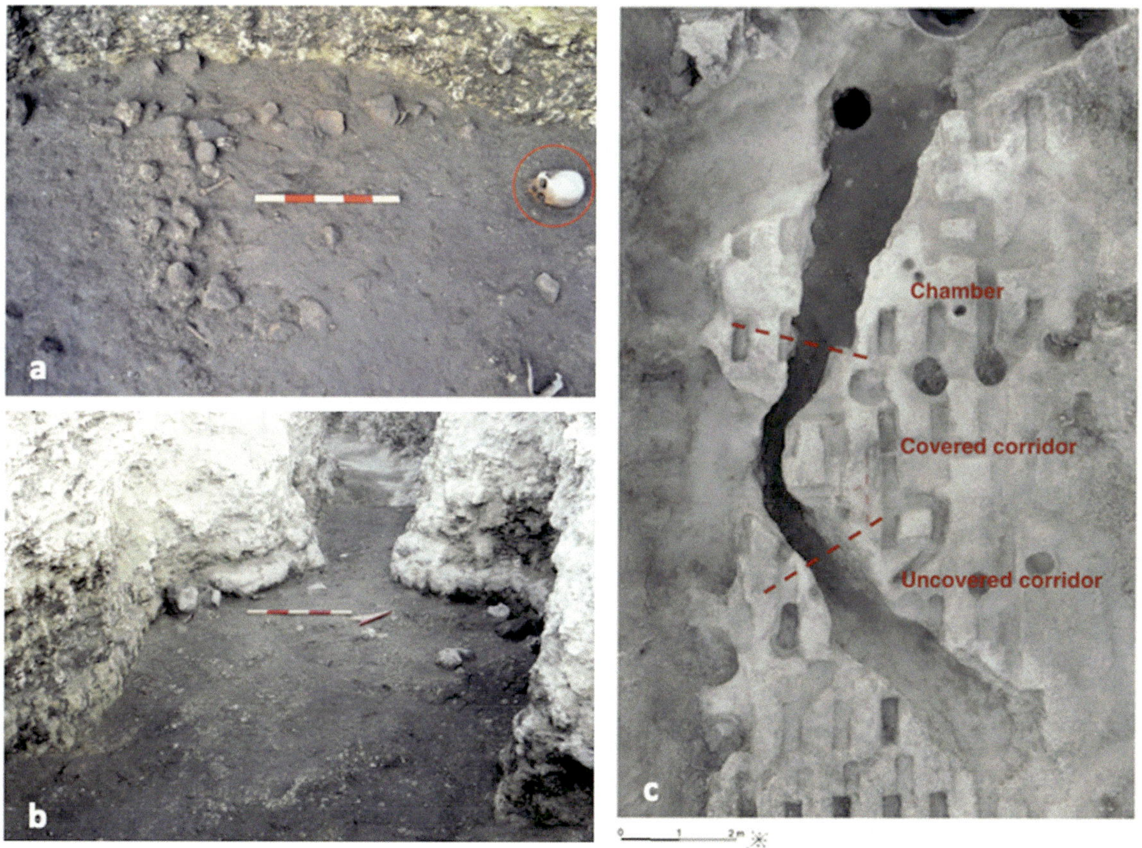


Figure 2. One of the archaeological levels (US 53) with abundant faunal material and remains of human skulls (a); one of the archaeological levels (US 58) with evidence of abandonment (b). Zenithal photo of the hypogeum during excavation (c) (After Tunzi *et al.* 2017).

1 the most frequent mammal is red deer represented by 7 antler portions all belonging to different individuals; on the whole this species is followed by roe deer, foxes and hares; however, the latter two species may probably be considered as intrusive, not intentionally introduced by humans in the archeological deposit. Furthermore, there are also some fish remains (gilthead bream and flathead grey mullet), tortoise elements, and bird bone fragments.

## 4 The dogs from Phase 2

Phase 2 is characterised by numerous dog specimens (NISP 221) that, as mentioned before, are totally absent from the rest of the archaeological sequence in this hypogeum. The remains are concentrated in the front and central portion of the Chamber (so-called Camera A and B, cf. Modesto 2019), but during the excavation no patterned spatial distribution has been recognised.

The analysis of the age at death indicates the presence of a minimum of 3 sub-adult individuals: at least one is perinatal, one is a very young puppy (between 2 and 6 months) and the latter is young. No adult animals have been identified so far.

The preliminary assessment of body part representation shows that, although with variable frequencies due to fragmentation, most of the skeletal elements are present for individual n. 1 (2–6 months) and n. 2 (perinatal), as shown in Figure 3 (a and b), although, probably because of sin- and post-depositional processes, the skeletal elements have not been found in anatomical connection.

The third individual, the young one, is represented only by very few fragments, in particular, portions of cranium.

This could suggest, especially for individuals 1 and 2, that very likely the whole carcass of these animals was originally transported into the hypogeum; however, it is not possible to exclude *a priori* the introduction of live animals that were then sacrificed in the underground structure during the ritual and were possibly subsequently consumed. In fact, the lack of cut-marks or other bone modifications on the dog elements may reflect either an actual absence, or may be due to the poor state of preservation of bone surfaces, considering that anthropogenic alterations are rare on the whole assemblage from this hypogeum.

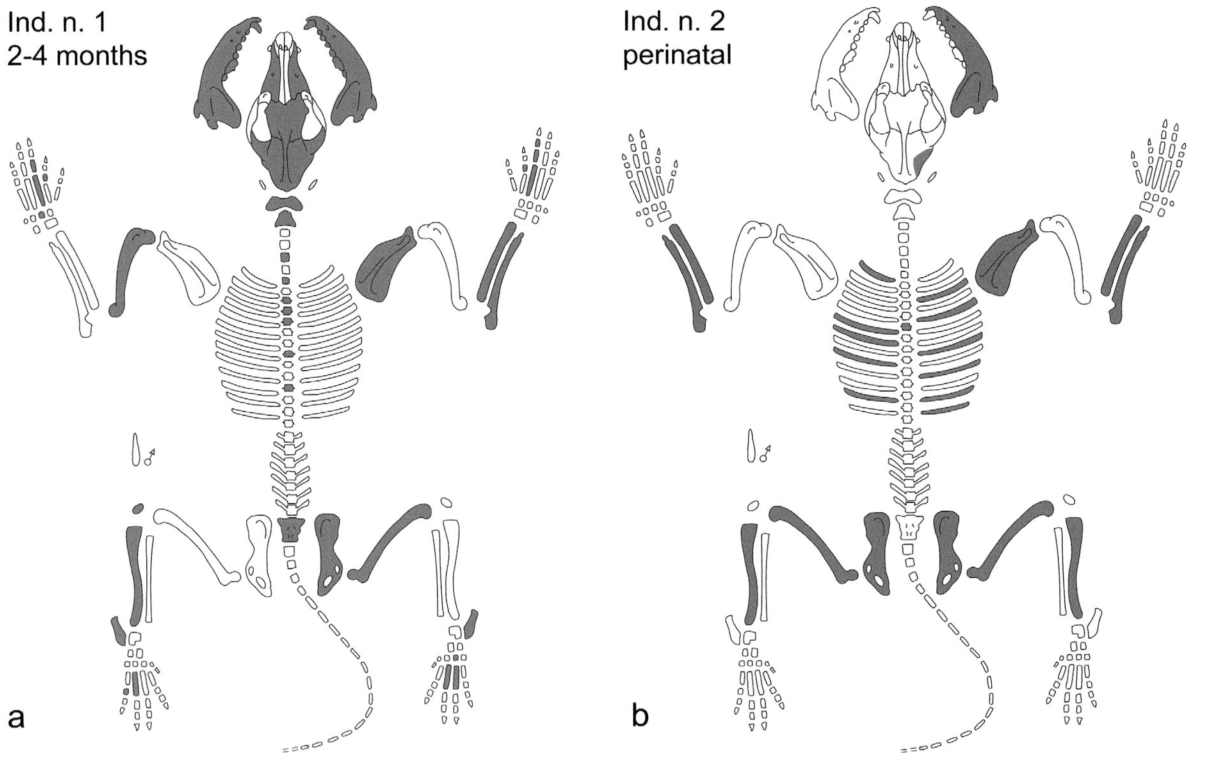


Figure 3. The body part representation of two individuals of *Canis familiaris* from Phase 2 (skeletons modified from ArcheoZoo.org/Michel Coutureau (Inrap) 2013).

## 5 Conclusions

The preliminary archeozoological data from *Ipogeo del Guardiano* indicate that the faunal composition, as well as other aspects of the animal assemblage (e.g., age selection), were significantly influenced by the ritual use of the hypogeum.

In particular the strong association in Phase 2 between human and dog remains points to a great significance of this species in cult activities. Furthermore, other *taxa* represented also by young animals, may have been relevant as well, especially ovicaprines and pigs that could be referred to some kind of ritual banquet or funerary offering.

The dog is known to have had a strong symbolic value and is often found in funerary contexts, from the Palaeolithic to the Roman age and beyond, precisely because of its leading role in life after death. For example, remains of dogs of perinatal age come from a funerary context coeval to the hypogeum, tomb 743 at Lavello (Basilicata) (Wilkens 1994).

The burial of at least one dog in a minor hypogeal structure at site 2 at the Diga del Rendina suggests a close relationship between hypogeism and dog burials (Cipolloni Sampò 1998b). Furthermore, sporadic remains of dogs have been found in Hypogeum 2 at Terra di Corte (San Ferdinando di Puglia; Oronzo 1995) and in the Fermatreccia Hypogeum (Minniti 2013–2014), also in the *Parco Archeologico degli Ipogei* at Trinitapoli.

Details on the rituals and their changes through time will be more deeply investigated and understood when the integration of archaeological, anthropological and faunal data will be completed.

## References

Cipolloni Sampò, M. 1986. La tomba tre dell'acropoli di Toppo Daguzzo (Potenza), elementi per uno studio preliminare. *Annali dell'Istituto Universitario Orientale di Napoli* VIII: 1–49.

Cipolloni Sampò, M. 1998a. Lavello: Ipogeo della Speranza, in *Scavi e ricerche archeologiche dell'Università di Roma 'La Sapienza'*, Catalogo della mostra: 190–192. Roma.

Cipolloni Sampò, M. 1998b. Ipogeismo funerario e cultuale nella Daunia meridionale, in A. Gravina, (ed.) *Atti del 19° Convegno sulla Preistoria, Protostoria e Storia della Daunia*, San Severo: 129–137.

Di Matteo, M., A.M. Tunzi, R. Modesto and F. Alhaique 2018. Primi risultati archeozoologici dall'Ipogeo del

Guardiano (Trinitapoli, BAT), poster presented at *9° Convegno Nazionale di Archeozoologia*, Ravenna, 28 novembre-1 dicembre 2018.

Minniti, C. 2013–2014. Resti animali dall'Ipogeo dei Fermatreccia, in R. Modesto, *Trinitapoli - Ipogeo dei Fermatreccia. Un Ipogeo dell'età del Bronzo*, unpublished dissertation School of Specialization in Archaeological Heritage, Sapienza University of Rome: 183- 186.

Modesto, R. 2019. La produzione ceramica degli ipogei dell'età del Bronzo di Trinitapoli e San Ferdinando di Puglia, unpublished PhD dissertation, Sapienza University of Rome.

Modesto, R., G. Eramo, I.M. Muntoni and A.M. Tunzi, A.M. 2020. Vasi interi o già rotti? Analisi morfometrica dei frammenti ceramici provenienti dagli Ipogei dell'età del Bronzo del Guardiano e dei Fermatreccia di Trinitapoli (BT), in A. Gravina (ed.) *Atti del 40° Convegno Nazionale sulla Preistoria, Protostoria e Storia della Daunia*: 227–240. San Severo.

Peroni, R., B. Barbaro and A. Vanzetti 2003. I materiali del nuovo ipogeo di Trinitapoli, in A. Gravina (ed.) *Atti del 23° Convegno sulla Preistoria, Protostoria e Storia della Daunia*, San Severo: 287–320.

Recchia, G. 1993. Grotta Manaccora (Peschici), considerazioni sulla Grotticella funeraria e sull'area antistante (scavi Rellini-Baumgärtel). *Origini*, XVII: 317- 401.

Recchia, G. 1999a. Rituale funerario e aspetti sociali a Grotta Manaccora e negli ipogei sepolcrali delle aree circostanti durante l'età del Bronzo, in A. Gravina (ed.) *Atti del 19° Convegno sulla Preistoria, Protostoria e Storia della Daunia*, San Severo: 21–50.

Recchia, G. 1999b. Una situazione di confronto per l'ipogeo dei Bronzi: il rituale funerario a Grotta Manaccora durante la media età del Bronzo, in A.M. Tunzi Sisto (ed.) *Ipogei della Daunia. Preistoria di un territorio*, Foggia: 281–283.

Recchia, G. and A.M. Tunzi Sisto 2003. Alcune note sull'articolazione interna di Grotta Manaccora durante l'Età del Bronzo, in A. Gravina (ed.) *Atti del 23° Convegno sulla Preistoria, Protostoria e Storia della Daunia*, San Severo: 339–348.

Oronzo, S. 1995. Analisi di un campione di resti faunistici dell'età del bronzo provenienti dall'ipogeo 2 in località Terra di Corte (S. Ferdinando di Puglia), in A. Gravina (ed.) *Atti del 16° Convegno sulla Preistoria, Protostoria e Storia della Daunia*, San Severo: 57–66.

Tunzi A.M., M. Lo Zupone, R. Modesto, V. Mironti, A. Zupancich, I. Caricola, C. Minniti and F.M. Martino 2016. *Living and Dead Underground: the case study of the 'Fermatreccia' Hypogeum of Trinitapoli (BT), Italy*, presented at 7th Conference of Italian Archaeology, Galway (Ireland).

Tunzi Sisto, A.M. 1990. L'ipogeo di San Ferdinando di Puglia, in A. Gravina (ed.) *Atti del 11° Convegno sulla Preistoria, Protostoria e Storia della Daunia*, San Severo: 129–137.

Tunzi Sisto, A.M. 1998a. Terra di Corte (San Ferdinando di Puglia, Foggia): l'ipogeo n. 2, in A. Gravina (ed.) *Atti del 16° Convegno sulla Preistoria, Protostoria e Storia della Daunia*, San Severo: 21–55.

Tunzi Sisto, A.M. 1998b. L'ipogeo dei bronzi di Trinitapoli, in A. Gravina (ed.) *Atti del 9° Convegno sulla Preistoria, Protostoria e Storia della Daunia*, San Severo: 77–86.

Tunzi Sisto, A.M. 1999a. San Ferdinando di Puglia, in A. M. Tunzi Sisto (ed.) *Ipogei della Daunia. Preistoria di un territorio*, Foggia.

Tunzi Sisto, A.M. 1999b. Il villaggio neolitico di Madonna di Loreto, in A.M. Tunzi Sisto (ed.), *Ipogei della Daunia. Preistoria di un territorio*, Foggia: 131–133.

Tunzi Sisto, A.M. 2001. L'ipogeo degli Avori di Trinitapoli,in A. Gravina (ed.) *Atti del 21° Convegno sulla Preistoria, Protostoria e Storia della Daunia*, San Severo: 253–274.

Tunzi Sisto, A.M. and M. Langella 1995. La grotticella trilobata di Madonna di Grottole. Taras XV, 2, Bari: 291–311.

Tunzi Sisto, A.M. 2005. *L'ipogeismo minore di Trinitapoli*, in A. Gravina (ed.) *Atti 25° Convegno sulla Preistoria, Protostoria e Storia della Daunia*, San Severo: 189–198.

Tunzi, A.M., R. Modesto, M. Lo Zupone and V. Mironti, V. 2017. L'Ipogeo del Guardiano (Trinitapoli, BT), *Notiziario di Preistoria e Protostoria*: 59–61.

Tunzi, A.M., R. Modesto, F. Alhaique, M. Di Matteo, M. Lo Zupone and V. Mironti, V. 2018. Nuove indagini nell'ipogeo dell'età del Bronzo del Guardiano (Trinitapoli, BT): considerazioni preliminari, in A. Gravina (ed.) *Atti del 38° Convegno sulla Preistoria, Protostoria e Storia della Daunia*, San Severo: 273–290.

Vanzetti, A. 1999. Combinazioni di corredo delle sepolture all'interno dell'ipogeo dei Bronzi di Trinitapoli, in A.M. Tunzi Sisto (ed.) *Ipogei della Daunia. Preistoria di un territorio*, Foggia: 222–226.

Wilkens, B. 1994. I resti faunistici della tomba 743 di Lavello (Basilicata), in *L'Ipogeismo nel Mediterraneo: origini, sviluppo, quadri culturali: atti del Congresso internazionale*, (23–28 maggio 1994, Sassari-Oristano, Italia). Sassari, Università degli studi di Sassari, Facoltà di Lettere e filosofia, Istituto di Antichità, arte e discipline etnodemologiche e Dipartimento di Scienze Umanistiche e dell'Antichità, 2: 667–670.

# 4.7 Four Dogs in the Road and Other Canine Oddities from Gabii (Rome, Italy)

Francesca Alhaique

Archaeozoology Division, Bioarchaeology Service, Museum of Civilisations, Piazza G. Marconi 14, Rome, Italy, francesca.alhaique@cultura.gov.it

**Abstract**

The ancient town of Gabii (Rome, Italy) was occupied from the 10th century BCE until the 3rd century CE. The rich faunal assemblage suggests that the economy was based on the three main domestic *taxa*, while equids and birds are rarely present; aquatic resources and wild mammals were only occasionally exploited. Dogs were recovered in many contexts and also include unusual findings such as depositions/burials, bones with cut marks and elements used for craft purposes. All these evidences indicate a special relationship, possibly not just utilitarian, between this species and the inhabitants of Gabii.

**Keywords:** Roman period, dog interments, rituals, human modifications, Central Italy.

## 1 Introduction

The ancient town of Gabii is located about 18 km E of Rome (Italy) along the *via Prenestina*, on the slopes of a former volcanic lake, *Lacus Gabinus,* later known as *Lago di Castiglione.* The site was occupied from at least the 10th century BCE until its decline in the 2nd and 3rd century CE.

The archaeological investigations (*Gabii Project*) in some areas of this settlement have been carried out since 2007 under the direction of Prof. Terrenato (University of Michigan) (see Becker *et al.* 2009; Mogetta, Becker 2014, and references therein).

## 2 The Gabii faunal assemblage

The excavations of the *Gabii Project* have yielded a very large faunal assemblage whose analysis is still in progress. The evidences collected so far indicate that, as expected, the proportions among species are variable within the settlement according to the time period and area (Moses and Alhaique 2022; for data on Area B and A see Alhaique 2016, 2021; for the Archaic phases of area D see Moses 2020).

In general, the economy at Gabii was based on the three main domestic *taxa* (pigs, ovicaprines, and cattle), the latter two were employed not only as a source of meat, but also for secondary products and as animal power. Equids, both horse and donkey, although rare, have also been identified. Dogs have also been recovered in almost every context or, when the actual bones were not found, their presence could be indirectly inferred from the presence of gnaw marks. Birds, especially chicken, that were probably reared locally, were a supplement to the diet, while aquatic, mainly marine, resources (mollusks, fish) (Alhaique *et al.* 2016) and wild mammals were exploited only occasionally. Furthermore, animal remains have sometimes been found as offerings in human burials (Motta *et al.* 2020).The rare occurrence of specimens belonging to unusual *taxa* (lion, leopard, bear, beaver, vulture) should also be mentioned; for the large wild carnivores, represented only by phalanges, the presence of pelts has been suggested, while cut marks on a beaver humerus indicate the occasional exploitation of this species as a food source (Alhaique 2019).

## 3 The Gabii dogs

### *3.1 The four dogs from Area F*

During the 2012 excavations, the partially articulated skeletal elements of at least four dogs were recovered within the fill, referable to the Imperial period, of a road of Republican age running along the NE wall of a large monumental public building in area F (Johnston *et al.* 2018).

The animals were of different ages and sizes (Figure 1): dog 1 was 9–10 months old with an estimated shoulder height of 55 cm, but still growing; dog 2 was 6–7 months old and of unknown size; dog 3 was 2–3 years old with a shoulder height of 61 cm; dog 4 was older than 3 years of age and had an estimated shoulder height of 44 cm.

The skeleton of dog 1 was relatively complete, while the other three are represented by fewer elements. These individuals were concentrated within a relatively small area (ca. 2 x 2 m) and were associated with some other

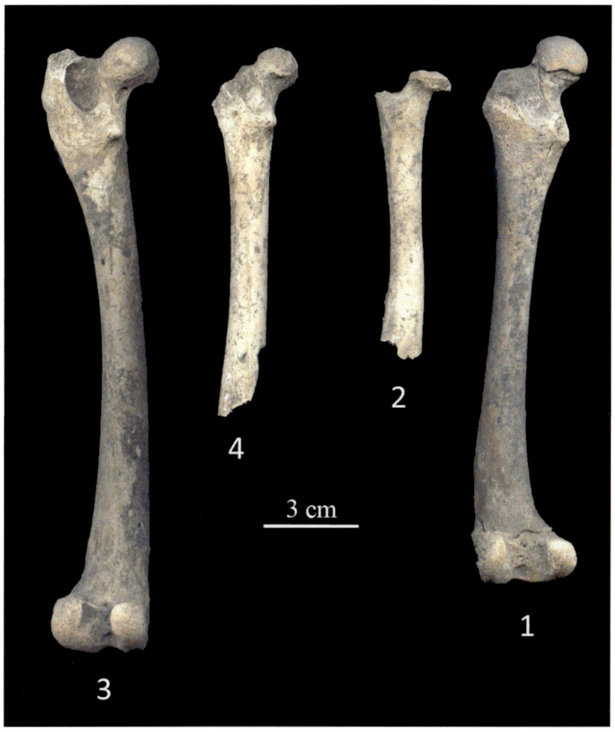


Figure 1. Area F dogs: left femurs of the four individuals identified (Photos by F. Alhaique).

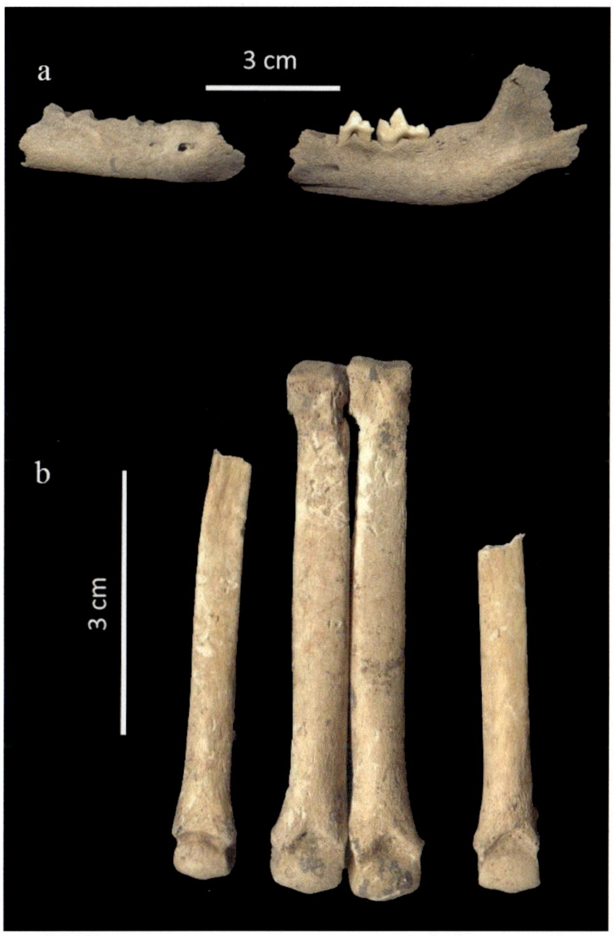


Figure 2. Area C House dogs. a, mandibles of the puppy. b, metacarpals of the adult individual, (Photos by F. Alhaique).

'more usual' domestic animal remains that represent food refuses or in a few cases crafting debris; the only exception is a fox humerus, which was probably a natural intrusive (i.e., not related to human exploitation) as supported by the lack of human modifications.

### *3.2 The dogs from the Area C House*

Some other dog remains were recovered in the *hortus* of the Area C House in layers related to the construction of this building (4th-3rd century BCE) (Mogetta and Opitz, forthcoming).

In this case a few specimens belonging to the puppy, about 2–4 months old (Figure 2a), and an adult dog, the latter about 51 cm at the shoulder (Figure 2b), were found in a hearth in association with an almost complete skeleton of a male lamb as well as with remains of very young ovicaprines and pigs (all unburnt) and other more common faunal elements.

### *3.3 Other Gabii dogs*

A dog burial, belonging to a 4–6 months old puppy, was found in Area A (Alhaique 2021) in a layer referable to the collapse and abandonment of the Republican house identified in this area.

The four dogs from Area F and those from the Area C and A do not display human modifications, but in other contexts of the site, a few dog specimens show cut-marks.

In Area C the traces on an occipital fragment (Figure 3a) indicate the disarticulation of the skull from the vertebral column, while other disarticulation marks were identified on a calcaneum from Area F (Figure 3b); in the first case the specimen was mixed with rubble in a layer referred to the construction of a road dated to the Republican period, in the second case the layer is within a structure in Area F and is dated, on the basis of the ceramic content, to the Republican-Early Imperial period.

Furthermore, occasional craft activities employing dog elements have been also documented at Gabii. Two pierced dog canines, used as pendants, have been recovered: one (Figure 3c) in Area B in a post-abandonment level referable to the Republican-Early Imperial period and another similar one in Area C in a collapse layer approximately dated to the Republican period. A dog proximal femur from Area A has been sawn off (Figure 3d) and therefore the shaft may have possibly been used as raw material for craft activities; this specimen, recovered in association with some other manufacturing debris (a cattle radius and

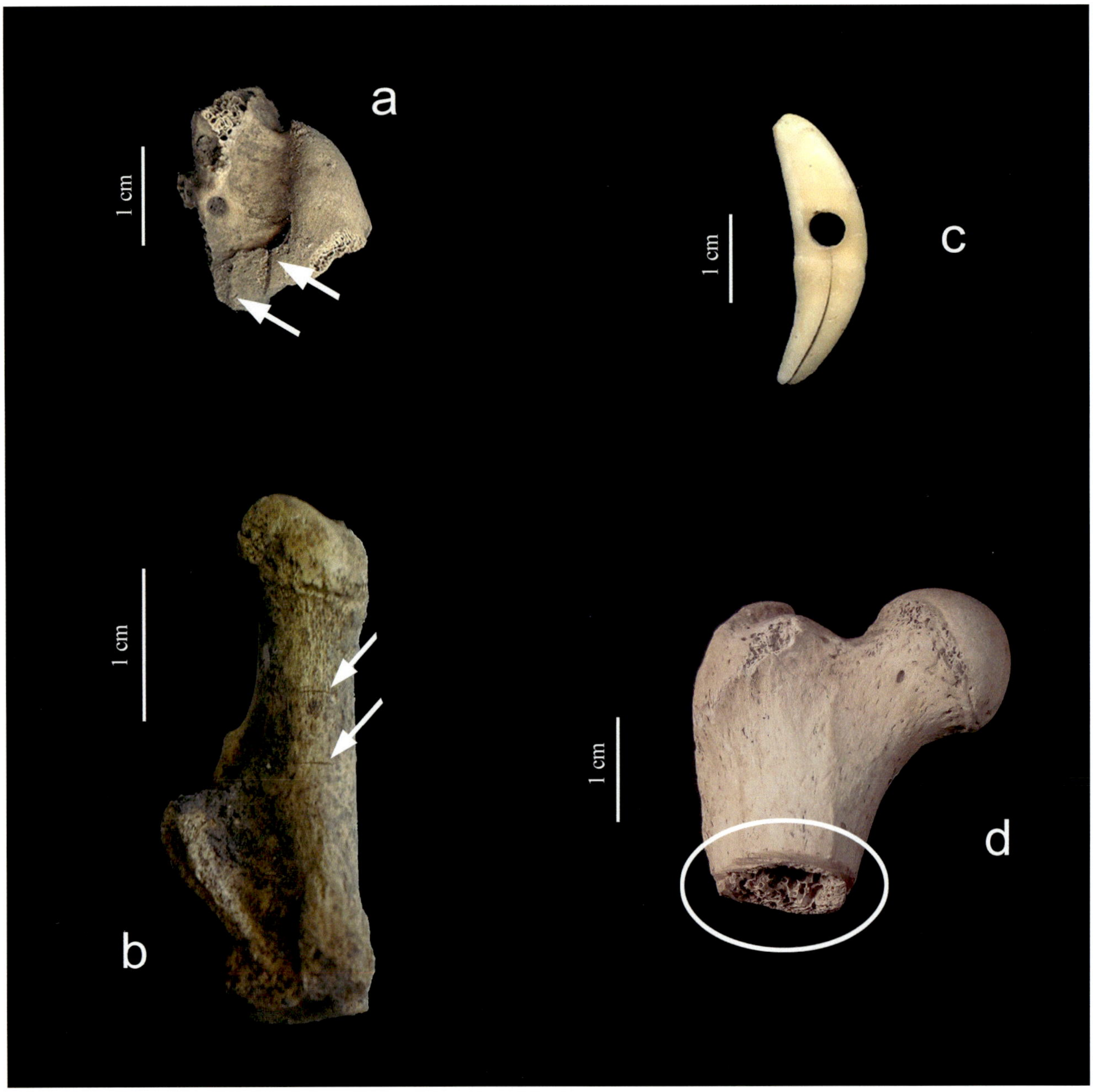


Figure 3. Dog specimens with human modifications. a, occipital condyle with cut marks (Area C). b, calcaneum with cut marks (Area F). c, canine pendant (Area B). d, proximal femur with saw marks (Area A), (Photos by F. Alhaique).

metapodial, Alhaique 2021), is possibly referred to the Imperial period.

**4 Discussion and conclusions**

The reason for the unusual concentration of dog remains near the corner of the monumental building in area F is not clear, especially since the layer in which the supposed depositions were cut represents an accumulation of rubble and debris when the street running along the wall of the building was no longer in use; it is possible that such interments represent just a way to get rid of the carcasses of these animals not a practice related to some kind of ritual or cult, although the concentration in such a small area of animals that are so different in size and age is certainly atypical. Similarly, the purpose of the Area A dog burial is still unclear, but in this case the anomaly may be represented by the young age of the animal.

The area C *hortus* context appears anomalous in itself for the presence in a hearth of the almost complete and unburnt skeleton of a lamb associated with the remains of other very young ovicaprines and pigs; the occurrence of the puppy may fit well within this scenario whose interpretation is however difficult. It is even more difficult to explain if the presence of the few adult dog remains is 'cultural' or accidental.

Regarding the few dog specimens with human modifications, the skeletal elements involved and the location of the marks suggest only disarticulation, but not meat removal therefore, at least so far, there is no positive evidence for the use of dogs as food. For the moment the data are too scanty to suggest the final purpose of such dismemberment, although possibly with the continuation of the analysis of the faunal assemblage of the site more evidences will emerge helping to clarify this issue.

The use of dog bones as raw material for craft purposes has, to my knowledge, not been documented in the Roman period; the Gabii example may indicate either a special meaning for the object obtained in the manufacturing process or, on the other hand, just be an occasional occurrence related to the exploitation of the material that was fit for the purpose and handy in that moment, since it is associated to other working debris, but of a larger animal; therefore in this latter case the specimen may just indicate the presence of a workshop nearby where all the available raw material was used regardless the species.

The use of teeth as pendants has been documented since the Palaeolithic, but it is rarer in historical times, although at Gabii, besides these two dog canines there are evidences of pierced pig female canines. Although in this case the ornamental purpose of the object is clear, the question of the deep meaning of the choice of this species is still unresolved. One possibility is their use as amulets since wolf and pig teeth have been used until recent times for children in order to develop strong and healthy dentition (see for example Bellucci 1881: 21); according to Pliny the Elder (Naturalis Historia ch. XXX) the longest tooth (possibly a canine) of a black dog could be used against quartan fevers (malaria).

Although rituals involving dogs are also well known in Roman times (e.g., De Grossi Mazzorin 2008), it is not possible to say if the four dogs in Area F, the ones from the Area C House or the butchered specimens, represented sacrificed animals or just occasional/ accidental occurrences.

Unusual dog finds although relatively rare are present in the ancient town of Gabii in different areas and periods, from the Republican to the Imperial age, suggesting a constant strict relationship between humans and dogs that cannot be explained only as their use as pets or 'co-workers' and indicating that for a long time this species had a special meaning in the culture of the people living in Gabii.

The 'odd' dog evidences from Gabii are actually raising more questions than answers but may represent a useful starting point for further investigating the variability of the long term relationship between this species and humans.

**Acknowledgements**

I am grateful to Nicola Terrenato, Marcello Mogetta, Anna Gallone, Laura Motta and all the members of the *Gabii Project* team for their support and useful discussions during the excavation and the analysis of the faunal remains from Gabii.

**References**

Alhaique, F. 2016. Zooarchaeological remains from the Tincu House at Gabii, in R. Opitz, M. Mogetta, N. Terrenato, A *mid-Republican House from Gabii*, (on line publication), Ann Arbor: University of Michigan Press. DOI: https://doi.org/10.3998/mpub.9231782.

Alhaique, F. 2019. The Gabii 'zoo': exotic and unusual animals from the Roman layers. *Atti 8° Convegno Nazionale di Archeozoologia*: 189–191.

Alhaique, F. 2021. Zooarchaeological remains from Area A, in L.M. Banducci and A. Gallone (eds) *A Cemetery and a Quarry from Imperial Gabii.* Gabii Project Reports 2. Ann Arbor: University of Michigan Press. DOI: https://doi.org/10.3998/mpub.11885571.

Alhaique, F., A. Crawford and L. Brancazi 2016. Preliminary data on the exploitation of aquatic resources at Gabii during the Roman period, presented at the 37th International Conference of the Association for Environmental Archaeology (Rome, September 29-October 1, 2016).

Becker, J.A., M. Mogetta and N. Terrenato 2009. A New Plan for an Ancient Italian City: Gabii Revealed, *American Journal of Archaeology*, 113(4): 629–642.

Bellucci, G. 1881. *Catalogo della collezione di amuleti inviata all'Esposizione Nazionale di Milano 1881*, Perugia: Tipografia Vincenzo Bartelli.

De Grossi Mazzorin, J. 2008. L'uso dei cani nel mondo antico nei riti di fondazione, purificazione e passaggio, in D'Andria F., De Grossi Mazzorin L., Fiorentino G. (eds), *Uomini, Piante e Animali Nella Dimensione del Sacro*: 71–81, Bari: Edipuglia.

Gallone, A. and M. Mogetta 2013. Gabii in età repubblicana: I rivestimenti pavimentali di alcune unità abitative, in C. Angelelli (ed.), *Atti del XVIII Colloquio AISCOM (Cremona, Italy, 14–17 March 2012)*: 717–725.

Johnston, A.C., M. Mogetta, L. Banducci, R. Opitz, A. Gallone, J. Farr, E. Casagrande Cicci and N. Terrenato 2018. A Monumental Mid- Republican Building Complex at Gabii. *Papers of the British School at Rome*, 86: 1–35.

Mogetta, M. and J.A. Becker 2014. Archaeological Research at Gabii, Italy: The Gabii Project Excavations 2009–2011, *American Journal of Archaeology*, 118 (1): 171–188.

Moses, V. 2020. The Zooarchaeology of Early Rome: Meat Production, Distribution, and Consumption in Public and Private Spaces (9th-5th cent. BCE). Unpublished Ph.D. dissertation, University of Arizona, Tucson.

Moses, V., Alhaique, F. 2022. Same Place, Changing Patterns? Animal Economy at Gabii (Latium, Central Italy) from the Early Iron Age through the Imperial Period. *Journal of Archaeological Science: Reports*, 46,103717 https://doi.org/10.1016/j.jasrep.2022.103717

Mogetta, M. and R. Opitz (eds) forthcoming. *A Domestic to Industrial Transition at Gabii*. Gabii Volume Reports 3. Ann Arbor: University of Michigan Press.

Motta, L., D. Fico, F. Alhaique and G. De Benedetto 2020. Offerings and rituals at the grave: insights from the macro- and micro-organic evidence, in M. Mogetta (ed.) *Élite burial practices and processes of urbanization at Gabii: the non-adult tombs from Area D of the Gabii Project excavations*, Journal of Roman Archaeology Suppl. 108: 115–124.

# 4.8 The Discovery of a Dog in the Excavations of the Rome Underground Line C in Largo Amba Aradam

Simona Morretta[1], Giovanni Ricci[2], Francesca Santini (†)[3]

[1]Soprintendenza Speciale Archeologia Belle Arti e Paesaggio di Roma, Piazza dei Cinquecento 67, 00185 Rome, ITALY, simona.morretta@cultura.gov.it

[2]Cooperativa Archeologia, Via Cairoli 88, 00185, Rome, Italy, gricci2012@libero.it

[3]Independent Researcher

Corresponding author: Simona Morretta, simona.morretta@cultura.gov.it

**Abstract**

During the archaeological excavations for the construction of the Rome underground Line C, in the Q15 Well 'Compensation grouting' of Largo Amba Aradam, some rooms of a building dating from the Hadrianic period building (early 2nd century AD) were found, probably pertinent to the ancient barracks found not far away (called 'Caserma di via Ipponio'). Beneath the wooden floor collapse which occurred as result of a fire that broke out in the middle of the 3rd century. AD, the almost complete skeleton of a dog was found. The dog is a young male of medium-large size, which by body structure can be assimilated to the modern breeds of the Setter, Pointer or Doberman and it can be traced back to a hunt or guard dog. The archaeozoological analysis and the archaeological investigation allowed the reconstruction of the events that occurred and the comparison with other remains found in similar situations.

**Keywords**: dog, chicken, domestication, biometric analysis, crushing-trauma.

## 1 Introduction

From December 2016 until the end of 2017 in Largo Amba Aradam the excavation of the Q15 well along the new Line C of the Rome Underground took place[1] (Figure 1). It is a 'compensation grouting' well, or a logistic structure to monitor the Aurelian Walls that are close by, to ensure their safety and stability during the underground tunnel excavation phases. Inside this structure, about 8 metres in diameter, archaeological excavation was carried out until the alluvial pre-anthropic depositions, which emerged at around 16 metres below the modern ground level.[2] This area includes the Southern slope of Celio Hill, which in the imperial age was the location of the top luxurious, aristocratic residences. Whereas, South of here a series of military buildings were located. Between these there were the famous military barracks recently unearthed in Via Ipponio during the excavation of the Amba Aradam Underground Station.[3]

In this occasion, taking into account the discussed topic, the focus will be on the area setting, mainly on the period between the Hadrian era and the second half of the 3rd century AD, which dates back to the animal bone remains.

## 2 The stratigraphic context

The two adjoining rooms (nos. 1–2) were built under Hadrian (first half of the 2nd century AD) (Figure 2). The Eastern wall in both rooms was built leaning against a previous wall in *opera reticolata*, which delimits a third room. Room 2 had a heating system and shows a mosaic floor made from white tesserae at 20,77 metres above sea level, in other words at around 9 metres deep. In Room 1 there is a mosaic with a black and white geometric motif featuring a double frame bordering a central emblem decorated with linked rings.[4] During the Antonine age (second half of the 2nd century AD) Room 3 undergoes substantial and significant modifications, affecting the ancient wall in *opera reticolata*. It was replaced by a partially redirected wall in *opera mista* (Figure 3).

[1] The excavation was carried out by Cooperativa Archeologia under the scientific supervision of Soprintendenza Speciale ABAP of Rome (Simona Morretta). S. Falzone handled the study of the wall paintings (*in situ* and fragmented ones) recovered during the archaeological investigations. F. Santini examined the animal bone remains. A special thanks needs to be made to Società Metro C S.p.A., to the Management Team Area and Roma Metropolitane for their constructive collaboration.

[2] Preliminary outputs in Falzone, Morretta, and Ricci 2019.

[3] Regarding the archaeological and topographic survey of Celio Hill see Colini 1994, Pavolini 2006. Concerning the new barracks in via Ipponio see Morretta, Rea 2018; Morretta, Rea 2020; Cardarelli *et alii* ongoing.

[4] The mosaic is currently being restored under the educational activities program of the Corso di Laurea in Conservazione e Restauro dei Beni Culturali of Tuscia University (Viterbo)/ Dipartimento per l'innovazione nei sistemi biologici, agroalimentari e forestali (DIBAF), following the agreement protocol between Soprintendenza and University.

Figure 1. Location of the excavation area (Photo from Google Earth, elaborated by A. Averini Cooperativa Archeologia).

Shortly afterwards (Severan age - first half of 3rd century AD), insubstantial, minor renovations were made in the neighbouring rooms 1–2, not leading to radical changes from the original building plan (Figure 4).

In Room 1 the ancient mosaic floor that was partially degraded has been restored using two different techniques. The perimeter band was patched up with *cocciopesto*, instead the gap on the central emblem and in some portions of the double row of frame boarding were repaired with coarse and unrefined mosaic integrations. At the same time, the entrance size along the western side was modified. A new N-S orientated wall placed at 1.80 metres from the northern jamb redefined the passage aspect, which was part of the original project design. Regarding Room 2, the restorations were exclusively focused on a limited portion of the western wall of which the facing and the plaster were rebuilt. Around the middle/second half of the 3rd century AD, Room 1 was filled with a huge amount of collapsed deposit, including burnt wooden slab elements (Figure 5), large fragments of the mosaics may be pertinent to the upper rooms or terraces,[5] mixed with plaster fragments from the walls and roof. All of these elements were in the primary deposition as they were linked to the architectural furniture in the room where they were recovered.

## 3 The archaeozoological analysis

The animal bone remains were recovered beneath this exceptional size of collapsed deposit. The articulated skeletons of two animals, a dog (*Canis familiaris* L., 1758) and a domestic fowl (*Gallus gallus* L., 1758) were found in the primary position placed close to the door communicating with Room 2, whereas disarticulated long bones of a hen were found in the opposite S-E corner of the same room (Figure 6).

### *3.1 Dog remains*

The dog skeleton is complete, in anatomical connection and is well preserved and largely intact. It was found in a rather twisted position.[6]

The special taphonomic relevance of these animals is that both the skeletons remained largely intact revealing the primary deposition, and probably traces and signs of the causes of death, attributable to the peri-mortem phase (Figure 7).

In effect, the skull is the skeletal portion that was the most heavily damaged, partly due to the intrinsic brittleness of the bone structure and partly due to the crushing-trauma caused by the floor collapsing and the

[5] The preliminary data about the precious mosaic restoration from the Tuscia University (see note 4) confirms that they belong to two different floors, covered or terraced rooms.

[6] The forward portion laid on the right side with retracted fore limbs and kept below the chest, and the rear portion crouched down flat on its belly to the ground, with spread hind limbs cross arranged one over the other and the tail between them.

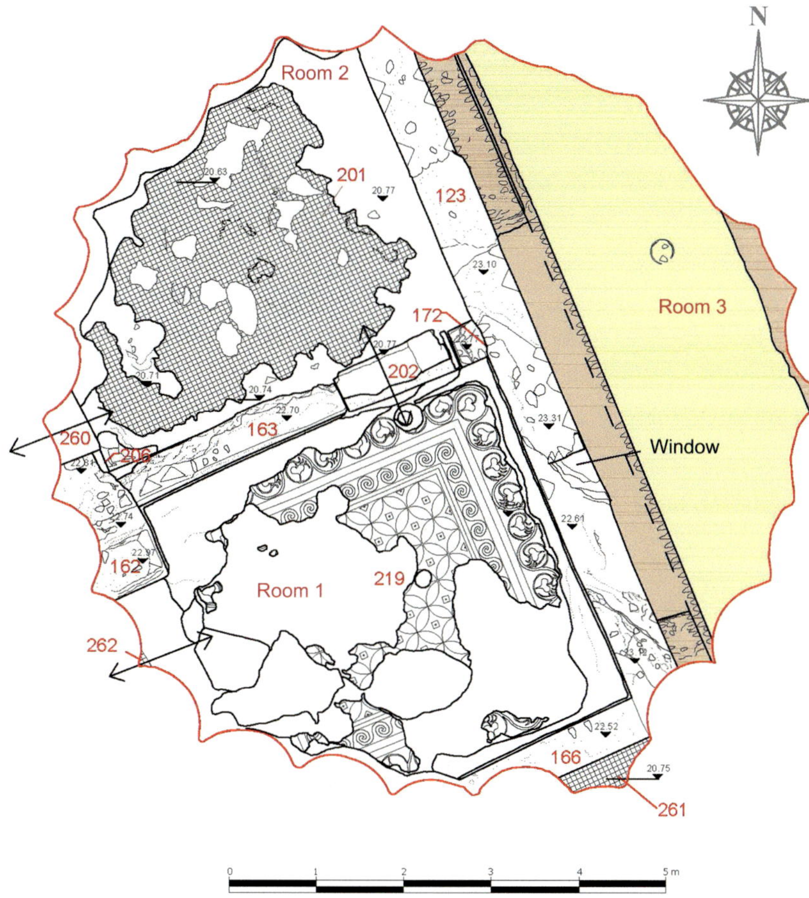


Figure 2. Hadrian building plan. In brown: the re-used walls, in yellow: the re-used walk-on floors (by A. Averini Cooperativa Archeologia).

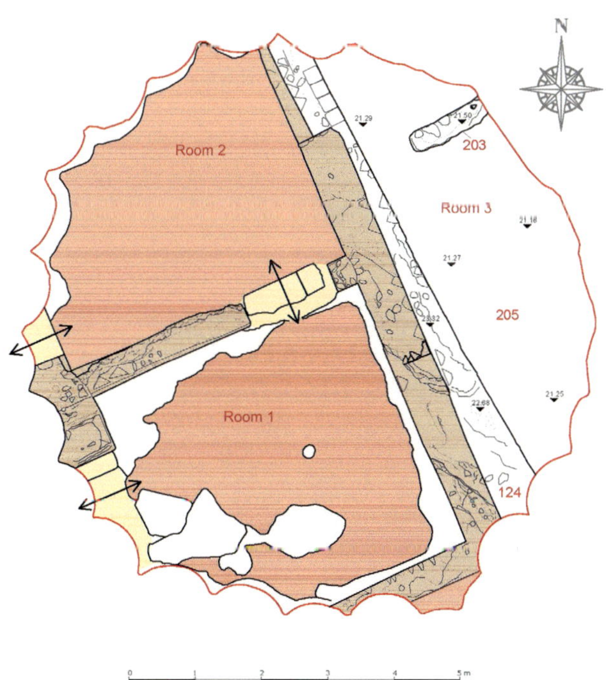


Figure 3. Antonine building plan. In brown: the re-used walls, in sand: the re-used thresholds, in pink: the re-used floors (by A. Averini Cooperativa Archeologia).

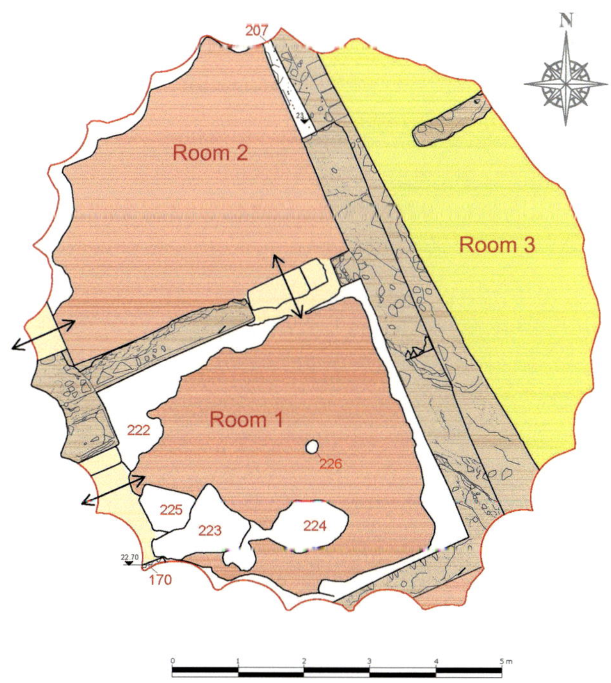


Figure 4. Severan renovations plan. In brown: the re-used walls, in sand: the re-used thresholds, in pink: the re-used floors, in yellow: the reused walk-on floors (by A. Averini Cooperativa Archeologia).

Figure 5. Details of the wooden burnt slab collapse (Photo by A. Scortecci Cooperativa Archeologia).

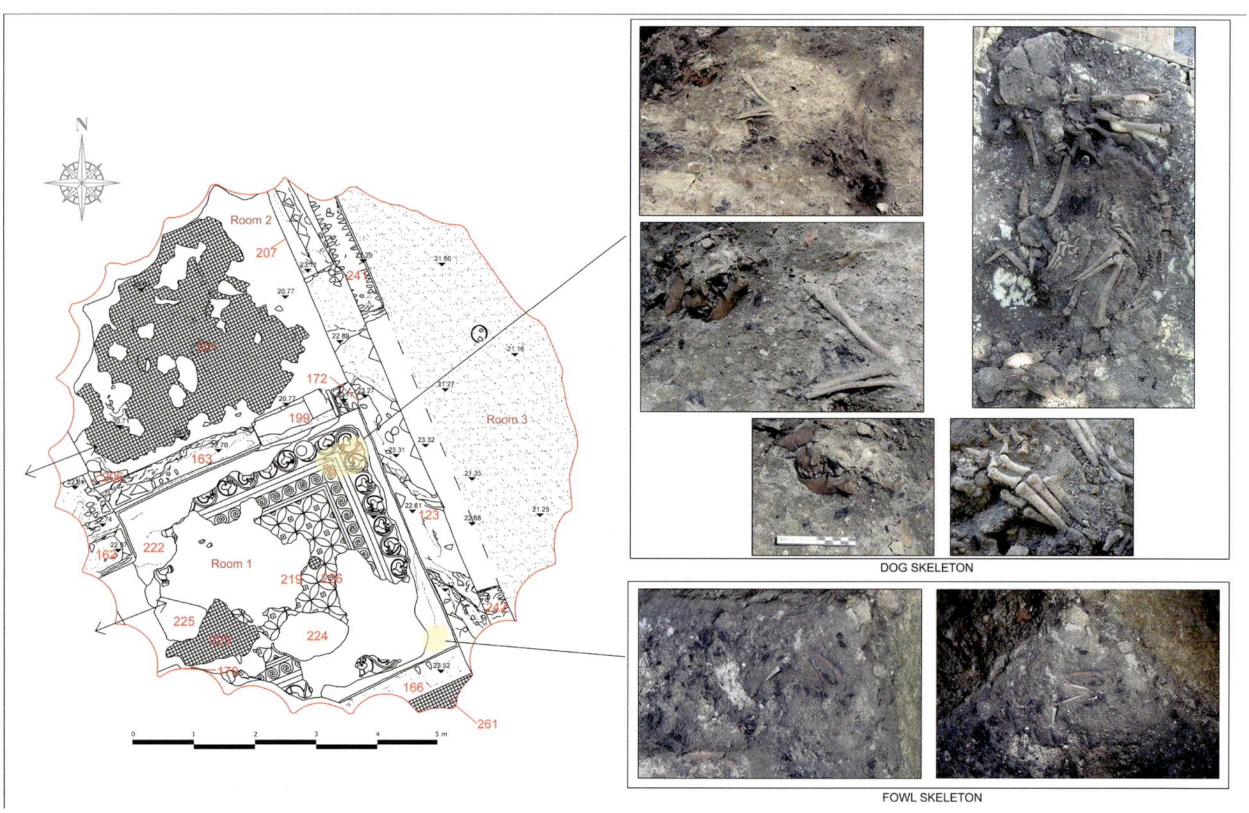


Figure 6. Positioning of the animal bone remains and details of dog and fowl skeletons (Photo by A. Averini Cooperativa Archeologia, F. Santini).

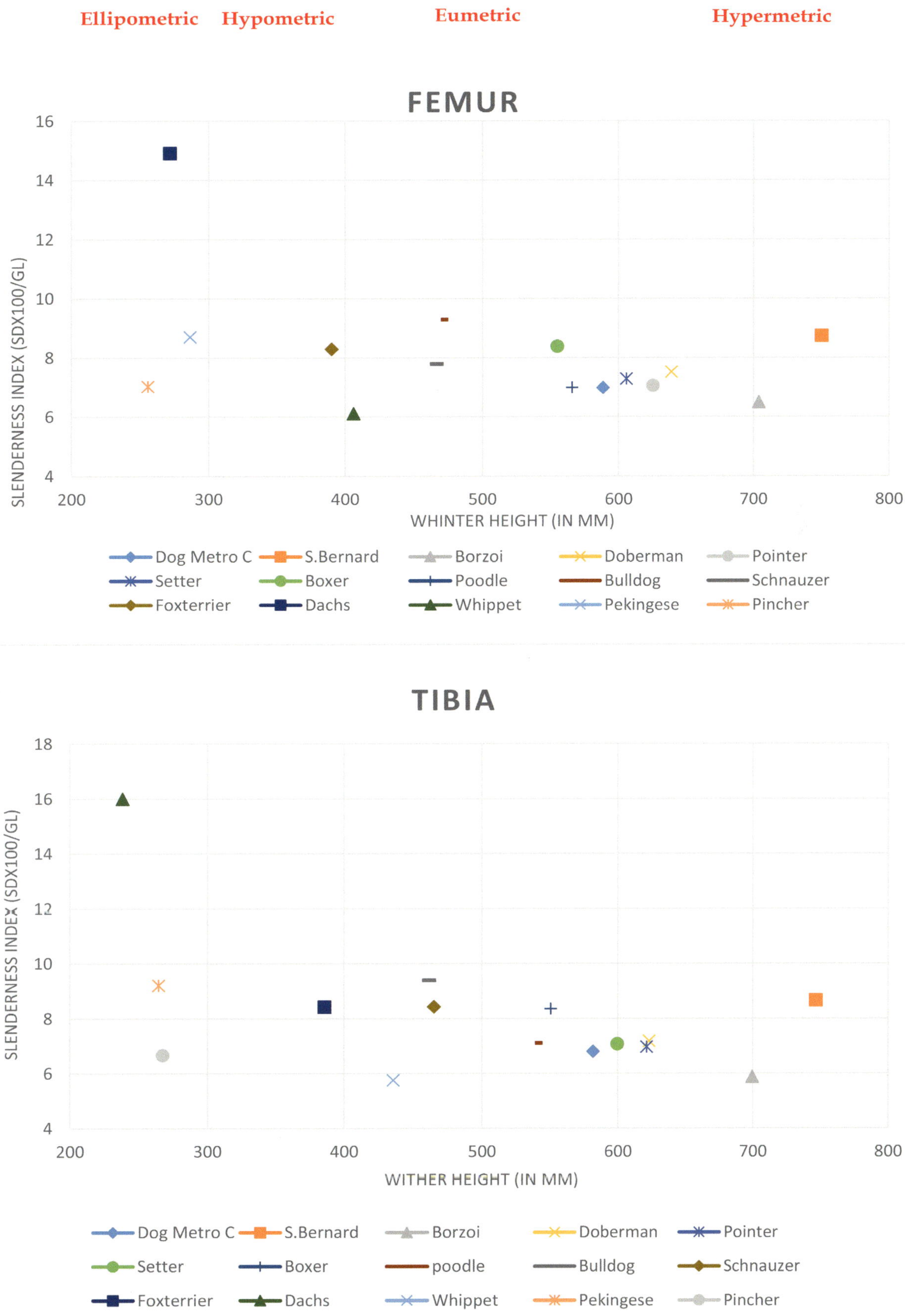


Figure 7. Scatter plot comparing withers height/slenderness index between the dog of Metro C and modern breeds (by F. Santini).

effect of the fire and/or hot ash.[7] The same marks were also registered on the long bones.[8]

The vertebral column, as well as the ribs and sternum, are very well preserved, in total anatomical and physiological connection, showing a tail with only five caudal vertebrae. This fact could suggest that this dog was an innately short-tailed breed or the tail had deliberately cut when it was a puppy.[9]

From the archaeozoological analysis, it emerges that this dog is a healthy young male[10] specimen, aged not more than 15 months.[11]

When long bones are completely preserved, as is the case with this specimen, using biometric analysis[12] the withers height may be estimated,[13] which fell between 56,24 - 58,65 mm (mean 57,74 mm), placing it within the eumetric group of dogs.

From the morphometric investigation, the specimen could be placed amongst the medium-tall sized dogs with slender, lean bones. In order to delve further into the morphological analysis, the withers height was correlated with the slenderness index and compared with modern breeds, bearing in mind that it is only an approximation because most of the modern dogs are products of artificial selection. The results show that this dog is anatomically similar to the modern Setter with a slightly slenderer tibia and with generally slimmer limbs.[14] In this case the skull shape seems to be related to a brachycephalic,[15] such as the Boxer breed easily recognisable by its short-snouted, but this outcome is not certain due to the poor preservation of the skull of this specimen.

Moreover, the dog of Metro C was plotted against dogs of different periods. It seems to fall well within the range of variability of the Bronze and Iron Age dogs, and close to the upper sized Roman dogs.[16]

The dog seems to have a well-defined morphological characteristics close to the large-size dog groups of the Roman period,[17] used maybe as a watch- or shepherd-dog.

### *3.2 Fowl remains*

Concerning the domestic fowl skeletons, the first specimen was recovered below the belly of the dog on the floor.[18] Its skeletal elements, like those of the dog, show fractures due to the crushing-trauma. The skeleton is almost complete and in anatomical connection and refers to an adult specimen.[19] The second individual includes only some skeletal elements, recovered in the South-Eastern corner of Room 1. Despite its partial preservation, it can be affirmed that it is a hen.[20]

Both the domestic fowls were compared with other ones and they can be included within the smaller Roman specimens.[21]

Noteworthy is that there are many interesting and well-known mosaics and wall paintings showing dogs and

[7] The anterior part of the snout, or the splanchnocranium, and of the jaws display clear typical fracture marks and slow combustion signs.

[8] On the femurs, ulnae and metapodials signs of fire and hot ashes were noted.

[9] After all, this was an ancient practice already well known since the Roman times. Pliny in the *Naturalis Historia* (VIII, 63, 153) mentioned '*Columella auctor est, si XL die quam sit natus castretur morsu cauda summusque eius articulus auferatur spinae nervo exempto, nec caudam crescere nec canes rabidos fieri*' about tail cutting for deworming. Columella reported in the *De re rustica* (VII, 12, 14) '*Catulorum caudas post diem quadragensimum, quam sintediti, sic castrare conveniet. Nervus est, qui per articulos spinae prorepit usque ad ultimam partem caudae; is mordicus conprehensus et aliquatenus eductus abrumpitur, quo facto neque in longitudinem cauda foedum capitin crementum, et, ut plurimi pastores adfirmant, rabies arcetur, letifer morbus huic generi*.' Moreover, it seems that in the following centuries this practice was also used on hunting and war dogs to avoid the tail being an obstacle.

[10] Sexual dimorphism is quite common among most mammals and the domestic dog is no exception. This specimen possessed a *baculum* or *os penis* allowing to unequivocally determinate its sex.

[11] The age was assigned, based on skull, post-cranial epiphyseal data (Barone 1998; Silver 1969). Moreover, it was possible to relate the first molar wear-stage with a deeper age estimation following the recent study of Horard-Herbin 2000. This scholar proposed a new age estimation method based on the occlusal surface attrition patterns of the first mandibular molar suggesting a series of age-specific wear stages. Using this criterion, the occlusal attrition of our dog was assigned as Stage C, defined not more than 15 months, classified in the young age-class.

[12] Measurements were taken following von den Driesch (1976) referring to the adult animals.

[13] By multiplying the greatest length (GL) of each type of long bone with the coefficients calculated from the skeletal proportions of modern individuals and skeletal measurements, using as models similar researches conducted on dogs, such as Clark 1995, Harcourt 1974, Koudelka 1884. See also De Grossi Mazzorin 2008, De Grossi Mazzorin and Tagliacozzo 2000.

[14] The comparison between the dog of Metro C and modern breeds is based on indices and logarithmic aggregations in Wagner's work (Wagner 1930).

[15] The shape of the skull is the most important criterion determining the breed of dogs. Usually three terms are frequently used to describe head shapes: dolicocephalic, mesocephalic and brachycephalic (Evans 1993).

[16] The comparison analysis summarised in a dataset includes skeletal dog remains taken from numerous literature reports and studies on dogs in North and Central Italy from the Bronze Age to the Late Roman Period. Its skull and mandible showed that this dog could be included in the Bronze and Iron Age cluster. The same holds true when post-cranial elements are compared.

[17] The two large-sized dog groups described and cited by Columella in *De re rustica* could be *canes villatici* or *canes pastorales* (Zedda 2006).

[18] The fowl was recovered under the dog skeleton facing up with open wings and spread legs, crushed against the floor by the dog and the floor weight.

[19] Due to the observed fusion of the long bones it could be considered an adult individual. It was impossible to establish the sex of this specimen because the distal portion of tarsus-metatarsus is not preserved.

[20] The long bones recovered demonstrate that this individual is a female domestic fowl adult thanks to the absence of the spur scar on the tarsus-metatarsus.

[21] These bone remains were compared with other fowl material of many reports and they seem to fall within the smaller Roman cluster when there was a great range of variability (Benecke 1993, De Grossi Mazzorin 2000).

domestic fowls in Roman daily life reinforcing also the specific anatomic characteristics of the two animals.[22]

## 4 Conclusions

In order to try to understand why the presence of these animals within this room and their singular relationship, the analysis of the main features of the surrounding context could be appropriate.

According to the archaeological and archaeozoological evidences, the dog may have trapped one of the fowls below itself in a panic and moment of excitement whilst looking for a safe place. Moreover, their deaths seem reasonably linked to the wooden slab collapse, which created huge harm to the stability to the upper floor.

Indeed, it is evident that this discovery is far from capturing a private domestic and usual situation. On the contrary, it seems to testify an extreme hazard and a highly confusing situation linked to a catastrophic, exceptional and sudden event where the animals have become trapped and died in their doomed attempt to escape and save themselves.

In further support of this argument we can provide some elements regarding the walls and mosaics not only of the room 1, but also in the adjacent Rooms (nos 2–3).

The walls display clear traces of cornerstones junction ejection, the mosaics show pronounced waviness mainly driven by differential structural failure clockwise rotation. All these elements might otherwise strongly suggest the effects of an earthquake causing also the onset of fire.[23]

Essentially, the presence of these animals could depend upon external factors, triggering conditions of extreme hazard and driving the animals to try to find a safe place, which have caused this unpredictable and accidental meeting probably in a room which could not have been their common ordinary residence.[24]

In conclusion, the present work could be in concordance with a number of ancient sources and artistic representations, archaeological reports and works, illustrating the features of dogs and chickens in the Roman Age. In effect, literature resources mentioning dog' bone remains reported them in similar finding situations, especially in Pompeian area.[25] Noteworthy is the dog skeleton found in the villa N. Popidi Narcissi Maioris (De' Spagnolis 2002)[26] and the plaster cast of a guard dog discovered in the House of *M. Vesonius Primus* (King 2002).[27]

## References

### *Ancient sources*

Columella, *De re rustica* (*On Rural Affairs*)
Pliny, *Naturalis Historia*, VIII, 63, 153

### *Modern sources*

Barone, R. 1998. *Anatomia comparata dei Mammiferi domestici*, Vol. 2. Bologna: Edagricole.

Benecke, N. 1993. On the utilization of the domestic fowl in Central Europe from the Iron Age up to the Middle Ages. *Archaeofauna 2*: 21–31.

Berry, J. 1998. *Unpeeling Pompeii. Studies in Region I of Pompeii*: 62–63. Milano: Electa.

Carandini, A. 1989. La villa romana e la piantagione schiavista, in *Storia di Roma IV*: 163–178. Torino: Giulio Einaudi Editore.

Cardarelli, V., F. Coletti, F. Failli, M. Galli, I. Montali and S. Morretta 2020. Tra archeologia e archeometria. Analisi preliminare degli indicatori di attività tessile e conciaria dagli scavi della Metropolitana C per la stazione Amba Aradam (Roma). *Atti di VII Purpureae vestes, International Symposium, Redefining Textile handcraft, structures, tools and production processes*: 341–350. Granada.

Clark, K. M. 1995. The later prehistoric and protohistoric dog: the emergence of canine diversity. *Archaeozoologia 7*: 9–32.

Colini, A.M. 1944. *Storia e topografia del Celio*. Città del Vaticano: Tipografia Poliglotta Vaticana.

De Grossi Mazzorin, J. 2000. Introduzione e diffusione del pollame in Italia ed evoluzione delle sue forme di allevamento fino al Medioevo, in I. Fiore, G.

[22] The most famous mosaics are the mosaic of Pompeian *villae*, such as the dog in the *vestibulum* of the House of the Tragic Poet (VI, 8.8) marked with '*CAVE CANEM*', the dog in the *vestibulum* of the House of *Paquius Proculus* (I, 7.1), the dog in the entrance of the House of *Caecilius Jucundus* (V, 1.26) (Jashemski 1979, King 2002). The most well-known painting is the dog on the wall of *caupona* of *Sotericus* (King 2002). Fowls were also depicted, like the wall painting in the atrium of the House of the *Vettii* and the mosaic in the House of the Labyrinth (Watson 2002).

[23] The data of Istituto Nazionale di Geofisica e Vulcanologia (Fabrizio Galadini) support this interpretation.

[24] The room where the animal bone remains were recovered seems to have already fallen into disuse as a representation room, or anyway as a precious room, to a secondary one due to the gaps which haven't been restored in the mosaic floor. The room probably already had a different use.

[25] We report the discovery of dog and domestic fowl skeletons in the Roman villa of Pisanella at Boscoreale (Carandini 1989, Della Corte 1903, Pasqui 1897), a dog skeleton in association with a mule or donkey were found in the House of *Amarantus* at Pompeii (Berry 1998, Fulford, Wallace 1995), a dog skeleton of a dog in the House of *Menander* (King 2002).

[26] At the villa of *N. Popidius Narcissus Maior* the skeletons of a dog and an equine were found under a below a roof collapse, placed on the floor unable to escape as they were tied up by chain (De' Spagnolis 2002).

[27] One of the most famous dog of Pompeii is the plaster cast of a guard dog left chained, found in a twisted position probably trying to save itself (King 2002).

Malerba and S. Chilardi (eds) *Atti del 3° Convegno Nazionale di Archeozoologia*: 351–361. Roma: Istituto Poligrafico e Zecca dello Stato.

De Grossi Mazzorin, J. 2008. *Archeozoologia. Lo studio dei resti animali in archeologia*: 99–103. Roma: Laterza Editore.

De Grossi Mazzorin, J. and A. Tagliacozzo 2000. Morphological and osteological changes in the dog from the Neolithic to the Roman Period in Italy, in S.J. Crockford (ed.) *Dogs through time: An archaeological perspective, Proceedings of the 1st ICAZ Symposium on the History of the domestic dog* (British Archaeological Reports International Series 889): 141–161. Oxford: Archaeopress.

Della Corte, M. 1903. Villa Rustica esplorata del Sig.re Ferruccio de Prisco nel fondo D'Acunzo immediatamente a mezzogiorno del piazzale della Stazione ferroviaria di Boscoreale (Ferrovie dello Stato), l'anno 1903. *Notizie Scavi* 1921: 436–42.

De' Spagnolis, M. 2002. *La villa N. Popidi Narcissi Maioris, in Scafati, Suburbio orientale di Pompei*: 81–88. Roma: L'Erma di Bretschneider.

Driesch von den, A. 1976. A guide to the measurement of animal bones from archaeological sites. *Peabody Museum Bulletins* 1: 1–148.

Evans, H. E. 1993. *Miller's Anatomy of the Dog*, Philadelphia: W.B. Saunders Eds.

Falzone, S., S. Morretta and G. Ricci 2019. Un cantiere di pittori al centro di Roma. Nuovi dati dallo scavo della linea C, Largo Amba Aradam. *Atti II Colloquio AIRPA, Sistemi decorative della pittura antica: funzioni e contesto*: 55–62. Roma: Edizioni Quasar.

Fulford, M. and A. Wallace-Hadrill 1995. The House of Amarantus at Pompeii (I, 9, 11–12), An interim Report on survey and excavations in 1995–96. *Rivista di Studi Pompeiani VII, 1995–96*: 77–113.

Harcourt, R.A. 1974. The dog in Prehistoric and Early Historic Britain. *Journal of Archaeological Science* 1: 151–175.

Horard-Herbin, M.-P. 2000. Dog management and use in the Late Iron Age: The evidence from the Gallic site of Levroux (France), in S.J. Crockford (ed.) *Dogs through time: An archaeological perspective, Proceedings of the 1st ICAZ Symposium on the History of the domestic dog* (British Archaeological Reports International Series 889): 115–121. Oxford: Archaepress.

Jashmeski, W. F. 1979. *The Garden of Pompeii, Hercolaneum and the Villas destroyed by Vesuvius Vol I*, 102–103. New York: Caratzas Brothers.

King, A. 2002. Mammals. Evidence from wall paintings, sculpture, mosaics, faunal remains, and ancient literary sources, in W.F. Jashemski and F.G. Meyer (eds) *The Natural History of Pompeii*: 410–14. Cambridge: Cambridge University Press.

Koudelka, F. 1884. Das Verhaltnis des Ossa longa zur Skeletthöhe bei deh Säugetieren, in Verhanndlung des Naturforschung Vereins. *Brünn 24*: 127–153.

Morretta, S. and R. Rea 2020. Una caserma alle pendici del Celio (II sec. d.C.): gli alloggi dei soldati, la domus del comandante, il giardino e l'edificio di servizio, *Atti del 7e Congrès international de Lyon sur l'armée romaine. Corps du chef et gardes du corps dans l'armée romaine*: 387–407. Université Jean Moulin – Lyon 3: Lyon.

Morretta, S. and R. Rea 2018. Una nuova caserma alle pendici meridionali del Celio, in A. D'Alessio, C. Panella and R. Rea (eds) *Roma Universalis. L'impero e la dinastia venuta dall'Africa*: 190–199. Milano: Electa.

Pasqui, A. 1897. La villa pompeiana della Pisanella presso Boscoreale. *Monumenti Antichi VII*: 397–554.

Pavolini, C. 2006. *Archeologia e topografia della Regione II (Celio). Un aggiornamento sessant'anni dopo Colini.* Roma: Quasar.

Silver, I.A. 1969. The ageing of domestic animals, in D. Brothwell and E. Higgs (eds) *Science in Archaeology*: 283–302. New York: Thames and Hudson.

Wagner, K. 1930. *Rezente Hunderassen. Eine osteologische Untersuchung*, Mit 36. Oslo: Jacob Dywad Kommission.

Watson, E.G. 2002. Birds. Evidence from wall paintings, sculpture, mosaics, faunal remains, and ancient authors, in W.F. Jashemski and F.G. Meyer (eds) *The Natural History of Pompeii*: 380–81. Cambridge: Cambridge University Press.

Zedda, M., P. Manca, V. Chisu, S. Gadau, G. Lepore, A. Genovese and V. Farina 2006. Ancient Pompeian dogs. Morphological and Morphometric evidence for different canine populations. *Anatomia Histologia Embryologia* 35: 319–324.

# 4.9 Dog and Human Sepultures at *Peltuinum* (L'Aquila, Italy)

Ivana Fiore[1,2], Luisa Migliorati[3], Antonella Pansini[3],
Tiziana Sgrulloni[3], Alessandra Sperduti[2]

[1]PhD Programme in Environmental and Evolutionary Biology, Dipartimento di Biologia Ambientale, Sapienza University of Rome, P.le A. Moro, 5 Rome, Italy, iva.fiore@gmail.com

[2]Bioarchaeology Service, Museo delle Civiltà, Piazza G. Marconi 14, Rome, Italy, alessandra.sperduti@cultura.gov.it

[3]Dep. Ancient World Studies, Sapienza University of Rome, P.le A. Moro, 5 Rome, Italy, luisa.migliorati@uniroma1.it, tizianasgrulloni@yahoo.it, antopansini@hotmail.com

Corresponding author: Ivana Fiore, iva.fiore@gmail.com

**Abstract**

*Peltuinum* is a Roman town in the central Apennines founded in the mid-1st century BC and abandoned in the 5th century AD when struck by violent earthquakes. The archaeological campaigns led to the discovery of three atypical funerary contexts in abandoned theatre structures (five shafts and a sewer) and of the city walls (a tower). All three sites show a strong association of human, dog, and other faunal remains. Still, they differ by the modality of deposition, the age at death profile of the human individuals, the number of associated dogs, and the presence/proportion of other domestic fauna, suggesting different interpretative scenarios. In this paper we describe the cases focusing on the dogs' remains in relation to their symbolic role. The analysis focused mostly on the dog remains from the shafts to understand whether there was a choice in the type of dog to be sacrificed based on certain characteristics such as sex, size and age.

**Keywords**: Central Italy, theatre, human and dog depositions, breeds, dogs sacrifice.

## 1 Introduction

The excavations conducted in the ancient city of *Peltuinum* concentrated on the area of the ancient theatre, whose construction history and layout have been clarified. The building, like the entire city, was abandoned in the second half of the 5th century AD due to a series of violent earthquakes. In the following centuries it was mostly used as a place to obtain building material. However, the entire northern sector of the theatre, which took advantage of the hillside, was covered by building drains, while, between the 11th and the 12th centuries, during the period of encastellation in the Abruzzi, the southern radial compartments of the theatre were incorporated into a fort. At some later time, the end of the 13th–beginning of the 14th centuries, the area between these units was stripped to be occupied by a 'workers' district, i.e., a series of small rooms used for reworking material in order to rebuild a nearby church (Tulipani 1996, Migliorati, 2014, 2015; Migliorati, *et al.* 2021). The excavations brought to light three atypical funerary contexts which are five shafts and a sewer in the disused theatre structures and a tower of the city walls. The three contexts show a strong association between humans and dogs. In the complex and diversified phenomenon of associated burials of dogs and humans across time and space, the site of *Peltuinum* provides three relevant funerary contexts (Fiore *et al.* 2018, Figure 1).

## 2 Methods

The animal remains were determined using the animal osteological collection of the Bioarchaeology Service of the Museo delle Civiltà, in Rome and by consulting the atlases of Schmid (1972) and Barone (1981). For the scientific nomenclature of domestic animals, the reference was Gentry, Clutton-Brock and Groves (2004). Age-at-death diagnosis was based on stages of teeth formation, eruption and wear according to Payne (1973) and Bull and Payne (1982) and on the fusion of the articular epiphyses of the long bones (Silver 1969, Bullock and Rackham 1982, Bull and Payne 1982). The minimum number of individuals (MNI) was calculated according to Bökönyi (1970), Chaplin (1971) and Cruz Uribe and Klein (1984). Osteometric data were collected according to the methodology of von den Driesch (1976).

The bones of canids recovered from the shafts were studied in detail. The material was analysed from an osteological, taphonomic and biometric point of view. The age-at-death was estimated according to Barone

(1981, 1995). The withers height was calculated following Koudelka (1884) and Harcourt (1974). Sex assessment, besides the presence or absence of *os penis*, was hypothesised on the morphological characteristics of the skull proposed by Colline Brassard, Cecile Callou (2020, Fig. 1). The taphonomic analysis was carried out macroscopically using a magnifying lens 10x, and microscopically using a stereomicroscope (Nikon SMZ 1000, 8x80). Statistical analyses were performed through the PAST (PAlaeontological STatistics) 3.20 software version (Hammer *et al.* 2001).

### 3 The shafts

The archaeological excavation of the theatre led to the discovery of seven shafts (85 x 55 cm and 400 cm average depth) located at the foot of the *pulpitum* wall and used in ancient times to house the poles of the *aulaeum* (Figures 1–2).

Shaft I (counting from north) was emptied during previous excavations and therefore no data are available on its filling, while shaft V was not investigated because of the 'workers' district superposition (13th - beginning of the 14th centuries). Shafts II-III-IV were brought to light below an accumulation of building materials, which can be dated, through the discovery of catacomb lamps, to the 5th cent. AD.

The excavation of the fill, which took place between 2011 and 2013, yielded, from the crest throughout the depth of the shafts, countless osteological finds related to human fetuses and newborns. Moreover, these findings were associated with dogs and some individual horses and other fauna, mixed with building material from the collapse of the *scaenae frons*.

On the other hand, the sixth shaft had partially been emptied already between the 13th and 14th centuries, to a depth of about 2.50 m, and the bone remains, found at the bottom, appeared to be sealed with a thick

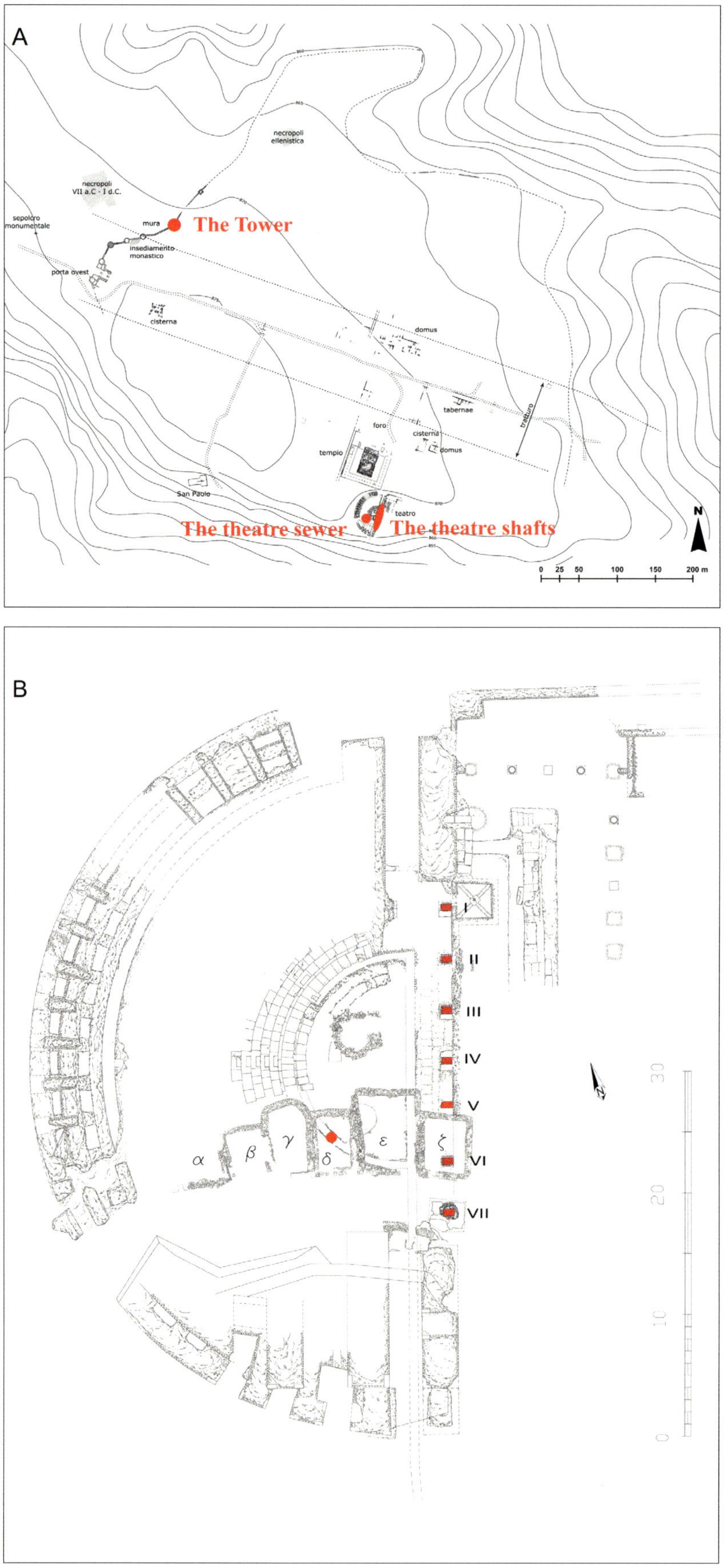


Figure 1. Plan with the location of the three contexts (a) and detail of the theatre with the shafts and sewer in room δ (b) (Migliorati *et al.* 2018).

layer of the local silty soil. The shaft was then backfilled with building waste mixed with earth and finally obliterated by the floor of a room of the 'workers complex'.

The most southern shaft, VII, was destroyed to a depth of approximately 3.00 m and completely emptied. It plausibly happened due to the construction of the fortress.

Figure 2. The shaft VII in the excavation phase (Photo by G. Calandra).

### 3.1 *Faunal bone remains from the shafts*

All the shafts contain faunal bone remains that can be attributed to a total of 89 individuals, predominantly dogs, the other domestic animals being represented by rare remains of horse, cattle, pig, sheep/goat.

(Fiore and Salvadei 2014, Migliorati *et al.* 2018, Sperduti *et al.* 2018). The remains of horse, cat and sheep/goat refer to whole skeletons or large portions (Tables 1–4). The shafts included different amounts of animal bone remains, three of them contained more than 1000 finds, III only 663. This is also reflected by the estimation of the Minimum Number of individuals, ranging from 30 (shaft II) to 12 (shaft III). Due to post-depositional phenomena the dog elements were not found in anatomical connection, except in rare cases, and it has become difficult to reassociate the elements to individuals.

Shaft VII, the last one investigated, had been emptied in antiquity and filled with building material; bone remains are rare and the largest ones were found near the walls, where patches of the previous filling were preserved. Careful sifting enabled the recovery of several small fragments of human and dog bone remains. The dog remains refer to limb bones of an

Table 1. NR and NMI of individual animals present in the different shafts.

| *Taxa* | Shaft II | Shaft III | Shaft IV | Shaft VI | Shaft VII | TOTAL |
|---|---|---|---|---|---|---|
| *Equus* sp. | 4 | 4 | 60 | 35 | | 103 |
| *Bos taurus* | 11 | 6 | 7 | 12 | 1 | 37 |
| *Sus domesticus* | 3 | | | 2 | 1 | 6 |
| *Ovis/Capra* | | | 2 | 15 | | 17 |
| *Canis familiaris* | 1192 | 614 | 1100 | 1200 | 8 | 4114 |
| *Felis catus* | 3 | | | 21 | | 24 |
| Indet. | 121 | 39 | 80 | 65 | 58 | 363 |
| **Total Remains** | **1334** | **663** | **1249** | **1350** | **68** | **4664** |
| *Equus* sp. | 2 | 2 | 3 | 2 | | 9 |
| *Bos taurus* | 2 | 1 | 2 | 2 | 1 | 8 |
| *Sus domesticus* | 1 | | | 2 | 1 | 4 |
| *Ovis/Capra* | | | 1 | 1 | | 2 |
| *Canis familiaris* | 24 | 9 | 18 | 15 | 2 | 68 |
| *Felis catus* | 1 | | | 1 | | 2 |
| **Total MNI** | **30** | **12** | **24** | **23** | **4** | **93** |

Table 2. NR and NMI of individual animals present in various contexts.

| | TOTAL SHAFTS | | SEWER - US 1263 | | TOWER | | TOTAL | |
|---|---|---|---|---|---|---|---|---|
| *Taxa* | NR | MNI | NR | MNI | NR | MNI | NR | MNI |
| *Equus* sp. | 103 | 9 | 1 | 1 | 25 | 4 | 129 | 14 |
| *Bos taurus* | 37 | 8 | 2 | 1 | 56 | 6 | 95 | 15 |
| *Sus domesticus* | 6 | 4 | 2 | 1 | 43 | 7 | 51 | 12 |
| *Ovis/Capra* | 17 | 2 | | | 71 | 9 | 88 | 11 |
| *Canis familiaris* | 4114 | 68 | 34 | 1 | 102 | 13 | 4250 | 82 |
| *Felis catus* | 24 | 2 | | | 3 | 1 | 27 | 3 |
| *Canis lupus* | | | | | 4 | 1 | 4 | 1 |
| Aves | | | 15 | | 2 | | 17 | |
| Indet. | 363 | | 163 | | 150 | | 981 | |
| **Total NR** | **4664** | **93** | **217** | **4** | **456** | **41** | **5537** | **138** |

Table 3. Percentages of NR and NMI of individual animals present in various contexts.

| *Taxa* | TOTAL SHAFTS | SEWER - US 1263 | TOWER | TOTAL |
|---|---|---|---|---|
| *Equus* sp. | 9.7 | 25.0 | 10.0 | 10.1 |
| *Bos taurus* | 8.6 | 25.0 | 15.0 | 10.8 |
| *Sus domesticus* | 4.3 | 25.0 | 17.5 | 18.7 |
| *Ovis/Capra* | 2.2 | | 22.5 | 8.0 |
| *Canis familiaris* | 73.1 | 25.0 | 30.0 | 59.4 |
| *Felis catus* | 2.2 | | 2.5 | 2.2 |
| *Canis lupus* | | | 2.5 | 0.8 |
| **Total MNI** | **93** | **4** | **41** | **138** |

Table 4. NMI and age of dogs present in various contexts.

| Dog | Foetus/ New Born | Young | Adult | MNI |
|---|---|---|---|---|
| Shaft II | 7 | 3 | 14 | 24 |
| Shaft III | 1 | 3 | 5 | 9 |
| Shaft IV | 9 | 1 | 8 | 18 |
| Shaft VI | 9 | 1 | 5 | 15 |
| Shaft VII | 1 | | 1 | 2 |
| Sewer US 1263 | | | 1 | 1 |
| Tower | 3 | 5 | 5 | 13 |
| **Total MNI** | **30** | **13** | **39** | **82** |

adult individual and those of a puppy. Cattle (fragment of ulna), pig (third phalanx) and probable *Ovis/Capra* (diaphysis of long bone) are rare.

In the VII shaft bone remains are rare, further careful excavation operations have collected evidence that this shaft was also used for the deposition of children and dogs, even though the original sediments were removed in ancient times.

In overall, the remains of 68 dogs were found from the shafts, mostly adults or aged 0–1 month (newborns/ fetuses). It was discovered that young dog remains (aged 2–12 months) are rare and older individuals are absent (Table 4).

On the evidence of several *os penis* and fetuses we recognised the presence of both males and females.

The height at withers shows that small to medium-sized dogs predominate.

### *3.2 Human bone remains from the shaft*

Nearly 2,500 human skeletal elements from 87 individuals were found in the shaft, unfortunately post-depositional processes altered the original positions, and it was not possible to recognise and isolate individual skeletal features (Sperduti *et al.* 2018).

Infants in the shafts are predominantly individuals who died at birth, including cases of miscarriage and premature birth. The age includes individuals from 22 fetal weeks to three years of age with a predominance of individuals of perinatal age (38–40 weeks of gestation). The human skeletal sample from *Peltuinum* seems to represent a particular case of collective burial of infants outside the formal burial ground of the community.

### *3.3 Methods of deposition*

The reconstruction of the deposition patterns of the finds reveals the complexity of the use of the shafts. Shaft IV contained two complete foals and one leg end of *Ovis/Capra*, all very young, and the right leg of an adult horse from the femur to hoof. In addition to these skeletons complete or large portions in anatomic connection there were 18 dogs of different ages. The animal remains of shaft IV were found only 70 cm thick from the top. They consisted of two whole foals, a whole leg of an adult horse, 9 puppies, 1 juvenile and 8 adult dogs and 17 infants. Based on the size of the shafts and the estimated footprint of the carcasses, we calculated that if they had been laid at the same time, they would have reached at least 2 m thick. This fact supports the hypothesis that the carcasses were probably placed at different times.

There are other elements, in addition to the constant presence of human infants and fetuses always associated with dogs, that make the *Peltuinum* shafts an accumulation to be read in ritual function (Migliorati *et al.* 2018). Dogs and horses were laid whole; dogs are in considerable quantity and among them there are also puppies and pregnant female dogs, some show traces of voluntary killing, finally a dog was laid at the closure of one of the shafts.

## 4 The sewer

Archaeological investigations in room δ, carried out in 2016, have brought to light a section of the semi-annular sewer channel for the disposal of rainwater from the theatre (Figure 1, B). A wall separating two different fills was identified inside the canal. The wall appears to be built against the ground with the curtain facing the north-west sector. From the excavation data and from the analysis of the structures of the workers' quarter, it can be deduced that the canal was excavated for the sector that could have presented problems for the support of a series of walls that were connected just above the canal; the fill soil would not have been sufficiently compact and consistent to support the structures. This sector was therefore emptied of earth and material from the collapse of the theatre (which occurred in the 5th century due to more than one strong earthquake) and filled with silt and limestone chips of various sizes to make the soil inside the canal compact and solid. The fill is dated by fragments of archaic faience. Since it was not useful to excavate the entire sector of the canal inside the room δ, the south-eastern part was not touched and the excavation did in fact reveal the materials from the collapse of the theatre and the sewer›s parapets: architectural elements, curved tiles, roof tiles and limestone elements. Within this sector, particularly in contact with the upstream part of the sewer partition wall, several human bone fragments belonging to two individuals were found (Fiore *et al.* 2018, Migliorati *et al.* 2021).

### *4.1 Human bone remains from the sewer*

The human bone remains consisted of small cranial fragments and a mid-proximal tibia; two zygomatic fragments allowed the identification of 2 male individuals over 30 years of age.

### *4.2 Animal bone remains from the sewer*

The bone remains recovered in this site (US 1236) include 34 fragments of dogs attributable to a single adult individual (Tables 2–3). Rare remains referring to domestic ungulates were also recovered: cattle (a fragment of radius and semilunar of an adult individual), pig (two fragments of teeth of a very young individual), a phalanx of an adult equid and one of an adult sheep/goat. Lagomorphs remains (tibiae, metapodial and maxilla) were found in this unit, rare bird and amphibian remains are also present.

The fragmentary dog remains refer to the different portions of the carcass.

The skull is documented by fragments of the mandible and lower teeth. The forelimb is represented by fragments of radius and ulna. The hind limb is represented by a fragment of the tibia, fibula, and tarsus bone. The axial skeleton consists of several thoracic, lumbar, and caudal vertebrae and several fragments of ribs. The extremities of the limbs consist of some metapodials and phalanges.

Half of the remains were found in a restricted area and at an altitude between 100 and 120 cm; radius and ulna, several vertebrae and the ribs were slightly misplaced.

In summary, the representation of all the anatomical parts of the 34 elements (all referable to one adult individual) suggests the presence of a complete dog body that was subsequently disrupted.

## 5 The Tower

Surveys and excavations have shown the city walls were bound to the shape of the plateau and, moreover, the northwestern side was protected by a few towers. The recent excavation of the northern section of the walls brought to light a new tower. The filling of the inside, below the plowed soil, consisted of a layer about 1.30 m thick, characterised by the presence of numerous animal and human bone remains, numerous squared blocks of medium and large dimensions and various ceramic fragments, falling within a very broad chronology (4th/3rd cent. BC - 4th cent. AD) (Migliorati *et al.* 2018).

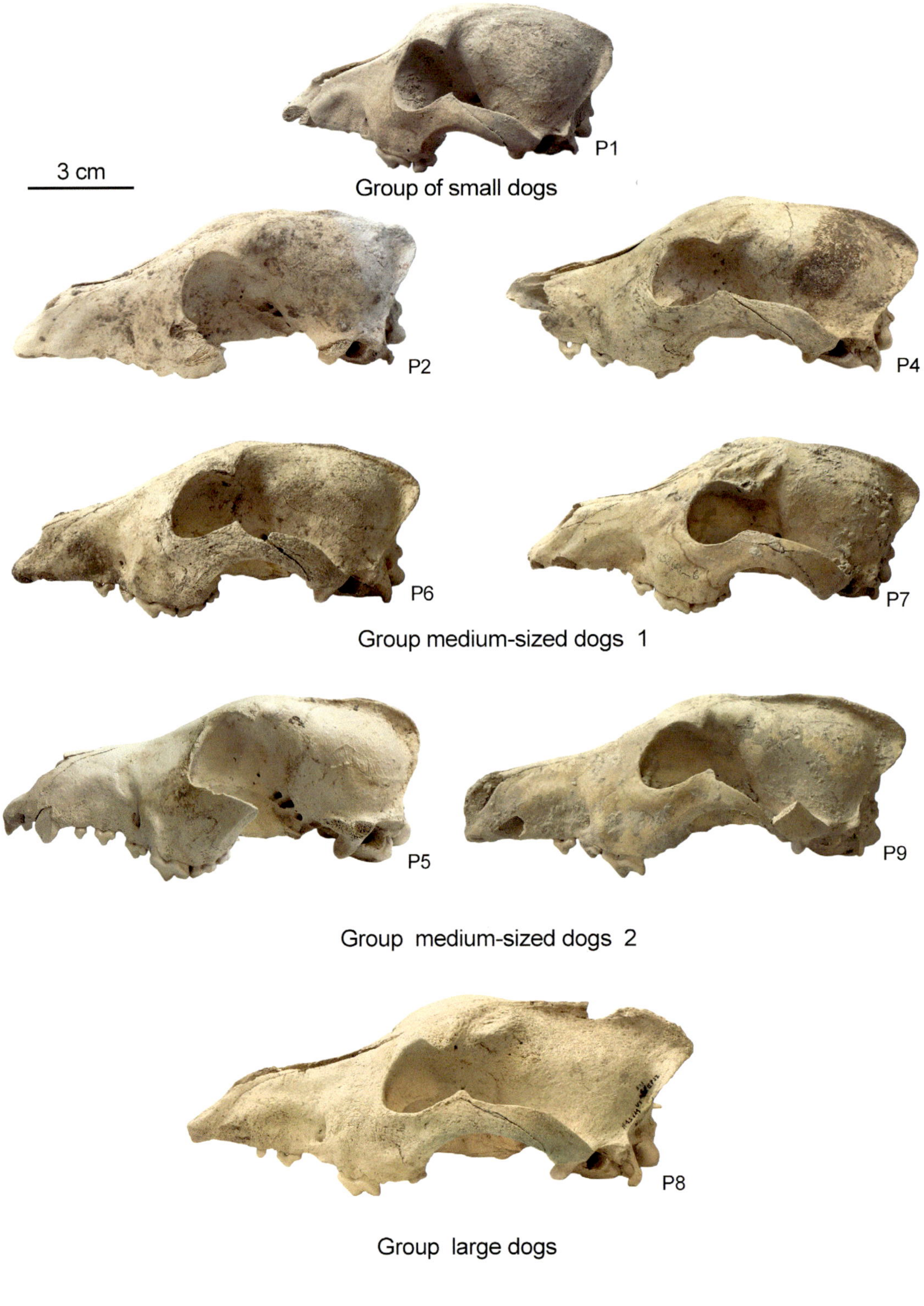


Figure 3. Skulls dog in lateral view, for descriptions and details, see paragraph 6.1 for description and details. (Photo by I. Fiore).

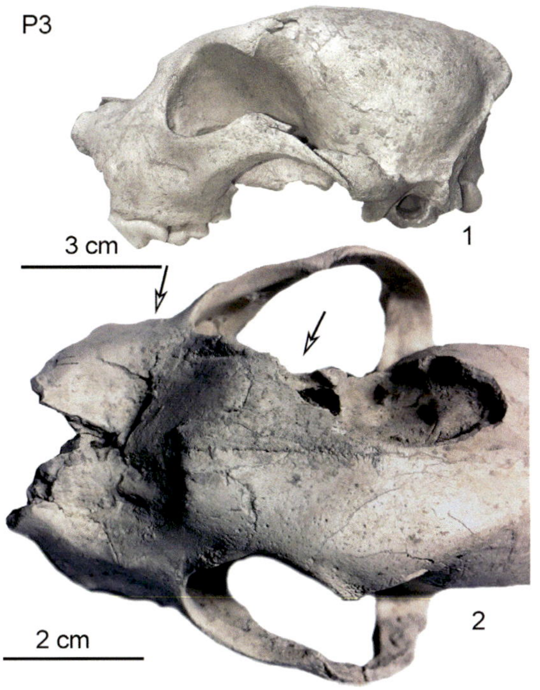


Figure 4. Skull dog (P3) in lateral view and details the blows, see paragraph 6.1 for description and details (Photo by I. Fiore).

Commingled disarticulated human material corresponds to dozens of individuals, both adults and subadults, associated with the remains of domestic animal species including at least two dogs in partial anatomical connection.

### 5.1 Human bone remains from the Tower

The human remains recovered from Tower number 984 relate to 35 individuals of different ages, mostly adults. No anatomically connected remains were recognised, and the bones are fragmented. Surface modifications and fractures suggest that the remains are in secondary deposition.

### 5.2 Animal bone remains from the Tower

A total of 456 animal bone remains from the tower were recovered and analysed, of which 304 have been specifically determined. The sample almost exclusively consists of domestic animals; the dog prevails both in the number of remains and in the MNI of individuals, followed in order by sheep/goat, ox, pig, and equids, with the presence of cats being sporadic (Tables 2–3).

The bone remains of ungulates are generally fragmentary and only in rare cases whole; they refer to cranial elements and limb ends, while long bones are less represented. Some remains show traces of cuts or cleft fractures from slaughter. Mortality profiles fall into those reflecting meat exploitation activities, with animals slaughtered at young adult/adult age (pigs, oxen, sheep/goats) and working and breeding activities, with animals kept alive until productive activity was exhausted (oxen, equids, and sheep/goats of adult/senile age).

The dog is represented by all parts of the body and in some cases, it is possible to reconstruct entire portions of the same individual. Anatomical elements are generally whole except for the ribs and skull, which are more susceptible to post-depositional fracture. All age groups from newborn to adult are represented. Some elements, two humeri and a radius of considerable size would testify to the presence of an adult wolf in the tower.

### 5.3 Mode of deposition

The bone remains show in some cases different degrees of preservation, perhaps due to longer exposure times before complete burial.

Bone remains of ungulates can be traced back to meal scraps or slaughter activity; the dogs, showing no traces of anthropogenic modifications, could be dead animals 'thrown inside the tower'. Therefore, an activity may be part of the cleaning of the area from carrion and slaughter waste. This interpretation may be plausible also considering that animal bones are associated with hundreds of human remains no longer in connection, and in secondary deposition and ceramic material covering a chronological span from the 4th BC to the 4th AD (Migliorati *et al.* 2018). Actually, near the tower, a monastery was built in the 7th cent. AD, remaining in use until the 17th cent. (Tulipani 1996). Archival documents testify that an agricultural area was associated with the monastery, in accordance with customary usage. In consequence, the tower may indeed be a 'deposit' of heterogeneous material, coming from a sort of reclamation of the area used centuries earlier as a necropolis and designated by the monastery for agricultural use.

## 6 Analysis of dogs from the shafts

The discovery of eight complete and of one almost complete (P3) dog skulls allows to analyse their morphology and to reconstruct the variety of dogs introduced into the shafts and in some cases also to trace the cause of death and the state of health of the animals (Figures 3–4).

Below is the description of the nine skulls that were discovered.

### *6.1 Skulls*

P1 - Skull almost complete, missing the left incisor area and some teeth. Head of small size (GL 138.5), short and wide, compared to the other skulls, but it is still within the dolichocephalic forms (Cephalic Index (CI) 68.8). The teeth are slightly worn; cranial sutures are evident. The teeth are in line, but very close together. Sagittal crest is not very evident. The foramen magnum has a keyhole shape. The sex is uncertain, probably female due to of the morphological features of the skull (POP, SC, BO = post-orbital process; sagittal crest; basioccipital) (Colline Brassard, Cecile Callou 2020, Fig. 1).

P2 - Skull almost complete, missing the left zygomatic and some teeth. Head of medium size (GL 172.6), slightly elongated and falling within dolichocephalic forms (CI 54.8). Teeth little worn and cranial sutures are evident. The teeth are in line, but very close together. Sagittal crest present. Shape of the *foramen magnum* nearly a keyhole. The sex is uncertain, probably female because of the morphological features of the skull (POP, SC, BO). Frontal depression and on right orbit possibly from blow or healed wound.

P3 - Incomplete skull, missing a large fragment of the maxillae. Head small in size (estimated GL about 155/160 mm), slightly elongated head falling within dolichocephalic forms (CI 54.0). Teeth are slightly worn and cranial sutures are not very evident. Sagittal crest is not very evident. Shape of the *foramen magnum* is nearly a keyhole. The sex is uncertain, probably female because of the morphological features of the skull (POP, SC, BO). There are blows at the level of the left orbit with breakage of the skull and around the snout evidence for crushing from blows.

P4 - Incomplete skull, missing a small fragment of the maxillae. Head of medium size (estimated GL about 190 mm), slightly elongated head falling within dolichocephalic forms (CI 40 ca.). The teeth are little worn and the cranial sutures evident. The teeth are in line, but very close together. Sagittal crest is present. Shape of the *foramen magnum* is nearly a keyhole. The sex is uncertain, probably cf. female because of the morphological features of the skull (POP, SC, BO). It shows blows at the level of the left orbit with a break in the skull.

P5 - Incomplete skull, missing the zygomatics and some teeth. Head of medium size (GL 188.7 mm), slightly elongated head which falls within the dolichocephalic forms (CI 54). The teeth are little worn and the cranial sutures evident. The teeth are in line, but very close together. Sagittal crest is very pronounced. Shape of the *foramen magnum* is almost oval. The sex is uncertain, probably cf. female because of the morphological features of the skull (POP, SC, BO). The canines are very wide compared to the other individuals. The left canine is broken at the waist and shows around the alveoli a halo of osteolysis on a reactive basis. The animal has probably suffered from an infection.

P6 - Incomplete skull, missing zygomatics and some teeth. Head of medium size (GL 169.0 mm), slightly elongated head falling within dolichocephalic forms (CI 55). Teeth are little worn and cranial sutures are moderately evident. Teeth are not in line, P4 is transverse. Sagittal crest present. Shape of the *foramen magnum* oval. The sex is uncertain, probably male because of the morphological features of the skull (POP, SC, BO).

P7 - Complete skull missing some teeth. Head of medium size (GL 168.4 mm), head not very elongated and falling within the dolichocephalic forms (CI 65). The teeth are little worn, and the cranial sutures are not evident. Sagittal crest is present. Shape of the *foramen magnum* is a keyhole. The sex is uncertain, probably male because of the morphological features of the skull (POP, SC, BO). The teeth are in line, but very close together. It shows irregularities in the contour of the upper margin of the right orbit, compatible with a previously healed fracture.

P8 - Incomplete skull missing the left zygomatic, nasal bones and some teeth. Head large (GL 221.3 mm), elongated head falling within dolichocephalic forms (CI 50). Teeth are worn, incisors and some premolars have been lost in life and alveoli have been or are being resorbed. Teeth are in line, but very close together. There is a very pronounced sagittal crest. Shape of the *foramen magnum* is almost oval. The sex is uncertain, probably male because of the morphological features of the skull (POP, SC, BO). It shows osteolysis on a reactive basis on the frontal, the animal has probably suffered from an infection.

P9 - Skull complete, missing some teeth. Head of medium size (GL 187.8 mm), elongated head falling within dolichocephalic forms (CI 45.5). The teeth are little worn, and the cranial sutures are not evident. The teeth are in line, but very close together. Sagittal crest is very pronounced. Shape of the *foramen magnum* is almost oval. The sex is uncertain, probably male because of the morphological features of the skull (POP, SC, BO).

#### *6.1.1 Dogs. Morphology, breeds, sex, age, health and death*

The nine skulls came from shafts II and IV. The morphometric data show skulls of varied sizes ranging from 138.5 mm to 221.3 mm with a mean of 175.1 mm and a high standard deviation of 22.8 (Table 5). In terms of size morphology, the dogs in shaft II show more

Table 5. Skull indices and ratios of dogs of *Peltuinum*. SI=Skull index, MI=Muzzle index, MWI=Muzzle width index.

| Dog Skulls | ID | SI | MI | MWI | Skull Length |
|---|---|---|---|---|---|
| P 1 - US 859 II Shaft | P1 | 62.8 | 39.1 | 38.8 | 138.5 |
| P 8 - US 859 II Shaft | P8 | 50.5 | 35.6 | 38.8 | 221.3 |
| P 9 - US 859 II Shaft | P9 | 45.5 | 40.0 | 38.5 | 187.8 |
| P 5 - US 881 II Shaft | P5 | | 44.3 | 34.5 | 188.7 |
| P 3 - US 881 II Shaft | P3 | 54.7 | | | 160.0 |
| P 6 - US 840 IV Shaft | P6 | 53.8 | 40.8 | 38.1 | 169.0 |
| P 4 - US 860 IV Shaft | P4 | 54.3 | | | 170.0 |
| P 2 - US 860 IV Shaft | P2 | | 43.2 | 34.0 | 172.6 |
| P 7 - US 860 IV Shaft | P7 | 55.9 | 40.7 | 39.3 | 168.4 |

| | SI | MI | MWI | Skull Length |
|---|---|---|---|---|
| N | 7 | 7 | 7 | 9 |
| Min | 50.5 | 35.6 | 34.0 | 138.5 |
| Max | 62.8 | 44.3 | 39.3 | 221.3 |
| Mean | 55.3 | 40.5 | 37.4 | 175.1 |
| Dev | 4.1 | 2.8 | 2.2 | 22.8 |

variability, in fact in this shaft, there is both the smallest and the largest dog skull with high standard deviation values (length 138.5–221.3; mean 187.8 and dev.St 31). The dogs in shaft IV illustrate less variability, falling within the range of medium-sized dogs, the length of the skulls is similar, in fact, the standard deviation is 1.9 (length 168.4–172.6; mean 170 and dev.St 1.9).

Cranial sutures are very evident in the skulls (P1, P2, P4, P5), moderately visible in P3, P6 and P8, and not visible in P7 and P9.

Overall, the teeth are a little worn, indicating that there are adult individuals of different ages, however not senile. The teeth are close together and only in one case, there is one 'out of line' with overcrowding and rotation of a premolar.

The morphological traits evaluated for sex determination (BO: basioccipital; POP: post-orbital process; SC: sagittal crest; Colline Brassard, Cecile Callou 2020, Fig. 1) do not always provide certain data, we have however assigned the sex when two or three characteristics are present. Since there is no exact correlation, we have preferred to record the data as probable. There are five dogs, probably females (P1, P2, P3, P4, P5) and four males (P6, P7, P8, P9).

The sagittal crest in the individuals P1, P3 is not very evident; it is very marked in P5, P8, P9 these are probably three male dogs defined also by other morphological traits (BO, POP, SC); the crest is present, nevertheless not very marked in the remaining skulls.

The shape of the occipital hole is oval in four cases (P5, P6, P8, P9), in two cases it is keyhole shaped (P1, P7), in three other cases it is close to keyhole shape (P2, P3, P4). The smallest skull has the *foramen magnum* in the shape of a key-hole, indeed it is a characteristic of 'Toy' dogs. Although the size of the *Peltuinum* skull is not so small, in fact from the height at withers of the long bones there are no individuals smaller than 25 cm. If one refers to the explanation given by Evans (2013) in his analysis of modern dogs, it could indicate that the other four (P2, P3, P4, P7) which have a key-hole notch, being of medium to large size are not purebred dogs, but mongrels.

The evaluation of the profile (nose-frontal jump or stop) was carried out considering the angle between the line of the frontal bones and the upper line of the muzzle. None of the skulls has a very marked stop, the angle measurement was carried out digitally on photos of the profile of the skulls, showing angle values between 148°-168° ca. The numbers fall within the ultra-long linear values; therefore, it is seen with a not very marked stop. In two cases, P1, the smallest skull, and P8 have the narrowest angle values (150° ca), indicating a relatively more marked stop than the others, while in the cases of P2 and P6 the profile is continuous without a stop, and in the remaining cases, the values are between 158°-168°. These differences suggest that there are at least four different morphologies or types of dogs as morphometric analysis verified.

Breeds were defined based on skull morphology and morphometrics compared to those of current dog breeds (Phillips *et al.* 2009, Knoest 2015), pending genetic analysis. Nine skulls were whole or could be measured for morphometric analysis (Table 5, Figures 5–6). The data confirm those obtained from height at withers, there is a predominance of medium-sized dogs with two exceptions: one particularly small and one large. The analysed record is predominantly composed of two slightly diverse groups of medium-sized dogs P2-P4-P6-P7 and P3-P5-P9 with P3 being quite close to the first set, while P1, which is particularly small, and P8, which is particularly large, diverge from the two groups and from each other. The skulls in the first group all come from the IV shaft, this could suggest a choice of animals of a particular breed or size.

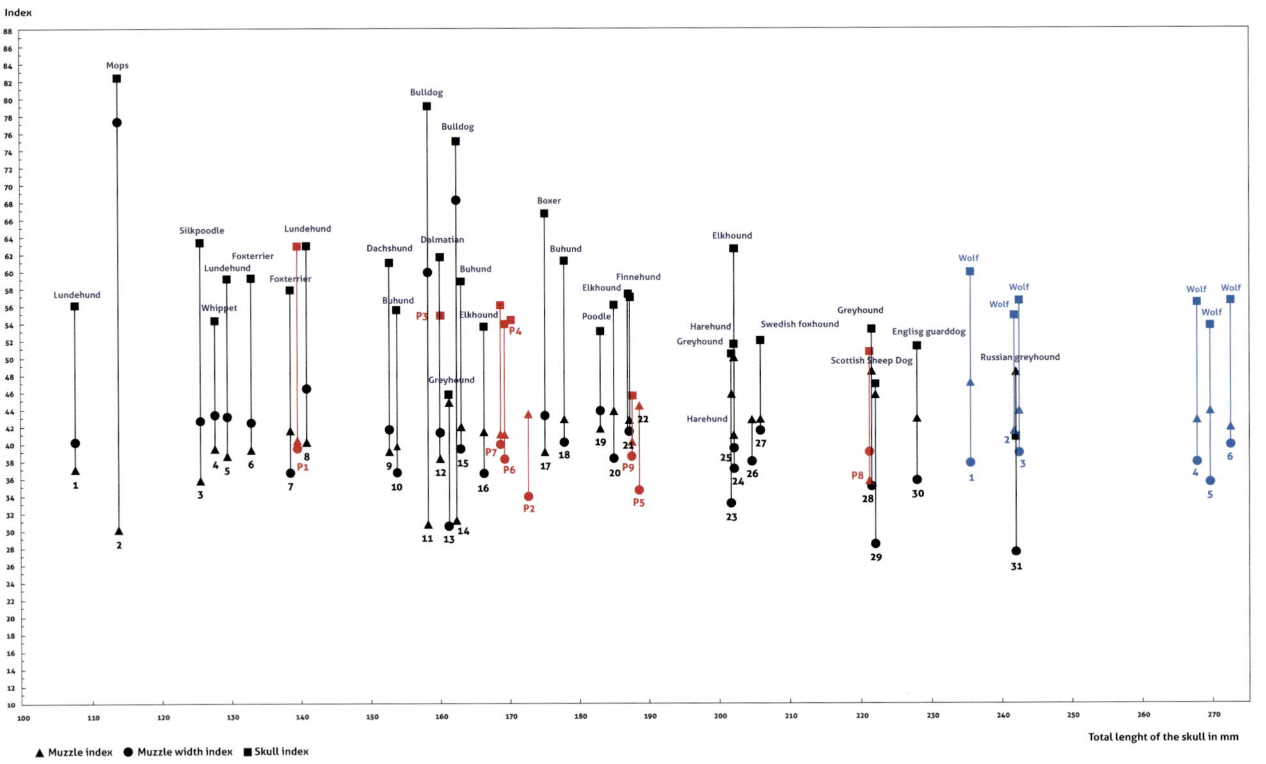


Figure 5. The skull, muzzle length and muzzle width indices for the 9 skulls of the *Peltuinum* dogs (red), the 31 skulls from modern dogs (black) and 6 skulls from modern wolf (blue), (Modified after Knoest 2015: 36).

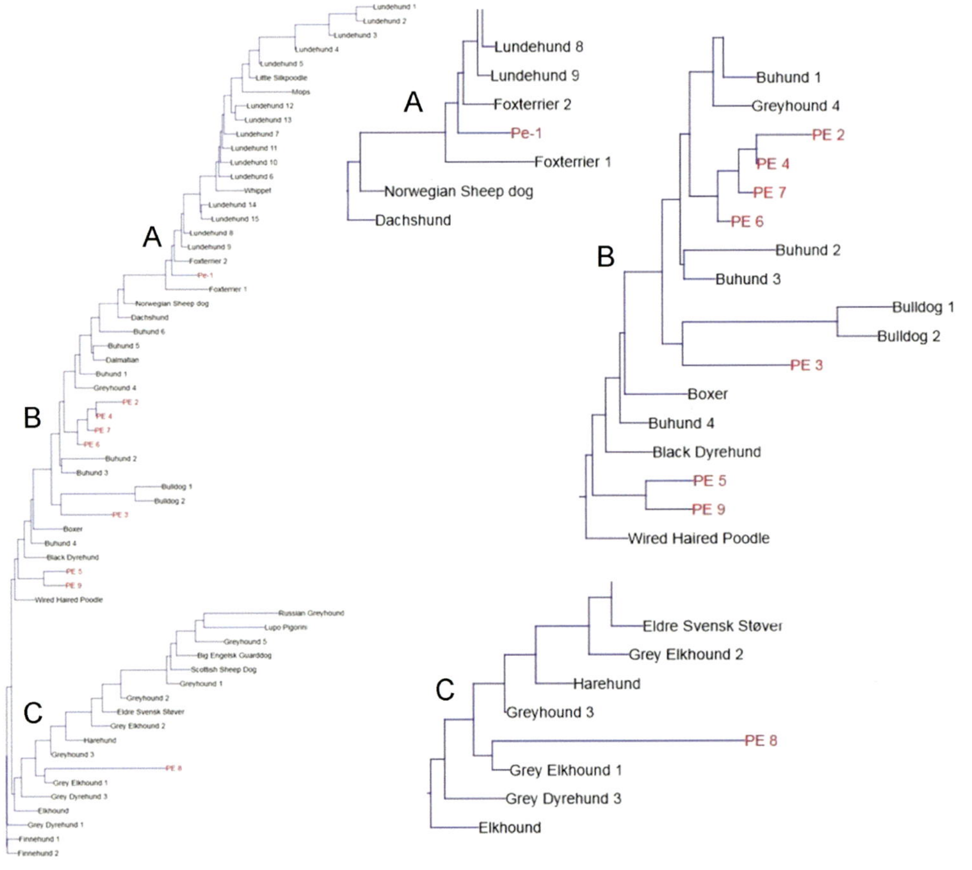


Figure 6. Neighbour joining clustering tree of selected skull measurements of some modern breeds and of the *Peltuinum* dogs. A-C details of the positions of the *Peltuinum* dogs (Data of modern breeds from Knoest 2015).

Although we believe that the comparison with current breeds should be made with prudence, we think it might be useful to have a reference to corroborate the differences found (Knoest 2015). The smaller individual (P1) with a short skull and muzzle is similar to the Terrier group (Federation Cynologique Internationale, Group 3), and resembles the Fox Terrier. The largest dog (P8), with the elongated muzzle, appears similar to the Greyhound group (FCI Group 10), with size and morphology comparable to the Greyhound. Most of the medium-sized skulls are related to the Spitz group (FCI Group 5), with size and morphology comparable to the Nordic Spitz for guarding and herding. Overall, the data seems to indicate the presence in prevalence of utility dogs in the shafts: herding, guarding, and hunting dogs and not companion dogs.

Traumatic and infectious pathologies are present. The analysis of the traces of modifications revealed two individuals with clear traces of blows on the skull (skull P3 and P4). On three skulls there were deformations on the orbits due to healed wounds (P2, P7, P8) probably due to blows in life (Figure 4).

Pathologies of infectious origin were found near the left dental arch of P8 where the left canine is broken and a halo of osteolysis on a reactive basis can be seen around the alveoli. The animal had probably suffered from an infection of dentaria. P5 skull shows rarefaction on the frontal bone possibly due to an unidentified pathology (Figure 3).

### *6.2 Limb long bones*

The femurs are the best preserved of the long limb bones and were used for size analysis of the dogs of *Peltuinum* with some evidence also taken from the other long bones. The bones are well preserved, mostly referable to adult but not senile. Measurements of 25 whole femurs came from theatre shafts, some certainly belonging to the same individual. There is considerable variability in the size of the femurs, ranging from 103–195.5 mm in length with an average of 147 mm (Tables 6–7).

This diversity is reflected in the height at withers which ranges between 31.2–58.8 cm with an average of 44.3 cm according to the Koudelka index (1885); and between 31.3–60.1 cm with an average of 44.9 cm following the Harcourt index (1974). There is no significant variation between the two indices.

The GL length and proximal width Bp of *Peltuinum* femurs were compared with those of 42 femurs belonging to dogs of different breeds analysed by Knoest (2015, p. 70). Comparison with actual breeds should always be considered with special caution, however, it is useful to have a reference of likely size and build. The data

Table 6. Femur measurements and height at withers of *Peltuinum* dogs (measurements: von den Driesch 1976).

| Femur of dogs | GL | Bp | SD | Bd | 3.01 * GL (Koudelka 1885) | -12,96 + 3.14 * GL K7Z (Harcourt 1974) |
|---|---|---|---|---|---|---|
| N | 25 | 24 | 25 | 25 | 25 | 25 |
| Min | 103.70 | 23.50 | 9.00 | 20.30 | 312.1 | 312.7 |
| Max | 195.50 | 39.60 | 14.80 | 34.30 | 588.5 | 600.9 |
| Mean | 147.14 | 31.87 | 11.82 | 26.63 | 442.9 | 449.1 |
| Dev | 24.66 | 4.71 | 1.71 | 3.76 | 74.2 | 77.4 |

Table 7. Humerus measurements and height at withers of *Peltuinum* dogs (measurements: von den Driesch 1976).

| Humerus | GL | Bp | Bd | SD | Bd | Dd | 3.37* GL (Koudelka 1885) | -26.54 + 3.37 * (Harcourt 1974) |
|---|---|---|---|---|---|---|---|---|
| N | 14 | 14 | 14 | 14 | 14 | 14 | 14 | 14 |
| Min | 98.2 | 19.0 | 23.0 | 8.2 | 22.3 | 15.4 | 320.2 | 293.6 |
| Max | 182.0 | 33.2 | 42.9 | 14.2 | 35.1 | 28.0 | 596.5 | 570.0 |
| Mean | 140.8 | 26.0 | 33.7 | 11.1 | 28.2 | 21.6 | 463.5 | 437.0 |
| Dev | 22.0 | 3.5 | 5.1 | 1.6 | 3.6 | 2.9 | 74.3 | 74.3 |

represented in the graph (Figure 7), shows that very small dogs are missing, for the rest, there are animals of different sizes. Although it is a constant that for the same length *Peltuinum* femurs have lower Bp and are therefore less robust.

The graph in Figure 8 basically shows that *Peltuinum* dogs may be distributed in three main groups.

- The first group of dogs, which are the smallest of *Peltuinum*, consists of three subgroups: a) small size 31–33 cm; b) medium-small size 38–40 cm; c) and 44 cm.
- The second comprises the group of medium-sized dogs (46–49 cm), where the largest number of measured elements is concentrated; only one element is located in the group of medium-large dogs with a height of 51–52 cm.

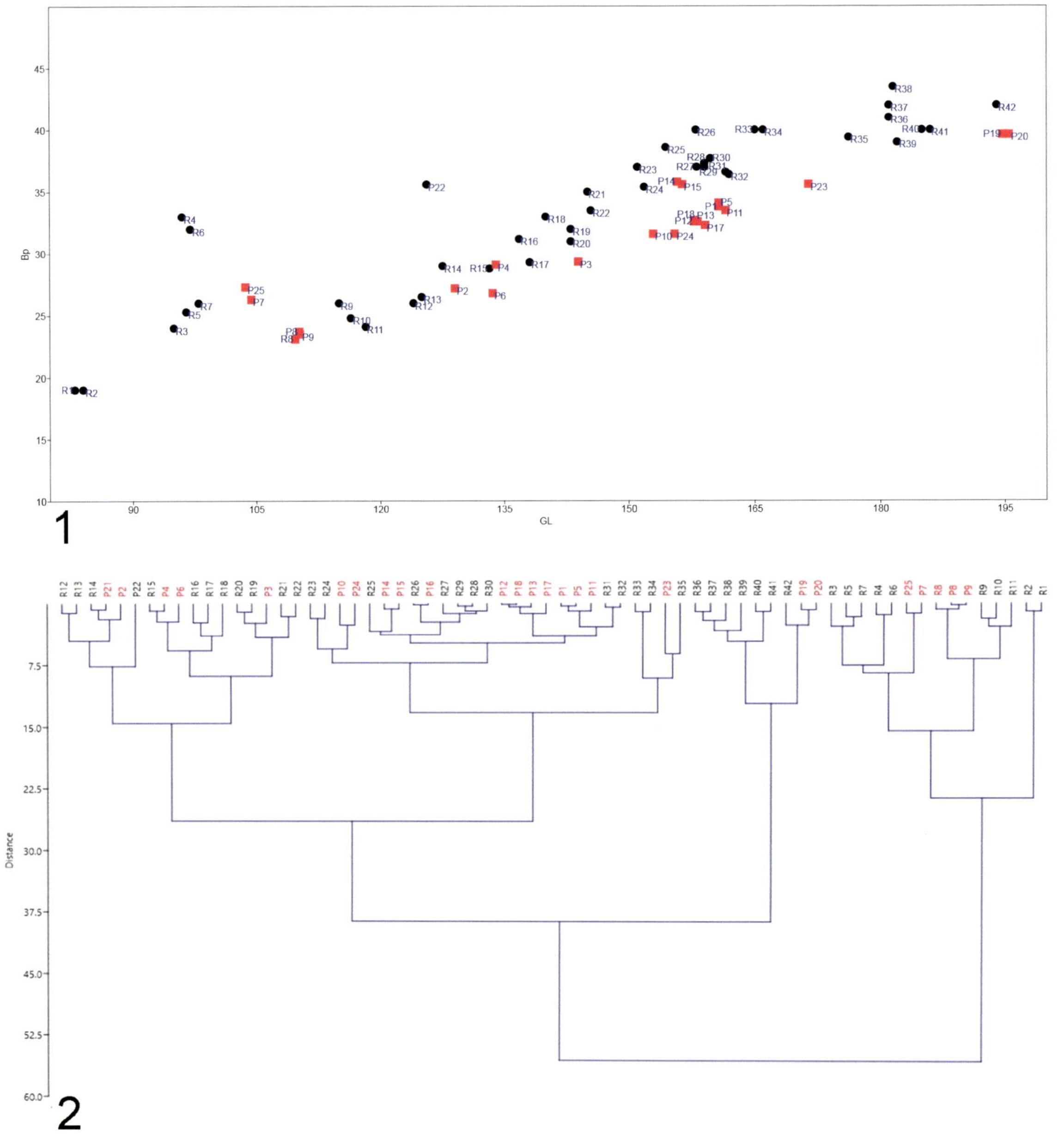


Figure 7. 1, scatterplot of the Bp (Breadth of the proximal end) of the femur against the GL (Greatest Length) of the *Peltuinum* femurs and modern breeds. 2, cluster analysis dendrogram of femurs of *Peltuinum* and of some modern breeds (Data of modern breeds from Knoest 2015).

- The third group consists of large dogs with a height of 59–60 cm.

The comparison with the dimensions of the humeri and ulnae confirms the variability found in the femurs. We note that one individual is smaller with a height at withers of 28.5–30 cm, probably the same as the smallest femur that gave values of about 31 cm. We found variability of 1–2 cm depending on the considered bones and the method of height estimation.

There are differences in the distribution of dogs in the shafts according to size particularly the dogs in shaft II have a great variability (between 31 and 60 cm), with differences also between US. In shaft II, in effect, small and medium dogs (31–50 cm) are found in the US 881 and small and medium-large dogs (31–58/60 cm) in US 859. The femurs from shaft IV, on the other hand, are more homogenous with minimal variability in height ranging from 45–48/50 cm. This difference was already evident from the analysis of the nine skulls.

### 6.3 The dogs of *Peltuinum*

Considering only the height at withers estimated from the femurs of *Peltuinum* and comparing it with those of

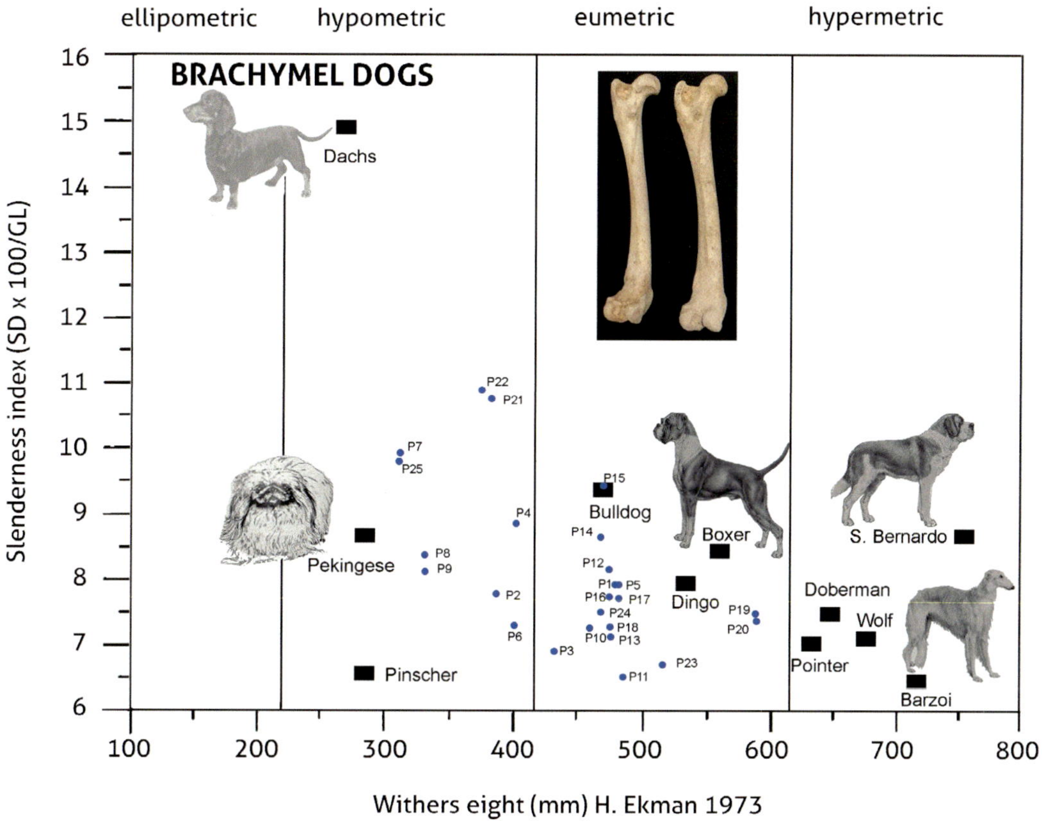


Figure 8. Graph of heights at withers from femurs of *Peltuinum* (After De Grossi Mazzorin and Tagliacozzo 2000).

Vindolandia (Bennett and Timm 2016 II, p. 106), if we exclude the absence of very small dogs (under 25/30 cm) and very large dogs (over 70 cm) the remainder fall within the variability of late antique dogs. Similarly, if one considers the data from some Roman Age sites analysed by De Grossi Mazzorin and Tagliacozzo (2000, fig. 10), the wither heights of *Peltuinum* fall within the size range of early and late Roman Age dogs, even if the six dogs from *Peltuinum* show greater variability. The variability in size observed in the femurs of *Peltuinum* may in some specimens reflect gender rather than breed diversity, in fact, both the morphological data of the skulls and the presence of os penis and fetuses (probable insertion of pregnant females) in the shafts show that dogs of both sexes are present.

The dog populations of the Roman period and *Peltuinum* show a great variation in size and many morphotypes and, as considered by various authors, this is found especially in urban contexts (De Grossi and Tagliacozzo, 2000). The variability found at *Peltuinum* is confirmed by the data known from ancient sources, in fact, the Romans had and bred a wide variety of dog types. The Romans classified dogs as *pastoricus* (shepherds), *villaticus* (guard dogs) and *venaticus* (hunting dogs); some dogs were also used in warfare. Marcus Terentius Varro (*De Re Rustica*) describes the characteristics of the shepherd dog as having an intermediate body between the agile hunting dog and the robust guard dog. Hunting dogs were further classified according to their aptitude and morphological characteristics: *seguges* (bloodhounds), with a particularly sensitive sense of smell, for sniffing out game; *celeres* (greyhounds), slender and fast, for chasing game; *pugnaces* (molossoids), strong and robust, for attacking game.

In addition to the dogs owned and used for specific purposes, we cannot exclude the presence of stray dogs that lived in urban areas or in the immediate vicinity where they could easily find food and shelter. Obviously, it is not easy to find exact correlations between the types of dogs found in *Peltuinum* and those known from ancient Roman sources. The hypothesis of the presence of dog-breeding in Roman times since morphological and size variability can also be confirmed by the dogs from *Peltuinum*.

The analysis of the traces of modifications suggests the cause of death of at least two individuals with clear traces of damage of the skull (skulls P3 and P4 ). Conversely, three skulls show depression from healed wounds (P2, P7, P8) probably due to blows. These wounds appear to be the result of blows and do not appear to be the result of aggression between dogs or other animals such as

Table 8. Dogs and Humans overview of the three different contexts.

| | SHAFTS | TOWER | SEWER - US 1263 |
|---|---|---|---|
| DOGS MNI | 68 | 12 | 1 |
| Complete individuals | yes | yes | No |
| Human MNI | 87 | 35 | 2 |
| Fetal/Newborns MNI | 84 | 8 | 0 |
| Interpretation | Ritual deposition of dogs and newborn | Retrieval and re-deposition from a nearby funerary area | Disturbed deposition |

in hunting accidents. The presence of healed wounds could suggest that the animals were cared for by their owners, but another possibility could be that they were village or street dogs and therefore nuisances, badly tolerated, beaten, and driven out of urban areas.

Even if on the other skulls there are no evident traces of killings, it is not excluded that other methods, e.g., strangulation, may have been used, leaving less evident marks.

### 7 Conclusions

This analysis allowed to better define the characteristics of the adult dogs that were placed in the *Peltuinum* shafts to accompany the dead infants on their afterlife journey. They were mainly medium-sized dogs, whose morphological analysis shows that their appearance is akin to that of the Spitz group. There is also a small dog that resembles a Fox Terrier and another resembling a large Greyhound, and the study of the long bones confirms the sizes corresponding to this type of dog (Figures 5–6). The dogs are adults of different ages, but not senile; they include males and possibly pregnant females. Perhaps a choice was made in the type of dog to be included in the shafts: in shaft II there is variability in size and morphology (small, medium, and large), in shaft IV only medium-sized dogs of similar morphology (Figures 7–8).

The morphological analysis and the present comparison seem to indicate the prevalence of utility dogs in the shafts: herding, guarding, and hunting dogs; missing very small dogs (under 25 cm, pet dogs) and very large dogs (over 70 cm).

The traumas on the skulls indicate some dogs may have been maltreated during life; while in two cases there is a clear evidence of intentional killing (Figure 4).

The pathologies found at the skeletal level are rare (except perhaps one case with unidentified pathology and one with inflammation of a broken tooth and probable osteoarthritis on the long bones of a medium-sized individual), but the traumas found at the cranial level show maltreated dogs. The wounds have healed, which may indicate that perhaps the dogs were cared for and therefore had someone caring for them, however, it is doubtful whether they were street dogs who had been beaten and chased away. Two skulls show clear traces of killing. When it comes to the choice of dog to be sacrificed, the greatest doubt remains as to whether these were animals owned by the relatives of the dead infants or whether they were actually street dogs caught and killed to accompany the infants to the world of the dead.

In antiquity, the dog played various roles: it was regarded as a companion and guide in the afterlife journey, sacrificed in special purification rites to female deities of the chthonic world, buried in a ritual closure of sacred areas, and immolated as a guardian in foundation rites of structures or city-walls; data are attested both in archaeological contexts (Camp 1986; Osanna 1993; Snyder 1999; De Grossi Mazzorin and Minniti 2001, 2006; Chenal-Venarde 2006; Bourbou and Themelis 2010; Fiore and Salvadei 2014; Pedrucci 2014; Fiore 2016; Trantalidou 2016; Migliorati *et al.* 2018; Sperduti *et al.* 2018) and in Literary Sources (Plut., Quaest. Rom. 52, 277[1]; Plin., Nat. Hist. XXIX, 58[2]).

The funerary association of human infants and dogs is documented in Greece in the Agora of Messene in the 3rd cent. BC (Bourbou and Themelis 2010) and in a well from *Kolonos Agoraios* of Athens in levels dated 2nd BC (Liston *et al.* 2018).

The shaft in Athens contained the remains of 450 children, mostly infants or full-term fetuses, and those

[1] Plutarch, Roman Questions by published in Vol. IV of the Loeb Classical Library edition, 1936. https://penelope.uchicago.edu/Thayer/e/roman/texts/plutarch/moralia/roman_questions*/c.html (11.01.2021)

[2] Pliny the Elder (Translated by H. Rackham, W.H.S. Jones and D. E. Eichholz, 1868–1963). *Natural History*. London: William Heinemann

of at least 150 dogs. There are other similarities between the two contexts:

- remains of other animal species with traces of butchery,
- prevalence of medium-sized dogs,
- some dog skulls retain traces of blows with skull crushing,
- cases of healed infectious and traumatic diseases (predominantly traumatic diseases in *Peltuinum*).

In Athens healed wounds were interpreted as evidence of free-ranging urban dogs. Animals are kept at a distance and therefore not given special consideration in life.

Apart from the different chronology, the main difference is the location of the depositions. In *Peltuinum* infants and animals were laid in the few small rectangular shafts (cm 85 x 55) related to the Roman theatre, while in Athens they were deposited in one single circular sacred well (one m diameter) associated with the sanctuary of Afrodite Urania (Liston *et al.* 2018). This feature shows the intention to put the infants under the protection of the goddess, while in *Peltuinum* there appears to be no such purpose (but the different chronology has to be taken into account). Moreover, in *Peltuinum* there are numerous newborn puppies and foetuses and two whole newborn foals which are not present in the well in Athens.

The dog-infant association has been interpreted as evidence of a purification sacrifice for the premature death of infants (Osanna 1993; Little 1999, De Grossi Mazzorin and Minniti 2006, Migliorati *et al.* 2018). A similar association was also found in the infant necropolis of Lugnano in Teverina, a case closer to *Peltuinum* in relation to time and space than the Kolonos Agoraios in Athens. The 47 infant tombs and the skeletons of 12 puppies under 6 months of age and a single subadult dog (Soren *et al.* 1995, 1999) testify the spread of this type of ritual in central Italy in the 5th century AD, increasingly corroborating the hypothesis of the intentional deposition of dogs and infants. Also in this case, we must underline the differences. Apart from the absence of foals, the Lugnano infants' tombs are single depositions and this feature places the case among the typical depositions of humans associated to dogs.

It is important to point out the presence of various contexts with the human-dogs association at *Peltuinum* (Table 8). Besides the deposition in the shafts, the workers' room δ (sewer) suggest a non-intact burial. The tower, with food waste, ceramics, human bones and abundant dog remains, would suggest the identification of a dumping area linked to a land reclamation. In the latter case it could have been an ossuary and perhaps at the time of the reburial, rituals involving dogs could have been performed.

## Acknowledgements

We are grateful to Eugenio Cerilli for assistance in preparing the statistical analysis and to Ana Pavlova for the first revision of the English text.

## References

Barone, R. 1981. *Anatomia comparata dei mammiferi domestici. 3, Splancnologia. (Italian edition by R. Bortolani).* Bologna: Edagricole.

Barone, R. 1995. *Anatomia comparata dei mammiferi domestici. 1, Osteologia* (Italian edition by R. Bortolani, E. Callegari). Bologna: Edagricole.

Bennett, D.G. and R.M. Timm 2016. The dogs of Roman Vindolanda, Part II: Time-stratigraphic occurrence, ethnographic comparisons, and biotype reconstruction. *Archaeofauna* 26: 93–112.

Bökönyi, S. 1970. A new method for the determination of the number of individuals in animal bone material. *American Journal of Archaeology* 74: 291–292.

Bourbou, C. and P. Themelis 2010. Child Burials at Ancient Messene, in A-M. Guimier-Sorbets, Y. Morizot (eds), *L'Enfant et la mort dans l'Antiquité: Nouvelles recherches dans les nécropoles grecques. Le signalement des tombes d'enfants*, Actes de la table ronde internationale (Athènes 2008), Athènes: 111–128.

Brassard, C. and C. Callou 2020. Sex determination of archaeological dogs using the skull: evaluation of morphological and metric traits on various modern breeds. *Journal of Archaeological Science: Reports* 31, 102294, https://doi.org/10.1016/j.jasrep.2020.102294.

Bull, G. and S. Payne 1982. Tooth eruption and epiphysial fusion in pigs and wild boar, in B. Wilson, C. Grigson, S. Payne (eds), *Ageing and Sexing Animal Bones from Archaeological Sites* (British Archaeological Reports, British Series 109): 55–72.

Bullock, D. and J. Rackam 1982. Epiphysial fusion and tooth eruption of feral goats from Moffatdale, Dumfories and Galloway, in B. Wilson, C. Grigson, S. Payne (eds), *Ageing and Sexing Animal Bones from Archaeological Sites* (British Archaeological Reports, British Series 109): 73–80. Oxford: BAR.

Camp, J.M. 1986. *The Athenian agora: excavations in the heart of classical Athens*, London: Thames and Hudson.

Chaplin, R.E.C. 1971. *The study of Animal Bones from Archaeological Sites.* London: Seminar Press.

Chenal-Velarde, I. 2006. Food, rituals? The exploitation of dogs from Eretria (Greece) during the Helladic and Hellenistic periods, in L. Snyder and E.A. Moore (eds), *Dogs and People in Social, Working, Economic or*

*Symbolic Interaction, Proceedings of the 9th Conference of the International.*

Cruz-Uribe, K. and R.G. Klein 1984. *The analysis of animal bones.* Chicago London: The University of Chicago press.

De Grossi Mazzorin, J. and C. Minniti 2001. Caratterizzazione archeozoologica: le sepolture di cani, in P. di Manzano (ed.), *ad deverticulum. Scavi archeologici lungo la bretella Nomentana-GRA*, Roma: 81–93.

De Grossi Mazzorin, J. and C. Minniti 2006. Dog Sacrifice in the Ancient World: A Ritual Passage?, in L.M. Snyder and E.A. Moore (eds), *Dogs and People in Social, Working, Economic or Symbolic Interaction*, Proceedings of the 9th Conference of the International Council of ArchaeoZoology, Durham 23–28 agosto 2002: 62–66.

De Grossi Mazzorin, J. and A. Tagliacozzo 2000. Morphological and osteological changes in the dog from the Neolithic to the Roman period in Italy, in *Dogs through time: an archaeological perspective*: 141–161. Oxford: Archaeopress.

Driesch von den, A. 1976. A guide to the measurement of animal bones from archaeological sites. *Peabody Museum Bulletins* 1: 1–148.

Evans, H.E. 2013. *Miller's Anatomy of the Dog.* Saunders: Philadelphia.

Fiore, I. 2016. Analisi archeozoologica dei resti ossei animali da alcune tombe della necropoli del Picentino, proprietà Colucci. S. De Natale, D'Agostino B. and Gastaldi B. (eds) *Pontecagnano. II 7. La necropoli del Picentino. Tombe della Prima Età del Ferro dalla proprietà Colucci*: 139–143. Naples: Centre Jean Bérard.

Fiore, I., L. Migliorati, A. Pansini, T. Sgrulloni and A. Sperduti 2018. The tower, the sewer, the shafts. Dog and human sepultures at Peltuinum (L'Aquila, Italy). in I. Fiore and F. Lugli (eds), '*Dogs, Past and Present: an Interdisciplinary Perspective*', Annali dell'Università di Ferrara. Sezione: Museologia Scientifica e Naturalistica 14: 32-33. Ferrara: Università degli Studi di Ferrara.

Fiore, I. and L. Salvadei 2014. I resti ossei di cani e neonati rinvenuti nei pozzetti II e III del teatro romano di *Peltuinum*: analisi preliminari. *Rendiconti della Pontificia Accademia* 84, 2011–12: 387–402.

Gentry, A., J. Clutton-Brock and C.P. Groves 2004. The naming of wild animal species and their domestic derivatives. *Journal of Archaeological Science* 31: 645–651.

Hammer, Ø., D.A.T. Harper and P.D. Ryan 2001. PAST: Paleontological statistics software package for education and data analysis. *Palaeontologia Electronica* 4 (1).

Harcourt, R.A. 1974. The Dog in Prehistoric and Early Historic Britain. *Journal of Archaeological Science* 1: 151–175.

Klein, R.G and K. Kruz-Uribe 1984. *The Analysis of Animal Bones from Archeological Sites*: 266. Chicago and London: University of Chicago Press.

Knoest, J.J.T. 2015. On the morphology of the domesticated dog in medieval Norway. Master thesis, The University of Bergen.

Koudelka, F. 1885. Das Verhältnis der ossa longa zur Skeletthohebei den Saugetieren. Verhandlung des Naturforschung, *Vereines in Brunn* 24: 127–153.

Liston, M.A., S.I. Rotroff and L.M. Snyder 2018. *The Agora bone well* (Hesperia Supplement 50). Athens: ASCSA.

Little, L.M. 1999. Babies in Well G5:3: preliminary results and future analysis. *American Journal of Archaeology* 103: 284.

Migliorati, L. 2014. *La città e il territorio*, in AA.VV., *Trent'anni di ricerche a Peltuinum. Rendiconti della Pontificia Accademia* LXXXIV, 2011–12: 351–386.

Migliorati, L. 2014. Piceno meridionale e Sannio vestino: gli insediamenti preromani e le opzioni di Roma, in P.L. Dall'Aglio, C. Franceschelli, L. Maganzani (ed.), *Atti IV Convegno Internazionale Studi Veleiati* (*Veleia-Lugagnano Val d'Arda* 2013): 313–330. Bologna.

Migliorati, L. 2015. *Peltuinum: una città sul Tratturo Magno*, in L. Pani Ermini (ed.), *Abruzzo sul Tratturo Magno*, Roma: 163–173.

Migliorati, L., T. Sgrulloni, G. Casazza 2018. Nuove indagini sulle fortificazioni di *Peltuinum*. *Scienze dell'Antichità* 24, 1: 295–308.

Migliorati, L., T. Sgrulloni, A. Vecchione 2021. Peltuinum (Prata d'Ansidonia, L'Aquila). Il teatro romano e il cantiere tardo-medievale di demolizione: lavori in corso. *Notizie Scavi* n.s. I, I: 107–144.

Osanna, M. 1993. Il problema topografico del santuario di Afrodite Urania ad Atene. *Annuario della Scuola Archeologica di Atene* LXVI-LXVII: 73–95.

Payne, S. 1973. Kill-off patterns in sheep and goats: the mandibles from Asvan Kale. *Anatolian Studies* 33: 281–303.

Payne, S. 1985. Morphological Distinctions between the Mandibular Teeth of Young Sheep, Ovis, and, Goats, Capra. *Journal of Archaeological Science* 12: 139–147.

Pedrucci, G. 2014. Cuccioli d' uomo, cuccioli di cane. Nuove proposte per l'interpretazione del materiale proveniente dalla necropoli di Lugnano in Teverina, in C. Terranova (eds), *La presenza dei bambini nelle religioni del mediterraneo antico*: 185–216. Roma: Aracne.

Phillips, C., I.L. Baxter and M. Nussbaumer 2009. The application of discriminant function analysis to archaeological dog remains as an aid to the elucidation of possible affinities with modern breeds. *Archaeofauna* 18: 51–64.

Schmid, E. 1972. *Atlas of Animal Bones (Tierknochenatlas).* New York and Amsterdam: Elsevier.

Silver, I.A. 1969. The Ageing of Domestic Animals. *Science in Archaeology: A Comprehensive Survey of Progress and Research* (revised edition): 250. London: Thames and Hudson.

Snyder, L.M. 1999. The animal bones from Well G5:3: domestic debris, industrial debris, and possible evidence for the sacrifice of domestic dogs in Late Hellenistic Athens. *American Journal of Archaeology* 103: 284.

Soren, D., D. Fenton and W. Birkby 1995. The Later Roman Infant Cemetery near Lugnano in Teverina, Italy: some implications. *Journal of Palaeopathology* VII, 1: 13–42.

Soren, D. and N. Soren 1999. *A roman Villa and a Late roman Infant Cemetery. Excavation at Poggio Gramignano (Lugnano in Teverina)* (Bibliotheca Archaeologica 23). Rome: L'Erma di Bretschneider.

Sperduti, A., L. Migliorati, A. Pansini, T. Sgrulloni, P.F. Rossi, V. Vaccari and I. Fiore 2018. Differential burial treatment of newborn infants from Late Roman Age. Children and dogs depositions at Peltuinum, in V. Nizzo (ed.), *Antropologia e archeologia a confronto: Archeologia e antropologia della morte*: 299–315. Rome: Romarché.

Trantalidou, K. 2016. Companions from the Oldest Times: Dogs in Ancient Greek Literature, Iconography and Osteological Testimony, in L.M. Snyder, E.A. Moore, (eds) *Dogs and people in social, working, economic or symbolic interaction. Proceedings of the 9th ICAZ Conference, Durham*: 96–120. Oxford: Oxbow Books.

Tulipani, L. 1996. Da *Peltuinum* alla *Civita Ansidonia*, in A. Campanelli (ed.), *Peltuinum. Antica città sul tratturo*: 50–61. Pescara.

# 4.10 The Dog as a Companion in Life and Death: The Case Study of Dog Burials in a Human Grave (VII - VI BC) Loc. Collina dei Gelsi - Poggio Sommavilla (RI)

Francesca Santini (†)

Independent Researcher
Corresponding: iva.fiore@gmail.com and luglifrance@gmail.com

**Abstract**

The purpose of this study is to clarify the use and presence of dogs in funerary contexts, starting from the case study of the chamber tomb in Loc. Collina dei Gelsi (Poggio Sommavilla, Collevecchio - RI) in a site in Central Italy, dating back to the 7th - 6th century BC, in which five dog skeletons were recovered in a multiple human grave. I will consider the role of this animal in daily life, its closeness to humans as a precious companion in hunting-sporting activities and war, as a guardian and a pet. By drawing on and examining different pieces of evidence, and by taking into consideration archaeological remains it is possible to depict a complete scenario of the preferential relationship between humans and dog which leads to an interpretation of the dog burials.

**Keywords**: dog burials, animal sacrifice, domestication.

## 1 Introduction

In this work I attempt to understand the use of dogs in funerary contexts and to interpret their presence with the human remains, taking into consideration the role of this animal in daily human life.

Compared to other domesticated animals, the dog has a preferential relationship with humans which involves participation in every aspect of social life, dogs being close to humans even in death.

During a rescue excavation in 1989[1] five dog skeletons were unearthed, buried with humans in a chamber tomb[2] dating from the 7th to 6th century BC[3], located in Central Italy, in Località Collina dei Gelsi (RI), and probably connected to the nearby Necropolis of Poggio Sommavilla (Figure 1).

In the grave, 3 human skeletons[4] were interred with five dogs[5], all of them oriented N-S except for the dog E which was oriented E-W (Figure 2).

Due to the site's emergency state, four soil blocks were realised in order to preserve the integrity of the dog skeleton remains, which could be then examined in the Lab. Only after several years, in 2014, was it possible to follow micro excavation procedures and fulfill the archaeozoological study[6] (Figure 3).

## 2 Material and archaeozoological investigation

From the position of the skeletons could be deduced that these animals had been carefully laid, rather than thrown, into the tomb. In fact, the taphonomic analysis revealed neither human modifications nor pathologies on the bones.

[1] The rescue excavation was headed by the then Soprintendenza per i Beni Archeologici del Lazio.

[2] The chamber tomb was entirely carved into volcanic rock (the typical geological formation of this area known in Italy as *tufo*). It was found with the ceiling collapsed and destroyed probably due to agricultural works. This tomb includes an entrance and a long corridor (called *dromos*) leading to a single rectangular-shaped room with three platforms. This chamber tomb was characterised by rich funeral furnishings, including numerous pottery objects (such as *oinochoai*, Sabine amphoras, *kantharoi*), bronze and iron objects including interesting bronze footwear (called *calzari* in Italy) (Alvino 1997).

[3] See Alvino 1997.

[4] The main inhumation was complete and interred on the Eastern platform, some long bones of the second one were uncovered during the excavations; some bones of the third young individual were recovered during the micro-excavation of the ground blocks of the dogs, in the same block as dog D (Santini 2014).

[5] The dogs are named with alphabetic capital letters A-E. In details, dog A was buried on the Northern platform above the main inhumated head; three dogs B, C and D were found on the Eastern platform close to each other; dog E was recovered isolated on the Southern platform (Santini 2014).

[6] I studied the five dog skeletons in order to carry out my dissertation for the MSc degree in Osteoarchaeology at the University of Sheffield. The micro-excavation was realised thanks to the cooperation between the then Soprintendenza per I Beni archeologici del Lazio (now SABAP-FR LT RI) (G. Alvino) and the Protohistoric Department of Palazzo Altemps-Soprintendenza Speciale Archeologia, Belle Arti e Paesaggio di Roma (A. Capodiferro).

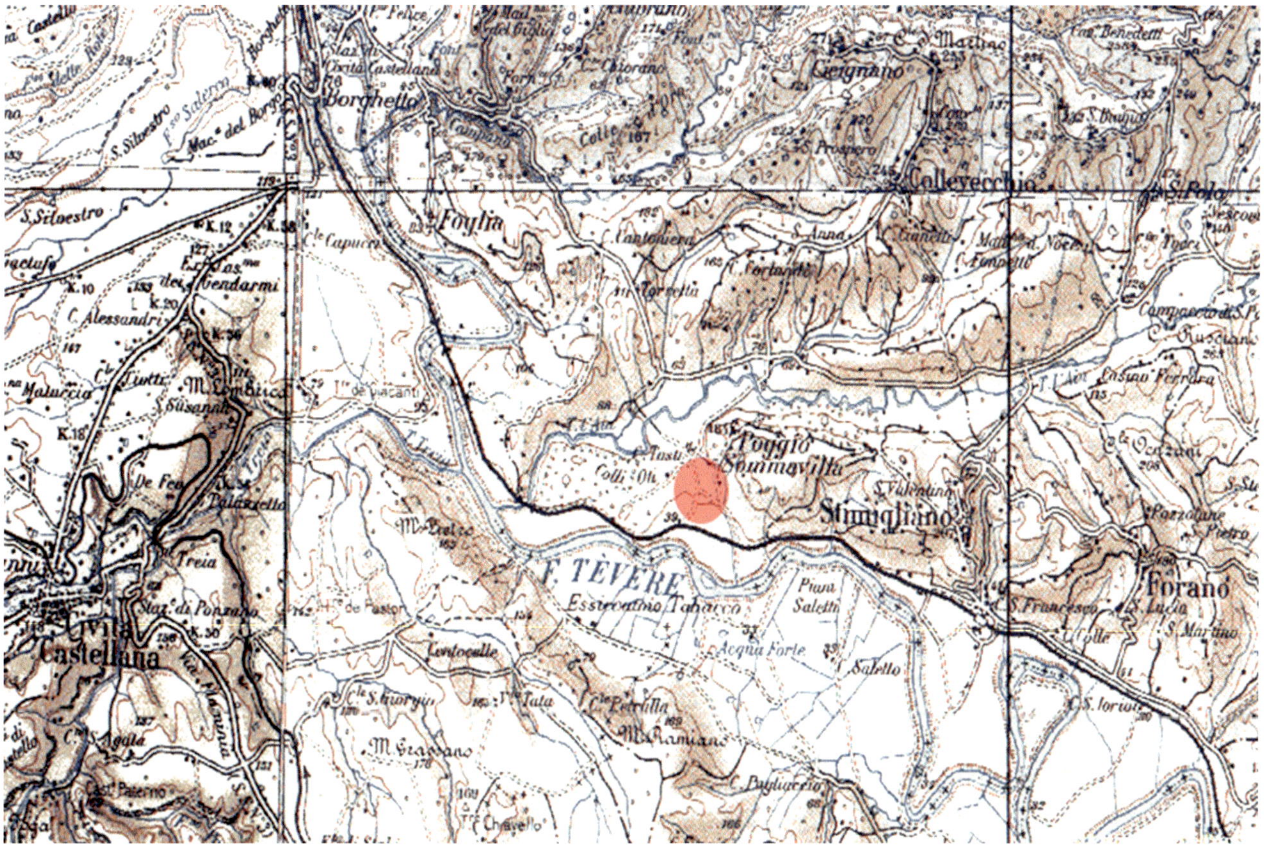


Figure 1. Topographical position of site Collina dei Gelsi (Poggio Sommavilla, Collevecchio - RI) (IGM 100.000).

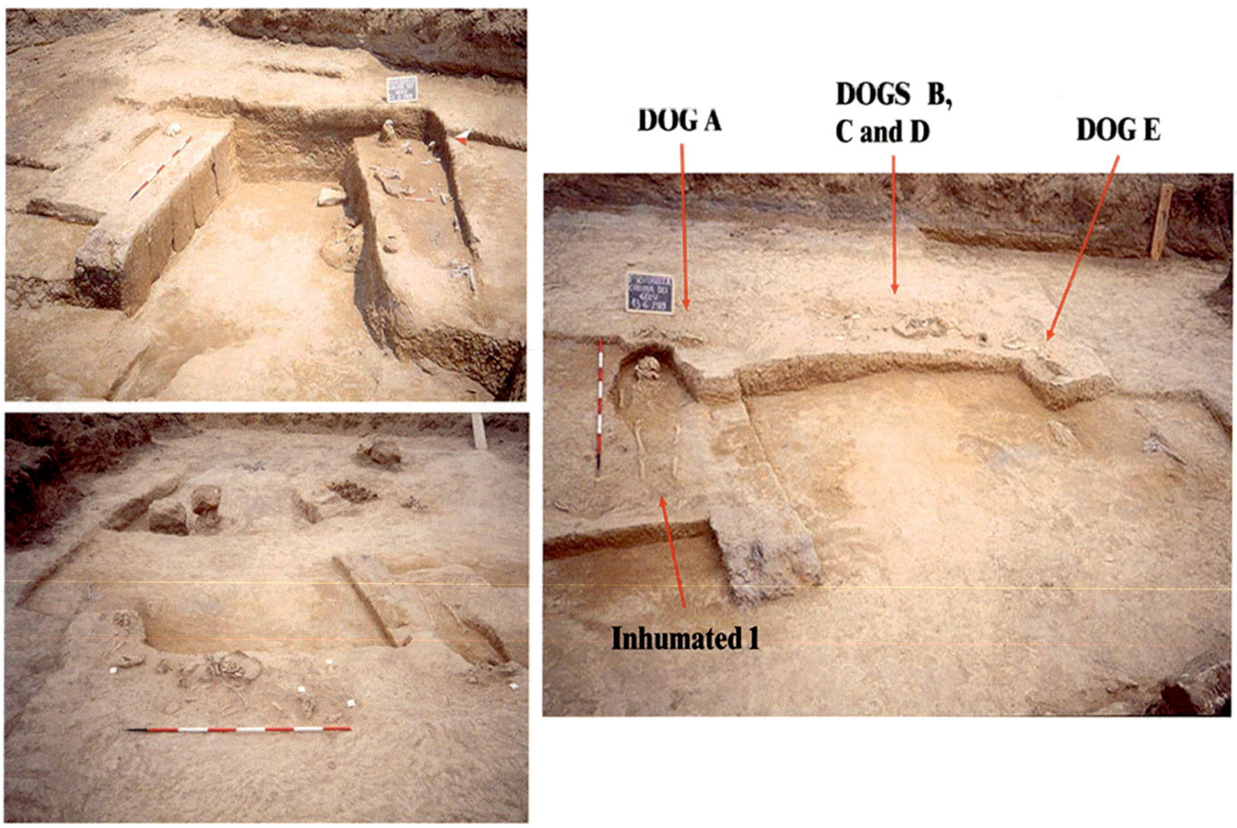


Figure 2. Chamber Tomb of Collina dei Gelsi with human and dog skeletal remains, (by Archives of SBAL).

Figure 3. The ground blocks containing the five dog skeletons (by F. Santini).

Despite the imperfect state of preservation of some of them, it was possible to subject the dog skeletons to morphometric and morphological analysis[7]. The investigation lead to the following results: dog A shows an incomplete skeleton, the sex is uncertain and the age at death could be established as not less than 6 years[8]; dog B[9] has a partially preserved skeleton (only the anterior portion), its sex was impossible to establish, and the age at death could be established as not less than 5 years; dog C shows an almost complete skeleton of a female, age between 2–4 years[10]; dog D was partially preserved, unestablished sex, and with an age not greater than 12–18 months[11], close to the dog an infant skull[12] was found; the skeleton of dog E was almost intact, it is a female over 6 years of age[13].

[7] The five dog skeletons were identified (Barone 1998, Schmid 1972) and analysed to establish sex (Grigson 1978) and age at death (Barone 1998; Cornevin, Lesbre 1894; Horard-Herbin 2000; Silver 1969). The measurements were collected and processed by hand, using electronic calliper and by measuring box, following the Driesch von den method (Driesch 1976).

[8] The fore limbs are retracted underneath the body, and although the hind ones are not completely preserved it was possible to confirm that they were also bent. The absence of the *baculum* or *os penis* could not indicate with certainty that the individual was a female because of the incomplete state of the skeleton. One of the ribs shows a healed simple fracture, with evident *callus* formation as a result of a direct traumatic event probably due to an impact or blow (Santini 2014).

[9] The skeleton of this dog was partially preserved and appeared to have been cut in half probably as consequence of the ground block during the excavation in 1989. Only the anterior portion was removed, including the axial skeleton up to the 3rd lumbar vertebra, and leaving the rest of the skeleton. It was lying on the right side with its forelimbs retracted underneath the body and partially below the head. Due to the poor preservation of this individual it was impossible to establish its sex. Based especially on the tooth wear stages, it could be assumed that this specimen was not less than 5/-5 and half years old (Santini 2014).

[10] The skeleton is complete. The dog lay on the left side, as in the natural rest position, with its right forelimbs markedly flexed and the left extended under the first one, and the hind limbs retracted. Moreover, during the micro-excavation a bowl of reddish impasto was recovered in correspondence with the final cervical and early thoracic vertebrae. Therefore, the dog seems to be placed just above this bowl, causing dislocation at the vertebral column. The *baculum* was not recovered, which means that this dog is presumably a female. The tooth wear stages suggest an age between 2–4 years. In the sediment underlying this skeleton, some ceramic vessel remains were found (Santini 2014).

[11] This dog skeleton only has the anterior portion in anatomical connection, including the two emi-mandibles joining to the skull found near to the dog B skeleton in its block. Because of the bad preservation it was impossible to establish the original anatomical position. The tooth wear stages indicate an age at death between 12–18 months (Santini 2014).

[12] Close to the humerus of the dog the upper portion of a skull and some teeth of an infant have been found, revealing an age between 3–5 years (Santini 2014).

[13] The skeleton is complete, and it appears to have been crouching down, lying on its belly with the forelimbs below the thoracic cage and the head, with the head lay on the right side. Conversely, the hind limbs position was unclear, probably as it had been subjected to dislocation and re-placement due to farming works (in particular to plough action) and because of having been outdoors for many years. Due to the completeness of this individual the absence of the *baculum*

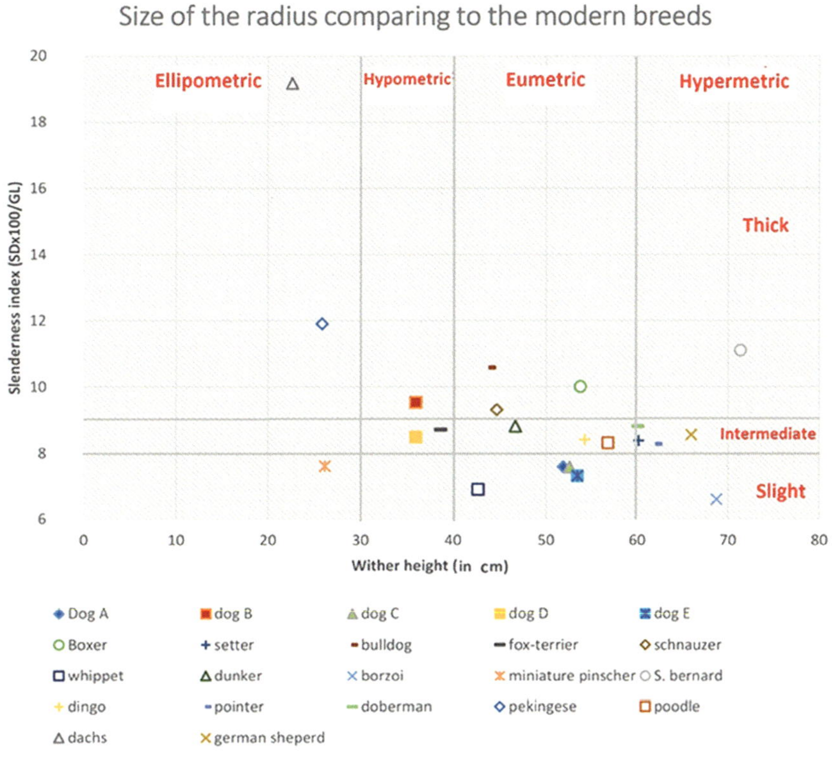


Figure 4. Scatter plot comparing wither height/ slenderness index between the five dogs and modern breeds (by F. Santini).

By comparing their withers heights with those of several modern breeds[14], it could be assumed that dogs B and D[15] had the same height as the modern Fox-Terrier; on the other hand, dogs A, C and E are very similar to the modern Boxer and Poodle (Figure 4).

## 3 Results

### *3.1 Dogs as companions in life*

Regarding the presence of dogs in archaeological sites, dog remains are relatively abundant in most archaeological contexts[16]. This shows an extraordinary interdependence between humans and dogs throughout the past 12000 years, when their friendship began with the domestication of dogs, which can be dated at least to the Mesolithic[17]. The relationship between man and dog can be defined as a long and intensive symbiosis, with visible benefits on both sides. Several studies offer important insights into the way in which this animal achieved a crucial role in both prehistoric and historical societies[18]. This close relationship probably started when the antagonism, favoured by the Great Neolithic Revolution, was not yet present, and hunters and wolves/first dogs cooperated in hunting activities.

Hence, domestication was also promoted by the wolf social structure, which is very similar to the human one[19], thus predisposing it both genetically and ethologically.

In order to depict a clear scenario of the interpretation of dog burials, the role of this animal in daily life must first be considered, where it is very close to humans as a useful companion, and then move towards the ritual sphere.

The association between man and dog concerns several fields: hunting, sporting activities, guarding and defence, war, the dog's role as a pet and in game time,

suggests that it was a female with an age of more than 6 years (Santini 2014).

[14] Regarding the post-cranial measurements, mainly the GL, Bp and Bd of femur, humerus, tibia and radius were used to calculate their wither heights (Clark 1995, Harcourt 1974, Koudelka 1884) in order to compare them with modern breeds and dogs of other chronological periods (Santini 2014).

[15] It was taken into consideration despite its young age.

[16] See Santini 2014.

[17] For more details on the domestication of the dog see Verginelli *et al.* 2006, De Grossi Mazzorin 2008a, Masseti 2008.

[18] See Benecke 1987, Verginelli *et al.* 2006, De Grossi Mazzorin 2008b.

[19] The wolf, being a carnivore, developed one of the highest levels of social behaviour showing solid group cohesion based on: strong social links, low aggressiveness between members of the same pack, a strictly hierarchical, family size system (dominant and subordinated individuals of both sexes), a social structure very similar to that of human societies.

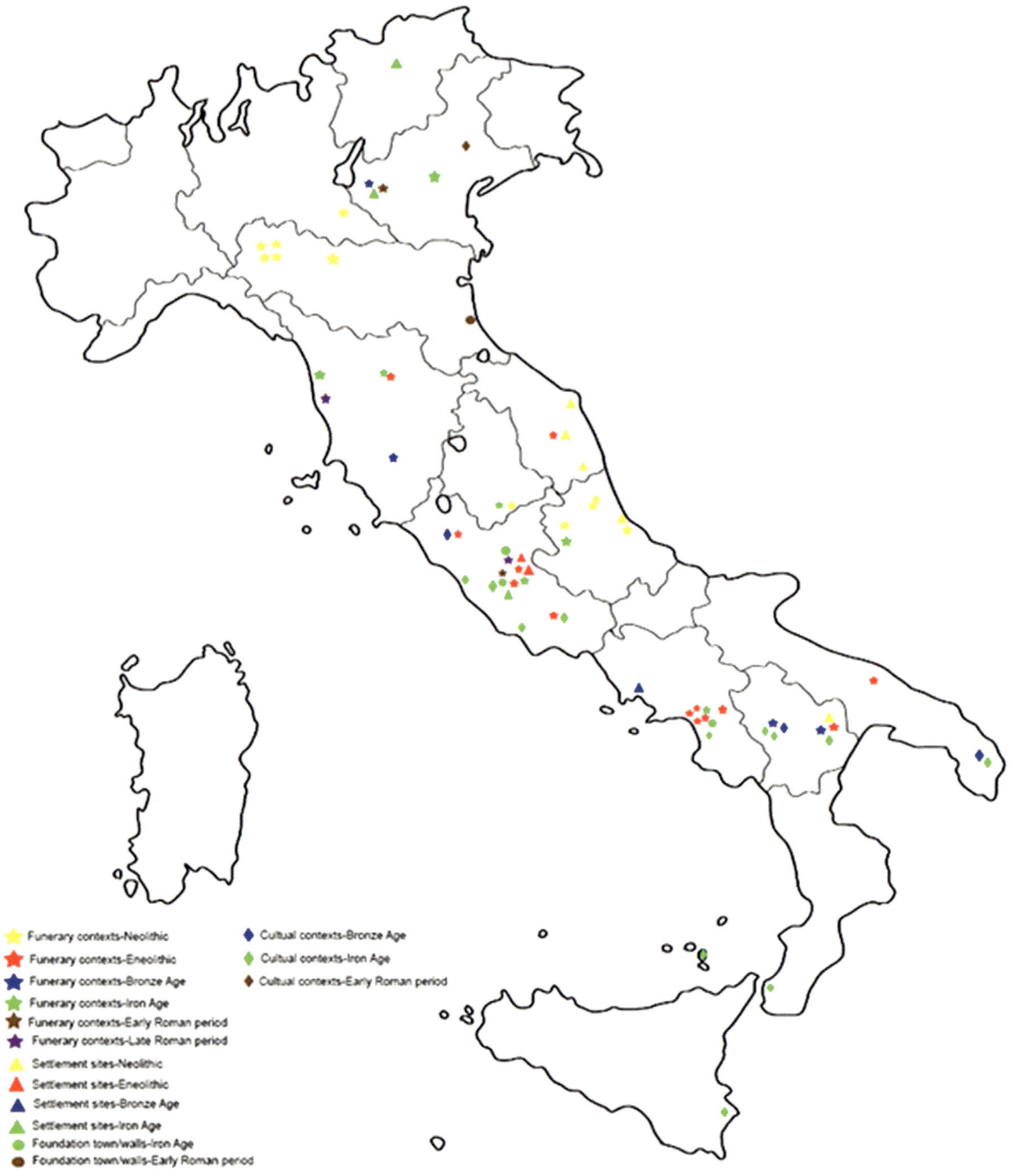


Figure 5. Map of Localisation of archaeological sites including dog remains in Italy, (by F. Santini).

etc. It is indeed a complex relationship, since humans and dogs even share living areas[20].

The dog is a unique animal that, for example in ancient Greece, could be present at symposiums and at feasts. It was seen not only as a participant but as a real individual, as part of the human community[21].

By looking closely at how the ancient Greek society valued the dog, a twofold, ambivalent picture emerges. Dogs were regarded as close and similar to man, yet not identical; they were perceived as beings straddling the bestial and human realms. They evoked feelings of proximity and friendship, but also impulses of hostility and cynical contempt, for dogs can be both savage and tame. They were considered loyal as well as treacherous, intelligent and stupid, vigilant and inattentive[22]. According to this perception, the word

[20] See Franco 2008.
[21] See Franco 2008

[22] See Menache 1997.

'*kyon*' was used as an insult, behind which there are different meanings, not least the meaning related to its metonymic nature[23].

Furthermore, whilst Plutarch and Pausanias noted that dogs were not allowed access into the temples, and that they were not sacrificed during divinatory rituals, Plutarch also refers to the purposeful spread of the dog sacrifice, in both the Greek and Roman worlds[24].

According to most scholars, the use of dog sacrifice in ancient times could be interpreted in different ways[25]. The ritual use of the dog is associated with two main general customs: 1) the sacrifice in funerary contexts, in which the dog is either buried close to a human in a grave or in the adjacent areas; 2) the sacrifice in non-funerary settings, including purification and rites of passage, agricultural rituals and foundation offerings.

In this respect, examining the site of Collina dei Gelsi, I have recorded the funerary practices combining human and dog remains in different types of sites located in Italy from the Neolithic to Roman period[26] (Figure 5).

It is possible to conclude that, since the Neolithic, dogs are the most common animal amongst domestic species. The use of the dog in a funerary context seems to be practised already in the Neolithic, throughout the Italic environment, and until the classical (Greek and Roman) worlds[27]. On the other hand, there are also cases in the Mesolithic, in a Natufian tomb, when the close relationship between man and dog starts to develop.

Concerning the arrangement of the dog burials in relation to human depositions, these kind of depositions reflect a clear intentional action, and the analysis of the spatial location of dog depositions can fall into two general categories, revealing two different meanings: when a dog was buried outside the human burials, its role could have been that of a guardian[28]; whereas the interment of a dog within a grave and together with the deceased (or, rather, with its owner), probably implied a different function, such as that of a faithful companion in daily life as in the afterlife[29]. Consequently, the site of Collina dei Gelsi can surely be seen as belonging to the second category.

Moreover, historical sources frequently mention the immolation of dogs on the graves of their owners, for example Homer in the Iliad[30]. Some authors report the same sacrifice, such as Plutarch in *Vitae Parallelae*[31], and *Claudius Aelianus* in *De Natura Animalium*[32]and in *Historia Varia*[33].

### 3.2 Dogs as companions in death

The role of the dog in the Underworld gains increasing importance in several Indo-European cultures[34]. Consequently, in the classical world there is a general recurring trend of linking the dog with the hereafter, in the role of a psychopomp animal (as escort of the soul of the dead in the afterlife). On the one hand the dog represents the human passage towards the afterlife and it removes, in the apotropaic meaning, death's evil forces. It has been represented in several funeral monuments and gravestones as a final tribute to the owner, accompanying him in the hereafter. On the other hand, dogs can also feature as monsters who guard and display the most terrifying and devouring aspect of death[35].

In order to give a full picture, it is possible to assume that the impurity associated with the dog turns their sacrifice into a cathartic act and a purifying gesture. The dog accompanied ceremonies related to the rituals of passage, reflecting its role as purifier[36].

In a woman's life, passages such as puberty had the same symbolic meaning, for they were linked to purification, pertaining to fertility and procreation. Other rituals involved the presence, and consequently the sacrifice, of the dog as protection. As in the case of ancient pagan worship concerning the agricultural cults such as *Rubigalia* and *Augurium Canarium*, feasts were linked to the protection of crops[37]. Another ritual to consider is known as *Porta Catularia*, during which a puppy was sacrificed close to one of the Gates of Rome[38].

Unfortunately, there are no historical sources on foundation rites, whereas the *Lupercalia* have been written by ancient authors such as Plutarch in his

23 See Mainoldi 1981; Franco 2008.

24 See De Grossi Mazzorin and Minniti 2002; Amoroso *et al.* 2005.

25 See Morris 2011.

26 See Santini 2014.

27 De Grossi Mazzorin and Tagliacozzo 2000; De Grossi Mazzorin 2008b.

28 Some examples are the sites: Madonna del Piano (FI) (Cencetti *et al.* 2006), Necropolis of Via Nomentana-GRA (RM) (De Grossi Mazzorin 2001a). See Santini 2014.

29 Some examples are the sites: Necropolis of Amelia (TR) (Salari *et al.* 2002), *Fidenae* Necropolis (RM) (De Grossi Mazzorin 2001b). See Curci and Muntoni 2008; Santini 2014.

30 Homer describes the sacrifice of two dogs during the funeral ceremonies of Patroclus (Homer, Iliad XXIII, 173–174).

31 Plutarch, *Vitae Parallelae*, Temistocles, X, 27–30.

32 *Claudius Aelianus*, *De Natura Animalium*, VII, 40–50.

33 *Claudius Aelianus*, *Historia Varia*, VIII, 4.

34 See Menache 1997.

35 See Mainoldi 1981.

36 See Menache 1997; De Grossi Mazzorin 2008b.

37 See Smith 1996, Amoroso *et al.* 2005, De Grossi Mazzorin 2008b.

38 The dog remains recovered under the walls or defensive structures of the towns seem to conform exactly to this ritual. Moving from the protection of the town/city to that of the house, the dogs found in the settlements sites could refer to the *Lares praestites*, often depicted with the dog skin and with a dog at their feet. They protected houses, but also the cities, the walls and their inhabitants. See De Grossi Mazzorin 2008b.

Figure 6. Funeral monument in Highgate Cemetery, London (Photo by F. Santini).

*Quaestiones Romanae*[39]. It seems that this rite was alluded to the *sulcus primigenius* done by Romulus, which refers to the transition from a tribal to an urban community[40].

**4 Conclusions**

Extrapolating the data and according to the results of this investigation, throughout time and thanks to the archaeological and historical data, a thin red line can be identified, a guiding thread that has driven past human communities to choose this animal at specific events in its life and to use the dog because of its primary characteristics.

The outcome is that, compared to other domesticated animals, the dog is an animal with a preferential relationship with humans, which involved participation in every aspect of social life.

In conclusion, the practice of dog sacrifices could lead us to believe in their liminality, giving the idea of two distinctive and contrasting worlds, intrinsically and extrinsically connected to each other.

The dog is close to boundaries, it embodies the point of contact and the transition state between human and non-human, between the world of the living and the underworld, but also between safety and danger, the inside and the outside, between a civilised and a primitive environment.

This point of view reflects the ambivalent nature of the dog in symbolic terms. Dogs are the only domesticated animals in which their original wild nature coexists with a domesticated one. This, in turn, underlines an unsolved relationship with humans. On the whole, however, the dog is a protective animal in life as in death.

[39] Plutarch, *Quaestiones Romanae*, 68, 280b-c. Some researchers affirm that this passage dates back to Rome's foundation, where the *luperci* ran around the Palatine Hill striking with tongs, suggesting that primitive humans inhabited this hill before Rome became a city, and the legendary she-wolf who nursed *Romulus* and *Remus*.

[40] The meaning of this ritual could be simplified in terms of countryside *versus* urban side, primitive *versus* civilised.

In light of all these findings, this case study reflects an emphasis on the role of the dog as a faithful companion both in daily life and in the afterlife.

In fact, its presence is strong to this day and its role is well testified in modern cemeteries as it was in ancient times (Figure 6).

### References

#### *Ancient sources*


Homer, *Iliad* XXIII, 173–174.
Claudius Aelianus, *Historia Varia*, VIII, 4.
Claudius Aelianus, *De Natura Animalium*, VII, 40–50.
Plutarch, *Quaestiones Romanae*, 68, 280b-c.


#### *Modern sources*


Alvino, G. 1997. *I Sabini. La vita, la morte, gli dei*. Catalogo della Mostra (Rieti 30 Ottobre - 15 Dicembre 1997). Roma: Armando Editore.

Amoroso, A., J. De Grossi Mazzorin and F. di Gennaro 2005. Sepoltura di cane (IX-VIII sec. a.C.) nell'area perimetrale dell'antica *Fidenae* (Roma), I. Fiore, G. Malerba and S. Chilardi (eds) *Atti del 3° Convegno Nazionale di Archeozoologia*: 311–327. Roma: Istituto Poligrafico e Zecca dello Stato.

Barone, R. 1998. *Anatomia comparata dei Mammiferi domestici* Vol. 2. Bologna: Edagricole.

Benecke, N. 1987. Studies on Early dog remains from Northern Europe. *Journal of Archaeological Sciences* 14: 31–49.

Cencetti, S., P. Mazza, F. Chilleri and F. Cozzini 2006. The Madonna del Piano (Sesto Fiorentino, Florence, Central Italy) ox and dog: a case of intentional Iron Age inhumation. *Geobios 39*: 328–336.

Clark, K.M. 1995. The later prehistoric and protohistoric dog: the emergence of canine diversity. *Archaeozoologica 7*: 9–32.

Cornevin, C.H. and X. Lesbre 1894. *Traité de l'Age des animaux domestiques d'après les dent et les productions épidermiques*. Paris: Libraire J.-B. Bailliere et Fils.

Cremonesi, G. 1965. Il villaggio di Ripoli alla luce dei recenti scavi. *Rivista di Scienze Preistoriche XX (1)*: 85–155.

De Grossi Mazzorin, J. 2000. Le sepolture dei cani della Necropoli di età imperiale di Fidene-via Radicofani (Roma): Alcune considerazioni sul loro seppellimento nell'antichità. *Atti del 2° Convegno Nazionale di Archeozoologia*: 387–398. Forlì: Abaco Edizioni.

De Grossi Mazzorin, J. 2001a. Caratterizzazione archeozoologica: le sepolture di cani, in P. di Manzano (ed.) *Ad deverticulum. Scavi lungo la bretella Nomentana - GRA*: 81–93. Roma: Arti Grafiche G. Cossidente.

De Grossi Mazzorin, J. 2001b. L'uso dei cani nei riti funerari. Il caso della necropoli di età imperiale a Fidene-via Radicofani, in M. Heinzelmann, J. Ortalli, P. Fasaldo and M. Witteyer (eds) *Culto dei morti e costumi funerari romani. Roma, Italia settentrinale e province nord-occidentali dalla tarda Repubblica all'età imperiale, Internationales Kolloquium (Rome 1-3 April 1998)*: 77–82.

De Grossi Mazzorin, J. 2008a. *Archeozoologia. Lo studio dei resti animali in archeologia*: 99–103. Roma: Laterza Editore.

De Grossi Mazzorin J., 2008b. L'uso dei cani nel mondo antico nei riti di fondazione, purificazione e passaggio, in F. D'Andria, J. De Grossi Mazzorin and G. Fiorentino (eds) *Uomini, piante e animali nella dimensione del sacro*, Beni Archeologici-Conoscenza e Tecnologie Quad.6: 71–81, Bari: Edipuglia.

De Grossi Mazzorin, J. and C. Minniti 2002. Dog Sacrifice in the Ancient World: A Ritual Passage? in L.M. Snyder and E.A. Moore (eds) *Dogs and People in Social, Working, Economic or Symbolic Interaction*, 9th *ICAZ Conference*: 62–66. Oxford: Oxbow Books.

De Grossi Mazzorin, J. and A. Tagliacozzo 2000. Morphological and osteological changes in the dog from the Neolithic to the Roman Period in Italy, in S.J. Crockford (ed.) *Dogs through time: An archaeological perspective, Proceedings of the 1st ICAZ Symposium on the History of the domestic dog* (British Archaeological Reports International Series 889): 141–161. Oxford: Archaeopress.

Driesch von den, A. 1976. *A guide to measurement of animal bones from archaeological sites*. Peabody Museum Bulletins 1: 1–148.

Franco, C. 2008. Cani e porci. Temi etnozoologici dal mondo antico, in C. Franco (ed.) *Gli animali e i loro uomini*: 45–51. Siena: Protagon Editore.

Grigson, C. 1978. Towards a blueprint for animal bone reports in archaeology, in D.R. Brotwell, K.D. Thomas and J. Clutton Brock (eds) *Research Problems in Zooarchaeology*: 121–128. London: Whitsabale Litho Ltd.

Harcourt, R.A. 1974. The dog in Prehistoric and Early Historic Britain. *Journal of Archaeological Science 1*: 151–175.

Horard-Herbin, M.-P. 2000. Dog management and use in the Late Iron Age: The evidence from the Gallic site of Levroux (France), in S.J. Crockford (ed.), *Dogs through time: An archaeological perspective, Proceedings of the 1st ICAZ Symposium on the History of the domestic dog* (British Archaeological Reports International Series 889): 115–121. Oxford: Archaeopress.

Koudelka, F. 1884. Das Verhaltnis des Ossa longa zur Skeletthöhe bei deh Säugetieren, in Verhanndlung des Naturforschung Vereins. *Brünn 24*: 127–153.

Mainoldi, C. 1981. Cani mitici e rituali tra il regno dei morti e il mondo dei viventi. *Quaderni Urbinati di Cultura Classica* 8: 7–41.

Masseti, M. 2008. *Uomini e (non solo) topi. Gli animali domestici e la fauna antropocora*: 33–43. Firenze: Firenze University Press.

Menache, S. 1997. Dogs: god's worst enemies? *Society and Animals* 5 (1): 23–44.

Morris, J. (ed.) 2011. *Investigating Animal Burials. Ritual, mundane and beyond* (British Archaeological Reports International Series 535): 1–15. Oxford: Archaeopress.

Salari, L., R. Sardella, E. Squazzini, A. Lisciarelli and T. Suadoni 2006. Il cane della Necropoli di Amelia (Terni, Umbria), in A. Curci, D. Vitali (eds) *Animali tra uomini e Dei. Archeozoologia del mondo preromano*: 179–191. Imola: Grafiche Baroncini Imolagrafiche.

Santini, F. 2014. The dog in the funerary practices: the case study of the human grave comprising the dog burials from Loc. Collina dei Gelsi - Poggio Sommavilla (Collevecchio, RI) dating back to the 7th-6th centuries BC. Unpublished MSc dissertation. University of Sheffield.

Schmid, E. 1972. *Atlas of animal bones for prehistorians, Archaeologista and Quaternary geologists*. Amsterdam: Elsevier.

Silver I.A., 1969. The ageing of domestic animals, in D. Brothwell and E. Higgs (eds), *Science in Archaeology*: 283–302. New York: Thames and Hudson.

Smith, C. 1996. Dead dogs and rattles. Time, space and ritual sacrifice in Iron Age Latium, in J.B. Wilkins (ed.) *Approaches to the Study of Ritual Italy and the Ancient Mediterranean* 2: 73–89. London: Accordia Research Centre, University of London.

Wagner, K. 1930. *Rezente Hunderassen. Eine osteologische Untersuchung*, Mit 36. Oslo: Jacob Dywad Kommission.

# 4.11 The Role of Dogs in the Xiongnu Society

Evgeniy S. Bogdanov

Institute of Archaeology and Ethnography, Siberian Branch, Russian Academy of Sciences, Pr. Akademika Lavrentieva 17, Novosibirsk, 630090 Russia. bogdanov@archaeology.nsc.ru

**Abstract**

The paper analyses different sources reflecting the role of the dog in the Xiongnu society. These are 'Records of the Grand Historian' by the Chinese scientist Sima Qian (around 94 BC) and materials from archaeological excavations in the territory of Transbaikalia and Northern China. The data obtained suggest that war and hunting were the main occupations of the nomadic elite and ordinary members of the Xiongnu society. In such hunts, dogs (*Canis familiaris*) were necessarily used. How they looked, can be demonstrated by the petroglyphs, by the images on metal buckle and bone remains found in the burials of Xiongnu and from settlements. The type of dogs for hunting can be conditionally called 'huskies' and 'steppe dogs'. Another type are guard dogs, large individuals whose descendants are still preserved in the territory of Mongolia (*khonchnokhoi* and *bankhar*).

**Keywords:** Xiongnu, Transbaikalia, Northern China, dogs, images on metal buckle, bone remains.

Figure 1. Map (and collage) of the location of peoples and Empires in Eurasia (Modified by E.S. Bogdanov).

## 1 Introduction

The Xiongnu Empire was formed in the end of the third century BC. It was one of the most significant events in the history of Central Asia (Figure 1). Chinese written sources and archaeological evidence are the most important information sources about the past of these populations. Sima Qian described the Xiongnu economy in the 'Records of the Grand Historian' (Shi Ji): 'They move about in search of water and pasture and have no walled cities or fixed dwellings, nor do they engage in any kind of agriculture. Their lands, however, are divided into regions under the control of various leaders... The little boys start out by learning to ride sheep and shoot birds and rats with a bow and arrow, and when they get a little older, they shoot foxes and hares, which are used for food. Thus, all the young men are able to use a bow and act as armed cavalry in times of war. It is their custom to herd their flocks in times of peace and make their living by hunting, but in periods of crisis they take up arms and go off on plundering and marauding expeditions...' (Records 1993: 129).

Based on current evidence, it can be assumed that hunting and war were the main activities for the Xiongnu elite and ordinary members of the society. Wars were outstanding events; hunting was an almost daily activity. Beating was entertainment, as well as the military method to prepare warriors. Hunting would be considered successful depending on how fast and organised the hunters' actions were. Consequently, young warriors learned tactical techniques that could be used in wars.

Hunting was often transformed into military raids on border territories of the Han dynasty. Chinese historians mentioned that those events included tens of thousands of Xiongnu horsemen. Xiongnu 'follow their livestock, and practice hunting in fields. ... All people gather together to go around forests in autumn when horses become fat' (Rudenko 1962: 49–50). There were particular hunting locations - for example, the Yin Mountains mentioned in the 'Book of Han' (Han Shu): 'There are many forests and much grass, many birds and animals in these mountains. *Shanyu* Modè settled down here, prepared his bows and arrows, and had his raids from here. It was his menagerie' (Rudenko 1962: 49–50). Summer was an unfavourable time of the year for round-up hunting. The hunts began in late autumn, with the onset of frost. In cold weather, meat was preserved better, and the skins were of better quality. For roundups in the primeval forest and in the mountains, good orientation on the ground and special training were necessary. With constant exercise, young people learned to act correctly and quickly, not to get confused in the deep forest, which was an important condition for a successful hunt. Discipline, dexterity and courage were necessary during the hunt, as they had to deal with dangerous large animals: bears, tigers, wild boars, wolverines, lynxes and wolves. Boars were especially dangerous for horses. Following the Chinese aristocracy, the Xiongnu began using hunting dogs. It was possible to collect some evidence of this in the analysis of art sources.

This paper will consider the following questions:

1. What types of dogs were depicted on various decorations in the culture of China?
2. What types of dogs were depicted on petroglyphs and on belt plates of the Xiongnu?
3. What information about the role of dogs in a nomadic society was obtained as a result of excavations of burial complexes and settlements?

## 2 Xiongnu and China. Animal art

Nomads had a maximum of 1.5 million people (this approximately corresponds to the population of one Han district), while the Han Empire reached almost 60 million people. However, for nearly 300 years, nomads have been the greatest danger to farmers.

Nomads could not have survived solely on animal products: agricultural products, silk, weapons, handicrafts, and jewellery were also necessary. It was one of the reasons to attack populations settled nearby. The Han dynasty of China conducted military operations as well as a special strategy called 'five temptations'. This strategy included:

1. to give nomads expensive textiles and chariots in order to damage nomads' eyes;
2. to give nomads tasty food in order to close their mouths;
3. to play wonderful music in order to close nomads' ears;
4. to construct outstanding buildings, storage places for cereal,and give nomads slaves in order to calm down their stomachs;
5. to donate expensive presents (Materialy 1968: 42, 44, 49, 73).

These actions were not just a simple shipment of products and Chinese goods including significant examples like Chinese chariots. This process caused the transformation of nomad society due to the impact of the Chinese culture. Accordingly, Xiongnu built tombs for their leaders in the same way as the tombs for Chinese noblemen were built. Besides, Xiongnu wore silk clothes, and organised round-up hunting, just like the Chinese aristocracy did. Dogs (*Canis familiaris*) could have been used in such hunting processes.

It should be recognised that the cult of dogs has existed in China since the Bronze Age. The dog was a loyal assistant to man during his lifetime and also accompanied him along the road to the 'world of the dead'. Therefore, the images of these animals are painted on the walls of tombs and cast in metal (Figures 2–4). This topic is too wide for this paper, so the focus will only be on materials related to nomads. Chinese craftsmen mass-produced and replicated bronze objects for Xiongnu based on their preferences. These objects mainly included decoration elements for costumes such as belts. Due to the popularity of the Scythian Siberian animal style, dogs were often represented quite similar to wolves and other predators (Figures 5–6). Nevertheless, we can identify dogs based on the presence of their rolled-up tails. The proportions of dogs and the forms of their tails allow us to make the preliminary conclusion that craftsmen

Figure 2. Wall painting in the Tomb in China. National Museum of Beijing (Photo by E.S. Bogdanov).

Figure 3. Bronze mirror and fragments scene with dogs. Wanquhou's tomb at Bojishan (2nd century BC) (Photo by E.S. Bogdanov).

Figure 4. Bronze decoration, random find, collection J.J. Lally and Co. (After So and Bunker 1995).

Figure 5. Bronze decoration, random find, Calonda collection (After So and Bunker 1995).

Figure 6. Bronze belt buckle, random find, collection Eskenazi Ltd. (After Bunker 2002).

represented husky types of dogs. There are many belt plates with motives of attacks (Bogdanov 2006: pl. LIV) (Figures 7–9). Scenes of peaceful life are very rare in the Xiongnu plastic art. For example, one belt plate represents a happy dog due to its master returning (Figure 10). Interestingly, there are practically no small figures of this type of dog in Xiongnu animal plastic art. Only figurines of deer, goats, or predators.

### 3 Archaeological evidence

Archaeological evidence from Xiongnu hillforts demonstrates that this population preyed on red deer, elks, mountain sheep, antelopes, wild boar, foxes, sables, bears, badgers, hares, and weasels. Bows and arrows, spears, and bear-spears were the main types of hunting weapons (Davydova 1995: 49–50). As already mentioned above, dogs were needed for successful round-up hunting. Graphic sources (art)

Figure 7. Bronze belt buckle, random find, collection C.T. Loo (After Rostovtzeff 1929).

Figure 8. Bronze belt buckle, random find, collection Eskenazi Ltd (After Bunker 2002).

Figure 9. Bronze belt buckle, random find, collection Arthur M. Sackler Gallery (After Bunker 1997).

provide a generalised image of this animal. More accurate data from archaeological excavations can be obtained. It is not only possible to see the physical appearance of dogs that lived next to the nomads, but also their relationship during life and after death can be revealed. Burials of symbolic herds were discovered near heads of buried Xiongnu. These symbolic herds were located on a special stage and included skulls of wild and farm animals. Dogs have also been discovered in these groups of skulls. Steppe dog sand huskies were excavated in burials of Xiongnu in *Cheremuhovaya Pad* and *Ilmovaya Pad* (the Republic of Buryatia, Russia)

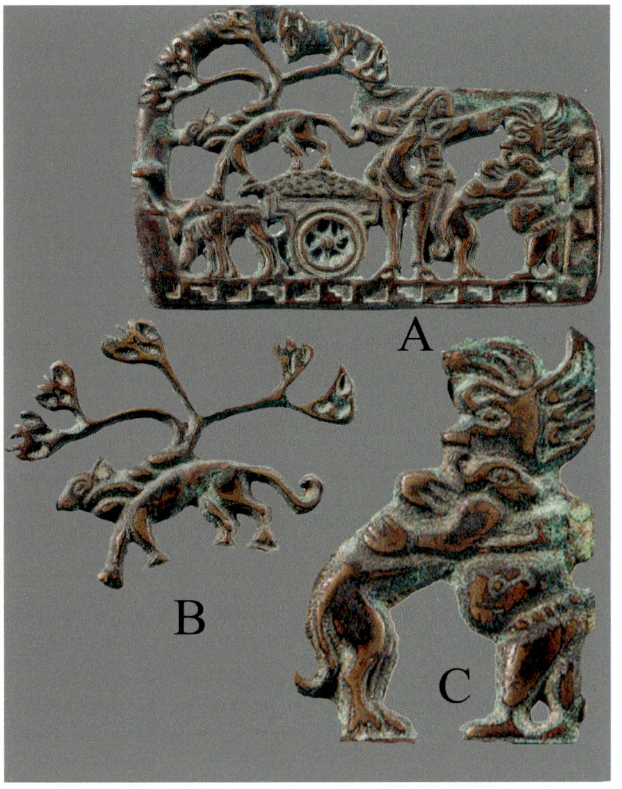


Figure 10. Bronze belt buckle (A) and fragments scene with dogs (B, C), random find, collection Shelby White and Leon Levy (After Bunker 2002).

(Konovalov 1976: 164–165; pl. 11). Most often, animals were laid at the feet of people. Sometimes dogs were isolated, and outlined by stone slabs (Konovalov 1976: 163). This fact indicates the special role of the dog in the worldviews of pastoralists. Xianbei and Wuhuan nomadic populations (eastern neighbours Xiongnu) believed that dogs guided the souls of dead people. In Han Shu, it says, 'Wuhuans take one fattened dog and lead it on a coloured cord; they also take the horse on which the dead man rode, his robe and things; then they burn everything and carry it behind the coffin. After the funeral, the soul of the dog accompanies the soul of the deceased to Mount Chishan' (Bichurin 1950: 143–144).

Animal husbandry was the main economic activity for Xiongnu, who practiced a nomadic lifestyle. In order to organise their needs, this population may have used big guard dogs. Their descendants still exist in Mongolia (*khonchnokhoi* and *bankhar*). Guard (herding) dogs were the most valuable types and were bred specifically. In order to fight predators with dignity, protect cattle, and, moreover, to win, a particularly developed respiratory system was needed. For that reason, an abnormal expansion of the chest is so noticeable in such dogs. It is this race of dogs that can be see non ancient Petroglyphs in Central Asia. They are all large-headed, broad-shouldered, with a short neck, with a voluminous, well-defined scruff (withers), with an expanded chest. Their important role in the life of cattle breeders is evidenced by the fact of the placement of tribal *tamga* Xiongnu along side the dogs. These knocks on the rocks can be seen not only in Mongolia, but also in China (Bogdanov 2017: 8).

It is important to remember that many mixed-breed dogs lived near settlements. Their origin was identified based on the analysis of dog remains from *Ivolginskoe* settlements (archaeological site in the Transbaikalia).[1] These dogs did not have particular relevance. Ethnographic data indicates that these dogs were not fed, and during hunger periods they were eaten. The huge number of dog bones within the boundaries of such settlements could be related to this data (Davydova 1995: 50).

## 4 Conclusions

Thus, archaeological data, graphic and written sources indicate that the selection of dogs began with nomads (like in China). New species appeared: for hunting and for protecting the herd. At the same time, most of the dogs lived near the nomadic settlements in the wild. Our sources indicate that it was not only the Xiongnu who lived in these settlements, but refugees from the territory of China did too. Not only did they bring specific tools and objects of art, but also a different worldview. That is why in the funeral rites of the nomads we find the idea of the role of guide dogs for the souls of dead people.

South Siberia, East Turkestan, Zhetysu, and Central Asia were also the main directions of the Xiongnu expansion. Many populations migrated West due to military conflicts. These significant movements (the Migration Period) caused changes in ethnicity in steppe regions in Eastern Europe. Dogs also migrated with their masters to the West. As a consequence, crossbreeding caused the emergence of new breeds. Some of them still exist nowadays. However, this question is not in the field of archaeology; this problem refers to genetics.

## References


Bichurin, N. Ya. (Iakinf) 1950. *Sobraniya svedeniy o narodakh, obitavshikh v Sredney Azii v drevniye vremena*. Vol. 1. Moskow- Leningrad: Izdatel'stvo Academii nauk SSSR. Бичурин, Н.Я. (Иакинф). 1950. Собрания сведений о народах, обитавших в Средней Азии в древние времена. Т. 1. Москва-Ленинград: Изд-во АН СССР.

Bogdanov, E.S. 2006. *Obraz khishchnika v plasticheskom iskusstve kochevykh narodov Tsentral'noy Azii (skifo-sibirskaya khudozhestvennaya traditsiya).* Novosibirsk: Izdatel'stvo IAET SO RAN. Богданов, Е.С. 2006. Образ хищника в пластическом искусстве кочевых народов Центральной Азии (скифо-сибирская художественная традиция). Новосибирск: Издательство ИАЭТ СО РАН.

Bogdanov, E.S. 2017. Proiskhozhdeniye tamg Xiongnu. *The Lower Volga. Archaeological Bulletin*, 16 (2): 5–32. Богданов Е.С. 2017. Происхождение тамг хунну, Нижневолжский археологический вестник. Т. 16, № 2. С. 5–32.

Bunker, E. 1997. *Ancient bronzes of the eastern Eurasian steppes from the Arthur M. Sackler collection.* New York: Arthur M. Sackler Foundation.

Bunker, E. 2002. *Nomadic art of the Eastern Eurasian Steppes. The Eugene V. Thaw and other New York Collection.* New Haven; London; New York: The Metropolitan Museum of Art, Yale University Press.

Davydova, A.V. 1995. *Ivolginskiy arkheologicheskiy kompleks.* Vol. 1. *Ivolginskoye gorodishche.* Sankt-Petersburg: Fond 'AziatIKA'. Давыдова, А.В. 1995. Иволгинский археологический комплекс. Т. 1: Иволгинское городище. СПб: Фонд «АзиатИКА».

Konovalov, P.B. 1976. *Khunnuv Zabaykal'ye (Pogrebal'nyye pamyatniki).* Ulan-Ude: Buryatskoe. Knijnoe Izdatel'stvo. Коновалов, П.Б. 1976. Хунну в Забайкалье (Погребальные памятники). Улан-Удэ: Бурят.кн. изд-во.

*Materialy po istorii syunnu (po kitayskim istochnikam).* 1968. Vol.1. V.S. Taskin. *(translator).* Moskow: Nauka. Glavnayaredaktsiyavostochnoyliteratury. Материалы по исто рии сюнну (по китайским источникам). 1968. Вып. 1. Предисловие, перевод и примечания В.С. Таскина. Москва: Наука. Главная редакция восточной литературы.

Qian Sima (W. Burton translator). 1993. *Records of the Grand Historian: Han Dynasty.* Vol. 1. Hong Kong: Renditions - Columbia University Press.

Rostovtzeff, M. 1929. *The animal style in South Russia and China.* Princeton - New York.: Princeton Univ. press.

Rudenko, S.I. 1962. *Kul'turakhunnovinoinulinskiyekurgany.* Moskow - Leningrad: Izdatel'stvo Academii nauk SSSR Руденко С.И. 1962. Культура хуннов иноинулинские курганы. Москва-Ленинград: Издательство АН СССР.

So, J. and E. Bunker 1995. *Traders and raiders on China's northern frontiers.* Seattle: Arthur M. Sackler Gallery, Smithsonian Institution in association with University of Washington Press.


[1] This is the only large Xiongnu settlement known to date. See (Davydova 1995: 47–50).

# 4.12 Dog Burial at the Ust-Voikarskoe-1 Settlement and its Interpretation Issues

Andrey V. Novikov[1] and Yuri N. Garkusha[1]

[1]Institute of Archaeology and Ethnography of the Siberian Branch of the Russian Academy of Sciences, 17, prospect Akad. Lavrentyeva, Novosibirsk, 630090, Russia. novikov@archaeology.nsc.ru; yunga1971@ngs.ru

Corresponding author: Andrey V. Novikov, novikov@archaeology.nsc.ru

**Abstract**

This article analyses the local accumulations of complete sets of canine osteological remains, which were found during the archeological study of the fortified settlement of Ust-Voykarskoe (Voykarsky fortified settlement) dated to the Late Middle Ages-Early Modern Period, conducted from 2012 to 2015. According to the current administrative division of the Russian Federation, the site is located on the territory of the Yamal-Nenets Autonomous Okrug, in the subpolar region of Western Siberia, and has a frozen cultural layer. Dendrochronological dating of the buildings where the remains of dogs were found, made it possible to determine the time of their emergence as the mid-18th century. On the basis of stratigraphic observations it was found that dwellings abandoned by the residents were used for the purpose of burials. The north-taiga zone of the Lower Ob region has historically been the home to the Northern Khanty, an ethnic group of the Ob Ugrians who have both Ugric and Samoyed components in their ethnic origins. The representatives of the Northern Khanty group are the Voykarsky Khanty who are associated with the functioning of the Voykarsky fortified settlement. Special respect for the dog is known both among the Samoyeds and the Ob Ugrians, but the form of reverence for dogs, manifested in the ritual practices of these ethnic groups, is sometimes expressed in the opposite way. Due to their ethnic multicomponent basis, the Northern Khanty reveal an ambivalent attitude toward the dog, which combines both traditions. It can be argued that in this case it is possible to speak about the tradition of intentional burial of dogs as sacrificial animals during the ritual of abandoning a dwelling among the Voykarsky Khanty.

**Keywords**: Lower Ob region, Voikarsky fortified settlement, 15th-19th centuries, Northern Khanty, burials of dogs, dwelling space.

## 1 Introduction

Archaeological studies in the north of Western Siberia with permafrost in the cultural layers make profound study of various spheres of activity of the ancient population of the region possible, since organic findings and remains are well preserved in them. Studies of one such archaeological sites with permafrost, the Ust-Voikarskoe-1 settlement, made it possible to obtain new evidence of the dog ‘participation’ in rituals of the aboriginal population of the Subpolar zone of Western Siberia.

Archaeological studies of this settlement, which is identified with the Voikarsky town known from written sources, were conducted from 2003 to 2008 under the direction of A. G. Brusnitsyna and N. V. Fedorova (Brusnitsyna 2003; Fedorova 2006); and from 2012 to 2016 under the direction of A. V. Novikov (Novikov and Garkusha 2017). The object is located in the Shuryshkarsky district of Yamalo-Nenets Autonomous Okrug, 2 km northeast from Ust-Voikary village (Subpolar zone of Western Siberia). The site location is confined to the southern end of the peninsula, limited by the floodplain of the left bank of the Gornaya Ob anabranch. The word ‘Voikar’ comes from the language of Komi-Zyryans and means ‘night town’. The Khanty called it ‘Aivozh pai’ - ‘small town’.

Modern landscape features of the object include a hill, up to 9 m high, about 100 m long, elongated from the north to the south and up to 50 m wide. The upper area of the hill is uneven, about 60 m long and 15–20 m wide. Archaeological studies have covered the northwestern part of the hill, including the area of the slope with the foot of the hill and along the western boundary of the upper area. This upland consisted mainly of organic cultural sediments formed on the basis of woodchips and other wood processing remains, containing permafrost formations.

Wood samples taken from structures located at the foot of the hill were dated back to the 14th and 15th centuries, and on the top of the hill from the17th to the end of the19th centuries. (Gurskaia 2008). According to dendrochronological data, the construction time

Figure 1. Settlement Ust-Voikarskoe-1. Photo of the burial of dog No. 1 (view from the South).

of the buildings found in 2012–2016 dates back to the last third of the 15th century to the middle of the 18th century. Wood buildings became dominant by the 1740s. In general, the middle of the 18th century is probably the end of active construction in the settlement. Thus, a chronological gap is formed between the capital development of the middle of the 18th century and single buildings of the early 19th century. Apparently, such late construction activity was no longer directly related to the history of the 'settlement' (Garkusha 2019).

Ethnic attribution of the site. The northern taiga zone of the Lower Ob region was, and still is, the place of residence of the Northern Khanty - ethnographic group of the Ob Ugrians. The Northern Khanty were formed through contacts between various groups of the population belonging to the Samoyed and Ugric language families (Perevalova 2004: 117–118; 216; 250). Based on the Samoyed substratum and the Ugric population migrations a new ethnic group was formed in the Voikar River basin - the Voikar Khanty (Perevalova 2004, p. 118–119; Martynova, 2005). The population of Voikarsky town was also preliminary characterised as Ugric Samoyed (Fedorova 2006: 17).

## 2 Material and method

### *2.1 Description of dog burial*

During the field studies in 2012–2015 the remains of three intentionally buried animals were found at the site[1]. They were accumulations of bone remains of dogs (*Canis lupus familiaris* - hereinafter referred to as 'dogs'), which preserved anatomical integrity. A peculiar feature of the animal remains location is their confinement to the interior of residential buildings. Animal skeletons were found at 0.2–0.25 m from the daylight surface.

Burial No. 1 (Figure 1). The dog's remains were located in the central part of the log building No. 1[2], at the level of the second - third crowns. The skeleton was located in a deepening made in the upper soil layer of the building's filling and covered with small pieces of wood. The animal was on the left side with a west-east orientation (along the spine); the head was directed to the west. A large fragment of a reindeer horn was found under the dog[3].

Burial No. 2 (Figure 2). The dog's remains were located 4 m to the northeast of burial No. 1, in the central part

Figure 2. Settlement Ust-Voikarskoe-1. Photo of the burial of dog No. 2 (view from the North).

of the site occupied by layers of two residential building ruins (No. 2, 2A), which were successively erected within the same area. Stratigraphic research showed that the bone remains were located in the area of a later construction - building No. 2. The insignificant thickness of the soil layer overlying the skeleton allows us to assume that it dates back to the time following the building's destruction.

The skeleton of a large animal was in anatomical order and located on the right side with a northeast-southwest orientation (along the spine); the head was directed to the southwest. The distal parts of the forelimbs were damaged. The lower jaw was missing. The dog's corpse was located on a large fragment of a birch bark sheet with a number of stitching marks. The fore paws were tied with a braided fibre rope in the bend region. There was a relatively long piece of a leather strap with stitching marks along the long sides in the neck region. A miniature birch bark box with conifer seeds was placed at the fore paws of the dog. A large fragment of a deer horn was found directly under the dog's skull, but due to the fact that horns were distributed throughout the studied territory of the settlement (Bachura, Nekrasov 2010: 208), it is difficult to relate them to the burial of the dog, as in the case of the first burial. Based upon the condition of buildings No. 1 and 2, they were partially dismantled before the burials appeared.

Burial No. 3 (Figure 3). The dog's remains were found in the central part of the large building No. 9 (No. 6 - according to N. V. Fedorova (Fedorova 2006: 14)). The condition of this building suggests that it was not deliberately destroyed, but abandoned and destroyed gradually.

The remains of an adult dog were located inside a rectangular wooden hearth structure, on top of the ash. It had been disturbed (part of the area occupied by the remains was affected by a later digging), but some parts of the skeleton preserved anatomical integrity; there were no traces of biting or cutting.

The presented materials continue a series of similar burials found by previous researchers in Voikarsky town. In the upper part of the filling of building No. 8 (according to N.V. Fedorova) there were skeletons of two dogs - in the centre and corner (Fedorova 2006: 14). It is important to stress that two dog burials found by N. V. Fedorova were also located in the area of dwellings that were no longer functional.

Figure 3. Settlement Ust-Voikarskoe-1. Photo of the burial of dog No. 3 (view from the South-South East).

It should be noted that apart from burials, single dog bone remains were found everywhere in the cultural layer of the settlement, but they occupy an insignificant share among remains of other mammals. Most of them are confined to the layers of the 18th- early 19th centuries (Bachura and Nekrasov 2010: 209).

## 3 Results

### *3.1 Analogies*

Anatomically integral dog remains have also been found at other settlement complexes of the second half of 2000 AD in the Lower Ob region. In the Nadymsky town, a dog skeleton was found accompanied by the remains of puppies (Kardash 2009: 32), two skeletons of adult animals were found in the Poluysky cape town (Kardash 2013: 119). However, at the abovementioned objects the dog remains were found outside residential buildings and are confined to horizons dated no later than the first third of the 18th century (Kardash 2009: 89, fig. 2.42; 90, fig. 2.43; Kardash 2013: 155, fig. 2.48). According to O. V. Kardash and T.V. Lobanova, they may represent 'ritual burial sacrifices associated with the construction of a house and, apparently, with the foundation of the village' (Kardash 2009: 282; Kardash, 2013: 119; Lobanova, Kardash, 2014: 71), however, no evidence is presented by the authors.

### *3.2 Dog burials relative chronology and dating issues*

Dendrochronological data and stratigraphic research are the basis for assumptions about the dating of dog burials. The main scope of dates of the wood used in the construction of building No. 8 (according to N.V. Fedorova), dates back to the middle - second half of the 17th century. (Gurskaya 2008: 219–220). Apparently, this building could have been used at the very beginning of the second third of the 18th century, judging by the dendrodates of wood samples found in the filling. This assumption is possible if the accidental ingress of wood harvested at such a late period onto the dwelling area would be excluded. Thus, it can be assumed that dog burials in this building could have appeared no earlier than the beginning - middle of the 18th century. The upper limit is determined by the construction of building No. 2 (according to N.V. Fedorova); it partially covers the area of the first building and, based on archaeological materials, dates back to the beginning of the 19th century (Fedorova 2006: 16).

Building No. 9 (No. 6 - according to N.V. Fedorova), judging by the dendrodates, continued to be used at the beginning of the 18th century (Gurskaya 2008: 219).

Dendrochronological dating indicates the time of construction of the building No. 1 (according to the numbering in force in 2012–2016) to the turn of the 1680's/90's.

During the destruction process, the building ruins were buried under the soil. External signs of small-sized buildings (No. 1, 2), where skeletons of dogs were found, were practically not visible. The finding of the remains inside large-sized buildings with a clear stratigraphy and certain external signs on the surface indicates a deliberate choice of abandoned buildings for burials. Certainly, the dog remains inside residential buildings could have appeared only after they were left by people and are partially destroyed.

Thus, a tradition was noted at the Ust-Voikarskoe-1 settlement, which is archaeologically manifested in the facts of finding anatomically integral dog remains. These facts have the following common features:

1. Dog burials were intentional (which, apparently, assumes dog killing);
2. All of them are located inside dwellings;
3. Dog burials were carried out in non-functional dwellings.

**4 Discussion**

***4.1 Dog burial interpretation issues***

The relationship between a human and a dog is a complex historical and cultural phenomenon which has lasted for many thousands of years. It has a huge volume of recorded manifestations of the use and presence of dogs in various forms of people's every day and ritual activities, and world view systems. The available studies on the role of dogs in the traditional culture of the aboriginal population of Western Siberia (see, e.g. Moshinskaia and Lukina 1982; Lukina 1983; Zhelvis 1984; Kosarev 1988; Perevalova 1996) rely mainly on ethnographic sources. The evidence-based interpretation of archaeological sources is difficult, since it must be based on a methodologically complex research procedure that involves conducting archaeological and ethnographic comparisons. Conducting methodologically correct archaeological and ethnographic comparisons has compulsory methodological limitations. The interpretation of the recorded facts should be based on a clear localisation of each dog bone finding in a specific archaeological context and conducted within the framework of a very specific ethnic and cultural tradition.

The classification of contexts of dog bone findings in the places of residence of the ancient people provides the distinction of several different types (Novikov 1997: 175–179; Novikov, 2001: 73–76). In this case, relatively complete skeletons of animals were found inside dwellings, in a special deepening, or with accompanying equipment, which makes it possible to assume the intentional burial of an animal. The combination of these common features also presupposes the interpretation of dog burials recorded at the Ust-Voikarskoe-1 settlement as traces of ritual activity after the intentional killing of an animal.

***4.2 Intentional dog killings in traditional culture of the Ugrians and Samoyeds***

The special attitude to a dog, which is based on mythological ideas, is known among the indigenous population of Western Siberia - the Khanty (belonging to the Ugric language group), Nenets and Selkups (belonging to the Samoyed language group). However, the basic principles of dog honouring are common among various ethnic groups in the region and have, at times, the opposite expression.

The ideas of the Ugric language group peoples (particularly, the Khanty) are based on the prohibition of bloody dog sacrifices, which can be performed only in exceptional cases (Moshinskaia and Lukina 1982: 55; Fedorova 2000: 85), at the same time, the ritual activity of the Samoyed language group peoples (particularly, the Selkups, Nenets) is characterised by the use of a dog as a victim.

Violation of the prohibition on killing a dog among the Khanty caused blame and led to various punishments (Startsev 1928: 98–99; Moshinskaia and Lukina, 1982: 55). One of the legends of the Khanty, living on theYugan River, describes the punishment of the hero who killed a dog (*Materialy po fol'kloru hantov* 1978: 43). For the Khanty, non-acceptance of the bloody dog sacrifice is one of the ways of ethnic self-identification. According to the observations of K.F. Karjalainen (1995: 101), this tradition appears in the Khanty's stories as a distinctive one At the same time, the possibility of the bloody dog sacrifice is recorded in the folklore of the Nenets (Patkanov 1999: 64; Perevalova 2004: 121).

***4.3 Dog killing in burial and memorial ritualism***

The role that the forest Nenets and Khanty, living in the same territory, in the Agan River basin, assign to a dog in the performance of the burial rite is an example of a combination of different traditions. The Nenets, when burying a human, along with the reindeer slaughter, pierced the dog with a stick and laid it next to the deceased (Perevalova and Karacharov 2006: 306), while the Khanty killed a dog in exceptional cases - 'if it grieves' (Perevalova and Karacharov 2006: 301). At the same time, among the Khanty, living on the Yugan, Kazym and Lyamin Rivers there was a custom to kill a hunter's dog and leave it at the grave of the buried (Kulemzin 1984: 142, 160). A dog could be sacrificed if the body of a frozen or drowned owner was not found;

at the same time, a dog was chosen as a victim according to the sex of the deceased (Vakh River) (Kulemzin 1984: 152). However, other information confirms the prohibition of killing a dog among these territorial groups of the Khanty (ibid., p. 160). The Khanty, living on the Lyamin River, allowed the killing of a dog if a human anticipates his/her death. In this case, he/she had the right to strangle a dog, i.e. 'take it with self' (Kulemzina and Lukinoy 1978: 177). An ambivalent attitude to a dog was noted among the Khanty, living on the Synya River, when the traditions of one surname allowed killing a dog (including as a burial sacrifice), and the traditions of another prohibited it (Sokolova 2011: 191). However, if killing a dog was inevitable due to some circumstances, then the supporters of the prohibition could invite a representative of another nationality to perform the sentence; and bloodless killing was performed (Sokolova 2011: 173, 191).

According to the observations of Perevalova (2004: 292–293), the Lower Ob region was the territory where both traditions were most prominently represented - killing (Samoyed) and prohibition (Ugric). This situation could not but affect the peculiar features of dog honouring by the Northern Khanty For example, until recently, the Khanty, living on the Voikar River, had a ritual dedicated to family spirits, during which a dog was sacrificed (ibid: 235–236). Christianisation of the aboriginal population led to the emergence of surrogate forms of pagan rituals, which also affected sacrifices associated with killing animals, particularly dogs. In the reports of missionary trips to the Obdorskaya volost at the end of the 19th century, cases of the use of a cross for hanging sacrificed dogs by the Ostyaks were recorded (Perevalova 2000: 170).

*4.4 Intentional killing of dogs for fur*

There is numerous evidence that wearing clothes made of dog skins was a widespread phenomenon among the aboriginal population of the Lower Ob region (Perevalova 2004: 292), which implies intentional animal killing. Some groups of the Khanty, living on the Pitlyar and Kunnovat Rivers did not prohibit wearing clothes made of dog skins, which also contradicts the prohibition of killing them (Perevalova 2004: 289, 291). According to Gondatti, the aborigines killed old, sick or crippled dogs; and their skins with long, white, fluffy fur were used to trim various parts of winter clothes (Gondatti 2000: 118). Breeding of dogs for further slaughter and economic use or trade and exchange operations was not something unusual (Perevalova 2004: 293).

Among the Mansi, whose negative attitude to killing and sacrificing dogs is noted by researchers (see. e.g. Nosilov 1904: 228–229), there are also examples of the use of dog corpses both for ritual purposes and for fur (Nosilov 1904: 117). In addition, dog skins could be used for ritual purposes. Gondatti (1888: 12) reports on the Mansi shaman tambourines covered in dog skin.

*4.5 Intentional dog killing for selection purposes*

The traditions of the Ob Ugrians allowed the deliberate killing of dogs in order to preserve and accumulate proven hunting qualities. The non-pedigree dogs were killed; puppies whose future as good hunting dogs was questioned were also killed as, in particular, black puppies, since they could not be used for hunting in winter. A. A. Dunin-Gorkavich notes that 'from the puppies born, the Ostyak chooses only good ones, whom he feeds, while the others are killed in order to avoid unnecessary and unproductive waste of food and save the mother's strength, which the extra puppies could drain' (Novikov 1999: 87). A dog was killed if it could not stop barking at domesticated deer; unnecessary and sick animals were also killed (Kulemzin and Lukina 1977: 29–30; 85; Fedorova 2000; Lukina 2004: 85; Sokolova 2011: 189). The Mansi allowed for old dogs to be strangled with a special strangler in order to help them to 'move to another world' (Moshinskaia and Lukina 1982: 57).

*4.6 Intentional dog burial*

The intentional burial of dogs, as an expression of a particular, respectful attitude, was most firmly entrenched among the Mansi as a separate tradition. The information about observance of the corresponding rituals is recorded throughout the history of ethnographic study of the Mansi (Ides and Brand 1967: 72, 77; Moshinskaia and Lukina 1982: 56–57). At the same time, there is no complete, detailed description of the rite itself. Only different ways of arranging burials are noted: individual burials made in pits, on a tree (in the hollow of a tree), on the surface with a dog's corpse covered with branches (burying in the snow), or in wooden structures above the ground. According to the ideas of the Mansi, living on the Ob and Nizhnyaya Sosva Rivers, a dead dog had to be buried in a certain place (Fedorova 1994: 86).

Ethnographers have recorded local manifestations of some rituals associated with dog burials, mainly in the territory inhabited by the Eastern Khanty. At the same time, this ethnographic group had the idea that dogs should not be buried in the ground at all, since 'burying a dog in the ground is like burying oneself' (Moshinskaia and Lukina, 1982: 58). It should be noted that a 'dog-human' association is a very ancient concept in the most archaic protoforms of the animal epic among peoples all over the world, where many animals first appear as people and only later, for various reasons, acquire animal characteristics (Kostyukhin

1987: 25–31). However, the Khanty, living on the Vakh River, sometimes buried dead dogs in the ground (Kulemzin 1984: 160). Before burial, a red cloth was tied to one forepaw, and a black cloth to the other (Kulemzin 1984: 160; Lukina 2005: 277). The Khanty, living on the Lyamin River, could bury a dog by wrapping it in a cloth and leaving it on the surface, covered with branches; sometimes red and black ribbons were tied to their paws (Moshinskaia and Lukina 1982: 57, 59). The Khanty on the Kazym River 'bury (cover with branches) a dead dog, tying a cloth to a paw' (Perevalova 2004: 291). The Khanty, living on the Yugan River, tied a piece of fur to one of the forepaws of a dead dog (Fedorova 2000: 85).

Information about the choice of burial sites is fragmentary. There is evidence that they could have been upstream of the river from the place where the village stands (Kulemzin 1984: 160); among the Khanty, living on the Yugan River, such a place could be located to the west of the settlement (Fedorova 2000: 85).

It is unlikely that a significant number of single dog bones found in settlement complexes can be associated with the constant performance of any rituals accompanied by sacrifices. Ethnographic materials indicate that there were special places outside the settlements for performing such rituals.

## 5 Conclusions. Interpretation of dog burials at the Ust-Voikarskoe-1 settlement

The contexts of finds of dog burials at the Ust-Voikarskoe-1 settlement do not allow interpreting them as the result of construction sacrifices. In addition, ethnographic materials show the underdevelopment of the construction sacrifice tradition among the Ob Ugrians, at least in this form. In the narrative sources of the Ob Ugrians, this sphere of myth-ritual representations is almost not reflected (Morozov 1993), at least regarding the development of the territory during the period of settlement organisation and the initial stage of constructions.

At the same time, special ideas among the Ob Ugrians are associated with the territory of the abandoned settlements, up to the endowment of a status of a sacred place. These ideas assumed various ritual activities on the area of abandoned settlements (as well as on the area of separate dwellings) and restrictions on the use of these lands (Kulemzin, Lukina 1977: 146; Karjalainen 1995: 65, 67). Various archaeological manifestations of dwelling abandonment rituals have been discussed in the literature (see, e.g. Novikova and Nesterova 2010: 218–220), including the Khanty, living on the Kazym River (Novikov 2008: 262–263, 270; Molodin *et al.* 2018: 87). The analysed facts of burial of dogs on the territory of abandoned dwellings of the village of Ust-Voikarskoe-1 indicate that the northern Khanty, living on the Voikar River, had a tradition of deliberately killing dogs with subsequent burial in dwellings. These could be rituals associated with the abandonment and further veneration of dwellings.

## Acknowledgments

This study was performed under the IAET SB RAS R&D Project 'Studies of Archaeological and Ethnographic Sites in Siberia during the Period of the Russian State' (FWZG-2022-0005).

## References

Bachura, O.P. i A.E. Nekrasov 2010. Promyslovye I domashnie zhivotnye v khozyajstvennoj deyatel'nosti naseleniya gorodishcha Ust'-Vojkarskij (XIV-XIXvv.). *Vestnik arheologii, antropologii I etnografii.* 2 (13): 206–213. Бачура, О.П. и А.Е. Некрасов. 2010. Промысловые и домашние животные в хозяйственной деятельности населения городища Усть-Войкарский (XIV-XIX вв.). *Вестник археологии, антропологии и этнографии* 2 (13): 206–213.

Brusnitsyna, A.G. 2003. Gorodishche Ust'-Vojkarskoe. Nachalo izucheniya, in A.V. Neskorov (ed.) *Ugry. Materialy VI-go Sibirskogo simpoziuma 'Kul'turnoe nasledie narodov Zapadnoj Sibiri'*: 45–52. Tobol'sk: Tobol'skij gosudarstvennyj istoriko-arhitekturnyj muzej-zapovednik. Брусницына, А.Г. 2003. Городище Усть-Войкарское. Начало изучения, в: Нескоров, А.В (ред.) *Угры. Материалы VI-го Сибирского симпозиума «Культурное наследие народов Западной Сибири»*: С. 45–52. Тобольск: Тобольский гос. ист.-арх. музей-заповедник.

Fedorova, E.G. 1994. Istoriko-etnograficheskie ocherki material'noj kul'tury mansi. Saint Petersburg: MAE RAN. Федорова, Е.Г. 1994. Историко-этнографические очерки материальной культуры манси. СПб.: МАЭ РАН.

Fedorova, E.G. 2000. Rybolovy i okhotniki basseyna Obi: problemy formirovaniya kul'tury khantov i mansi. Saint Petersburg: Evropeyskiy Dom. Федорова, Е.Г. 2000. Рыболовы и охотники бассейна Оби: проблемы формирования культуры хантов и манси. СПб.: Европейский Дом.

Fedorova, N.V. 2006. Voykarskiy gorodok. Itogi raskopok 2003–2005 gg. *Nauchnyy vestnik Yamalo-Nenetskogo avtonomnogo okruga* 4: 11–17. Федорова, Н.В. 2006. Войкарский городок. Итоги раскопок 2003–2005 гг. *Научный вестник Ямало-Ненецкого автономного округа* 4: 11–17.

Garkusha, Ju. N. 2019. Dendrokhronologija gorodishha Ust'-Vojkarskoe: novye rezul'taty (po materialam rabot 2012–2016 godov), in N.M. Chairkina (ed.) *V Severnyj arkheologicheskij kongress: tezisy dokladov:*

221–222. Hanty-Mansijsk; Ekaterinburg: OOO Universal'naja tipografija 'Al'fa-Print'. Гаркуша Ю.Н. 2019. Дендрохронология городища Усть-Войкарское: новые результаты (по материалам работ 2012–2016 годов, в: Чаиркина, Н.М. (ред.) *V Северный археологический конгресс: тезисы докладов*: 221–222. Ханты-Мансийск; Екатеринбург: ООО Универсальная типография «Альфа-Принт».

Gondatti, N.L. 1888. *Sledy yazychestva u inorodtsev Severo-Zapadnoi Sibiri. (Iz VIII knigi Trudov Etnograficheskogo Otdela)*. Moscow: Tipografiya E.G. Potapova. Гондатти, Н.Л. 1888. Следы язычества у инородцев Северо-Западной Сибири. (Из VIII книги Трудов Этнографического Отдела). Москва: Типография Е.Г. Потапова.

Gondatti, N.L. 2000. Predvaritel'nyi otchet o poezdke v Severo-Zapadnuyu Sibir'. *Lukich* 4: 96–144. Гондатти, Н.Л. 2000. Предварительный отчет о поездке в Северо-Западную Сибирь. *Лукич* 4: 96–144.

Gurskaya, M.A. 2008. Dendrokhronologicheskaja datirovka arheologicheskih obrazcov drevesiny gorodishha Ust'-Vojkarskogo (Severo-Zapadnaja Sibir'): 212–231, in P.A. Kosintsev (ed.) *Fauny I flory Severnoj Evrazii v pozdnem kajnozoe*. Ekaterinburg; Cheljabinsk: Rifey. Гурская, М.А. 2008. Дендрохронологическая датировка археологических образцов древесины городища Усть-Войкарского (Северо-Западная Сибирь), в: Косинцев, П.А. (ред.) *Фауны и флоры Северной Евразии в позднем кайнозое*: 212–231. Екатеринбург; Челябинск: Рифей.

Ides, I. and A. Brand 1967. Zapiski o russkom posol'stve v Kitay (1692–1695). Moscow: Nauka. Идес, И. и А. Бранд 1967. Записки о русском посольстве в Китай (1692–1695). Москва: Наука.

Kardash, O.V. 2009. Nadymskij gorodok v kontse XVI - pervoj treti XVIII vv. Istoriya I material'naya kul'tura. Yekaterinburg; Nefteyugansk: Izd-vo 'Magellan'. Кардаш, О.В. 2009. Надымский городок в конце XVI — первой трети XVIII вв. История и материальная культура. Екатеринбург; Нефтеюганск: Изд-во «Магеллан».

Kardash, O.V. 2013. *Polujskij mysovoj gorodok knyazej Tajshinykh*. Yekaterinburg; Neftyugansk: Izd-vo 'Magellan'. Кардаш, О.В. 2013. *Полуйский мысовой городок князей Тайшиных*. Екатеринбург; Нефтюганск: Изд-во «Магеллан».

Karjalainen, K.F. 1995. *Religiya yugorskikh narodov*. Vol. II. Tomsk: Izdatel'stvo Tomskogo universiteta. Карьялайнен, К.Ф. 1995. *Религия югорских народов*. Т. II. Томск: Издательство Томского университета.

Kosarev, M.F. 1988. Chelovek I priroda v svete sibirskikh etnograficheskikh i arkheologicheskih materialov, in M.A. Devlet (ed.) *Nekotorye problemy sibirskoj arkheologii*: 84–113. Moscow: Institut arkheologii AN SSSR. Косарев, М.Ф. 1988. Человек и природа в свете сибирских этнографических и археологических материалов, в: М.А. Дэвлет (отв. ред.) *Некоторые проблемы сибирской археологии*: 84–113. Москва: Институт археологии АН СССР.

Kostyukhin, E.A. 1987. *Tipy I formy zhivotnogo eposa*. Moscow: Nauka. Костюхин, Е.А. 1987. *Типы и формы животного эпоса*. Москва: Наука.

Kulemzin, V.M. 1984. *Chelovek i priroda v verovanijah hantov*. Tomsk: Izdatel'stvo Tomskogo universiteta. Кулемзин, В.М. 1984. *Человек и природа в верованиях хантов*. - Томск: Издательство Томского университета.

Kulemzin, V.M. and N.V. Lukina 1977. *Vasjugansko-vahovskie hanty v konce XIX - nachale XX vv.* Tomsk: Izdatel'stvo Tomskogo universiteta. Кулемзин, В.М. и Н.В. Лукина 1977. *Васюганско-ваховские ханты в конце XIX - начале XX вв.* Томск: Издательство Томского университета.

Lobanova, T.V. and O.V. Kardash 2014. Khozyajstvennye, bytovye I ritual'nye aspekty zhiznedeyatel'nosti naseleniya Poluyskogo mysovogo gorodka (po rezul'tatam analiza arkheozoologicheskoy kollektsii). *Archaeology, Ethnology and Anthropology of Eurasia* 3: 66–79. Лобанова, Т.В. и О.В. Кардаш 2014. Хозяйственные, бытовые и ритуальные аспекты жизнедеятельности населения Полуйского мысового городка (по результатам анализа археозоологической коллекции). *Археология, этнография и антропология Евразии* 3: 66–79.

Lukina, N.V. 1983. Formy pochitaniya sobaki u narodov Severnoj Azii, in N.I. Tolstoy (ed.) *Areal'nye issledovaniya v yazykoznanii I etnografii*: 226 233. Leningrad: Nauka. Лукина, Н.В. 1983. Формы почитания собаки у народов Северной Азии, в: Н.И. Толстой (отв. ред.) *Ареальные исследования в языкознании и этнографии (язык и этнос)*: 226–233. Ленинград: Наука.

Lukina, N.V. 2004. Khanty ot Vasyugan'ya do Zapolyar'ya. Istochniki po etnografii. Vol. 1: Vasyugan. Tomsk; Yekaterinburg: TSU. Лукина, Н.В. 2004. Ханты от Васюганья до Заполярья. Источники по этнографии. Т. 1: Васюган. Томск; Екатеринбург: ТГУ.

Lukina, N.V. 2005. Khanty ot Vasyugan'ya do Zapolyar'ya. Istochniki po etnografii. Vol. 2: Srednyaya Ob'. Vach, iss. 1. Tomsk: TSU. Лукина, Н.В. 2005. Ханты от Васюганья до Заполярья. Источники по этнографии. Т. 2: Средняя Обь. Вах. Кн. 1. Томск: ТГУ.

Martynova, E.P. 2005. Vojkarskie khanty: kto oni? in *Kul'turnoe nasledie narodov Sibiri i Severa: Materialy Shestykh Sibirskikh chteniy*: 121–126. Saint Petersburg: MAE RAN. Мартынова, Е.П. 2005. Войкарские ханты: кто они?, в: *Культурное наследие народов Сибири и Севера: Материалы Шестых Сибирских чтений*: 121–126. СПб.: МАЭ РАН.

V.M. Kulemzina and N.V. Lukinoy. 1978. *Materialy po fol'kloru hantov*. Zapis', predislovie i primechanie Tomsk: Izdatel'stvo Tomskogo universiteta. *Материалы по фольклору хантов*. Запись, предисловие и примечание В.М. Кулемзина и Н.В. Лукиной. 1978. Томск: Издательство Томского университета.

Molodin, V.I., A.V. Novikov, A.V. Kenig, V.N. Dobzhanskiy, A.V.Vybornov, G.P. Vedmid', V.S. Myglan, E.A. Zayceva, A.Yu. Maynicheva and A.A. Shil' 2018. *Kazymskiy arkheologo-etnograficheskiy kompleks*. Novosibirsk: IAET SB RAS Publ. Молодин, В.И., А.В. Новиков, А.В. Кениг, В.Н. Добжанский, А.В. Выборнов, Г.П. Ведмидь, В.С. Мыглан, Е.А. Зайцева, А.Ю. Майничева и А.А. Шиль 2018. *Казымский археолого-этнографический комплекс*. Новосибирск: Издательство ИАЭТ СО РАН.

Morozov, V.M. 1993. Domostroenie u obskikh ugrov (po dannym razlichnykh istochnikov), in Koryakova, L.N. (ed.) *Znanie I navyki ural'skogo naseleniya v drevnosti I srednevekov'e*: 192–203. Yekaterinburg: Nauka.Морозов, В.М. 1993. Домостроение у обских угров (по данным различных источников), в: Корякова, Л.Н. (отв. ред.) *Знания и навыки уральского населения в древности и средневековье*: 192–203. Екатеринбург: Наука.

Moshinskaya, V.I. and N.V. Lukina 1982. O nekotorykh osobennostyakh v otnoshenii k sobake u obskikh ugrov, in N.V. Lukina (ed.) *Arkheologiya i etnografiya Priob'ya*: 46–60. Tomsk: Izdatel'stvo Tomskogo universiteta. Мошинская, В.И., Лукина, Н.В. 1982. О некоторых особенностях в отношении к собаке у обских угров, в: Лукина, Н.В. (ред.) *Археология и этнография Приобья*: 46–60. Томск: Издательство Томского университета.

Novikov, A.V. 1997. Sobaki v poselencheskikh kompleksakh lesostepnogoiy uzhno-taezhnogo Ob'-Irtysh'ya, in Kiryushin, Yu.F. and A.B. Shamshin (eds) *Sotsialno-ekonomicheskie struktury drevnikh obshchestv Zapadnoy Sibiri. Materialy Vserossiiskoy nauchnoy konferentsii*: 175–179. Barnaul: Izdatel'stvo ASU. Новиков, А.В. 1997. Собаки в поселенческих комплексах лесостепного и южно-таежного Обь-Иртышья, в: Кирюшин, Ю.Ф. и А.Б. Шамшин (отв. ред.) *Социально-экономические структуры древних обществ Западной Сибири. Материалы Всероссийской научной конференции*: 175–179. Барнаул. Издательство АГУ.

Novikov, A.V. 1999. A.A. Dunin-Gorkavich - issledovatel' traditsionnogo sobakovodstva aborigennogo naseleniya Zapadnoi Sibiri, in N.L. Krivonosova (ed.) *III Bersovskie chteniya. K 95-letiyu A.A. Bersa i 90-letiyu E.M. Bers. Materialy nauchno-prakticheskoi konferentsii. g. Ekaterinburg, sentyabr' 1997*: 85–89. Yekaterinburg: Bank kul'turnoi informatsii. Новиков, А.В. 1999. А.А. Дунин-Горкавич — исследователь традиционного собаководства аборигенного населения Западной Сибири, в: Н.Л. Кривоносова (ред.) *III Берсовские чтения. К 95-летию А.А. Берса и 90-летию Е.М. Берс. Материалы научно-практической конференции. г. Екатеринбург, сентябрь 1997*: 85–89. Екатеринбург. Банк культурной информации.

Novikov, A.V. 2001. Sobaki v mirovozzreni I iritual'noi praktike drevnego naseleniya lesostepnoi I yuzhno-taezhnoi zon Zapadnoi Sibiri. *Archaeology, Ethnology and Anthropology of Eurasia* 1 (5): 72–84. Новиков, А.В. 2001. Собаки в мировоззрении и ритуальной практике древнего населения лесостепной и южно-таежной зон Западной Сибири. *Археология, этнография и антропология Евразии* 1 (5): 72–84.

Novikov, A.V. 2001. Dogs in the system of views and ritual practice of ancient populations of west siberian forest-steppe and soud-taiga zones. *Archaeology, Ethnology and Anthropology of Eurasia* 1 (5): 72–84.

Novikov, A.V. 2008. Etnokul'turnoe vzaimodeistvie kazymskikh khantov I vostochnoslavyansko go naseleniya v XVIII-XIX vekakh (na primere domostroitel'nykh traditsii), in A.Ya. Trufanov (ed.) *Barsova Gora: drevnosti taezhnogo Priob'ya*: 262–274. Yekaterinburg; Surgut: Ural'skoe izdatel'stvo. Новиков, А.В. 2008. Этнокультурное взаимодействие казымских хантов и восточнославянского населения в XVIII - XIX веках (на примере домостроительных традиций), в: А.Я. Труфанов (отв. ред.) *Барсова Гора: древности таежного Приобья*: 262–274. Екатеринбург; Сургут: Уральское издательство.

Novikov, A.V. and Yu.N. Garkusha 2017. Predvaritel'nye rezul'taty polevyh issledovanij gorodishha Ust'-Vojkarskoe-1 (Pripoljarnaja zona Zapadnoj Sibiri) v 2012 - 2016 godah. *Vestnik Rossijskogo fonda fundamental'nyh issledovanij. Gumanitarnye I obshhestvennye nauki* 3 (88): 141–149. Новиков, А.В. и Ю.Н. Гаркуша, 2017. Предварительные результаты полевых исследований городища Усть-Войкарское-1 (Приполярная зона Западной Сибири) в 2012 - 2016 годах. *Вестник Российского фонда фундаментальных исследований. Гуманитарные и общественные науки* 3 (88): 141–149.

Novikova, O.I. and M.S. Nesterova 2010. Arkheologicheskie svidetel'stva ritualov ostavleniya zhilishcha, in Chernaya, M.P. (ed.) Kul'tura kak sistema v istoricheskom kontekste: opyt Zapadno-sibirskikh arkheologo-etnograficheskikh soveshchanii: materialy XV Mezhdunarodnoy Zapadno-Sibirskoy arkheologo-etnograficheskoy konferentsii, Tomsk, 19–21 maya 2010 g.: 218–220. Tomsk: Agraf-Press. Новикова, О.И. и М.С. Нестерова 2010. Археологические свидетельства ритуалов оставления жилища, в: Черная, М.П. (отв. ред.) Культура как система в историческом контексте: Опыт Западно-Сибирских археолого-этнографических совещаний: материалы XV

Международной Западно-Сибирской археолого-этнографической конференции, Томск, 19–21 мая 2010 г.: 218–220. Томск: Аграф-Пресс.

Nosilov, K.D. 1904. *U vogulov. Ocherki i nabroski*. St. Petersburg: Izdanie A.S. Suvorina. Носилов, К.Д. 1904. *У вогулов. Очерки и наброски*. СПб.: Издание А.С. Суворина.

Patkanov, S.K. 1999. Tip ostjackogobogatyrja po ostjackim bylinam I groicheskim skazaniyam, in S.K. Patkanov *Collected Works in 2 vols. Vol. 2*: 21–111. Tyumen': Izdatel'stvo Yu. Mandriki. Патканов, С.К. 1999. Тип остяцкого богатыря по остяцким былинам и героическим сказаниям, в: С.К. Патканов *Сочинения в двух томах*. Т. 2: 21–111. Тюмень: Издательство Ю. Мандрики.

Perevalova, E.V. 1996. Dve traditsii v sakral'nom otnoshenii k sobake u nizhneobskikh khantov, in V.I. Sobolev (ed.) Integratsiya arkheologicheskikh I etnograficheskikh issledovanij. Materialy IV Vserossijskogo nauchnogo seminara, posvyashchennogo 60-letiyu so dnya rozhdeniya V.I. Vasil'eva. Pt. 2: 83–87. Novosibirsk; Omsk: NSPU Publ. Перевалова, Е.В. 1996. Две традиции в сакральном отношении к собаке у нижнеобских хантов, в В.И. Соболев, (ред.) Интеграция археологических и этнографических исследований. Материалы IV Всероссийского научного семинара, посвященного 60-летию со дня рождения В.И.Васильева. Ч. II: 83–87. Новосибирск; Омск.: Издательство НГПУ.

Perevalova, E.V. 2000. Obdorskie knjaz'ja Tajshiny (istoriko-jetnograficheskij ocherk), in A.V. Golovnev (ed.) *Drevnosti Yamala. Iss. 1*: 152–190. Yekaterinburg; Salekhard: UrO RAN. Перевалова, Е.В. 2000. Обдорские князья Тайшины (историко-этнографический очерк), в: А.В. Головнев (ред.) *Древности Ямала. Вып. 1*: 152–190. Екатеринбург; Салехард: УрО РАН.

Perevalova, E.V. 2004. Severnye khanty: etnicheskaya istoriya. Ekaterinburg: UrO RAN. Перевалова, Е.В. 2004. *Северные ханты: этническая история*. Екатеринбург: УрО РАН.

Perevalova, E.V. and G.K. Karacharov 2006. *Reka Agan i eyo obitateli*. Ekaterinburg; Nizhnevartovsk: UrO RAN; Studiya 'GRAFO'. Перевалова, Е.В. и Г.К. Карачаров 2006. *Река Аган и её обитатели*. Екатеринбург; Нижневартовск: УрО РАН; Студия «ГРАФО».

Sokolova, Z.P. 2011. *Severnye khanty: (polevye dnevniki)*. Moscow: Institut etnologii i antropologii RAN. Соколова, З.П. 2011. *Северные ханты: (полевые дневники)*. Москва: Институт этнологии и антропологии РАН.

Startsev, G. 1928. Ostyaki: Social'no-jetnograficheskij ocherk. Leningrad: Priboy. Старцев, Г. 1928. Остяки: Социально-этнографический очерк. Ленинград: Прибой.

Zhel'vis, V.I. 1984. Chelovek i sobaka (vospriyatie sobaki v raznykh etnokul'turnykh traditsiyakh). *Sovetskaya etnografiya* 3: 135–143. Жельвис, В.И. 1984. Человек и собака (восприятие собаки в разных этнокультурных традициях). *Советская этнография* 3: 35–143.

# 4.13 The Dog in the Castle: A Dog Skeleton from the Castle of Santa Severa (Latium, Italy)

Eugenio Cerilli[1] and Marco Fatucci[2]

[1] Independent researcher, via Torquato Taramelli 14, 00197 Roma, Italy. cerillieugenio@gmail.com
[2] Independent researcher, via Sagripanti 24, 00052 Cerveteri (Roma), Italy. marcofatucci@inwind.it
Corresponding author: Eugenio Cerilli, cerillieugenio@gmail.com

**Abstract**

Between 2003 and 2010, several excavations were carried out in various sectors of the citadel and the village within the Castle of Santa Severa. In particular, the remains of an early Christian church were identified in the Piazza della Rocca under the modern pavement. Two human burials and one of a dog (*Canis familiaris*) were identified inside the church. The radiocarbon dates indicate that the burials occurred between 1380 and 1450. The skeleton of the dog was analysed from an anatomical, morphological, biometric, and taphonomic point of view, in the attempt to reconstruct the burial processes and to identify the canine breed.

**Keywords**: dog, castle, dog breeds, artistic representations, Late Middle Ages.

## 1 Introduction

The Santa Severa Castle is located on the Tyrrhenian coast, about 50 km north of Rome. This area has been inhabited by humans for a long time, since the Palaeolithic (Enei, 2013a, and references therein), but it became especially important during the Etruscan period.

The settlement of *Pyrgi* was founded in the 7th century BC; it was a port-emporium, through which goods and people from many areas of the Mediterranean passed, but also an important religious centre, as indicated by the three sanctuaries (Baglione *et al.* 2017; Michetti and Belelli Marchesini, 2018; and references therein).

In the 3rd century BC *Pyrgi* became a Roman colony (Enei, 2013b) and a *Castrum* protected by polygonal walls was founded. Its structures overlapped those of the Etruscan period, exploiting and fortifying as well as expanding them; the importance of the port was preserved, also because of the import of minerals from the Elba island. In the imperial period, the town became the home of villas with a widely used port area.

After the martyrdom of the young Severa (AD298) the site became an important place of Christian worship.

The port remained in full activity even in the Late Antiquity and in the Middle Ages, with the increase of the village around the castle and the Saracen tower.

During the 11th century the village of Santa Severa was property of the powerful Abbey of Farfa and a fortress with a tower was built, with the subsequent expansion, transformation and development of the fortified village.

At the beginning of the 12th century, Santa Severa became part of the properties of the Benedictine monks of the Monastery of St. Paul Outside the Walls, bound to the Apostolic Chamber and its possessions. During the 14th and 15th centuries, a series of important noble families alternated in the ownership of the village or portions of it: Tiniosi, Bonaventura, Venturini, Di Vico, and Anguillara. When the castle was owned by the Bonaventura and Venturini families (second half of the 14th century) it underwent a series of renovations, which changed several internal areas.

From 1482, for five centuries, the village was the property of the Religious Order of the Holy Spirit, until the 20th century when it passed to the municipality of Santa Marinella, who in 2003 began a series of restoration and renovation works.

Between 2003 and 2010, during the restoration of the settlement, a series of surveys and excavations were carried out in various sectors of the village within the Castle of Santa Severa (Enei 2013c). In an open space inside the castle, called Piazza della Rocca, the remains of an early Christian church, partially hypogeal, were identified under the modern pavement. The church was built in the second half of the 5th century and was in use, with several transformations, until the mid 14th century. In the 9th century, a large cemetery area developed around the church, remaining in use until the end of the 14th century and beyond. In the 14th century the church was abandoned, probably because of a devastating fire that almost destroyed the structure. This event could be potentially related to the disastrous consequences of the plague that swept through Italy and Europe between 1348 and 1350, resulting in the decline, several collapses, fires, and improper uses of the internal structures of the castle.

During the second half of the 14th century, when the castle was owned by the powerful noble Roman families of Trastevere, Bonaventura and Venturini, the refurbishment of the castle was undertaken, so the walls of the church were partially levelled and the interior was filled with several layers of the backfill from the discard pits present in the area, containing archaeological and osteological materials that allow a reconstruction of the history and daily life of the castle between the 13th and the mid-14th centuries (Fatucci and Cerilli 2013, 2015, 2016, 2019). At the end of the 14th century, the remaining portions of the walls of the church were razed.

During the investigations inside the church and along its northern wall, along the same alignment and within the same depositional horizon, two burials of adult men and one of an adult dog (*Canis familiaris*) were brought to light (Figure 1). A third adult man was found buried inside the church, close to the nave. The direct radiocarbon dates on the human skeleton without legs discovered near the dog skeleton, indicate that the burials occurred between 1380 and 1450 (Enei, 2013d), just before the levelling of the walls of the church and the completion of the refurbishment of the castle. If the upper limit of the dating range for this skeleton indicates the end of the demolition and filling of the church, these could also have been completed at the same time as the acquisition of the property by the powerful Di Vico Roman family.

Figure 1. Santa Severa, the two adult men and the dog buried inside the church, along the northern wall, during the excavations (After Enei 2013c: 86, modified).

## 2 Material and methods

The skeleton of the dog was in almost complete anatomical connection, laid on its right side with its limbs bent. Almost all the skeletal elements are present, except for a few bones that are missing because of ancient works of rearrangement of the area.

The age of the dog at its death was estimated according to Barone (1981, 1995). The measurements were taken according to von den Driesch (1976). The withers height was calculated according to Koudelka (1885) and Harcourt (1974). The weight of the individual was established according to Hamblin (1984), Wing (1978) and Anyonge (1993).

The taphonomic analysis was carried out macroscopically using a magnifying lens 10x, and microscopically using a digital microscope (Celestron® Handheld Digital Microscope Pro) at different magnifications.

Statistical analyses have been executed by means of the PAST (PAlaeontological STatistics) 3.20 version software (Hammer *et al.* 2001).

## 3 Results

### *3.1 Age at death*

The complete eruption with a low degree of wear of the teeth, and the state of fusion of the axial and appendicular skeletal elements, indicate that the dog died around one year of age (Barone 1981, 1995).

### *3.2 Sex*

The sex determination remains uncertain because the baculum was missing, perhaps because of post-depositional events. According to The and Trouth (1976, p. 3) an anatomical difference between male and female can be established by the surface markings of the basioccipital morphology analysis, showing three main typologies. However, the surface markings of the basioccipital in the dog of Santa Severa has very few markings, similar to Type II that The and Trouth (1976, p. 3) attribute to females.

### *3.3 Withers height*

The withers height, calculated on the humerus, radius, ulna, femur and tibia of both sides, is between 61.52 and 69.32 cm, with a mean of 65.94 cm and a standard deviation of 2.89, if calculated according to Koudelka (1885), and between 64.67 and 69.92 cm, with a mean of 66.91 and a standard deviation of 1.83, if calculated according to Harcourt (1974). Therefore, the withers height of the dog of Santa Severa is between about 62 and 70 cm, with a mean of all calculated withers heights of about 66.5 cm.

### *3.4. Body weight*

The estimated weight of the dog varies according to the formulae used.

Hamblin (1984) and Wing (1978) proposed formulae involves the mandible height (mm) measured on the labial side at the centre of the lower carnassial tooth. According to Hamblin (1984) the weight of the dog of Santa Severa is between 20.511 - 20.701 kg, while according to Wing (1978) the weight is between 23.227 - 23.388 kg, considering the two measurements of the left and right mandibles.

Anyonge (1993) proposed two formulae involving the stylopodial bones. From the formula involving the humerus the weight is 25.924 kg (both humeri have the same greatest length), while from the formula involving the femur the weight is between 24.551 - 24.770 kg, considering the two measurements of the left and right femurs.

Therefore, the dog weight is attested at a range between approximately 21 and 26 kg, with a mean of all calculated weights of 23.3 kg.

### *3.5 Slenderness*

The ratio between the slenderness index of the femur (left femur = 6.53; right femur = 6.42; mean = 6.476) and the withers height (considering the mean = 66.5 cm) places the Santa Severa dog in the hypermetric dogs field of the graph proposed by De Grossi Mazzorin and Tagliacozzo (2000).

All these data indicate that the Santa Severa dog was an individual of about one year of age, quite tall, but slender and relatively light in weight, who was perhaps a female.

### *3.6 Taphonomy*

On the extremities of the limbs several cut marks are present.

Thin cut marks that are often deep are localised on the anterior side of the distal portion of the diaphysis of the second, third, and forth left metacarpals, and on the anterior-lateral side of the surface of the distal portion of the diaphysis of the fifth left metacarpal. Other repeated, thin, deep cut marks are localised on the medial side of the distal portion of the diaphysis of the second right metacarpal, on the lateral side of the central portion of the diaphysis of the fourth right metacarpal, and on the anterior lateral side of the central portion of the diaphysis of the fifth right metacarpal. One thin cut mark is localised on the medial side of the proximal end of the diaphysis of the fifth left metatarsal and another on the anterior side of the distal end of the diaphysis of the third right metatarsal. Other cut marks, even if less marked, are present on some phalanges.

These cut marks were inflicted most probably in relation to the skinning of the animal, with a thin blade by cutting transversely to the bones and in a circular way from the lateral side to the medial side of the left anterior autopodium, and perhaps with a double circular movement from the medial to the lateral side of the right anterior autopodium (Figure 2).

## 4 Discussion

The question to be defined is why this dog seems to have been buried so carefully and apparently in connection with the two human burials. One approach to answer this question would be to try defining which of the modern dog breeds could be the most similar to the dog buried in Santa Severa.

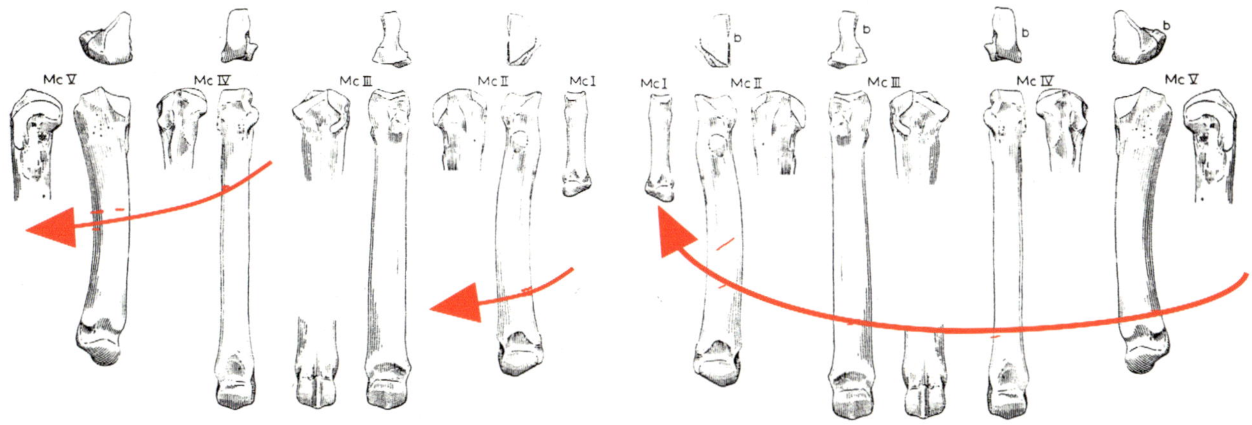


Figure 2. Sketch of cut marks on the metacarpals of the Santa Severa dog. (Basic drawing from von den Driesch 1976, modified).

Figure 3. Cranium of the Santa Severa dog.
Clockwise from top-left: left lateral view, right lateral view, cranial view, buccal view.

### *4.1 Breed*

Whilst waiting for possible future genetic analysis, a morphological approach was attempted (see e.g., Phillips *et al.* 2009), which focused the analysis on the cranial morphology and morphometry, mainly because of the scarcity of comparative data in the literature regarding the measurements of post-cranial elements of modern breeds.

The cranium of the Santa Severa dog is relatively elongated, the snout is of medium length and not particularly narrow (Figure 3).

Following the available comparative data, only some measurements (1÷9, 12÷14, 18, 23÷28, 30, 32, 33, 36÷38, 40; according to von den Driesch, 1976: 42–43) were selected to perform the biometric analysis. The measurements of the crania in the modern dog collection in the museum of the University of Bergen in Norway and reported in the master thesis by Knoest (2015) were used as the main source of biometric comparative data.

The bivariate analysis of the relationship between the condylo basal length (measurement nr. 2, according to von den Driesch, 1976) and the median palatal length (measurement nr. 13, according to von den Driesch, 1976), which defines the relationship between the total length of the skull and the length of the snout, indicates preliminarily that the cranium of the Santa Severa dog is placed near some types of Greyhound, a Scottish Sheep Dog (Border Collie) and a Swedish Foxhound. The ratio between the median palatal length (measurement

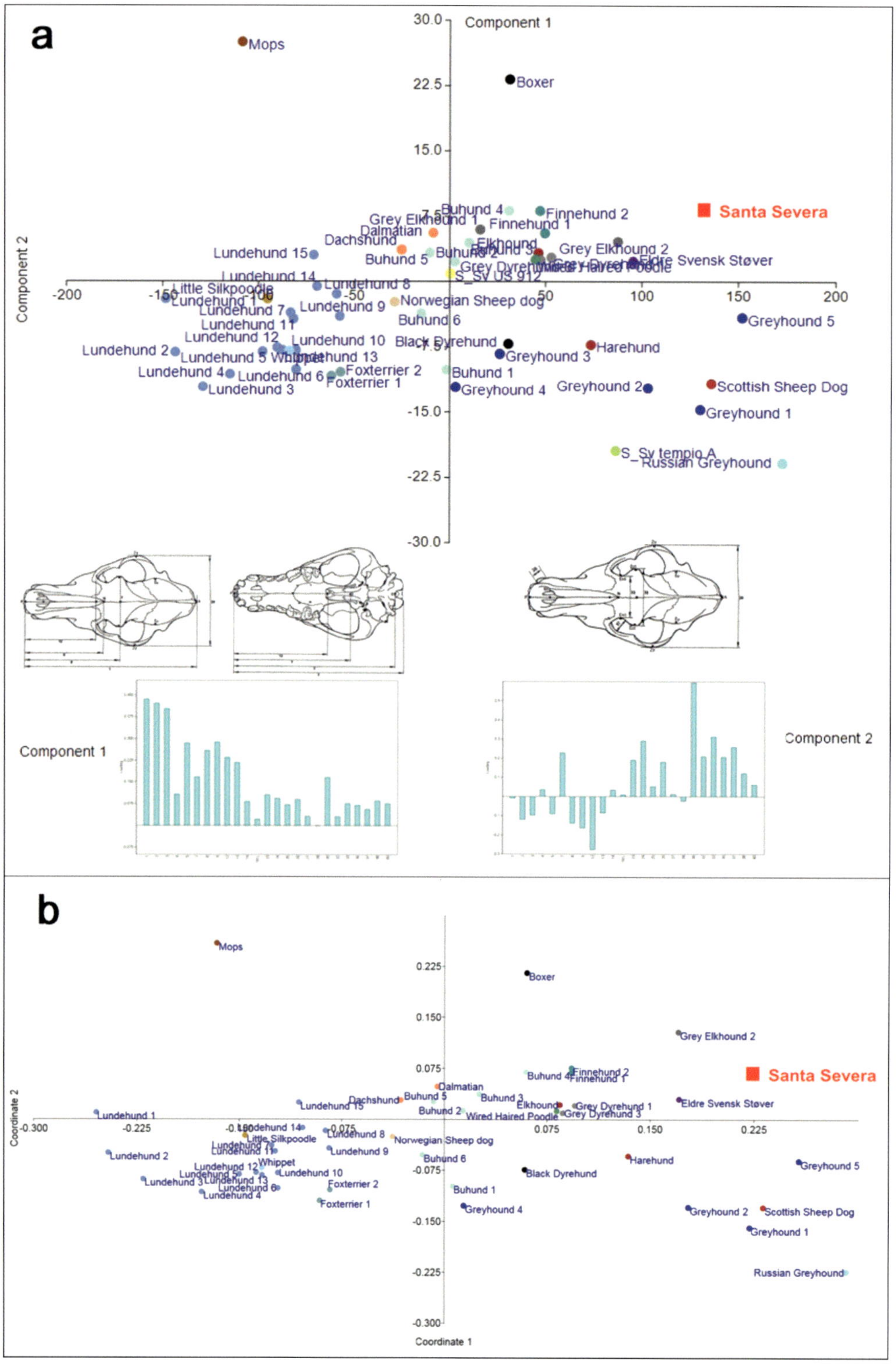


Figure 4. Scatter plot of the first and the second components of the PCA of selected skull measurements of some modern breeds and of the Santa Severa dog (a). Scatter plot of the first and the second coordinates of the PCoA of selected skull measurements of some modern breeds and of the Santa Severa dog (b) (Data of modern breeds from Knoest 2015).

nr. 13, according to von den Driesch, 1976) and the breadth of the canine alveoli (measurement nr. 36, according to von den Driesch, 1976), which defines the proportion of the snout (slender or stocky), indicates that the cranium of the Santa Severa dog has a snout that is wider than the previously mentioned breeds.

In order to better compare the measurements and proportions of the cranium of the Santa Severa dog with those of some modern breeds, we proceeded to the multivariate statistical analysis of the entire biometric matrix.

Regarding the Principal Component Analysis (matrix: variance-covariance), the scatter plot of the first (Eigenvalue = 6951.12; variance = 91.921%), to which the longitudinal measurements of the skull contribute most, and the second (Eigenvalue = 214.524; variance

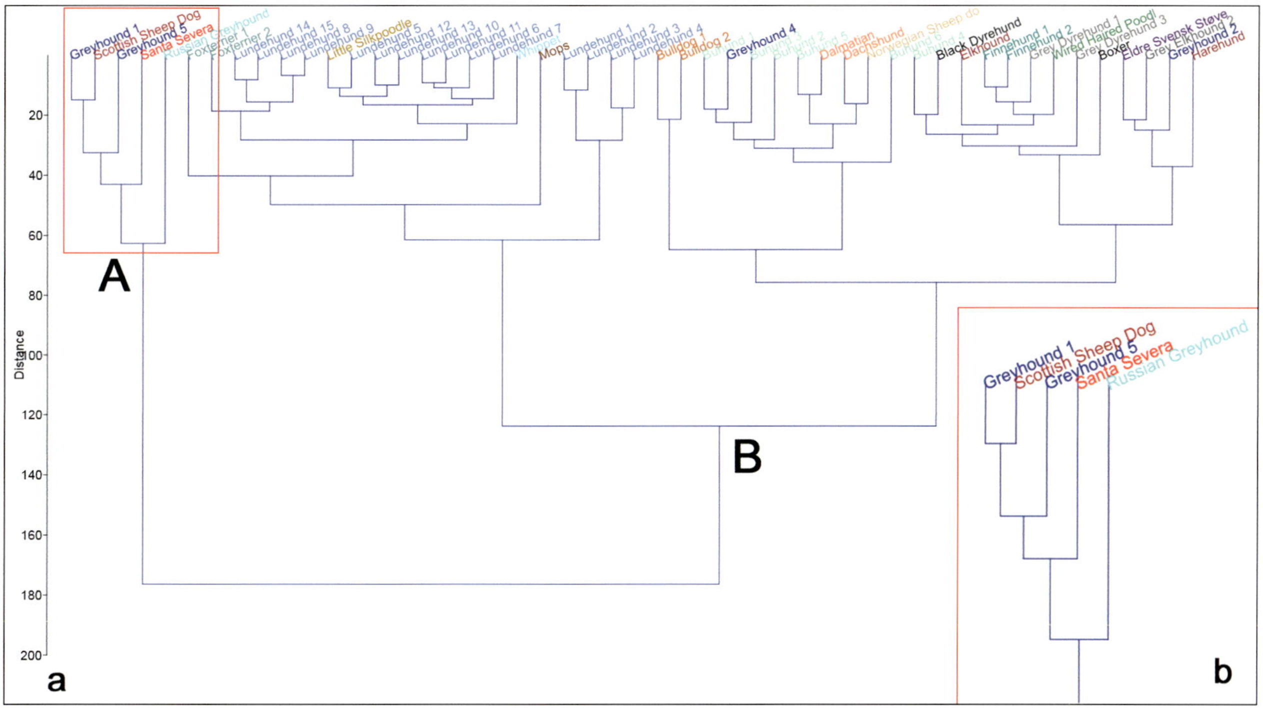


Figure. 5. Cluster analysis dendrogram of selected skull measurements of some modern breeds and of the Santa Severa dog (Data of modern breeds from Knoest 2015).

= 2.8368%), to which the longitudinal measurements of the skull contribute most, does not place the Santa Severa dog close to the position of the breeds mentioned above (Figure 4a). This distance increases in the scatter plot of the first and the third (Eigenvalue = 97.1449; variance = 1.2846%) components; the third components however provide less information and the length of the neurocranium mainly contributes to it, but other measurements contribute significantly less.

With the Principal Coordinates Analysis (Euclidean similarity index; transformation exponent c=2) this distance decreases slightly, as it is possible to see in the scatter plot of the first (Eigenvalue = 3.7652E05; percent = 91.462) and the second (Eigenvalue = 10824; percent = 2.6292) coordinates (Figure 4b).

In the similarity analysis (using the Unweighted Pair Group Method with Arithmetic Mean algorithm, Euclidean similarity index, final branch root) the dog skulls used for comparison are distributed in two well separated asymmetrical clusters as far as the quantity of canine breeds included. The Greyhounds are located both in clusters and in different rami, due to the wide variability of cranial dimensions, withers height and weight, which characterise this canine breed (see below). The Santa Severa dog is located in cluster A, in close similarity with two specimens of Greyhounds, as in the bivariate scatter plot between the condylo basal length and the median palatal length discussed above, together with a Russian Greyhound, and a Scottish Sheep Dog (Figure 5).

The hierarchical analysis performed applying the Neighbour joining clustering method (Euclidean similarity index, final branch root) places the Santa Severa dog together with three types of Greyhound, a Russian Greyhound, a Scottish Sheep Dog, and a Swedish Foxhound (Figure 6).

To summarise, the analysis of the data from statistical analysis would restrict the possibilities to four breeds: Greyhound (Federation Cynologique Internationale standard nr. 158), Border Collie (FCI 297), Swedish Foxhound (FCI 132), and Russian Greyhound (Borzoi) (FCI 193).

The latter breed tends to be heavier (male 34–38 kg, female 26–41 kg) and taller (male 75–85 cm, female 68–78 cm) than the Santa Severa dog. The Swedish Foxhound has more recent origins than the dog of Santa Severa, but it can also be considered as representative of a similar hunting dog, such as the Italian Segugio (FCI 337), which has an older origin, a similar weight (male 18–28 kg, female 18–28 kg) but a lower height (male 52–58 cm, female 48–46 cm) than the Santa Severa dog. The Border Collie is significantly smaller (weight: male 14–20 kg, female 12–19 kg; height: male 48–56 cm, female 46–53 cm) than the Santa Severa dog.

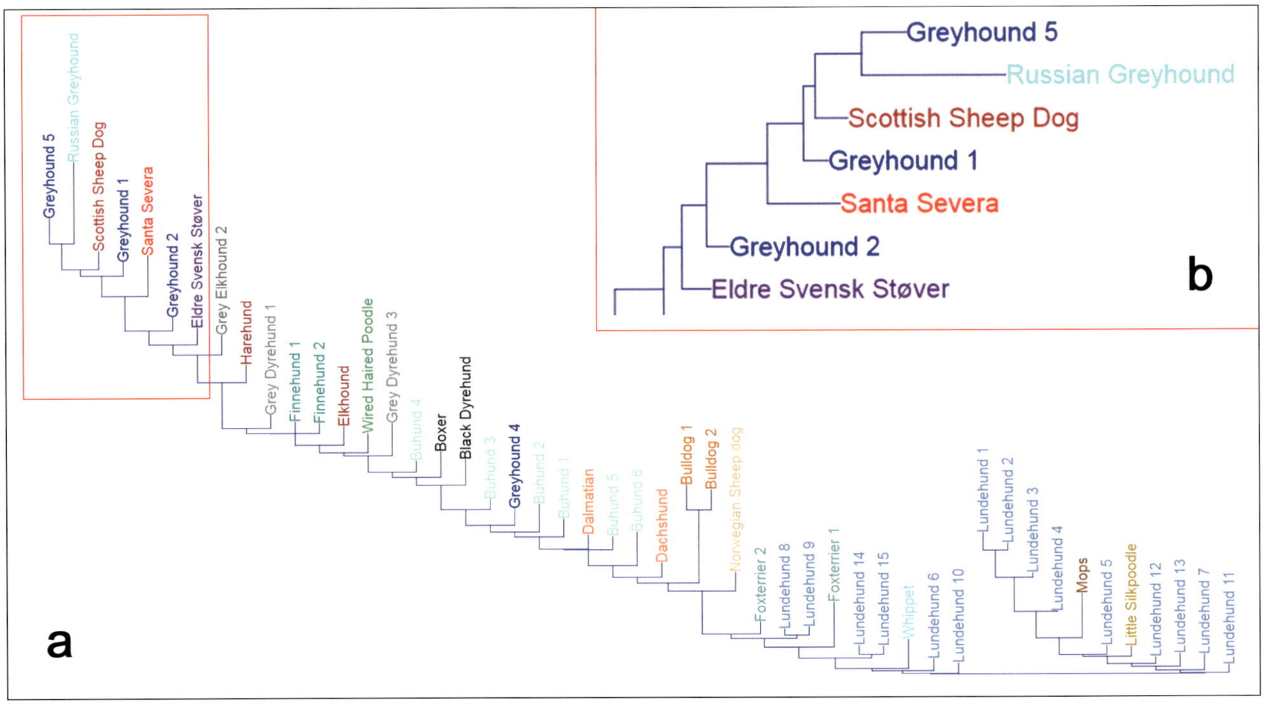


Figure 6. Neighbour joining clustering tree of selected skull measurements of some modern breeds and of the Santa Severa dog (Data of modern breeds from Knoest 2015).

From the comparison with these four breeds and considering the height (62–70 cm, mean 66.5 cm) and weight (about 21–26 kg, mean 23.3 kg) of the dog of Santa Severa, our animal appears closer to the Greyhound (height: male 71–76, female 69–71; weight: male 27–40 kg, female 27–34 kg). We could therefore suggest that the dog of Santa Severa resembled a Greyhound, perhaps a prime-adult female, even if it does not have a particularly elongated and slender snout like the modern specimens.

### *4.2 Greyhounds in history*

The group of Sighthounds (or Gazehounds), hounds that hunt more by sight and speed than by scent and endurance, include 25 recognised breeds (Afghan Hound, Azawakh, Borzoi, Chart Polski, Galgo Español, Greyhound, Hortaya borzaya, Irish Wolfhound, Italian Greyhound, Magyar agár, Rajapalayam, Rampur Greyhound, Saluki, Sloughi, Whippet; Chippiparai, Combai, Kanni, Khalag Tazi, Mudhol Hound, Old Croatian Sighthound, Scottish Deerhound, Silken Windhound, Taigan, Xigou). The first 15 breeds are considered more strictly as Greyhounds and are characterised by a wide range of cranial dimensions and proportions, withers height and weight, as evidenced by the dispersion in the previously illustrated diagrams of the greyhound specimens preserved in the collection of the museum at the University of Bergen in Norway (Knoest 2015).

The smallest representative of the greyhounds is the little Italian Sighthound (FCI 200; height 23–38 cm; weight 3–5 kg), the biggest is the Irish Wolfhound (FCI 160; height 71–78.5 cm; weight 40.5–54 kg). The greyhounds are also a group whose origins/selection (see below) are widely dispersed in time (from 5th-4th millennium BC to 19th century AD) and space (from Morocco to India and from Mali to Russia).

On their origin there is no unanimity in the scientific literature. The ancient forms, including two forms (Afghan Hound, FCI 228, and Saluki, FCI 269), are considered to be basal dogs (Parker *et al. 2004*; Larson *et al.* 2012), the recent forms are re-developed or developed *ex novo* in the 19th century.

Archaeozoological evidence of the presence of greyhounds is very rare. A probable example is given by the finding of two limb bones of a greyhound-like dog dated to AD 800 ± 60 (AMS radiocarbon dating) at the acropolis of the Chotěbuz-Podobora hillfort (Czech Republic), a settlement inhabited from the 10th-8th century BC to the first half of the 11th century AD. The calculated height of this dog is 70 cm according to Harcourt’s method (1974) and the ratio between the maximum length and the minimum width of the radius falls into the interval of the ratio for greyhounds, mainly that of the Polish greyhound (Fišáková 2010). The genetic analysis has confirmed this attribution (Svobodová 2015) and the isotopic analysis indicated that this specimen could have been imported into the Czech settlement from Poland, and represented a luxury commodity reflecting the elite status of its owner (Fišáková 2010).

Since it is not possible to rely on statistically significant and accurate samples of archaeozoological evidences, in order to trace the history of greyhounds, a brief mention will be made of the evidence that derives from artistic or literary representations in antiquity. Distinctive features of the greyhound are the height combined with the extreme slenderness, marked by a strong narrowing of the abdomen before the thighs, and the elongated and tendentially pointed snout, although in some forms the snout may be slightly squatter.

Dogs that look like greyhounds with long slender bodies, are depicted in temple drawings dating from 6000 BC at Çatalhöyük (Turkey), a possible indication of a near-eastern origin of greyhounds, and on the rock paintings in the Tassili (Algeria), dated between 5000–2000 BC, evidences indicating that the greyhound is indeed one of the oldest dog breeds (Svobodová 2015).

In the mastaba of Ptahhotep (Egypt; 5th dynasty, 25th-24th century BC) a slender dog with an elongated snout, straight erect ears, and a curled tail is represented. Although in Egyptian iconography this cranial morphology could also represent a jackal, the hunting context shown in the bas-relief and the curled tail would make it more plausible that the canid represented is a dog.

A similar dog is represented on a relief fragment from Giza (Egypt), now preserved in the Walter Art Museum (Baltimora, USA), dating back to 2400 BC. In the tomb of Anteff II (11th dinasty, *c.* 2072 BC) another dog is depicted with the same characteristics described above. A golden flabellum from the treasure of the Pharaoh Tutankhamun (1341–1323 BC) represents a scene of ostrich hunting with a slender dog characterised by a pointed snout, straight erect ears, and a curled tail. A similar dog is carved on the wall of the tomb of Sarenput I (12th dynasty, 20th-19th century BC) at the site of Qubbet el-Hawa (Egypt).

This type of dog, which corresponds to the Tesem (*tsm*) hieroglyph, has long been considered as the ancestor of today's Pharaoh Hound, native of Malta, but following Parker *et al.* (2004) this breed has been recreated in more recent times from the combinations of other breeds, because its genome does not match that of the ancient Egyptian sighthounds, even if its appearance is very similar.

A red-figured krater from Cuma (Pan painter, 530–470 BC) preserved in the Museum of Fine Arts (Boston, USA) shows *Aktaion* being killed by his dogs which look like greyhounds (inv. 10.185). A similar scene with the same subjects is depicted on a Poseidonian skyphos dated to 400–350 BC (inv. 76/106), kept in the Baden State Museum (Germany). A greyhound-like dog is also with *Ermes* on a black-figured amphora dated to 520 BC in the collections (inv. 06.1021.68a) of the Metropolitan Museum of Art (New York, USA) (Trantalidou 2016).

The head of a dog with the appearance of a greyhound, or a Cirneco dell'Etna, is depicted in a rython by the painter of Patera's Workshop (late Apulian red-figure pottery, *c.* 340–320 AD) (Pellegris 2004), preserved in the Poldi Pezzoli Museum (Milan, Italy), an iconography which is very represented in this ceramic typology.

A pair of Greyhounds playing are represented in a marble sculpture from Monte Cagnolo near Lanuvio (Rome, Italy) dated to the 2nd century BC, and now exhibited in the Museo Pio-Clementino (Sala degli Animali, Vatican Museums, Vatican City).

A greyhound is depicted alongside the bed of the funerary monument in polychrome terracotta representing dying Adonis, dated between 250 and 100 BC and exhibited in the Museo Gregoriano Etrusco (Vatican Museums, Vatican City).

A relatively slender running dog with an elongated snout, albeit not pointed, is depicted on the *verso* of the silver denarius of the *Gens Postumia* (73–74 BC), while on the *recto* the head of Diana is shown in profile.

Four little Greyhounds with pointed elongated snouts, are sculptured at the corners of a base in Luni marble (c. 27 BC - 197 AD). They were discovered in the *Horti Maccenatiani* (Esquiline Hill, Rome, Italy), and are now located in the Marcus Aurelius Exedra of the Palazzo dei Conservatori (Rome, Italy).

A slender dog with a pointed elongated snout, a greyhound, accompanies Endymion who contemplates the beloved Selene, who comes down to him covered with a cloak dark, depicted in a fresco dated before AD 79 from the Dioscuri House (Pompei, Italy), now exhibited in the National Museum of Naples (Italy).

The earliest European textual reference to sighthounds is reported in Arrian's *Kynēgetikos* (2nd century AD) (Phillips and Willcock 1999; Trantalidou 2006).

Several dogs are depicted on the so-called Bayeux Tapestry, an embroidered cloth dated to the 11th century (a few years after the battle of Hastings of 1066), some of them have a slender form, with a strong narrowing of the abdomen before the thighs, a more or less elongated snout, and usually represented as they run. Considering the iconography of the artwork it is not certain that these dogs are exactly greyhounds, but it is still possible to imagine that they are at least sighthounds and perhaps even greyhounds.

Albertus Magnus (Albert the Great; before 1200–1280), a German Catholic Dominican friar and bishop, portrayed the greyhound in one of the earliest descriptions of this breed, extolling its forms (Kitchell and Resnik 1999).

In the debate on the history of greyhounds, a large space is taken by the possible Celtic origin of this breed. In many Irish myths and legends, several dogs which may have corresponded to the Greyhound, Irish Greyhound, or Scottish Deerhound, are described. On the Hilton of Cadboll Stone, a Pictish stone dated to the end of the 8th century AD and discovered in the Scottish Highlands, on exhibit in the Museum of Scotland, three dogs resembling greyhounds accompany two knights and a lady during a deer hunt. So, it is possible that some specimens of greyhounds have been brought to Ireland and Scotland and that they have been bred and further selected in these regions. The 10th century laws of Hywel Dda, or Hywel the Good, of South Wales describe greyhounds and others dog (Clutton-Brock 1976). By the thirteenth century a cult has emerged in France to venerate Saint Guinefort, a greyhound, who was held to be a special protector of young children. Ralph Neville, first Earl of Westmorland (1364–1425), was buried in his tomb in the collegiate church of Saint Mary at Staindrop (Durham, UK) with his greyhound (Friedman 2016).

The proof of the consideration that the greyhound had in the Anglo-Saxon world is its quotation in the holy book for antonomasia. The King James Bible's Version (1611) reads: '*There be three things which go well, yea, four are comely in going: A lion which is strongest among beasts, and turneth not away for any; A greyhound; an he goat also; and a king, against whom there is no rising up.*' (Proverbs 30: 29–31), interpreting the original Hebrew expression *zarzìr mothnàyim* (Proverbs 30:31; Vigini 2002: 166), in its literal sense, that is 'girded of waist', with direct reference to the abdominal narrowing characteristic of the body of greyhounds. There is no unanimity on this translation however, and it remains an interpretative enigma. Consequently, there is a great variety of hypotheses on the identification of the animal and its prerogatives, so in many versions the translation is 'strutting rooster' (International Standard Version 2020) or 'strutting cock' (New American Bible 2002).

#### *4.3 Greyhounds in art*

To hypothesise the reason for such great care in the burial of the Santa Severa dog, assuming that it was a greyhound based on the morphological and morphometric evidence discussed above, we can investigate the role of greyhounds during the late Middle Ages and Renaissance. The greyhound, with its sharp sight and speed (it reaches a speed of 17 metres per second; Hudson *et al.* 2012), is particularly suitable for pursuit hunting small furred and feathered game such as hares (the latin name *canis leporarius* derived from *lepus*: hare), foxes, and bustards. It is also a dog with an elegant bearing, which makes it an ideal companion for the upper social classes. For these reasons, the greyhound is a dog breed which was originally bred only for monarchs and the social elite (Kholová 1987, Stuchlý and Císařovský 1991) and used in hunting activities or as a pet. Therefore, we would expect to find it represented in life scenes of aristocratic or socially elevated classes, rather than in lower class contexts. This type of research can also be effectively conducted through the analysis of artistic representations of this canine form. For this reason, 79 artworks representing greyhounds of the late Middle Ages, Renaissance and the first half of the 17thcentury, have been selected (Table 1).

Table 1 shows the relevant information regarding the selected artworks: the century in which they were produced, the author/authors, their dates of birth and death, the main geographical area of activity or the location in which the artwork was produced, the title, the date of execution, the place of conservation/exhibition, the type of work of art (fresco, painting, drawing, engraving, tapestry, sculpture, illuminated manuscript, ceramics, or coat of arms), the matter and technique, the subject/context that is represented (mythological; Roman history; religious: Bible Old Testament, Bible New Testament, Saints, nativity scene/Magi, nativity scene/shepherds, noble ceremony, allegorical, hunting, noble banquet, court daily life, court meeting, popular daily life, policy/travel, battle, portrait, or dog), the types of dogs depicted (only greyhound, or greyhound with another dog), the number of greyhounds depicted (1,2,3, or more than 3), the coat colour (white, merle/piebald, brown, or black/dark), and the morphology of the snout (stocky/not slender, medium slender and pointed, slender and pointed, not slender/slender and pointed when different types are present).

As for the appearance of the greyhounds represented, they mostly appear to be of marked height, with a generally elongated skull and pointed snout. However, in some cases, greyhounds with a tendentially less elongated snout are present. For example in the fresco *St. George Liberating the Princess of Trebizond*, painted by Antonio Puccio Pisano (Pisanello) between 1433 and 1438, and located in the Pellegrini Chapel of the church of Sant'Anastasia in Verona (Italy), a greyhound is depicted, its snout is not extremely elongated, as it is possible to see in some of the preparatory studies of greyhounds contained in the *Vallardi Codex* (1433–1438) preserved in the Louvre Museum (Paris, France). In the painting the *Vision of Saint Eustace,* again by Pisanello (1436–1438), exhibited in the National Gallery of London (UK), some dogs similar to greyhounds, with not such a slender snout, are represented together with other types of dogs. Another greyhound without a slender snout is represented in the fresco *Sigismondo Pandolfo Malatesta in prayer in front of San Sigismondo* by Piero della Francesca (1451) located in the chapel of the Conception of the Malatesta Temple in Rimini (Italy).

Numerous greyhounds without elongated snouts are represented together with other dogs in the *Boar and Bear Hunt* tapestry, part of The Devonshire Hunting Tapestries (Victoria and Albert Museum, London, UK), woven by an anonymous French or Flemish artist around 1420, and depicting a hunting scene. The greyhound painted in the *Adoration of Magi* by Gentile da Fabriano (1423), on exhibit in the Uffizi Gallery (Florence, Italy), does not have an elongated snout either. In another *Adoration of Magi* painted by Il Sassetta (Stefano di Giovanni di Consolo) around 1430, there is another greyhound represented without an elongated snout. The final greyhound without an elongated snout accompanies *Enea Silvio Piccolomini leaving for the Council of Basel* as represented by Bernardino di Betto Betti (Pinturicchio) in a fresco located in the Piccolomini Library, Siena Cathedral, and painted at the beginning of the 16th century (1502–1507).

As for the colour of the coat, white greyhounds are most frequently represented throughout the chronological period considered. Of the 56 works depicting white greyhounds, 31 contain only one specimen and in the other 11 there are two or more white greyhounds, in some cases they are associated with brown or merle/piebald specimens. Brown greyhounds, mostly singles, are represented in 13 works; while the merle or piebald ones are present only in 7 representations.

Most of the scenes depicted concern hunting (17), followed by religious scenes concerning the saints (13) or taken from the Old Testament (6) and the New Testament (13). From the latter category, for their peculiarity, the Nativity scenes stand out (10): five represent the arrival/adoration of the Magi and five the adoration of the shepherds. The mythological scenes (8), the scenes of banquets (7) and of 'political' meetings, such as triumphs or travels (6), are well represented. The other types of representation are less abundant. It should be noted that in all cases, except for the scenes of adoration of the shepherds and the only defined one of daily popular life (which then refers to the collection of the mandrake, an activity that mostly concerns doctors or their workers), the characters and scenes represented fall within a noble context or at least of high census classes. This testifies how greyhounds, with their elegance and haughty bearing, are companions of adoption of aristocrats and wealthy classes (see above) both as pets and in hunting activities; in the latter case by virtue of their acute sight, effective speed in pursuit and innate courage. As for the Nativity scenes, if the presence of greyhounds associated with the Magi falls within the reasons mentioned above, the association with the shepherds can be, apparently, contradictory, not being the greyhounds suitable for the tasks of keeping and guiding the flocks, but it is probable that the greyhounds depicted are not associated with the shepherds, but rather with the presence of Christ, Son of the Most High made Man, undoubtedly considered by the artists, and their clients, belonging to the upper classes.

The representations of greyhounds in the artworks of the centuries following those considered, which have not been taken into consideration in detail in this paper, increase considerably, also because of the increase in the economic availability of the bourgeois class which gradually established itself from the 17th century onwards, taking its place side by side with the aristocracy in the possession of power. The greyhound, while maintaining its utilitarian function as a spotter and pursuer in hunting, increasingly assumes the role of elegant companion to show off as a distinctive sign of the achievement and consolidation of a high level on the social scale by those who owned it.

Consequently, its breeding and selection of new forms increased and gradually greyhounds become a subject increasingly present in artistic representations.

## 5 Conclusion

During the investigations inside the church located in the Caste of Santa Severa, the skeleton of an adult dog (*C. familiaris*) was found in the same alignment and within the same depositional horizon of two burials of adult men. The direct radiocarbon dating on the human skeleton placed near the dog, indicates that these burials occurred in a range between 1380 and 1450. The skeleton of the dog is almost completely preserved and belongs to a prime-adult individual, perhaps a female, with a withers height between 62 and 70 cm (mean 66.5 cm, slender proportions and a weight between 21 and 26 kg (mean 23.3 kg). Some cut marks produced by a metal tool have been identified on some metacarpals and metatarsals, as well as on some phalanges, suggesting that the animal had been skinned before the burial. The bivariate and multivariate analysis on the biometric parameters of the cranium compared to those of numerous modern canine breeds suggest that the dog was probably similar to a greyhound. Artistic representations of dogs over the centuries straddling the dog burial date, have shown that the greyhound breed was well known and used in various human activities.

In conclusion, whilst awaiting desirable future genetic analyses to confirm the hypothesis, it is probable that the dog of Santa Severa was a hunting dog, possibly a Greyhound, whose close relationship with the adult men buried in the same stratigraphic horizon made it deserve a similar burial, possibly after the preservation of the coat in its remembrance.

### References


Anyonge, W. 1993. Body mass in large extant and extinct carnivores. *Journal of Zoology* 231 (2): 339–350.

Baglione, M.P., B. Belelli Marchesini, C. Carlucci and L.M. Michetti 2017. Pyrgi, harbour and sanctuary of Caere: landscape, urbanistic planning and architectural features. *Archeologia e calcolatori* 28(2): 201–210.

Barone, R. 1981. *Anatomia comparata dei mammiferi domestici. 3, Splancnologia. (Italian edition by R. Bortolani).* Bologna: Edagricole.

Barone, R. 1995. *Anatomia comparata dei mammiferi domestici. 1, Osteologia. (Italian edition by R. Bortolani, E. Callegari).* Bologna: Edagricole.

Clutton-Brock, J. 1976. The animal resources, in D.M. Wilson (ed.) *The Archaeology of Anglo-Saxon England*: 373–392. London: Methuen.

De Grossi Mazzorin, J. and A. Tagliacozzo 2000. Morphological and osteological changes in the dog from the Neolithic to the Roman period in Italy, in *Dogs through time: an archaeological perspective*: 141–161. Oxford: Archaeopress.

Driesch von den, A. 1976. *A guide to the measurement of animal bones from archaeological sites.* Peabody Museum Bulletins 1: 1–148.

Enei, F. (ed.) 2013a. *Santa Severa tra leggenda e verità storica. Pyrgi e il Castello di Santa Severa alla luce delle recenti scoperte. Scavi 2003-2009.* Grotte di Castro: Ceccarelli editore.

Enei, F. 2013b. Il punto sulle conoscenze, in F. Enei (ed.) *Santa Severa tra leggenda e verità storica. Pyrgi e il Castello di Santa Severa alla luce delle recenti scoperte. Scavi 2003-2009*: 313–387. Grotte di Castro: Ceccarelli editore.

Enei, F. 2013c. Lo scavo della Piazza della Rocca: la scoperta della chiesa paleocristiana di Santa Severa, in F. Enei (ed.) *Santa Severa tra leggenda e verità storica. Pyrgi e il Castello di Santa Severa alla luce delle recenti scoperte. Scavi 2003-2009*: 51–123. Grotte di Castro: Ceccarelli editore.

Enei, F. 2013d. Le analisi al radiocarbonio degli scheletri dalla Casa del Nostromo, dalla Piazza della Rocca e dalla Chiesa, in F. Enei (ed.) *Santa Severa tra leggenda e verità storica. Pyrgi e il Castello di Santa Severa alla luce delle recenti scoperte. Scavi 2003-2009*: 214–219. Grotte di Castro: Ceccarelli editore.

Fatucci, M. and E. Cerilli 2013. Primi risultati dallo studio dei resti faunistici rinvenuti nello scavo effettuato al castello di Santa Severa, 'Piazza della Rocca', in F. Enei (ed.) *Santa Severa tra leggenda e verità storica. Pyrgi e il Castello di Santa Severa alla luce delle recenti scoperte. Scavi 2003-2009*: 224–233. Grotte di Castro: Ceccarelli editore.

Fatucci, M. and E. Cerilli 2015. Gli uccelli nel castello di Santa Severa (Roma) durante il Basso Medioevo (XIII-XIV secolo): significato paleoeconomico e ambientale, dati preliminari, in U. Thun Hohenstein, M. Cangemi, I. Fiore, J. De Grossi Mazzorin (eds) *Atti del 7° Convegno Nazionale di Archeozoologia, Ferrara 22-23 november 2012, Annali dell'Università degli Studi di Ferrara, Museologia Scientifica e Naturalistica* 11(2): 123–126.

Fatucci, M. and E. Cerilli 2016. Allevamento e caccia al castello di Santa Severa (Roma) durante il basso medioevo: mense signorili e popolari dal XIII al XIV secolo, in U. Thun Hohenstein, M. Cangemi, I. Fiore, J. De Grossi Mazzorin (eds) *Atti del 7° Convegno Nazionale di Archeozoologia, Ferrara 22-23 november 2012, Annali dell'Università degli Studi di Ferrara, Museologia Scientifica e Naturalistica* 12(1): 235–242.

Fatucci, M. and E. Cerilli 2019. Analisi tafonomica del campione faunistico dal riempimento basso-medievale della chiesa del castello di Santa Severa,in J. De Grossi Mazzorin, I. Fiore, C. Minniti (eds) *Atti dell'8° Convegno Nazionale di Archeozoologia, Lecce 11-14 novembre 2015*: 203–210.

Friedman, J.B. 2016. Dogs in the Identity Formation and Moral Teaching Offered in Some Fifteenth-Century Flemish Manuscript Miniatures, in L.D. Gelfland (ed.) *Our Dogs, Our Selves. Dogs in Medieval and Early Modern Art, Literature, and Society*: 325–362. Boston: Brill.

Fišáková, M.N. 2010. The Local Elite at Chotěbuz-Podobora (Czech Republic) from zooarchaeological and archaeological perspectives, in A. Pluskowski, G. Kunst, M. Kucera, M. Bietak, I. Hein (eds) Using animal to construct human identities in medieval Europe animal as material culture in Middle Ages 3, *VIAVIAS Vienna Institute for Archaeological Science* 3: 31–38.

Hamblin, N.L. 1984. *Animal Use by the Cozumel Maya.* Tucson: University of Arizona Press,.

Hammer, Ø., D.A.T. Harper and P.D. Ryan 2001. PAST: Paleontological statistics software package for education and data analysis. *Palaeontologia Electronica* 4(1).

Harcourt, R.A. 1974. The Dog in Prehistoric and Early Historic Britain. *Journal of Archaeological Science* 1: 151–175.

Hudson, P.E., S.A. Corr and A.M. Wilson 2012. High speed galloping in the cheetah (*Acinonyx jubatus*) and the racing greyhound (*Canis familiaris*): spatiotemporal and kinetic characteristics. *Journal of Experimental Biology* 215(14): 2425–2434.

International Standard Version, 2020, Proverbs 30, viewed 06 February 2020, https://biblehub.com/isv/proverbs/30.htm

Kholová, H. 1987. *Historie psího rodu*. Praha: Práce.

King James Bible's Version, 1611, Proverbs 30:29–31., viewed 06 February 2020, https://www.kingjamesbibleonline.org/Proverbs-Chapter-30/#31.

Kitchell, K.F. and I.M. Resnik 1999. *Albertus Magnus, On Animals. A Medieval Summa Zoologica*. Translated and Annotated by K.F. Kitchell Jr. and I.M. Resnick, 2 Vols. Baltimore: John Hopkins University Press.

Knoest, J.J.T. 2015. On the morphology of the domesticated dog in medieval Norway. Master thesis, The University of Bergen.

Koudelka, F. 1885. Das Verhältnis der Ossa Longa zur Skeletthöhebei den Säugetieren. *Verhandlungen Naturforschenden Vereins Brünn* 24(1): 127–153.

Larson, G., E.K. Karlsson, A. Perri, M.T. Webster, S.Y. Ho, J. Peters, P.W. Stahl, P.J. Piper, F. Lingaas, M. Fredholm, K.E. Comstock, J.M. Modiano, C. Schelling, A.I. Agoulnik, P.A. Leegwater, K. Dobney, J.-D. Vigne, C. Vilà, L. Andersson and K. Lindblad-Toh 2012. Rethinking dog domestication by integrating genetics, archeology, and biogeography. *Proceedings of the National Academy of Sciences* 109(23): 8878–8883.

Michetti, L.M. and B. Belelli Marchesini 2018. Pyrgi, porto e santuario di Caere: tra conoscenze acquisite e ricerche in corso. *Annali della Fondazione per il Museo 'Claudio Faina'* 20: 245–280.

New America Bible 2002. Proverbs 30, viewed 06 February 2020, http://www.vatican.va/archive/ENG0839/_PKT.HTM

Parker, H.G., L.V. Kim, N.B. Sutter, S. Carlson, T.D. Lorentzen, T.B. Malek, G.S. Johnson, H.B. DeFrance, E.A. Ostrander and L. Kruglyak 2004. Genetic Structure of the Purebred Domestic Dog. *Science* 304: 1160–1164.

Pellegris, C. 2004. *Rhyton* apulo a figure rosse, in G. Sena Chiesa, E.A. Arslan (eds) Miti Greci. *Archeologia e pittura dalla Magna Grecia al collezionismo*: 72, n. 35. Milano: Electa.

Phillips, A.A. and M.M. Willock 1999. *Xenophon & Arrian, On Hunting (Kynēgetikos)*. Warminster: Arris and Phillips.

Phillips, C., I.L. Baxter and M. Nussbaumer 2009. The application of discriminant function analysis to archaeological dog remains as an aid to the elucidation of possible affinities with modern breeds. *Archaeofauna* 18: 51–64.

Schmitt, J.-C. 1983. *The Holy Greyhound: Guinefort, Healer of Children since the Thirteenth Century* (Studies in Oral and Literate Culture 6). Cambridge: Cambridge University Press.

Stuchlý, I. and M. Císařovský 1991. *Chrti. Canis*, Praha.

Svobodová, H., M. Bartoś, M.N. Fišáková and P. Kouřil 2015. Genetic analysis of the possibly the oldest greyhound remains within the territory of the Czech Republic as proof of a local elite presence at Chotěbuz-Podoborahillfort in the 8th-9th century AD. *Acta Musei Nationalis Pragae, Series B, Historia Naturalis* 71(1–2): 17–24.

The, T.L. and C.O. Trouth 1976. Sexual dimorphism in the basilar part of the occipital bone of the dog (*Canis familiaris*). *Acta Anatomica* 95(4): 565–571.

Trantalidou, K. 2016. Companions from the Oldest Times: Dogs in Ancient Greek Literature, Iconography and Osteological Testimony, in L.M. Snyder, E.A. Moore, (eds) *Dogs and people in social, working, economic or symbolic interaction. Proceedings of the 9th ICAZ Conference*, 96–120, Durham 2002, Oxbow Books.

Vigini, G. 2002. *L'Antico Testamento, Proverbi*. Milan Paoline Editoriale libri.

Wing, E.S. 1978. Use of dogs for food: An adaptation to the coastal environment, in B.L. Stark and B. Voorhies (eds) *Prehistoric Coastal Adaptations: The Economy and Ecology of Maritime Middle America*: 29–41. New York: Academic Press.

Table 1 - List of selected artworks representing greyhounds between the 14th and 17th century.

| Century | Author | Made in / Region of activity | Title | Date | Collection | Artwork type | Matter technique | Subject | Dog type | Greyhound number | Coat colour | Snout morphology |
|---|---|---|---|---|---|---|---|---|---|---|---|---|
| 13th | Azzo di Masetto (?) / Memmo di Gilippuccio (?) (13th cent.) | Central Italy | Scene of a tournament and hunting | *c.* 1291 | Palazzo del Popolo (San Gimignano, Italy) | fresco | | hunting | greyhounds with other dogs | 2 | white, dark | slender/pointed |
| 14th | Bernardo Daddi (*c.* 1280–1348) | Central Italy | Nativity (Polyptich of Santa Reparata) | 1340–1345 | Uffizi Gallery (Florence, Italy) | painting | oil on wood | religious: nativity scene/ shepherds | greyhounds with other dogs | 1 | white | slender/pointed |
| 14th | Buonamico Buffalmacco (Buonamico di Martino) (1262–1340) | Central Italy | | 1336–1341 | Campo Santo (Pisa, Italy) | fresco | | religious: allegorical | only greyhounds | 1 | brown | slender/pointed |
| 14th | Gaston III, Count of Foix (1331–1391) | France | Livre de La Chasse Ms. 616–619 | 1387–1389 | National Library of France (Paris, France) | illuminated manuscript | | hunting | greyhounds with other dogs | more than 3 | white, brown | not sleder, slender/pointed |
| 14th-15th | Tacuina Sanitatis | | Hunting (codex 4182) | late 14th | Biblioteca Casanatense (Rome, Italy) | illuminated manuscript | | hunting | greyhounds with other dogs | more than 3 | white, merle/ piebald | slender/pointed |
| 14th-15th | Tacuina Sanitatis | | Fructus Mandragorae, (Cod. Vindob. ser. nov. 2644) | second half or 14th cent. - *c.* 1450 | Vienna (Austria) | illuminated manuscript | | popular daily life | only greyhounds | 1 | merle/ piebald | slender/pointed |
| 15th | Anonymous | Florence (Italy) | Ameto's Discovery of the Nymphs and Contest between the Shepherds Alcesto and Acaten | *c.* 1410 | Metropolitan Museum of Art (New York, USA) | desco da parto (Childbirth tray) | tempera on wood | mythological | only greyhounds | more than 3 | white | slender/pointed |
| 15th | Anonymous | northern France or Flanders | Otter and Swan Hunt | late 1420s | The Devonshire Hunting Tapestries, Victoria and Albert Museum (London, UK) | tapestry | | mythological/court meeting | greyhounds with other dogs | 2 | brown | medium slender/ pointed |
| 15th | Anonymous | France or Flanders | Falconry | late 1420s | The Devonshire Hunting Tapestries, Victoria and Albert Museum (London, UK) | tapestry | | hunting | greyhounds with other dogs | 1 | white | slender/pointed |
| 15th | Anonymous | France or Flanders | Deer Hunt | late 1420s | The Devonshire Hunting Tapestries, Victoria and Albert Museum (London, UK) | tapestry | | hunting | only greyhounds | more than 3 | white | slender/pointed |
| 15th | Anonymous | France or Flanders | Boar and Bear Hunt | late 1420s | The Devonshire Hunting Tapestries, Victoria and Albert Museum (London, UK) | tapestry | | hunting | greyhounds with other dogs | more than 3 | white | not slender |
| 15th | Anonymous | France | Livres des Merveilles du Monde, 'Ethiopia' M.461, fol. 26v | *c.* 1460 | Pierpont Morgan Library (New York, USA) | illuminated manuscript | | hunting | greyhounds with other dogs | 2 | white | slender/pointed |
| 15th | Anonymous | France | Horologe of Sapience from Henri Suso (Ms. IV.111 f, 16v) | 1455–1460 | Bibliothèque Royale (Brussels, Belgium) | illuminated manuscript | | religious: allegorical | greyhounds with other dogs | 1 | white | slender/pointed |
| 15th | Anonymous | Flanders (Belgium) | Hunters' Picnic | late 15th | Louvre Museum (Paris, France) | tapestry | wood and silk | hunting | only greyhounds | 2 | white, merle/ piebald | slender/pointed |
| 15th | Anonymous (1480–1520) | Germany | St. Ursula Announces to her Father her Departure on a Pilgrimage to Rome | 1490s | Louvre Museum (Paris, France) | painting | canvas | religious: Saints | only greyhounds | 1 | merle/ piebald | slender/pointed |

| Century | Author | Made in / Region of activity | Title | Date | Collection | Artwork type | Matter technique | Subject | Dog type | Greyhound number | Coat colour | Snout morphology |
|---|---|---|---|---|---|---|---|---|---|---|---|---|
| 15th | Antonio Vivarini (Antonio da Murano) (*c.* 1418–1476,1484) | Venice (Italy) | Adoration of the Magi | 1445–1447 | Gemäldegalerie (Berlin, Germany) | painting | | religious: nativity scene/Magi | only greyhounds | 1 | white | slender/pointed |
| 15th | Benozzo Gozzoli (*c.* 1420–1497) | Central Italy | The Procession of the Three Wise Men, with Giuliano de' Medici and Joseph, Patriarch of Constantinople | 1459–1464 | Cappella dei Magi, Palazzo Medici Riccardi (Florence, Italy) | fresco | | policy/travel | greyhounds with other dogs | 3 | white, brown | slender/pointed |
| 15th | | Central Italy | Procession of the Youngest King, with Lorenzo, Piero e Giovanni de' Medici | 1459–1464 | Cappella dei Magi, Palazzo Medici Riccardi (Florence, Italy) | fresco | | policy/travel | greyhounds with other dogs | 1 | brown | slender/pointed |
| 15th | Benvenuto di Giovanni (Benvenuto di Giovanni di Meo del Guasta) (1436-*c.* 1518) | Central Italy | The Adoration of the Magi | *c.* 1470–1475 | National Gallery Of Art (Washington, USA) | painting | tempera on wood | religious: nativity scene/Magi | only greyhounds | 1 | white | slender/pointed |
| 15th | Biagio d'Antonio Tucci (1446–1516) | Central Italy | The Triumph of Camillus | *c.* 1470–1475 | National Gallery Of Art (Washington, USA) | painting | tempera on wood | policy/travel | greyhounds with other dogs | 3 | white | slender/pointed |
| 15th | | Central Italy | The Story of Joseph | 1476–1504 | Metropolitan Museum of Art (New York, USA) | painting | tempera on wood | religious: Bible Old Testament | only greyhounds | 3 | white | slender/pointed |
| 15th | Biagio d'Antonio Tucci (?) (1446–1516) | Central Italy | Scenes with Stories of Argonauts | *c.* 1465 | Metropolitan Museum of Art (New York, USA) | painting | tempera on wood | mythological | greyhounds with other dogs | more than 3 | white, brown, dark | slender/pointed |
| 15th | Bicci di Lorenzo (1373–1452) | Central Italy | Nativity | 1435 | Chiesa di S. Giovannino dei Cavalieri (Florence, Italy) | painting | | religious: nativity scene/ shepherds | only greyhounds | 1 | white | slender/pointed |
| 15th | Fra Carnevale (Bartolomeo di Giovanni Corradini) (*c.* 1420–1425–1484) | Central Italy | The Presentation of the Virgin in the Temple (?) | 1467 | Museum of Fine Art (Boston, USA) | painting | tempera and oil on wood | religious: Bible New Testament | only greyhounds | 1 | white | slender/pointed |
| 15th | Gentile da Fabriano (*c.* 1370–1427) | Central Italy | Adoration of the Magi | 1423 | Uffizi Gallery (Florence, Italy) | painting | tempera on wood | religious: Bible New Testament | only greyhounds | 1 | white | not slender |

| Century | Author | Made in / Region of activity | Title | Date | Collection | Artwork type | Matter technique | Subject | Dog type | Greyhound number | Coat colour | Snout morphology |
|---|---|---|---|---|---|---|---|---|---|---|---|---|
| 15th | Hans Memling (*c.* 1430–1494) | Flanders (Belgium) | The Martyrdom in Cologne, part of the Saint Ursula Shrine | 1489 | St. John's Hospital (Bruges, Belgium) | painting | oil on wood | religious: Saints | only greyhounds | 1 | white | slender/pointed |
| 15th | il Sassetta (Stefano di Giovanni di Consolo) (*c.* 1392–1450,1451) | Central Italy | Adoration of the Magi | *c.* 1430 | Palazzo Chigi-Saracini (Siena, Italy) | painting | oil on wood | religious: nativity scene/Magi | greyhounds with other dogs | 1 | white | not slender |
| 15th | Jacopo di Arcangelo (Jacopo del Sellaio) (?) (*c.* 1441–1493) | Central Italy | Scenes with Stories of Argonauts | *c.* 1465 | Metropolitan Museum of Art (New York, USA) | painting | tempera on wood | mythological | greyhounds with other dogs | more than 3 | white, brown, dark | slender/pointed |
| 15th | Jean Fouquet and Associates (1420–1481) | France | Greyhound with motto and anagram initials of Simon de Varie (Ms. 7 f. 2v) | *c.* 1455 | Jean Paul Getty Museum (Malibu, USA) | illuminated manuscript | | coat of arms | only greyhounds | 1 | white | slender/pointed |
| 15th | Limbourg brothers (Jean) (15th cent.) | France | Très Riches Heures du Duc de Berry, January, Ms.65 f.1v | *c.* 1412–1416 | Condé Museum (Château de Chantilly, France) | illuminated manuscript | tempera on vellum | noble banquet | only greyhounds | 1 | white | slender/pointed |
| 15th | Lorenzo Ghiberti (1378–1455) | Central Italy | Isaac sends Esau to hunt, from the Gates of Paradise | 1425–1452 | Doors of Baptistry of Florence (Florence, Italy) | sculpure | bronze | religious: Bible Old Testament | only greyhounds | 2 | | slender/pointed |
| 15th | Lorenzo Monaco (Piero di Giovanni) (*c.* 1370-*c.* 1425) | Central Italy | Adoration of the Magi | 1420–1422 | Uffizi Gallery (Florence, Italy) | painting | tempera on wood | religious: nativity scene/Magi | only greyhounds | 1 | white | slender/pointed |
| 15th | Paolo Uccello (Paolo di Dono) (1347–1475) | Italy | Battle of San Romano | *c.* 1438 | Uffizi Gallery (Florence, Italy) | painting | oil on wood | battle | only greyhounds | 1 | white | slender/pointed |
| 15th | | Italy | The Hunt in the Forest | *c.* 1470 | Asmolean Museum (Oxford, UK) | painting | tempera on wood | hunting | greyhounds with other dogs | more than 3 | white, brown, dark | slender/pointed |
| 15th | Piero della Francesca (Piero Benedetto de' Franceschi) (*c.* 1416–1492) | Central Italy | Sigismondo Pandolfo Malatesta in prayer in front on Saint Sigismondo | 1451 | Cappella della Concezione Tempio Malatestiano (Cattedrale di Santa Colomba, Rimini, Italy) | fresco | | religious: Saints | only greyhounds | 1 | white | not slender |

| Century | Author | Made in / Region of activity | Title | Date | Collection | Artwork type | Matter technique | Subject | Dog type | Greyhound number | Coat colour | Snout morphology |
|---|---|---|---|---|---|---|---|---|---|---|---|---|
| 15th | Pisanello (Antonio Puccio Pisano) (1390–1455) | Central Italy | St. George Liberating the Princess of Trebizond | 1433–1438 | Pellegrini Chapel, Churc of Sant'Anastasia (verona, Italy) | fresco | | religious: Saints | greyhounds with other dogs | 1 | white | not slender |
| 15th | | Central Italy | St. George Liberating the Princess of Trebizond, preparatory study of greyhound (Vallardi Codex, 2334) | 1433–1438 | Louvre Museum (Paris, France) | drawing | pen and brown ink | dog | only greyhounds | 1 | white | not slender |
| 15th | | Central Italy | eight study of greyhound (Vallardi Codex, 2427–2433, 2435) | 1433–1438 | Louvre Museum (Paris, France) | drawing | brown pencil and watercolour | dog | only greyhounds | 8 | nr. 2 white, nr. 2 brown, nr. 4 merle/ piebald | nr. 3 not slender, nr. 5 slender/ pointed |
| 15th | | Central Italy | The Vision of Saint Eustace | 1436–1438 | National Gallery (London, UK) | painting | egg tempera on wood | religious: Saints | greyhounds with other dogs | 1 | white, brown | not slender |
| 15th | Rogier van der Weyden (1399–1400–1464) | Holland | Saint Columba Altarpiece (central panel) | c. 1455 | Alte Pinakothek (Munich, Germany) | painting | | religious: nativity scene/Magi | only greyhounds | 1 | white | slender/pointed |
| 15th | | Holland | Baptism, Confirmation, Penance | 1455–1450 | Royal Museum of Fine Arts (Antwerp, Holland) | painting | | religious: noble ceremony | only greyhounds | 1 | brown | slender/pointed |
| 15th | Robinet Testard (activity 1471–1531) | France | Les Grandes Chroniques de France (Ms. fr. 2609, folio 150) | | Bibliothèque Nationale de France (Paris, France) | illuminated manuscript | | policy/travel | only greyhounds | 2 | white | not slender |
| 15th | Vittore Carpaccio (*c.* 1465–1525,1526) | Venice (Italy) | Saint George Baptises the Selenites | 1507 | Scuola di San Giorgio degli Schiavoni (Venice, Italy) | painting | tempera on wood | religious: Saints | greyhounds with other dogs | 1 | white | slender/pointed |
| 15th | | Venice (Italy) | Saint Ursula Legend, arrival of the English ambassadors to court of the king of Brittany | *c.* 1495 | Gallerie dell'Accademia (venice, Italy) | painting | oil on canvas | policy/travel | only greyhounds | 1 | brown | slender/pointed |
| 16th | Albrecht Dürer (1471–1528) | Europe | Saint Eustace | *c.* 1501 | Metropolitan Museum of Art (New York, USA) | engraving | copper | religious: Saints | only greyhounds | more than 3 | white | slender/pointed |
| 16th | Anonymous | | The Unicorn is Attacked | 1495–1505 | The Hunt of the Unicorn, The Cloister (New York, USA) | tapestry | | hunting | greyhounds with other dogs | more than 3 | white, brown | slender/pointed |
| 16th | Anonymous | France or Flanders | Rustic Sports | 1500–1515 | Victoria and Albert Museum (London, UK) | tapestry | | hunting | greyhounds with other dogs | 2 | white | slender/pointed |
| 16th | Anonymous | Brussel (Belgium) | Ecouen, Story of David and Bathesheba, part 9 | 1510–1520 | Château d'Écouen (Ecoen, France) | tapestry | wool, silk, silver and gold | religious: Bible Old Testament | only greyhounds | 1 | white | slender/pointed |
| 16th | Anonymous | France or Flanders | The Elephant Hunt | 1530 | Louvre Museum (Paris, France) | tapestry | | hunting | greyhounds with other dogs | 2 | white | slender/pointed |

| Century | Author | Made in / Region of activity | Title | Date | Collection | Artwork type | Matter technique | Subject | Dog type | Greyhound number | Coat colour | Snout morphology |
|---|---|---|---|---|---|---|---|---|---|---|---|---|
| 16th | Bernard van Orley ( *c.* 1491–1541) | Brussel (Belgium) | The Battle of Pavia | 1528–1531 | National Museum of Capodimonte (Naples, Italy) | tapestry | wool, silk, silver and gold | battle | only greyhounds | 1 | white | slender/pointed |
| 16th | Bernard van Orley ( *c.* 1491–1541) | Brussel (Belgium) | The Month of September | 1531 | The Hunt of Maximilian, Louvre Museum (Paris, France) | tapestry | | hunting | greyhounds with other dogs | 1 | merle/ piebald | slender/pointed |
| 16th | | Brussel (Belgium) | The Month of June | 1531 | The Hunt of Maximilian, Louvre Museum (Paris, France) | tapestry | | hunting | greyhounds with other dogs | 2 | white, merle/ piebald | slender/pointed |
| 16th | | Brussel (Belgium) | The Month of July | 1531 | The Hunt of Maximilian series , Louvre Museum (Paris, France) | tapestry | | hunting | greyhounds with other dogs | 2 | white, merle/ piebald | slender/pointed |
| 16th | Dosso Dossi (Giovanni Francesco di Niccolò Luteri) (*c.* 1468–1542) | Central Italy | Circe and her Lovers in a Landscape | *c.* 1520 | National Gallery Of Art (Washington, USA) | painting | oil on canvas | mythological | only greyhounds | 1 | brown | slender/pointed |
| 16th | Gaspar Dias (*c.* 1560–1591) | Portugal | The Appearance of the Angel to St. Roch | 1584 | Museu de São Roque (Lisbon, Portugal) | painting | oil on canvas | religious: Saints | only greyhounds | 1 | merle/ piebald | slender/pointed |
| 16th | | Holland | The Virgin and Child with Saints and Donor | 1505–1510 | National Gallery (London, UK) | painting | oil on wood | religious: Bible New Testament | only greyhounds | 1 | brown | slender/pointed |
| 16th | Giovanni Bellini (*c.* 1430–1516) | Venice (Italy) | An Episode from the Life of Publius Cornelius Scipio | after 1506 | National Gallery Of Art (Washington, USA) | painting | oil on canvas | Roman history | only greyhounds | 1 | white | slender/pointed |
| 16th | Jan Matsys (*c.* 1510–1575) | Flanders (Belgium) | David and Bathsheba | 1562–1563 | Louvre Museum (Paris, France) | painting | oil on wood | religious: Bible Old Testament | greyhounds with other dogs | 1 | white | slender/pointed |
| 16th | Jan Mostaert (*c.* 1475–1555,1556) | Holland | Portrait of the knight Abel von Coulster | 1500–1525 | Royal Museum of Fine Art of Belgium (Brussels, Bergium) | painting | oil on wood | portrait | only greyhounds | 1 | white | slender/pointed |
| 16th | Lorenzo Lotto (*c.* 1480–1556,1557) | Nothern Italy | Story of Santa Barbara | 1524 | Cappella Suardi (Trescore Balneario, Bergamo, Italia) | fresco | | religious: Saints | greyhounds with other dogs | 1 | white | slender/pointed |
| 16th | Luca Penni (School of Fontainebleau) (*c.* 1500,1504–1556) | France | Diana the Huntress | 1550–1560 | Louvre Museum (Paris, France) | painting | oil on canvas | mythological | only greyhounds | 1 | white | slender/pointed |
| 16th | Lucas Cranach the Elder (*c.* 1472–1553) | Germany | The Stag-Hunt | 1506 | Metropolitan Museum of Art (New York, USA) | engraving | woodcut | hunting | greyhounds with other dogs | more than 3 | | slender/pointed |
| 16th | | Germany | A Stag Hunt with Frederick III, Elector of Saxony | 1529 | Kunsthistorisches Museum (Vienna, Austria) | painting | oil on lime | hunting | greyhounds with other dogs | more than 3 | white, brown | slender/pointed |
| 16th | | Germany | Garden of Eden | 1530 | Kunsthistorisches Museum (Vienna, Austria) | painting | oil on lime | religious: Bible Old Testament | only greyhounds | 1 | white | slender/pointed |

| Century | Author | Made in / Region of activity | Title | Date | Collection | Artwork type | Matter technique | Subject | Dog type | Greyhound number | Coat colour | Snout morphology |
|---|---|---|---|---|---|---|---|---|---|---|---|---|
| 16th | Paolo Veronese (Paolo Caliari) (1528–1588) | Venice (Italy) | Presentation in the Temple | 1560 | Church of Saint Sebastian (Venice, Italy) | painting | oil on canvas | religious: Bible New Testament | only greyhounds | 1 | brown | slender/pointed |
| 16th | | Venice (Italy) | Actaeon Watching Diana and her Nymphs Bathing | *c.* 1560 | Museum of Fine Art (Boston, USA) | painting | oil on canvas | mythological | only greyhounds | 2 | white | slender/pointed |
| 16th | | Venice (Italy) | The Wedding at Cana | 1562–1563 | Louvre Museum (Paris, France) | painting | oil on canvas | noble banquet | greyhounds with other dogs | 2 | white | slender/pointed |
| 16th | | Venice (Italy) | Moses saved from the waters | 1570–1580 | Museum of Fine Art (Lyon, France) | painting | | religious: Bible Old Testament | only greyhounds | 2 | white, brown | slender/pointed |
| 16th | | Venice (Italy) | Boy with a Greyhound | *c.* 1570s | Metropolitan Museum of Art (New York, USA) | painting | oil on canvas | portrait | only greyhounds | 1 | white | slender/pointed |
| 16th | Pinturicchio (Bernardino di Petto Petti) (*c.* 1452–1513) | Central Italy | Bernardino brings to life a dead man found under a tree | 1473 | National Gallery of Umbria (Perugia, Italy) | painting | oil on wood | religious: Saints | only greyhounds | 1 | brown | slender/pointed |
| 16th | | Central Italy | Nativity, Adoration of Shepherds | 1501 | Cappella Eaglioni (Spello, Italy) | fresco | | religious: nativity scene/ shepherds | only greyhounds | 1 | white | slender/pointed |
| 16th | | Central Italy | St. Catherine's Disputation | 1492–1494 | Borgia Apartments, Vatican Museums (Vatican City) | fresco | | religious: Saints | only greyhounds | 1 | brown | slender/pointed |
| 16th | | Central Italy | Enea Silvio Piccolomini leaving for the Council of Basel | 1502–1507 | Piccolomini Library, Siena Cathedral (Siena, Italy) | fresco | | policy/travel | only greyhounds | 1 | brown | not slender |
| 16th | Vincenzo Catena (*c.* 1480–1531) | Venice (Italy) | The Adoration of the Shepherds | after 1520 | Metropolitan Museum of Art (New York, USA) | painting | oil on canvas | religious: nativity scene/ shepherds | only greyhounds | 1 | merle/ piebald | slender/pointed |
| 17th | Abraham Bloemaert (1566–1651) | Holland | Venus and Adonis | 1632 | Statens Museum for Kunst (Copenhagen, Denmark) | painting | oil on canvas | mythological | greyhounds with other dogs | 1 | brown | slender/pointed |
| 17th | Anonymous | France | Daphne Choosing Diana as Her Ideal | 17th cent. | Virginia Museum of Fine Arts (Richmond, Virginia, USA) | tapestry | wool and silk | mythological | | 1 | brown | slender/pointed |
| 17th | Anonymous (School of Raphael) | Brussel (Belgium) | Adoration of the Shepherds | 1624–1630 | Scuola Nuova, Vatican Museums (Rome, Italy) | tapestry | | religious: nativity scene/ shepherds | only greyhounds | 1 | white | slender/pointed |
| 17th | Jan Brueghel the Elder (1568–1625) | Flanders (Belgium) | The Archdukes Albert and Isabella Visiting a Collector's Cabinet | 1621–1623 | Walters Art Museum (Mount Vernon-Belvedere, Baltimore, USA) | painting | oil on wood | court daily life | greyhounds with other dogs | 1 | merle/ piebald | slender/pointed |
| 17th | Hieronymous Francken II (1578–1623) | Flanders (Belgium) | | | | | | | | | | |
| 17th | Jan Brueghel the Elder (1568–1625) | Flanders (Belgium) | The Vision of St Hubert | 1615–1630 | Museo del Prado (Madrid, Spain) | painting | oil on wood | religious: Saints | greyhounds with other dogs | 3 | merle/ piebald | slender/pointed |
| 17th | Peter Paul Rubens (1577–1640) | Flanders (Belgium) | | | | | | | | | | |

etail from Devienne fig.4 (chapter 5.8). East wall of an isolated stone slab. Songshan, axiang district; Upper panel: Dongwangong (King-Father-of-the-East), hybrids and cloud wirls with dog's head; Lower panel: Kitchen scene with a dog slaughter at a well.

# Section 5
# Representation of Dogs in Different Cultures

# 5.1 *Lupus in Fabula*: The Representation of the Wolf (*Canis lupus*) in European Palaeolithic Art


Gianpiero Di Maida[1], Margherita Mussi[2],
Alberto Lombo Montañés[3], Manuel Bea[3]

[1] Niedersächsisches Landesamt für Denkmalpflege, Scharnhorststraße 1, 30175 Hannover, Germany. gianpiero.dimaida@nld.niedersachsen.de.

[2] ISMEO - Associazione Internazionale di Studi sul Mediterraneo e l'Oriente, Corso Vittorio Emanuele II 244, 00186 Roma margherita.mussi@fondazione.uniroma1.it

[3] Department of Sciences of Antiquity, University of Zaragoza, C/ Corona de Aragón 42, Edificio Cervantes 50009 Zaragoza Spain. Research Group P3A; IPH. manubea@unizar.es, albertolommon@hotmail.com

Corresponding author: Gianpiero Di Maida, gianpiero.dimaida@nld.niedersachsen.de



**Abstract**

The representations of wolves in the Palaeolithic art record of Europe reported in the literature are presented here, and re-considered taking into account other species that might be easily misinterpreted for wolves, and finally re-assessed into three different classes of probable attribution.



**Keywords**: Upper Palaeolithic, Europe, art, wolf, Canidae


## 1 Introduction

Ever since the very first discoveries of Paleolithic art - at La Madeleine first, and later at Altamira and at other sites, like La Mouthe, Marsoulas etc. - archaeologists and the public as a whole understood that the focus was on animal representations. Anthropomorphic figures were and are the exception, and even the figurines later to be called 'Venuses' were not accepted without controversies (e.g. De Mortillet 1898). However, the represented animals are nothing but a fraction of the Pleistocene fauna. Leroi-Gourhan and Laming-Emperaire both underlined the prevalence of equids, bovids and caprids. This selective process has been perfectly summarised by Leroi-Gourhan when he wrote that 'nobody drew a frieze of lions and storks surrounded by hyenas and eagles' (Leroi-Gourhan 1972, cited by Bahn 2016: 307). Within this selection, carnivores are a small percentage, and among those wolves and other canids are even less frequent: '*la quasi-absence du loup est frappante*', in Leroi-Gourhan's words (1992: 377).

The structuralist approach of Leroi-Gourhan made clear that the represented animals were not necessarily linked to the food supply, as initially notably proposed by the hunting magic theory. It is more likely that they had a symbolic role in the cosmologies of the human groups. This is aptly summarised in the well-known sentence of Levi-Strauss (1962) '*les espèces sont choisies non comme bonnes à manger, mais comme bonnes à penser*'. Given the relevance of the wolf in the life and cosmology of hunter-gatherers in Eurasia (Otte 2012), which makes the wolf an animal that is 'good to think', its scarcity in the art record is not easily explained.

Here we focus on the wolf, and provide a catalogue of the rare representations of this carnivore in the record of the European Upper Palaeolithic. For an overall discussion we also include the other canids of the Late Pleistocene, as listed by Sommer and Benecke (2005), i.e. the red fox, *Vulpes vulpes* and the dhole, *Cuon alpinus* which is limited today to South-East Asia and the Indian subcontinent, but was present in Western Europe until the early Holocene (Ripoll *et al.* 2010). In the case of foxes, no attempt is made to distinguish *Vulpes vulpes* from *Vulpes corsac* and *Vulpes lagopus*, the arctic fox, both of which are much less frequent in the faunal record. We add another medium-sized carnivore, the wolverine (*Gulo gulo*), a large mustelid whose overall shape might resemble that of a canid. It is often difficult to determine the exact species when dealing with representations, which may only be a few centimetres long, or fragmentary, or just schematic. Figures that are clearly unrealistic (like theriomorphic figures or *monstra* in general) have been excluded from the present work.

## 2 Methods

As a first step, we researched and listed the most characteristic and most easily recognisable anatomical traits of the wolf and of the other middle-sized carnivores listed above which can be mistaken for wolves.

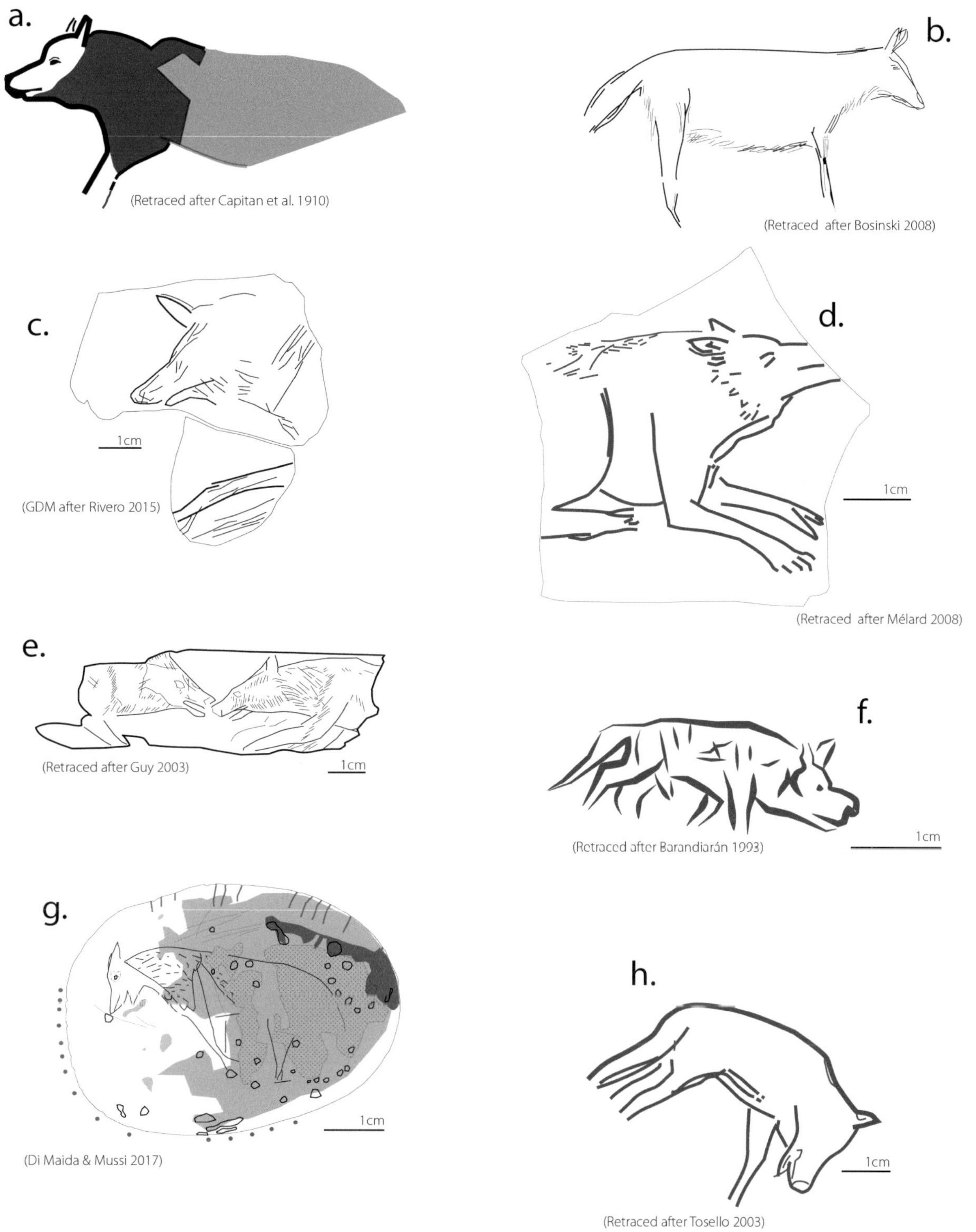


Figure 1. Reliable figures of wolves. a, Font de Gaume. b, Gönnersdorf A. c, Isturitz. d. La Marche. e, La Vache. f, Les Eyzies. g, Polesini. h, Rochereil. All the figures have been traced in vector by GDM.

The wolf is the largest member of the Canidae family, and if compared with other carnivores it is characterised by less pointed features (ears and the muzzle). In Europe it has co-existed with humans since an advanced phase of the Middle Pleistocene (Anzidei *et al.* 1999; Boudadi-Maligne 2012).

The fox has a more flattened skull, upright triangular ears, a pointed, slightly upturned snout, and a long bushy tail. The frequently quoted Palaeolithic representation of a fox from Altxerri (Altuna and Appellainz 1976, Sieveking 1979) was identified mainly thanks to the thick tail.

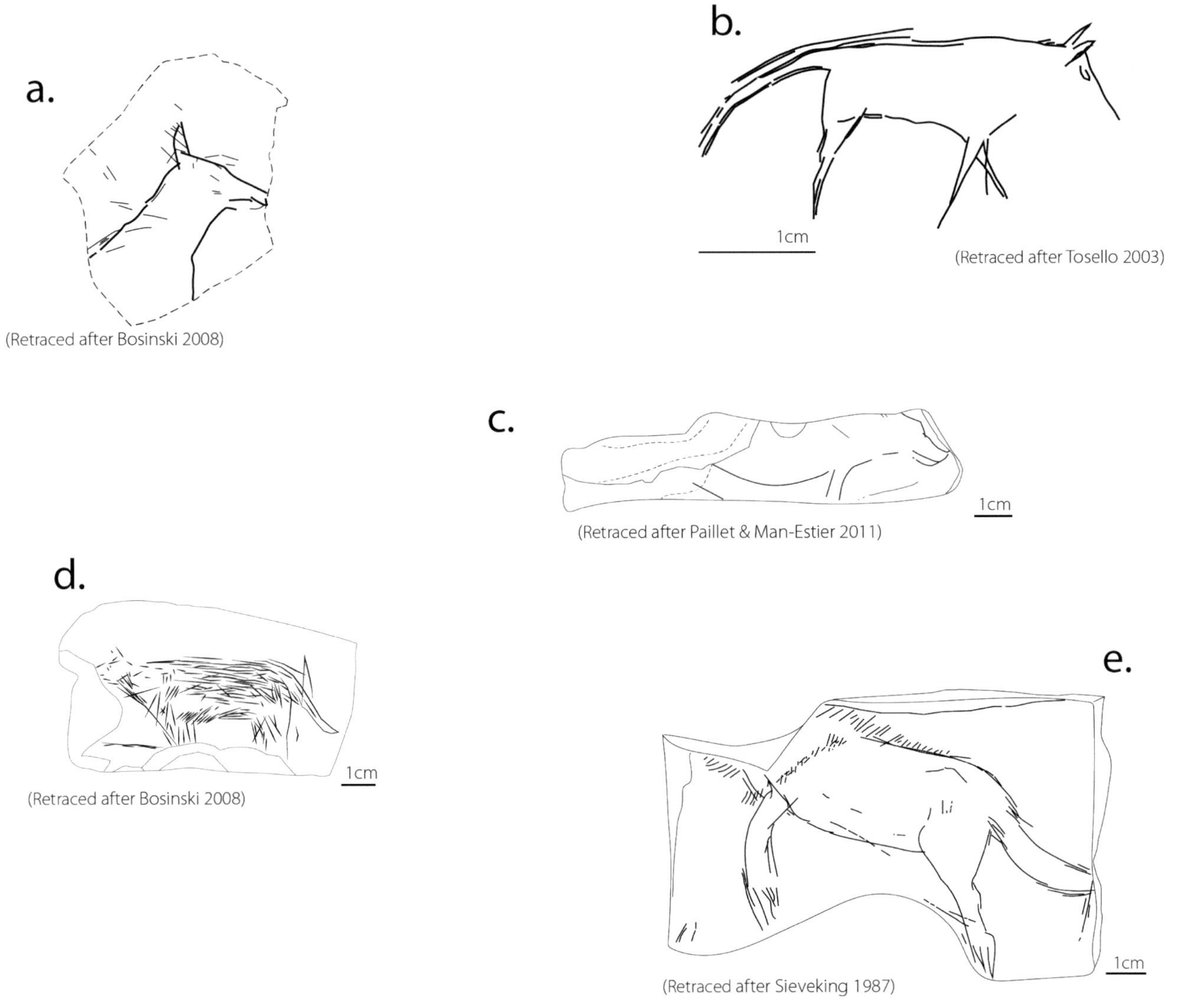


Figure 2. Possible figures of wolves. a, Gönnersdorf. b, La Madaleine. c, Laugerie-Basse. d, Marsoulas. e, Montastruc. All the figures have been traced in vector by GDM.

In the case of the dhole, the profile of the skull is characteristically convex, without a concave depression between the snout and the front.

The wolverine is a sturdy and muscular animal, renowned for its ferocity and tenacity, that more closely resembles a small bear or a large skunk than other members of its own family. It has been recognised in some depictions of Paleolithic age (see Barandiarán 1974a).

Then, as a second step, we collected from the literature the representations of wolves and other middle-sized carnivores, isolating the wolves, positively recognised as such thanks to specific anatomic traits, from the doubtful ones that cannot be definitely assessed. In between these two categories, a third, intermediate one, encompasses figures left with an open interpretation: 'possible' figures of wolves (see Lombo Montañés 2018).

We list below the representations belonging to the above-mentioned three groups, with the name of the site and the main reference, followed by a brief motivation of the placement in whichever group.

### 3 Catalogue (in alphabetical order)

Reliable figures of wolves (Figure 1):

- Font-de-Gaume (Capitan *et al* .1910): the snout, the mane, the ears and the overall shape, all suggest that this is a wolf.
- Gönnersdorf A (Bosinski 2008): we accept Bosinski's interpretation that this is a wolf, based on the overall proportions, the tail, the hair, and partially the ears. The face, a major characteristic, is so reduced and schematic that it does not really help.

- Isturitz (Rivero 2015): despite being fragmentary, the rendering of the snout and ears, as well as the slender front leg(s) are all consistent with a wolf.
- La Marche (Mélard 2008): the mane, ears and snout (albeit only partially visible) allow a rather safe attribution to a wolf.
- La Vache (Guy 2003): ears and snouts are consistent with those of wolves. The confrontation between two individuals fits with wolf behaviour within the complex pack structure of this species. However, dholes, which also have a highly social pack structure, cannot be totally ruled out.
- Les Eyzies (Barandarián 1993): despite the high degree of simplification, as in the case of Rochereil (see below), the tail, slightly hunched back, ears, snout and overall proportions definitely suggest a wolf.
- Polesini (Di Maida and Mussi 2017): the profile of a mammal is accurately engraved on a small pebble. The surface is damaged by water alteration, which mostly affected the hindquarters of the figure. The following anatomical elements suggest a wolf: the snout (albeit pointed); the mane and the pelage, carefully detailed by tiny sub-parallel lines from the neck to the back of the shoulder.
- Rochereil (Tosello 2003): see Les Eyzies.

Possible figures of wolves (Figure 2):

- Gönnersdorf B (Bosinski 2008): this figure closely resembles Gönnersdorf A, included in the previous category. This one however is quite incomplete and does not allow for a clear interpretation.
- La Madeleine (Tosello 2003): the lack of head compounded with the schematic rendering makes an attribution very difficult. Even if it was a canid, it might be another species, such as a fox.
- Laugerie-Basse (Paillet and Man-Estier 2011): in this case, despite the figure being incomplete, there are possibly more solid elements for the attribution to a wolf, even if we prefer to stay on the safe side: the outline of the silhouette, the snout, the proportions, all are compatible with a canid.
- Marsoulas (Bosinski 2008): unfortunately, the support is fractured and most of the head lost. Otherwise the posture, proportions, details (albeit faint) of the tail and body, all point to a wolf. We leave this interpretation open.
- Montastruc (Sieveking 1987): this figure is incomplete and headless, and furthermore scarcely realistic in the overall conception. However, the mane, tail, and overall proportions are compatible with those of a wolf.

Dubious figures of wolves (Figure 3):

- Alkerdi (Barandarián 1974b): this figure (together with the one from Les Combarelles B) shows some characteristics attributable to a canid. But the fragmentary status and the scarcity of details do not allow for a clear attribution.
- Arancou (Barandarián 2003): the overall proportions and the snout are compatible with those of a smaller canid (maybe a fox?). But the representation does not provide strong evidences allowing for the positive attribution to a species.
- Les Combarelles A (Barrière 1997): the remarkably long neck suggests that this is not a canid.
- Les Combarelles B (Barrière 1997): this figure is too fragmentary to be safely identified. While remaining doubtful, however, it is the one with the most characteristics of a wolf (the snout and the few lines possibly referring to a mane).
- Parpallò (Villaverde 1994): the overall shape of this figure is too generic and defies any identification with a canid (or any other species).
- Roc-la-Tour (Rozoy 1997): like the figure from Les Combarelles A, this one is rather unrealistic, making any attribution to a species difficult.

All the figures listed above are directly or indirectly associated with the final phases of the Upper Palaeolithic (Magdalenian or Epigravettian). Accordingly, they are all of Lateglacial age, possibly with the exception of Parpallò (for a detailed discussion of the Parpallò engraving cfr. Lombo Montañés 2018).

## 4 Conclusions

Overall, the Palaeolithic artists rarely depicted large carnivores. All the same, Fritz *et al.* (2011) record some hundreds of bears and lions, often from the early phases of the Upper Palaeolithic. Wolves are much rarer, and generally later in age. This is in contrast with a substantial faunal record, that shows a constant and well-distributed presence of wolves all over the Euro-Asian continent and spanning from the last pleniglacial conditions (~75 ka BP) well into historical times (e.g. Sommer and Benecke 2005).

Of the relevance of wolves in the spiritual and cultural world of European Palaeolithic hunter-gatherers, we have limited but safe proof in the archaeological record, for instance in the use of wolves' perforated teeth as elements of parures and body ornaments, also in a burial context (Moreau 2003; Otte 2012; Vanhaeren and

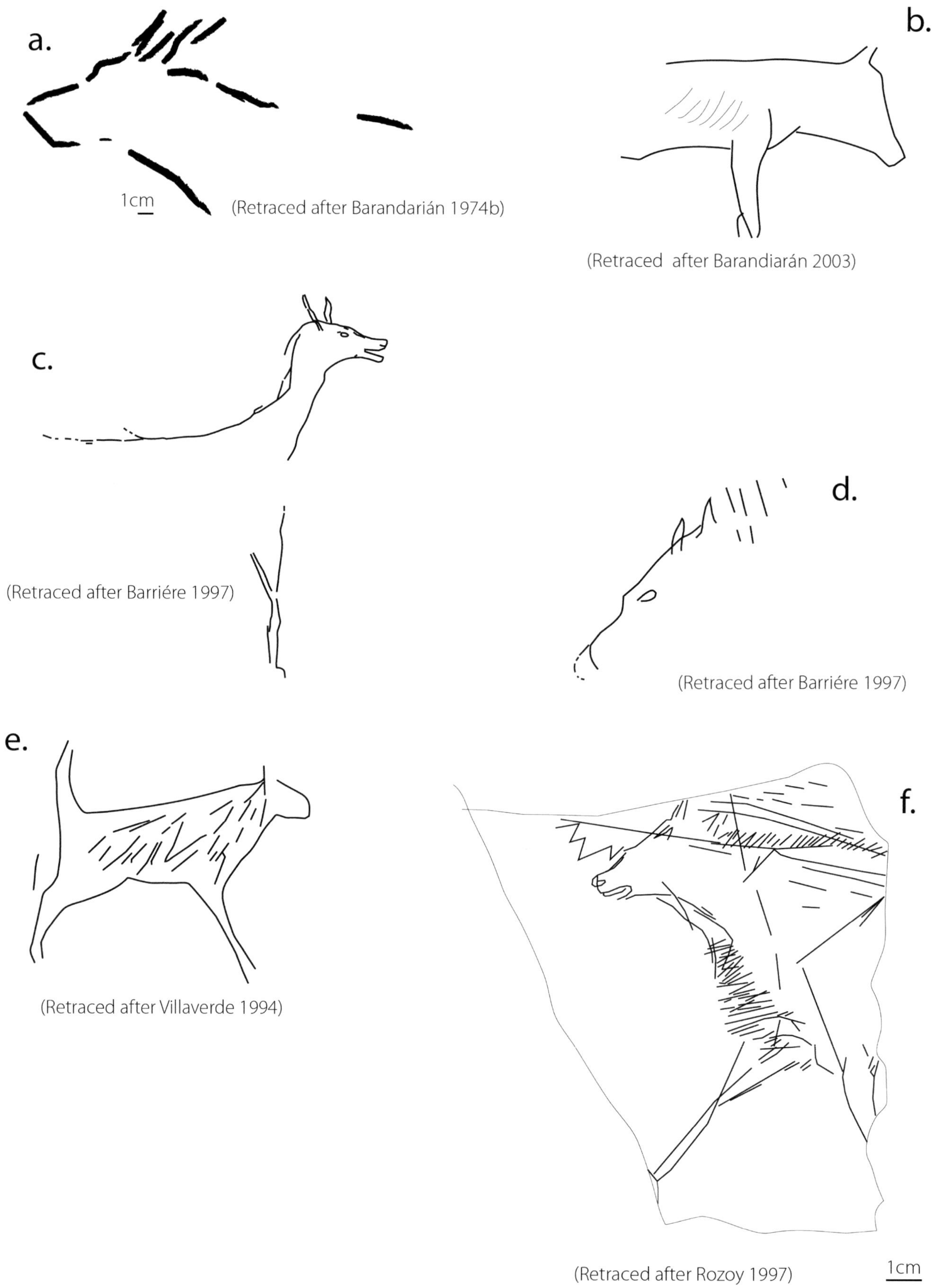


Figure 3. Dubious figures of wolves. a, Alkerdi. b, Arancou. c, Les Combarelles A. d, Les Combarelles B. e, Parpallò. f, Roc-la-Tour. All the figures have been traced in vector by GDM.

D'Errico 2001). Additionally, seeing the amply testified role that the wolf has played in the imagination of human kind of all times (with its feral nature often contraposed to its domesticated opposite - the dog; its similar-to-men habits in raising its offspring, in hunting, in the social structure of the pack, e.g. Ronnberg 2010), it seems generally safe to assume that the wolf must have played a central role in the folklore and mythology of Palaeolithic human groups too.

Of this role though, the current known art record of the European Palaeolithic bears no undisputedly clear trace and thus, we conclude that, for whichever reason, possibly wolves were 'good to think, but not good to draw'.

### References


Altuna, J. and J.M. Appellainz 1976. Las figuras rupestres paleoliticas de la Cueva de Altxerri (Guipuzcoa). *Munibe* XXVIII, 1–3.

Anzidei A.P., A. Arnoldus-Huyzendveld, L. Caloi., M. Palombo and C. Lemorini C. 1999. Two Middle Pleistocene sites near Rome (Italy): La Polledrara di Cecanibbio and Rebibbia - Casal de' Pazzi. In: *The role of early humans in the accumulation of European Lower and Middle Palaeolithic bone assemblages*, 173–195. Mainz: RGZM.

Bahn, P.G. 2016. *Images of the Ice Age*. Oxford: OUP.

Barandiarán, I. 1974a. El Glotón (*Gulo gulo L*) en el Arte Paleolítico. *Zephyrus* XXV, 177–196.

Barandiarán, I. 1974b. Arte paleolítico en Navarra. Las cuevas de Urdax. *Príncipe de Viana* 134–135: 9–47.

Barandiarán, I. 1993. El lobo feroz: La vacuidad de un cuento magdaleniense. *Veleia* 10: 7–38.

Barandiarán, I. 2003. *Grupos homoespecíficos en elimaginario mobiliar magdaleniense. Retratos de familia y cuadros de género*. Vitoria: Universidad del País Vasco.

Barrière, C. 1997. *L´art pariétal des grottes les Combarelles*. Angoulême: AMPRA/PALEO.

Bosinski, G. 2008. *Tierdarstellungen von Gönnersdorf: Nachträge zu Mammut und Pferd sowie die übrigen Tierdarstellungen*. Mainz: RGZM.

Boudadi-Maligne, M., 2010. Les Canis pléistocènes du Sud de la France: approche biosystématique, évolutive et biochronologique. Unpublished PhD dissertation, Univ. Bordeaux 1.

Capitan, L., H. Breuil and D. Peyrony 1910. *La caverne de Font-de-Gaume aux Eyzies (Dordogne)*. Monaco: Impr. Vve A. Chéne.

De Mortillet, G. 1898. Statuette fausse des Baoussé-Roussé. *Bulletins et Mémoires de la Société d'Anthropologie de Paris* 9: 146–153.

Di Maida, G. and M. Mussi 2017. Cry wolf! The engraved pebble of Grotta Polesini (central Italy), in D. Wojtczak, N. Al Najjar, R. Jagher, H. Elsuede, F. Wegmüller and M. Otte (eds) *Vocation préhistoire : hommage à Jean-Marie Le Tensorer*: 99–108. Liège: Éditions ERAUL. Études et Recherches Archéologiques de l'Université de Liège.

Fritz, C., Ph. Fosse, G. Tosello, G. Sauvet and M. Azema 2011. Ours et lion: réflexion sur la place des carnivores dans l'art paléolithique, in J.P. Brugal, A. Gardeisen and A. Zucker (eds) *Prédateurs dans tous leurs états. Evolution, biodiversité, interactions, mythes, symboles. XXXIe rencontres internationales d'archéologie et d'histoire d'Antibes*: 299–318. Antibes: Éd. APDCA.

Guy, E. 2003. Côte, in J. Clottes and H. Delporte (eds) *La grotte de La Vache (Ariège). Fouilles Romain Robert. II. L´art mobilier*. Paris: Éditions de la Réunion des Musées Nationaux.

Leroi-Gourhan, A. 1992. *L'art pariétal. Langage de la préhistoire*. Grenoble: Editions Jérôme Millon.

Levi-Strauss, C. 1962. *Le Totémisme aujourd'hui*. Paris: PUF.

Lombo Montañés, A. 2018. Los cánidos en las manifestaciones gráficas paleolíticas. *Munibe Antropologia-Arkeologia* 69. https://doi.org/10.21630/maa.2018.69.10

Mélard, N. 2008. Pierres gravées de la Marche à Lussac-Les-Châteaux (Vienne). Techniques, technologie et interprétations. *Gallia Préhistoire* 50: 143–268.

Moreau, L. 2003. Les éléments de parure au Paléolithique supérieur en Belgique. *L'anthropologie* 107: 603–614.

Otte, M. 2012. Les premiers loups, les premiers hommes, in Ô Loup ! De nos campagnes à nos imaginaires: 16–21. Mariemont: Musée Royal de Mariemont.

Paillet, P. and E. Man-Estier 2011. Oeuvres d'art méconnues de Laugerie-Basse (Dordogne). Collection Capitaine Maurice Bourlon - Institut de paléontologie humaine, Paris. *L'Anthropologie* 115: 505–521.

Ripoll, M.P., J.V. Morales Pérez, A. Sanchis Serra, J.E. Aura Tortosa and I.S. Montañana 2010. Presence of the genus *Cuon* in upper Pleistocene and initial Holocene sites of the Iberian Peninsula: new remains identified in archaeological contexts of the Mediterranean region. *Journal of Archaeological Science* 37: 437–450.

Rivero O. 2015. *Art mobilier des chasseurs magdaléniens à la façade atlantique*. Liège: Éditions ERAUL. Études et Recherches Archéologiques de l'Université de Liège.

Ronnberg, A. (ed.) 2010. *The Book of Symbols*. Köln: Taschen.

Rozoy, J-G. 1997. Les capacités mentales des artistes paléolithiques de l'Ardenne. *L'Anthropologie* 101 (1): 83–113.

Serpell, J. (ed.) 1995. *The Domestic Dog: Its Evolution, Behaviour and Interactions with People*. Cambridge: CUP.

Sieveking, A. 1979. *The Cave Artists*. London and New York: Thames and Hudson.

Sieveking, A. 1987. *A Catalogue of Paleolithic Art in the British Museum*. London: The Trustees of the British Museum by British Museum Publications.

Sommer, R. and N. Benecke 2005. Late-Pleistocene and early Holocene history of the canid fauna of Europe (Canidae). *Mammalian Biology* 70: 227–241.

Tosello, G. 2003. *Pierres gravées du Périgord magdalénien. Art, symboles, territoires*. XXXVe Supplément à Gallia Préhistoire. Paris: C.N.R.S.

Vanhaeren, N. and F. D'Errico 2001. La parure de l'enfant de la Madeleine (fouilles Peyrony). Un nouveau regard sur l'enfance au paléolithique supérieur. *Paleo* 13: 201–240.

Villaverde, V. 1994. *Arte paleolítico de la Cova del Parpalló. Estudio de la colección de plaquetas y cantos grabados y pintados*. Valencia: Servei d'Investigació Prehistórica.

# 5.2 At the Beginning of a Beautiful Friendship. Canid Representations in Levantine Rock Art

Manuel Bea[1], Alberto Lombo[1], Gianpiero Di Maida[2], Margherita Mussi[3]

[1]Department of Sciences of Antiquity, University of Zaragoza, C/ Corona de Aragón 42 Edificio Cervantes 50009 Zaragoza, Spain. Research Group P3A; IPH. manubea@unizar.es, albertolommon@hotmail.com

[2]Niedersächsisches Landesamt für Denkmalpflege, Scharnhorststraße 1, 30175 Hannover, Germany. gianpiero.dimaida@nld.niedersachsen.de

[3]ISMEO - Associazione Internazionale di Studi sul Mediterraneo e l'Oriente, Corso Vittorio Emanuele II 244, 00186 Roma margherita.mussi@fondazione.uniroma1.it

Corresponding author: Manuel Bea, manubea@unizar.es

**Abstract**

Referring to Levantine rock art there are still many open questions which are most relevant in the case of rarely depicted species, such as the canids. In contrast to most-frequently represented animals, the depictions of 'minor' species, were less realistic, smaller and with few anatomical details. Therefore, it is virtually impossible to make a distinction between wolves, foxes and dogs and even to identify the canid species. For this reason, the focus of this study is on the activity depicted, on the association with other animals and humans, and on the stylistic phase. In any case, canids and humans are the only predatory species represented in Levantine rock art. Previous approaches are discussed, known figures re-examined and assessed, and newly-described ones added.

**Keywords**: Post-Palaeolithic, Levantine Rock Art, Canid motifs, Spain.

## 1 Introduction

The late prehistoric rock-art sites of the eastern Iberian Peninsula form an exceptionally large cluster. The way of life during a critical phase of human development is vividly depicted in paintings whose style and subject matter are unique in records worldwide[1]. This is why Levantine rock art was included in the World Heritage List on 1998 by the general assembly of UNESCO, in Kyoto. It is characterised by stylistic, thematic, technical and geographical features that define it as a major prehistoric corpus (Domingo 2012; Hernández 2012; Villaverde *et al.* 2012) (Figure 1). Ever since archaeologists came across this artistic cycle, they have attempted to provide a global definition as well as stylistic and chronological classifications. Nevertheless, the studies on Levantine rock art leave many open questions (García *et al.* 2012).

One of the main characteristics is the naturalistic style of animal depictions, including anatomical details, exact proportions and the use of a correct perspective. This is the case with deer, bovids and ibexes, which are the most commonly represented species. The canids, however, are quite different. In almost every case, their identification is dubious as they are small-sized and the rendering is scarcely naturalistic. Furthermore, they are part of scenes, which can vary in their level of complexity; can be cumulative or not, and which include other animals as well as humans.

Probable wolves, foxes, dogs, jackals, and felines have all been tentatively identified (Breuil *et al.* 1912; Hernández-Pacheco 1924: 94; Ripoll 1961; Beltrán 1968, 1982; Jordá 1974; Dams 1984), but sometimes simply canids (Sanchidrián 2000: 388) or even carnivores (Porcar *et al.* 1935: 45–46; Alonso and Grimal 1996; Mateo Saura 1999; Mateo Saura and Carreño 2003; Soria and Zorrilla 2019) are mentioned. Overall, Canidae figures seem merely anecdotic. The limited records and less naturalistic style make it extremely difficult to positively recognise them and nearly impossible to make a distinction between wild and domesticated animals, as already pointed out by Jiménez and Ayala (2006).

In the following paragraphs we provide an updated overview of Canidae representations, including a discussion on the domestication of the dog and the geographical extension of Levantine rock art (Figure 1).

## 2 Towards a definition of canid motifs in Levantine rock art

As mentioned above, all identifications of the Canidae in Levantine Rock Art are dubious. The careless morphological characterisation and stylistic patterns are quite different from those of bovids, deer and ibexes

[1] https://whc.unesco.org/en/list/874/

Figure 1. Distribution of Levantine rock art sites.

(Figure 2). It is just impossible to take into account morpho-skeletal characteristics or specific anatomical traits.

The rare figures possibly representing wolves, foxes, dogs or generic carnivores are generally stylised or semi-naturalistic, mostly ca. 4–5 cm-long. Frequently there is a long snout, standing ears, a long horizontal or raised tail and short legs. Only in a couple of cases have some details been added: raised hair and a howling attitude at Barranc de la Palla (Alicante) (Figure 3.6) and a big fox-like tail at Cueva de la Araña (Bicorp) (Hernández-Pacheco 1924) (Figure 3.3). Canidae are never the main or only theme represented in a shelter, unlike the case of other animal species. They are always found in panels with other animals and humans, but are not always synchronous with them.

Figure 2. Levantine art. Exemplification of conventions and stylistic patterns in animal figures : bovid, deer and ibex motifs used to follow much more realistic conventions while canidae motifs were represented in a more stylised way (the canid is after Breuil *et al.* 1912).

Figure 3. Probable canids in Levantine rock art. 1 and 13, Cueva del Polvorín. 2–3, Cueva de La Araña. 4, Abrigo de La Viñuala. 5 and 9–12, Cueva de La Vieja. 6–8, Barranc de la Palla.14, La Vacada. 15–16, Cortijo de Sorbas. 17, Hoya de Navarejos III. 18, Torcal de las Bojadillas. (After Hernández-Pacheco 1924; Breuil 1912; Alonso and Grimal 1996; Bea 2009, 2017).

Figure 4. Fox representation from the Cueva de la Araña (Hernández-Pacheco 1924, Lam. XX).

In Palaeolithic art, isolated canids are represented, sometimes in detail (Di Maida *et al.* in this volume). In Levantine sites, canids are instead part of a scene, or at least share the same space as humans and herbivores. The extant record is interpreted as documenting either domestic dogs collaborating with humans during the hunt, or packs of wild animals hunting on their own.

As mentioned above, there are very few canid identifications, even including the dubious ones. The best example is definitely that from Cueva de la Araña II, where just a single fox is depicted. Despite its small size (it is 7 cm long) it is represented with the typical thick tail in a natural attitude, as if suddenly stopping with erect ears in an alert posture (Hernández-Pacheco 1924: 94) (Figure 4). The good preservation, the style and the expression all make it unique (Beltrán 1968: 39; Dams 1984; 136 and 249; Hernández-Pacheco 1924: 94).

Despite this fox depiction and a few more well-defined canids, the stylistic features rarely allow the species to be identified. To be more specific, in any decorated panel we rather the focus on the location and relationship with other elements. In the case of Cueva de la Vieja, a group of eight ibexes in a row, apparently passing through a narrow pathway, are driven by a possible canid that apparently pushes them towards a group of archers. However, the panel includes more figures and possible canids, and has been variously interpreted as depicting a hunting pack of wolves (Breuil and Lantier 1952), jackals or coyotes (Breuil and Obermaier 1927) or even African wild dogs (Eickstedt 1952). The various hypotheses emphasise the differences in the rendering of the tail: an animal with a narrow tail is possibly a wolf, while one with a wider tail might correspond to a jackal or coyote (Breuil *et al.* 1912: 547) (Figure 5). The morphology is similar to that of a canid from Fuente del Sabuco defined as a wolf by Dams (1984: 251, fig. 204.8).

In discussing the Cueva de la Vieja, Dams (1984: 151) rather suggests a 'composed scene' because of the different colours of ibexes and canids: accordingly, the canids would have been depicted later. In any case, given that collaboration between archers and canids cannot be ruled out, this beautiful and vivid scene might represent domesticated dogs included in a complex hunting strategy. Other canids (dogs or wolves) may be just part of the landscape, accompanying some schematic human figures (Dams 1984: 149) or simply filling empty spaces left among larger animals. The presence of domestic dogs in this site had already been suggested by Breuil 'Ce qui nous a, plus que toute chose, porté à soupçonner que peut-être ces 'Chacals" étaient domestiques, est la découverte de deux dessins où des

Figure 5. Different canid species from Cueva de La Vieja (After Breuil *et al.* 1912: 546, fig. 7). Breuil classified them as wolves or jackals/coyotes according to the shape of the tail.

animaux analogues, mais moins reconnaissables, sont tenus ou accompagnés par des chasseurs portant arc et flèches sous un bras, et paraissant, de l'autre, soit tenir à l'encolure, soit exciter l'animal' (Breuil *et al.* 1912: 548) and 'en examinant ces dessins, une idée vient qu'on ne doit pas rejeter sans examen: ne pourrait-on songer à des chiens?' (Breuil *et al.* 1912: 547). According to Dams (1984: 249), a few rather undefined figures in this panel could also be dogs.

In another scene, at El Polvorín, a possible canid, identified as a wolf, seems to face two wild boars (Beltrán 1982: 33; Dams 1984: 249). However, the panel could also depict a hunting party led by humans, with a dog driving the game, running after the prey and helping the hunters in many ways. No human figure is actually preserved, but a linear stroke on the rear of a wild boar has been interpreted as a rope or broken lasso, indirectly suggesting human hunters (Dams 1984: 41–42).

Other panels at El Polvorín are less clear, as the in the well-known case of a human figure holding a long rope coiled around the neck of an animal. The latter is an indeterminate quadruped according to Dams (1984: 41); an ibex taken by a lasso according to Beltrán (1968) and Viñas *et al.* (2019); or even a dog according to Vilaseca (1947).

With the exception of the fox at Cueva de la Araña II and of a few more figures from Cueva de la Vieja, Barranc de la Palla and Torcal de las Bojadillas, the remaining canids in the Levantine area are of dubious interpretation. Some cannot be accepted at all, for example a couple of motifs mistakenly identified as Levantine at Os de Balaguer (Jiménez and Ayala 2003: 168).

The small size and stylised conventions hamper any straightforward interpretation. Some, originally defined as probable canids, have recently been re-interpreted as undefined quadrupeds or just colour spots, such as the supposed canid from Mas d'en Josep (Beltrán 1985: 138; Domingo *et al.* 2003); the 'wolf' from El Polvorín (Beltrán 1985: 125); the '*canidé probable*' from Cingle de la Ermita del Barranc Fondo (Dams 1984: 103); the dubious '*chien-loup*' from Torcal de las Bojadillas (Dams 1984: 249); the 'canid' in association with an archer from La Vacada (Martínez-Bea 2005: 114 and 158; 2009: 75); and some poorly detailed ones from Covacho Ahumado (Almagro 1956; Beltrán and Royo 1998: 19) and Hoya de Navarejos III (Bea 2017).

### 3 Domesticated animals?

Chronology is a major pitfall in Levantine rock art studies. Despite some attempts (Ruiz *et al.* 2006, 2009, 2012; Viñas *et al.* 2016), it has so far been impossible to provide any absolute dating for this artistic cycle and its phases. Overall, however, there is general agreement that the chronology is post-Palaeolithic (Jordá 1966; Beltrán 1968; Ripoll 1968, 1970; Dams 1984; Llavori de Micheo 1988; Martí and Hernández 1988; Alonso and Grimal 1994; Utrilla and Calvo 1999; Martí and Juan Cabanilles 2002; Mateo Saura 2002, 2009; Villaverde and Martínez-Valle 2002; Molina *et al.* 2003; Martínez-Bea

2005; Domingo 2006; Villaverde *et al.* 2006; Hernández 2012; Mas *et al.* 2012).

It has been hypothesised that some scenes in Levantine rock art represent domesticated canids, opening up a discussion on the domestication of the dog and its use for hunting (Beltrán 1968: 39; 1982: 53). However, even accepting as proven that there is positive evidence of domestication (Beltrán 1982: 33), this does not help in establishing the chronology of Levantine rock art, especially for the earliest phases.

Overall, there is a general consensus that, not including the intriguing evidence from some Aurignacian sites in Belgium, in Western Eurasia and the Middle East the process leading to the domestication of the dog started at least 14,000 years ago, with examples of deliberate burials of dead dogs dating back 12,000 to 14,000 years (Davids and Valla 1978; Clutton-Brock 1980, 2000; Tchernov and Valla 1997; Pionnier-Capitan *et al.* 2011; Larson and Fuller 2014). Relevant discoveries have been made at Palegawra (Iraq), as well as at sites across Central and Eastern Europe such as Mezin, Eliseevichi, Kniegrotte, Oelknitz, Teufelsbrücke, Hauterive-Champréveyre, Saint-Thibaud-de-Couze or Pont d'Ambon (Vigne 2005–2006). There have also been discoveries dating to the Mesolithic in England and Northern Europe (Degerböl 1961; Detry and Cardoso 2010). Evidence of the individual burial or ritual disposal of dead dogs and humans with dogs in different chrono-cultural complexes has been discovered worldwide, some examples being very early, such as those from Natufian sites (Hayonim or Mallah-Eynan) (11,000–12,000 BP) or from southwestern North America (Koster, Dust Cave or Modoc Rock Shelter) (ca. 7000–8400 BP) (Morey 2006). In any case, domestic dogs are well documented during the Neolithic and increase in number during the Chalcolithic.

In the Iberian Peninsula, it has been pointed out that the dog is the earliest domesticated animal (Altuna 1980: 75, note 1). A *Canis* humerus, similar in size to the humerus of a dog, was found in a Magdalenian level at Erralla (Altuna *et al.* 1985: 110; Vigne 2005–2006). More remains of similar age were discovered at Abauntz, Marizulo, Arenaza and Nerja. The dogs increase in number later as the Neolithic develops (Catagnano 2016; Olivier *et al.* 2018).

In both Palaeolithic and Levantine art, the canids are very few and, furthermore, small-sized. There are differences, however, as during the Palaeolithic they are mostly engraved on mobile supports, while in Levantine art they are always painted on rockshelter walls. Furthermore, the Levantine canids are represented in groups, sometimes associated with humans in a lively attitude, possibly participating in the hunt. Could the differences be the result of domestication?

## 4 Conclusions

The canids in Levantine Rock Art have scarcely been studied. Those that have been discovered display very little naturalistic style (compared to other Levantine animals such as bovid, deer or ibex, and even horse or wild boar) and are small in size. The lack of detail makes it impossible to distinguish the species and difficult even to define them as canids, unlike the case, with Palaeolithic art. Dams mentions a concentration of these animals in just one site: '*Parmi les 9 figurations qui pourraient être interprétées comme des canidés, 6 sont concentrées à la Cueva de la Vieja*' (Dams 1984: 249), pointing out that canids represent only 0.57% of the animals found in Levantine art (Dams 1984: 316). Even if the number of both probable and dubious canids has since increased (Alonso and Grimal 1996: 76), they remain unquestionably rare, despite the different opinion of some authors (Jiménez and Ayala 2006: 165).

The most important animals, i.e. those represented most frequently, in a more naturalistic way and at a larger scale, are the species providing the highest calorie intake (i.e. bovids, deer, ibexes, equids and wild-boars). Apart from humans, the canids are the only predators found in Levantine art to date, although some doubtful felines and bears have been tentatively identified (Cabré 1915: 136; Ripoll 1961: 22; Alonso and Grimal 1996; Mesado *et al.* 1997: 9; Ruiz and Royo 2016: 28). Canid representations were clearly less important than those of other species, at least from a quantitative point of view. They are found at just a few sites dispersed throughout the Levantine territory (Figure 1), further evidence that they were not the focus of art. However, taking into account their scarcity, small size, lack of naturalism, location in marginal or empty spaces, Jordán (2001–2002: 41) has suggested an alternative symbolic explanation: a hypothetical sacred nature, especially in the case of wolves, with some kind of taboo not allowing them to be depicted.

One of the authors of the present work (Bea 2017) recently suggested that the so-called Levantine Art is one of several post-Palaeolithic rock-art styles, synchronic or not, with a naturalistic tendency. From a stylistic point of view the canids correspond to the most recent phases of Levantine art, associated with stylised human figures in hunting activities, or possibly added to such scenes. Jiménez and Ayala (2006) underline the importance of other aspects, notably body language: ear and tail position; and body, head, legs, tail and ears posture. The bad preservation of rock art found in shelters with a wide opening rarely allows detailed observation. All the same, the canids discussed above were consistently represented standing on their four legs, leaning slightly forward, with a straight tail and rigid limbs in an alert or attentive attitude.

The small size of our sample makes it difficult to know whether any differences in rendering was due to varying artistic ability or attempts to represent species diversity. In any case, we would remark that the archaeological record suggests marked differences from the dogs of the Chalcolithic (Catagnano 2016), when *Canis familiaris* became more abundant in the Iberian Peninsula. Could the different patterns of canid representations be related to the chronology and to the time when *C. familiaris* became more abundant and diverse? Or do they represent an actual development of the later phases of Levantine art? Or could they be related to artistic cycles different from the Levantine one, but also with a naturalistic tendency?

Forthcoming research and new discoveries will hopefully provide a better understanding of the start of the beautiful friendship between humans and dogs.

## Acknowledgements

Manuel Bea is member of the Project '*Gaps and Sites: Vacíos y ocupaciones en la Prehistoria de la Cuenca del Ebro*' (HAR2017-85023-P) and Instituto Universitario de Investigación en Patrimonio y Humanidades (IPH-Universidad de Zaragoza); Alberto Lombo is a research collaborator with the Area of Prehistory (University of Zaragoza).

We are also grateful to Julian Chancellor for the English revision of the text.

## References

Almagro, M. 1956. Las pinturas rupestres en el Bajo Aragón, in M. Almagro, A. Beltrán and E. Ripoll. *Prehistoria del Bajo Aragón*: 66–90. Zaragoza: Instituto de Estudios Turolenses.

Alonso, A. and A. Grimal 1994. El Arte Levantino o el 'trasiego' cronológico de un arte prehistórico. *Pyrenae* 25: 51–70.

Alonso, A. and A. Grimal (eds) 1996. *El arte rupestre prehistórico de la Cuenca del río Taibilla (Albacete, Murcia): Nuevos planteamientos para el estudio del Arte Levantino.* 2 vols. Barcelona: Alonso & Grimal.

Alonso, A. and A. Grimal 1999. *Introducción al arte levantino a través de una estación singular. La cueva de la Vieja (Alpera, Albacete).* Albacete: Asoc. Cult. Malecón.

Altuna, J. 1980. Historia de la domesticación animal en el País Vasco, desde sus orígenes hasta la romanización. *Munibe* 33: 75–81.

Altuna, J., A. Baldeón. and K. Mariezkurrena 1985. Bases de subsistencia de los pobladores de Erralla: macromamíferos, in J. Altuna, A. Baldeón and K. Mariezkurrena (eds) *Cazadores magdalenienses en Erralla (Cestona, País Vasco). Munibe* 37: 87–117.

Bea, M. 2017. El abrigo de Hoya de Navarejos III (Tormón, Teruel). Nuevas perspectivas para el análisis del arte levantino interior. *Complutum* 28 (1): 37–50.

Beltrán, A. 1968. *Arte rupestre levantino.* Monografías Arqueológicas No. 4. Zaragoza: Universidad de Zaragoza.

Beltrán, A. 1982. *El arte rupestre del Levante español.* Madrid: Ediciones Encuentro.

Breuil, H. and R. Lantier 1952. *Les homes de la pierre ancienne (Paléolitique et Mésolithique).* Paris: Payot.

Breuil, H. and H. Obermaier 1927. Las pinturas rupestres de los alrededores de Tormón. *Boletín de la Real Academia de la Historia* XC: 511–531.

Breuil, H., P. Serrano and J. Cabré 1912. Les peintures rupestres d'Espagne IV. Les Abris del Bosque a Alpéra (Albacete). *L'Anthropologie* XXIII: 34–36.

Cabré, J. 1915. *Arte rupestre en España.* Madrid: Comisión de Investigaciones Paleontológicas y Prehistóricas, Mem. No. 1.

Catagnano, V. 2016. *Aproximación morfométrica y paleogenética al estudio de la variabilidad de Canis l. familiaris.* Barcelona: Universidad de Barcelona.

Clutton-Brock, J. 1980. Los orígenes del perro, in J. Clutton-Brock and M. Hills (eds) *Ciencia en Arqueología*: 310–316. Madrid: Fondo de Cultura Económica.

Clutton-Brock, J. 2000. *Dogs through time: an archaeological perspective* (British Archaeological Reports International Series 889). Oxford: Archaeopress.

Dams, L. 1984. *Les peintures rupestres du Levant Espagnol.* Paris: Picard.

Davids, S.J.M. and F.R. Valle 1978. Evidence for domestication of the dog 12.000 years ago in the Natufian of Israel. *Nature* 276: 608–610. https://doi.org/10.1038/276608a0

Degerböl, M. 1961. On a find of a preboreal domestic dog (*Canis familiaris* L.) from Star Carr, Yorkshire, with remarks on other Mesolithic dogs. *Proceedings of the Prehistoric Society* 21: 35–55. https://doi.org/10.1017/S0079497X0001598X

Detry, C. and J.L. Cardoso 2010. On some remains of dog (*Canis familiaris*) from the Mesolithic shell-middens of Muge, Portugal. *Journal of Archaeological Science* 37 (11): 2762–2774. https://doi.org/10.1016/j.jas.2010.06.011

Di Maida, G., Mussi, M., A. Lombo and M. Bea (this volume). Lupus in fabula. The representations of Wolf (*Canis familiaris*) in the European Palaeolithic art, in I. Fiore and F. Lugli (eds) *Dogs, Past and Present. An interdisciplinary perspective.* Oxford: Archaeopress.

Domingo, I. 2006. La figura humana, paradigma de continuidad y cambio en el arte rupestre Levantino. *Archivo de Prehistoria Levantina* 26: 161–192.

Domingo, I. 2012. Human figure, techniques and territory: towards a technical redefinition of Levantine rock art, in J.J. García, H. Collado and G. Nash (eds) *The Levantine Question: Post-Palaeolithic rock art in the Iberian Peninsula*: 117–144. Cáceres-Budapest: Archaeolingua.

Domingo Sanz, I., E. López-Montalvo, V. Villaverde Bonilla, P. Guillem Calatayud and R. Martínez-Valle 2003. Las pinturas rupestres del Cingle de Mas d'en Josep (Tírig, Castelló). Consideraciones sobre la territorialización del arte levantino a partir del análisis de las figuras de bóvidos y jabalíes. *Saguntum* 35: 9–49.

Eickstedt, E. 1952. Die Sauhartz von Valltorta. Ein Beitrag zur Biohistorie von Altiberien. *Homo* 3: 123–130.

García, J.J., H. Collado and G. Nash (eds) 2012. *The Levantine Question: Post-Palaeolithic rock art in the Iberian Peninsula*. Cáceres-Budapest: Archaeolingua.

Hernández-Pacheco, E. 1924. *Las pinturas prehistóricas de las Cuevas de La Araña (Valencia)*. Comisión de Investigaciones Paleontológicas y Prehistóricas, Mem. No. 34, Serie Prehistórica. Madrid: Museo Nacional de Ciencias Naturales.

Hernández, M.S. 2012. Defining a Neolithic rock art: Levantine, Macroschematic and Schematic arts in the Mediterranean basin, in J.J. García, H. Collado and G. Nash (eds) *The Levantine Question: Post-Palaeolithic rock art in the Iberian Peninsula*: 145–166. Cáceres-Budapest: Archaeolingua.

Horard-Herbin, M.-P., A. Tresset and J.-D. Vigne 2014. Domestication and uses of the dog in western Europe from the Paleolithic to the Iron Age. *Animal Frontiers* 4 (3): 23–31. https://doi.org/10.2527/af.2014-0018

Irving-Pease, E., H. Ryan, A. Jamieson, E.A. Dimopoulos, G. Larson and L. Frantz 2018. Paleogenomics of animal domestication, in C. Linqvist and O. Rajora (eds) *Paleogenomics. Population genomics*: 225–272. Springer, Cham. https://doi.org/10.1007/13836_2018_55

Jiménez Lorente, S. and M.M. Ayala Juan 2006. Avance al estudio de la representación del *Canis familiaris* en la pintura rupestre postpaleolítica. *Cuadernos de Arte Rupestre* 3: 161–184.

Jordá, F. 1966. Notas para una revisión de la cronología del arte rupestre levantino. *Zephyrus* XVII: 47–76.

Jordá, F. 1974. Formas de vida económica en el arte rupestre levantino. *Zephyrus* XXV: 209–223.

Jordán, J.F. 2001–2002. Los animales en el arte rupestre postpaleolítico de la Península Ibérica. Emblemas, alegorías, epifanías y ausencias. *AnMurcia* 16–17: 37–52.

Jordana, J., A. Sánchez and J. Piedrafita 1990. *Los cánidos españoles. Orígenes y clasificación*. Barcelona: Purina.

Larson, G. and D.Q. Fuller 2014. The evolution of animal domestication. *Annual Review of Ecology, Evolution, and Systematics* 45: 115–136. https://doi.org/10.1146/annurev-ecolsys-110512-135813

Larson, G. E.K. Karlsson, A. Perri, M.T. Webster, S.Y.W. Ho, J. Peters, P.W. Stahl, P.J. Piper, F. Lingaas, M. Fredholm, K.E. Comstock, J.F. Modiano, C. Schelling, A.I. Agoulnik, P.A. Leegwater, K. Dobney, J.D. Vigne, C.Vilà, L. Andersson and K. Lindblad-Toh 2012. Rethinking dog domestication by integrating genetics, archaeology and biogeography. *PNAS* 109 (23): 8878–8883. https://doi.org/10.1073/pnas.1203005109

Llavorí de Micheo, R. 1988–1989. El arte postpaleolítico levantino de la península Ibñerica. Una aproximación sociocultural al problema de sus orígenes. *Ars Praehistorica* VII-VIII: 145–156.

López, M.G., M. Soria and D. Zorrilla 2009. *El arte rupestre en las sierras giennenses. Patrimonio de la Humanidad. Sierra Morena Oriental*. Jaén: Instituto de Estudios Giennenses, Diputación Provincial de Jaén.

Martí, B. and M.S. Hernández 1988. *El Neolític Valencià. Art rupestre i Cultura Material*. Valencia: Servei d'Investigació Prehistòrica.

Martí, B. and J. Cabanilles 2002. La decoració de les ceràmiques neolítiques i la seua relació amb les pintures rupestres dels abrics de La Sarga, in M.S. Hernández and J.M. Segura (eds) *La Sarga. Arte rupestre y territorio*: 147–170. Alcoy: Ayuntamiento de Alcoy, Caja de Ahorros del Mediterráneo.

Martínez-Bea, M. 2005. Variabilidad estilística de distribución territorial del arte rupestre levantino en Aragón: el ejemplo de La Vacada (Castellote, Teruel). PhD dissertation, Universidad de Zaragoza.

Martínez-Bea, M. 2009. *Las pinturas rupestres del abrigo de La Vacada (Castellote, Teruel)*. Monografías Arqueológicas No. 49. Zaragoza: Universidad de Zaragoza.

Mas, M., R. Maura and M. Solís 2012. Absolute chronologies and relative chronologies. On the initial sequences of post-Palaeolithic rock art in the Mediterranean arc, in J.J. García, H. Collado and G. Nash (eds) 2012. *The Levantine Question: Post-Palaeolithic rock art in the Iberian Peninsula*: 187–208. Cáceres-Budapest: Archaeolingua.

Mateo Saura, M.A. 1999. *Arte rupestre en Murcia. Noroeste y Tierras Altas de Lorca*. Murcia: Editorial KR.

Mateo Saura, M.A. 2002. La llamada fase 'pre-levantina' y la cronología del arte rupestre levantino. Una visión crítica. *Trabajos de Prehistoria* 59 (1): 49–64.

Mateo Saura, M.A. 2009. *Arte rupestre levantino. Cuestiones de cronología y adscripción cultural*. Murcia: Tabularium.

Mateo Saura, M.A. 2012. Del arte paleolítico al arte levantino. ¿Continuidad o ruptura?, in J.J. García, H. Collado, H. and G. Nash (eds) 2012. *The Levantine Question: Post-Palaeolithic rock art in the Iberian Peninsula*: 167–186. Cáceres-Budapest: Archaeolingua.

Mateo Saura, M.A. and A. Carreño 2003. Nuevos yacimientos con arte rupestre en Albacete: los abrigos del Cortijo de Sorbas III (Letur), Barranco de los Buitres (Nerpio) y Arroyo de los Covachos II (Nerpio). *Al-Basit* 47: 5–40.

Morey, D.F. 2006. Burying key evidence: the social bond between dogs and people. *Journal of Archaeological Science* 33: 158–175. https://doi.org/10.1016/j.jas.2005.07.009

Mesado, N., E. Barreda and J. Andrés 1997. Las pinturas rupestres del abrigo del Mas de Barberà (Forcall, Castellón). *Archivo de Prehistoria Levantina* XXII: 117–137.

Olivier, M., A. Tresser, L.A.F. Frantz, S. Bréhanrd, A. Balasescu, M. Mashkour, A. Boroneant, M. Pionnier-Capitan, O. Lebrasseur, R.M. Arbogast, L. Bartosiewicz, K. Debue, R. Rabinovich, M.V. Sablin, G. Larson, G. Hänni, C. Hitte and J.D. Vigne 2018. Dogs accompanied humans during the Neolithic expansion into Europe. *Biology Letters* 14: 20180286. http://dx.doi.org/10.1098/rsbl.2018.0286

Pionnier-Capitan, M., C. Bemilli, P. Bodu, G. Célérier, J.F. Ferrié, P. Fosse, M. Garcià and J.D. Vigne 2011. New evidence for Upper Palaeolithic small domestic dogs in South-Western Europe. *Journal of Archaeological Science* 38: 2123–2140. https://doi.org/10.1016/j.jas.2011.02.028

Ripoll, E. 1961: *Los abrigos pintados de los alrededores de Santolea (Teruel)* (Monografías de Arte Rupestre. Arte Levantino 1). Barcelona:Instituto de Prehistoria y Arquelogía.

Ripoll, E. 1968. Cuestiones en torno a la cronología del arte rupestre postpaleolítico en la Península Ibérica, in *Simposio Internacional de Arte Rupestre*: 165–192. Barcelona: Diputación Provincial de Barcelona.

Ripoll, E. 1970. Acerca del problema de los orígenes del arte levantino, in *Valcamonica Symposium*: 57–67. Capo di Monte.

Rodríguez, L., J.M. Torrado, M. Villalba, J. González, A. Domínguez, E. García, H. Collado, N. Sala, M. Algaba, J.L. Arsuaga, J.J. García, I. Domínguez, L. Nobre, M. Mas and M. Solís 2015. Nuevos enclaves con arte rupestre en la Meseta castellano-leonesa. Las representaciones pintadas y grabadas en las cuevas de la Zarzamora y El Portalón (Perogordo, Segovia, España), in M.A. Medina, A.J. Romero, R.M. Ruiz-Márquez and J.L. Sanchidrián (eds) *Sobre rocas y huesos: las sociedades prehistóricas y sus manifestaciones plásticas*: 216–227. Córdoba: Fundación Cueva de Nerja y Universidad de Córdoba.

Rubio, M. 1995. Aproximación al estudio de las figuras zoomorfas representadas en el Arte Rupestre Levantino. *Recerques del Museu d'Alcoi* 4: 103–109.

Ruiz, J.F. and J. Royo 2016. *Guía de la Cañada de Marco. Alcaine (Teruel)*. Teruel: Ayuntamiento de Alcaine.

Ruiz, J.F., A. Hernanz, R.A. Armitage, M.W. Rowe, R. Viñas, J.M. Gavira and A. Rubio. 2012 Calcium oxalate AMS $^{14}C$ dating and chronology of post-Palaeolithic rock paintings in the Iberian Peninsula. Two dates from Abrigo de los Oculados (Henarejos, Cuenca, Spain). *Journal of Archaeological Science* 39: 2655–2667. https://doi.org/10.1016/j.jas.2012.02.038

Ruiz, J.F., M. Mas, A. Hernanz, M.W. Rowe, K.L. Steelman and J.M. Gavira 2006. First radiocarbon dating f oxalate crusts over Spanish prehistoric rock art. *INORA* 46: 1–5.

Sanchidrián, J.L. 2018. *Manual de arte prehistórico.* Barcelona: Ariel.

Soria, M. and D. Zorrilla 2019. Arte levantino y arte esquemático en Andalucía oriental, in R. Viñas (coord) *I Jornades Internacionals d'Art Rupestre de l'Arc Mediterrani de la Península Ibèrica. XXè Aniversari de la Declaració de Patrimoni Mundial*: 91–128. Tarragona: Museu Comarcal de la Conca de Barberà, Museu Arqueòlogic de Catalunya (MAC), Ajuntament de Montblanc.

Tchernov, E. and F.F. Valla 1997. Two new dogs, and other Natufian dogs, from the Southern Levant. *Journal of Archaeological Science* 24 (1): 65–95. https://doi.org/10.1006/jasc.1995.0096

Utrilla, P. and M.J. Calvo 1999. Cultura material y arte rupestre 'levantino': la aportación de los yacimientos aragoneses a la cuestión cronológica. Una revisión del tema en el año 2000. *Bolskan* 16: 39–70.

Utrilla, P. and M. Bea 2018. El arte Levantino, in J. Soler, R. Pérez and V. Barciela (eds) *Rupestre. Los primeros santuarios. Arte prehistórico en Alicante*: 126–139. Alicante: MARQ.

Vigne, J.D. 2005. L'humérus de chien magdalénien de Erralla (Gipuzkoa, Espagne) et la domestication tardiglaciaire du loup en Europe. *Munibe* 57 (1): 279–287.

Vilaseca, S. 1947. *Las pinturas rupestres de la Cueva del Polvorín*. Madrid: Comisión General de Excavaciones Arqueológicas. Memoria No. 17.

Villaverde, V. and R. Martínez Valle. 2002: Consideraciones finales, in R. Martínez-Valle and V. Villaverde (eds) *La Cova dels Cavalls en el Barranc de la Valltorta*, Monografías del Instituto de Arte Rupestre I: 191–202. Valencia: Instituto de Arte Rupestre de la Valltorta.

Villaverde, V., Martínez-Valle, R., P. Guillem, E. López-Montalvo and I. Domingo 2012. What do we mean by Levantine rock art?, in J.J. García, H. Collado, H. and G. Nash (eds) 2012. *The Levantine Question: Post-Palaeolithic rock art in the Iberian Peninsula*: 81–106. Cáceres-Budapest: Archaeolingua.

Viñas, R., A. Rubio, J.F. Ruiz, M. Vaquero, J. Vallverdú, M.W. Rowe and N. Santos 2016. Investigación cronoestratigráfica en el conjunto rupestre de a Sierra de la Pietat: abrigos de Ermites I y IV (Ulldecona, Tarragona, Catalunya). *Revista de Arte Prehistórico* 2: 70–85.

# 5.3 Dog Images in the Altai Rock Art

Dmitry V. Cheremisin

Department of the Bronze and Iron Ages in Siberia, Institute of Archaeology and Ethnography, Siberian Branch, Lavrentiev St, 17, Novosibirsk 630090, Russia. topsya@bk.ru

**Abstract**

The article presents the images of the dog in the rock art of Altai from the Bronze Age to the present day. Based on the extensive material obtained by the author in his expeditionary research, various subjects involving dogs are shown. These are scenes of hunting and chasing wild animals, deer, and mountain goats. The difficulties in determining the characters of the rock art - whether a wolf or a dog are indicated, and identification criteria are determined. An attempt is made to distinguish different breeds of dogs, as they are depicted in the rock art of Altai. The different semantics of the plots of petroglyphs depicting dogs for different eras and cultures from the Bronze Age to the present are noted.

**Keywords:** Altai, petroglyphs, image identification, hunting with a dog, dog breeds.

## 1 Introduction

Petroglyphs are the most numerous archaeological sites in Altai. They are a part of rock surfaces, which are natural objects – and cannot independently move in space and therefore directly indicate the presence of certain peoples who have a certain style in art and left these marks on the rocks. The study of these types of engravings enables us to follow the ancient lines of communication, and to clarify the cultural links between ancient populations. The materials of rock images are considered an informative and relevant source of archeological data about the occupations of the ancient population, about the type of economy, about interactions with nature, in particular, with the animal world. Rock Art compositions in which dogs are depicted make it possible to reveal these cultural and historical aspects of the life of the ancient population of Altai. The article is based on the author's many years of research on the petroglyphs of the Russian and Mongolian Altai.

## 2 Methods

The author's methodology is based on the principle of determining the dating and cultural affiliation of petroglyphs. Determination of the style, cultural realities, cases of overlapping of earlier images with later ones, the practice of including ancient petroglyphs in new compositions, have all used these positions. The context of the compositions in which the dogs are placed was taken into account. Dated petroglyphs were interpreted in a historical and cultural approximation.

## 3 Materials and discussion

The dog image in the rock and folk art of Southern Siberia and Central Asia in the diachronic aspect did not previously attract great attention of researchers The images of canine animals, i.e. the wolf and the dog, look quite similar especially in metal works, carved wood, ivory and horn as well as in old toureutics and sculptures. Recently, Russian archaeologists V.V. Bobrov and N.A. Moor (2019) released an article on the rock art of Central Asia and Southern Siberia and identified dog images among other heroes on the basis of the analysis of various published data. These researchers identified certain iconographic features, proposed identifications of the portrayed species, and classified popular compositions with dogs. They also executed statistical analysis and defined distinctive features between the images of dogs and other representatives of the *Canis* family in the rock art of the Bronze and early Iron Age periods. The researchers pointed out that the reliable identification of the images is only possible in Rock art in contrast to works of applied art (toreutics, wood and stone carvings). They saw the major difficulty in the identification of images due to the similarity in the appearances of ancient dogs and wolves rather than the chronological attribution of the petroglyphs (Bobrov and Moor 2019).

The paper publishes some incredibly exciting some most exciting compositions with dog images dating from the Bronze Age to modernity that were found by the present author in his field studies in the Altai. The earliest images of representatives of *Canis* in the rock art of the Sayan-Altai Range date back to the early and

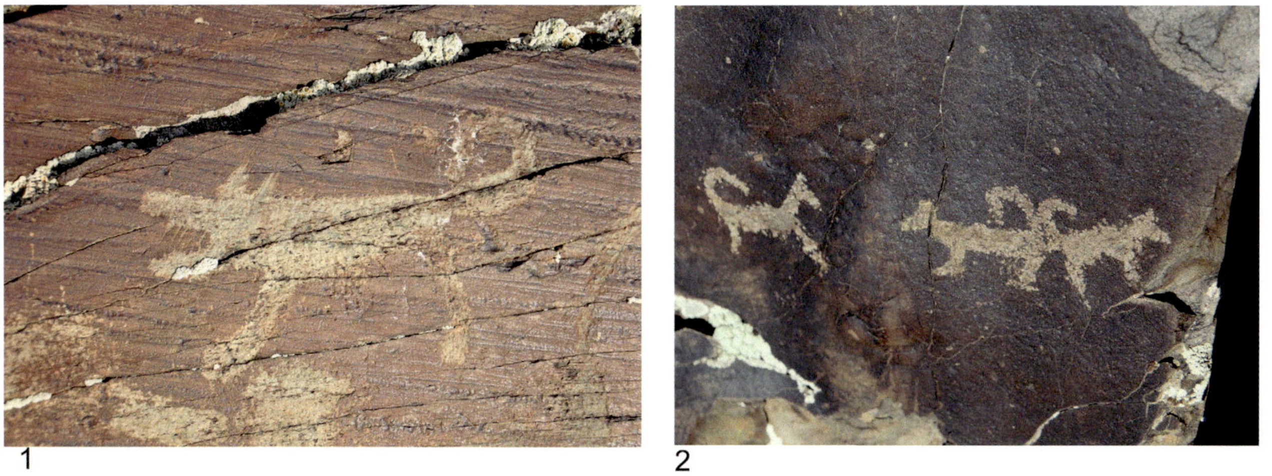

1 2

Figure 1. One of the main features making it possible to identify dog images is the tail that is shown either upright or turned to the back, straight or curved. 1, Elangash river valley, southeastern Russian Altai; 2, Chagan river valley, southeastern Russian Altai (Photo by D.V. Cheremisin).

Figure 2. Landscape of the Chagan river valley (Photo by D.V. Cheremisin).

Middle Bronze Age periods (the 3rd - 2nd millennia BC). The pecked out and engraved petroglyphs show isolated images and multi-figured compositions with wolves and dogs. It should be noted that one of the main features making it possible to identify dog images is the tail that is shown either upright or turned to the back, straight or curved (Bobrov and Moor 2019: 129) (Figure 1, n. 1–2). The other reason for the identification of an image as that of a dog is the inclusion of this image into the multi-figured composition and the direct participation of this hero in the economic activities of humans, mostly in the most numerous hunting scenes.

The set of Bronze petroglyphs that was found by the present author in the Chagan River valley in the south of the Russian Altai shows exactly these iconographic features in both isolated images and the compositions showing the dogs chasing and attacking ibexes (Figures

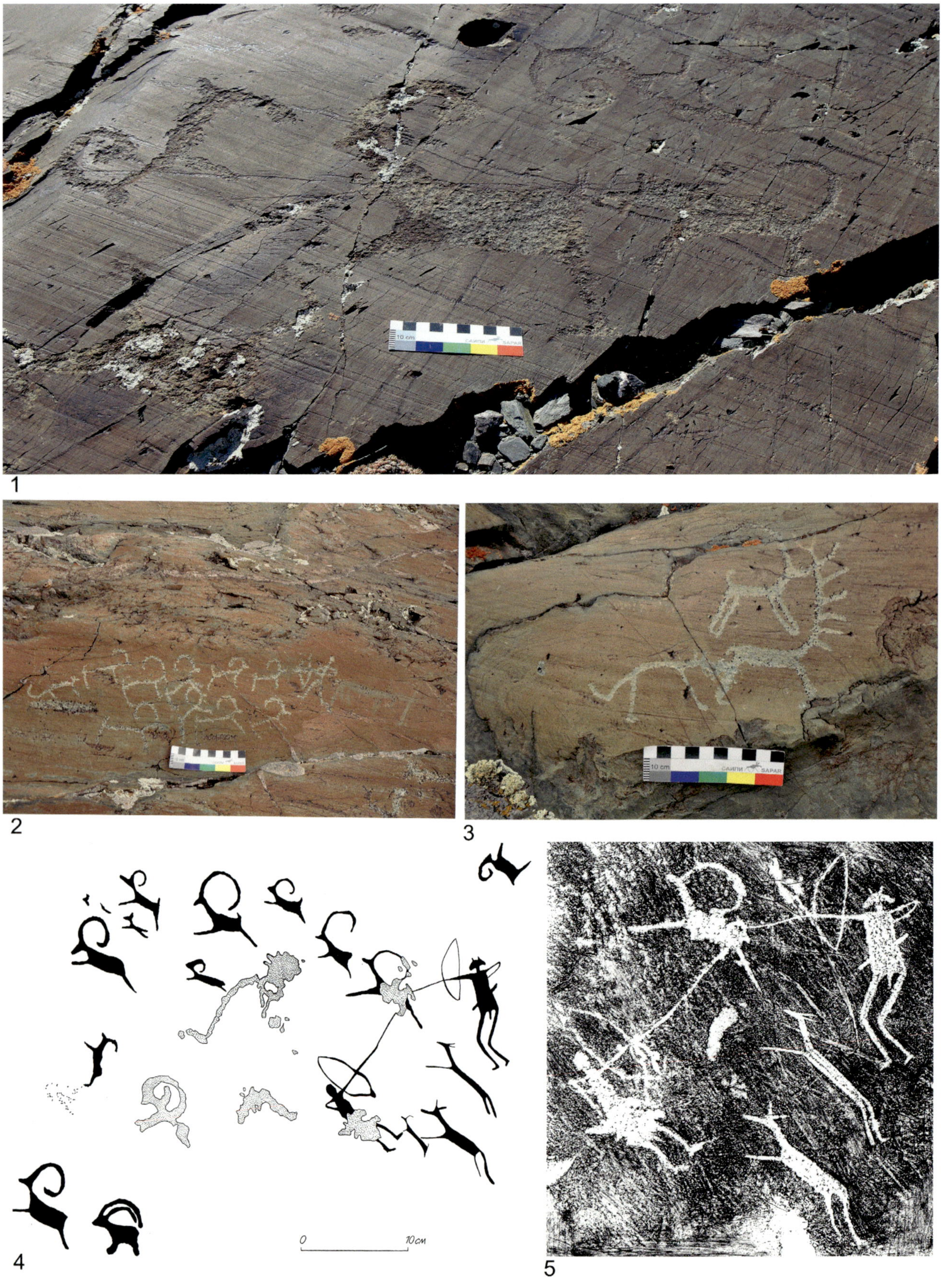


Figure 3. The other reason for the identification of an image as that of a dog is the inclusion of this image into the multi-figured composition and the direct participation of this hero in the economic activities of humans. 1, Chagan river valley, southeastern Russian Altai; 2–3, Elangash river valley, southeastern Russian Altai; 4–5, Kara-Chad river valley, Ukok plateau, southern Russian Altai (Graphic drawing and tracing by D.V. Cheremisin).

Figure 4. A hunter keeping a dog on a lead attached to his waist. Baga-Oigur river valley, northwest of Mongolian Altay (Photo by D.V. Cheremisin).

3–4). Similar features are noted on the composition from the Mongolian Altai presenting a hunter keeping a dog on a lead attached to his waist (Figure 4). Dogs with curled tails are shown in the scene of elk chasing (Figure 5). The elk hunting composition shows 27 (!) dogs, two of them are kept on the lead (one lead is attached to the waist, the other is held in the hunter's hands) have been found at the Tsagaan-Salaa II site in the Mongolian Altai (Jacobson *et al.* 2001: 173, fig. 130).

The rock art composition on the cliffs on the right side of the Chagan represents wild boar hunting: five dogs chase the wild boar and drive it to the hunter attacking the animal from various sides, while the hunter shoots the prey with bow and arrow (Figure 6). Apparently, the composition renders the typical features of boar hunting, which is a considerably dangerous animal or the hunter. Hunting wild boar with several dogs or with a pack of hounds was easier and less dangerous. The boar hunting composition with a pack of hounds was one of the significant scenes in the rock art repeated in a number of Bronze Age petroglyphs in the Russian and Mongolian Altai; this composition was also typical for Scythian-Siberian art. For instance, this composition is rendered on the gold plate of the second part of the 1st millennium BC found in Tuva. It shows the hunter hitting the wild boar with a dagger; the boar is attacked from behind by a dog with the tail turned up on the back (Grach 1980: 81, fig. 117). Apparently, this composition played an important role in the art and ideology of the population of Southern Siberia and Central Asia and was linked with the mythology and heroic epic stories of the cattlemen in the Bronze and early Iron Ages. The compositions with several dogs are also widely spread over the vast territory. In all appearances, these compositions represent chasing and driving methods of hunting wild boar, deer and elks by a large group of hunters and dogs.

Quite a few compositions are known with Bronze Age wheeled vehicles attended by dogs. One of such compositions was found by the present author on the cliffs in the Chagan valley; it shows four anthropomorphic images in swift motion, a wheeled carriage and gracile dogs. (Cheremisin 2003).

Similar compositions with dogs chasing, hunting and rendering hoofed animals, and dogs present in animal pastures are known from the rock art of the early Iron

1

2

Figure 5. Dogs with curled tails are shown in the scene of elk chasing (Chagan river valley, southeastern Russian Altai) (Photo by D.V. Cheremisin).

Figure 6. 1, Wild boar hunting: five dogs chase the wild boar and drive it to the hunter attacking the animal from various sides, while the hunter shoots the prey with bow and arrow (Chagan river valley, southeastern Russian Altai) (Photo by D.V. Cheremisin); 2, several rock engravings representing a single story including the composition of hunting the maral red deer were found in the Chagan valley (southeastern Russian Altai) (Photo by D.V. Cheremisin); 3, a hunter pointing his arrow to a wolf or a dog with a thick tail (Chagan river valley, southeastern Russian Altai) (Graphic drawing and tracing by D. Cheremisin); 4–5, guns with supports are shown in many hunting scenes including.

Age. These scenes were highly important for the tribes of early pastoralists. The images of feline and wolf-like carnivores are well presented in the applied art of the early Iron Age. The Scythan-Siberian Animal Style was typical for the epoch and was widely spread over the vast Eurasian territory. The predominant images of this style are wild animals. The Wolf as a representative of *Canis* was the leading hero in the art of the tribes of the Scythian period populating Eurasian steppe and forest-steppe ecozones. Feline and wolf-like carnivores are the typical heroes of the specific torques artefacts of the Pazyryk culture. However, even the images with the tail turned up at the back are traditionally perceived as the mythological wolf image rather than that of the dog (Kubarev and Cheremisin 1987).

The broad spread of the technique of fine engraving among the Turkic population of the Eurasian steppes in the early medieval period (the 6th - 9th centuries) is represented by the rock images of armed horsemen, scenes of military confrontation as well as traditional scenes of hunting, chasing deer and ibexes by mounted hunters armed with bows and arrows. The present author found several rock engravings representing a single story including the composition of hunting the maral red deer in the Chagan valley (Figure 6, n. 2). The features of the dog image suggest that it is a tazy hound, a hero from the heroic epic poem of the Turkic people. Another synchronous rock art composition executed through fine engraving in the same style shows the hunter pointing his arrow to a wolf or a dog with a thick tail (Figure 6, n. 3). The scene cannot be interpreted in a single way; it either shows a hunter with his dog; or the animal image can be interpreted as the game being the wolf.

The southern part of the Russian Altai represents a locus where the Telengit indigenous population still practices the rock art tradition. The local folk art is characterised by its own style and realism in rendering the features of modernity. The compositions related to the traditional economic activities are traditional too. One of such traditional compositions is the scene of hunting wild animals with guns. Guns with supports are shown in many hunting scenes including hunting with dogs (Figure 6, n. 4–5) les have changed through time, but the traditional subjects related to the culture of the Altai have survived. The dog is one of the essential elements of their culture which has survived from the remote past till modernity. Petroglyphs represent the important role of the dog in the everyday life of the Altai herdsmen.

## References

Bobrov V.V. and N.N. Moor 2019. Obraz sobaki v naskal'nom iskusstve narodov Yuzhnoy Sibiri I Mongolii (epoha bronzy I skifskoe vremya). *Problemy istorii, filologii I kul'tury* 2: 128–139. Бобров В.В. и Н.Н. Моор. 2019. Образ собаки в наскальном искусстве народов Южной Сибири и Монголии (эпоха бронзы и скифское время). *Проблемы истории, филологии и культуры* 2: 128–139.

Cheremisin D.V. 2003. Rock art composition with a chariot and 'dancers' from Chaganka (Kara-Oyuk) in the Altai. *Archaeology, Ethnology, & Anthropology of Eurasia*, 4 - (16) 57–63.

Grach, A.D. 1980. Drevnie kochevniki v tsentre Azii (Ancient Nomads in the Center of Asia). Moscow: Nauka. Грач А.Д. 1980. *Древние кочевники в центре Азии*. Москва: Наука

Jacobson E., V. Kubarev and D. Tseevendorj 2001. *Répertoire des pétroglyphes d'Asie Centrale*. T. 5, fasc. 6: Mongolie du nord-ouest. Tsagaan-Salaa/Baga Oigor. - 132., taf. 346, photogr. 399. Paris: Diffusion de Boccard.

Kubarev, V.D. and D.V. Cheremisin 1987. Volk v iskusstve i verovaniyakh kochevnikov Tsentral'noy Azii, in: I.N. Gemuev and A.M. Salagaev (eds) Traditsionnye verovaniya i byt narodov Sibiri. 98–117. Novosibirsk: Nauka. Кубарев, В.Д. и Д.В Черемисин 1987. Волк в искусстве и верованиях кочевников Центральной Азии, И.Н. Гемуев и А.М. Сагалаев (ред.) *Традиционные верования и быт народов Сибири*. 98–117 Новосибирск: Наука.

# 5.4 Representations of Dogs in Attic Funerary Monuments: A Question of Symbolism?

Francesco Tanganelli

Dipartimento di Scienze Umane (DiSU), Università della Basilicata, via Nazario Sauro 85, 85100 Potenza, Italia, ftanganelli87@gmail.com

**Abstract**

The iconographic repertoire of Attic funerary monuments (*sémata*) of Classical age is particularly rich in zoomorphic representations, among which dogs appear to be very common. However, Attic productions attest only three of the numerous dog typologies recorded by literary sources and artistic expressions: the Spitz-type dog, the 'greyhound' and the Molossian dog. Based on the analysis of a substantial group of *sémata*, the frequency and the choice of a specific type of dog seems to be due not only to its attitude, but to the gender and age of the deceased as well, revealing the extraordinary social value of this animal.

**Keywords**: Classical Greece, Attic sculpture, funerary art, iconography, animal symbolism.

## 1 Introduction and methodology

In ancient times, Athens and Attica knew a very rich and important tradition of funerary monuments (in ancient Greek, *sémata*), carved in precious marble, in different sizes and shapes, and attested in particular between the second half of the 5th century and the last decade of the 4th century BC (Clairmont 1993). Along with the deceased and their relatives, Attic artists often used to represent different kinds of animals (Woysch-Méautis 1982), among which one of the most common was, without doubt, the dog (Zlotogorska 1997).

The present paper aims to reconsider the iconography and meaning of different dog types presented by Classical Attic tombstones. Starting from a general review of the importance and the role of dogs in ancient Greek society, this analysis takes into consideration a remarkable sample of 159 funerary *sémata*, inclusive of both relief slabs (*stèlai*) and decorative statues. According to academic custom, several authors in the past have examined these two categories of materials individually, and never together, thus missing the opportunity to gain an overview of their production and implications. Funerary statuary, in particular, seems to be more problematic: in fact, it is rather difficult to gather a large number of original funerary dog statues nowadays, mainly due to the practice of copying and reusing Greek originals for decorative purposes in the Roman period. Another puzzling problem also concerns the modern habit of antiquities collection and restoration, which often caused the loss of original characters, contexts and meanings.

For all these reasons, this study requires a multidisciplinary approach, which considers not only the literary, epigraphic, artistic and archaeological data, but the phenotypes and ethology of all the identified typologies as well. The concept of dog typology, based on the analysis of phenotypic characters, is preferable to breed definition: in fact, despite the great number of ancient breeds recorded by Greek and Latin texts, nearly 50 names (Hull 1964: 28–30), it is almost impossibile today (with very few exceptions) to recognise them precisely in ancient sculptures and paintings, due to the widespread absence of connection between names and images.

## 2 Dogs among Classical Attic *sémata*: typologies, frequencies, associations

Greek artistic tradition, from the Mycenaean age to the Roman period, offers representations of different dog typologies, among which we can notice small lap dogs, greyhounds, molossers and mastiffs, fox-like and wolf-like dogs (Tanganelli 2013: 55–57). This evidence shows the great variability of these domestic animals throughout the Aegean area during Antiquity. However, if we examine the well-known Attic production of funerary *sémata* of Classical age, surprisingly we find only the presence of a very small number of recurring dog types.

The first one - the most common - is the Spitz-type dog. A small size characterises this *kynídion* - the 'lap dog' of the ancient Greeks - along with triangular-shaped ears, a tapered muzzle, thick fur and a curled, upturned tail. Thanks to the epigraphic evidence of a lost red-figured amphora from the Etruscan town of Vulci (Figure 1), it is possible to identify this dog with the renowned Melitaean dog (Keller 1909: 93), so small that Aristotle thought to compare it to a marten (*Historia animalium*

Figure 1. Drawing of an Attic red-figured amphora with a young man talking to his Melitaean dog. From Vulci (Italy), ca. 500 BC. The vase is currently lost (After Keller 1909: Fig. 34).

612b, 10). The Melitaean dog traditionally took its name from the island of *Melítē*, identified with Malta by Strabo (*Geographica* 6, 2, 11) and with Mljet by Pliny the Elder (*Naturalis historia* 3, 152). Although the provenience of Melitaean dog from Malta seems to be most likely (Busuttil 1969: 207–208), it is rather difficult to connect this ancient breed with the modern Maltese dog, which clearly shows different phenotypic characteristics. Despite the numerous depictions of the Melitaean dogs in ancient Greek art, Attic *sémata* seems to show this dog only in relief monuments, where it is mostly presented in association with children of both sexes (45% males; 16.7% females). Besides these, it can also be seen in the company of male teenagers (11.7%) and girls (6.7%), and yet at the side of young women (3.3%) or under the chairs of mature women (10%), while it seems to be very rarely portrayed with ephebes (males of 18–20 years old, still doing military service) and adult men. A widespread iconographic solution presents the Melitaean dog pointing to a bird, usually held by its deceased master: although difficult to understand in its deep, symbolic meaning, this image seems to recall, at least, a specific attitude of the Spitz dog (especially, the Finnish one), considered suitable for the capture of small-sized birds (Dennis-Bryan, Clutton-Brock 1988: 72).

The second recorded dog type consists of a sort of sighthound, here referred to as 'greyhound' (according to the terminology adopted by K. Dennis-Bryan and J. Clutton-Brock), but more similar to a modern Podenco dog, or - better yet - to the smaller Italian Cirneco dell'Etna (Dennis-Bryan, Clutton-Brock 1988: 23–24). In fact, this dog type shows a sharp muzzle with a big, pointed nose; the ears are long and straight; the body is, at the same time, slim and strong, and the tail appears thin and long. One Attic tombstone, dedicated to the brothers Apollodoros and Lakon, sons of Lakon, shows the representation of one dog with Podenco phenotypes, depicted alone, smelling the ground: this evidence led some authors to identify it with the famous *Lakonikós kýōn* (the 'Laconic dog'), assuming the existence of a word pun between the name of the breed and the anthroponym Lakon (Freyer-Schauenburg 1970). Indeed, Laconic dogs were famous for their smell, to the point that Sophocles compares Odysseus's intuition to the nose of a Laconic bitch (*Ajax*, 7–8). However, it is quite difficult to verify this word pun theory on a larger scale, due to the physiognomy and behaviour of this dog, which partly seems to be common to other ancient dog breeds, i.e. the Cretan and the Carian (Tanganelli 2013: 21–22). In Attic funerary reliefs, this greyhound-type mostly follows ephebes (50%), but is at times also visible in the company of male teenagers and adult men (15.6% for both). In some funerary slabs, the deceased master holds a *lagobólon*, the typical hunting stick, which seems to characterise him - along with the dog - as a heroic hunter (Figure 2). Only in very few cases Attic artists chose to also represent this dog as a decorative statue (Vedder 1985: T59), used to adorn funerary precincts (*períboloi*).

The third and last dog type identified on Classical Attic *sémata* is the Molossian dog. Often confused with the mastiff, the Molossian type must be considered closer to the modern Italian Cane Corso, known also as 'the Catch-and-Hold Dog' because of its powerful bite (Breber 2014). The Molossian dog shows a big, muscular body, a wide neck with evident skin flaps, a short muzzle, a big nose and abundant, pendulous lips. Among the representations considered here, there is evidence that Greeks and Romans cut the ears of these dogs, eventually in order to prevent their loss in fights. Referring to the Molossian dog, in the 1st century AD, Columella uses the Latin expression '*canis villaticus*' (*De re rustica* 7, 12, 2–4), most likely alluding to its guardian attitude, and tells us that the best coat colour for this dog is black, so as to make it invisible to thieves in the darkness. Julius Pollux (*Onomasticum* 5, 39) assigns a mythical origin to the first specimen of this dog, claiming that Hephaestus would have created it with the bronze of Demonesus as a gift for his father Zeus. The presence of Molossian dogs in Classical Attic *stèlai* seems to be unusual: in fact, the only funerary slab showing a Molossian dog, seated at the feet of its young master - the so-called 'Grottaferrata relief' -, has been recently classified as a non-Attic masterpiece, carved in white Paros marble and assigned to the hand of an insular or Ionic sculptor (Ghisellini 2007). Unlike Spitz-dogs and greyhounds, Attic artists apparently

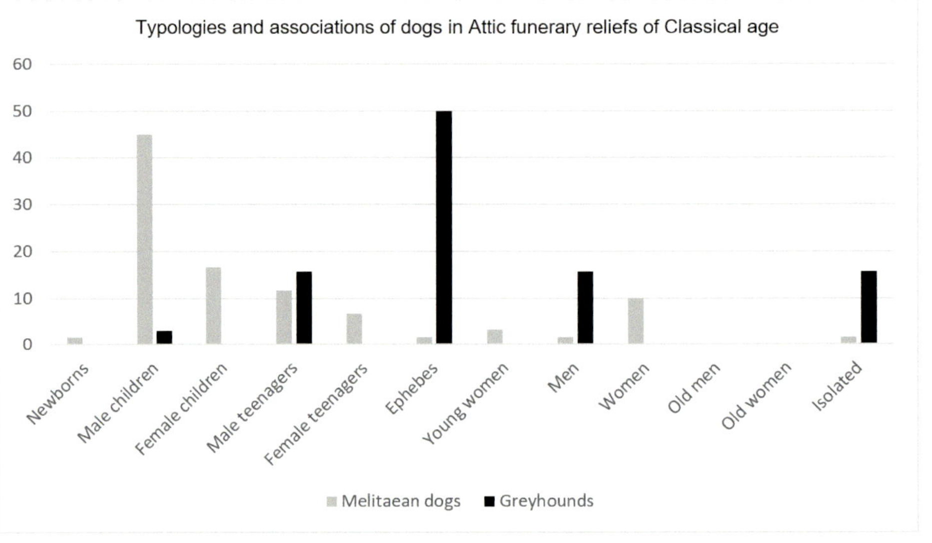


Table 1. Frequencies and associations of Melitaean dogs and greyhounds in Classical Attic funerary reliefs (Table by F. Tanganelli).

Figure 2. So-called 'Ilissos Stele' (NM 869). From the bank of the Ilissos river in Athens, 4th century BC. Athens, National Archaeological Museum , © Hellenic Ministry of Culture and Sports / Hellenic Organisation of Cultural Resources Development, (Photo by F. Tanganelli).

chose to employ images of Molossian dogs only as decorative statues, often displayed in pairs at the sides of rich family *períboloi*, whereas some modern authors occasionally mistook them for fierce lions or panthers (Vermeule 1972: 58). The few and certain funerary Molossian dogs, recorded by different scholars, seem to come principally from the Piraeus necropolis (Figure 3), the Kerameikos excavations and the Acropolis slopes (Tanganelli 2013: 97–99).

### 3 Dog typologies in funerary art: a question of symbolism?

The review of a very large number of Classical Attic *sémata* shows that, despite the wide variability of dog typologies handed down by ancient Greek art, Attic sculptors have adopted only three of these types. Referring to this evidence, a question arises: is there any logic in the use of these canine images in funerary art? What was their meaning?

If we consider the Melitaean dog, we can notice that this lap dog appears mostly in association with children, girls, and women of every age; on the other hand, greyhounds are usually portrayed along with ephebes and adult men (Table 1). This difference seems to become significant if we consider the role and the social positions of these persons: in fact, according to Xenophon (*Oeconomicus* 7, 30), we must imagine women (and obviously children) confined to closed home spaces (òikos), while ephebes and men were free to move and travel in the outdoor world. In this sense, the lap dog appears to be regarded as a tender playfellow for children and faithful companion in domestic life. On the other hand, greyhounds are more properly associated with adult male subjects, capable of going

Figure 3. Funerary statue in the shape of a Molossian dog (NM 3574). From the Piraeus' necropolis, 4th century BC. Athens, National Archaeological Museum, © Hellenic Ministry of Culture and Sports / Hellenic Organisation of Cultural Resources Development, (Photo by F. Tanganelli).

hunting in the woods or simply free to take walks and to attend meetings in public squares (*agoràì*). There is still only one human category that significantly appears with both Melitaean dogs and greyhounds: the male teenage group (12–17 years old). Most likely due to their particular age - which made them no longer children, but still not adults - these subjects were portrayed in some cases with a greyhound by their sides, and in other cases with a small Melitaean dog. The occasional presence of the Melitaean dog with boys showing gym tools (strigil, *arýballos*) also seems to recall the words of Athenaeus of Naucratis (*Deipnosophistae* 12, 16, 6), who records that the inhabitants of the Greek town of Sybaris would bring these lap dogs in the gymnasia.

Quite different is the evidence of the Molossian dog, never used in funerary reliefs, but always represented as decorative statues, often crouched or seated on their hind legs. One of the most famous funerary Molossian dogs adorned the tomb of Lysimachides in the Kerameikos necropolis (Knigge 1991: 126). Lysimachides, son of Lysimachos, from the *dêmos* of Acharnai, had been ruler (árchon) of Athens in 339/338 BC. A connection between the two subjects - the dog and the ruler - seems to lie in a passage of Plato (*Res publica* 2, 375e-376a), where the philosopher compares the rulers of the Athenian state (*archontēs*) to guardian dogs, usually friendly with their citizens and guarded with strangers. Nevertheless, as happens with the slabs of Apollodoros and Lakon, this one supposition cannot be considered as sufficient enough in explaining the meaning of all the other Molossian dog statues recorded by archaeological research. Indeed, it is perhaps more probable to think of these canine representations as funerary guardians, also due to their physical positions, which seem to recall the attitude of the Molossian dog in defending its master's house and properties.

## 4 Conclusions

The composition and nature of dog images recognisable in Classical Attic *sémata* - inclusive of both relief monuments and decorative statues - apparently imply a compelling logic in the use of different dog typologies. In fact, the data emerging from this review seems to indicate that Attic sculptors most likely adopted the Melitaean dog and the greyhound as markers of the social condition of deceased people. In funerary reliefs, Melitaean lap dogs have always been associated with 'domestic subjects', such as children (of both sexes), girls and women, while greyhounds usually appeared at the side of ephebes and men. Male teenagers represent one of the most interesting human categories, since they were represented in association with both these canine typologies, likely due to their transitional age and uncertain status. Much different is the condition of Molossian dogs, never employed in funerary reliefs, but adopted only as decorative statues in rich grave precincts: the preference accorded to these dogs in this particular context could be due to their attitude as loyal guardians, often used by their masters to defend houses and cattle against thieves.

## Acknowledgements

I would like to express my deep gratitude and appreciation to Marco Masseti, for our enjoyable and stimulating discussions about dogs in ancient and modern times.

## References

### *Ancient sources*

Aristotle, *Historia animalium* 612b, 10.
Athenaeus of Naucratis, *Deipnosophistae* 12, 16, 6.
Columellauses, *De re rustica* 7, 12, 2–4.
Julius Pollux, *Onomasticon* 5, 39.
Plato, *Res publica* 2, 375e-376a.
Pliny, *Naturalis historia* 3, 152.
Strabo, *Geographica* 6, 2, 11.
Xenophon, *Oeconomicus* 7, 30.

### *Modern sources*

Breber, P. 2014. *The Catch-and-Hold Dog in Italy (Il Cane da Corso)*. Borgoricco: Websterpress.
Busuttil, J. 1969. The Maltese dog. *Greece & Rome* 16, 2: 205–208.
Clairmont, C.W. 1993. *Classical Attic Tombstones*. Kilchberg: Akanthus.

Dennis-Bryan, K. and J. Clutton-Brock 1988. *Dogs of the last hundred years at the British Museum (Natural History)*. London: The Museum Edition.

Freyer-Schauenburg, B. 1970. KUÔN LAKÔNOS - KUÔN LAKAINA. *Antike Kunst* 13: 95–100.

Ghisellini, E. 2007. La stele funeraria greca del Museo dell'Abbazia di Grottaferrata. *Bollettino d'Arte* 92, 139: 19–58.

Hull, D.B. 1964. *Hounds and hunting in Ancient Greece*. Chicago: The University of Chicago Press.

Keller, O. 1909. *Die Antike Tierwelt*. Leipzig: Verlag von Wilhelm Engelmann.

Knigge, U. 1991. *The Athenian Kerameikos. History - monuments - excavations*. Athens: Krene

Tanganelli, F. 2013. The dog in the art and literature of Classical Greece, with particular reference to Attic funerary monuments. Unpublished dissertation, University of Florence.

Vedder, U. 1985. *Untersuchungen zur plastischen Ausstattung attischer Grabanlagen des 4. Jhs v. Chr.* Frankfurt am Main: Lang Edition.

Vermeule, C. 1972. Greek funerary animals, 450–300 B.C. *American Journal of Archaeology* 76, 1: 49–59.

Woysch-Méautis, D. 1982. *La représentation des animaux et des êtres fabuleux sur les monuments funéraires grecs: de l'époque archaïque à la fin du IVe siècle av. J.-C.* Lausanne: Cahiers d'Archeologie Romande.

Zlotogorska, M. 1997. *Darstellungen von Hunden auf griechischen Grabreliefs. Von der Archaik bis in die römischen Kaiserzeit* (Antiquates, Band 12). Hamburg: Verlag Dr. Kovač.

# 5.5 'Do Not Laugh, I Beg of You, for This Is a Dog's Grave': The Human-Canine Bond in the Ancient Greek World

Liubov Eliseeva[1] and Eugenia Andreeva[2]

[1]Institute of World History, Russian Academy of Sciences. Leninsky Prospekt 32A, 119334, Moscow, Russia; State Academic University for the Humanities, Maronovskiy pereulok, 26, 119049, Moscow, Russia; Lomonosov Moscow State University, Leninskiye gory 1, GSP-1, 119991, Moscow, Russia; liubovgeliseeva@gmail.com

[2]Institute of World History, Russian Academy of Sciences. Leninsky Prospekt 32A, 119334, Moscow, Russia; State Academic University for the Humanities, Maronovskiy pereulok, 26, 119049, Moscow, Russia; aenik@yandex.ru

Corresponding author: Eugenia Andreeva, aenik@yandex.ru

**Abstract**

The paper focuses on Greek epitaphs set up for dogs and attempts to demonstrate the kind of data that can be extracted from such texts. It analyses the texts of canine epitaphs and compares them not only to similar Latin inscriptions and literary evidence, but also to Greek funerary inscriptions set up for humans, as the structure of the canine epitaphs is often similar to the structure of human ones. Topics such as dog names and 'professions' are also touched upon in the paper, and it is concluded that dog owners seem to have named their dogs in a similar manner to how they have named their slaves and have adopted a pattern of expressing their grief over the loss of a pet from the epitaphs commemorating the deaths of dependent members of their households.

**Keywords:** dog burials, epitaphs, epigraphy, Greek and Latin inscriptions, human-canine bond.

## 1 Introduction

In the most general terms, it can be said that the attitude towards dogs in the ancient Greek and Roman culture was ambiguous. On the one hand, dogs were viewed as 'impure' or chtonic creatures[1] and thus were despised. For instance, the findings of dog skeletons in ancient settlements of the Cimmerian Bosporus might be interpreted as a ritual purification, sacrifice for protection or sacrifice to the chthonic deities.[2] In the Homeric epic the word 'dog' is often used as a curse word or an insult, one that could be applied to a person who was greedy, cowardly, treacherous, irritating or vulgar.[3] Insults of the same kind can also be found in epigraphical sources, e.g., in a profane inscription from the Roman Agora in Athens, cut out on the steps on the northeast side of the Tower of the Winds,[4] an unknown person is called μωρὸς κύων. This insult might not be interpreted only literally, i.e. simply 'a stupid dog': according to Hesychius of Alexandria (*s.v.* κύων), the word 'dog' in this context could also be used to signify male genitalia. The epic poems evoke dogs in threats of outrageous treatment of human remains. A dog in Homer is a necrophagic, i.e. corpse-eating animal: a hero sometimes threatens to smite his enemy down and leave his corpse to the beasts (in most cases - the dogs).[5] This trope occurs in inscriptions, too, e.g., the imprecation on the gravestone of T. Puficius Rufus that dates back to the Roman Imperial period threatens anyone who would bury another corpse in the grave, saying that this person then would be doomed to remain unburied and glutted by 'dogs and birds'.[6]

On the other hand, in some sanctuaries dogs were not only tolerated but even venerated as belonging to god and as a source of healing.[7] For instance, the inscriptions (4th century BC) from the sanctuary of Asclepius in Epidaurus mention the names of the patients, the impairments from which they suffered, and the specific treatments applied. A blind child named Hermioneus was treated by one of the dogs from the sanctuary.[8] A child named Aiginatas, who had some kind of a tumour in his throat, was also 'healed' by one of the dogs from the sanctuary: it 'made him healthy using its tongue'.[9] It is noteworthy that dogs helped these two children who presumably had no fear of them, whereas a dumb girl was scared by a serpent from the sacred grove, and ran away calling for her mother and father.[10]

[1] Scholz 1937: 7; Franco 2019: 44.
[2] Zhuravlev *et al.* 2016: 34–37; cf. Franco 2019: 46f.
[3] Franco 2014: 7–8, see also *LSJ s.v.*
[4] Oikonomides 1987: 37–42; *SEG* XXXVII: 203.
[5] Franco 2014: 7–8; 54–74.
[6] *SEG* XLII: 1156, found near the ancient city of Sinope (present-day Turkey). Cf. *Il.* 1.4f.
[7] Franco 2014: 8.
[8] *IG* IV$^2$,1 121.
[9] *IG* IV$^2$,1 122.
[10] *IG* IV$^2$,1 123. So it seems to be the shock that helped her recover from her condition. For a possible explanation of this two-fold attitude towards dogs among the Greeks see Franco 2014: 7–8, 189 with note 6; 2019: 34, 44. However, one should keep in mind that this paradoxical attitude towards dogs is by no means unique to ancient Greek culture.

In our research, however, we will not focus on the positive and negative aspects of the public's perception of dogs, but on the existing evidence of emotional connections between dogs and their owners in an attempt to highlight the personal dimension of the issue.

## 2 Material and methods

In order to analyse the wide range of relationships that existed between humans and dogs in the ancient Greek society we draw on epigraphic evidence, namely funerary inscriptions. For the convenience of further discussion, we can categorise the analysed epitaphs as follows:

### *2.1 Literary evidence*

Some Greek epitaphs composed for animals are known to us from the literary tradition. These epitaphs are all attributed by their sources to specific authors, and the attributions define a chronological arc extending from the 4th century BC to the 1st century AD: the epitaphs of dogs named Lycas,[11] Locris,[12] Tauros,[13] and Lampon.[14]

### *2.2 Epigraphic evidence*

Greek epitaphs that were cut out on stone, dated mostly from the 4th century BC to the 2nd century AD: an epitaph of a 'bewept' dog;[15] an epitaph of a dog on a gravestone made by her owner, Balbos;[16] and epitaphs of dogs named Theia,[17] Tyrannos,[18] Parthenope,[19] Philokynegos,[20] and Stephanos, the latter on the dog's sarcophagus.[21] Some stones were lost, and the inscriptions are known to us only in copies.

### *2.3 Mentions of pets in human epitaphs*

The third group consists of Greek epitaphs - both from the literary tradition and found on actual gravestones or in copies - of humans that mention their pets: an epitaph of a soldier Hippaemo mentioning his dog Lethargos;[22] an epitaph of a girl named Bassa mentioning her dog Ounion;[23] an epitaph of an athlete Antonian mentioning his dog Paregoris;[24] and an epitaph of a gladiator Autolycos mentioning his dog Epiodis.[25] Usually it is only the name of the dog that is mentioned in this type of inscriptions. It does not necessarily mean that the dog was buried with its master, the pet could have had a tomb of its own, however it seems to have been important to refer to the dog in the epitaph or to depict it on the funerary relief.

### *2.4 Latin sources*

The Latin-speaking world has provided a considerable number of dog epitaphs (e.g., epitaphs of dogs named Margarita,[26] Patrice,[27] Myia,[28] Helena,[29] Aeolis[30] and an epitaph of a dog who guarded chariots),[31] and these texts will be used as comparative material.

The parlance of the epitaphs - regardless of whether they are human or canine - should be considered to be a specific type of language. One of the most significant features of the epigraphic texts is the use of *formulae*, i.e., well-established expressions used in such texts. The key method that we applied in our research is the analysis of the formulaic language of the inscriptions in order to further define the nature of the human-canine bond. To put this in perspective we also draw on human epitaphs as well as the evidence from literary sources.

## 3 Results and discussion

### *3.1 'Professions' of dogs reflected in their epitaphs*

It seems that in Classical culture dogs were primarily classified by their functions and people of certain social standing were supposed to own certain types of dogs: a herdsman was expected to have dogs to protect his flocks, or an aristocrat was expected to have hunting dogs or even lapdogs whose only 'job' was to catch one's fancy.[32] Commemoration of a dog's virtues expressed in the epitaphs reflects this classification: lapdogs are praised for their affection, good behaviour and playfulness; owners of guard dogs or hunting dogs are grateful for the fulfilment of their duties. For instance, owners praised 'huntress Lycas', 'hunter Lampon', 'a servant and a companion in many sea travels' (epitaph set up by Balbos); 'guardian of the chariots' (in Latin),

[11] Poll. *Onom.* V, 48; see also Garulli 2014: 30–31, No. 1.
[12] Poll. *Onom.* V, 48; see also *GVI* 1463; Garulli 2014: 31, No. 2.
[13] *Anth. Pal.* VII, 211; see also Garulli 2014: 31, No. 3.
[14] *Anth. Pal.* IX, 417; see also Garulli 2014: 32, No. 4.
[15] *IG* XIV, 2128; see also *Epigr. Gr.* 627; *GVI* 1365; Geist 1969: 150, No. 398; Garulli 2014: 33–34, No. 7.
[16] *IG* XII,2 458; *Epigr. Gr.* 329; see also *GVI* 309; also *SEG* XL: 5199; XLVIII: 2103 and Garulli 2014: 34, No. 8.
[17] *IG* XIV 1647; see *Epigr. Gr.* 626; IGUR III 1230; *SEG* XL: 1599, XLIV: 1692; Garulli 2014: 33, No. 6.
[18] Dobias-Lalou, Gwaider 1997: 28–29; see *SEG* XLVII: 2176; Chamoux 2001: 1310; Garulli 2014: 32–33, No. 5.
[19] *IG* XII,2 459; see *GVI* 691; *SEG* XLIV: 1692; Garulli 2014: 34–35, No. 9.
[20] *CIG* 2, 3559; see *Epigr. Gr.* 332; *GVI* 1032; Garulli 2014: 35, No. 10.
[21] See *SEG* XL: 1599; XLI: 1283; Garulli 2014: 35–36, No. 11.
[22] *Anth. Pal.* VII, 304; see Chamoux 1990: 118; Robert, Robert 1976: 207, with note 215.
[23] *IK Ephesos* 2231; see *BÉ* 1979: 16; Chamoux 1990: 117.
[24] See Chamoux 1990: 116; *SEG* XX: 752.
[25] *IG* XII,2 644; but see mainly Robert 1940: 223–225, No. 285; also *AE* 1973: 521; *BÉ* 1974: 459; Bean 1973: 408–409, No. 43; Carter 1999: 327–328, No. 150.
[26] *CIL* VI 4, 29896; *CLE* 1175; see Geist 1969: 151, No. 400; Booms 2016: 92–93, No. 23.
[27] *CLE* 1176; see Geist 1969: 151–152, No. 401.
[28] *CLE* 1512; see Geist 1969: 152, No. 402.
[29] *CIL* VI, 3 19190; see Booms 2016: No. 24; Erpetti 2017.
[30] See *AE* 1994: 348.
[31] *CIL* IX 5785 (1); *CLE* 1174; see Geist 1969: 150, No. 399.
[32] Brewer, Clark, Phillips 2001: 64.

Figure. 1. Grave stele of Helena (© The J. Paul Getty Museum)

and Margarita in a Latin epitaph was 'trained to run boldly through strange forests and to hunt out furry wild beasts in the hills'.[33] In the case of lapdogs, we can notice multiple descriptions of a dog's good heart and kind nature.[34] A little dog Theia was 'joy' to her owner; with another dog, Parthenope, her owner used to play; Stephanos was 'pleasant' to his owners. Latin epitaphs of lapdogs Patrice and Myia demonstrate the same pattern. Sometimes we can deduce a dog's functions from the context: the gravestone of a soldier Hippaemo mentions his dog and it is reasonable to say this is the case of a 'chien de guerre'.[35] There are some traces of dogs' professional duties even in their names, e.g., Philokynegos translates to 'He who loves the hunt'.

The rare case when it is possible to talk about an indication of a specific breed, not 'profession', is possibly the epitaph of Tauros, which declares that this dog derives from the Melitian field, and that breed is described by both Strabo (*Geogr.* 6.2.11) and Pliny (*NH* 3.152) as a small dog (κυνίδιον; *catulus*). Some images could also be indicative of this breed's appearance: the dog Helena, as can be judged by the relief on her tombstone, was of the same breed - a small fluffy dog (Figure 1). However, in the epitaph Tauros (i.e., 'Bull'), is described as 'the most loyal guard of Eumelos' and his voice is denoted through the word φθέγμα - the same word is used to describe the bellow of a bull. So in this case a lap dog is described as 'the guard' ironically.[36] We know of another dog named Taurōn - a hunting dog of Zenon, 'Taurōn the Indian' who hunted boars,[37] and in this case the name is absolutely justified.

### *3.2 Dog names*

We have already noted that dogs' names often represented their duties. However, according to ancient authors, the main purpose of a dog's name was to be catchy and expressive. Xenophon suggests a range of names for dogs - from his point of view, the owners should use a short name in order to be able to call dogs easily. These names speak for themselves: Psyche - 'Soul', Phylax - 'Guard', Phonax - 'Killer', Chara - 'Joy', etc. (*Cyn.* 7.5). Columella agrees with Xenophon and adds some Greek names to the list (e.g., Skylax -'Pup'; *cf.* Petron. 64), and also gives some specific examples for naming female dogs: Spoude - 'Speed', Alke - 'Courage'; and some Latin names: Ferox ('Fierce'), Lupa ('She-wolf'), Tigris ('Tiger') (*Rust.* 7.12, 13).

Some dog names found in the epitaphs listed above seem to be similar to names like 'Killer' or 'Guard': the dog's name Epiodis can be interpreted as 'Roadster' (and it is comparatively short). However, most of these dogs had more complicated names than ancient authors advised; these names reflect their 'duties' or personal traits: Philokynegos - 'he who loves the hunt', Paregoris - 'someone soothing'. Simple translation of the name of the 'war dog' Lethargos - 'lethargic' - seems to be a little odd, but it could be an indication of his calm nature; however, researchers have suggested that his name was actually Laithargos - 'biting secretly'.[38] Two names, Ounion and Margarita, have the same

[33] The British Museum, Collection online, viewed 10 April 2020 <https://research.britishmuseum.org/research/collection_online/collection_object_details.aspx?objectId=459488&page=1&partId=1&subject=16699>.

[34] However, the virtue of gentleness is not only reserved for lapdogs: the hunting dog Margarita liked to 'lie on the soft lap of her master and mistress' in the same manner as the lapdog Patrice. This is reflected by the literary tradition: Xenophon advises treating dogs with kindness rather than force - they should be coaxed, encouraged, called by name (*Cyn.* 6.10, 17.22, 17.25). Arrian says that dogs become fond of humans who show affection to them and beside whom they sleep no less than of humans who give them food (*Cyn.* 9.1, 10).

[35] Robert, Robert 1976: 207, with note 215.

[36] E.L. Hicks even thinks it is 'a clumsy forgery', not even entertaining the idea that this name could be given to a small dog as a joke (Hicks 1882: 130, with note 1).

[37] *P. Cairo Zen.* 4 59532, 256–246 BC; Lloyd-Jones, Parsons 1983: 489–490, No. 977; Pepper 2007: 605–622.

[38] Masson 1962: 139; Zajcev 1996: 140; Ivantchik 2005: 121.

Figure. 2. Grave stele of Tyrannos, tracing (After Chamoux 2001: 1310).

meaning - 'pearl',[39] - and seem to point to the colour and/or the general beautiful appearance of the dogs. Some names seem to have a jocose undertone. To the already discussed name Tauros ('Bull'; see note 36) for a small dog, the name Tyrannos can also be added (Figure 2): it was probably some sort of a pun on the contrast between the 'great toil' endured by the dog (πολλὰ πονήσας) while serving his master and the meaning of the name - 'Ruler', 'Prince'. On the other hand, this name could also reflect this dog's dominant position within the pack.[40] The name Myia - 'Fly' - in a Latin epitaph could probably sound humorous and endearing at the same time.

This does not mean, however, that dogs could not have had human names. Among the names in the listed epitaphs there are quite ordinary human personal names like Stephanos (and his human name corresponds to the statement that he 'was buried like a human'), Helena, and Margarita. Also, such 'self-explanatory' names as Philokynegos, Tauros and Tyrannos have also been attested as human names (see: *LGPN* I 429; III.A 437; IV 347). Some names, even though unattested directly among humans, seem to be very close to known human names: Lycas (cf. Lycos, *LGPN* I 290–291, or Lycās, *LGPN* III.B 262), Theia (cf. Theios, *LGPN* I 211).

What could be said about people with these kind of names? It seems that the suggestion that these names belonged to people of dependent positions is quite obvious. However, before the comparison of slaves' and dogs' names takes place, it should be emphasised that this analysis requires a lot of preparation. In the most general sense, we can assume that some dog and human names have the same structure: the name contains a hint at the most significant and distinctive feature of a person or a dog. We can specify the following features of a personal name that is characteristic of dependent social status: the name could point to the person's duties, (e.g., Phylax - 'Guard' - is both a dog and a slave name; one Philokynegos from the 2nd century AD was a gladiator, a net fighter, which almost certainly means that he was a slave or a freedman);[41] it could describe some positive characteristics required of the person or some abstract good qualities (cf. slave name Agathos and dog's name Theia); it could also derive from myths, historical narrations and religious practices (e.g., Parthenope and Helen are both dog and slave names).[42]

In this context it is noteworthy that the dog names from the above-mentioned Latin epitaphs are of Greek origin (cf. dog named Lydia in Martial's epigram 2.69). This could have two, not necessarily mutually exclusive, reasons: firstly, the owners could have tried to pick a reasonably unusual and beautiful sounding name (cf. Ounion in a Greek epitaph); and secondly, the people with Greek names in Italy at the time were usually slaves or freedmen.

### *3.3 The structure of dogs' epitaphs as a reflection of the human-canine relationship*

By analysing dogs' epitaphs we can notice that all the epitaphs have a similar structure. A dog's life is described in terms of its qualities and its actions, and it is sometimes stated that the master had erected the tomb as a reward for the dog's noble qualities and glorious deeds. This structure is shown in its purest form in the epitaph of Parthenope: she has got a reward - a tomb - for her love and loyalty that were expressed in her deeds.

[39] Here we see an interesting reversal: the Latin epitaph has the name Margarita, Greek in origin (μαργαρίτης), while the Greek one has Ounion, Latin in origin (*unio*).

[40] Chamoux 2001: 1308–1311. Fr. Chamoux also suggested that this name could point to the playful nature of the dog.

[41] *SEG* XXXIX: 1340.

[42] Masson 1973.

A common feature of the epitaphs of humans and dogs is that sometimes poets were recruited in order to compose an epitaph. Of course, not all funerary inscriptions are in verse, but some are, which seems to indicate a higher status of the family. Moreover, the names of some poets are attributed with more or less accuracy (e.g., Simmias of Rhodes, Anyte of Tegea, Antipater of Thessalonica, Peisander of Camirus). Sometimes the customer was not satisfied by one poem and had the poet compose two or more for one epitaph: in the case of Taurōn the Indian, two epitaphs were made for him (see note 37). In the texts of some epitaphs it is also possible to notice references to literary traditions, which in some cases also 'humanise' the pet. The deceased dog Tauros is described as ἀργός, and this word is used by Homer in the same sense to describe 'swift' dogs (*Il.* 1.50, 18.283, etc). It seems that authors of Latin epitaphs found inspiration in poetic language, e.g., in the poems by Catullus, Vergil and Ovid. It is impossible not to compare the phrase in Patrice's epitaph - 'no more will you give a thousand kisses' (*non dabis oscula mille*) - to a verse from poem Catull. 5 'give me a thousand kisses' (*da mihi basia mille*). One more reference to Catullus is contained in another Latin epitaph - this time of Myia - where a direct quote from Catull can be found. 3: *o factum male* - 'oh evil deed!'.[43]

The authors of dogs' epitaphs also used standard expressions of grief, e.g., by mentioning that dogs are bewept, or expressing hopes for some kind of posthumous prosperity (for instance, Balbos wished that the ground be weightless for his dog - and it is quite a frequent expression).[44] The epitaphs of Tyrannos, Philokynegos, Stephanos, Margarita, as well as the one of the 'bewept' dog demonstrate another trope characteristic of human epitaphs: in some cases the text of the inscription is composed in the first person, i.e., it is as if the buried addresses the reader.[45]

In a funerary inscription a dog may be described as young, which might also be a rhetoric trope, e.g., Balbos prays that the ground be weightless for 'the young pup who in now underground' (τῇ κατὰ γῆς σκύλακι); Stephanos is described as dearly departed. Probably, it should not be literally perceived (however, they could have indeed died young) as the young age makes the description of a dog's notable qualities and deeds even more extraordinary. We can classify the references to a dog's 'puppyhood' as a similar rhetoric trope: Locris was the fastest among other puppies; Philokynegos loved hunt from the very beginning of his life. This trope has the same aim - to demonstrate the dogs' good qualities as being even more outstanding. Even though some epitaphs could sound ironic (e.g., those of Tauros and Tyrannos), it does not belittle the owner's sadness and grief for the deceased dog.

All the aforesaid, however, do not mean that the line between human and dog tombs was being blurred in any way. On the contrary, authors of the epitaphs were fully aware of this borderline. One of them wrote 'laugh not, I pray, though it is a dog's grave';[46] the owner of Stephanos, a girl named Rodopa, buried her dog 'as a human' (i.e., the author of the text realised that there was a distinction to be breached here); and Balbos says that in a way for the earth (γῆ) there was no difference between a canine and a human grave (again, implying that there was one in the eyes of people). Nevertheless, it is interesting that only in one case is this distinction accentuated in a somewhat negative way ('laugh not, I pray, though it is a dog's grave');[47] other listed epitaphs have direct indications that it is a dog's grave, but it is not presented as something strange or amusing.

### *3.4 Epigraphic formulae and social hierarchies*

Let us compare actual *formulae* of dog epitaphs to human ones. As a rule, Greek epitaphs listed the name of the father of the deceased in the Genitive case. In the case of men there are usually no words that explicitly express this parental connection, which is supposed to have been obvious. In the epitaphs of women, however, such words are much more common, as the male name in the Genitive case could be not only her father's but her husband's, brother's or son's as well, so, if the word 'father', 'husband' (or 'daughter', 'wife'), etc. was not inserted into the text, it might have sounded ambiguous. Such a difference is most likely to point to the dependent status of women in ancient Greek society, and some analogies to this formulaic expression of dependence in texts pertaining to dogs can be found. Probably the most famous dog of Ancient Greece, Argos, the dog of Odysseus, is described with the master's name in the Genitive case so it determines the animal's identity like a high-sounding patronymic:[48] he is 'Argos of long-suffering Odysseus' (*Od.* 17.292). We can notice a similar use of the Genitive case in the epitaph of Lampon, who is 'the dog of Midos'. The difference here is that this Genitive does not connect the two names directly.

The dog owner's name in our texts may be classified by the noun ἄναξ ('lord') or δεσπότης ('master') or their synonyms, which usually describe a master, a lord of a household (οἶκος), especially when a dog had a male owner: the tombstone of the 'bewept' dog was

[43] Some researchers suggest, however, that this epitaph is a charming parody (cf. Walters 1976 with bibliography; it is also interesting that its lines are used to restore the corrupt line in the *Veronensis MS* of Catull. 3).

[44] Cf., e.g., *IG* XII,1 151 from Rhodes; *GVI* 475; *MAMA* X 63; etc.

[45] Cf. *CIRB* 610; *IG* I$^3$ 1503; *IG* IV$^2$,1 735; etc.

[46] The epitaph of the 'bewept' dog; Garulli 2014: 33–34, No. 7.

[47] Cf. Arr. *Cyn.*: 'I don't think I should feel ashamed to immortalise the name of this dog' (see Franco 2014: 122–123; 2019: 48).

[48] Franco 2014: 37, see also Frisch 2017: 7–18 with bibliography.

erected 'by the hands of its master' (χεῖρες... ἄνακτος); Tyrannos has served his master (addressed as 'δέσποτα') well; in the epitaph of Parthenope her master is also her τροφεύς - 'a foster father'. Generally speaking, the word choice 'lord/master' to signify the dog's owner is not a coincidence. It may be noted here that the regular term in Homer for a dog's master is ἄναξ (e.g, *Il.* 23.173). It is used to signify 'lord' as head of the people and the army but also in the sense 'the lord of the household' (*Od.* 1.397, cf. *LSJ s.v.*). A variant δεσπότης also has the meaning 'the master of the household'. This finds a vivid visual analogy in reliefs on funerary steles. Depictions of a horse, a dog and a servant are in fact the elements used to identify a person of free status, a master of an οἶκος.[49] To own a dog was perceived as a 'badge' of distinction and dogs were in fact a kind of status symbol. There is an epitaph of a soldier that represents this pattern perfectly: 'The name of the man was Hippaemo, of the horse - Podargos, of the dog - Lethargos (or Laithargos), of the servant-squire - Babes' (note the order in which they are listed). Other epitaphs that contain only a name of a dog (an epitaph of the athlete Antonian mentioning his dog Paregoris; an epitaph of the gladiator Autolycos mentioning his dog Epiodis) can be included into the same group. The same idea can be extrapolated on the epitaphs of Midos's dog Lampon and of Tauros, 'guard of Eumeles'. The same is apparently true for female tombstones: an epitaph of a girl named Bass mentions the name of her dog - Ounion.[50] We can find direct reflections of this depiction practice in the literary tradition: one of the characters of Satyricon by Petronius, Trimalchio, specifically ordered in his will to depict his dog at the foot of his funerary statue (Petron. 71), which could be a pun (one of many) on his vanity and desire to indicate his elevated social status.

If the words ἄναξ or δεσπότης or their synonyms are not explicitly used, in some dog epitaphs there are other *formulae* that point to this 'master-slave' hierarchy and express the same idea in reverse, i.e., the dog is identified as a servant or slave. The dog that was buried by Balbos is described as δουλίς - 'a female slave'. It could be said that these words - 'master' and 'slave' - are almost interchangeable. Presumably, it is possible to compare this epitaph to the epitaph of Helena, where she is called *alumna*. This word might mean that Helena could have been perceived by her owner either as 'a foster child' or as 'a slave brought up in the house' (in the light of the epitaph set up by Balbos the latter seems to be more likely). In the epitaph of the dog named Theia erected by her female owner there is an indication of the same specific dog-owner relationship - a dog is a τροφίμη ('foster daughter').[51] There are a considerable number of Greek epitaphs mentioning 'foster-children' (a common term is θρεπτός, however, other words deriving from the same root, such as τρόφιμος/θρεπτός are also used), either commemorating a deceased 'foundling' or erected by them for their 'parents', and in the majority of cases it seems that these 'adopted' children did not enjoy the same social status as their 'parents', but were dependent members of the household.[52] It should be noted however, that τροφεύς in the context of dog epitaphs could be interpreted in a more down-to-earth sense as a 'breeder'.

The names of the dogs' owners could be mentioned in the part of the text indicating the person who set up the funeral monument. This part of the inscription *formula* is also quite common in human epitaphs. A relative or a community quite often erected tombstones *in memoriam* of the deceased, and such epitaphs included the names of the sponsors or creators as well. For instance, the abovementioned epitaph of the net fighter Philokynegos has a specific formula very common in Greek epitaphs, which indicates that the tombstone was erected by his wife 'at her own expense in order to honour his memory' (ἐκ τῶν ἰδίων αὐτοῦ μνήμης χάριν).[53] Very similarly, the owner of Theia erected a gravestone for her in order to have 'an accurate memory' (μνῆστιν ἔχουσα ἀτρεκῆ) about her. The owner of the 'bewept' dog from Florence made the epitaph and put the remains in the grave himself. In the same way Balbos buried his dog in the earth of Lesbos. The nature of the relationship between Stephanos and his mistress Rhodope is not expressed in any technical terms, however, it is specifically indicated that it was Rhodope who ordered the construction of the tomb. All these mentions of a master may be considered not only as a specific formula for dog epitaphs adopted from human ones, but also as another indication of the hierarchical relationships between dogs and their masters: the very fact that one erects a tombstone for a dog means that they were its master.

According to Cr. Franco, dog was rarely depicted as a slave: it was rather assimilated to a member of a family - 'a wife, a daughter or a servant born and raised in the house'[54] - however, all of these social positions can be described as 'dependent' or 'lower-ranked' in Greek social hierarchy. Quite fittingly dog owners grieving the loss of their pets seem to have adopted a pattern of expressing that grief from the epitaphs commemorating the deaths of dependent members of households - slaves, freedmen, foster children. In

[49] Franco 2014: 73.
[50] For more information about dog images on women's tombstones see Franco 2014: 83–85.
[51] It is noteworthy, however, that Xenophon uses this word - τροφίμη- to talk of a dog sleeping inside the house (*Hell.* 5.3.9).
[52] See Ricl 2009 for further references and information.
[53] *SEG* XXXIX: 1340 (2nd cent. AD).
[54] Franco 2019: 38f.

addition, they named their pets in a similar manner to how they named their slaves (see 3.2 above).

## 4 Conclusions

Having studied the epitaphs listed in the beginning we can draw the following conclusions. The structure of the dogs' epitaphs can be compared to the structure of human ones. The most important part of a dog's epitaph is the reference to the owner who raised and buried it. In the case of male owners, we can see this indication not as a mere reflection of affection towards the pet, but as a status symbol, as an affirmation of his sovereignty over all members of the household, including dogs. It is quite possible that the owner could have used the epitaph praising his dog's outstanding deeds to reaffirm his own achievements both as a capable dog-trainer and as a master of a successful and thriving household. Indeed, the very fact that the owner had the means to erect a costly stone monument, or even pay a poet to compose an epigram for a dog is by itself an indication of an above-average wealth.[55] This, however, does not deprecate the emotional bond between the dog and the owner, which is plainly expressed in our source material.

As for the dogs' names, they do not appear to be strictly short and functional, as Xenophon and his followers would have liked it, but rather express the attitude of the owner towards the dog or the qualities and characteristics that the owner saw (or would have liked to see) in his dog.

The dogs' names and images could have been put on the graves of the owners, men and women, as if they symbolically accompanied their masters even to the underworld. It is worth adding that sometimes dogs are depicted on funerary reliefs on gravestones. Possibly, the aim of a dog's image was to render the image of the deceased more complete. The positive value of a dog for the deceased goes even further if we assume that it is guarding the tomb: for instance, the epitaph of the chariot guard dog plainly states that his 'spirit now guards his ashes' (*cineres vindicat umbra suos*).

While dogs are not the only pets whose memory might be honoured with a tomb and/or an inscribed gravestone, proportionally it is the most commemorated animal by far. This commemoration, unlike the Egyptian tradition of mummification of religiously venerated animals, such as crocodiles, cats, or ibises, which continued on through the Hellenistic and Roman times, seems to derive not from the general ubiquitous status of a dog as a 'sacred' or contrarily 'impure' animal, but from singular personal stories of human-dog emotional bonds.

## Acknowledgements

We would like to thank Dr. Elena V. Lyapustina, Dr. Alexey V. Belousov, and Prof. Dr. Askold I. Ivantchik for their help and guidance in some aspects of this work. The research of L. Eliseeva is supported by the Russian Foundation for Basic Research, project no. 20-09-00386.

## References

### *Ancient sources*

Abbreviations after OCD (https://oxfordre.com/classics/page/3993)
*Anth. Pal.* - *Anthologia Palatina* (*The Palatine Anthology*).
Arr. *Cyn.* - Arrian, *Cynegeticos* (*Hunting with Dogs*).
Catull. - Catullus, *Carmina.*
*Char.* - Theophrastus, *Characteres.*
*Geogr.* - Strabo, *Geographica* (*Geography*)
*Il.* - Homer, *Iliad.*
*NH* - G. Plinius Secundus, *Naturalis Historia* (Pliny the Elder, *Natural History*).
*Od.* - Homer, *Odyssey.*
Petron. - Petronius, *Satyricon.*
Poll. *Onom.* - Pollux, *Onomasticon.*
*Rust.* - Columella, *De re rustica* (*On Rural Affairs*)
Xen. *Cyn.* - Xenophon, *Cynegeticos* (*Hunting with Dogs*).
Xen. *Hell.* - Xenophon, *Hellenica* (*Greek History*).

### *Abbreviations*

AE - *L'Année épigraphique.* Paris.
BÉ - *Bulletin épigraphique.* (Suppl. Revue des Études Grecque). Paris.
CIL - *Corpus Inscriptionum Latinarum.* Berolini.
CIRB - Struve, V. *et al.* (eds) *Corpus inscriptiorum regni Bosporani.* Moscow-Leningrad, 1965.
CLE - Riese, A., F. Bücheler and E. Lommatzsch (eds). *Carmina Latina Epigraphica.* Vol. II. *Anthologia Latina sive Poesis Latinae supplementum.* Lipsiae, 1868–1926.
*Epigr. Gr.* - Kaibel, G. *Epigrammata graeca ex lapidibus coniecta.* Berolini, 1878.
GVI - Peek, W. *Griechische Vers-Inschriften. I. Grab-Epigramme.* Berlin, 1955.
IG - *Inscriptiones Graecae.* Berolini.
IGUR - Moretti, L. *Inscriptiones graecae urbis Romae.* Vol. I–IV. Rome 1968–1990.
IK Ephesos - *Die Inschriften von Ephesos.* Vol. I-VIII. Bonn, 1979–1984.
LSJ - Liddell, H.G., R. Scott, and H.St. Jones (eds), *Ancient Greek Lexicon With a Revised Supplement.* Oxford, 1996
LGPN - *A Lexicon of Greek Personal Names.* Oxford.
MAMA X- Levick, B., St. Mitchell, J. Potter, and M. Waelkens (eds), *Monumenta Asiae Minoris Antiqua.*

[55] Theophrastus (*Char.* 21.9), for instance, considers setting up an epitaph for one's dog as one of the sure signs of petty ambition.

Vol. X. *Monuments from the Upper Tembris Valley, Cotiaeum, Cadi, Synaus, Ancyra, and Tiberiopolis recorded by C.W.M. Cox, A. Cameron, and J. Cullen.* London, 1993.

P. Cair. Zen. IV - Edgar, C.C. (ed.) *Zenon Papyri, Catalogue général des antiquités égyptiennes du Musée du Caire.* IV. (Cat. 90). Nos. 59532—59800. Cairo, 1931.

SEG - *Supplementum Epigraphicum Graecum.* Leiden-Boston.

### Modern sources

Bean, G.E. 1973. 'New Inscriptions', in J.M. Cook (ed.), *The Troad: An Archaeological and Topographical Study*: 395–418. Oxford: Study Clarendon Press.

Booms, D. 2016. *Latin Inscriptions. Ancient Scripts.* Los Angeles: J. Paul Getty Museum.

Brewer, D.J., T. Clark and A. Phillips 2001. *Dogs in Antiquity. Anubis to Cerberus: The Origins of the Domestic Dog.* Warminster: Aris & Phillips.

Carter, M.J.D. 1999. *The Presentation of Gladiatorial Spectacles in the Greek East: Roman Culture and Greek Identity.* Unpublished PhD diss., McMaster University. Viewed 4 April 2020 <http://hdl.handle.net/11375/6583> .

Chamoux, Fr. 1990. 'Une stèle funéraire de Cyrène'. *Bulletin de la Société nationale des Antiquaires de France*, 1988: 113–120.

Chamoux, Fr. 2001. 'Chiens cyrénéens'. *Comptes rendus des séances de l'Académie des Inscriptions et Belles-Lettres*, 3: 1307–1313.

Dobias-Lalou, C. and R.A. Gwaider 1997. From the cemeteries of Cyrene. *Libya Antiqua. Nuova Serie* 3: 25–30.

Garulli, V. 2014. Gli epitafi greci per animali. Fra tradizione epigrafica e letteraria, in A. Pistellato (ed.), *Memoria poetica e poesia della memoria. La versificazione epigrafica dall'antichità all'umanesimo*: 27–64. Venezia: Edizioni Ca' Foscari.

Geist, H. 1969. *Römische Grabinschriften.* München: Heimeran.

Franco, Cr. 2014. *Shameless: The Canine and the Feminine in Ancient Greece.* Berkeley-Los Angeles: University of California Press.

Franco, Cr. 2019. 'Dogs and Humans in Ancient Greece and Rome: Towards a Definition of Extended Appropriate Interaction', in J. Sorenson and A. Matsuoka (eds) *Dog's Best Friend? Rethinking Canid-Human Relations*: 33–58. Montreal: McGill-Queen's University Press.

Frisch, M. 2017. 'ἦ μάλα θαῦμα κύων ὅδε κεῖτ' ἐνὶ κόπρῳ: The Anagnorisis of Odysseus and His Dog Argos (Hom. Od. 17, 290–327)'. *Literatūra*, 59/3: 7–18.

Hicks, E.L. 1882. 'On the *Characters* of Theophrastus'. *Journal of Hellenic Studies,* 3: 128–143.

Erpetti, M. 2017. 'Dirk Booms, Latin Inscriptions. Ancient Scripts. Los Angeles: J. Paul Getty Museum'. *Bryn Mawr Classical Review* 2017.02.14. Viewed 1 April 2020 <https://bmcr.brynmawr.edu/2017/2017.02.14>.

Ivantchik, A.I. 2005. 'L'épigramme VII, 304 de l'Anthologie Palatine', in A. Kolde, A. Lukinovich, and A.-L. Rey (eds) *Κορυφαίῳ ἀνδρί. Mélanges offerts à André Hurst*: 118–124. Genève: Droz.

Masson, O. 1962. *Les fragments du poète Hipponax.* Paris: Klincksieck.

Masson, O. 1973. Les noms des esclaves dans la Grèce antique. *Actes du Groupe de Recherches sur l'Esclavage depuis l'Antiquité* 2: 9–23.

Lloyd-Jones, H. and P.J. Parsons 1983. *Supplementum Hellenisticum.* Berlin and New York: De Gruyter.

Oikonomides, Al.N. 1987. 'Ancient Inscriptions Recording the Restoration of Greco-Roman Shrines by the Emperor Flavius Claudius Julianus (361–363 AD)'. *The Ancient World: a Scholarly Journal for the Study of Antiquity*, XV/1–2: 37–42.

Pepper, T.W. 2007. 'A Patron and a Companion: Two Animal Epitaphs for Zenon of Caunos', in Tr. Gagos (ed.) *Proceedings of the 25th International Congress of Papyrology*: 605–622. Ann Arbor: Scholarly Publishing Office, The University of Michigan Library.

Ricl, M. 2009. 'Legal and Social Status of *threptoi* and Related Categories in Narrative and Documentary Sources', in H.M. Cotton, R.G. Hoyland, J.J. Price and D.J. Wasserstein (eds) *From Hellenism to Islam. Cultural and Linguistic Change in the Roman Near East*: 93–114. Cambridge University Press.

Robert, L. and J. Robert 1976. 'Une inscription grecque de Téos en Ionie. L'union de Téos et de Kyrbissos'. *Journal des savants*, 3–4: 153–235.

Robert, L. 1940. *Les gladiateurs dans l'Orient grec.* Paris: E. Champion.

Scholz, H. 1937. *Der Hund in der griechisch-römischen Magie und Religion.* Berlin: Triltsch & Huther.

Walters, K.R. 1976. Catullan Echoes in the Second Century A.D.: CEL 1512. *The Classical World* 69/6: 353–359.

Zajcev, A.I. 1996. 'L'épigramme Anth. Pal. VII 304'. *Hyperboreus,* 2/1: 139–150. Зайцев, А.И. 1996. Эпиграмма «Палатинской антологии» VII 304. *Hyperboreus*, 2/1: 139–150.

Zhuravlev, D.V., Sablin, M.V. and A.A. Strokov 2016. 'Dog burials on the Golubitskaya 2 settlement', in D. Zhuravlevand and U. Schlotzhauer (eds) *Asian Bosporus and Kuban Region in Pre-Roman Time. Materials of the International Round Table, June 7-8 2016.* Moscow: Sam Poligrafist. Журавлев, Д.В., М.В. Саблин и А.А. Строков 2016. Захоронения собак напоселении Голубицкая 2. В: Д. Журавлев и У. Шлотцауер (ред.), Азиатский Боспор и Прикубанье в доримское время. Материалы Международного круглого стола *7-8 июня 2016 г.* 34–37. Москва: Сам Полиграфист.

# 5.6 The Image of the Dog on Ancient Coins in the Mediterranean Area

Alessandra Bottari

Università degli Studi di Messina. Polo Annunziata 98168 Messina, Italy.
alessandrabottari@yahoo.it

**Abstract**

In ancient times, the dog image is very widespread in the coin typologies of the Mediterranean area and shows very varied iconographic patterns. From the fifth century BC, it appears in Segesta as a principal form coupled with the personification of the city, where it is the protagonist of the legend of the town's foundation. It is subsequently adopted by the Elymian towns in western Sicily to then spread to the eastern part of the island in the Hellenistic age. In the remaining Greek and Roman worlds, the animal is more often the companion of gods and mythical or historical figures, especially of the goddess Artemis as both goddess of the hunt and of the moon.

**Keywords**: dog, iconography, symbolism, coins.

## 1 Introduction

The study of coin documents enables the reconstruction of the meaning of the dog image, exceeding the mere depiction of the faunistic reality. In some cases explicit references to symbolic meanings can be recognised, which partly derive from the behavioural nature of the animal, but above all from the mythical-ritual traditions of the peoples who minted the coins. The image of the dog is characterised by a great variety of representations: the money *typoi* differ first of all based on the position the animal has within the flan, but also based on its stance. It can therefore represent the main isolated half-length (head, protome) or full-length figure in several positions (standing, standing while looking back, sniffing, crouched, seated, sleeping, moving, following a smell, raising a forepaw, jumping or running). It can be part of the main type as a companion of the gods (Artemis/Diana; Asklepios; Dionysos; Dioscuri; Lares; or Victoria) or of a mythical or historical figure (Androclus; archer; hunter; horseman; Krimisos; Kydon; Iolaus; or Ulixes). The dog may also appear with another animal (deer; fawn; dolphin; hare; sea-moth; or tuna). It can also represent a secondary element as a symbol (head or full-length, standing, raising a forepaw or running whilst looking back).

The great variety of representations and meanings, which characterise the dog, cannot be completely addressed here. Nonetheless, an overarching theme can be retrieved, which links the most widespread typologies and raises the most important values of this animal.

## 2 Main isolated dog type

### *2.1 Origin and spread in western Sicily*

The role of the main isolated type firstly reveals an important link between the dog and the people who mint the coins, the area and sometimes with historical-political events. When observing the distribution map of the coins showing the image of a dog as a main isolated type (Figure 1), one can observe a concentration in the western Mediterranean area, especially in Sicily. The first evidence is dated at the beginning of the first quarter of the fifth century BC in Segesta. The dog is connected to this town and this is demonstrated by the legend of the town's foundation, as revealed by the literary texts (even if later in relation to the coins).[1] In accordance with these texts, the river Krimisos, in the resemblance of a dog, would have coupled with the nymph Egesta, who would have given birth to the *oikistes* hero Egestes. The importance of the dog on the coins and its presence in the legend of the foundation of Segesta have given rise to a question among scholars about the existence of a proper cult of the animal or the river as a god in zoomorphic appearance. Nonetheless, in 1991 A. Dubourdieu rejected this hypothesis and interpreted the legend of the coupling between Segesta and the dog as the mythical transposition of initial rituals characterised by an animal disguise. On the other hand this does not explain the several behaviours of the dog on the coins if they did make reference to a divine epiphany, it would not be possible to explain

[1] Cf. Serv. *Ad Aen.* I, 550, *Schol. Lycophr.* 952 and 964.

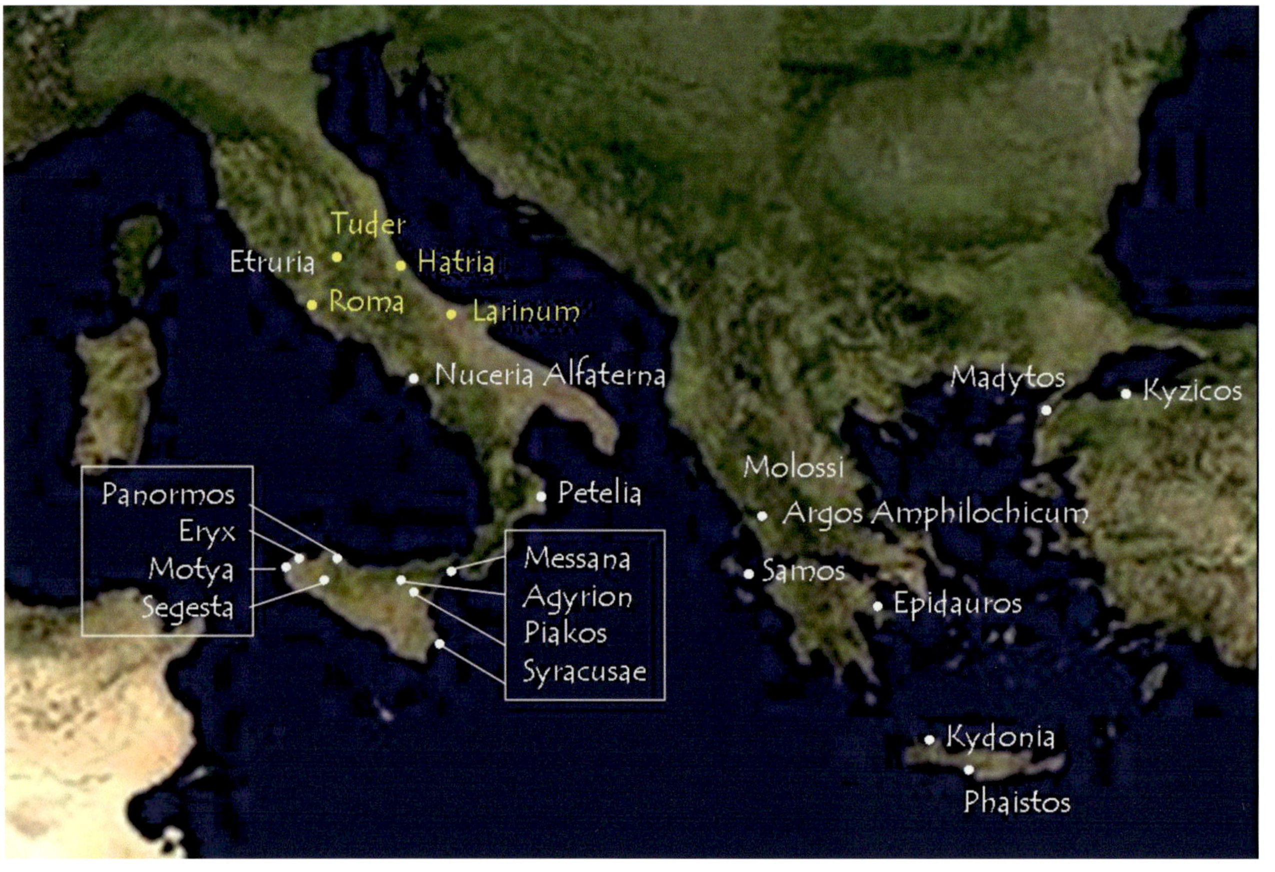


Figure 1. Distribution map of the main isolated dog type (by A. Bottari).

Figure 2. Segesta, AR, didrachma 455–450 – 445–440 BC (Gerhard Hirsch Nachfolger. Auction 355. Lot number: 1608).

them (Marconi 1997: 1085–1087). The first coin typology minted by the mint of Segesta in the years between 475–470 and 455–450 BC shows a dog[2] sniffing the ground (Hurter 2008: table 1, 1–12.); between the years 455–450 and 445–440 BC the minted image of the dog is standing (Figure 2) (Hurter 2008: table 6, 63–96).[3] Both the images appear on the front of the didrachmas, on whose reverse is depicted the head of the eponymous nymph Egesta. These years are those immediately following the Battle of Himera, when the Greeks of Sicily lead by Syracuse and Agrigento prevailed on Carthage, which exerted its influence on the western area of the island. From 460 BC Erice[4] mints the same typologies of Segesta on didrachmas and that of the dog looking back on litrae. At the same time this last typology was also chosen by Segesta. This choice shows a clear link between these two Elymian towns.[5] According to S. Cataldi, the option of minting the dog type - Elymian symbol and possible reference to Punic cults - is justified by the will of Segesta to state its belonging to 'another' *ethnos* other than the Greek climate of Syracuse, which would imply a link with Carthage. On the contrary, in his research on the Elymian-Punic relationships[6] L. Gallo believes that

[2] Regarding the dog of Segesta, it is possible to speak of an exact breed which his unanimously recognised as the Cirneco dell'Etna (Manganaro 2004).

[3] The traditional repertoires give a higher chronology to this series of didrachmas with the standing dog, from 470 BC, which would be contemporary to the sniffing dog. Cf. e.g. Grose (1923: 2533–2541, table 86. 7–14).

[4] Earlier, the town showed typologies referring to Selinunte in 480 BC, in Agrigento after this date and until the fall of the tyranny of Phalaris (Cutroni Tusa 1997: 417).

[5] The divisional coins of Segesta and Erice in the middle of the fifth century show, as noted by A. Cutroni Tusa (1992; 1997) in two contributions to the International Days of Study on the Elymian Area, the close relationship between the two towns. In particular, the scholar underlines that these coins had to facilitate a common and undistinguished circulation of money in order to help the reciprocity of exchanges. These coins are distinguished as 'parallel' (i.e. characterised by the same type but with different *ethnos*) and 'common' (i.e. with double *ethnos*) (Cutroni Tusa 1997: 419).

[6] Cataldi (1992: 75) talks about the dog cult of Segesta and suggests a relationship between this animal and the Punic religiousness referring to an item of news from Pompey Trogue in Justin, XIX, 1, 10. In accordance with him, the Punic people were used to offering human beings as a sacrifice and eating dog meat.

the Elymian town shows - in moments of Carthaginian weakness and disengagement, that is after 480 BC - the intention of freeing itself from the heavy influence of the African town.[7]

From 430 BC the dog image also appears in Panormos and Motya (Jenkins 1990: part. 1, table 2, 25–17). The dog of Motya is standing still and is in the act of sniffing the ground, while Panormos presents two dog typologies. One (Jenkins 1972: 35, Z 1) is similar to the dog of Motya, the other (Jenkins 1990: 50, table 6, 1) shows a dog that seems to follow a smell. At the same time this iconography also appears in Segesta. According to L. Gallo, Segesta would have again expanded its area of influence damaging the areas around Carthage (Gallo 1992: 322–323), and probably following an armed conflict against Motya[8].

The last fifteen years of the fifth century BC was a period of great development of the dog typologies in western Sicily. Once again in parallel Segesta and Erice, mint the first series in bronze showing the standing and the looking back types. These series continue until the first ten years of the fourth century BC. One of the three bronze series,[9] which L. Lazzarini first attributes to the mint of Alikai, seems to be dated from 415 BC. The iconography of the front is very particular, showing a nymph sacrificing what appears to be a dog or a wolf on an altar (the head of the animal is not clearly recognisable). A *tetras* (Hurter 2008: table 29, 3) belongs to the third series. This coin shows a dog following a scent on its reverse. From this period a Segestan litra (Hurter 2008: table 25, 29–30) shows a standing dog with the symbols of a shell and a *gorgoneion* (or solar head) on its back, while a three-quarter nymph's head within a laurel crown is represented on its front. The beautiful litrae of Erice are coeval. These coins show a dog proudly standing, which is paired with a full-length Aphrodite seated on a chair who holds a dove (Grose 1923: 2234, table 72. 7) in her right hand or stretches out her hand to Eros, who crowns her (Zodda 1989: 3–26, 15–16, table III, 35).

Between 412/410 and 400 BC Segesta chooses to represent a dog following a scent (Hurter 2008: table 16, 187–197), while Motya represents a standing dog (Jenkins 1990: part. 1 table 3, 20; 22; 24; 27; 29). In this period, other dog typologies appear. One typology is the attacking dog or carrying a prey, in particular the head of a deer on didrachms[10] of Segesta and Motya (Jenkins 1990: part. 1 table 4, 31), which is minted with perfectly corresponding typologies. The other typology represents a hare as a prey on smaller nominal value coins of Segesta[11] and Erice.[12] The hunting context referenced by the presence of a prey could represent an aggressive attitude of the towns of western Sicily, in particular of Segesta, which during this period is involved in a territorial dispute with Selinunte. This attitude is certainly also referred to on another typology minted in this period by Segesta: the typology of the hunter/Krimisos accompanied by the dog, which will be discussed below.

The cases described above do not represent true hunting scenes, but rather the end of the hunting act, when the dog displays and offers its prey. The hare (and overall the game) frequently appears offered as a love symbol on pottery with black figures, where the dog is sometimes also a gift to the *eromenos* by the *erastès* (Schnapp 1997: 247–257) The hare is also represented in different iconographic patterns on pottery with red figures (Schnapp 1997: 325–354) and on some *lekythoi* with a white background (Schnapp 1997: 320). The interpretation of the game as a love symbol could well connect with the female figure on the other side of the coin, which, if not identified in all cases as the goddess Aphrodite, certainly possesses her characteristics. The pairing with a female figure (head or full-length, eponymous nymph or Aphrodite) on the other side of the coin is almost exclusively a standard feature in the dog coin typologies of western Sicily. This fact is very important, because the mints in eastern Sicily change the dog type, but pair it with male gods and figures.

### *2.2 The dog type spreads to eastern Sicily*

Two coins of AE in Piakos are dated between 420 and 400 BC. One is characterised by a dog jumping (Grose 1926: 4644, table 174. 12), while the other shows a beautiful *unicum* (Jenkins 1972: 410) of a dog attacking a fawn. Both the dogs are paired with a young head of a river god on the front of the coin.

[7] L. Gallo (1992: 315–323) through his reconstruction of the relationship between Carthage and Segesta demonstrates that the the shift to the Rome side by Segesta during the Punic Wars, justified by the Trojan common origins, was not a sudden U-turn to the historical ally Carthage, but rather a predictable choice. The Scholar recognises an imbalance of strength which from time to time gives to one or the other a position of supremacy.

[8] If one can interpret in his sense the passage of Diodorus (11, 86, 2) which recalls the conflict between Segestans and *Lilybaitai* (Gallo 1992: 322).

[9] With exception are the two litrae attributed to the second series from around 400 BC; the third series places itself between 390 and 370 BC. (Lazzarini 2005: 21–23).

[10] On the front the dog is depicted biting a deer head between its paws; the back of the coin shows the dog paired with a nymph head on the right with her hair gathered on the back of her neck (Hurter 2008: table 16, 185). The same typology can be found on uncias from between 390 and 380 BC (Hurter 2008: table 29, 16).

[11] Segesta chooses to represent the dog dragging along a hare on silver fractions: on litrae (Hurter 2008: table 25, 28) and on *hemiobolon* (SNG Cop. Sicily 2: 242); on bronze fractions the dog is depicted as still and sniffing the hare, which lies under the dog's paws (Hurter 2008: table 29, 15).

[12] The types of Erice are characterised by the dog looking back on a litra from 410–400 BC (SNG Cop. Sicily 2: 246) and on an uncia from between 400 and 380 BC (SNG ANS: 1328). In Piakos a coin shows a dog attacking a prey, which will be addressed later.

The bronze coin (Grose 1923: 2125, table 68. 2) of Agyrion is dated between 345 and 300 BC, which shows a dog intent on following a scent, paired with the head of Apollo. This is followed on a *hemilitron* (Grose 1923: 2127–8, table 68. 4) dated after 241 BC by the dog accompanying the hero Iolaus crowned by Nike. During the age of Timoleon, the dog appears in Syracuse, where S. Cataldi thinks Timoleon could have chosen the animal as a symbol of his politics of liberation from the oppressor Hicetas (Cataldi 1992: 79–81). These types are new in relation to the ones considered above. One type is a dog with its right paw raised (later it will appear on a Roman bronze coin from the republican age),[13] and which will later appear on two bronze coins paired with Zeus Eleutherios (Calciati 1983: 84/2) or with Asklepios (Calciati 1986: 193, 84). Another type shows a crouched dog (SNG Cop. Sicily 2: 744) turned to the left, which looks back straining its ears and putting its rightforepaw on the ground, as if ready to stand up if necessary. The last mint to assume the dog type is Messana at the time of the Mamertines, who clearly refers to the Siculian god Adranus[14] given his esteem as a warrior. As a result, the god's head appears on the front of the coin (Särström 1940: 137–146). The Mamertines probably used this iconography to try to dampen the animosity of the Siculians, who suffered from their hegemony (Cataldi 1992: 82). Therefore, a symbol of 'other' origin could have been used in particular moments to bear witness to a new politics, or even a politics of liberation.

## 3 Main dog type with other subjects

### *3.1 The dog as a companion of several characters*

The tetradrachms have been mentioned among the coins of Segesta. These coins were minted from 415 BC, during the period of some territorial disputes with Selinunte, and they refer to a cynegetic context. In particular, the tetradrachm dated 415 BC or between 415 (Marconi 1997: 1071) and 412–410 BC (Hurter 2008: 26, table 1) shows the typology of the moving quadriga on the front of the coin, turned to the right and ridden by Segesta, who is crowned by a flying Nike. The reverse of the coin shows a young naked hunter turned to the right, with a *pileus* on the back of his neck, his left foot on a rock, a *chlamys* on his left arm put on his knee and a javelin. This hunter is accompanied by a dog, which is sniffing the ground. Other tetradrachms are from 410 BC (Figure 3) (Marconi 1997: table CCXVII, 2). A character very similar to the hunter mentioned above appears on the front of the coin and little horns on his forehead make him a personification of the river Crimiso. This character has got two javelins with him and is in the company of a dog, which together with its owner scans the horizon. Also, in this case, the nymph Egesta is depicted on the back of the coin, while she is crowned by a flying Nike; but this time the nymph is in the act of making a sacrifice on an altar. The same hunter typology appears on the front of tetradrachms (Hurter 2008: table 26, 6, 7) from 405–400 BC, on whose reverse the head of the nymph Egesta is shown. In these two last cases, two dogs appear: one smells the ground, while the other looks ahead towards an ithyphallic herm. The previously mentioned in-depth analysis by C. Marconi (1997: 1081–1082) refers to the military interpretation of these typologies. The scholar has identified the male figure characterised by the javelin and the chlamys hanging from his arm with Egestos, explaining that the Segestans way of fighting would have been similar to hunters. It is important to underline the role of the dog, which accompanies the hunter and shares his patient and cautious behaviour in a land characterised by the presence of the herm as a boundary in relation to the *chora* of the town. Naturally, the dog plays the role of the land guardian and of man's faithful companion. Nonetheless, the setting of the scene in a boundary area brings to mind the relationship that this animal has with the gods of the places and the transition stages. In 2008, in his very interesting analysis on the dog sacrifices made among Greeks, Etruscans, Italic peoples and Romans, J.-C. Lacam (2008: 29–80) emphasised that the sacrifice is offered in all cases to both celestial and chthonian gods, although each ethnic group's rituals are totally independent. This setting is linked to the transitions and the contact between different worlds: the civil and the wild worlds, the world of the living and that of the dead. In accordance with the ancient sources, in the Hephaestus (Ael. *N. A.* XI, 3) and in the Sicilian sanctuaries of Adranus (Ael. N. A. XI, 20), Sicilian dogs played the role of guardians following the transition between the sacred and the profane.

In light of these considerations, in my opinion it is fundamental to reconsider the previously mentioned analysis by A. Dubourdieu concerning the rites of passage characterised by dog disguises. These refer to goddesses, to whom the characteristics of Artemis, Aphrodite and the Phoenician Astarte can be attributed. With regards to this point, J. C. Lacam (2008: 32–34) refers to the remains of eighty-five dogs found in a well north-east of the *Hephaisteion* in the Athens

[13] A dog with its forepaw raised appears on two bronze coins with nominal value from 230 BC: on a *quadrans* the dog is on the front of the coin, while it is paired with a six-spoked wheel on the reverse of the coin (Rutter 2001: 239); on a *hemilitron* the same dog appears on the back of the coin, while on the front it is paired with the head of Rome wearing an elm (Rutter 2001: 309).

[14] Adranus, war god of the Mamertines, is depicted (Canciani 1981: 229–230, s.v. *Adranos I*, LIMC) as a native Sicilian god. His origin is probably volcanic, and he is identified by the Greeks with Hephaistos. He is considered the father of Palikoi and a sanctuary on the slopes of Etna is dedicated to this god in the homonymous town. Another term within the Lexicon is dedicated to the river god Adranus, edited by Arnold-Biucchi (1981: 230), which underlines 'à ne pas confondre avec Adranos I', and where he comments on a coin showing on the front the head of the river god with small taurine horns. The author connects this head with the river Simeto, in which Adranus (if a river with this name exists) is a branch.

Figure 3. Segesta, AR, tetradrachm; 410 BC (http://www.magnagraecia.nl/coins/).

Figure 4. Pandosia. AR, drachma; 400–336 BC (Numismatica Ars Classica. Auction 52. Lot number: 51).

agora. These remains accompanied the remains of one hundred and seventy-five infants, probably victims of an epidemic, together with a sword and a herm surmounted by a female head.[15] The remains of dog sacrifices and other animals have also been found in Locri, at Centocamere, in the sanctuary of Aphrodite.[16] In addition, a chthonian value is given to the remains of dog sacrifices found in a *bothroi* from the Lucanian age in the sanctuary at the mouth of the Sele (Ferrara 2008: 106) and to the remains found in the sanctuary of Lucania in Torre di Satriano (Osanna and Sica 2005: 137–138). This once again underlines the recurrence of rites of passage overseen by a patron goddess. In these last cases, the goddess belongs to suburban sanctuaries, of the *limen*. On the other hand, other gods are famously linked to places of transition, and are depicted in the company of a dog on coins.[17] Pan can be considered as a patron of rites of passage, who is identified on a series of silver coins of Pandosia dated between 400 and 336 BC as a young male figure seated on a rock in the company of a dog. This figure is paired with the head of Hera Lacinia on the other side of the coin (Figure 4). The god is characterised by a javelin; this tool refers to Pan as a god of hunting and war and alludes to the requirement that he should defend the *chora* from external attacks.[18] An iconographic typology comparable to Pan of Pandosia appears on Medma bronze coins (Rutter 2001: numbers 2427–2430), paired with a head which is attributed to Persephone; however, no weapons are represented on these coins. G. Salamone (2012: 98–103) does not exclude that the typology of Medma could be Dionysus in 'his definition as the god of life and death' connected to Persephone, with whom he shares the relationship with nature and with plant cyclicity. The scholar reconstructs, in relation to the analysis of the religious context of Medma, the function of this god as a male element in an ideal of a divine couple, who can assure fertility and prosperity to the *polis*. In this sense, exceeding the limit of an exact denomination of the gods in question, this reasoning can also be extended to the couple Hera Lacinia (god of the town) and Pan[19] on the Pandosia coins, for whom the reference to the area is more easily readable.[20] Moreover, the same reasoning can also be widened to the nymph couple Egesta and the personification of the river Krimisos (Caccamo Caltabiano 1992: 135–137), depicted first as a dog, then - on a tetradrachm from 410 BC - as a male figure in the company of one or more dogs comparable to the figure of a hunter, but identifiable as a river god thanks to the presence of two small horns. In the case of Segesta, it has already been emphasised that this divine couple is considered as the one giving birth to the founder of the town, and therefore to the town itself.

### *3.2 The dog as a companion of the gods*

Among the gods depicted in the company of a dog, only Artemis/Diana is considerably widespread throughout time and civilisations. The dogs, as a feature of the hunter goddess, are described in poetry[21] and are shown with her on several materials, in particular in the scene showing the murder of Actaeon.[22] Nonetheless,

[15] It is not important for our reasoning if she is Aphrodite, as thought by Lacam (2008: 43) considering the indication of Paus. I, 14, 7, who mentions the temple of Aphrodite Ourania (or Artemis as read in Kahil 1984: 630) close to that of Hephaistos.

[16] According to Torelli (1987: 599), based on the inscription from the end of the seventh century-the beginning of the sixth century BC found in the sanctuary, the sacrifice was offered to the Asiatic and Lydian Cybele, to whom the dog sacrifice was traditional and who was identified by the Ionic tradition with Aphrodite.

[17] The *Lares* on Roman coins from the republican age dated at 112–110 BC can be defined as such (Crawford 1974: 298). According to Lacam (2008: 35–55), dog sacrifices are offered to these gods, who have an infernal value. At the same time, the horn of abundance characterises these gods, who are patrons of borders and boundaries, and of places of transition.

[18] From Dionisio and the Lucans (Taliercio Mensitieri 1998: 360).

[19] The pairing of the town god (in this case as charioteer of a biga of mules/Pan seated on a rock, with a hare in his hand) had already appeared on tetradrachms of Messana in 420 BC. Furthermore, Herakles seated on a rock appears paired with the head of Hera Lacinia in Crotone in around 390 BC (SNG ANS: 375, 383).

[20] According to Salamone (2012: 96). The figure of Pan would refer to the toponym of the town, whilst the typology of Hera Lacinia on the front of the coin would refer to a federal dimension.

[21] In Call. Hymn. In D. v. 87 ff., Pan gives two white and black dogs, and two reddish dogs to Artemis, which are even able to kill and bring lions. He also gives her a pack of seven Laconian dogs faster than the wind to chase fast preys (Bormann 1968: 45–49).

[22] Beautiful slim, quick, and muscular hounds appear in this scene, which is often depicted on pottery with red figures. An example is the bell crater by the Pan painter preserved in the Museum of Fine Arts, Boston, where the hounds are depicted savaging their owner, who collapses under the hounds' bites and raises his right arm to the sky, as if begging for a divine aid, while the goddess draws her bow to kill Actaeon. The same scene, characterised by the same dramatic force, appears on the famous metope in the E temple in Selinunte and on a small plate reconstructed from fragments, which belongs to the Melos reliefs and is preserved in the Museo Archeologico Nazionale of Naples. In certain craters of Italiot origin the scene appears less vivid, because of the almost distant attitude of Artemis. It is the hounds

it is important to underline that the dog is not a fixed feature of the goddess and of the coins. The quiver and the bow can be considered as 'denoting' features which are necessary for a certain identification of this divine figure, because they generally appear behind Artemis's shoulder even when only her head is depicted. As a result, the presence of the dog next to the goddess on the coin typologies can be considered more significant. Artemis in the company of a dog first recurs in Greece. In Orchomenos, in Arcadia, the dog is depicted seated next to the kneeling goddess on the front of a bronze coin[23] dated between 370 and 340 BC, while on the reverse the dog appears paired with the figure of Callisto. In Kydonia, at the end of the third century BC, the dog is depicted seated next to Artemis on the back of a tetradrachm (Grose 1926: 7090). The goddess is standing and wearing a short tunic and *endromides*, holding a long lit torch in her right hand. The head of Apollo appears on the front of the coin. In Kabyle, in Thrace, the same pairing of Apollo's head/Artemis appears on a bronze coin dated between 270 and 230 BC. In this case, the goddess is depicted standing with a *patera* in her right hand and a torch in her left hand (SNG BM Black Sea, 193). During the Second Punic War, a certain diffusion of the type 'Artemis with dog' is recorded in the Italic area. In particular by the Brettians the dog accompanies a standing Artemis characterised by a torch and a *patera*. This scene is depicted on the back of a drachma (Grose 1923: 1508, table 47. 23; 1509) linked to a coin with a higher nominal value with the Hera/Zeus types, while on the back of the coin the goddess is paired with Apollo's head (Figure 5), but sometimes also with a young river god standing and in the act of crowning himself, together with the head of Nike (Grose 1923: 1496, table 47. 17 Arslan 1989: table X, 102). In the same years, in Rhegion the dog accompanies Artemis with a bow in her right hand and again with along torch in her left hand (Grose 1923: 1931, 1932, table 61. 12). In Syracuse, on the coins (SNG Cop. Sicily2, table 19, 876–879) attributed to the V Republic, the dog is depicted while dashing towards the goddess, who is going to shoot an arrow. Artemis appears to be going to the right with a torch brandished like a weapon and accompanied by a dog dashing to the right on a Mamertine bronze coin (Carroccio 2004: 13) of Messana minted between 215 and 212 BC.A bronze coin (Carroccio 2004: 2) with the same iconography of Abacaenum is dated after 212 BC. Finally, in Tauromenion (Carroccio 2004: 18) Artemis is depicted standing in the company of a dog on a bronze coin of uncertain dating.[24] On the other hand, also during

Figure 5. Bruttians, AR, drachm; 217–203 BC (http://www.magnagraecia.nl/coins/).

Figure 6. Syracusa, AR, litra, 215–212 BC (Classical Numismatic Group. Auction 114. Lot number: 81).

the Second Punic War the paring of Artemis's head/running dog recurs on a semuncia of Larinum (Grose 1923: 166) and on a bronze coin of Petelia (Grose 1923: 1846, table 58. 22). These towns are remembered by the sources for their constant fidelity (at least initially) to Rome. Nonetheless, it is possible to think that a split between pro-Romans and pro-Punics occurred within them. The research by M. Caccamo Caltabiano has demonstrated this fact (Caccamo Caltabiano 1977: 53–54). In Petelia, celebrated for its fidelity to Rome by Livy,[25] a division occurred between a small group of Optimates, who later escaped to Rome, and the people's party, which had the intention to agree with Hannibal from the beginning. A similar situation occurs in the events of Syracuse. Although there were those who had already plotted an alliance with Carthage,[26] when the old King Hiero II, who had been faithful to the alliance with the Roman power[27] died, the town suffered a moment of disorder leading to the murder of Hieronymus and the foundation of the V Republic, whose money sustained the expenses of the people of Syracuse (Burnett 1995: 397), and who intended to defend their autonomy from Rome. Also, in connection with Rhegion, Livy[28] talks about the resistance to every attempt of Carthaginian penetration, supported by the forces sent by Appius Claudius Pulcher the propraetor of Sicily. As for Larinum, it is known that in 217 BC, near the area of its territory, the Roman army tried to

which give life to the scene running from different points towards Actaeon, who tries to repel the attack with weaponry.

[23] In BCD Peloponnesos: 1575, the dog appears seated next to the kneeling goddess, paired with the figure of Callisto on the reverse of the coin.

[24] Carroccio (2004: 182) singles out another bronze coin, attributed to the mint of Paropos and dated after 211 or between 204 and 190 BC, where the god accompanied by a dog is not identifiable.

[25] Liv. XXIII, 20, 4–10.

[26] This refers to Gelon, son of Hiero ( Liv. XXIII, 30, 10–12).

[27] Liv. XXIII, 21, 1–5.

[28] Liv. XXIII, 30, 9; 1, 11.

oppose the Carthaginian troops, who had camped by the Gereonium.[29]

In all these cases, the recurrence of the dog stands out, in particular together with Artemis, on the Hannibalic coins of Southern Italy, and even in Syracuse during the years of the rebellion against the Roman power and of the alliance with Carthage. Rome chose this symbol, after regaining its power, for a series dated after 216–215 BC,[30] within typologies that were specific to the Italic mints, allied with Hannibal and then under Rome's command.[31] It is even more interesting to notice that the attitude of the dog changes on the basis that the value of the goddess assumes. It takes part in the, so to speak, cynegetic action on the coins of Syracuse, whose iconography transmits a message that fits perfectly with the war spirit of the town, which the exponents of the V Republic should have had, as they intended to keep the attained autonomy. In the other cases, the dog turns its muzzle and paw to the goddess - as if to make contact - in its salvific acceptation, which is expressed by the epithet of *Soteira*, connecting both the symbol of the eight-pointed star on some Bruttian drachmas with the pairing of the Apollo typologies of the front of the coin.[32]

The salvific and liminal value of the dog is well matched with its purifying properties. For example, the purifying role can refer to the presence of the dog next to Asklepios, the healing god of physical and spiritual infirmities. This animal is depicted next to Asklepios in Rhegion, where the cult of this god was very strongly felt, on tetradrachms (Franke-Hirmer 1964: table 99, 285) dated at 435–425 BC and in Epidauro on drachmas[33] dated between 350 and 325 BC. According to the literary sources, the dog played an active role in the temple of Asklepios in Epidauro, thanks to the curative proprieties of its saliva, and according to the testimony of Pausania, the animal was depicted crouched under the god's throne in the cult statue made by Thrasymedes.[34] The only iconographic evidence is the reproduction on the previously mentioned drachmas.

It is evident that, as previously noted, it is not possible to have a unique reading of the dog figure on coins, nor to examine all the received attestations.[35] However, this close examination has tried to rebuild the versatile meaning that the animal has, both as land guardian, and as man's companion in several situations: in the transitions characterising human life, and in the extreme passage from life to death. In this sense, it is possible to understand the many statements of dog sacrifices which were widespread among the Italic people and in Rome, but also in Greece. The liminal value of the dog also goes well with its purifying properties, which have been emphasised in relation to Asklepios and rivers. This value also explains the reference by literary sources (Mainoldi 1981) to purifying rituals, which are characterised by the passing of the Macedonian army across the two halves of a dog during lustral rites.

## Acknowledgements

My sincere thanks go to the organisers of the conference, where it was possible to reflect on a topic, which when observed from different points of view shed light on various aspects and implications. A special thanks to Professor M. Caccamo Caltabiano for her inspiration about the examined subject and her continuous support, which has always been constant.

## References

### *Ancient sources*

For the citation of classical works, reference is made to the Thesaurus Linguae Latinae, LIDDELL-SCOTT-JONES Oxford 1968, LAMPE.

Aelianus, *De Natura Animalium libri XVII.*
Callimachus, *Hymnus in Dianam.*
Diodorus Siculus, *Bibliotheca Historica.*
Livius, *Ab urbe condita libri.*
Iustinus, *Epitoma Historiarum Philippicarum Pompeii Trogi.*
Pausanias, *Graeciae descriptio.*
*Scholia in Lycophronis Alexandram.*
Servius, *Commentarii in Vergilii Aeneidos libros.*

### *Abbreviations of classical works*

Ael. N. A. = Aelianus, *De Natura Animalium libri XVII.*
Call. Hymn. In D. = Callimachus, *Hymnus in Dianam.*
Liv. = Livius, *Ab urbe condita libri.*

[29] Liv. XXIII, 18, 7; 23, 9.

[30] The series consists of a victoriatus (Crawford 1974: 122/1), where a little dog is at the feet of the personification of Victory, who crowns a trophy; of a didrachm (Crawford 1974: 122/2) showing Dioscuri, who rides galloping horses, and the standing dog; and of bronze coins where the dog is a small symbol on the bow of a ship (Crawford 1974: 219/2–6).

[31] Artemis in the company of a dog also appears on other later Roman coins: on a denarius from 71 BC three dogs accompany Artemis depicted on a deer biga; this iconography recurs in the provinces during the imperial age, but without dogs. In addition, the dog sometimes accompanies Artemis on aurei and denari of the series marked with the inscription SICIL to commemorate the victory of Augustus against Sextus Pompeius in Nauloco. During the imperial age Artemis typologies increase. Artemis Laphria recurs more frequently with the dog; this iconography appears in the mints of the Peloponnese, but also in several other mints of the eastern provinces.

[32] The Herakles's head paired with Artemis of Messana is an exception and appears on the bronze coin of Mamertines, to which the inscription of the back refers.

[33] Grose 1926: 6882, 6883, table 233. 8, 9.

[34] Paus. II, 27, 2.

[35] For a review of the coin typologies with the dog (Bottari 2011; Bottari, A., I tipi monetali del cane e del lupo. Iconografia e simbolismo, Unpublished Ph Ddissertation, Università di Messina 2009–2010.)

Paus. = Pausanias, *Graeciae descriptio*
Schol. Lycophr. = Scholia in Lycophronis Alexandram.
Serv. Ad Aen. = Commentarii in Vergilii Aeneidos libros.

### *Modern sources*


Arnold Biucchi, C. 1981. S.v. Adranos II, in *Lexicon Iconographicum Mythologiae Classicae* I: 230. Zürich-München: Artemis & Winkler Verlag.

Arslan, E. A. 1989. *Monetazione aurea e argentea dei Brettii*, Milano: Milano Edizioni ennerreSrl.

BCD Peloponnesos. *LHS Numismatics. Coins of Peloponnesos. The BCD Collection.* Catalog of public auction 96, 8–9 May 2006. Zurich.

BMC Italy. Poole, R. S. 1873. *A Catalogue of the Greek Coins in the British Museum. Italy.* London: Head and Gardner.

Bornmann, F. 1968. *Hymnus in Dianam*. Firenze: la Nuova Italia.

Bottari, A. 2011. Tipi del cane e del lupo sulle monete del Mediterraneo antico, in N. Holmes (ed.) *Proceedings of the XIV International Numismatics Congress, Glascow 2009*: 1247–1253. Glasgow: International Numismatica Council, London: Spink & Son.

Caccamo Caltabiano, M. 1977. *Una città del Sud tra Roma e Annibale. La monetazione di Petelia*. Palermo: Sophia.

Caccamo Caltabiano, M. 1992. s.v. Krimisos, in *Lexicon Iconographicum Mythologiae Classicae* VI: 135–137. Zürich-München: Artemis & WinklerVerlag.

Caccamo Caltabiano, M. 2007. *Il significato delle immagini. Codice e immaginario della moneta antica*. Reggio Calabria: Falzea editore.

Calciati, R. 1983. *Corpus Nummorum Siculorum. La monetazione di bronzo* I. Milano: Edizioni G.M.

Calciati, R. 1986. *Corpus Nummorum Sicolorum. La monetazione di bronzo* II. Milano: Edizioni I.P.

Canciani, F. 1981. v. Adranos I, in *Lexicon Iconographicum Mythologiae Classicae* I: 229–230. Zürich-München: Artemis & WinklerVerlag.

Carroccio, B. 2004. *Dal* basileus *Agatocle a Roma. Le monetazioni siciliane d'età ellenistica (cronologia-iconografia-metrologia)*. Pelorias 10. Messina: DICAM.

Cataldi, S. 1992. Popoli e città del lupo e del cane in Italia meridionale e in Sicilia tra realtà e immagine, in M. Sordi (ed.) *Autocoscienza e rappresentazione dei popoli nell'antichità*: 55–82. Milano: Vita e pensiero.

Crawford, M.H. 1974. *Roman Republican Coinage*. London: Cambridge University Press.

Cutroni Tusa, A. 1992. *Le emissioni frazionarie di argento di Segesta in Giornate internazionali di Studi sull'area elima, Atti di Convegno (Gibellina, 19–22 settembre 1991)*: 647–669. Pisa-Gibellina: LSA.

Cutroni Tusa, A. 1997. Un incisore monetale a Segesta nell'ultimo ventennio del V sec a.C., in *Terze giornate internazionali di studi sull'area elima, Atti di Convegno (Gibellina, 22–26 ottobre 1994)*: 415–428. Pisa- Gibellina: LSA.

Doubourdieu, A. 1990–1991.Le chien de Segeste. *Kokalos* XXXVI-XXXVII: 53–81.

Ferrara, B. 2008. Il sistema dei doni votivi nei *bothroi* del santuario di Hera alla foce del Sele, in G. Greco, B. Ferrara (eds) *Doni agli dei. Il sistema dei doni votivi nei santuarii, Atti del Seminario di Studi (Napoli 21 aprile 2006)*: 77–111. Napoli: Naus Editoria.

Franke, P.R. and M. Hirmer 1964. *Die griechische Münze*. München: Hirmer.

Gallo, L. 1992. Alcune considerazioni sui rapporti elimo-punici, in *Giornate internazionali di studi sull'area elima, Atti di Convegno (Gibellina, 19–22 settembre 1991)*: 315–340. Pisa-Gibellina: LSA.

Grose, S.W. 1923. *Catalogue of the Mc Clean Collection of Greek Coins I*. Cambridge: University press.

Grose, S.W. 1926. *Catalogue of the Mc Clean Collection of Greek Coins II*. Cambridge: University press.

Hurter, S.M. 2008. *Die Didrachmen Prägung von Segesta*. Biel: Schweizerische Numismatische Gesellschaft.

Jenkins, G.K. 1972. *Ancient Greek Coins*. London: Barrie and Jenkins.

Jenkins G.K. 1990. *Ancient Greek Coins 2*. London:Seaby.

Kahil L. 1984. *s.v. Artemis*, in *Lexicon Iconographicum Mythologiae Classicae* I. Zürich-München: 618–753. Artemis & Winkler Verlag.

Lacam, J.C. 2008. Le sacrifice du chien dans les communautés grecques, étrusques, italiques et romaines. Approche comparatiste. *Mélanges de l'école française de Rome* 120/1: 29–80.

Lazzarini, L. 2005. La monetazione e il sito di Halikiai (Alicie), città della Sicilia occidentale, *Schweizerische Numismatische Rundschau* 84: 15–25.

Mainoldi, C. 1981. Cani mitici e rituali tra il regno dei morti e il mondo dei viventi. *Quaderni urbinati di cultura classica* XXXVII, n.s. 8: 7–41.

Manganaro, G. 2004. Il cane cirneco, Aigestes e Aktaion: realtà antiche e moderne in documenti letterari, figurativi e numismatici. *Quaderni catanesi di studi antichi e medievali* 3, Gennaio-Dicembre: 47–59.

Marconi, C. 1997. Storie di caccia in Sicilia occidentale, in *Seconde giornate internazionali di studi sull'area elima, Atti II (Gibellina 22–26 ottobre 1994)*: 1071–1120. Pisa-Gibellina: LSA.

Osanna, M. and M. Sica 2005. Articolazione dello spazio e pratiche rituali nel santuario lucano di Torre di Satriano, in *Lo spazio del rito*: 125–140. Venosa: Osanna edizioni.

Rutter, N.K. 2001. *Historia Numorum Italy.* London: British Museum Press.

Salamone, G. 2012. *'Una' e 'molteplice': la Ninfa eponima di città. Iconografie monetali e semantica.* Reggio Calabria: Falzea editore.

Särström, M. 1940. *A Study of the Coinage of the Mamertines*. Lund: V.K. Gleerups Forlag.

Schnapp, A. 1979. Images et programme: les figurations archaïques de la chasse au sanglier. *Revue Archéologique*: 195–218.

Schnapp, A. 1997. *Le chasseur et la cité. Chasse et érotique dans la Grèce ancienne*. Paris: Albin Michel.

SNG ANS Sylloge Nummorum Graecorum 1975, *3. Bruttium - Sicily I: Abacaenum-Eryx.* New York: The Collection of the American Numismatics Society.

SNG BM Black Sea Sylloge Nummorum Graecorum 1993. *The Black Sea.* 1, IX, London: The Collection of the British Museum.

SNG Cop Sylloge Nummorum Graecorum. 1942. *The Royal Collection of Coins and medals Danish National Museum*, 3: Italy 3: Lucania - Bruttium, Copenhagen.

SNG Cop Sylloge Nummorum Graecorum 1942. *The Royal Collection of Coins and medals Danish National Museum*, 5: Sicily II: Segesta-Sardinia, Copenhagen.

Taliercio Mensitieri, M. 1998. Monete di Pandosia, in *Mito e Storia in Magna Grecia: Atti del trentaseiesimo Convegno di studi sulla Magna Grecia :* Taranto, 4–7 ottobre 1996, 357–365. Taranto: Istituto per la Storia e l'Archeologia della Magna Grecia.

Torelli, M. 1987. I culti in S. Settis (ed.) *Storia della Calabria, I. La Calabria antica.* Roma: 589–612.

Zaganiaris, N.J. 1980. Le chien dans la mythologie et la littérature gréco-latines. Platon 32: 52–87.

Zodda, D. 1989. Contributo alla storia della monetazione di Erice nel V secolo a.C. *Rivista Italiana di Numismatica* XCI: 3–26.

# 5.7 The Numismatist's Best Friend. Images of Dogs on Roman Coins

Alessandro Crispino

Department of Humanities (DISUM), University of Bari Aldo Moro, Piazza Umberto I, 1, Bari, Italy. ale.crispino@libero.it

**Abstract**

Latin literary sources and Roman art have provided significant testimony to man's best friend in frescoes, sculptures, mosaics and even coins. Indeed, from a study of Roman coins, we can learn much about the various breeds of dogs present in the Mediterranean area during the period, the role they played in ancient society, and the link between dogs and the gods. The dog in the Roman world is only used for utility purposes. It was exploited in herding, in surveillance, in hunting, and in war. Roman coins confirm these dog roles in Roman society between the republican and early imperial ages.

**Keywords**: coins, breeds, pastoralism, war, hunting.

## 1 Introduction

The dog was always an important figure in ancient history. There are many literary and archaeological documents which are testimony to its presence. The aim of this work however, is to shed light on the role which dogs played in ancient Roman society. The paper first focuses on literary sources before going on to concentrate on another fundamental historical source rich in information, coins.

Coins are sources of primary interest to historians as they can provide missing information or, in some cases, confirm information deriving from other sources. An experienced historian should use all available sources including: written historical sources, inscriptions, archaeological finds, and coins. Archaeology and numismatics form some of the documentary basis of ancient history. Ancient currency is an archaeological monument. It is one of the smallest monuments of Antiquity, a 'miniature' monument containing a wealth of information and just as relevant as other sources, it is necessary to interpret and reconstruct the past. Interestingly, many Greek and Roman coins feature images of dogs.

Here the focus will be only on coins minted in Rome between the third century BC (the period in which the first images of dogs on coins appear) and the fourth century AD (the period in which the last dog is found).

From a study of Roman coins, it is possible to see that there were a number of different dog breeds present in the Mediterranean area: Braccoids, Molossers and images of other dogs similar to the Cirneco. It is also possible to understand from each coin the specific role played by the dog in the Roman age and to which god it was connected.

Figure 1. *Denarius* of *Caius Mamilius Limetanus* (Crawford 1974, 362/1, 82 BC).

## 2 Results of research

### *2.1 Literary sources*

Many literary sources of both the Greek and Latin ancient world refer to dogs. Some regard the animal as man's best friend, while others state the exact opposite.

In the Odyssey (17, 291–327) (Privitera 1991: 521–523) the faithful Argo happily dies after recognising his beloved owner, who has returned to Ithaca after a twenty-year absence.

In 82 BC *Caius Mamilius Limetanus* put Mercury's face on the obverse of his coin (Crawford 1974: 362/1) (Figure 1) while on the other side of the coin there is an image of Odysseus (Ulysses) who has just returned home and is reunited with Argo who seems happy to finally see his master again. Odysseus is not gleefully embracing his faithful hound, but looking over the top of him, past him, and trying to settle him down. This

Dogs, Past and Present (Archaeopress 2023): 356–362

makes perfect sense: Odysseus was in disguise at this point (part of what makes the image so touching is that the dog recognises his master despite his disguise); he is trying not to let his dog give the game away. Here, the artist has softened the rather upsetting imagery of the dying flea-bag on a dung-heap, replacing it with the more palatable image of a happy dog. The dog on the coin could hardly look further from death. In what is a lovely touch, the dog has his rear paws excitedly resting upon the curve of the dotted border.

On the reverse, the coin magistrate chose this image because *Mamilia's* family came from *Tusculum*, and was considered a descendant of Odysseus. *Mamilia* was the daughter of Telegone, who was in turn the daughter of Odysseus and Circe. This coin represents the divine bloodline of *gens*. The obverse shows Mercury as an uncle of the ancestor Odysseus (Luce 1968: 25–39).

In an episode from his *Satyricon* (64, 7–9), whilst at a sumptuous dinner held by Trimalchione, *Petronius Nigrus* (Aragosti 1995: 284–287) illustrates how these animals participated in banquets organised by their master and his friends, and how they delighted dinner guests with their virtuosity. Furthermore, we learn of the affection that Trimalchione held for his dog *Scylax*, saying that no one in his home loved him more. The ladies of the house were more familiar with dogs, and of the various breeds, preferred small dogs. These household dogs often came from Gaul (France) and Malta, and were in great demand in high roman society, particularly during the *Saturnalia* celebrations when they were given as gifts along with epigrams.

Other episodes of canine fidelity are reported by other illustrious classical authors, although they are not enough to make up for the generally negative opinion that the Greeks and Romans had of the animal.

After years of conflict in Troy, Elena realises how many deaths her escape with Paris has caused. Merciless towards herself, Elena talks about her 'dog eyes' (*Iliad*, 3, 180) (Franco 2003).

Actaeon, on the other hand, was eaten by his own dogs. Artemis, who had been seen by Actaeon while bathing at a spring, turned him into a deer and ordered the fifty dogs to attack their master, as they did not recognise him.

The Romans too did not generally value the dog highly: indeed, if Pliny the Elder[1] asserts that dog is man's most faithful friend (*Naturalis Historia 8, 61, 142)*, at the same time, and without any pity for the fate of the poor animals, he remembers that every year dogs were hung up alive from an elder tree between the Temples of Youth and *Summanus*. Clearly, the species had never been forgiven for neglecting to defend the Capitoline from the Gauls led by Brennus (*Naturalis Historia 29, 14, 57)*.

Virgil defines the dog as *obscenae* (*Georgica 1, 470)*, horrible, foul; Horace however believes the dog is *immundus* (*Epistulae 1, 2, 26)*, dirty (Cantarella 2018: 219–221).

Regarding Roman women, who often went for walks with their dogs in their arms, Plutarch remembers that Caesar blamed and accused them of no longer being able to procreate.

In his *Saturae*, Iunus Iuvenal comments with disdain on the customs of these times. In his sixth Satire, he speaks of the corruption of women, accusing them of preferring the death of their husbands over the death of their dogs.

With the 'Christians' we read similar vituperations: for John Chrysostom dogs are *'the vilest animals'*, *'which live in irremediable disdain and have no hope for redemption'* (*Homiliae. 10, 3)*. For Augustine, the dog is *'despicable and ignoble'*, *'the last of men and beasts'* (*Quaestiones in Heptateuchum. 6, 7; 7, 73)*.

Due probably to this bad reputation, the dog, along with the snake, rooster and the monkey, was locked up in the sack (*culleus*) of *Poenacullei* (from the Latin, 'penalty of the sack', the death penalty imposed on a subject who had been found guilty of parricide).

### *2.2 The role of dogs in the Roman age*

In ancient Rome there was great demand for dogs and, specialised officers called *procurators cinogiae* were sent to the provinces of the Empire to search for and collect high quality breeding dogs, which were then transported to Rome for training and reproduction.

The Romans classified dogs into *pastoricus* (shepherd's), *villaticus* (guard dogs of farms, houses, fields and camps) and *venaticus* (hunting dogs).

There is also evidence that dogs were used during war.

#### *2.2.1 The Pastoricus*

For Romans, the dog was a real and irreplaceable working animal which was employed particularly in pastoral farming, one of the foundations of their economy. The *canis pastoricus* was raised with great care. Virgil recommended a whey-based diet and trimming fur from its tail and above all the ears, in order to protect the dog from the bites of wolves and foxes.

[1] Pliny the Elder focuses on dogs in *Naturalis Historia* (VIII 142–153) (Giannarelli 1983: 233–239).

In *De Re Rustica*, *Marcus Terenzius Varro*, describes the characteristics of dogs best suited to guarding livestock, and provides useful information on their required diet and how to protect against wolves. Talking of pastoral dogs, he says they are characterised by an intermediate body between the light and agile '*canis venaticus*' and the imposing and muscular *canis villaticus* (the Molosser).

Specific references to dogs' physical aspects, and in particular to their coats, can be found in *De Re Rustica* by *Moderatus Columella*. In its precepts, the author gives advice on how to choose a house watch dog. The animal should ideally be, robust with a large head and bright eyes, possessing a thunderous bark, and have a black coat which will instil fear at night as it merges with the darkness. As for the protection and defence of a flock, *Columella* recommends that the animal be equally robust and strong so as to intimidate wolves and follow them if they flee with their prey but should also have a white coat so as not to be confused with an attacker at night. The colour of the dog's eyes was also important. Yellow was preferred, although the most terrifying were silver, like the eyes of a ghost, which creates great fear.

The pastoral dog was produced by crossing the ferocious dog of the legionaries and some local lupoid breeds (also Sighthound); this is how the breed is found in the Abruzzese, Sicilian, Silano and Fonnese mastiffs. This kind of dog was selected because it had no predatory instinct towards the flock and due to its aggressive nature and ability to fight off predators. Therefore, when it was necessary to fight the 'wolf' and defend the flock, the Romans used the 'domestic wolf'; when it was necessary to fight for other purposes, they used the *pugnaces*.

It is no coincidence that the first roman and bronze coins of the third century BC featured a picture of the *molossus* dog[2] with 'ROMA' written underneath (Figure 2),[3] while on some other bronze examples [4] we find the two-faced *Janus* on the obverse side of the coin while on the opposite side we see the prow of a ship, the symbol of the city, with above a large dog above it. In these coins the dogs are detailed signs of Roman mint which produced them. One of these signs is the Molosser, symbol of the conquering strength of Rome.

*2.2.2 The Villaticus*

In ancient Rome dogs were commonly used to guard houses. Indeed, in some homes, depictions of dogs have been found with various aggressive postures, near the entrances of houses symbolically seeming to scare off intruders and burglars.

A mosaic in the House of the Tragic Poet (Pompei VI 8,5), located on the floor of the *fauces* and another in the House of *Paquius Proculo* (Pompei I 7,1), depicted in the *vestibulum*, show two guard dogs in an aggressive pose, although represented in different ways (Pelagalli and Di Gerio 2017). The image shows a large dog with a black coat and white spots all over its body. The animal's massive body, tied to a chain with a red collar, has its muscles contracted, in harmony with the tension displayed in the ears and eyes, and snarling with a half open mouth. As a whole the image provides a representation of a subject ready to attack any intruder with brute force and anger. In the lower part of the mosaic in the Tragic Poet's house, the inscription *cave canem* is clearly visible informing outsiders entering the home of the presence of a guard dog. However, the image of the dog depicted in the vestibule of the house of *Paquius Proculo* is very different. The artist has not shown the animal as aggressive or wanting to attack a possible intruder but has instead portrayed him lying calmly on the ground. The image depicts a large, slender dog with black fur, tied to a chain which is the same colour as the coat. It has a rather special red collar. Also of interest, due to its special features and docile appearance, is the dog depicted in mosaics in the vestibule of the House of Venosio the First (Pompei VI, 14, 20), restrained by a red leash attached to its collar. The host, as the primary guardian of the home did not choose a fearful guard dog but instead his pet dog and likely companion on walks and hunting trips. This pet is portrayed in the hall, restrained not by a chain but by a leash, and ready to welcome visitors as a friend rather than an intruder. The dog is a large specimen with a black coat and a particular red collar and the physical features of a wolf-like dog.

This is the same dwelling in which a plaster casting returned the three-dimensional form of a dog tied to a chain, who clearly died in a state of convulsion. The animal has an open mouth, showing the throat which is clearly visible and is wearing a large collar fitted with two rings, with which it was tied to a chain.

A dog used to guard a house is also shown on a silver coin minted in 112–111 BC. The *Caesia* family placed upon the obverse of this coin an image of Apollo seen from behind, with his head turned to the left and with a thunderbolt in his right hand. On the reverse we find the *Lares*, the owners of the house, seated facing each other, with a dog between them, each holding a staff in the left hand; above there is a bust of Vulcan with tongs over his shoulder. One of the *Lares* is petting the dog, protector of streets and borders of fields (Figure 3).[5]

[2] Crawford 1974: 24/6a; 24/6b (265–242 BC); 26/5; 26/6; 26/7; 26/8 (234–231 BC);293/3 (113–112 BC).

[3] Crawford 1974: 26/4 (234–231 BC).

[4] Crawford 1974: 122/4; 122/5; 122/6; 122/7 (206–195 BC).

[5] Crawford 1974: 298/1 (112–111 BC).

Figure 2. *Half Litra* (anonymous) (Crawford 1974, 26/4, 234–231 BC).

Figure 3. *Denarius* of *Lucius Caesius* (Crawford 1974,298/1, 112–111 BC).

*2.2.3 The Venaticus*

For hunting too, the sons of *Romulus* handed down much information regarding how dogs were selected. The Romans were the first to classify dogs according to their use in hunting. These were:

- *Seguges*: (bloodhounds) which were able to track game thanks to their extremely sensitive sense of smell;
- *Celeres*: (greyhounds) which chased game at great speed;
- *Pugnaces*: (Molossers) which attacked game.

The Romans enjoyed hunting in large numbers, on horseback if the terrain allowed: Molossers, surrounded the prey which had been first found by the *seguges* and then pursued and trapped by the *celeres* until the hunters arrival. When the dogs were recalled they would face the javelin at close range to show their courage.

The popularity of hunting is widely documented on sarcophagus engraved with mythological scenes which, since the age of Hadrian, became precious funeral monuments for wealthier Romans. The symbolism of the decorations equates success in hunting with victory in war and with triumph over death.

The land-owning classes passion for hunting became widespread in the second century BC and its popularity increased further still during the imperial period. The *venation* was a hunt for mammals of any size, with activities carried out by either individuals or groups. During the late republican and imperial periods, the custom of organising large wild boar hunts became widespread.

In artistic depictions and on coins, hunting dogs are portrayed mainly as chasing or confronting their prey, like modern day hunting hounds. The Segugi only knew how to hunt and attack game. They are not depicted indicating or retrieving hunted game killed by the hunter. Like all ancient peoples, the Romans were not interested in pointer or retriever dogs which are modern creations of man, a result of the radical change in weapons used in hunting today.

The dogs engaged in hunting activities had their own names. We know from sources that even in the ancient world it was customary to give a name to a dog. In *Kynegetikos*, his treatise on hunting (Labiano 2012.), Xenophon recommends short names, which are easy to pronounce, and give urgency to a master's orders. The author cites 47 names, including: *Psychè*, *Thymòs*, *Bìa*, *Aktìs* and *Hèba*. Most of the hunting dogs depicted resemble greyhounds (Levrieri), fast animals with slender bodies and long legs.

There are also many paintings in which there are large hound dogs with drooping ears and a short coat. The strongest dogs depicted are the wolf-like canines and mastiffs: the first breed being similar to wolves, while the latter have a large head and short snout.

In mythology and in the artistic representations of the past, as well as on coins from the Roman age, the dog often accompanies the goddess Diana who moved silently through the woods. It is clear that the dog offers her protection. The dogs defended her and obeyed her orders. In many artistic expressions, the dog that accompanies the goddess is docile and obedient. In some images, Artemis' dog is a greyhound.

During the republican age, the magistrates responsible for minting coins, chose to put images of dogs in hunting scenes with the Goddess Diana.

a. In 79 BC *Lucius Papius* placed a dog under a griffin, used as a sign to indicate which mint produced the coin mint.
b. In 77 BC *Publius Satrienus* placed a beautiful she-wolf on the reverse.
c. While in 74 BC *Caius Postumius* put on the obverse of a coin the beautiful head of Diana with a quiver, arrows and hair tied up while the reverse

Figure 4. *Denarius* of *Caius Postumius* (Crawford 1974,394/1, 74 BC).

Figure 5. *Denarius* of *Caius Hosidius Geta* (Crawford 1974,407/2, 68 BC).

of the coin features a beautiful greyhound running (Figure 4).[6]

d. In 71 BC *Lucius Axius Naso* chose two gods: Mars on the obverse and Diana on a chariot pulled by two elks on the reverse. She is depicted followed by two greyhounds while another hound precedes her in the race.
e. In 68 BC *Caius Hosidius Geta* chose an image of Diana on the obverse side, here, embellished with diadem and earrings, while on the reverse side we find a boar hunt, one of the most popular artistic scenes of the period, and very much a status symbol for noble and aristocratic families. Here, the boar is represented as being hit by a lance and almost bitten by what looks like a small but courageous greyhound. The choice of putting images of dogs on coins is also the result of an artistic elaboration in which the coiners had the opportunity to show off their great technical ability (Figure 5).[7]

Coins from the Imperial age are in fact very different to those from the first to third century AD where the Goddess Diana is always represented with a single dog. The scene results as fixed and standard. Indeed, the Goddess is represented standing up with hunting equipment in the coins of Augustus, Nerva, of Faustina for Antonino Pio, Caracalla and Giulia Domna and of Gallienus.

*2.2.4 The dog of war*

Three centuries before Christ, Alexander the Great used the Molosser in battle, spreading panic among the enemy. Indeed, his favourite Molosser *Periles,* died in battle.

When Alexander returned from India, he gave the mastiff to King *Porus* as a gift and in this way mastiff breeding began. Later, the breed became more widespread in Rome and was given the name Italian Molosser. Most dog breeds came from the ancient Assyrian-Babylonian mastiff and then spread throughout the world thanks to the Phoenicians who sailed all over the known seas. Like the Britons, the Romans bread Molossers to fight in war. In battle, dogs were used in large numbers and as brave guards too, in order to aid the men on horseback Indeed, dogs have always been involved in war, and right from the beginning, ancient Molossers were created exclusively for this purpose.

Pliny the Elder reports that dogs '*were the most faithful and cheapest assistants*'. In 231 BC, with their help, the legionaries of *Marcus Pomponius Matho* resolved the problem of *Pelitis* in Sardinia (*Fonnese* mastiffs are indeed the descendants of these dogs).

The Imperial Roman Army valued Molosser dogs highly. The *procurator cinegeti* selected them according to their performance in the arena and in battle. The Molosser was used as a combatant in war, for fighting against beasts and gladiators in the arenas and for big-game hunting. The Roman Molossian was a functionally complete dog, and in the lands conquered by Roman legions it engendered similar dogs, which were trained for similar purposes. In Spain, the catching Perro and in France the Dogue de Bordeaux became widespread.

The presence of these dogs within the Empire and in breeding centres can be traced to the arrival of merchants and territorial conquests. The courage, strength, and temperament of these *Bellator* or *Pugnator* dogs who fought alongside soldiers increasingly amazed those who selected them based on how trustworthy and effective they proved to be.

The war Molosser, with its collar covered with iron spikes and blades on its back, was trained to attack enemies, knock them down and dig its fangs into their throats. A similar war dog derived from the wolf, since the Romans used domesticated wolves. The Romans found in the wild animal's bloodline, all the qualities sought in a four-legged soldier: fearlessness, determination in combat and the drive to conquer territory.

[6] Crawford 1974: 394/1.
[7] Crawford 1974: 407/1.

This highlights the symbolic relationship between the wolves and some warlike peoples. On the Italian peninsula, for instance, the Lucanian tribe drew its name from *Aukol*, the wolf. The legend of the Roman foundation only took on the form we know today in the latter times when Romans extended their hegemony over the whole peninsula. They chose the wild canine, 'the she-wolf', as a totemic animal and as an image of courage and the city used it also on the legionary banners.

The *Bellatores*, *Pugnatores* or *Pugnaces* (as the Greek historian *Strabo* calls fighting dogs) fought alongside soldiers, watched over their sleep and, when trained, were also used in communications. The courage, strength and temperament of these dogs amazed the Roman Legionaries, who designated them to different groups according to the military structure in which they were used:

- *Cohors*: combat unit consisting of six companies (including 100 or more men);
- *Praetoria Cohors*: General bodyguards;
- *Cohortes*: auxiliary troops.

Ferocity, also outside of warfare, defined the evolution of the ancient Molossers, which were also used as killers. Individuals who had committed specific crimes were thrown into pits where they were torn to pieces by dogs which had been kept without food. Indeed, the expression 'thrown to the wolves' derives from this gruesome custom. The ferocity instilled in Molossers by man was also employed in the games of ancient Rome. In the *ludi gladiatori*, wild animals, Molossers and gladiators engaged in mortal combat. Thus, the ancestor of the Cane Corso was forced to fight against bears, lions, and bulls. Many Roman emperors became famous because of these cruel circus games. In the Roman amphitheatres, during the *ludi gladiatori*, wild beasts, Molossers and gladiators would fight. Thousands of bears were captured and used in the arenas. In the first century BC the emperor Caligula organised a ferocious fight of 400 bears against a group of gladiators accompanied by their dogs. Strabo writes that four Molossers were required to face a lion. The Molossers' fighting skills also emerged during bullfighting shows, where bulls fought among themselves, against other animals or even against men. Widespread throughout the Mediterranean area, in Rome, the *Theatrum Tauri* was destined for taurine hunting (*venationes*). Then there were the fights between men and beasts, where the former were victorious; '*venatores*' or '*bestiari*', were trained in schools similar to those for the training of the gladiators, but they were not held in high esteem by the spectators, they wore short tunics with sleeves and bands on their legs. Their weapons were a lance and a leather whip and they were always accompanied by a group of dogs. Among the *domini factionum* who managed the organisation of the performances, we might recall a curious episode involving *Aulus Fabritius*, a praetor of Nerone, who tried to counter the excessive requests of these administrators by presenting wagons pulled by previously trained Molossers.

Figure 6. *Denarius of Caius Antestius Labeo* (Crawford 1974,219/1e, 146 BC).

In 146 BC, *Caius Antestius Labeo* minted a series of coins with the Roman helmeted head on one side and Castor and Pollux riding upon two horses on the other (Figure 6). On one example in particular we find the horses underneath, as an extra element added by the mint, a large dog wearing a heavy collar around his neck like that worn by dogs during battle. It is no coincidence that it was *Caius Antestius Labeo* himself who used a dog jumping onto the usual bow of a ship on some bronze specimens.[8] With more aggressive features here is Cerberus, another famous mythological dog who had the role of protecting the door to the underworld, preventing entrance to humans while at the same time preventing the dead from escaping from the kingdom of Hades. In many images this dog is a Molosser. Examples are the coins of Caracalla, Elagabalus, Gordianus and Probus.

In the fourth century after Christ, the dog only appears in the coinage of Alexandria of Egypt, related to the empire's last years of paganism. Indeed, we find Serapis or the Goddess Isis sitting on the God Sothis who here has the form of a dog.

Therefore, with the exception of these unique specimens of the fourth century AD, in which an Egyptian divinity with the appearance of a dog is represented, this animal is attested until the third century AD, after which it disappears completely, perhaps a symptom of the radical change which would involve the empire in late antiquity or possibly the result of the dog's bad reputation which also continued with Christianity.

[8] Crawford 1974: 219/2–4 (146 BC).

Later, images on Roman coins became simpler and more standardised, losing the realism which in art heralds the advent of the Middle Ages.

**3 Conclusions**

The research sought to analyse how the dog' was seen in the Roman world. From literary sources the dog does not appear to be man's best friend. On the contrary, this animal lent itself to being involved in many negative elaborations, to of an odious inclination to transgression. In many societies where the dog is considered an intelligent, communicative, valid helper and trusted collaborator, in which it is allowed to be present in organised spaces of culture and even in human homes, the word 'dog' is an insult or recurs in disparaging expressions.

The affection which is shown towards dogs today was almost unknown for the ancients, who had altogether more practical uses for the animal.

The dog has certainly been the subject of many wildly contrasting opinions in historical sources. Some consider the animal as a symbol of loyalty whereas others accuse it of a wide range of transgressions, from simple bad behaviour to violence and deceit. Dogs evidently played such an integral part in the community that they were almost conceived of in human terms.

Dogs were expected to be docile and obedient, loyal and supportive, grateful and collaborative but due to an occasional inability to control their instincts, were not always able to be this. It is man's disappointment in this inability which is at the root of the large number of negative stereotypes linked to the dog and the 'dog's face'. The traitor, the blind executioner and the bloodthirsty madman are all figures born from the fear of an animal that man has always trusted greatly.

The ancient Roman coins, however, seem to rehabilitate the figure of the dog. In fact, on all the specimens it is never represented negatively but rather it is the faithful dog Argo, or the helper always present next to the Goddess Diana, as well as by the Lari Gods caressing a dog's head.

It is the symbol of an aristocratic society engaged in hunting or recalls the origins of Rome which has always been linked to sheep farming and agriculture. After all, the city of Rome did not choose a dog as its symbol, but a she-wolf, an ancient' cousin' of the dog.

**References**


Aragosti, A. (ed.) 1995. *Petronio Arbitro. Satyricon.* Milano: BUR.

Cantarella, E. 2018. *I supplizi capitali. Origine e funzioni delle pene di morte in Grecia e a Roma.* Milano: Feltrinelli.

Crawford, M.H. (ed.) 1974. *Roman Republican Coinage.* Cambridge: Cambridge University Press.

Giannarelli, E. (ed.) 1983. *Gaio Plinio Secondo. Storia Naturale.* Torino: Einaudi.

Franco, C. 2003. *Senza ritegno. Il cane e la donna nell'immaginario della Grecia antica.* Bologna: Il Mulino.

Labiano, M. 2012. *The problem of autorship of Xenophon's* Cynegeticus, in J. Martinez (ed.) *Mundus vult decipi. Estudios interdisciplinares sobre falsificación textual y literaria*: 171–183. Madrid: Ediciones Clásicas.

Luce, T.J. 1968. Political Propaganda on Roman Republican Coins: circa 92–82 BC. *American Journal of Archaeology* 72.1: 25–39.

Privitera, G.A. (ed.) 1991. *Omero. Odissea.* Milano: BUR.

Pelagalli, G.V. and M. Di Gerio 2017. *Il cane nell'arte pompeiana.* Napoli: Valtrend editore.

# 5.8 Dogs in Early Imperial China: Anthropo-Zoological Reading of Iconographic Sources from the Han Dynasty (206 BC-AD 220)

Frédéric Devienne

Centre de recherche sur les civilisations de l'Asie orientale (CRCAO) - UFR Langues et civilisations de l'Asie orientale (LCAO). Université de Paris. Case 7009. 5 rue Thomas Mann. 75205 Paris cedex 13. France. frd.dvn@gmail.com

**Abstract**

During the Han dynasty, at the end of Chinese Antiquity, dogs gained visibility in art, to the extent that the period can be seen as a true golden age of canine representation in China. In its analysis of iconographic materials taken from archaeological sources and mainly from funerary art, the multifocal approach adopted here aims to highlight the extent of the domestic condition of dogs, and to reveal their supposed role in beliefs during that period in China.

**Keywords:** dogs, China, iconography, anthropo-zoological aspects.

## 1 Introduction

The study of dogs in ancient China based on various cultural sources - texts, images, artefacts - is not new, either in the West or in China. However, the subject is worthy of renewed attention following the discovery of a large number of archaeological material in recent decades. While Chinese authors have been prompt to report these new discoveries, they have so far only done so on a case-by-case basis, in the limited context of detailed excavation reports.[1] Moreover, the question of the dog remains subordinate to the general animal theme. Once overlooked, the subject of the animal has begun to gain momentum, especially since the mid-1990s.[2] Nevertheless, it remains very limited and confined to the realm of zooarchaeologists.

Western sinology, for its part, has three major studies on the question of dogs in China, but these date from the first half of the 20th century.[3] Among them, those of Laufer and Erkes are undoubtedly the most significant, in their innovative efforts to initiate canine studies in China, the former from a perspective of cultural history in relation to the theme of canine varieties, and the latter focusing on the religious aspect, including its ritual, folkloric and magical dimensions.

In the present study, the purpose is to take both these approaches further, with a focus on iconographic sources. Moreover, the perspective chosen here is consistent with the current conception of animal studies in social sciences and humanities. This reflection begins with a few methodological considerations.

## 2 Method, materials and challenges

Following a presentation of the adopted method and corpus, some issues will be raised.

### *2.1 For an anthropo-zoological approach to animal iconography*

The study of zoography, or animal representation and description by man, is a vast field of research. Apart from the strictly zoological point of view of what could be called 'scientific zoography', illustrated by scientific imagery and natural description of animals in science since the 18th century, we face a wide range of cultural testimonies on animals achieved by man since ancient times, during all the history he has shared with animals. Less objective, sometimes totally symbolic and unrealistic, this type of animal representation and description established by man, through art, folklore, myths, literature, beliefs, rituals, and any knowledge of the humanities, belong to what we could call 'cultural zoography'. According to the sources considered, materials from different nature appear, oral or immaterial, textual, and visual. In this study, it is clearly the last type of material, and particularly the pictorial representations in visual arts, that constitute the main sources of documentation. As far as the dog is concerned, the zoography of the general scope becomes a 'cynography' when all the materials collected are strictly limited to this animal.

The visual nature of the 'cynographic' materials allows to document, by making them visible, the concrete

[1] For example, Liu (1986), Wei and Tian (1991) and Zhang and Chi (1997), Wang (2006), for the study of *terra cotta* statuettes discovered in large quantities in Nanyang (Henan) since the 1980's.
[2] Elisseeff (1992; 1998).
[3] Chronologically, Laufer (1909), Collier (1921) and Erkes (1944).

and spiritual manifestations of what constitutes the heart of the subject here: the human domination over animals. Choosing the dog makes it particularly relevant, as in the case of man's closest companion, anthropic pressure is particularly strong. By focusing on the iconographic sources, such an approach falls de facto in the field of the cultural study of animals and more precisely, with regards to the past, in the field of their cultural history. This notion is not new, in that it was initiated by O. Keller, between 1887 and 1913, and also practiced, in the case of the dog, by zoologists of the era of scholarly cynology (1860–1930) up to Max Hilzheimer.

In the 1980s, a new perspective on the relationship between man and animals emerged, in particular the recognition of animal contribution to human civilisation. Under the impetus of innovative writings (Delort 1984, the journals *Anthropozoologica* and *Ethnozootechnie*), the trend was accentuated over the following decade (Digard 1990, works edited by Bodson since 1988) leading to the emergence, particularly in French and Belgian university circles, of an independent branch of zoology: anthropo-zoology. The aim now is to study animals in their historical, cultural and religious relationship with man and no longer from the sole perspective of natural history. The extent of publications on the dog in recent years has made it possible to consider the subject as a branch in its own right.

The thematic filter used here to sift through the iconographic material - the morphology of dogs, their functions in reality and their emotional, spiritual and symbolic understanding by the Chinese of the time - is therefore in keeping with this perspective.

### *2.2 The corpus*

As far as representativeness of sources is concerned, the decision to limit the study to the Han dynasty was a natural one.

#### *2.2.1 Period*

Representations of a domestic animal such as the dog were rare in China before the end of antiquity (Table 1).[4] In contrast to their relative invisibility in pre-imperial art, dogs became abundantly represented under the Han (206 BC-AD 220), especially during the second period, the Eastern Han (AD 25–220) (Figure 1). Of varied appearance, these dogs are depicted in a variety of supports and techniques: pictorial wall carvings on stone and stone slabs (*huaxiangshi*) for funerary structures such as tombs, sarcophagi, commemorative shrines and offering chambers; brick reliefs and hollow bricks (*huaxiangzhuan*, *kongxinzhuan*), stamped clay tiles (*wadang*); murals; funerary *terracotta* statuettes (*taoyong*); decors and ornaments of vases and artefacts such as glazed and unglazed earthenware, bronzes, lacquerware; paintings on silk; woven fabrics, etc.

Table 1. Representations of dog motifs in art up to early imperial China.

| **Late Neolithic** (4500 – *c.* 2000 BC) | **From the Bronze Age to the Qin dynasty** (*c.* 2000 – 206 BC) | **Han dynasty** (206 BC – AD 220) |
|---|---|---|
| 80 motifs | 50 motifs | More than 375 motifs |

#### *2.2.2 Qualitative and quantitative factors*

In formal terms, the techniques of the Han period stemmed from the modular manufacturing processes of pre-imperial art.[5] The highly stereotypical character of the images, the dog motif being one of them, went hand-in-hand with their production in large quantities using stencils, layers, templates, matrices or molds. This factor needs to be taken into consideration when counting the occurrences of the dog motifs in the corpus (Table 2).[6]

### *2.3 Points of discussion*

From a material and concrete perspective, and in view of the new concept of domestication, in that it is now understood as a continuous and gradual process of permanent interaction between the domesticated animal and man, the aim of this study is to determine the degree of specialisation in the use of dogs in Han China. Enthusiasm for dogs at that time can thus be measured according to physical and technical criteria, by the degree of morphological differentiation of dogs, by socio-cultural and practical criteria, by the degree of diversification of their activities for and with humans or, on the contrary, by their versatility.

From the perspective of ideas and beliefs, it needs to be seen whether the phenomenon is accompanied by a shift in mentality in favour of the dog. This requires the examination of more intangible criteria such as the degree of autonomy and integration of dogs with man in his private sphere, and the degree of sacralisation of the dog in religion.

The results of this iconological investigation are broken down into the following three points that will be examined and discussed in relation to one another.

[4] Figures in Tables 1 and 2 are rounded off, according to my iconographic database on the 9th August 2019.

[5] Ledderose (2000).

[6] Table 2 updates the data of Masumitsu (2004: 164, tab. 1).

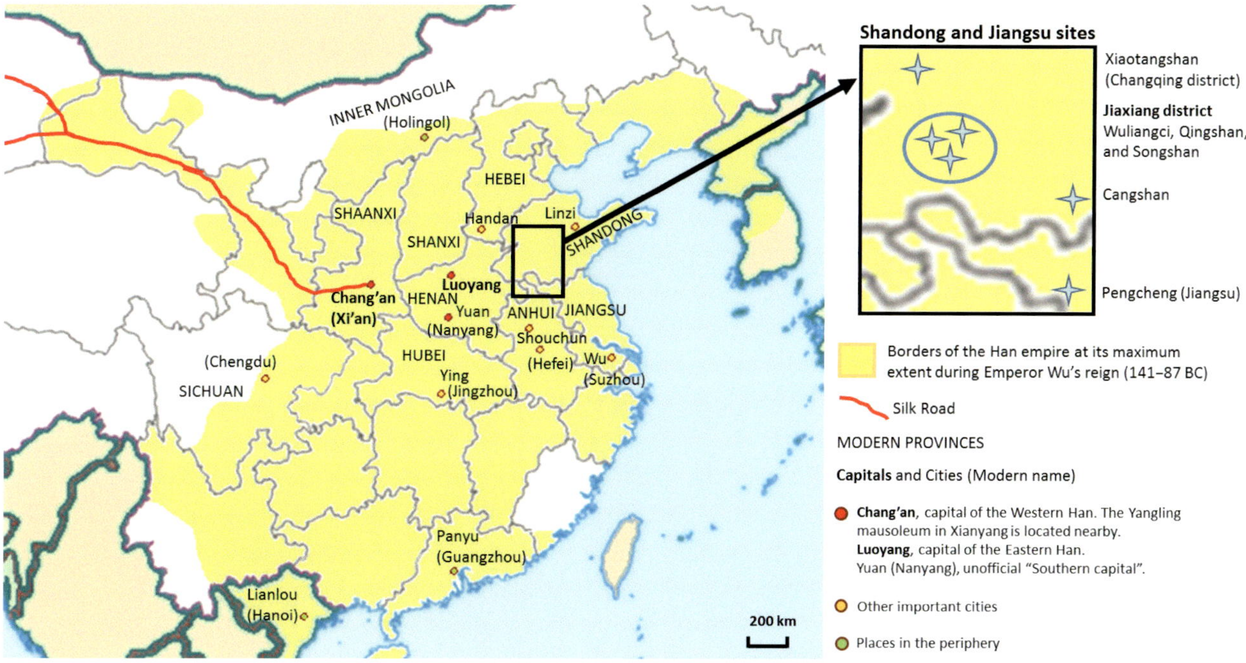


Figure 1. Map of Han dynasty, China.

Table 2. Dog in Han art: scale of magnitude in absolute and relative quantities.

| **Pictorial patterns on architectural structures** (Stone with relief and incised carving, brick reliefs and stamped tiles, murals...) | | **Small-scale patterns on pottery, vases, objects** (Lacquerware, earthenware, stamped ceramics, bronzes...) | | **Statuettes, figurines, monumental statuary** | | **Total** | |
|---|---|---|---|---|---|---|---|
| Duplicates | Single motif | Duplicates | Single | Duplicates | Single | Duplicates | Single |
| 450 | 160 | 250 | 65 | 3000 | 150 | 1000 | 375 |

Before focusing on the Eastern Han, their predecessors will first be examined, the Western Han.

## 3 The morphology of dogs in Western Han art

The visibility of dogs in the decorative and figurative arts of the Han period was not fully achieved until the second half of the dynasty. Under the Western Han, modular techniques still adopted methods of representation that suggest an entirely symbolic representation of reality. In relation to the animal world, this amounts to representing the generic traits of the species or breed under consideration, or even attributing to a particular dog the general characteristics of the entire species.

### *3.1 Multi-functionality and modularity*

The first example shown here is the large dog present in the historied ornamentation of stamped hollow bricks (*huaxiang kongxin zhuan*) of the Western Han in Luoyang (Figure 2A).[7] This unique pattern represents a specimen of a large hunting dog. Morphologically situated between the wolfdog and the greyhound, it has some of the features of the former; pointed ears, conically tapering muzzle, triangular head, and others of the latter; short hair, with a slender athletic body.

This particular motif, being a replicable matrix and therefore versatile by definition, is used to depict various representatives of the canine species in a variety of contexts. Isolated and non-contextualised, it depicts a large hunting dog in a 'pointing' stance.[8] Then depicted at a distance behind a deer, we can imagine it chasing its prey. We then see it catching a hare. Behind a man with a stick, this same dog becomes

[7] The *intaglio* design on the hollow brick is made with a stamp on the still damp grey clay. The matrix stamp itself figures an embossed version of the same design, in reverse. (White 1938; Fairbank 1942; Huang 1982; Zhou *et al.* 1985).

[8] Hunting specific term. In French, 'marquer l'arrêt'.

Table 3. Distribution of the 1391 animal statuettes from the Yangling Mausoleum pit 13.

| **Caprids** (Caprinae) | | **Canids** (Canidae) | | **Pigs** (Suidae) | |
|---|---|---|---|---|---|
| Goats | Sheep | Wolfdogs (type A) | Mastiffs or basic dogs (type B) | Adult pigs | Piglets |
| 235 | 189 | 458 (16 individual types according to colour, to gender, and to morphological differentiation) | | 455 | 54 |

Figure 2. Representation of dogs during the Western Han (206 BC-AD 23):
A) Large hound. A pattern and its variations. Stamped hollow bricks. Luoyang (Henan, 3rd-1rst century BC). Matrix stamp in relief (reconstitution) and rubbings of the *intaglio* designs. Modified from Huang 1982 and others; B) Wolfdog (or wolf type hound) and mastiff type dog (or basic dog). Front and side view. Funerary statuettes. Yangling mausoleum (Shaanxi, 141 BC). Hand drawing, modified from *Wenwu* 1994.6: 14, fig. 20; C) Hunting dogs - Rooster fight. Murals. Tomb M1 of Houtun, Dongping district (Shandong, AD 9–23). Hand drawing, modified from *Dongping Houtun Handai bihua mu* 2010: 30, fig. 21.

Table 4. Muzzle length criterion: assumptions and representation.

| **Context** \ **Muzzle length** | | **Very long** | **Long (pointed)** | **Mid long (and thick)** | **Short and crushed** |
|---|---|---|---|---|---|
| **Chinese generic name** | | *Quan* | | *Gou* | |
| **Morphotype (primary type of Mégnin)** | | Greyhound (graioid) | Wolfdog (lupoid) | Pseudo-braque (pre-braccoid) | Mastiff or Dog (molossoid) |
| **Function** | | Sight hound | Sight / Scent | Scent hound | Guard dog, domestic dog |
| | | Hound, hunting dog | | | |
| | | *Zouquan* | | *Zougou* (?) | *Shougou, chugou* |
| **Chinese specific name** | | *Xian* | | *Xiexiao* | *Ao* |
| **Site** | **Yangling (fig. 2B)** | | A type | Re-use? ← | ← B type |
| | **Luoyang (fig. 2A)** | A+ type | | | |
| | **Houtun (Fig. 2C)** | A++ type | | A/B mixt type | |

a traveler's companion.[9] In front of an armed soldier, he turns into a garrison dog. Only the context and the accompanying motifs allow us to appreciate the versatility of this dog, whose appearance and attitude remain unchanged from one scene to another.

### *3.2 Morphological duality: a case of functional dimorphism?*

In the comparison of two generic types, as shown in the second set of examples (Figure 2B), a further step is taken in the process of representing dogs. In the funerary context of a meat reserve for the afterlife, the zoomorphic statuettes in the accompanying pit no.13 of Emperor Jingdi's Yangling Mausoleum highlight a marked morphological duality between the following two profiles.[10]

- Type A of a wolfdog (lupoid morphotype): triangular head, pointed muzzle, marked furrow in the middle of the forehead (cf. the sagittal crest, prominent in wolves), oblique, close, round eyes, pointed ears located on the side of the skull, slightly protruding jaws, tuft of hair from the neck down to the cheeks (like wolves), a straight hanging bushy tail, thin and relatively high members, and with a slim overall body, smaller than type B.
- Type B of a mastiff or a basic dog (molossoid morphotype): square head, flat muzzle, rounded forehead, eyes set wide apart on the side, bulges around the eyes, ears rounded and located at the top of the skull, massive prominent jaws, thin tail, straightened in an arc and set on the croup, short large limbs with a powerful heavy body. The tuft of hair on the cheeks is also present in these dogs.

These statuettes were unearthed from a limited selection of pair-figured domestic animals, whose function as animals for slaughter is clear, according to the editors of the excavation report. There are other variations such as gender and coat colour. Such modularity is expressed on two subsequent levels: that of variety and that of the individual (Table 3).[11]

Beyond their immediate purpose as a meat reserve, the dogs of the Yangling Mausoleum raise the question of the specificity and origin of their morphological differentiation. It can be assumed that the two morphotypes in question, lupoid and molossoid, must have represented the two dominant morphological types of dogs at the time in China, or perhaps the two main strains of the existing breeding channels (Figure 3). Table 4 provides a hypothetical overview of the situation.[12]

[9] The scene possibly depicts the hypothetic meeting between Confucius and Laozi, as mentioned in *Shiji*, ch. 63. The traveler with the stick may be Confucius and the dog behind him, his companion, evokes a rambling homelessness. The grieving family dog (*sangjia zhi gou*) is indeed a metaphor of Confucius himself, in his exile, as a political advisor without assignment, nor kingdom. (Cf. Han texts *Hanshi waizhuan*, ch. 9; *Kongzi jiayu*, ch. 22; *Shiji*, ch. 47).

[10] Emperor Jindi (r. 157–141 BC, posthumous name Xiaojing). Mausoleum discovered in 1990 at Xianyang, Shaanxi. Dog statuettes dimensions: H. = 19.9 cm, L. = 31.2 cm, l. = 8.5 cm. Excavation reports in *Wenwu* 1992.4 and 1994.6.

[11] *Wenwu* 1994.6: 4–23 and 30.

[12] Table 4. For the Chinese names, see the definitions in the *Shuowen jiezi* and *Erya*. For the distinction between *xian* (long muzzle dog) and *xiexiao* (short muzzle dog), see also the commentaries of the *Shijing*, I.XI.2 (no.127), Sitie.

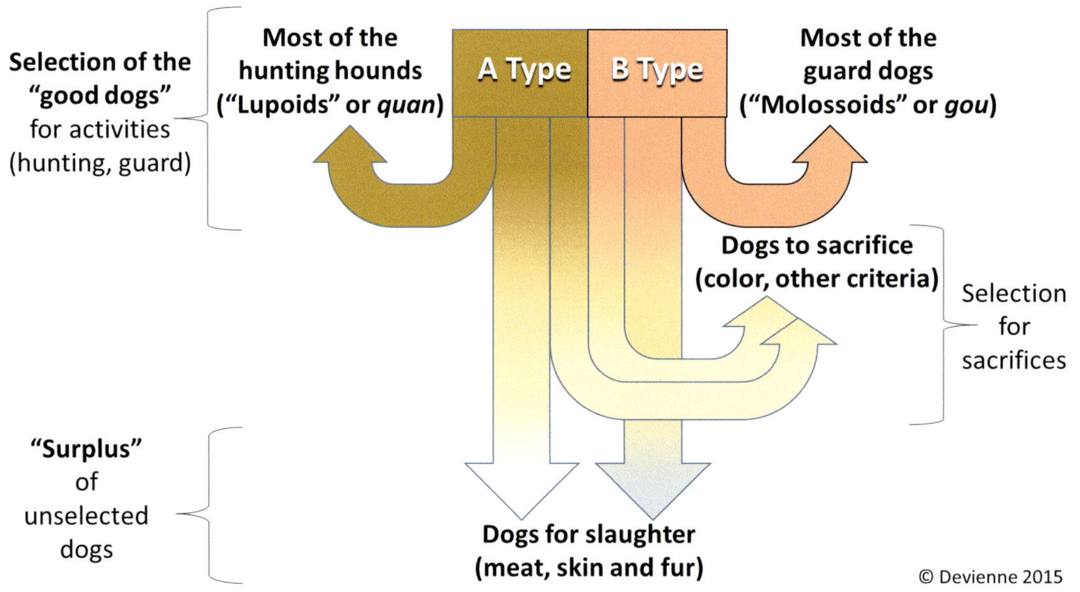


Figure 3. Dog supply chain hypothesis.

Subsequently, from the Eastern Han period onwards, the description of canine morphology in art becomes inseparable from the functional context in which dogs evolve.

## 4 The role of dogs in reality according to the 'contextualised' art of the Eastern Han

The traditional interpretation of the Confucian classical texts assigns to the dogs of Chinese Antiquity the three main roles of guard dog, hunting dog and slaughter dog.[13] This limited conception does not reflect the complex reality. That said, the most frequently illustrated cases do indeed fall under the three canonical functions: butchery, hunting and guarding. The outline of the following presentation also reflects this tri-partition, which needs however to be refined. The following examples all date, for the most part, from the Eastern Han.

### *4.1 Kitchen imagery. Slaughter and skinning of dogs amidst their peers*

In Shandong during the Eastern Han, the imprint of Confucian ideology on funerary art is reflected in the staging of ritual animal sacrifices in the bas-reliefs. In this context, the dogs that are sacrificed or eaten, and whose fate is clearly shown, are not the only ones present in the kitchens. On the contrary, this place of domesticity reveals differences in treatment of the canine population.

#### *4.1.1 The differentiated status of dogs revealed by the iconography of kitchens*

When examining the iconography (Table 5), it is possible to distinguish the status of dogs as follows:

- As the **dog of the house**, its duties include the task of watching the kitchen (Figure 4D).[14] As such, he is not tied but left free. Sometimes wearing a collar, it is a well-trained and behaved dog, responsible for chasing away rodents and other undesirable animals, or for guarding against theft. Its depiction with a turned head testifies to his increased vigilance. When the dog is out with his master or perhaps offered as a present, he can be tied up in the kitchen of the host as shown in Figure 4E.
- **Slaughter dogs** for butchery or sacrifice are slaughtered by caning (Figure 4B) or by hanging and strangulation at the gallows of a well (Figure 4A-inf.). The two methods are first shown together on the Xiaotangshan stone slabs (AD 76–88).[15] It is this depiction of the dog being skinned at the well that gains prominence at the offering chambers in Wuliangci and Songshan.[16] These practices still exist today in China and Korea.[17]
- **Village dogs or street dogs.** Undesirable and a nuisance, they are persecuted or repelled. One of the rare depictions is at the Wuliangci, in a scene

[13] Cf. the commentaries of the *Liji* (ch. 17, Shaoyi) by Kong Yingda (*Wujing zhengyi*, *juan* 35) at the beginning of the 7th century AD and the commentaries of the *Zhouli* by Jia Gongyan *c.* 650 AD (*Zhouli yishu*, ch. 36, Quanren).

[14] In total, this pattern has more than a dozen occurrences, including *Zhongguo huaxiangshi quanji* (*ZHQ*) 1.123.

[15] *ZHQ* 1.42.

[16] About fifteen occurrences including *ZHQ* 1.42, 1.50, 1.58, 1.77, 1.90, 2.134, 3.9, 3.140.

[17] Orange (1992).

Figure 4. Dog in kitchens. Pictorial stone carvings from the Eastern Han (Shandong, unless otherwise stated):
A) East wall of an isolated stone slab. Songshan, Jiaxiang district; Upper panel: Dongwangong (King-Father-of-the-East), hybrids and cloud swirls with dog's head; Lower panel: Kitchen scene with a dog slaughter at a well; Rubbing from *ZHQ* 2.98.
B) Caning of a dog led to slaughter: B1) Xiaotangshan; B2) Liangtai, Zhucheng district; B3) Guhexiang, Sichuan; Rubbings from *ZHQ* 1.42 and 7.105. Hand drawing from *Wenwu* 1981.10: 19, fig. 7.
C) Undesirable dogs repelled at the entrance of the kitchens. Wuliangci. Rubbing from *ZHQ* 1.88.
D) Dog watching the kitchens: D1) Greyhound, Wuliangci; D2) Lupoid, Jiangsu; Rubbing from Gu 1997: 236. Photograph from *ZHQ* 4.14.
E) Tied mastiff devouring a hanging leg. Baizhuang, at Linyi. Rubbing from *ZHQ* 3.9.

Table 5. The intervention of dogs in kitchens after the scenes engraved on stone in Shandong.

| Mixed utility (active and social): **Dogs of the house.** A dog with a name (and often with a collar) | | Passive utility: **Slaughter dogs** for butchery or sacrifice. A nameless dog | | | | Latent utility: **Stray dogs or free-ranging dogs** |
|---|---|---|---|---|---|---|
| Housedog watching the kitchens, chasing away rodents | Guard dog accompanying his master outside or offered to the host. | After death (dog as meat) | | The act of killing the dog | | **Village dogs, street dogs** (latent function of scavenging) |
| | | Above the stove | Hanging on a butcher's hook | | | |
| Not tied and left free dog in the middle of the kitchens (Fig. 4D1 and D2) | A large mastiff tied in the kitchens (Fig. 4E) | Dog's head (Fig. 4A-inf., left) | Dog's body (Fig. 4B3, left) | Dog led to slaughter: tied on a leash, pummelled by cudgel (Fig. 4B) | Flaying at a well (Fig. 4A-inf., right) | Kitchen clerks kicking the dogs (Fig. 4C) |

Table 6. The six domestic animals (*liuchu*). * Introduced in China after the sheep, goats are assimilated to the same inclusive category of *yang* (Caprinae). ** Pirazzoli-t'Serstevens and Bujard 2017: 436.

| Animals | Five domestic animals for alimentary or sacrifice purpose (*wuchu*) | | | | | Horse |
|---|---|---|---|---|---|---|
| | Cattle | Pig | Sheep (and goat)* | Poultry | Dog | |
| Sacrifices | Sacrificial triad (Chinese *suovetaurile*, *tailao*): Bull, pig, ram. | | | Small sacrifices | Small sacrifices | Formerly, deposit in pit |
| | / | Sacrificial dyad: Pig and ram (some rituals, as the *shaolao*) | | | Meat for sacrifice (taboo?) | |
| Used as food common / (rare) | (Meat) (Milk) | Meat | Meat (Milk) | Meat, eggs | → becomes a common dish and an object of sacrifice | (Meat) Milk |
| Other uses (animal labour) | Transport, pack, ploughing, dry dung for fuel | Scavenging | Pasture, weeding, wool (passive) | Limited (alert, wake up) | Multiple (hunt, guard, sled...) | Transport, pack, riding, dung for fuel |
| Cost (coins)** | 1200 to 3750 | 300 to 900 | 150 to 500 | 70 | 100 | 5 to 10 000 |

representing the return from a hunt with game being brought into the kitchens (Figure 4C). The kick given to the dog is unambiguous: the dog does not belong to the household; it is rather an intruder that must be chased away and should not have access to the kitchens.

*4.1.2 Ritual cooking and regulated use of dog meat*

Although each kitchen scene is unique in its composition (Figure 4A-inf.), a limited number of modular elements are found in each of them, in a specific combination and spatial distribution. Each scene has its particularity. Apart from the modulation in the arrangement of the decorative elements (stove, well, kitchen furniture: hanging gallows, cutting tables, jars, barrels), it is above all the modulation within the group of five edible domestic animals (Table 6) that is particularly instructive.

The extent of animal killing, the choice of animals and the funeral context make the case for ritual cuisine, in which the sacrifice of animals offered to ancestors, or even consumed, is an integral part of the ceremony (Chang 1977; Sterckx 2011). For such commemorative or processional banquets, it is the ceremonial framework

Figure 5. Morphological diversity of dogs under the Eastern Han. From pictorial stone carvings, statuettes, etc.: A) Hare hunt, with a leashed greyhound, mastiff type tracking hounds, and hunting-nets carriers. Qingshan (formerly Ciyun) temple, near Jiaocheng village. Jiaxiang district, Shandong. Xylographic reproduction from *Jinshi suo* 1821, Shisuo 4: 27b-28a. B1) Henan hound. Nanyang, Henan; B2–4) Hound, 'fox-dog', and giant mouth wolfdog. Xiaotangshan shrine. Changqing district, Shandong; B5–6) Wolfdog and large greyhound. Miao Yu tomb at Pengcheng, Jiangsu; B7) High-sized mastiff type hound. Cangshan, Shandong. Rubbings from *ZHQ* 6.153, 1.43 and 3.114. Hand drawing from *Wenwu* 1984.8: 25, fig. 8. C1) Shepherd. Mural painting, Holingol (Inner Mongolia). From *Han Tang bihua* 1974: pl. 36; C2) Ao, large mastiff. Brick relief, Zhengzhou (Henan). Rubbing from Zhou *et al.* 1985: no.102; C3) Farm wolfdog. Carved stone slab, Suide (Shaanxi). Rubbing from *ZHQ* 5.139; C4) Chained large mastiff. Large brick relief, Fanji (Xinye, Henan). Rubbing from *Kaogu xuebao*, 1990.4: 501, fig. 1; C5–8) Guard dogs and/or companion dog. Funerary statuettes: C5) Jiangsu. Metropolitan Museum of Art, New York. From Bower 2005: 434, cat. no.53; C6) Shaanxi. British Museum; C7–8) Nanyang, Henan. From Zhang and Chi 1997: 35, fig. 1 and Nanyang Municipal Museum. D) Re-employement of greyhounds, as pageantry and prestige dogs. Wuliangci and Songshan, Shandong. Rubbings from *ZHQ* 1.81 and 2.100. E) Zhao Dun attacked by an Ao mastiff. Wuliangci. Rubbing from *ZHQ* 1.216.

rigorously defined in the *Book of Etiquette and Ceremonial* (Yili) and the *Books of Rites* (Liji), that applies. These texts teach us two points. Firstly, while explicit references appear in the Han period commentaries, dog meat is never explicitly mentioned in the canonical sources themselves. This implies a possible taboo around canine meat in ancient times or, failing that, a reinterpretation of the Han in that regard. Secondly, the commentaries based on these two books place great emphasis on the predominant use of dog meat in lesser rituals of sacrifice, which fits well with the humble and virtuous character attributed to dogs under the Han. Finally, as at the Yangling Mausoleum, the slaughtered dogs do not seem to be of a particular variety. This is demonstrated by the use of undifferentiated dogs, and the similarities between dogs sacrificed at the well and house dogs watching the kitchens.[18]

[18] *ZHQ* 3.140.

#### 4.2 The imagery of the hunt. Highlighting the prominent role of dogs

Equally fundamental is the contribution of Han iconography to a better, and visually explicit, knowledge of hunting practices. Indeed, visual sources emphasise the major contribution of dogs to hunting, whereas the texts barely mention them. Even the writings of court poets and scholars such as Sima Xiangru, Yang Xiong and Zhang Heng, who nevertheless excelled at describing these hunts, contain little reference to the dog.[19] While certain morphological differences in hunting dogs date back to at least the Eastern Zhou (771–256 BC),[20] it is indeed under the Han that we may gain a fuller appreciation of their role.

##### *4.2.1 The differentiated composition of dog packs*

Inherited from High Antiquity, the great imperial and royal hunts were transformed by Emperor Han Wudi (r. 141–87 BC) into veritable processions to legitimise imperial and military power. Although imperial hunts declined under the Eastern Han, hunting continued to flourish in some provinces. Smaller, more individual hunting, which first appeared under the Western Han among the new elites of enriched commoners, notables and merchants, became common amongst town folk of the suburbs, in the form of hare racing.[21] The following three emblematic sites illustrate these different modes of hunting:

- The shrine or stone chamber at Xiaotangshan (AD 76–88) illustrates the great imperial hunts at their apogee, after the revival of Shanglin Park.[22] The hunting parties were exceptionally large, with hundreds of men-at-arms, including lancers, crossbowmen, horsemen, falconers and net bearers. It is here that we find the most diversified packs of hounds, with at least four different varieties of hounds, each with their specific assignment.
- The tomb of Miao Yu (†AD 150) at Pengcheng (present-day Jiangsu)[23] presents another configuration, probably depicting hunting of a high local dignitary, with at most three distinct varieties of hounds: wolfdogs for stalking and attack, majestic greyhounds in the rear and a third specimen with an uncertain appearance.
- Although more diverse, the small individual hunts possess their own model of representation. One of the most characteristic is the scene represented on one of the stone slabs found at Qingshan Temple (formerly Ciyun Temple), near the village of Jiaocheng (AD 25–88, Figure 5A), in Shandong and kept in the collections of the University of Tokyo. Taking the form of an illustrated canon, the scene appears in multiple variations until the AD 140s, following the same general model: a greyhound on a leash, ready to pounce, facing a pack of howling dogs that chase the prey towards him. The scene can be interpreted as a shortened version of the final assault on prey in the great imperial hunts.[24]

##### *4.2.2 Morphological differentiation adapted to specific tasks*

For hunting, it is very clearly the profile of a large, slender, pseudo-lupoid hound which dominates. Used alone or in the packs with his fellows, it is absent from composite packs. Its versatility allows it to pursue and attack all types of prey, from the largest to the smallest. In Henan, it even replaces greyhounds.[25] It is therefore assumed that it also runs very fast and hunts by sight as well as by scent. There are two variants of this dog, depending on the region:

- In central and eastern regions, these large hounds have very high muscular legs and barring exceptions, a short, black coat. Under the Western Han and up to the very beginning of the Eastern Han, they were similar to the Luoyang type described above (Figure 2A). Under the Eastern Han, they were replaced by the Henan standard; a dog with a flared head, wider jaws and especially long, floppy, tapering ears (Figure 5B1).[26] Its black coat makes it possible to identify it with the Hanlu mentioned in the texts.[27] The Baiquan dogs, north of Henan, are another later variation, with beige hair and white legs.[28]
- In the western regions, on the other hand, it is a real wolfdog, clearly of the lupoid type, with a red, white or black coat, which replaces it.[29] This is also the case in Jiangsu (Figure 5B5).

In contrast to the general-purpose dogs, specialised hounds have more prominent morphotypes and, except for greyhounds, are more useful whilst operating in packs:

---

[19] Cf. the literary genre of *fu* in the *Wenxuan*, an anthology of Chinese poetry. Translation by Knechtges 1982 and 1987.

[20] The existence of distinct profiles - dogs with long snout (*xian*) and short snout (*xiexiao*) - is assumed in the princely hunts of Duke Xiang of Qin (r. 777–766 BC). See the comments on the poem 'Sitie', already cited. The polymorphism of hunting dogs is also evident in the art of the Warring States period (453–221 BC). Cf. Weber 1966–1968.

[21] For the elite families enjoying the pleasures of hunting, see *Yantielun*, II.9. For hare hunting as a distraction for humble people, see *Shiji*, ch. 101, ed. Zhonghua shuju (Zhsj) 1982, 8: 2744.

[22] *ZHQ* 1.43.

[23] *Wenwu* 1984.8: 22–29; *ZHQ* 4.139.

[24] Zhang Pingzi, *The Capital of the West* (Xijing fu), *Wenxuan*, II.2b. Traduction by Knechtges 1982: 219–221.

[25] *ZHQ* 6.153.

[26] See also *ZHQ* 6.79.

[27] *Shijing*, I.VIII.8 (no.103), Luling. *Zhanguoce*, Qice 3. *Xinxu*, Xinshi 5. *Hanshi waizhuan*, ch. 7. Etc.

[28] Ding 1970: 63.

[29] See *ZHQ*, vol. 5, devoted to Shanxi and Shaanxi. For the coat colour, see the scene of *Hare Hunt with Dogs and Birds of Prey*, represented on a painted brick from the Gansu (3rd century AD), Jiayuguan Museum.

- The greyhound (graioid type), a slender, very swift dog, which hunts by sight and quickly becomes exhausted, is depicted held on a leash, ready to pounce. It is found in Shandong (Qi Hanlu), Jiaocheng and as far as Wuliangci. It can be considered the ultimate thoroughbred, developed from the standard hound by selecting the fastest specimens. Indeed, the iconography of the scenes where the greyhound comes into play between the years AD 25–140 in Shandong (Figure 5A and 5B6, for example), depicts a hesitant and sometimes divergent morphology from one site to another, until the greyhound finally appears in Wuliangci in the years AD 140–180 (Figure 5D, among others).
- The tracking hounds (mixed type, pseudo-molossoid) have a more imposing appearance: a tubular body, a large tucked in neck, capable of supporting a large head. They hunt by scent and their loud bark is used to frighten game (Figure 5A). The Xiaotangshan variety with short hair seems almost to herald the braccoid type, in that it presents a thick rectangular muzzle as well as a sharp nasal angle (Figure 5B2). It differs from the long-haired variety of Houtun (Figure 2C).
- Large mastiffs, more closely representative of the molossoid type, also play the role of trackers in Shandong and Henan (Figure 5B7).
- Atypical morphotypes still appear in the Xiaotangshan stone chamber. A fox-headed dog on a leash could perform the function of tracker (Figure 5B3). The large, stocky, lupoid dog with a giant mouth (Figure 5B4) is also found in Juxian, where it is used to kill prey.[30]

### *4.3 Imagery of the watchdog at home. Reconsideration of bias in the texts*

The polymorphism of dogs is further illustrated in their guard duties. Apart from the significant re-use of wolfdogs, greyhounds and undifferentiated dogs, it is here the molossoid morphotype, that of mastiffs or domestic dogs in the broad sense, which dominates in the function of watching and protection (Figure 5C). This type, which includes about fifteen varieties, constitutes the most pronounced morphological disparity under the Han period. For these dogs, the images reveal two uses.

#### *4.3.1 The active use of guard dogs*

Whereas the official texts, in continuity with the pre-imperial tradition, favour the moral scope of anecdotes,[31] the iconography of the Han demonstrates the material conditions of the life of almost all dogs, and the varying degrees of autonomy that they enjoyed:

a. **House dogs** respond to utilitarian security tasks (protection of property and people, vigilance) but as ratters, they are also guarantors of hygiene by keeping rodents and other carriers of disease and insalubrity away. They are usually accepted inside the house:
   - **Tied dogs.** Posted in front of dwellings and at front doors, in niches or cavities, they are chained ferocious medium or large-sized[32] imposing mastiffs (Figure 5C4). Others are placed in strategic locations, such as under warning drums.[33]
   - **Semi-free dogs.** Free to come and go within the enclosure of a house,[34] they watch over the kitchens as previously mentioned; they are the guardians of grain silos and more generally, act as ratters, sometimes shown hunting rats[35] or devouring them.[36]
   - **Dogs with total freedom.** Untied and unhindered, they still depend on a home but can enter and leave the enclosure of the house at will, through a small hatch door specifically made for them.[37] Such freedom is granted to dogs in the New Fengyi town on the outskirts of Chang'an.[38] In a rural context, the same liberty is given to farm or barnyard dogs and wolfdogs that watch over livestock, in Shaanxi (Figure 5C3 among others)[39] and in Jiangsu.[40]

Most of these dogs wear collars, and some of them, the most ferocious or the most powerful, have a single or double harness. In Sichuan, large bulldogs are still to be found.[41]

b. **Herding dogs.** Just as absent in iconography as in texts, their obvious under-representation is perhaps due to the fact that pastoral activity

[30] *ZHQ* 3.139.

[31] See the literary cliché in texts of the ferocious guard dog (*menggou*), a metaphor for the clumsy, cowardly or stupid advisers of a sovereign, in the writings of Han Feizi (ch. 34, a wine merchant's dog who scares his customers) and of Guanzi (ch. Jie, snarling dogs, ready to bite all passers-by). In Han time entertainment literature, for example the *Miscellaneous Notes on the Western Capital* (Xijing zaji), canine anecdotes provide more material examples.

[32] *ZHQ*, 7.14. See also the dog statuettes and the miniature towers and dwellings with a dog posted at the entrance.

[33] Greyhound under a drum, Tai'an. *Zhongguo tapian*: 70, no.2130.

[34] Household spaces open to dogs appear on a Sichuan brick. (Zhang 1982: 17, no.3).

[35] *ZHQ* 1.222.

[36] *ZHQ* 7.35.

[37] *Yanzi chunqiu*, Zapian (xia). Out of contempt, Emissary Yanzi is forced to use the small side door reserved for dogs.

[38] Cf. accustom brought from Jiangsu by the first Han emperor, so that his father, native from there, would not feel out of place when settled in the capital. It consisted in letting dogs, cattle and poultry come in and out alone. *Xijing zaji*, II.40.

[39] *ZHQ* 5.39, 5.106, 5.142, 5.174, 5.196.

[40] *ZHQ* 4.106.

[41] Statuettes from the sites Tianfu and Tianhuishan in Sichuan. Provincial Museum, Chengdu.

takes place in non-Chinese regions: the Steppes or Central Asia. The murals from the tombs of Holingol are one of the rare examples.[42] Dogs roam there amid herds of sheep, horses and large long-haired cattle. The first type of dog has the physiognomy of a large long-haired mastiff, similar to that of large woolly-maned dogs, which, being of Central Asian or Himalayan origin, constitute a probable strain of molossoids (Figure 5C1). The sheep seem to be guarded by other dogs, of an uncertain, possibly lupoid, appearance.

c. **Companion dogs** accompanying travelers, butchers, acrobats, etc. In the iconography, these dogs are sometimes difficult to dissociate from street or village dogs that are either stray or abandoned.[43]

#### *4.3.2 A social dimension: pageantry, prestige and company*

In addition to their guarding or hunting activity, dogs from wealthy houses may have other, less tangible, functions. Often of a social and affective nature, not mentioned in texts but associated to a much-decried inactivity[44], this aspect stands out in the images:

d. **Pageantry and prestige dogs**. In social functions, dogs assist or represent their master to mark his status and hierarchical rank. In the Jiaxiang district (Figure 5D), prestige dogs are beautiful greyhounds, and wolfdogs in Feicheng. These dogs stand at the feet of their master, at the threshold of the entrance door, during ceremonial receptions of visitors.[45] Tolerated in the interior spaces of the house, they are often kept at bay, confined under the stairs (Figure 5D2). There are also escort dogs, used as bodyguards of convoys[46] or people.[47] Similarly, attack dogs are represented by mastiffs (Figure 5E).[48]

e. **Recreational and companion dogs**. Without any specific activity, they may serve as guards and protectors, but their specificity lies rather in the affective bond, comfort and presence they provide their master. They are primarily small (Figure 5C6 and 5C7), most likely the *bai* dogs of *Shuowen jiezi*.[49] Among them, some pugs show their fangs, others look more like peaceful bichons. However, it is evident that some dogs of medium and large stature belong to this category too.

### 5 Attitudes towards the dog according to Han art

The sociability of dogs towards humans, as previously mentioned, raises the question of its reciprocity. The measurement of the degree of esteem that the ancient Chinese could show their dogs will be discussed first, before approaching the strictly spiritual point of view.

#### ***5.1 Consideration of dogs in the private sphere***

The ambivalence regarding dogs is apparent from the comparison between the various Han materials.

On the one hand, the effigies and representations of pleasure, luxury and prestige dogs, specific to the richest tombs, highlight an infatuation with dogs. Beyond their utilitarian requirements, the exuberance of certain physical traits of dogs suggests the emergence of purely aesthetic criteria for their selection. The tendency to particularisation, originality and curiosity is particularly striking in Nanyang, one of the main manufacturing centres of these canine statuettes.[50] When we see the variety of postures; waking, growling or threatening, barking, sudden awakening and the accentuation of the features to the point of caricature - big head, bulging eyes, overly large ears and jaws, disproportionate neck - we understand that these are the watchdogs of a tomb (Figure 5C5 to 5C8). However, behind the stylistic deformations, there are also individual dogs to whom their owner wished to pay homage. In the Anhui and Jiangsu regions, the ornate fabrics and blankets on the dog's back also attest to a certain consideration (Figure 5C5).[51] Archaeological evidence still attests to the symbolic presence of the dog even in the most modest tombs; these were small,

---

[42] *Han Tang bihua* 1974: plates 33 and 36.

[43] *ZHQ* 1.150 and 2.88.

[44] In the historiography of pre-imperial China, the inactivity of exotic luxury dogs is symbolised by the Ao mastiffs from the country of Lü. These dogs were sent to the court of one of the early Western Zhou kings (around 1040 BC), who was finally compelled to refuse them, on moral grounds (*Shangshu*, ch. Lü'ao). Contrarily, the Han emperor Lindi's (AD 156–189) excessive affection for his dogs, to whom he granted titles and disguised as humans, was considered a model of depravation (*Hou Hanshu*, ch. 8, ed. Zhsj 1965, 2: 346). Similarly, the Roman emperor Caligula tried to promote his favourite horse, Incitatus, consul, according to Suetonius, *De vita Caesarum* (Lives of the Twelve Caesars), book IV: Caligula, chapter 55.

[45] *ZHQ* 1.91 and 2.19.

[46] *ZHQ* 2.19 and embossed decor on a brick from Sichuan, *Zhongguo tapian*: 173, no.23.

[47] *Zuozhuan*, X: Zhaogong, 23 (518 BC). Imprisoned with his entourage - a few servants and a beautiful guard dog (*feigou*) - Minister Shusun Chuo waits to be released before killing his dog to offer it as food to his jailers who had demanded it whilst he was under lock and key. His intransigence goes hand in hand with his lack of affection for the dog.

[48] Zhao Dun attacked by an Ao mastiff, according to *Gongyang zhuan*, Xuangong 6. See also *Zuozhuan*, VII: Xuangong 2.

[49] Erkes 1944: 202. *Shuowen*, Quan 12. Formerly, dwarf dogs were thought to appear in China as a result of the introduction of the Maltese dogs from Constantinople, via the Silk Road in the 7th century AD. (Collier 1921: 127; Laufer 1909: 278).

[50] Designated capital of the south, this city enriched by the state monopoly of salt and iron during the Western Han, is also the stronghold of Emperor Guangwu, founder of the Eastern Han, and his principal ministers. See above, note 1.

[51] Described in *Han Feizi*, ch. 9, the phenomenon has existed since the time of the Warring States.

Figure 6. The dog: symbols and religious beliefs. From pictorial stone carvings, Eastern Han.
A) Archery scene, with two dogs. South wall niche of Stone Chamber no.1, Wuliangci. Rubbing from *ZHQ* 1.66.
B) Solar dog and Three-legged Raven in the sun disk, carried by the Red Bird. Tengzhou, Shandong. Rubbing from *ZHQ* 2.165 and close-up. C) Distraught house dog of a grieving family, tied to a tree. Beishan, Anhui. Rubbing from *ZHQ* 4.196.
D) Orphan Ding Lan, his father's statue, a neighbour and a dog. Wuliangci. Xylographic reproduction from *Jinshi suo* 1821, Shisuo 3: 24–25, with the dog added.

crude figurines, representing rather large hardy dogs.[52] This shows the extent of the canine phenomenon.

On the other hand, there are moral ideological aspects, with strong reservations about dogs, even in the private sphere. In Confucian texts, the dog must remain confined to its utilitarian dimension as a domestic animal, despite the progressive recognition and benevolence it receives. A Confucian text reconstituted during the Han like the *Liji* thus provide recommendations on the care with which the master Confucius himself should bury his own dog; by wrapping it in a chariot cover (*gai*) or, failing that, in a simple mat.[53] In the 'Neize' chapter of the *Liji*, statements by Zengzi encourage people to continue to care for the dogs and horses of their deceased parents.[54] Under the Han, the sanctification of these ancient precepts legitimised the recasting of the servant dog's status, inseparable from that of the horse.

[52] For example at the Dawan site in Fengdu, Chongqing.

[53] Zhongni zhi xugou (the dog fed by Confucius) from *Liji*: II.4, ch. Tan Gong (xia) and *shougou* (guard dog) from *Kongzi jiayu*, ch. Zigong wen.

[54] *Liji*, VI.12, ch. Neizi. Conversely, Confucius deplored the excessive attention paid to dogs and horses, to the detriment of parents. *Lunyu*, II.7, ch. Weizheng: *Jin zhi xiaozhe, wei neng yang. Zhiyu quanma, jie neng you yang. Bujing, heyi bie hu?*

The two animals have indeed become the paragons of good servants, but who exhaust themselves in their tasks.[55] Recognition of legitimate rewards for their chores is however only relative as it only applies after their death, in the care given to their burial.[56] The merit of dogs is even compared to that of birds of prey, in that their skill and promptness are a result of their training by good masters.[57]

Furthermore, the moral sacrifice imposed on the dog has no limits. Its condition means that it can at any time pass from life to death[58] and see its status relegated to the lowest level of the canine hierarchy, namely that of slaughter dog. The expression, 'When the prey runs out, the good hunting dog goes to the pan', illustrates this. This is the punishment of any dog that becomes useless, and metaphorically, the risk that any adviser to a king incurs, once his task is accomplished.[59] The reversibility of status that applies to dogs of excellence testifies to the precariousness of their condition.[60] It also sets them up as a model of selflessness and humility.

### 5.2 *A religion of the dog*

Beyond the transposition from guardian of the home to that of the tomb, the almost systematic presence of canine figures in the funeral context in Han China can also be explained by spiritual motivations. Some have already been pointed out by Western sinologists: to purify deadly breath, repel miasmas and demons and to refresh the air.[61] However, the iconography of the Han renders other beliefs visible.

#### *5.2.1 Paths to Heaven and the afterlife*

In the pictorial cosmography of the Han, canids play the role of intermediary between the two worlds, as in other religions.[62] Rather benevolent despite their fierce appearance, these fabulous beings share several functional or morphological traits with dogs:

- **The Cerberean type wild beasts**, belonging to the statuary of mythological guardians, protect, in pairs, the sacred path leading to the tomb.[63] In the pictorial representations, they guard the successive levels that mark the ascent to Heaven. With a few exceptions,[64] these large polymorphic beasts do not have the appearance of dogs but have the function of guardians and the posture of faithful attendants. In the later periods of Imperial China, these beasts are replaced by the Foo dogs or 'lion dogs', as guardians of temples, gradually assuming the physique of real dogs[65].
- **The winged canids** may correspond to the *chailang* (wild dogs and wolves) of the texts; ferocious beasts dwelling on the confines of the universe that are believed to devour lost souls or push them into the depths of hell.[66] In Shandong, these creatures adopt dancing or seated anthropomorphic postures.[67]
- **Celestial dogs** appear to have the role of psychopomps. Among them, the Red Dog inhabits the sun alongside the Three-legged Raven. Believed to devour the sun during an eclipse, this solar dog's daily task is also to guide it in its journey from east to west, which is also the path followed by souls in their ascent to Heaven, towards the afterlife.[68] In the pictorial carvings in Shandong, the Red Dog is enthroned in the middle of the sun disk, carried like a shield by the mythical Fuxi[69] or by the Red Bird, emblem of the South (Figure 6B).[70] In a way, it prefigures the Nine-tails Fox, another solar symbol. Then we find the dogs that watch over the gusts and the winds. The scrolls of clouds, from which appear the heads of dogs and one of their front paws (Figure 4A-sup.),[71] as well as the suggested effect of velocity and direction, lead us to interpret the dog slaughter depicted nearby, at the lower earthly levels (Figure 4A-inf.), as a sacrifice to the Winds or an act of purification.[72]

---

[55] Cf. the Chinese expressions *quanma* and *quanma chi*. Dozens of occurrences in *Shiji*, *Hanshu* and *Hou Hanshu*.

[56] *Hanshu* ch. 70, ed. Zhsj 1962, 9: 3021: *Fu quanma you lao yu ren, shang jia weigai zhi bao*.

[57] Cf. the Chinese expressions *yingquan zhi gong* and *yingquan zhi cai*. *Hou Hanshu* ch. 74A, ed. Zhsj 1965, 9: 2384 and 9: 2393. Similarly, in *Shiji* chapter 53 (ed. Zhsj 1982, 6: 2015), hunting dogs are considered to be mere servants.

[58] For example, a dog, victim of the poisoning perpetrated by Lady Li Ji, in *Zuozhuan*, V: Xigong 4. *ZHQ* 1.91 and 2.101.

[59] The anecdote refers to the warning to the loyal ministers of King Goujian of Yue (r. 496–465 BC), Wen Zhong and Fan Li, in *Han Feizi*, *Wu Yue chunqiu* and *Lunheng*. Under the Han, the metaphor is applied to General Han Xin (†197 BC), Marquis of Huaiyin, in *Shiji* and *Lunheng*. Out of any context, read the *Wenzi*, ch. 6 and the *Huainanzi*, ch. 17.

[60] Such is the case of the valuable dogs that end in stew for reasons of moral conduct. One example is the good hunting dog (*zougou*) of Duke Jing of Qi (r. 547–490 BC), for whom, after his death, his master intended, but failed, to grant the honour of a burial in a coffin (*Yanzi chunqiu*, Neipian, ch. Lian (xia)). Another example is the guard dog of the emissary Shusun Chuo (*supra*, note 47).

[61] Erkes, in his works from 1930–1932 and 1944, was one of the first authors to emphasise the apotropaic aspects of dogs. For their role of protection against evil spirits, read Harper 1985: 497 (*shengou*). On the question of cooling, read Collier 1921, ch. 2 and particularly p. 21–22.

[62] Leach 1961.

[63] Probable reference of the fantastical animals, which guard, in pairs, the two sides of an entrance or doors of temples throughout the ancient East: griffins, sphinx, *kebalim* and other Cerberus. Li 2001.

[64] Particularly in the Xuzhou region, in Jiangsu. *ZHQ* 2.226, 4.6, 4.9 and 4.83.

[65] Mitchell 1991. See also the stone dogs in Leizhou.

[66] Erkes 1944: 193–197.

[67] *ZHQ* 1.234, 1.235, 2.14 and 2.15.

[68] Granet 1994: 376–381, 528 and 537–538.

[69] *ZHQ* 3.19.

[70] See also *ZHQ* 1.137, 1.153, 2.145 and 3.210.

[71] See also *ZHQ* 1.64, 1.73, 2.95 and 2.109.

[72] Erkes 1930–1932 and 1944: 217–221.

*5.2.2 Postmortem use of dogs*

Dogs reveal yet other aspects of the religious syncretism of the Eastern Han. The role of the two dogs represented on the south wall of the Wuliangci Stone Chamber no.1 is particularly instructive (Figure 6A).[73] The first dog bears witness to the beliefs of Chinese folklore that associates the flight of souls to Heaven with the symbolism of archery. The archery indicates the gradual ascent of the sun to its Zenith, path to Heaven, and then its descent and total disappearance in the Far West, pathway to the Beyond.[74] The primordial suns are represented here by the flock of birds that nest on the mythical Fusang mulberry, a kind of *axis mundi* between the world of the living and that of the dead. In its undifferentiated physiognomy, the dog that is ready to pounce on the birds is similar to the dog sacrificed at the well shown on the eastern walls of offering chambers throughout the district of Jiaxiang (Figure 4A-inf.). It could therefore be the transposition of the same dog into another state, after death.[75] The second dog refers to the Confucian virtue of honour and memory for the deceased.[76] The beautiful greyhound, alive and well, represents the distraught housedog after the disappearance of his late master.[77] It could also be an evocation of the magical ceremony of the recall of souls, in that the dog, at the feet of the tree of life, and even attached to it in certain representations (Figure 6C), serves as a vector for the return of the dead souls, down to earth. As one of the closest ties to the deceased, the good and faithful surviving dog is a legitimate link between his passed-away master and the living word. The offering chambers are indeed a place of exchange between descendants and their deceased ancestors.

**6 Discussion**

The foray into the world of images has made it possible to re-establish certain important aspects, partly concealed and distorted by texts, of the material living conditions of dogs at the end of Chinese Antiquity. This approach reveals some elements linked to beliefs surrounding them. It remains to be seen whether the accumulation of canine images throughout the Han dynasty was accompanied by a real change of mentality in favour of dogs and whether the iconography actually reflects this change. Even if the answer remains unclear with regard to the dogs themselves, the iconographic sources have provided undeniable documentary value. As visual documents, they testify to numerous advances in what could be considered as the principal branches - technical, religious and socio-cultural - of anthropozoology as a field of research:

- In terms of selection and breeding techniques, a relatively pronounced morphological differentiation of dogs is manifested at several levels. This differentiation is essentially functional for dogs of active employment (hunting, guarding), but for specimens of social utility (prestige, pageantry, companionship), where appearance and originality are valued, an aesthetic dimension begins to emerge. Crossbreeding or selection from two strains or dominant profiles, lupoid and molossoid, which perhaps are at the origin of the linguistic distinction between *quan* and *gou*, seem to have given rise to around thirty sub-varieties. These include dwarf dogs, a hyper-type of hound built for racing, and an emerging pseudo-braccoid type. As for the passive utility of slaughter dogs, there seems to be no morphological specificity, either in separate breeding or in the elimination of the specimens unfit for activity. The scavenging function is barely mentioned. Finally, it should be noted that in their uses, dogs form very heterogeneous groups and that, as it has been shown, there is no unequivocal correlation between a morphotype and a given activity, but rather for certain specimens, a specific task, and for others, re-employment.
- In the religious sphere, paradoxically, it is the positive consideration towards the dog in the beliefs, that keeps this animal in a state of absolute sacrifice. Its apotropaic effects indeed imply a bloody sacrifice. As for the link it allows between the living and the deceased in Confucian ceremony, if it can take place by keeping the dog of the deceased alive, it also presupposes the sacred use of canine meat to invoke the ancestors and honour their memory.
- In the private sphere, the situation is very ambivalent. In theory, on reading the available sources, both texts and images, Confucian ideology and its visual manifestations in funerary art constitute a major obstacle to the emancipation of dogs. Thus, the possibility of ending its life as slaughter meat, *a priori* applies to all dogs, despite some improvements (status of the dog as servant, survival of the dog of the deceased, etc.). At the same time, the emergence of companion dogs as a widespread

[73] See also *ZHQ* 1.50, 1.58, 1.63, 1.77, 1.88, 1.90, 1.134, 2.98 and 2.131. Bower 2005; Brashier 2005; Wu 1989.
[74] In reference to the gesture of the mythical archer Yi, who shoots at the Ten Suns to leave only one, but also in reference to the eclipses, to the dog devouring the sun and the purification of the deadly breaths. Cf. note 68.
[75] Its condition of a dead dog is shown by the contact of its paws with a chariot cover, on which it stands in balance. It is indeed the material used to bury dogs. Cf. textual sources mentioned in note 53.
[76] Cf. the anecdote of the orphan Ding Lan who makes a statue of his father after the death of his parents (*Xiaozi zhuan*), illustrated in the Wuliangci with a dog (fig. 6C), possibly that of the deceased for whom the son continues to care. Wu 1989: 282–285.
[77] Cf. the concept of *sangjiagou* to which Confucius is compared, as an unappointed counsellor. Cf. notes 9 and 53.

phenomenon is undeniable, particularly in the Nanyang region. Some dogs therefore seem to have freed themselves, at least partially, from the strictly utilitarian yoke.

## 7 Conclusions

By focusing on the visual arts at the end of Chinese Antiquity, and analyzing their visual testimonials from a new perspective, this study has made it possible to highlight the contribution of dogs to the society of the time, both in material and symbolic terms. Despite the utilitarian and ideological servitude, hardly favourable to the emancipation of the dog, we can nevertheless observe an underlying but widespread canine infatuation unprecedented in the entire history of imperial China.

## References

Bodson, L. (ed.) 2002. *D'os, d'images et de mots. Contribution à la réflexion sur les sources de l'histoire des connaissances zoologiques*. Liège: Université de Liège.

Bower, V. 2005. Notice of catalogue no.53. Figure of a dog, in C.Y. Liu, M. Nylan and A. Barbieri-Low (eds) *Recarving China's Past*: 433–439. New Haven and London: Princeton University Art Museum.

Brashier, K.E. 2005. Symbolic discourse in eastern Han memorial art: the case of the birch leaf pear. *Harvard Journal of Asian Studies* 65: 281–310.

Chang, K.C. (ed.) 1977. *Food in Chinese culture*. New Haven: Yale University Press.

Chavannes, É. 1893. *La sculpture sur pierre en Chine au temps des deux dynasties Han*. Paris: Leroux.

Collier, V.W.F. 1921. *Dogs of China and Japan in Nature and Art*. New York: Strokes.

Delort, R. 1984. *Les animaux ont une histoire*. Paris: Le Seuil.

Digard, J.-P. 1990. *L'homme et les animaux domestiques*. Paris: Fayard.

Ding X. 1970. *Zhonghua guobao*, vol. 2. Taibei: Guoji chubanshe.

丁星五 1970. 中華國寶, 第二集. 臺北: 國際出版社.

*Dongping Houtun Handai bihua mu* 2010. Beijing: Wenwu chubanshe.

東平後屯漢代壁畫墓 2010. 北京: 文物出版社.

Elisseeff, D. 1992. Vers une zoohistoire chinoise. *Revue bibliographique de sinologie* X: 171–176.

Elisseeff, D. 1998. L'anthropozoologie: un thème nouveau dans quelques revues chinoises. *Revue bibliographique de sinologie* XVI: 273–281.

Erkes, E. 1930–1932. Strohhund und Regendrache. Ein Beitrag zur altchinesischen Ikonographie. *Artibus Asiae* 4.4: 205–212.

Erkes, E. 1944. Der Hund im alten China. *T'oung-pao* 37.5: 186–225.

*Ethnozootechnie* 78 (2006). *Le chien*.

Fairbank, W. 1942. A structural key to Han mural art. *Harvard Journal of Asiatic Studies* 7.1: 52–88.

Granet, M. 1994. *Danses et légendes de la Chine ancienne*. Paris: PUF.

Gu S. 1997. *Zhongguo Hanhua tudian*. Hangzhou: Zhejiang sheying.

顧森 1997. 中國漢畫圖典. 杭州: 浙江摄影.

*Han Tang bihua* 1974. Beijing: Waiwen chubanshe.

漢唐壁畫 1974. 北京: 外文出版社.

Harper, D. 1985. A Chinese demonography of the third century BC. *Harvard Journal of Asiatic Studies* 45.2: 459–498.

Huang M. 1982. *Luoyang Xihan huaxiang kongxinzhuan*. Beijing: Renmin meishu.

黃明蘭 1982. 洛陽西漢畫像空心磚. 北京: 人民美術.

Keller, O. 1905. Hunderassen im Altertum. *Jahreshefte des Österreichischen Archäologischen Institutes in Wien* VIII: 242–269.

Knechtges, D. 1982–1987. *Wen Xuan or Selections of Refined Literature*, vol. I-II. Princeton: Princeton University Press.

Laufer, B. 1909. Jottings on the races of dogs in ancient China, in *Chinese Pottery of the Han Dynasty*: 247–281. Leiden: Brill.

Leach, M. 1961. *God had a dog*. New Brunswick: Rutgers University Press.

Ledderose, L. 2000. *Ten Thousand Things*. Princeton: Princeton University Press.

Li L. 2001. Lun Zhongguo de youyi shenshou. *Zhongguo xueshu* 5: 62–134.

李零 2001. 論中國的有翼神獸. 中國學術 5: 62–134.

Liu W. 1986. Handai de taogou yu Zhongguo gudai de yanggou fengxi. *Sichuan wenwu* 3: 7–10.

劉文傑 1986. 漢代的陶狗與中國古代的養狗風習. 四川文物 3: 7–10.

Masumitsu Y. 2004. Inukara mita Chūgokukodai no shakai to bunka. *Tōyō bunka kenkyū* 6: 153–181.

益満義裕 2004. イヌから見た中国古代の社会と文化. 東洋文化研究 6: 153–181.

Mégnin, P. 1898. *Le chien et ses races*. Vincennes: Bureaux de l'éleveur.

Mitchell, E.P. 1991. *The Lion-Dog of Buddhist Asia*. New York: Fugaisha.

Orange, M. 1992. En Corée, le chien se mange surtout en été, in *Savourer, goûter*: 373–377. Paris: Presses de l'Université Paris-Sorbonne.

Pirazzoli-t'Serstevens and M. Bujard 2017. *Les Dynasties Qin et Han*. Paris: Les Belles Lettres.

Sterckx, R. 2011. *Food, Sacrifice and Sagehood in Early China*. New York: Cambridge University Press.

Wang W. 2006. Gounian shuo taogou. *Shanghai wenbo* 2: 44–49.

王蔚波 2006. 狗年說陶狗. 上海文博 2: 44–49.

Weber, C. 1966–1968. Chinese pictorial bronze vessels of the late Chou period. *Artibus Asiae* 28: 107–154 and 271–311, 29: 115–192 and 30: 145–236.

Wei R. and Tian Y. 1991. Nanyang Handai taogou diaosu yishu, in Tang W. (ed.) *Nanyangshi bowuguan jianguan 30 nian jinian wenxuan*: 160–168. Zhengzhou: Zhongzhou guji.
魏仁華, 田玉芳 1991. 南陽漢代陶狗雕塑藝術, 收入湯文興(編), 南陽市博物館建館30年紀念文選: 160–168. 鄭州: 中州古籍.
White, W. 1938. *Tomb Tile Pictures of Ancient China*. Toronto: University of Toronto Press.
Wu H. 1989. *The Wu Liang Shrine*. Stanford: Stanford University Press.
Zhang W. 1982. *Hanhua xuan*. Tianjin: Renmin meishu.
張萬夫 1982. 漢畫選. 天津: 人民美術.
Zhang X. and Chi Y. 1997. *Nanyang Handai taogou*. Zhengzhou: Zhongzhou guji.
張曉軍, 赤銀 1997. 南陽漢代陶狗. 鄭州: 中州古籍.
*ZHQ* = *Zhongguo huaxiangshi quanji* 2000. Jinan: Shangdong meishu and Zhengzhou: Henan meishu.
中國畫像石全集 2000. 濟南: 山東美術. 鄭州: 河南美術.
*Zhongguo tapian* 1985. Beijing: Zhongguo guoji shudian.
中國拓片 1985. 北京: 中國國際書店.
Zhou D., Lü P. and Tang W. 1985. *Henan Handai huaxiangzhuan*. Shanghai: Renmin meishu.
周到, 呂品, 湯文興 1985. 河南漢代畫像磚. 上海: 人民美術.

# 5.9 'Cobalt Greyhounds'. An Artistic Proof in Ceramics

Silvia Nutini[1] and Marino Marini[2]

[1]Coworker of National Museum of Bargello, via del Proconsolo 4, 50122, Florence, Italy, silvianutini2002@yahoo.it
[2]Curator of National Museum of Bargello, via del Proconsolo 4, 50122, Florence, Italy, marino.marini@beniculturali.it
Corresponding author: Silvia Nutini, silvianutini2002@yahoo.it

**Abstract**

This study analyses the technology and iconography of a jar of the first half of the fifteenth century with two handles characterised by the oriental relief blue decoration named 'zaffera a rilievo'. The jar is now exhibited in the majolica's room in the National Museum of Bargello in Florence. A true peculiarity can be found in the subjects represented: the rampant dogs. Their physical characteristics allow some reflections about the morphology and body structure of the breed. A type of 'ancient dog' is depicted, similar to the greyhound but with traits of the 'primitive dog' coming from Egypt: the 'Tesem'.

**Keywords**: jar, art, Renaissance, breed, greyhounds.

## 1 Introduction

Renaissance iconography has always been influenced by themes and subjects of a specific historical period and often recalls events or underlines the fashion of the moment; it has become a kind of proof of uses, customs, habits, everyday life or events of an era.

The representation on the jar preserved in the National Museum of Bargello and donated by the heirs of Luigi Pisa in 1933 (Conti 1971: n.511), does not show only an evidence of the European people's passion for those dog breeds imported from the East, which, during the Renaissance, in 15th-16th century, became protagonists of figurative arts because of their elegance and particularity; this dogs depiction can be also a comparison for a detailed analysis of 'ancient dogs' morphological characteristics.

## 2 The decoration

The decoration's detail of the jar allows us to set up the study on an iconographic analysis.

The ceramic is an artefact and so it can be examined in its composition, recognising the materials with which it was made, the stilistic elements and, where are the figurative representations, we can study the images.

The iconography, in fact, groups representations according to their subject, to explain what they intend to depict and to decipher those that are the contents and their evolution in a specific chronological range.

In the maiolica jug of the National Museum of Bargello (Figure 1) the thin silhouettes of pairs of rampant dogs, represented on each of the two faces, are highlighted, rendered with the typical intense blue of 'zaffera'on the white background of glaze; the remaining space is completely occupied by oak leaves and berries in a tight composition, as if the painter had been conditioned by a sort of so defined *horror vacui* (fill completely and finely the surface of a work with details).

The characteristic thickness of the paint is possible by the high percentage of lead present in the blue pigment.

'Zaffera' is the true first Florentine ceramic typology with 'Renaissance' connotations and proves to be an emblematic expression of the general fifteenth century figurative evolution in Florence; this is confirmed by the profiles of half-length human figures that have appeared more frequently on pottery since the beginning of the 15th century.

Among the figurations that appear on these vases we can recognise animal subjects, in pairs or singles, and still foliage and geometric decorations, letters in Gothic characters, religious symbols, noble emblems and signs borrowed from heraldry (as the Florentine lily).

The jars and 'albarelli' of various sizes for ointments, jugs and bottles for syrups and oily preparations, constitute the fundamental nucleus of the 'supplies' necessary for the pharmaceutical activity.

## 3 The iconography

The iconographic analysis of the animals represented on the Bargello jar has made clear some characteristic features, such as the dilatation of the chest, the narrowing of the abdomen, the elongated tail tending

Figure 1. Florentine manufacture, pharmacy jar with rampant dogs, first half of the fifteenth century, National Museum of Bargello (Archive photo from National Museum of Bargello).

to curl and the straight ear sallowing to attribute the morphology and structure to that coming from the Levant, specifically the greyhounds and the 'primitive dogs'.

This study is based on an article by Alberto Bertelli, whore constructed the evolutionary framework of these races, through artistic testimonies and anatomical comparisons with current animals (Bertelli 2014).

Bertelli mentions an important research by the Fred Hutchinson Cancer Institute in Seattle, carried out by comparing mitochondrial matrilineal DNA (Lynch and Madeoy 2004). The study identified ten families of dog breeds, with relative progeny and has affirmed that the greyhounds and the group called 'primitive dogs' are the first 'ancient dogs', those who first differentiated themselves from the wolf.

Despite the hypothesis of a monophyletic origin for these two types, evidenced by archaeological and artistic findings, the 'primitive dogs' and their descendants are immediately diversified by greyhounds, not only physically but also for the hunting technique.

Between the greyhound and primitive dog there are substantial phisically differences that are immediately evident at first sight (Figure 2).[1]

The greyhound is predisposed to run by nature; it is a fairly long, wide and square animal with raised hips, an extended and broad chest characterised by long ribs, well circled and developed backwards.

In the greyhound the fusion between length, shape and curve of the vertebral column, with the strong musculature which encapsulates and supports the area of the loins, creates that characteristic, prominent and conspicuous arching above the kidneys; these are breed characteristics, together with the width of the loins and the pelvis and the balance between the length of the femur and tibia.

The 'primitive dog' has a light constitution, with long and strong limbs, very similar to the greyhound although from the greyhound it differs in robustness.

The skull is flat, oval in a sagittal sense; the straight nasal bridge. The pointed muzzle has a length almost equal to that of the skull. The ears, a true breed emblem, are triangular, erect and rigid, with front opening. The tail, uniform throughout its length, is carried like a sabre when the dog is at rest, higher on the rump when it is attentive.

The greyhound would have originated in the vast low prairies of the Fertile Crescent (Przezdziecki 2001), the strip of land with, in the centre, the north of Arabia and in eastern end, the Persian Gulf, a habitat that had led to this animal an increase of sight first of all for its survival and than to perceive every slight movement of the preys, even in the distance.

The 'primitive dog', originated in the heights of North Africa, however, was certainly less fast than the greyhound bu the compensated this with an extraordinary agility even on inaccessibile terrain type sin which a greyhound would surely have been in difficulty; during the hunting activity, he wasable to use sight, hearing and smell at the same time.

Archaic depictions of 'primitive dogs', similar to the *Kritichos Ichnilatis*, the 'Cretan hound' mentioned by Aristotle in *Historia Animalium*[2] (Gomez Gane, 2014: 196), the oldest pure dog breed in the Mediterranean, are the cave paintings of the Algerian mountain massif named Tassili n'Ajjer, in the Sahara's desert, dating back to the Neolithic (Layoux, 1977).

[1] The site of Italian Authority for Canophilia named ENCI (https://www.enci.it/libro-genealogico/razze/) was consulted for the identification of morphological differences and breed peculiarities.
[2] Aristotle, *Hist. Anim.* VIII, 28

Figure 2. Greyhound and descendant of 'primitive dog' (Cirneco dell'Etna). Archive photos from the sites *iocaccio.it* and *clublevriero.org*. The greyhound is predisposed to run by nature; it is a fairly long, wide and square animal with raised hips, an extended and broad chest characterised by long ribs, well circled and developed backwards. The 'primitive dog' has a light constitution, with long and strong limbs, very similar to the greyhound although from the greyhound it differs in robustness. The ears, a true breed emblem, are triangular, erect and rigid, with front opening.

Figure 3. Small statue of Anubis, late period, 664–332 BC, National Archaeological Museum of Florence (Archive photo from National Archaeological Museum of Florence, Regional Directorate of Museums in Tuscany). The main representant of 'Tesem' breed is the god Anubis. The evocative Egyptian works of Anubis art depicting unmistakably remind us of the traits of Mediterranean 'primitive dogs', especially in the morphology of the head and ears and also the ancient Greeks were convinced that Anubis was a dog descended from the wolf.

On an iconographic level, however, it is worth considering a further species of 'ancient dog', come to us in artistic reproductions of the Egyptian people: the so-called 'Tesem' (Canton, 2018: 173–176).

The main representant of 'Tesem' breed is actually well known to us: it is the god Anubis, protector of necropolis and world of the dead in the Egyptian religion, which was depicted as a man with the head of dog or as a dog of not precise derivation, often attributed to *Canis lupus lupaster* (African wolf) or to *Canis aureus lupaster* (golden jackal).

Bertelli cites a further analisys (Rueness E.K 2011), based on both phenotypic and genetic comparisons, by studying the mitochondrial DNA of the wolf, jackal and other representatives of the *Canis, Cuon* and *Lycaon,* which revealed that the Egyptian jackal would actually be an African wolf . In fact, even the evocative Egyptian works of Anubis art depicting unmistakably remind us of the traits of Mediterranean 'primitive dogs' (Figure 3), especially in the morphology of the head and ears and also the ancient Greeks were convinced that Anubis was a dog descended from the wolf, ashanded down by Plato in the dialogue *Gorgia*[3].

Considering what has been shown up to this time and the eastern origin of the decoration of Bargello jar, it's clear that the representation of dogs on this object contains the essence of those who are the 'ancient dogs', collecting in the morphology of the body the salient features of the two races: the thoracic thinning of the greyhound and the expansion of the abdomen of the 'primitive dog', with some dominant traits of the 'Tesem', such as the straight ears and the curling tail.

## 4 Conclusions

During the 15th century, at the height of the Renaissance, the greyhound consolidated its role as 'king of hunt'; the most spectacular night hunting scene with greyhound is offered by Paolo Uccello in 1470, with the work *Night Hunting*, preserved at the Ashmolean Museum in Oxford: in the painting, deer and greyhounds blend into the darkness while people on horse back in the foreground are dedicated to hunting.

There are numerous attestations of the importance of the greyhound also in the minor arts and in sculpture: an example is the bronzebas-relief by Benvenuto Cellini which reproducts a greyhound ('Saluki' breed) always kept at National Museum of Bargello in Florence.

Greyhounds were also included in the medieval and renaissance ideal where noble families loved to surround themselves with 'strange' animals, especially for their beauty and rarity, to increase the prestige of the owner. It therefore seems natural that these animals could arouse great surprise and it was logical that the artists had repentely portrayed them in their works.

The little-known countries that supplied 'strange' animals were almost always Islamic lands, from Turkey to North Africa, and it was from these lands that the noble families, including the Medici of Florence, already, in the 15th century, imported, wild animals as Arabian horses, falcons, greyhounds and 'primitive dogs' and the cheetahs to be used in hunting (Spallanzani 1983: 359–366).

The dogs depicted on the National Museum of Bargello jar's surface are a 15th century ceramic proof of the role possessed by greyhound who, for elegance and extravagance, was often reproduced on artisticlevel; this hypothesis is also supported by the type of 'zaffera' decoration present on the object, of oriental derivation such as the 'ancient dogs' referred to.

However, the analysis and comparison with iconographic evidences in history suggests an additional assumption regarding the represented animals: despite the anatomical structure of these, they recall the greyhounds, as the main physical characteristics of the breed are confirmed but they also possess a morphology of the tail tending, in veterinary jargon, the 'question mark-tail' and the ears pointing up wards which suggest a representation attributable, to a greater extent, to the Egyptian 'Tesem' race, thus suggesting a probable artistic testimony related, more properly, to the 'primitive dogs'.

## References

### *Ancient sources*

Aristotle, *Historia Animalium,* VIII, 28

Plato, *Gorgia*, 482 b

### *Modern sources*

Bertelli A. 2014. Levrieri e cani primitivi nel bacino del mediterraneo, viewed 10 October 2019, <www.dicasamarziali.com>

Canton M. 2018. *Levrieri e segugi primitivi. Etnogeografia di tutte le razze canine del mondo che inseguono la preda a vista,* Antonio Crepaldi Editore.

Conti G. 1971. *Catalogo delle maioliche*, Museo nazionale del Bargello, Florence: Museo del Bargello.

Gomez Gane Y. 2015. Il 'Cirneco' tra Sicilia e Mediterraneo: saggio storico-linguistico in *Bollettino Centro di Studi Filologici e Linguisitici Siciliani*, XXV. Palermo: 193–222: Centro Studi Filologici e Linguistici Siciliani.

Layoux J.D. 1977. *Tassili n'Ajjer. Art rupestre du Sahara préhistorique,* Paris: Editions du Chêne.

[3] Plato, *Gorgia* 482 b

Lynch D. and J. Madeoy 2004. Genealogical map reveals 10 top dogs, *News Service*, viewed 12 September 2019,<www.newscientist.com>

Przezdziecki X. 2001. *Our Levriers: The Past, Present and Future of All Sighthounds*, Tremouilles, France: LesAmis de Xavier Przezdzieck.

Rueness E.K. 2011. *The cryptic african aolf: canis aureus lupaster is not a golden jackal and is not endemic to egypt*, *PLoS ONE*, January - Volume 6-Issue, viewed on 12 September 2019, <www.plosone.org>

Spallanzani M. 1983. Saluki alla corte dei Medici nei secoli XV- XVI, in '*Mitteilungen des kunsthistorischen Insitutes in Florenz*', XXVII. BAND: 359–366: HEFT.

Detail from Portillo Gómez fig. 4 (chapter 6.3).
Copy of a relief with a banquet scene. Museo della Civiltà Romana (Rome).
Original: Muséed'Art et Histoire (Geneva, Switzerland), first half of the Second Century AD.

# Section 6
## Dogs: Myth and Symbolism

# 6.1 'Implore Me Not, Dog'. The Dog in the Classical World: An Apotropaic View

Marco Giuman and Miriam Napolitano

Università degli Studi di Cagliari. Piazza Arsenale 1, Cittadella dei Musei, 09124 Cagliari, Italy, mgiuman@unica.it, miriam.napolitano@gmail.com

Corresponding author: Marco Giuman mgiuman@unica.it

**Abstract**

In the Classical world, the dog has polysemic meaning, as proved by the analysis of poetic and ethological ancient sources. As a symbol of absolute fidelity to its owner, the dog stands for a fundamental iconographic marker for the aristocratic self-representation, but it can also be interpreted in a negative sense. The ambiguity gives it a liminal meaning, in which the symbolic value is intensified by its relationship to the gods, connected to the concept of passage from one state to another. This double aspect contributes to project an apotropaic sense onto itself, which remained in the Roman world until Late Antiquity.

**Keywords**: dog, Greek art, Roman art, classical world, apotropaic value.

## 1 Introduction

In the Classical world the figure of the dog seems to assume a polysemic connotation, as can be clearly seen from the comparison between poetic and ancient ethological sources, above all from Aristoteles and Aelianus. In this perspective, the dog is primarily the symbol of total and eternal fidelity to its owner,[1] as reflected by the notorious Homeric episode of Argos,[2] who died only after seeing Odysseus again. But this animal also represents a fundamental iconographic sign for the self-representation of aristocratic status, particularly during the archaic age, as can be seen in the Greek artistic repertoires relating to hunting activities or to *symposium* scenes. This latter point is clearly shown in Argos's description given by Odysseus:[3] 'it is fine of form, but I do not clearly know whether it has speed of foot to match this beauty or whether it is merely as table-dogs are, which their masters keep for show'.

With regard to this, there are countless images that can help to focus on this symbolic value. It is the case, for instance, of the Attic Red-Figure *lekythos* from Gela, which is now in Boston, attributed by John Beazley to the Pan Painter and representing the hunter Kephalos and his dog.[4] It is of one of the most famous Greek vases in the world, also known as the François vase. It is now in Florence and represents a complex scene of the Calydonian Boar hunt (Figure 1).[5] Also for the world of the *symposium*, the images related to the dog are many, as we can see in an Attic Red-Figure *psykter*, which is now in Rome, and is attributed to the Achelous Painter or the Leagros Group.[6] In this image, we can also read the close ideological connection that, through the concept of protection of the owner and his house, links the dog to the idea of *Oikos*, seen as a domestic space, and particularly as a defence of the door. These symbolic mechanisms, emphasising the prophylactic sense of its meaning, turn the dog into a fundamental figure of funerary iconography, not only in iconographical terms, as we can see for example in a funerary stele from Thebes (Figure 2),[7] but also in the symbolic mediation of the myth, which translates into the figure of Cerberus, the three-headed dog who guards the gates of Hades.[8]

(MG)

## 2 Material and methods

According to C. Mainoldi,[9] the ambiguity attributed to the figure of the dog derives from two fundamental and complementary aspects that give it a liminal status.[10] The first of these identifies the domesticated dog, which plays the role of guardian and helper in the hunt.

[1] Plin. *H.N.* 8.61.40; 10.83.63.
[2] Homer, *Odyssey* 27.300–327.
[3] Homer, *Odyssey* 27.307–310.
[4] Boston, Museum of Fine Arts 13.198. *BAPD* 206356; Panvini and Giudice 2003: 334; Rizza and Giudice 1996: 77, fig. 14.
[5] Florence, Museo Archeologico Etrusco 4209 (François Vase). *BAPD* 300000. For this vase see Torelli 2007, with previous bibliography.
[6] Rome, Caltagirone Collection. *BAPD* 718.
[7] Thebes, Archaeological Museum (A88). Funerary stele with a dog dating at the middle of IV century BC.
[8] Paris, Musée du Louvre, Etruscan Black-figure *hydria* from Caere.
[9] Mainoldi 1984: 37–93.
[10] For this value and the interpretation of the figure of dog in Greek society, also related to homoerotic love see Kitchell 2004; Neils 2014.

Figure 1. Calydonian Boar hunt, detail on the François vase (from Torelli 2007: 93).

Figure 2. Funerary stele with a dog, unknown provenance (Photo by M. Giuman).

The second aspect is that of the wild animal, linked to the underworld, given its terrifying, contaminating and impure character: the frightening apparitions of monsters or infernal ghosts with canine features.[11] The chthonian appearance of the dog is also evident by its use in private ritual practices, offered as a sacrificial victim in private purification rituals.[12] These sacrifices reflect the important role of the dog as a mediator and *pharmakos*, which, by being possessed, can acquire the powers of evil, removing impurities from the individuals whom it is placed in contact with.[13] From a positive perspective, linked with its keeper's role, this animal was believed to be the guardian of the house. This aspect is provided by its apotropaic image on the mosaics of the vestibules in several Pompeian houses, the boundary space between the interior and exterior of the Roman *Domus*.[14] For example, the mosaic of a chained dog with the Latin inscription *CAVE CANEM* as 'Beware of the dog', occupies the House of the Tragic Poet's threshold, and its context suggests it was a useful instrument to ward off evil spirits that could lead to disease from the streets to the house.[15]

The same valence, translated in a symbolic sense, converts the dog into a guardian of the cycle of time and life, strictly related to Hecate, the goddess of the Underworld who watches over crossroads and frontiers,[16] as well as other goddesses, identified with her under the symbolic concept of procreation, birth,

[11] Levi 1941: 224, note 49: '*Hekate was accompanied during her wandering in the night by (...) the souls of dead, mostly of murdered men*'; see also Mainoldi 1981.

[12] Carboni 2017: 16–18; De Grossi Mazzorin 2008; Sassù 2016.

[13] Plutarch, *Quaest. Rom.* 68.280C.

[14] Battelli 1998; Wilburn 2018.

[15] Levi 1941: 224, note 53; Wilburn 2018: 108–111.

[16] Carboni 2017; Levi 1941: 224; Sassù 2016.

Figure 3. Nicolo intaglio in ring from Wroxeter (from Henig 1984: 244, fig. 1 c).

Figure 4. Bronze *Tintinnabulum* from Herculaneum (from Carcanis 1771: tav. XCV).

and development, like Leukothea-Eileithyia, Astarte, Uni, or the Italic *Genita Mana*,[17] the latter being linked to fertility and the menstrual cycle. An example is offered by some Greek-Roman rings with apotropaic function, in which there is a dog emerging from a sea-shell, which according to M. Henig,[18] is 'the *uterus* from which life comes, a symbol of fertility and rebirth' (Figure 3). However, many gems show the predatory dog attacking a hare: the image could suggest their interpretation both as love gifts and as amulets to attract the love of the beloved, manipulating real life according to the dynamics of the magic of love.[19] The hunting images reproduced on gems could also be donated or purchased in order to promote profits and good affairs, as well as symbolising their appearance in dreams according to the Greek philosopher Artemidorus,[20] who lived in the 2nd century AD.

The apotropaic value of the dog as a threatening predator is revealed by a bronze rattle, a Roman *tintinnabulum* dating from the 1st century BC to the 1st century AD, which adorned a room inside a Herculaneum house and now shows in the *Gabinetto Segreto* of Naples National Archaeological Museum (Figure 4).[21] This device depicts a dwarven gladiator fighting his own improbable monstrous phallus in the form of a dog and five suspended bells. According to C. Johns,[22] the motif shown could indicate the sexual sphere as the uncontrollable impulsiveness typical of animals. On the other hand, C.A. Barton[23] recognises the symbolic representation of the masculinity of the gladiator, fused in its main features demonstrated by sexuality and violence. But it seems more likely that, a comic subject like this may have been used to enjoy, laugh or scare away malignant forces. The metal material, the noise of the bells and the phallus are notoriously powerful magic and prophylactic charms against envy[24] and the dog is interpreted as a symbol of impudence.[25]

Many Greek and Roman objects depict the apotropaic representation of the *oculus malignus*, an eye surrounded by animals and various objects which menaced to attack it[26]. These are mosaics, medals, pendants and engraved gems (Figure 5).[27] Among these *probaskania*, namely antidotes against envy and negative influences, appears the dog, able to hurt and avoid the Evil Eye with its bark, and bite but above all its saliva.[28] According to the therapeutic meaning referred to by Pliny,[29] the prophylactic power of the dog's saliva is manifested in

[17] De Grossi Mazzorin and Minniti 2006: 63.

[18] Boardman 1970: 297, 232–233, pl. 698; Henig 1984: 244, fig. 1c (nicolo intaglio in ring from Wroxeter); Walters 1926, no. 2416.

[19] Molesworth, Henig 2011: 182, pls. 16a-b.

[20] Artemidorus 2.11.

[21] Museo Archeologico Nazionale, Naples, inv. 27853; Carcani 1771: 385–386, tav. XCV.

[22] Johns 1982: 68, fig. 14.

[23] Barton 1993: 73.

[24] Elliott 2016: 202, ill. 27; Parker 2018.

[25] Carcani 1771: 385–386.

[26] The most common animals are the crow, the scorpion, the dog, the lizard, the lion, the ibis and the snake; among the objects are often attested the phallus, the trident, the dagger, the spear, the nails and the thunderbolt: Elliott 2016, ills. 47–50.

[27] Elworthy 1895: 129–131, figg. 14–16, 19; Mastrocinque 2003: 418–420; Pirzio Biroli Stefanelli 1990: 67, no. 46.

[28] Levi 1941: 220–225.

[29] Plin. *H.N.* 7.3.

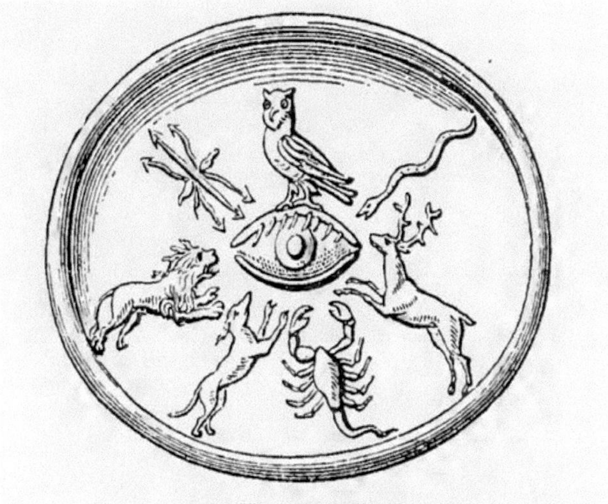


Figure 5. Engraved gem, representation of the evil eye under attack by animals and objects with apotropaic value (from Elworthy 1895: 131, fig. 19).

healing the venom of a snake, an animal that is often juxtaposed to the figure of Envy, to its personification, represented while strangling the envious, the *Phthoneros.*[30] We can see an analogue image in a floor mosaic at Antioch on the Orontes (Syria), where some creatures attack an ithyphallic hunchback, notoriously considered to be both bringer of bad luck and protector against evil.[31] This use continues during Late Antiquity which a bone plaque from Butrint demonstrates, dated between the 4th and the 5th century AD.[32] The bone plaque is carved with the representation of a hunting dog leaping over the evil eye; its finding in the construction levels of a house suggests that it is to be understood as a domestic amulet.

### 3 Results

The analysis of the literary sources and the material evidence shows that the dog is a humanised animal, that was integrated into the domestic environment a long time ago. Its nature appears twofold since it is part of both the world of man, as his companion and loyal guardian, and the non-human world, preserving its wild nature.[33] Its role as a guardian is also transferred to the Underworld, and the shape of the dog is used to imagine demons, and is interpreted as an evil omen. The power attributed to the goddess Hekate is transferred to the dog. With regard to this, both play the role of powerful *apotropaion*, giving protection against the spirits, demons and souls of the dead sent by the goddess, as well as the barking of a dog which was considered capable of scattering phantoms.[34] On the other hand, its representation in amulets indicates that the dog is considered a symbol of fertility, rebirth, and protection. It is well known that its blood and its body were offered to the gods in expiatory rites, with the intention of protecting and increasing the fertility of crops, but also for favouring and helping with the birth of children.[35] Because the dog as a being is considered to belong to the two worlds, *supernus* and *inferius*, it is endowed with supernatural powers and therefore associated with the magical sphere of dreams and omens.[36] It also has therapeutic and curative properties,[37] according to the magic precept that 'the one who hurt you will cure you'.[38] Ancient sources demonstrate that various parts of the dog were considered magical because they were able to heal diseases and illnesses:[39] the dog's eyes[40] and gall were used for eye-illnesses, its liver, lick[41] and teeth[42] for the healing of human beings. As we have already discussed, the saliva is considered by Pliny[43] as one of the most effective antidotes to protect from the venom of snakes, but it becomes a dangerous agent when it's produced by a rabid dog, a disease easily transmissible to humans through the contact with the bite of this animal.[44] With regard to this, it is important to point out that dog's saliva and spitting are considered useful elements to guard against any accidents from oneself as well as to protect from the actions of the evil eye,[45] as shown by Theocritus[46] in the famous episode of Polyphemus, who spits three times to avoid bad luck.

(MN)

### 4 Discussions

As we said at the beginning, in addition to the positive perception, strictly that of loyalty, the dog can assume a negative meaning. In this sense, we can identify the first basic step in an extreme version of the

[30] Giuman 2013: 130–131.
[31] Giuman and Napolitano (in press); Levi 1941: 220–225; Trentin 2015: 56–57.
[32] Mitchell 2007: 282–283, 294–296, fig. 11d; Wilburn 2018: 110.
[33] Sergis 2010.
[34] Levi 1941; 224.
[35] De Grossi Mazzorin and Minniti 2006: 65.
[36] Sergis 2010, pp. 64–65.
[37] Gourevitch 1968.
[38] Elliott 2016: 264.
[39] De Grossi Mazzorin and Minniti 2006: 64; Goebel and Peters 2014: 601.
[40] In the II Book of the Kyranides, the following magic remedy is reported: 'With the two eyes of a white dog, of the stone magnet, of the obsidian stone make a preparation like a dry eye drops and spread it on the lashes in the evening, you will see in the darkness everything that happens' see Macrì 2009: 146, note 171.
[41] The healing of children cured by the licking of a dog is remembered by inscriptions from the sanctuary of Asclepius at Epidaurus: Mainoldi 1981: 37.
[42] According to Pliny, ashes of dogs' teeth mixed with honey availed to help children who were slow in teething: Elliott 2016: 264.
[43] Plin. *H.N.* 7.3.
[44] Plin. *H.N.* 8.152; Aristotle, on the other hand, rules out its transfer to the man with the bite: Aristotle, *HA*. 8.22.
[45] Giuman 2013: 118.
[46] Theoc. 6.35–40.

Figure 6. Detail on the Statue-*rhython* from Kush (Sudan) attributed to Sotades. Boston, Museum of Fine Arts (graphic design by M. Giuman).

fidelity concept, understoodsas the idea of servility, submission, absolute denial of individual freedom; all attitudes that represent the well-known corollary by which the Greeks feel the Eastern world.[47] In this perspective, it may be useful to better understand this concept with an image related to a configurated *rhython* by the Sotades Painter. In it, the *episemon* of the pelta-shield[48] of a Persian warrior is decorated with a dog with its tail between its legs and its head bowed, an unequivocal proof of submission (Figure 6).[49] On the other hand, it is not by chance that the pelts held by oriental enemies and reproduced on the Greek vases never have a symbol. The *episemon*, in fact, represents a space that, already in the epic world, constitutes a true and proper paradigmatic projection of the moral and military virtues of the warrior who parades it;[50] such as, for instance, in the famous description of the Achilles shield.[51]

But this reading is not the only negative one, as we can clearly understand by the disdainful words with which Achilles replied to Hector, mortally wounded by Peleus' son. To the request made by the Trojan hero to return his dead body to his father Priam, Achilles replies with hard and hopeless words:[52] '*implore me not, dog*', says with ruthlessness the Achaean warrior, nobody will '*ward off the dogs*' from his body.

This Homeric image is confirmed by many passages of Greek literature, such as in the final *Chorus* of *Seven against Thebes*, during which to the unburied body of Polynices it is said '*prey to dogs*'.[53] Another example can

[47] Vegetti 1979: 133. Sassi 1988: 104 ff. about the way in which the Greeks perceive the Persians, especially after the Persian Wars, see Vlassopoulos 2013, with previous bibliography.

[48] For a preliminary study about *episemata* (that is the symbols of the Greek shields) in Attic ceramic repertoires see Chase 1902; Giuman 2000.

[49] Boston, Museum of Fine Arts 21.2286. *BAPD* 209548; Beazley 1963: 772; Bothmer 1957: 57, tav. 90, 1; Hoffmann 1997: 89 ff; Kahil 1972: 283, figs. 18–19. It seems interesting to note that, in the Persian world, the dog does not appear only as an animal generically connotated to the funerary field – and in this way consecrated to Ahuramazda, the deity of the dead – but it also seems to fulfill specific functional areas related to the ritual sphere of death. See also Herodotus 1.140.

[50] Giuman 2000: 39 '*L'area centrale occupata dall'episemon ci si presenta come l'unico spazio sufficientemente visibile sul quale poter collocare un qualche segno di riconoscimento che, nella calca e nel fragore assordante di un combattimento – proprio le 'fiere percosse di scudi ' ricordate da Tucidide (4, 96, 2) –, possa favorire un rapido riconoscimento dell'avversario o del compagno d'arme (Cfr. Vegezio, Epitome rei militaris 11.18)*'.

[51] Homer, *Iliad* 18.477–606.

[52] Homer, *Iliad* 22.345–348: 'μή με κύον γούνων γουνάζεο μὴ δὲ τοκήων· / αἳ γάρ πως αὐτόν με μένος καὶ θυμὸς ἀνήη / ὤμ' ἀποταμνόμενον κρέα ἔδμεναι, οἷα ἔοργας, / ὡς οὐκ ἔσθ' ὃς σῆς γε κύνας κεφαλῆς ἀπαλάλκοι, οὐδ' εἴ κεν δεκάκις τε καὶ εἰκοσινήριτ' ἄποινα /στήσωσ' ἐνθάδ' ἄγοντες, ὑπόσχωνται δὲ καὶ ἄλλα'. In this perspective, see also Homer, *Iliad* 1.4; 22.335; 24.409.

[53] Aeschylus, *Seven against Thebes* 1013–1017: 'τούτου δ' ἀδελφὸν τόνδε Πολυνείκους νεκρὸν / ἔξω βαλεῖν ἄθαπτον, ἁρπαγὴν κυσίν, / ὡς ὄντ' ἀναστατῆρα Καδμείων χθονός, / εἰ μὴ θεῶν τις ἐμποδὼν ἔστη δορὶ / τῷ τοῦδ'.

be found in a passage of *Hecuba* by Euripides, in which Polymestor threatens Trojan women after the murder of his sons:[54] *'where am I rushing, leaving my children unguarded for maenads of hell to mangle, to be murdered and ruthlessly cast forth upon the hills, a feast of blood for dogs?'*. It is precisely in this instance that it is imperative to note that, little further, Hecuba herself is turned into a ghost dog with eyes of fire.[55]

Therefore, it is in the symbolic idea of the dog as a stray carnivorous animal that feeds on unburied bodies, an attitude confirmed by ancient ethological sources, that we must read its negative connotation. In this respect, it could be useful to underline a significant matching in terminology: in ancient Greek, the verb *skylao* or *skyleo*, related in etymological terms to the substantive *skylax* ('puppy'), can extend its semantic meaning and indicate the deplorable action of robbing a cadaver[56] or, by analogy, the violent and bloody raid of an enemy city fallen after a siege.[57]

### 5 Conclusions

The image of the dog has been commonly used with a protective aim, especially because of its known role as keeper. Furthermore, it could represent a defence against the dangers which the dog itself personifies, following the ambivalent peculiarity of the superstition based on the ambiguity related to this figure.

(MG)

### Acknowledgments

M. Napolitano gratefully acknowledges Sardinia Regional Government for the financial support of her PhD scholarship in Storia, Beni Culturali e Studi Internazionali (P.O.R. Sardegna F.S.E Operational Programme of the Autonomous Region of Sardinia, European Social Found 2014–2020 – Axis III Instruction and Formation, Objective 10.5, Line of Activity 10.5.12).

### References

#### *Ancient sources*

Aeschylus (Translated by G. O. Hutchinson 2017). *Seven against Thebes*. Oxford: Oxford University Press.

Aristotle (Translated by D.M. Balme and A. Gotthelf 2002). *Historia animalium*. Cambridge: Cambridge University Press.

Artemidorus (Translated by D.E. Harris-McCoy 2012). *Artemidorus Oneirocritica: text, translation, and commentary*. Oxford: Oxford University Press.

Euripides (Translated by M. Mc Donald 2019). *Hecuba*. London: Nick Hern Books.

Herodotus (Translated by D. Grene 1987). *The history*. Chicago: University of Chicago Press.

Pliny the Elder (Translated by H. Rackham, W.H.S. Jones and D. E. Eichholz, 1868–1963). *Natural History*. London: William Heinemann.

Plutarch (Translated by F.C. Babbitt 1936). *Plutarch's Moralia*. Cambridge: MA. Harvard University Press. London. William Heinemann Ltd.

Polybius (Translated by R. Waterfield 2010). *The histories*. Oxford: Oxford University Press.

Theocritus (Translated by R. C. Trevelyan 1925). *The idylls of Theocritus*. London: The Casanova Society.

Thucydides (Translated by J. Mynott 2013). *The war of the Peloponnesians and the Athenians*. Cambridge: Cambridge University Press.

Vegetius (Translated by M. D. Reeve 2004). *Epitoma rei militaris*. Oxford: Oxford University Press.

#### *Modern sources*

Barton, C.A. 1993. *The Sorrows of the Ancient Romans. The Gladiator and the Monster*. Princeton: Princeton University Press.

Battelli, P. 1998. L'ingresso della domus come confine tra città e abitazione. *Studi Romani* 46: 281–301.

Beazley, J.D. 1963. *Attic Red-Figure Vase-Painters*. Oxford: Oxford University Press.

Bothmer, D. von 1957. *Amazons in Greek Art*. Oxford: Oxford University Press.

Carboni, R. 2017. '...come uno che ha gettato via gli oggetti delle purificazioni, senza volgere gli occhi'. Liminarità e impurità nel mondo greco. *OTIVM*, [S.l.], 2.

Carcani, P. 1771. *Delle antichità di Ercolano. De' bronzi di Ercolano e contorni*, II. Napoli: Regia Stamperia.

Chase, G.H. 1902. The Shield Devices of the Greeks. *Harvard Studies in Classical Philology* 13: 61–127.

De Grossi Mazzorin, J. and C. Minniti 2006. Dog Sacrifice in the Ancient World: A Ritual Passage?, in L.M. Snyder and E.A. Moore (eds) *Dogs and People in Social, Working, Economic or Symbolic Interaction. Proceedings of the 9th Conference of the International Council of Archaeozoology, Durham, August 2002*: 62–66. Exter: Oxbow Books.

De Grossi Mazzorin, J. 2008. L'uso dei cani nel mondo antico nei riti di fondazione, purificazione e passaggio, in F. D'Andria, J. De Grossi Mazzorin and G. Fiorentino (eds) *Uomini, piante e animali nella dimensione del sacro. Seminario di studi di Bioarcheologia (28–29 giugno 2002), Convento dei Domenicani-Cavallino (Lecce)*: 71–81. Bari: Edipuglia.

[54] Euripides, *Hecuba*, 1075–1078: ποῖ πᾶι φέρομαι τέκν' ἔρημα λιπὼν / Βάκχαις ῞Αιδα διαμοιρᾶσαι / σφακτά, κυσίν τε φοινίαν δαῖτ' ἀνή- / μερόν τ' ὄρειον ἐκβολάν'.

[55] Euripides, *Hecuba* 125 ff.

[56] So, for example, in Herodotus 1.82.

[57] Plb. 9, 10, 13.

Elliott, J.H. 2016. *Beware the Evil Eye. The Evil Eye in the Bible and the Ancient World. Greece and Rome*. Eugene: Cascade Books.

Elworthy, F.T. 1895. *The evil eye: An account of this ancient and widespread superstition*. London: J. Murray.

Giuman, M. 2000. Episemata e politica. Scudi e monete nell'Atene di VI secolo a.C. *Ostraka* 9, 1: 31–42.

Giuman, M. 2013. *Archeologia dello sguardo. Fascinazione e* baskania *nel mondo classico*. Roma: Giorgio Bretschneider Editore.

Giuman M. and M. Napolitano 2020 'Contra invidentium effascinationes'. Prophylaxis and evil-eye in some gems of the Republican Roman Era with grotesque subjects. *Acta Antiqua Academiae Scientiarum Hungaricae* 60, 3–4: 303–317.

Goebel, V. and J. Peters 2014. Veterinary Medicine, in G.L. Campbell (ed.), *The Oxford Handbook of Animals in Classical Thought and Life*: 589–606. Oxford: Oxford University Press.

Gourevitch, D. 1968. Le chien, de la therapeutique populaire aux cultes sanitaires. *Mélanges d'Archéologie et d'Histoire* 68 (1): 247–281.

Henig, M. 1984. The elephant and the sea-shell. *Oxford Journal of Archaeology* 3.2: 243–247.

Hoffmann, H. 1997. *Sotades. Symbols of Immortality on Greek Vases*. Oxford: Claredon Press.

Johns, C. 1982. *Sex or Symbol. Erotic Images of Greece and Rome*. London: British Museum Publications.

Kahil, L. 1972. Un nouveau vase plastique du potier Sotadès au Musée du Louvre. *Revue Archéologique* 2: 271–284.

Kitchell, K.F. 2004. Man's Best Friend? The Changing Role of the Dog in Greek Society, in B. Santillo Frizell (ed.) *Pecus: Man and Animal in Antiquity. Proceedings of the conference at the Swedish Institute in Rome, September 9-12, 2002* (The Swedish Institute in Rome. Projects and Seminars, 1): 177–182. Rome: Swedish Institute.

Levi, D. 1941. The evil eye and the lucky hunchback, in R. Stillwell (ed.) *Antioch on the Orontes III: The Excavations 1937-1939*: 220–232. Princeton (NJ): Princeton University Press.

Macrì, S. 2009. *Pietre Viventi. I minerali nell'immaginario del mondo antico*. Druento: UTET.

Mainoldi, C. 1981. Cani mitici e rituali tra il regno dei morti e il mondo dei viventi. *Quaderni Urbinati di Cultura Classica* 8: 7–41.

Mainoldi, C. 1984. *L'image du loup et du chien dans la Grèce ancienne*. Paris: Éditions Ophrys.

Mastrocinque, A. (ed.) 2003. *Sylloge Gemmarum Gnosticarum. Parte I*, (Bollettino di Numismatica 8.2.I), Roma: Istituto Poligrafico e Zecca dello Stato, Libreria dello Stato.

Mitchell, J. 2007. Keeping the Demons out of the House: The Archaeology of Apotropaic Strategy and Practice in Late Antique Butrint and Antigoneia, in L. Lavan, T. Putzeys and E. Swift (eds) *Objects in context, objects in use: material spatiality in late antiquity*: 273–310. Leiden-Boston: Brill.

Molesworth, H. and M. Henig 2011. Love and Passion. Personal Cameos in Late Antiquity from the Content Collection, in C. Entwistle and N. Adams (eds) *Gems of Heaven. Recent Research on Engraved Gemstones in Late Antiquity, c. AD 200-600* (British Museum Research Publication 177): 179–185. London: British Museum.

Neils, J. 2014. Hare and the Dog: Eros Tamed, in A. Avramidou and D. Demetriou (eds) *Approaching the Ancient Artifact. Representation, Narrative, and Function. A Festschrift in Honor of H. Alan Shapiro*: 311–318. Berlin-Boston: De Gruyter.

Panvini, R. and F. Giudice (eds) 2004. *Ta Attika. Veder greco a Gela. Ceramiche attiche figurate dell'antica colonia*. Roma: L'Erma di Bretschneider.

Parker, A. 2018. 'The Bells! The Bells!' Approaching *Tintinnabula* in Roman Britain and Beyond, in A. Parker and S. McKie (eds) *Material Approaches to Roman Magic. Occult Objects & Supernatural Substances* (Themes in Roman Archaeology 2): 57–68. Oxford: Oxbow Books.

Pirzio Biroli Stefanelli, L. 1990. Una raccolta per collezionisti – I gioielli, in M.C. Molinari (ed.) *Il tesoro di via Alessandrina*: 33–75. Cinisello Balsamo - Milano: Silvana.

Rizza, G. and F. Giudice (eds) 1996. *I vasi attici ed altre ceramiche coeve in Sicilia*. Atti del convegno internazionale, Catania, Camarina, Gela, Vittoria, 28 marzo - 1 aprile 1990 (Cronache di Archeologia 29). Catania: Consiglio Nazionale delle Ricerche.

Sassi, M.M. 1988. *La scienza dell'uomo nella Grecia antica*. Torino: Einaudi.

Sassù, A. 2016. Through impurity: a few remarks on the role of the dog in purification rituals of the Greek world, in P.A. Johnston, A. Mastrocinque and S. Papaioannou (eds) *Animals in Greek and Roman religion and myth. Proceedings of the symposium grumentinum Grumento Nova (Potenza) 5-7 June 2013*: 393–418. Newcastle upon Tyne (UK): Cambridge Scholars Publishing.

Sergis, M.G. 2010. Dog sacrifice in ancient and modern Greece: from the sacrifice ritual to dog torture (kynomartyrion), *Folklore. Folklore: Electronic Journal of Folklore* 45: 61–88.

Torelli, M. 2007. *Le strategie di Kleitias. Composizione e programma figurativo del vaso François*. Roma: Electa.

Trentin, L. 2015. *The Hunchback in Hellenistic and Roman Art*. London: Bloomsbury Academic.

Vegetti, M. 1979. *Il coltello e lo stilo. Animali, schiavi, barbari, donne, alle origini della razionalità scientifica*. Milano: Petite Plaisance.

Walters, H.B. 1926. *Catalogue of the engraved gems and cameos, Greek, Etruscan and Roman, in the British Museum*. London: Trustees of the British Museum.

Wilburn, A. 2018. The Archaeology of Ritual in the Domestic sphere: Case studies from Karanis and Pompei, in A. Parker and S. McKie (eds) *Material Approaches to Roman Magic. Occult Objects & Supernatural Substances* (Themes in Roman Archaeology 2): 103–114. Oxford: Oxbow Books.

***Website***

*BAPD*. Beazley Archive Pottery Database, University of Oxford, viewed 10 December 2019, <http://www.beazley.ox.ac.uk/pottery/default.htm>.

# 6.2 Dogs in Phoenician Culture

Giuseppe Minunno

Scuola di Specializzazione in Beni Archeologici, Università degli Studi di Firenze, Via Laura 48, 50121 Florence, Italy, giuseppe.minunno@unifi.it

**Abstract**

Evidence concerning dogs in Phoenician culture is provided by texts, artefacts, and archaeo-zoological findings. Although dogs certainly played a wide range of functions in everyday life, just as they did in the Greek and Roman world, this subject is not well documented. More data are available regarding the symbolic functions of dogs and ritual activities involving them, although geographical and chronological differences must be considered. There is evidence of cynophagy in some settlements, and several dog depositions have been found, the interpretation of which is still open to question.

**Keywords:** animal studies, dogs, Phoenicians, Carthage, cynophagy.

## 1 Phoenician culture and sources relating to dogs

The general term 'Phoenician culture' is used here to refer to the culture of the Phoenician cities on the Levantine coast as well as the local cultures which developed in the areas where Phoenician people settled (including, therefore, those usually termed Punic or Carthaginian). Since the evidence regarding dogs in Phoenician culture[1] is scattered from the eastern shores of the Mediterranean to the Atlantic Ocean, and spans over a millennium, caution is required in order to avoid undue generalisations.

As with many other aspects of Phoenician culture, direct textual sources on dogs are scanty. Furthermore, Greek and Latin writers, who often provide complementary data, did not, as a rule, consider the presence of dogs in these contexts worth mentioning, probably because most of the duties assigned to dogs were similar to those they carried out in the Greek and Roman world. Thus, for instance, sheepdogs are referred to as a component of a Punic shepherd's belongings in Silius' poem *Punica* (2.442–444: *omnia Poenum / armenti vigilem patrio de more secuntur / gaesaque latratorque Cydon tectumque focique*). This reference, which occurs in the description of a scene depicted on Hannibal's imaginary weaponry, is nothing more than a literary echo of Virgil (Verg. *G.* 3.343–345: *Omnia secum / armentarius Afer agit, tectumque laremque / armaque Amyclaeumque canem*). On the imaginary weapons, dogs are also supposedly employed in the hunt of Dido and Aeneas (Sil. *Pun.* 2.417–418), here again echoing Virgil (Verg. *Aen.* 4.132). While in the first case Virgil was speaking broadly of Libyan (not specifically Carthaginian) shepherds, the second reference is to a fictional episode in which the poet assumed that dogs were used in Carthaginian hunts just as they were in literary - as well as real - Greek and Roman hunts. Therefore, although dogs were certainly used in many practical activities, especially in hunting, breeding and as watchdogs, the literary evidence is meagre (deemed too obvious to be mentioned), vague, or even unreliable (it may simply have been assumed that their functions were the same as they were in Greek and Roman culture). Several representations of dogs are attested, especially from North Africa (Yazidi 2009: 143–148; D'Andrea 2018a: 204–210) but these provide little evidence concerning the use of dogs in ordinary tasks. As a rule 'Les scènes pastorales ne faisaient pas partie des thèmes de l'iconographie punique' (Yazidi 2009: 298). A terracotta stamp found in a third-second century BC Carthaginian *favissa* (Astruc 1959: 116–117; Yazidi 2009: 148) may depict a dog watching some poultry being threatened by a weasel (Yazidi 2009: 298). Zoo-archaeological evidence, however, is progressively being added to the literary and iconographic sources and increasing our knowledge concerning dogs in Phoenician culture.

## 2 The consumption of dog meat

One of the aspects that have recently emerged from zoo-archaeological evidence is the consumption of dog meat, which appears to have been practised, at least occasionally, at some Phoenician sites (cf. Campanella 2008: 70–71; D'Andrea 2018a: 191–193).[2] Cut marks on dog bones have been noted in the Iberian Peninsula (Rocha Branca: Cardoso 2000: 322, 325; Ibiza: Saña 1994); in Sardinia (Sulky: Wilkens 2008: 249–251) and in North Africa (Lixus: Iborra Eres 2005: 231; Ceuta:

[1] Cf. D'Andrea 2018a.

[2] According to Morales *et al.* 1994: 53, the small sample of dog bones from Castillo de Doña Blanca (8th–6th century BC) 'indirectly indicates that dogs were not items of consumption'.

Camarós and Estévez 2010: fig. 8; Utica: Cardoso *et al.* 2016: 319 fig. 4; Carthage: Weinstock 1995: 115). To date we have no evidence of the consumption of dog meat in Phoenicia and this might have been a western practice, perhaps resulting from cultural influences exerted by indigenous populations (cf. D'Andrea 2018a: 193). Cynophagy is well attested in the Maghreb from the Middle Ages on (Simoons 1981; Prévost 2006), and is apparently specifically tied to the cultural heritage of Berber peoples (Simoons 1981; Mansouri 2015: 129–134). Evidence of the consumption of dog meat prior to the Phoenician presence in the region, however, is lacking. In Utica, where the evidence in this regard dates back to the earlier Phoenician settlement (10th-9th century BC), the relevant archaeological context has also yielded indigenous ceramics (Cardoso *et al.* 2016: 315). Indeed, the Persian king Darius is said (Just. *Epit.* 19.1.10–13) to have asked the Carthaginians to abstain from consuming dog meat, but there is no claim of a similar request being made of the Phoenician cities of the Levant, which were at that time under Darius' dominion. The reason why Darius requested, or is presumed to have requested the Carthaginians to refrain from consuming dog meat might lie in Persian cultural norms in which dogs enjoyed high status as well as having funerary connotations (Bucci 1977: 450; Boyce 1982: 302; van Wickevoort Crommelin 2005: 20). According to our source, the Carthaginians would appear to have agreed to Darius' request. They were also requested to cremate their dead instead of burying them and to cease from performing human sacrifice (*edictum, quo Poeni humanas hostias immolare et canina vesci prohibebantur, mortuorum quoque corpora cremare potius quam terra obruere a rege iubebantur*). Due to this association with burial and sacrificial customs, one might wonder whether the consumption of dog meat was regarded as a ritual, rather than an alimentary, issue.

### 3 Dogs in ritual contexts

Although the Punic inscriptions known as 'sacrificial tariffs' do not testify to the practice of sacrificing dogs, it must be borne in mind that these tariffs only regulated the ritual practice of specific sanctuaries and the possibility that dogs were sacrificed in other sanctuaries cannot therefore be excluded. In fact, archaeological evidence hints at a consumption of dog meat in ritual contexts. In Utica, dog bones were found in the fill of a pit which was closed (about the 10th-9th century BC) 'possibly in a ritualistic way'. The presence of the dog bones could be associated with a collective banquet related to this ritual closure (Cardoso *et al.* 2016: 315, 319–320). Dog consumption in Sulky may also be connected with ritual activities (Carenti and Wilkens 2006: 176). Dog remains have also been found inside a *favissa* in Motya (Alhaique 2012). Some disarticulated dog bones which were found within shafts in the necropolis of Cadiz (Niveau de Villedary y Mariña and Ferrer Albelda 2004) have been interpreted as the possible remains of ritual meals, but new evidence suggests, instead, that these remains are to be related to the more or less complete dog skeletons deposited in underground structures of the same necropolis. These dogs were supposedly involved in ritual activities centred on their killing and deposition (Niveau de Villedary y Mariñas 2008). Dogs, therefore, appear to have been considered as a suitable material for cthonic rituals in third century BC Cadiz.[3] Dog bones, or even complete skeletons, are sometimes found in funerary contexts. In Malaga an adult dog, which had been burned but not consumed, was deposited next to the foundations of a sixth century BC tomb, near its entrance (Martín Ruiz *et al.* 2003: 151–152, 156–157). According to Niveau de Villedary y Mariñas (2008: 112), 'nos hallamos ante un sacrificio de 'holocausto', el animal es sacrificado y quemado'. In the necropolis of Puig des Molins (Ibiza) complete dogs were deposited in some of the hypogea (Morales Pérez 2008: 28; 27 fig. 10). At Sidon seven skulls of 'greyhounds' were found inside a sarcophagus in the necropolis of Ayaa (Hamdy Bey and Reinach 1892: 27).[4] At Carthage, in a tomb of Sainte-Monique, animal bones were found mingled with human skeletons (Bénichou-Safar 1982: 280). Among these bones at least seven skulls of dogs (but also one of a pig) could be identified (D'Andrea 2018a: 201). A dog's head was found in front of the door of a tomb in Sainte-Monique (Bénichou-Safar 1982: 279). In the pit of a tomb of the Odéon a niche contained the complete skeletons of an adult dog and a puppy (Gaillard 1938–1940: 331; Bénichou-Safar 1982: 279–280).[5] Bones of dogs (at least one male and one female) have been found in a Punic tomb in Villamar (Sardinia). These bones, which show no cut-marks, were found disarticulated but had probably been moved when the tomb was reopened (Pompianu 2017: 12). Some dog bones were also found in tombs at Villaricos (two bones, Castaños Ugarte 1994: 3), Sulky (a lower jaw, Guirguis and Unali 2012: 2015–2016) and on Malta (bones of a dog and a sheep in a possible ossuary:

[3] Dog bones (showing no cut marks) were found in the necropolis of the Senhor dos Mártires, Alcácer do Sal (Cardoso and Arruda 2016: 206).

[4] Since dogs are represented in the hunting scenes decorating the plinth of the sarcophagus (Fleischer 1983: 30–35, Taf. 12–17), it has been suggested that the deceased had been a hunter, and the skulls were those of his favourite hunting dogs (cf. Elayi 1988: 315). The hypothesis that the sarcophagus only held the remains of dogs whose master was buried in another sarcophagus (Del Medico 1957: 91 note 2) is disproved by the explicit assertion that 'les ossements du mort' were found together with the dog skulls (Hamdy Bey and Reinach 1892: 27).

[5] Bénichou-Safar (1982: 280) affirms that the niche contained, 'les squelettes complets d'un chien adulte et d'un jeune chiot et celui, complet, d'une chèvre ou d'un mouton', but Gaillard (1938–1940: 331) only mentioned 'deux os de membre d'un ruminant, chèvre ou mouton'. Probably, therefore, just some parts of other animals, but no other complete individual, had been deposited in the niche with the dogs.

Sagona 2002: 881; remains of two small dogs in a tomb: Sagona 2002: 966).[6]

In addition to the evidence of real dogs, images of dogs are also attested in funerary contexts. From the necropolis of Tuvixeddu (Cagliari) comes a terracotta figurine of a dog which could be ascribed to the 3rd century BC (Barreca 1986: 259, fig. 254). A dog ('sorte de lévrier' according to Delattre 1895: 282) wearing a collar is represented together with a riding figure on a terracotta disk from the necropolis of Douïmés. A similar item was found in a tomb at Utica (Yazidi 2009: 144). From a necropolis at Kerkouane there is a scarab depicting four animals, probably dogs, surrounding a corpse wrapped in a shroud (Redissi and Tillot 1995: 153; Pl. III, 9; Yazidi 2009: 148). To the wealth of iconographic evidence collected by D'Andrea (2018a: 204–207) can be added the picture of a dog in an unpublished tomb at Othoca (Stiglitz 1999: 78, 80–81). Dogs, and their images, could have played varying roles in the different funerary contexts in which they have been found. They could have been meant to guard the tomb, to watch over their masters in the afterlife; to follow them as pets or hunting dogs, to serve as food for the dead, or for their relatives during the burial rituals or be employed in other ritual activities for which they appeared to be an especially suitable, if not essential, element.[7]

Since dogs are well documented in funerary contexts, it could be worth noting that there is scarce evidence of dog remains from the so-called *tophet*, a ritual space the purpose and function of which is currently the object of heated debate amongst scholars (cf. Xella 2012–2013; D'Andrea 2018b). The traditional interpretation, which considers the *tophet* an area where children were deposited after having been ritually killed and burned, has been challenged by a different explanation which considers them instead to be basically burial grounds, mainly dedicated to very young children who had died before having properly acquired a full personality. It is to be noted that the claim that dog bones were present in the urns of the *tophet* of Motya (Whitaker 1921: 257) has not been confirmed by later research (Ciasca *et al.* 1996: 329). Some dog bones were recovered, but only outside the urns and these showed no trace of burning, like the skull of a dog found in the *tophet* of Tharros (Fedele 1977: 191–193). Since the *tophets* were open areas, the possibility of intrusive materials should be considered. Dog bones were also reportedly found in the *tophet* of Carthage, but this information cannot be confirmed, and the context is unclear (Bénichou-Safar 2004: 52; D'Andrea 2018a: 198). Representations of dogs on the stele of the *tophets* are also extremely rare. A Carthaginian stela supposedly representing a dog was mentioned by Babelon (1896: 69) but was later lost (Hours-Miedan 1951: 49).[8] It was perhaps one of those two or three anepigraphic stelae, whose images are kept in the Cabinet du Corpus and on which, according to Bénichou-Safar (2004: 52), dogs were represented. As for the dog represented on a clay figurine of Artemis found in the *tophet* of Motya (Bevilacqua 1972: 115–116, Tav. LXXXIX,2), it has no particular relevance to this topic since, as correctly pointed out by D'Andrea (2018a: 207 note 117), the dog belongs directly to the iconography of the Greek hunting goddess, who was possibly identified in Punic culture with Tinnit.[9]

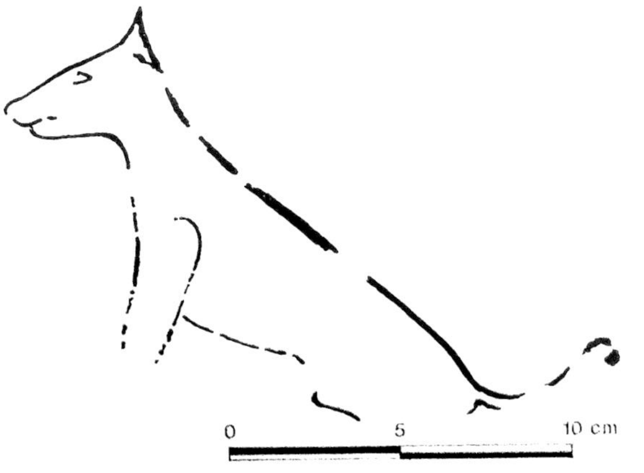

Figure. 1. Possible representation of a dog in the Grotta Regina (from Coacci Polselli *et al.* 1979, fig. 62).

In a Phoenician inscription from Kition (*CIS* I 86) *klbm* and *grm* are recorded as receiving payments in connection with a temple of Astarte. Although different interpretations are also possible, *klbm* is usually translated as 'dogs', while *grm* might mean 'whelps' (cf. Watson 1997: 93). Scholars accepting a translation of *klbm* as 'dogs' disagree, however, about whether the text deals with real dogs or with men, whilst those who interpret *klbm* as 'men' disagree as to whether they were male prostitutes or some other kind of (cultic?)

[6] It is unclear to me whether the tooth found in a tomb of Solus actually belonged to a dog (cf. D'Andrea 2018a: 202), since the phrase mentioning the tooth, 'schegge ossee combuste (non classificabili ad esclusione di un frammento di fibula e un dente canino)', Calascibetta 2009: 636, could refer to a human cuspid rather than implying that it belonged to a dog.

[7] At Ghyneh (south east of Byblos), two dogs appear on a rock relief (Renan 1864: 292–293; Pl. XXXVIII) dated to the Hellenistic (Parlasca 1982: 7) or the Roman period (de Jong 2017: 330). The dogs feature in a hunting scene, whose context is funerary (Seyrig 1940: 113–120).

[8] Vassel (1921: 46) supposed that the stela was kept at the Musée Guimet.

[9] At the Cerro do Castelo de Garvâo (Ourique) dog bones were found among the materials deposited in a ditch in the 3rd century BC. The ditch has been interpreted as a *favissa* or *bothros* pertaining to a sanctuary where Tinnit was supposedly worshipped (Varela Gomes and Tavares da Silva 1994; Tavares da Silva and Varela Gomes 2006). The dog bones represented 15% of the animal remains examined (Cardoso and Varela Gomes 1997: 105–107). The context shows strong indigenous features.

Figure 2. Terracottas from Santa Gilla (from Moscati 1991: 112).

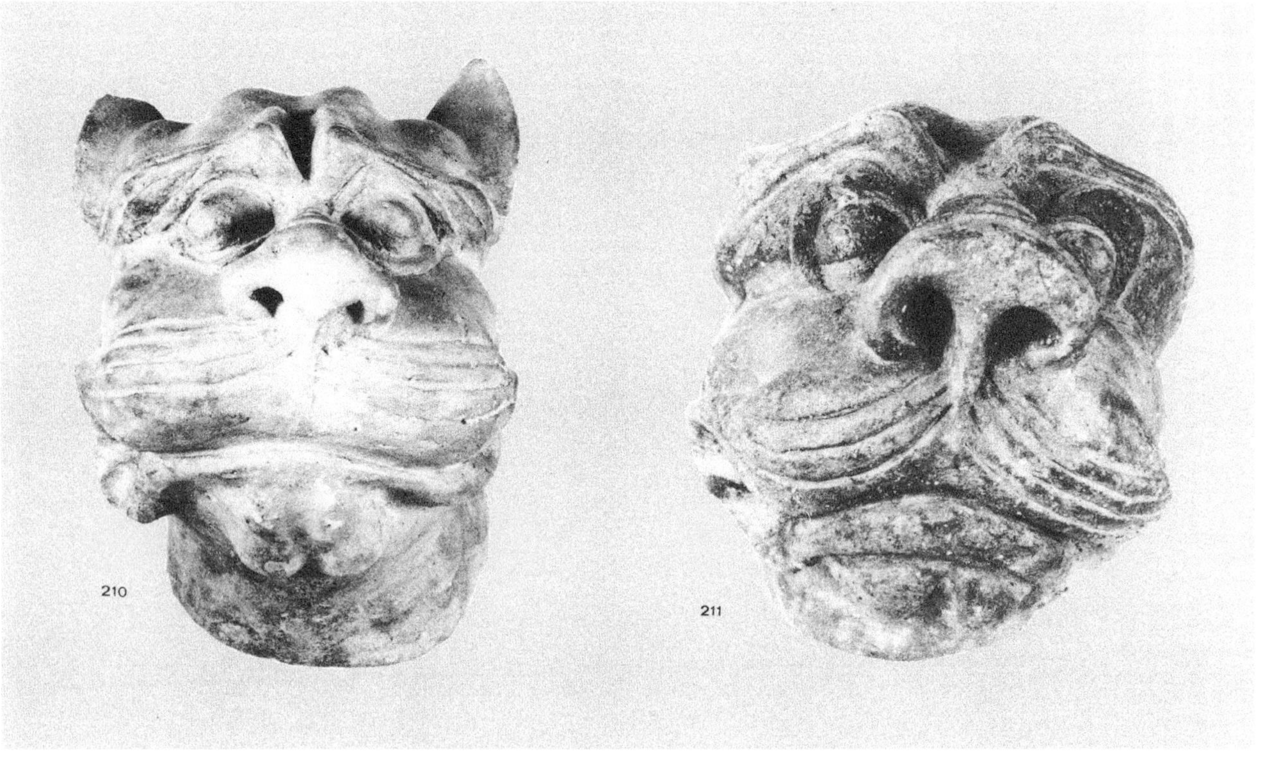


Figure 3. Terracottas from Santa Gilla (from Moscati 1991: 113).

personnel.[10] Also in the Old Testament (Deut. 23.18, 1 Kgs 15.12, 2 Kgs 23.7; cf. Rev. 22.15 and Phil. 3.2) *klbm* are mentioned in connection with the temple and are usually interpreted as cultic (male) prostitutes (cf. Brunet 1985; Day 2004), although an interpretation as real dogs has also been suggested (Goodfriend 1995). In the case of the Phoenician inscription, a difficulty in interpreting *klbm* as real dogs is that they receive payment like the men listed. A possible association of real dogs with the cult of Astarte has been suggested for a supposed Persian period 'neighbourhood temple' excavated in Beirut (area BEY 010). Inside this building dog bones were found, as well as terracotta figurines suggesting that the building might have been mainly devoted to the cult of Astarte (Elayi 2010: 166). Dogs, however, were probably also associated with other Phoenician deities. To the corpus of dog images collected by D'Andrea (2018a: 204–210) a possible representation of a dog can be added (Figure 1) painted on a wall of Grotta Regina (Coacci Polselli *et al.* 1979: 85; fig. 62; Tav. XXXIX; cf. Bisi *et al.* 1969: 60; fig. 35; Tav. XXXI), a cave near Palermo where the worship of the god Shadrapa is well attested, but was not exclusive. It is unknown if this image was related to the worship of Shadrapa, or of Isis (as suggested by Rocco 1969: 28). Other images of dogs occur in Eshmun's sanctuary at Bostan esh-Sheikh. A hunting dog appears on a frieze of the 'piscine du trône d'Astarté' (Stucky 2005: 170–172), with a hunter on horseback who might represent the god himself (who was εἰωθώς κυνηγετεῖν according to Dam. *Isid.* fr. 348; Dunand 1983; Will 1985: 113). A small dog or puppy is also represented, whilst another two small dogs or puppies appear on the decoration of the 'Bâtiment aux frises d'enfants' in the same sanctuary (Stucky 2005: 172–174). A group of terracottas which was found in the lagoon of Santa Gilla, near Cagliari (Vivanet 1892, 1893: Levi 1937, Moscati 1991), may also have related to the worship of Eshmun. The terracottas, which included twenty foreparts of 'greyhounds' (Figure 2; Moscati 1991, n° 190–209) and two heads of 'molossers' (Figure 3; Moscati 1991, n° 210–211), were possibly intended as votive material for a local sanctuary (Moscati 1991: 27), which might have been dedicated to Eshmun (or to Sid, according to Barreca 1986: 290). The terracottas are dated by scholars to either immediately before (Salvi 2004: 71–73) or shortly after the Roman conquest (3rd-2nd century BC). Foreparts of 'greyhounds' were later found in a different part of the lagoon (Fanari *et al.* 1988: 15, 20, 24 fig. 12), and another 'molosser' head was found in the area of the present-day harbour of Cagliari (Tore *et al.* 1992: 539; Tav. I, 2). On a razor from Carthage (Acquaro 1971: 34; fig. 16, 1; Tav. IV, 1; Yazidi 2009: 146) a dog is associated with Hermes, whose

[10] 'The *klbm* and *grm* (...) were probably temple servants masked (and clothed?) as dogs and lions. Both groups may have been singers and dancers, but there is no evidence that they were prostitutes - or at least that prostitution was their ritual function' (Peckham 1968: 317 note 4); 'The *klbm* and the *grm* are cult personnel dressed like dogs and lions' (Healey 1974: 56). Even if they were prostitutes, they were not necessarily personnel of the temple (Ribichini 2004: 60).

Figure 4. Razor from Carthage (from Acquaro 1971, fig. 38, 2).

possible Carthaginian *interpraetatio* is uncertain. On another razor from Carthage (Acquaro 1971: 69–70; fig. 38, 2; Tav. XXVI; Yazidi 2009: 145), a dog is associated with Melqart (Figure 4). The same god could be identified, according to T. Redissi (1995: 120) in the figure wrapped in a shroud and surrounded by four (probably) dogs on the aforementioned scarab from Kerkouane. Melqart is also involved in the mythical tradition, only attested for a relatively late age, which explains the discovery of purple dye through the story of a dog having been stained by murex on the shore near Tyre. Tyrian coins issued under the Roman Empire celebrate this discovery by depicting the dog and the murex (Naster 1985). As stated in a source (Poll. *Onom.* 1.45–46), this dog belonged to Heracles (ὁ Ἡράκλειος κύων), Heracles being the usual Greek *interpretatio* for the god Melqart. Other sources claim that Heracles only saw the dog and realised the value of the murex (Suda, η 476 Adler). According to some sources (Johannes Malalas, *Chronographia* 2.32; Georgius Cedrenus, *Historiarum compendium* 34; and a spurious passage[11] formerly attributed to Johannes Antiochenus, cf. Müller, *FGH* IV, p. 544) the dog seen by Heracles was a sheepdog, κύνα ποιμενικόν. Other sources only mention a sheepdog (or a bitch: Gregory of Nazianzus, *Orationes* 4.108, cf. Cassiod. *Var.* 1.2.7: *fame canis avida*) with no reference to Heracles. According to one version, the dog of a shepherd found the murex which had been thrown away by a fisherman (*Ach. Tat.* 2.11); perhaps mistaking the latter for the dog's master, one source (Nonnus, *Dion.* 40.304–310) labels the dog a 'fishing dog', κύων ἁλιεργὸς (hapax). The involvement of Heracles in this tradition is probably due to his being the patron of Tyre (cf. Bonnet 1988: 74–77), rather than to any particular relationship with dogs or to a role of his temples in the economy of purple dye (as suggested by Mazzucato 2002: 86, cf. Acquaro 1998: 104). On the other hand, the dog might have symbolised the *canicula*, the myth being connected with the behaviour of the murex during 'dog days' (Naster 1985). A symbolic value, although unknown, was presumably attributed to the glass pendants representing dogs' heads, one of which was found in Carthage (Seefried 1982: 144). Dogs are well represented on coins issued by Motya (Jenkins 1971, Pl. 1–2: 14–17; Pl. 3; Pl. 4: 31–36; Mani Hurter 2005: 8; Taf 2, 2–3) and Panormos (Jenkins 1971, Pl. 1–2: Z1-Z3; Pl. 6: 1, 7–8, 11; Pl. 7: 1–4).[12] Panormos was most probably the origin also of the coins carrying the legend *ṣyṣ*, on some of which a dog occurs (Jenkins 1971, Pl. 6: 9–10). These were, however, imitations of coinage from the Elymian city of Segesta (cf. Cutroni Tusa 1982). According to some scholars, the symbolism of the dog in western Sicily was a mark of eastern or specifically Phoenician

[11] Cf. S. Mariev, ed., *Ioannis Antiocheni fragmenta quae supersunt omnia*, Corpus Fontium Historiae Byzantinae 47, Berlin 2008, p. 591.
[12] Jenkins 1971, Pl. 6: 2 is a fake: Mani Hurter 2005: 7.

Figure 5. Dog deposition at Tel Burak (from Kamlah and Sader 2003: 152, fig. 8).

influence (cf. Holm 1870: 89; Movers 1850: 321 note 34; Pace 1945: 633; Schilling 1982: 237), whilst others assume that it belonged to the native, or specifically the Elymian tradition (Ciaceri 1911: 122–133; Dubourdieu 1990–1991; Cataldi 1992: 73–74). In fact, the breed represented on the coins has been linked to the present-day 'cirneco dell'Etna', a hound originally from Sicily (Ciaceri 1911: 129–130). Since it seems to have been typical of the Elymian region, this hound would have been chosen as a suitable symbol for Segestan coins (Marconi 1997: 1077–1078). Due to the close political and economic relations between the Elymians and the Phoenician settlements on Sicily, such dogs were presumably also physically present in the latter. On some of the coins of Panormos (Jenkins 1971, Pl. 6: 8; Jenkins 1971, Pl. 6: 9, carrying the legend *ṣyṣ*) the dog is associated with a shell, presumably a murex. A reference to the aforementioned myth of the discovery of the purple by a dog (Mazzucato 2003: 126; cf. Acquaro 1998: 102–104) cannot be ruled out, although the association of the shell and the dog on coins of Panormos does not occur earlier than 415 BC (cf. Mazzucato 2003: 125), while it is attested on Segestan coins of the years 455/50–445/40 BC (Mani Hurter 2008: 70–72; Taf. 6, 60–68). It is not clear why Segesta should have alluded on its coins to the production of purple dye, the 'spécialité des Phéniciens' as suggested by Dubourdieu (1990–1991: 77 note 141). Furthermore, as already mentioned, the myth is only attested by very late texts and coins (note that the shell featuring on Tyrian and Byblian coinage of the Persian period is not a *murex* but a *Charonia variegata*: Elayi and Elayi 2009: 272–273; Elayi and Elayi 2014: 46–47).

## 4 Dog depositions

Dog burials are attested at several Levantine coastal sites, especially during the Persian period, and some Phoenician sites have also furnished evidence of dog burials.[13] At Khalde, about 12 km south of Beirut, eight dog skeletons were found in the 1960's. Identified as 'desert greyhounds', they were found near a tenth-eighth century BC cemetery but also near buildings 'd'époque gréco-perse' (Saidah 1967: 166–167). At Tell Burak, a site located 9 km south of Sidon, dog burials were found in Persian period deposits (Figure 5; Kamlah and Sader 2003: 149; Sader and Kamlah 2010: 132; Çakirlar *et al.* 2013). Two pits, each containing the articulated skeleton of a dog, were found a few metres apart. It was not possible to ascertain whether a domestic building complex on the slope of the tell, with which both pits were associated, had already been abandoned at the moment when the burials were performed. One of the dogs, a puppy whose age-at-death can be estimated as six months or slightly younger, lay on its left side with its limbs folded. The corpse was covered with sherds of an amphora. The other skeleton was only partially preserved, and the sex could not be determined in either case. In Beirut, a dog cemetery dating to the Persian period was found in the area of the Iron Age glacis (Finkbeiner and Sader 1997: 130–132, 131 Fig. 7). At least ten dogs were buried in shallow pits, lying on their sides with the extremities folded under their bodies. Three of the skeletons were covered by large potsherds, just like the aforementioned cases in Tell Burak. A further 16 dog burials dating to the Persian period, found on the outskirts of ancient Beirut, have been carefully studied (Hourani 2018a). The dogs, some of which were surely male, were buried lain on their side, with their limbs either extended or flexed. No pits could be identified, and the dogs might have been deposited in perishable containers. At least one of the corpses had been covered by a fragment of an amphora, and another one by a stone (Hourani 2018a: 159, 178–180). The age of the animals gave no indication of any apparent human selection and they were medium to medium-large specimens. Later dog burials (2nd century BC) have also been found in Beirut (Hourani 2018b), within abandoned structures and in refuse dumps on the fringes of the settlement (BEY 198).

[13] At Quinta do Almaraz (Almada) dogs burials have also been reported (Correia 2015; cf. Cardoso and Varela Gomes 1997: 107–108). The dogs (whose skulls were apparently removed after the soft tissue had decayed) were buried in the town moat. The burials date back to a period between the 8th and the 4th century.

Although complete dog skeletons are also known from later as well as earlier contexts (e.g. Maher 2005: 286–288), it is in the Persian period that most of the dog burials occur in the Levant. There may be no single explanation for the phenomenon of Levantine dog depositions as a whole but, in attempting to furnish one, scholars have mainly focused on the case of Ashkelon, where at least 1250 dog burials have been excavated (Wapnish and Hesse 1993). The dogs were buried in pits, lying on their sides. Some scholars associate the custom with a Phoenician cultural influence. According to one explanation, the dogs had previously been revered as sacred animals, free to roam in a sanctuary that was probably linked to a cult of healing (Stager 1991). Alternatively, the dogs may have been associated with the worship of Eshmun (the conjecture being that 'it was in the Phoenician incarnation of Eshmun that Asklepios/Gula was worshiped' in Ashkelon, Halpern 2000: 141) or Astarte (Heltzer 1998). As already mentioned, 'dogs' (*klbm*) and *grm* (possibly 'whelps') are mentioned in connection with a temple of Astarte in a Phoenician inscription from Cyprus, and a supposed temple in Beirut where the same goddess may have been worshipped and where dog bones were found (but not in dog depositions). No evidence of a sanctuary has been found, however, in relation to the dog depositions of Ashkelon, nor those in Beirut, Khalde or Tell Burak. Furthermore, as observed by Helen Dixon (Dixon 2018: 20), the phenomenon of the deposition of dogs in the Levant is not limited to culturally Phoenician sites, therefore it is difficult to accept an explanation based solely on Phoenician cultic influences. The Philistine tradition should also be evaluated (Horwitz 2015: 148). One suggested explanation is that the dogs were not buried with reverence due to their association with a deity, but were deposited in specific areas after having been killed, possibly during some healing/purifying ritual and in a manner which left no traces in the archaeological record (Edrey 2008). In support of this hypothesis is a passage in Isaiah (66. 3), which might indicate that to break a dog's neck (it is not clear whether or not blood was to be shed, according to Sasson 1976: 201) was a heathen ritual practice. An economic explanation suggests that the Tyrians used Ashkelon as a hub for the trade in dogs, and that the dogs which died before being shipped were buried there (Smith 2015). This interpretation, however, does not explain the other occurrences of dog cemeteries. For Geoffrey Miller the most likely explanation 'is that the Phoenicians buried dogs to which they had some emotional attachment' (Miller 2008: 493), while Helen Dixon (2018) even suggests that such an attachment was not necessary, since dogs could have acquired the role of 'persons', therefore being buried according to the contemporary burial custom adopted for people of low social status.

## 5 Human 'dogs'

A symbolic relationship between dogs and human beings is attested in Phoenician culture by the onomastics, where personal names formed with the element *klb*, meaning 'dog' are well attested (cf. Benz 1972: 131–132, 331; Dixon 2018: 35 Table 4). These names could express the attitude of the devotee towards the divinity. Labelling someone, or oneself, 'a dog', in order to express an inferior position and imply deference and obedience is a well attested practice in the Levant, especially in the Amarna correspondence and in the Lachish letters (Winton Thomas 1960). Calling someone 'a dog' was not necessarily intended as an offence (cf. Galán 1993). A more (self)-debasing phrase was 'dead dog', also used in the Old Testament (1 Sam 24.15; 2 Sam 9.8; 16.9; the LXX has it in 2 Kgs 8.13) and ambivalence towards dogs is a common feature in many cultures. In *EA* 60–61 the sender calls himself a 'dog of the house' of his lord, clearly stressing his positive attributes of faithfulness and devotion to his master. The same positive value should be assumed for the element *klb* in the Phoenician personal names, 'dog (of god)' meaning 'obedient worshipper' (Krahmalkov 2000: 227). Such a meaning might also explain the use of the word *klb* to indicate some kind of temple personnel (if such is the meaning of *klb* in *CIS* I 86). New Testament readers will also remember the dialogue of Jesus with the Syrophoenician, or Canaanite woman (Mark 7: 24–30; Matthew 15: 21–28). When the woman and her daughter are compared to dogs, the woman, who was a Phoenician, answered that dogs may eat the scraps from their lord's table.

## 6 Conclusions

Although, for classical writers, sheepdogs, hunting dogs, and watchdogs were probably just too common a feature to deserve mentioning, the role of dogs in daily life was probably much wider than the available evidence can prove. Owing to the nature of most archaeological contexts whose animal remains have been comprehensively analysed, the symbolic and ritual functions of dogs are better known. The apparent diversity of those functions suggests that the dog was, in Phoenician culture, an animal that was very 'good to think with'. Such diversity might be partly explained by local and chronological peculiarities within what we call as a whole 'Phoenician culture'. Further research will surely shed more light on the life of a Phoenician's best friend.

## Acknowledgements

I wish to thank Bruno D'Andrea and Yasha Hourani for kindly providing me with a copy of their recent publications.

## References

### *Ancient sources*

Achilles Tatius, *Clitophon and Leucippe*
Cassiodorus, *Variae*
Damascius, *Vita Isidori*
Gregory of Nazianzus, *Orationes* 4.108
Justinus, *Epitome* (of Trogus)
Nonnus, *Dionysiaca*
Pollux, *Onomasticon*
Silius Italicus, *Punica*
P. Vergilius Maro, *Aeneis*
P. Vergilius Maro, *Georgica*

### *Abbreviations*

Abbreviations follow S. Hornblower, A. Spawforth, and E. Eidinow (eds), *The Oxford Classical Dictionary*, 4th Edition. Oxford/New York: Oxford University Press.

*Ach. Tat.* = *Achilles Tatius*
Cassiod. *Var.* = Cassiodorus, *Variae*
Dam. *Isid.* = Damascius, *Vita Isidori*
Just. = Justinus, *Epitome* (of Trogus)
Nonnus, *Dion.* = Nonnus, *Dionysiaca*
Poll. *Onom.* = Pollux, *Onomasticon*
Sil. *Pun.* = Silius Italicus, *Punica*
Verg. *Aen.* = P. Vergilius Maro, *Aeneis*
Verg. *G.* = P. Vergilius Maro, *Georgica*

### *Modern sources*


Acquaro, E. 1971. *I rasoi punici* (Studi Semitici 41). Roma: Consiglio Nazionale delle Ricerche.

Acquaro, E. 1998. I Fenici, Cartagine e l'archeologia della porpora, in O. Longo (ed.) *La porpora. Realtà e immaginario di un colore simbolico. Atti del Convegno di studio. Venezia, 24 e 25 ottobre 1996*: 99–110. Venezia: Istituto veneto di scienze, lettere ed arti.

Alhaique, F. 2012. Resti faunistici dalla Favissa F.2950, in L. Nigro and F. Spagnoli (eds) *Alle sorgenti del Kothon. Il rito a Mozia nell'Area sacra di Baal 'Addir-Poseidon. Lo scavo dei pozzi sacri nel Settore C Sud-Ovest (2006–2011)* (Quaderni di archeologia fenicio-punica/CM 02): 33. Roma: Università di Roma 'La Sapienza'.

Astruc, M. 1959. Empreintes et reliefs carthaginois de terre cuite. *Mélanges d'archéologie et d'histoire* 71: 107–134.

Babelon, E. 1896. *Carthage*. Paris: Ernest Leroux.

Barreca, F. 1986. *La civiltà fenicio-punica in Sardegna* (Studi e Monumenti 3). Sassari: Carlo Delfino.

Bénichou-Safar, H. 1982. *Les tombes puniques de Carthage. Topographie, structures, inscriptions et rites funéraires*. Paris: Éditions du Centre National de la Recherche Scientifique.

Bénichou-Safar, H. 2004. *Le tophet de Salammbô à Carthage. Essaie de reconstitution* (Collection de l'École Française de Rome 342). Paris: École Française de Rome.

Benz, F.L. 1972. *Personal Names in the Phoenician and Punic Inscriptions* (Studia Pohl 9). Roma: Biblical Institute Press.

Bevilacqua, F. 1972. Considerazioni sulle terrecotte a stampo, in F. Bevilacqua, A. Ciasca, G. Matthiae Scandone, S. Moscati, V. Tusa and A. Tusa Cutroni, *Mozia VII* (Studi Semitici 40): 113–117. Roma: Consiglio Nazionale delle Ricerche.

Bisi, A.M., M.G. Guzzo Amadasi and V. Tusa 1969. *Grotta Regina-I* (Studi Semitici 33). Roma: Consiglio Nazionale delle Ricerche.

Bonnet, C. 1988. *Melqart. Cultes et mythes de l'Héraclès tyrien en Méditerranée* (Studia Phoenicia 8). Namur/ Leuven: Peeters.

Boyce, M. 1982. *A History of Zoroastrianism*, II. Leiden/ Köln: Brill.

Brunet, G. 1985. L'hébreu *kèlèb*. *Vetus Testamentum* 35: 485–488.

Bucci, O. 1977. Una pagina dimenticata di storia: i rapporti tra Cartagine e l'impero dei Persiani. *Africa. Rivista trimestrale di studi e documentazione dell'Istituto italiano per l'Africa e l'Oriente* 32: 446–455.

Çakirlar, C., V. Amer, J. Kamlah and H. Sader 2013. Persian Period Dog Burials in the Levant: New Evidence from Tell el-Burak (Lebanon) and a Reconsideration of the Phenomenon, in B. De Cupere, V. Linseele and S. Hamilton-Dyer (eds) *Archaeozoology of the Near East X. Proceedings of the Tenth International Symposium on the Archaeozoology of South-Western Asia and Adjacent Areas* (Ancient Near Eastern Studies Supplement 44): 243–264. Leuven: Peeters.

Calascibetta, A.M.G. 2009. Nuovi dati sulla necropoli soluntina, in C. Ampolo (ed.) *Immagine e immagini della Sicilia e di altre isole del Mediterraneo antico. Atti delle Seste Giornate Internazionali di Studi sull'area elima e la Sicilia occidentale nel contesto mediterraneo (Erice, 12–16 ottobre 2006)*: 633–643. Pisa: Edizioni della Normale.

Camarós, E. and J. Estévez 2010. Los restos arqueozoológicos de mamíferos: gestión y explotación del recurso animal en los niveles del siglo VII a.C. de Plaza de la Catedral (Ceuta), in F. Villada Paredes, J. Ramon Torres and J. Padilla (eds) *El asentamiento protohistórico de Ceuta. Indígenas y Fenicios en la orilla norteafricana del estrecho de Gibraltar*: 383–405. Ceuta: Ciudad Autónoma de Ceuta, Archivo General.

Campanella, L. 2008. *Il cibo nel mondo fenicio e punico d'Occidente. Un'indagine sulle abitudini alimentari attraverso l'analisi di un deposito urbano di Sulky in Sardegna*. Pisa/Roma: Fabrizio Serra.

Cardoso, J.L. 2000. Fenícios e Indígenas em Rocha Branca, Abul, Alcácer do Sal, Almaraz e Santarém. Estudo comparado dos mamíferos, in M.E. Aubet Semmler and M. Barthélemy (eds) *Actas del IV Congreso Internacional de Estudios Fenicios y*

*Púnicos (Cádiz, 2-6 octubre 1995)*, I: 319–327. Cádiz: Universidad de Cádiz.

Cardoso, J.L. and A.M. Arruda 2016. Faunas domésticas e rituais funerários em Alcácer do Sal (Idade do Ferro), in R. Vilaça and M. Serra (eds) *Matar a fome, alimentar a alma, criar sociabilidades. Alimentação e comensalidade nas sociedades pré e proto-históricas*: 193–217. Coimbra: Universidade de Coimbra.

Cardoso, J.L., J.L. López Castro, A. Ferjaoui, A. Mederos Martín, V. Martínez Hahnmüller, and I. Ben Jerbania 2016. What the people of Utica (Tunisia) ate at a banquet in the 9th century BCE. Zooarchaeology of a North African early Phoenician settlement. *Journal of Archaeological Science: Reports* 8: 314–322.

Cardoso, J.L. and M. Varela Gomes 1997. O consumo de cão, em contextos fenício-púnicos, no território português. *Estudos Orientais* 6: 89–117.

Carenti, G. and B. Wilkens 2006. La colonizzazione fenicia e punica e il suo influsso sulla fauna sarda. *Sardinia, Corsica et Baleares antiquae* 4: 173–186.

Castaños Ugarte, P.M. 1994. Estudio de la fauna de la necrópolis de Villaricos (Almería). *Archaeofauna* 3: 1–12.

Cataldi, S. 1992. Popoli e città del lupo e del cane in Italia meridionale e in Sicilia tra realtà e immagine, in M. Sordi (ed.) *Autocoscienza e autorappresentazione dei popoli nell'antichità* (Contributi dell'Istituto di Storia antica 18): 55–82. Milano: Vita e Pensiero.

Ciaceri, E. 1911. *Culti e miti nella storia dell'antica Sicilia*. Catania: Battiato.

Ciasca, A., R. Di Salvo, M. Castellino and C. Di Patti 1996. Saggio preliminare sugli incinerati del tofet di Mozia. *Vicino Oriente* 10: 317–346.

Coacci Polselli, G, M.G. Guzzo Amadasi and V. Tusa 1979. *Grotta Regina-II. Le iscrizioni puniche* (Studi Semitici 52) Roma: Consiglio Nazionale delle Ricerche.

Correia, F. 2015. Enterramento de cães na Quinta do Almaraz (Almada, Portugal), in G. Branco, L. Rocha, C. Duarte, J. de Oliveira and P. Bueno Ramírez (eds) *Arqueologia de Transição: O Mundo Funerário. Actas do II Congresso Internacional Sobre Arqueologia de Transição (29 de Abril a 1 de Maio 2013)*: 113–124. Évora: Universidade de Évora, Centro de História de Arte e Investigação Artística.

Cutroni Tusa, A. 1982. Riflessioni sulla monetazione di Segesta ed Erice, in M.L. Gualandi, L. Massei and S. Settis (eds) *ΑΠΑΡΧΑΙ. Nuove ricerche e studi sulla Magna Grecia e la Sicilia antica in onore di Paolo Enrico Arias*, I: 239–244. Pisa: Giardini.

D'Andrea, B. 2018a. Le chien dans la religion et dans la vie quotidienne des communautés phéniciennes et puniques de la Méditerranée occidentale. *Mélanges de l'École Française de Rome* 130: 185–217.

D'Andrea, B. 2018b. *Bambini nel 'limbo'. Dati e proposte interpretative sui tofet fenici e punici* (Collection de l'École Française de Rome 552). Roma: École française de Rome.

Day, J. 2004. Does the Old Testament Refer to Sacred Prostitution and Did it Actually Exist in Ancient Israel?, in C. McCarthy and J.F. Healey (eds) *Biblical and Near Eastern Essays. Studies in honour of Kevin J. Cathcart* (Journal for the Study of the Old Testament Supplement Series 375): 2–21. London/New York: T&T Clark International.

de Jong, L. 2017. *The Archaeology of Death in Roman Syria. Burial, Commemoration, and Empire*. Cambridge: Cambridge University Press.

Delattre, A.-L. 1895. Lettre sur les fouilles de la nécropole punique du terrain Douïmès à Carthage, *Comptes rendus de l'Académie des Inscriptions et Belles-Lettres* 39: 281–285.

Del Medico, H.E. 1957. L'illustration du thrène d'Aqahat sur le sarcophage dit des 'Pleureuses' au Musée d'Istanbul, in Z.V. Togan (ed.), *Proceedings of the Twenty-Second Congress of Orientalists*, II: 91–94. Leiden: Brill.

Dixon, H. 2018. Late 1st-Millennium B.C.E. Levantine Dog Burials as an Extension of Human Mortuary Behavior. *Bulletin of the American Schools of Oriental Research* 379: 19–41.

Dubourdieu, A. 1990–1991. Le chien de Segeste. *ΚΩΚΑΛΟΣ* 36–37: 51–83.

Dunand, M. 1983. L'iconographie d'Echmoun dans son temple sidonien, in *Atti del I Congresso Internazionale di Studi Fenici e Punici (Roma, 5-10 novembre 1979)*: 515–519. Roma: Consiglio Nazionale delle Ricerche.

Edrey, M. 2008. The Dog Burials at Achaemenid Ashkelon Revisited. *Tel Aviv* 35: 267–282.

Elayi, J. 1988. Les sarcophages phéniciens d'époque perse. *Iranica Antiqua* 23: 275–322.

Elayi, J. 2010. An Unexpected Archaeological Treasure: The Phoenician Quarters in Beirut City Center. *Near Eastern Archaeology* 73: 156–168.

Elayi, J. and A.G. Elayi 2009. *The Coinage of the Phoenician City of Tyre in the Persian Period (5th-4th cent. BCE)* (Orientalia Lovaniensia Analecta 188, Studia Phoenicia 20). Leuven: Peeters.

Elayi, J. and A.G. Elayi 2014. *A Monetary and Political History of the Phoenician City of Byblos in the Fifth and Fourth Centuries B.C.E.* Winona Lake, IN: Eisenbrauns.

Fanari, F., G. Nieddu, E. Usai and R. Zucca 1988. *Santa Gilla e Marceddì. Prime ricerche d'archeologia subacquea lagunare*. Cagliari: Regione Autonoma della Sardegna.

Fedele, F. 1977. Antropologia fisica e paleoecologia di Tharros. Nota preliminare sugli scavi del *tofet*, campagna 1976. *Rivista di Studi Fenici* 5: 185–193.

Finkbeiner, U. and H. Sader 1997. Bey 020 Preliminary report of the excavations 1995. *Bulletin d'Archéologie et d'Architecture Libanaises* 2: 114–166.

Fleischer, R. 1983. *Der Klagefrauensarkophag aus Sidon* (Istanbuler Forschungen 34). Tübingen: Ernst Wasmuth.

Gaillard, L. 1938–1940. Réservoir à amphores et sépultures puniques de Carthage. *Bulletin Archéologique du Comité* des Travaux historiques et scientifiques 1938–1940: 327–333.

Galán, J.M. 1993. What is he, the dog? *Ugarit-Forschungen* 25: 173–180.

Goodfriend, E.A. 1995. Could *keleb* in Deuteronomy 23:19 Actually Refer to a Canine?, in D.P. Wright, D.N. Freedman and A. Hurvitz (eds) *Pomegranates and Golden Bells. Studies in Biblical, Jewish, and Near Eastern Ritual, Law, and Literature in Honor of Jacob Milgrom*: 381–397. Winona Lake: Eisenbrauns.

Guirguis, M., and A. Unali 2012. Ipogei sulcitani tra età punica e romana: la Tomba Steri 1, in M.B. Cocco, A. Gavini and A. Ibba (eds), *L'Africa Romana XIX. Trasformazione dei paesaggi del potere nell'Africa settentrionale fino alla fine del mondo antico, Sassari, 16–19 dicembre 2010*: 2011–2029. Roma: Carocci.

Hamdy Bey, O. and T. Reinach 1892. *Une nécropole royale à Sidon. Fouilles de Hamdy Bey*. Paris: Ernest Leroux.

Halpern, B. 2000. The Canine Conundrum of Ashkelon: A Classical Connection?, in L.E. Stager, J.A. Greene and M.D. Cogan (eds) *The Archaeology of Jordan and Beyond. Essays in Honor of James A Sauer*: 133–144. Winona Lake: Eisenbrauns.

Healey, J.P. 1974. The Kition Tariffs and the Phoenician Cursive Series. *Bulletin of the American Schools of Oriental Research* 216: 53–60.

Heltzer, M. 1998. On the Vth Century B.C.E. Dogs from Ashkelon. *Transeuphratène* 15: 149–152.

Holm, A. 1870. *Geschichte Siciliens im Alterthum*, I. Leipzig: Engelmann.

Horwitz, L.K. 2015. Dog Remains, in R. Kletter, I. Ziffer and W. Zwickel, *Yavneh II. The 'Temple Hill' Repository Pit* (Orbis Biblicus et Orientalis Series Archaeologica 36): 145–149. Fribourg: Academic Press/Göttingen: Vandenhoeck & Ruprecht.

Hourani, Y. 2018a. Persian Period Dog Burials of Beirut: Morphology, Health, Mortality and Mortuary Practices. In C. Çakırlar, J. Chahoud, R. Berthon and S. Pilaar Birch (eds) *Archaeozoology of the Near East XII. Proceedings of the 12th International Symposium of the ICAZ Archaeozoology of Southwest Asia and Adjacent Areas Working Group. Groningen, Institute of Archaeology, June 14–15 2015, University of Groningen, the Netherlands*: 153–184. Groningen: Barkhuis Publishing & University of Groningen.

Hourani, Y. 2018b. Congenital Anomalies and Traumatic Injuries in Dogs from *Laodicea in Canaan* (Hellenistic Beirut, Lebanon), in L. Bartosiewicz and E. Gál (eds) *Care or Neglect? Evidence of Animal Disease in Archaeology. Proceedings of the 6th meeting of the Animal Palaeopathology Working Group of the International Council for Archaeozoology (ICAZ), Budapest, Hungary, 2016*: 79–96. Oxford: Oxbow Books.

Hours-Miedan, M. 1951. Les représentations figurées sur les stèles de Carthage. *Cahiers de Byrsa* 1: 15–160.

Iborra Eres, P. 2005. El material faunístico, in C. Aranegui Gascó (ed.) *Lixus-2 Ladera sur. Excavationes arqueológicas marroco-españolas en la colonia fenicia. Campañas 2000-2003* (Sagvntvm. Papeles del Laboratorio de Arqueología de Valencia. Extra-6): 229–239. Valencia: Universitat de València.

Jenkins, G.K. 1971. Coins of Punic Sicily, Part I. *Schweizerische numismatische Rundschau* 50: 25–78.

Kamlah, J. and H. Sader 2003. The Tell el-Burak Archaeological Project Preliminary Report on the 2002 and 2003 Seasons. *Bulletin d'Archéologie et d'Architecture Libanaises* 7: 145–173.

Krahmalkov, C.R. 2000. *Phoenician-Punic Dictionary* (Studia Phoenicia XV. Orientalia Lovaniensia Analecta 90). Leuven: Peters.

Levi, D. 1937. Scavi e ricerche archeologiche della Reale Soprintendenza alle opere d'Antichità e d'arte della Sardegna (1935–1937). *Bollettino d'Arte 1937*: 193–210.

Maher, E.F. 2005. The Faunal Remains, in M. Dothan and D. Ben-Shlomo (eds), *Ashdod VI. The Excavations of Areas h and K (1968-1969)* (Israel Antiquities Authority Reports 24): 283–290. Jerusaem: Israel Antiquities Authority.

Mani Hurter, S. 2005. Addenda et corrigenda zu G.K. Jenkins, coins of punic Sicily, Part I. *Schweizerische numismatische Rundschau* 84: 5–14.

Mani Hurter, S. 2008. *Die Didrachmenprägung von Segesta* (Schweizer Studien zur Numismatik 1). Bern: Schweizerische Numismatische Gesellschaft.

Mansouri, M. 2015. Cynophagy, homosexuality and anthropophagy in medieval Islamic North Africa as signs of hospitality. *Journal of North African Studies* 20: 128–142.

Marconi, C. 1997. Storie di caccia in Sicilia occidentale, in *Atti delle seconde giornate internazionali di studi sull'area elima (Gibellina, 22-26 ottobre 1994)*: 1071–1120. Pisa: Scuola Normale Superiore di Pisa/ Gibellina: Comune di Gibellina; Centro Studi e Documentazione sull'Area Elima.

Martín Ruiz, J.A., A. Pérez-Malumbres Landa and J.R. García Carretero 2003. Tumba de cámara de la necrópolis fenicia de Gibralfaro (Málaga, España). *Rivista di studi fenici* 31: 139–160.

Mazzucato, C. 2002. L'industria della porpora: un'eredità fenicia, in M. Khanoussi, P. Ruggeri and C. Vismara (eds) *L'Africa romana. Lo spazio marittimo del Mediterraneo occidentale: geografia storica ed economica. Atti del XIV convegno di studio, Sassari, 7-10 dicembre 2000*: 83–96. Roma: Carocci.

Mazzucato, C. 2003. Il murice nelle monete fenicie e puniche. *Byrsa* 2: 121–140.

Miller, G.D. 2008. Attitudes toward Dogs in Ancient Israel: A Reassessment. *Journal for the Study of the Old Testament* 32: 487–500.

Morales, A., M.A. Cereijo, P. Brännstöm and C. Liesau 1994. The mammals, in E. Roselló and A. Morales (eds) *Castillo de Doña Blanca. Archaeo-environmental investigations in the Bay of Cádiz, Spain (750-500 B.C.)*

(British Archaeological Reports International Series 593): 37–69. Oxford: Tempus Reparatum.
Morales Pérez, J.V. 2008. Zooarqueología en un contexto ritual: posibilidades de estudio y ejemplos de aplicación en el Mediterráneo, in E. Ferrer Albelda, J. Mazuelos Pére and J.L. Escacena Carrasco Sevila (eds) *De dioses y bestias. Animales y religión en el Mundo Antiguo*: 13–32. Sevilla: Universidad de Sevilla.
Moscati, S. 1991. *Le terrecotte figurate di S. Gilla (Cagliari)*. Roma: Multigrafica Editrice.
Movers, F.C. 1850. *Das phönizische Alterthum*, II. 2. Berlin: Dümmler.
Naster, P. 1985. Le chien et le murex sur des monnaies impériales de Tyr. *Quaderni Ticinesi. Numismatica e antichità classiche* 14: 257–260.
Niveau de Villedary y Mariñas, A.M. 2008. ¿Compañero en la muerte o guía hacia el más allá? El perro en la liturgia funeraria púnica, in E. Ferrer Albelda, J. Mazuelos Pére and J.L. Escacena Carrasco Sevila (eds) *De dioses y bestias. Animales y religión en el Mundo Antiguo*: 97–141. Sevilla: Universidad de Sevilla.
Niveau de Villedary y Mariña, A.M. and Ferrer Albelda, E. 2004. Sacrificios de cánidos en la necrópolis púnica de Cádiz. *Huelva Arqueológica* 20: 63–88.
Pace, B. 1945. *Arte e civiltà della Sicilia antica*, III. Genova/Roma/Napoli/Città di Castello: Dante Alighieri.
Parlasca, K. 1982. *Syrische Grabreliefs hellenistischer und romischer Zeit. Fundgruppen und Probleme*. Mainz am Rhein: von Zabern.
Peckham, B. 1968. Notes on a Fifth-Century Phoenician Inscription from Kition, Cyprus (CIS 86). *Orientalia* 37: 304–324.
Pompianu, E. 2017. Nuovi scavi nella necropoli punica di Villamar (2013–2015). *The Journal of Fasti Online*: viewed 9 January 2020, www.fastionline.org/docs/FOLDER-it-2017-395.pdf
Prévost, V. 2006. Les ragoûts de chien: prophylaxie, beauté et gastronomie. *Horizons Maghrébins-Le droit à la mémoire* 55: 88–96.
Redissi, T. 1995. Étude des scarabées et scaraboïdes de Kerkouane. *Revue des Études Phéniciennes-Puniques et des Antiquités Libyques* 9: 115–146.
Redissi, T. and M. Tillot 1995. Catalogue des scarabées et scaraboïdes de Kerkouane. *Revue des Études Phéniciennes-Puniques et des Antiquités Libyques* 9: 147–188.
Renan, E. 1864. *Mission de Phénicie*. Paris: Imprimerie Impériale.
Ribichini, S. 2004. Al servizio di Astarte. Ierodulia e prostituzione sacra nei culti fenici e punici, in A. González Blanco, G. Tilla Séiquer and A. Egea Vivancos (eds) *II Congreso Internacional del mundo púnico* (Estudios Orientales 5–6): 55–68. Murcia: Universidad de Murcia.
Rocco, B. 1969. La Grotta di Monte Gallo (iscrizioni e disegni). *Sicilia Archeologica* 5: 18–29.
Sader, H. and J. Kamlah 2010. Tell el-Burak: A New Middle Bronze Age Site from Lebanon. *Near Eastern Archaeology* 73: 130–141.
Sagona, C. 2002. *The Archaeology of Punic Malta* (Ancient Near Eastern Studies Supplement 9). Leuven: Peeters.
Saidah, R. 1967. Fouilles de Khaldé. *Bulletin du Musée de Beyrouth* 20: 165–169.
Salvi, D. 2004. Attraccare sul passato: il giacimento archeologico del porto di Cagliari, in A. Benini and M. Giacobelli (eds) *Atti del II Convegno Nazionale di Archeologia subacquea, Castiglioncello, 7-9 settembre 2001*: 61–75. Bari: Edipuglia.
Saña, M. 1994. Análisis zooarqueológico del Pozo HX-1, in J. Ramon, *El pozo púnico del 'Hort d'en Xim' (Eivissa)* (Trabajos del Museo Arqueológico de Ibiza 32): 71–81. Ibiza: Museo Arqueológico de Ibiza.
Sasson, J.M. 1976. Isaiah LXVI 3–4a. *Vetus Testamentum* 26: 199–207.
Schilling, R. 1982. *La religion romaine de Vénus depuis les origines jusq'au temps d'Auguste* (2nd edition). Paris: De Boccard.
Seefried, M. 1982. *Les pendentifs en verre sur noyau des pays de la Méditerranée antique* (Collection de l'École française de Rome 57). Roma: École Française de Rome.
Seyrig, H. 1940. Antiquités syriennes. *Syria* 21: 113–122.
Simoons, F.J. 1981. Dogs as Human Food in Northwest Africa. *Appetite: Journal for Intake Research* 2: 253–266.
Smith, A.M. 2015. The Ashkelon Dog Cemetery Conundrum. *Journal for Semitics* 24: 93–108.
Stager, L.E. 1991. Why Were Hundreds of Dogs Buried at Ashkelon? *Biblical Archaeology Review* 17/3: 26–42.
Stiglitz, A. 1999. Osservazioni sulla pittura funeraria nella Sardegna punica. *Annali della Facoltà di Lettere e Filosofia dell'Università di Cagliari* 54: 75–110.
Stucky, R.A. 2005. *Das Eschmun-Heiligtum von Sidon. Architektur und Inschriften* (Antike Kunst Beiheft 19). Basel: Vereinigung der Freunde antiker Kunst.
Tavares da Silva, C. and M. Varela Gomes 2006. O santuário proto-histórico de Garvão. *Cadernos Culturais d'Ourique* 1: 43–53.
Tore, G., M.A. Amucano and P. Filigheddu 1992. *Notulae punicae Sardiniae*, in A. Mastino (ed.) *L'Africa romana. Atti del IX convegno di studio, Nuoro, 13-15 dicembre 1991*, II: 533–60. Sassari: Gallizzi.
van Wickevoort Crommelin, B. 2005. Zoroastrier in Karthago? Anmerkungen zu einer Gesandtschaft des Dareios I., in R. Wiegels, W. Spickermann, K. Matijević, and H.H. Steenken (eds) *Rom, Germanien und das Reich: Festschrift zu Ehren von Rainer Wiegels anlässlich seines 65. Geburtstages*: 1–29. St. Katharinen: Scripta Mercaturae.
Varela Gomes, M. and C. Tavares da Silva 1994. Garvão. Un sanctuaire protohistorique du sud du Portugal. *Les dossier d'archéologie* 198: 34–39.

Vassel, E. 1921. Les animaux exceptionnels des stèles de Carthage. *Revue de l'histoire des religions* 84: 36–76.

Vivanet, F. 1892. Avanzi di terrecotte votive ripescati nella laguna di Santa Gilla presso Cagliari. *Notizie degli Scavi di Antichità* 1892: 35.

Vivanet, F. 1893. Nuove terrecotte votive ripescate nella laguna di Santa Gilla presso la città. *Notizie degli Scavi di Antichità* 1893: 255–258.

Wapnish, P. and B. Hesse 1993. Pampered Pooches or Plain Pariahs? The Ashkelon Dog Burials. *The Biblical Archaeologist* 56: 55–80; Reprinted in: L.E. Stager, J.D. Schloen and D.M. Master (eds), *Ashkelon 1. Introduction and Overview (1985-2006)*: 541–564. Winona Lake: Eisenbrauns.

Watson, W.G.E. 1997. Comments on the Phoenician Tariff Inscriptions from Kition. *Die Welt des Orients* 28: 89–95.

Weinstock, J. 1995. Some bone remains from Carthago, 1991 excavation season, in H. Buitenhuis and H.P. Uerpmann (eds) *Archaeozoology of the Near East: Proceedings of the 2nd international symposium on the archaeozoology of southwestern Asia and adjacent areas*: 113–118. Leiden: Backhuys.

Whitaker, J.I.S. 1921. *Motya. A Phoenician Colony in Sicily*. London: G. Bell and sons.

Wilkens, B. 2008. I resti faunistici dell'US 500, in Campanella 2008: 249–259.

Will, E. 1985. Un problème d'*interpretatio graeca*: la pseudo-tribune d'Echmoun à Sidon. *Syria* 62: 105–124.

Winton Thomas, D. 1960. *Kelebh* 'dog': Its Origin and Some Usages of It in the Old Testament. *Vetus Testamentum* 10: 410–427.

Xella, P. 2012-2013. 'Tophet'. An Overall Interpretation. *Studi epigrafici e linguistici sul Vicino Oriente antico* 29–30: 259–281.

Yazidi, S.Z. 2009. *Le bestiaire dans l'imaginaire des puniques*. Tunis: Faculté des Lettres, des Arts et des Humanités de Manouba.

# 6.3 Dog in War, Hunting, Livestock Work and Everyday Life of Greco-Roman Society

Ana Portillo Gómez

Faculty of Philosophy and Letters, Department of Art History, Archaeology and Music, University of Cordoba. Plaza del Cárdenal Salazar, 3, 14071, Córdoba, Spain. aportillogomez@gmail.com

**Abstract**

The dog has been since ancient times one of the closest creatures to man, it was the first domesticated animal and therefore, within the study of Human Sciences, it was considered necessary to pay a small tribute by giving a brief overview of their ways of life, uses and functions during the Classical Period. As a partner for hunting, a guardian and custodian of the house or of the temples, a brave soldier, a guide in the transit to death or a faithful friend, the dog has had since Antiquity, an intimate relation with man, different from that of any other pet. In Europe the first to develop their own breeds, selecting the characteristics and appearance of the animal, were probably the Romans. At that time, different breeds that were produced can be distinguished following a functional pattern and it is possible to differentiate between hunting dogs, shepherd, partner or war dogs.

**Keywords**: dogs, Classical period, Greco-Roman life, Antiquity life, dog's uses.

## 1 Introduction

The relationship of ancient civilisations with animals doesn't differ that greatly from the situation nowadays. Some animal species were particularly valued, such as the dog and the horse (Brodrick 1972; Hyland 1990), especially among young people (Maspero 1997, 12). Ideologically, the Greek and Roman people shared some basic differences between men and animals (Castignone and Lanata 1994). This issue came to generate a genuine debate among some philosophical schools, such as the Stoic and the Epicurean (Isnardi 1989). The main difference was the possession of logos, that is, the knowledge of language and reasoning (Gasti and Romano 2003).

Another characteristic inherent to man is the commitment to honour and respect. Writers like Xenophon (425–430 BC to 355 BC ca.) shows this idea in his texts:<<*And it seems to me, oh, Hieron !, that is what makes man different to other animals, craving honours; because, in everything related to food or drink or sleep or the carnal, it seems that all animals enjoy the same ...* >> (Xen. *Hier.*, 7, 3).

During the classical period it can be considered that two ideologies or ways of thinking coexisted about the relationship established between animals and humans. The most widespread idea was the one that considered animals as creatures born to be at the service of man. The most similar philosophical current to this cause was the Stoic one (Arnim 2002), in such a way that famous philosophers such as the Greeks Crisipo or Posidonio and the Romans Seneca or Cicero, would advocate for this belief, reflecting it in numerous writings. In this quotation, for example, a brief paragraph of the work of Cicero *De Natura Deorum*, where he demonstrates this same conviction: << ... *the same animals, as we can see, were created for the benefit of men ...* >> (Cic. *Nat. D.*, II, 158).

Epicureans and Peripatetics, such as Carneades or Sextus Empiricus, were against this thought. They denied the creation of animals for human satisfaction and benefit (Dal Pra 1950, 162). The Stoics rejected the skill of reasoning and intelligence in animals (Isnardi 1989, vol. I, 532; Vauclair 1992), there is no legal or moral obligation on them and therefore, they are likely to be consumed as food or slaughtered (Maspero 1997, 14).

Plutarch, among others, masterfully fights against these arguments using epicurean teachings (Santese 1994). It is a conviction that he shows in the book XII of his reflections collected in 'Moral and customs works' (*Moralia*), where Ulysses establishes an intense dialogue with Grilo, a character whom the magician Circe transforms into a pig and that, temporarily, gives him the ability to communicate verbally (Plut. *Mor.*, 991F).

This lack of sensitivity with animals is evident in public shows, which implied a real torture for them, until their death (Mancioli 1987). Among these forms of entertainment, cockfighting can be specially highlighted ,which was very present among the Athenian and Roman people.[1]

During Antiquity no monographic treatise or writing were generated about the figure of the dog, therefore it is necessary to research the works of those authors who

[1] Aelianus in his *Varia Historia* (2,28) writes about the origin of these cockfights.

approached the animalistic theme in search of some information about this creature. In general, there are many Classical authors who address issues related to animals, however we could say that it will be Aristotle (384–322 BC) who establishes some of the most relevant on the basis of zoology. In his work (*History animalium*), he makes important observations about gender, anatomy and reproduction of animals, carrying out a fundamental classification of them in two large groups that enclose vertebrates and invertebrates.

His studies were copied and translated for hundreds of years by numerous authors without many new contributions. His main heir in zoology would be Theophrastus (370–287 BC), followed by some writers who had a strong dependence on Aristotelian texts such as Apollodorus of Alexandria (3rd century BC), Aristophanes of Byzantium (257–180) or Tryphoon of Alexandria (1st century BC) (Maspero 1997, 8–9).

Among the Roman academics who assimilate and continue the knowledge generated by the Greek authors, Pliny the Elder should be acknowledged (1st century AD). His work *Historia Naturalis* consists of thirty-seven books, of which four of them (VIII, IX, X and XI), are dedicated to the animal kingdom. Another famous Roman writer who dedicates a good part of his work to animals was Columella (1st century AD), author of one of the best written treatises on agriculture, *De Re Rustica*. However, perhaps the broader and most complete Roman-Era treaty on the animal world is owed to Claudius Aelianus (170–235 AD). It is a compendium of sixteen books entitled *De Natura Animalium*, focused on the study of the characteristics, behaviour and curiosities of animals. Saint Isidore of Seville (7th Century) takes this zoological knowledge from Classical Times and dedicates his twelfth book to the animal question, following the teachings of Pliny the Elder.

Therefore, returning to the figure of man's best friend, it can be appreciated the large number of writers who mention the inherent qualities of the dog, but of all of their qualities, faithfulness stands out. Who has not been excited to hear the story of the faithful Argo, who maintained the necessary forces until he saw his owner Ulysses return to Ithaca, dying happily after being the only creature that recognised him after twenty years? (Hom., *Od.*, XVII, 300 ss). Plutarch (*Mor.*, 969D ss) discusses it in a passage in his work where he addresses the issue of animal intelligence.

Columella also speaks about this faithfulness characteristic, and dedicates several paragraphs on the subject in his treatise on agriculture. Loyalty to his master, his qualities as watcher and guard are the attributes of canids that he highlights. He also makes a distinction between the different species of dogs according to the needs to which they should dedicate themselves, thus evidencing the morphological and behavioural qualities and characteristics of dogs destined for guarding, custody and defence of livestock, briefly mentioning the hunting dog. He provides a series of tips on the training and care of these animals, even indicating some remedies or treatments for the most common conditions (Columella. *Rust.*, VII, 12, 13).

## 2 The Hunt and the Shepherd

### *2.1. Agreytikós kýon/ Venaticus canis*

Hunting is one of the oldest activities in which the dog plays a fundamental role. In this practice, his alliance with man is strong and deep, creating a team when it comes to demonstrating the skills acquired in the art of hunting. The importance of canids in hunting in the ancient world, is revealed in the numerous sculptural, pictorial and literary passages dedicated to this animal in the development of these tasks.

There are many classical authors who highlight the role of the dog in this activity, Xenophon (430–355 BC), Seneca (4 BC–65 AD) or Flavius Arrianus (95–175 AD) are some of them. These authors provide interesting data, such as the diet or training that dogs intended for hunting should follow, with the intention of controlling the animal's predatory instinct and thus preventing them from devouring prey (Sen. *Thyest.*, 497–505).

They used to receive a diet based on boiled products and cereals, because they should not appreciate the taste of meat or blood (AA. VV. 2005: 13). For these dogs it was recommended to grant short names, so that they were easily recognisable to the animal, thus optimising the response time of the dogs to the orders they should receive.

Arrianus highlights some instinctive and psychological characteristics that every good hunting dog must possess, such as the nobility of spirit, the attitude and courage of the animal (Arr., *Cyn.*, 4, 2; 5, 9; 5, 11; 6, 2; 7, 7; Martin and Berti 2012, 391). In classical art a large number of representations of this theme can be found, in which, it is even possible to distinguish some of the breeds used in this activity. Basically, four different types of hunting dogs can be identified:

- *The Greyhound*. These dogs had stylised bodies, with elongated heads, pointed ears, fine joints and recessed bellies. Their fundamental characteristic is speed, for that reason, they were especially used for the hunt of fast prey like rabbits or hares.[2] There are numerous

[2] Part of the descendants of this typology of dog are known as Greyhounds, which are still today used for hunting this type of prey.

Figure 1. Copy of a mosaic with a hunting scene from Horti Liciniani, near Santa Bibiana's Church, Rome. Beginnings of the IV Century AD. Museo della Civiltà Romana (Rome).

representations of these canids in various paintings in Pompeian houses, such as in the atrium of Lucrezio Frontone's house (Reg. IV, ins. 4, n. 10), in the peristyle of the Caserma dei gladiatori (Reg. IV, ins. 5, n. 3), in the atrium of the Meander House (Reg. I, ins. 10, n. 4) or in a relief from Noviomagus, (Neumagen, Germany) preserved in the Museo della Civiltà Romana (Rome).

- *The Wolf -type*. These are dogs that bear a strong physical resemblance to wolves. They share an important genetic load with them, being created as hybrids between dogs and wolves. They have pyramidal heads, with tall straight ears and elongated snouts. Some of these dogs were used in shepherding.[3] Representations of these animals can also be found in pictorial scenes in Pompeian houses, such as in the atrium of the Casa dell'Ara Massima (Reg. VI, ins. 15, n. 16), in the peristyle of the Casa delle Nozze d'Argento (Reg. V, ins. 2), in the Casa della caccia (Reg. VII, ins, 14, n. 48) or in the *Thermopolium* (Reg. VI, ins. 10, n. 1). -The Molossian type. These were dogs characterised by their strong complexion and physical power, used in different activities[4] such as the work of big game, to catch large beasts such as bears. These dogs came from Molossia, a city in the region of Thessaly (Greece) where they were used to care for livestock or for war. These are very robust animals and usually have broad, rounded or cubic heads, small ears, short noses and long lips.

Again Pompeii becomes an inexhaustible source of representations where we can distinguish this breed in different scenes. They appear painted on the wall of a tavern (Reg. VI, ins. 10, n. 1), in the peristyle of the Casa del Meandro (Reg. I, ins. 10, n. 4), in the lobby of the Casa di Vetti (Reg. VI, ins. 15, n. 1) or in the *triclinium* of the Casa di Meleagro (Reg. VI, ins. 9, n. 2).

- *The Mongrel type*. In many occasions, some dogs do not have the characteristics of the general typologies mentioned above, so they have been included in the same group under the word 'mongrel' and they are very present in several iconographies. Some of them can be found in the *oecus* of the Casa del Sirico (Reg. VII, ins. I, n. 47), on the pavement of the lobby of the Casa del Cinghiale (Reg. VIII, ins. 3, n. 15) or in the *triclinium* of the Casa di Giulia Felice (Reg. II, ins. 4, n.3).

Hunting themes frequently appear on the fronts of the Roman sarcophagi, since hunting used to be one of the favourite activities of the *dominus* who, on many occasions, appears as a protagonist participating in the scene. An interesting specimen preserved in the Capitoline Museums of Rome shows up to three different breeds of dogs. At the front face of the sarcophagus, Calidón's wild boar hunt is depicted, perhaps the most famous hunt in classical mythology.

Dogs were used to hunt all kinds of prey, including large beasts such as bears (Merlen 1971). Two copies of the most illustrative examples are two mosaics preserved in different museums, but belonging to the same piece (Figure 1). Both of them reveal the process used to capture these huge animals. A perimeter was circled with the help of a net, thus limiting the transit space of the bears. The dogs threatened the beasts by directing

[3] Nowadays some of the breeds that descend from this typology of dog are the German Shepherd, the Siberian Husky and the Samoyed.
[4] These dogs were used in fights as war dogs and in some *ludi romani* such as *venationes*.

Figure 2. Detail of the sarcophagus with hunting scene. Inventory number: 837. Asian marble. First decades of the IV Century AD. Musei Capitolini, Centrale di Montemartini (Rome).

them to cages equipped with access ramps, where a servant, located at the top of them, waited patiently to close the door.

Some men participate in the scene Wearing some kind of protective gloves that were used to cover the entire arm, to avoid possible attacks. These men had the task of ensuring the capture, placing the nets, guiding the beasts towards the cages, that is, working together with the dogs to achieve the common task. The concern of humans for our most faithful friend is evident in a curious scene that takes place in a sarcophagus dating back to the first decades of the fourth century AD and preserved in the Centrale Montemartini of the Capitoline Museums of Rome.

On the front of the piece there is a scene in which a wild boar and a deer are hunted. In the heat of a battle, one of the beasts has injured a small dog, which shows a bleeding wound on its back. The restlessness of its owner or caretaker, is clearly manifested in the act of rescuing the animal, carrying it with him in his arms, wrapping it up and offering it comfort, in addition to his countenance, since he is evidently afflicted (Figure 2).

### *2.2. Poimenikós kýon / Pastoralis canis*

Dogs have also been used since ancient times in rugged terrains as guardians and custodians of cattle (Figure 3) For these tasks, dogs of great physical complexion, strong, long-haired and preferably light-coloured to distinguish them from aggressors who used to usually be dark-coloured (wolves, for example). This typology called *poimenikós kýon* or *canis pastoralis*, encompasses the so-called Molossians of the regions of Laconia, Salento and Umbria (AA. VV. 2005: 14–15) considered since ancient times, to be the best grazing dogs.

One of the main concerns that dog owners show is the care of diseases and injuries that make their task impossible. The attention dedicated to these animals, began in their juvenile state, because at birth, those puppies considered to be the strongest were selected, and they were provided with extra nutrition using goat's milk. The breeds remained with their mother for a period of six months, progressively introducing them to grazing tasks (Varro, *Rust.*, II).

Figure 3. Marble plate with grazing scene. Vatican Museums (Rome).

These dogs had to receive good food so that they would not get hungry and not be tempted to attack cattle. Their diet was based on bread dipped in milk, bean soup, cereals and some bones to fortify their teeth[5] (AA. VV. 2005: 15; Columella, *Rust.*, VII, 12). Ancient people treated, with special consideration, the health of these animals (Ov., *Fast.*, IV, 763–766), because living outside with the cattle, exposed them to numerous diseases, especially parasitic ones. Columella bequeathed a whole series of suggestions for the prevention of different diseases including the treatment of ticks and fleas, which must have been very common.

Columella recommended smearing an ointment on the dogs made out of tar and lard in order to make the ticks fall off on their own, instead of tearing them off. For fleas, it was advisable to rub them with ground cumin and vegetable water, both substances had a strong smell and disinfectant function[6] (Columella, *Rust.*, VII, 13).

### 3 The guardian dog: *Pylorós Kýon* / *Canis ostiarius*

The insecurity of the streets and the existence of thieves and criminals have been a reality since ancient times and have worried a big part of society. Protecting the home against any threat became a task that fell largely on the figure of the *pylorós kýon* or *oikurós kýon* or the *canis ostiarius, villaticus canis* or *canis catenarius.*[7] These guard dogs'a main task was to ensure the safety of the house, so, as advised by Columella (*Rust.*, VII, 12), large breeds should be chosen, with robust bodies, a fine sense of smell, a loud bark and if possible, coloured dark, so that they could not be visible in the gloom of the night and go unnoticed in the eyes of the assailant.

These animals were usually chained in the *ostium* of the houses where they used to sleep during the day, as they were busiest at night, when they had to remain alert. This fact is confirmed by archaeology with cases as well known as the remains of a *canis catenarius* found in Pompeii or the mosaics documented in the *ostium* of several houses in the city, where representations of these animals appear. One of the most famous is the mosaic of the House of the Tragic Poet (Reg. VI, ins. 8, n. 8), where the image of an imposing black Molossian, tied to a chain in a defensive attitude can be seen The famous inscription *cave canem* appears between its front legs warning us of the ferocity of this tenacious guardian. Another noteworthy example is the mosaic in the lobby of the *Paquius Proculus* house (Reg. I, ins. 7, n. 1), in which we can see a majestic black dog that awaits vigilantly lying on the ground. In this case, the animal is part of a symbolic language that transmits the defensive character, so important in the entrances, in which a series of weapons are represented (shields, spears and axes ), instruments that, like the canids, were used for the protection and custody of places that had value for a community.[8]

The Greeks entrusted the custody of temples, palaces and cities to these animals, employing for this purpose the Molossians from Epirus[9] (AA. VV. 2010). These dogs possessed a great complexion and strength (AA. VV.

[5] These foods provided them with starch and other carbohydrates, substances easily digestible for dogs, including the proteins of the beans and fatty acids.

[6] The purifying benefits of the vegetable water is the reason why it was recommended to mix it with lime to plaster the floors and walls of the stables and folds.

[7] Seneca refers to this animal as *canis catenarius*, while Columella calls it *villaticus canis.*

[8] These could be homes, temples, cities or even tombs.

[9] The descendants of this type of dogs are the Mastiffs. This name comes from the Latin word *mansuetus*, (tame), participle of the verb *mansuesco*, which also means domesticate. This verb is formed by the words *manus* (hand) and *suesco* (usual). That is, a mastiff is a tame dog that is used to the hand of its master but it doesn't mean that it is not brave. It is well known how ferocious these dogs can be.

2005: 16). There are numerous classic passages written by various authors that talk about the protection of sanctuaries and cities by these animals, such as the thousand guard dogs of the Sanctuary of Ádrano in Sicily. These dogs were ministers of this local divinity and kept their temple. They were kind to those who approached them with good intentions but turned into cruel beasts if the visitors intended to perform misdeeds or steal, tearing apart anyone who dared to do evil in a sacred place (Ael., *NA*, XI, 20).[10]

Plutarch also mentions an interesting episode about the dog Cáparo, the guardian of a temple dedicated to Asclepios[11] (Plut., *De soll.an.*, 969 E, F). One dark night, an unfortunate thief disturbed the house of the God by stealing several offerings and leaving in a hurry, convinced that he had gone unnoticed. However, the skillful guardian launched himself into the race after the bandit barking.The thief tried to bribe the animal with bread and cakes without any success, because the dog did not let up and. finally the Athenians arrested the thief and after interrogating him, he ended up confessing the fact and was punished. The animal was rewarded with the honour of being fed and cared for by the city.

The feat of the guardian *Soter* (saviour, in Greek) is also well-known, one of the fifty dogs that protected the city of Corinth from enemy attacks. One day during 581 BC, in the course of the celebration in honour of Aphrodite, Nauplia's troops took the opportunity to attack the city. The dogs responded by attacking the enemy but it was only the wise *Soter* who came to alert the citizens of the evil that was stalking them. Therefore, the dog was honoured with a silver necklace engraved with its name and earned eternal fame. Rome also had its guard dogs although, in this case, the episode had an unfortunate ending. In the year 381 BC the Gauls were preparing an assault on the city and the dogs that were guarding the Capitol did not warn of the threat, the alarm of the attack was raised by the geese that protected the temple of Juno Moneta. This unfortunate event caused that every anniversary of this attempted attack, dog sacrifices were made on this mountain.

Homer in his Odissey tells that the palace of Alcinous, King of the Phaeacians was decorated with numerous gold and silver statues of dogs, highlighting the guardian function of this animal.

### 4 Everyday life: *Oikurós kýon / Canis familiaris*

The dog was part of social and family life of the ancient Greeks and Romans. This is a historical fact that can be verified through different sources, such as classical literature, archaeology or iconography. This is what Petronius depicts in the central episode of Satyricon, the so-called 'Trimalchionis Dinner', where the little dog *Perla* and the Molossian *Squilace* appear as protagonists sharing the stage with the other guests (Petron., *Sat.*, 64, 6).

They were not only well accepted in the home environment, they also participated in banquets and other activities of the house. Homer (*Od.*, XVII, 309) and Oppian of Apamea (*C.*, I, 14) mention them as 'lap dogs'. The veneration for these animals is especially evident in Roman high society, especially by the females (Franco 2003) (Figure 4).

The passion for these little dogs, many of them coming from regions like Gaul, Sicily or Malta, was a constant. Many poets such us Juvenal, criticised this, telling us that these women 'almost prefer the death of their husbands before their dogs' (*Sat.,* VI, 653). These animals were included in the *apophoreta*[12] lists sent as gifts during the *Saturnalia* festivities (Mart., *Spect.* 14,1).

This type of dogs was considered to be authentic luxury dogs, and their presence has been verified in several sites located in the Iberian Peninsula, such as Santo Domingo (Lugo, Spain) and the Roman *villae* of Arellano (Navarra, Spain), finding dogs with heights between 26 and 31 cm (Altuna 1994; Oliver 2014). Again, the iconography shows numerous scenes that display the presence of this animal during everyday life in the classical community. The front of the sarcophagi are clear examples that demonstrate the characteristics and life of the deceased.

However, it is truly significant that this animal is devoted so much attention, to the point of getting a leading position in classical portraits, as a true protagonist of certain spaces, which not only are limited to the domestic sphere, but also surpass it by even crossing to the sacred sphere.

In this sense, we cannot forget the unique case of the so-called 'Cagna ferita' (injured dog) preserved in the Giovanni Barraco Museum (Rome) (Figure 5). It is an exceptional piece carved in pentelic marble, represented in the act of licking a wound on one of its hind legs. The figure is a Roman copy, from the first imperial era, of a Greek original in bronze, from the end of the fourth century BC, attributed to the sculptor Lisipo (Moreno 1981, 196–199). Pliny (*HN*,XXXIV, 38) claims to have seen her in the Juno cella of the Jupiter Capitoline temple before the fire and assault of the Vitellians. He describes it with great enthusiasm and amazement for the beauty

[10] There are other passages about enshrined dogs to divinities in the work by Aelianus. They are *NA.*, XI, 3, 5.

[11] Although the location of the temple is not specified, it could be supposed that this is the famous temple of Epidauro.

[12] These are presents of a festive nature delivered to the Roman houses during some celebrations. It is a custom that comes from Greece and it was included with pleasure in Roman life.

Figure 4. Copy of a relief with a banquet scene. Museo della Civiltà Romana (Rome). Original: Muséed'Art et Histoire (Geneva, Switzerland), first half of the Second Century AD.

Figure 5. Wounded dog. Roman copy signed by Sopatro of an original work by Lisippo from the end of the IV Century BC. Pentelic marble. Museo Giovanni Barracco (Rome).

and realism of the effigy, giving it an incalculable value, and such is so, that he speaks of a decree that was drawn up in which it was stipulated that his custodians should ensure the security of this piece and if something happened to it, they would pay for it with their heads.[13]

The sculpture of the injured dog must have enjoyed great fame and its model became widespread in the arts, since other copies of obvious similarity have been documented, following the same scheme, as evidenced in a piece preserved in the Archaeological Museum from Naples. The effigy of the canids was also represented in common objects such as crockery, of which countless specimens have been conserved. On many occasions these ceramic vessels became part of the funeral trousseau, from which most of these artefacts have been recovered.[14]

### 5 The dog of war: *Polemistés kýon / Canis pugnaces*

The dog has shared the war scene with man since very remote times, being employed for a large number of tasks, such as guards, guardians, messengers, fighters and carriers of ammunition, medication or food. To perform these roles, it was necessary to have a large number of animals with physical strength, a good sense of smell, and a resistant and combative spirit.

All these qualities masterfully meet in those known as Molossians, dogs that come from Epirus (Greece), from which the breeds of mastiffs and current bulldogs derive. Representations of this canine typology of ancestral civilisations have been conserved, such as those developed in ancient Mesopotamia, where they were sculpted in stone accompanying several soldiers of the Nineveh Palace, or as sculptures that adorned gardens and palaces. In Ptolemaic times, Pharaoh Ptolemy II (285–246 BC) paraded through the streets of Alexandria with a court of 2,400 Molossians as part of his army, which frightened the spectators with their enormous size (Carreras 2013). A very famous Molossian in history was *Péritas*, the dog of Alexander the Great (356–323 BC) with whom he fought as the most loyal of his soldiers on numerous occasions, until his death during the conquest of India.[15]

The *polemistés kýon* or *canis pugnaces*, highlighted by several classical authors (Ael., *NA*, III, 2; Arist., *Hist. an.*, 608–28,31), was employed as a true warrior by many civilisations from Antiquity[16] (Forster 1940- 41). According to Pliny the Elder (*HN.*, VIII, 61), cities in Asia Minor such as Colophon (Lidia) or Castabala (Cilicia) used soldier dogs as auxiliary troops. The Gauls also had an army of mastiffs that accompanied them to battle (Str., *Geogr.*, IV, 5, 2). Polyaenus (*Stratagemata*) tells us some of the tasks that dogs used to develop on the battlefield, showing us their use as a true instrument of war, used strategically to reduce the forces of the enemy or as an unfortunate messenger, because it had to ingest a metal tube with instructions, and once it reached its destination would be sacrificed in order for the instructions to be recovered.

As battle animals, their bodies were protected with leather breastplates and their heads with small leather helmets and were sent to the opposite ranks either as carriers of containers with fire to start fires, or as true soldiers to hurt enemies and their horses, in this case they were also equipped with spiked breastplates and collars (Carreras 2013, 2). As for defensive work, the Roman writer Flavius Vegetius Renatus (*De Re Militari*), recommended that these types of dogs were kept in watchtowers in order to sound the alarm whenever there were enemies in the vicinity (*Mil.*, IV, XXVI).

The Romans, amazed by the ferocity and temperament of these animals, used them in different public shows for the fun and training of society. These cruel *ludi* called *venationes*[17]took place both in the circuses and in the amphitheatres, where the stage was prepared with artificial bushes, small ponds and trees that mimicked a natural landscape (Cass. Dio., *Var.*, V, 42). Later, wild animals of all kinds were released, and those of an exotic nature that came from regions like Africa or India, such as bears, bulls, lions, tigers or elephants, were forced to fight each other until death (Sen., *Dial.*, III, 42,2).

Before jumping into the sand these animals had been held in the dark and went without food for a long time, to enhance and increase their appetite and ferocity. These fierce struggles could face animals or men against animals. The *venator* was the fighter who faced these beasts, he used to receive a similar training to

13 <<*Euecta supra humanam fidem ars est successu, mox et audacia. in argumentum successus unum exemplum adferam, nec deorum hominumue similitudinis expressae. aetas nostra uidit in Capitolio, priusquam id nouissime conflagraret a Vitellanis incensum, in cella Iunonis canem ex aere uolnus suum lambentem, cuius eximium miraculum et indiscreta ueri similitudo non eo solum intellegitur, quod ibi dicata fuerat, uerum et satisdatione; nam quoniam summa nulla par uidebatur, capite tutelarios cauere pro ea institutum publice fuit*>>. Plin. *HN*, XXXIV, 38.

14 The representation of dogs has also been used since ancient times as offerings to some divinities asking for protection. This is the function of a terracotta head of a dog located in Cádiz (Spain). This piece was part of a group of ritual objects related with the cult to the Goddess Astarté (Venus Marina), protector of the sailors. It was written about by Avieno in his *Ora Maritima.*

15 Plutarch (*Parallel lives)* and Pliny the Elder (*Natural History*) wrote about this passage in their works.

16 This use of the dog in wars was extended in certain times and places, being especially important during the conquer of America. This is addressed by Fray Bartolomé de las Casas in his chronicles. Also Theodor de Bry represented these activities in his engravings. Vid. Bueno, 2011

17 M. Fulvio Nobiliore offers for the first time this type of show for the Romans in the year 186 B.C., presenting lions and panthers as a celebration of the victory against the Ethiopians (Liv., *Ab Urbe Condita*, XXXIX 22, 2). It is noticed that these types of performance soccurred until the times of Totila (IV Century), being abolished in Italy during the VI Century (Colini and Cozza, 1962). The latin poet Martial writes about these shows in his *Liber De Spectaculis,* inside his work *Epigramas.*

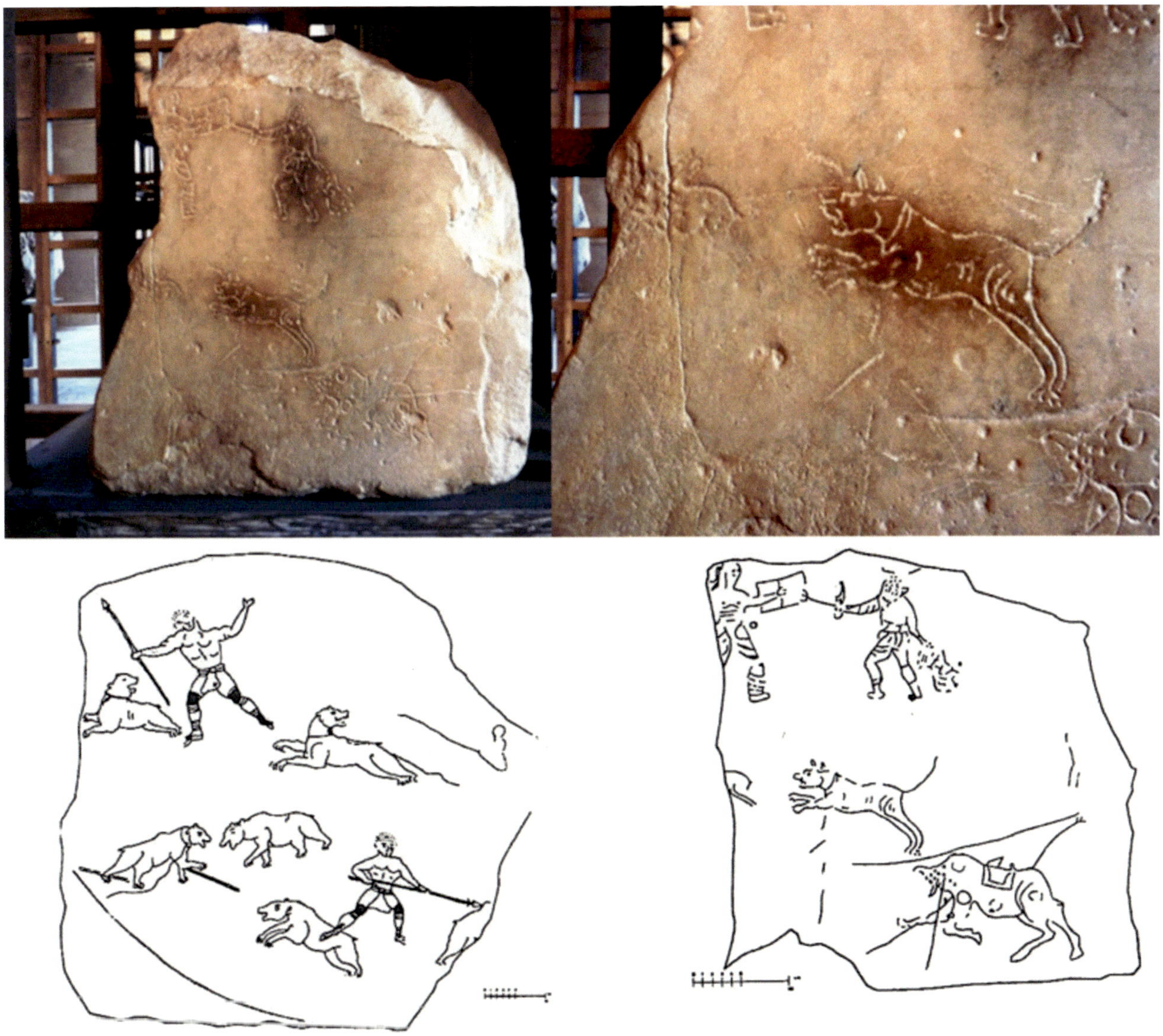

Figure 6. Upper section. Plate of marble with graphite that represents a *venatio* scene. It comes from the Colosseum (Rome), Inv. n. 375838. Lower part. On the left. Drawing of a part of the same piece. Inv. n. 375837. On the right. Drawing of the piece number 375838 (Rea, 1987, 84, 3 y 4).

that of the gladiators, but the venator had less rank and social consideration. As weapons, they were provided with a long spear, a whip and a pack of dogs (Mancioli 1987, 66–67). Together with the Molossians and after the conquest of Britannia, a type of British dog that could kill a bull, was incorporated in the games. So big was the enthusiasm for these types of dogs and for these activities in this region that an important figure was established, that of the *procurator cinegii*, a person in charge of selecting the best specimens of this breed of dogs for their shipment to Rome (AA. VV. 2005: 21).

In this sense, a passage from the work of Quintus Aurelius Symmachus is very significant, who writes, in one of his letters to Brother Flaviano, the following comment:

> *<I thank you for the gift of seven British dogs, which were presented at the circus games with great admiration and stupor of the Roman people, who could not believe they had been transferred to Rome in iron cages, such as tigers or lions, because of their fierceness>>* (*Ep.* II, LXXVII).

Emperor *Comodus* was a great fan of *venationes*, who trained himself with the best masters in javelin throwing (SHA., *Hist. Aug.*, *Hdt.* I, 15). He used to show his skills and abilities as a pitcher in the amphitheatre, but it was not an act of courage, but a banal display of his precision in the shot and a mere exaltation of his figure, because he only had the audacity to target some species of herbivores (always juveniles), such as deer or goats and never dared to face more dangerous beasts. A corridor or platform was built in the immediate vicinity of the arena, from where, with all the security that this position provided, he threw javelins to ferocious beasts, such as bears, lions or panthers (Rea 1987, 83). He also appeared in the amphitheatre as a great secutor-type gladiator, with absurd representations of fights with other gladiators in which he participated and, obviously, always emerged victorious, since no adversary would tempt against the emperor (SHA., *Hist. Aug.*, Com, XV; *Hdt*, I, 15).

There are numerous representations that show us scenes of this type of *ludi* that, due to their popularity, appear in all types of artistic pieces and objects, marble reliefs, skylights, dishes, etc. Among them, a series of '*engraved stone*' made by spectators are especially interesting, which appeared in different areas of the Flavius Amphitheatre (Figure 6). In them several fighting scenes can be observed, such as a battle between gladiators (*retiarius* and *secutor*) and two sequences of *venationes*, in which the presence of a huge dog that pursues a prey should be highlighted, while under it, there is an agonised bull pierced by a spear. In a plate made of *lunense* marble from the back of the *vomitoria* of the Colosseum, another example can be found which depicts these games in precise detail, where the image of a dog has been represented, well equipped with its collar, in full race and just at the moment of hunting an antelope.

## 6 Conclusions

As we have had the opportunity to verify, the dog has had, since ancient times, a wide variety of functions and meanings for man, both in the most primitive daily tasks such as hunting and grazing, as well as being a home guard, for military conflicts and spirituality. The dog has always proved to be a faithful ally and friend of his master. The importance and attachment that the classical civilisations gave to this animal, are translated into the rich iconographic, archaeological and literary passages that have been preserved about them.

This fact confirms the relevance and role of the animal in these societies and stands as one of the closest creatures to mankind since the origins of Humanity. This reality, which has an evident continuity to this day, dignifies and ennobles it, since, as an animal species, it is unparallelled with any other kind of living being, so we want to grant it with the nickname of 'companion animal', since no other creature other than the dog has provided more fidelity and selfless love to humans.

Archaeological research has provided a lot of information about the relationship between this animal and man, especially those related to its role in society, uses and functions in a practical way. However, we consider that it is necessary to address the issue, from a broader perspective, of the psychological aspects, the emotional ties and the symbolic values that the canines possessed during the classical period. The amount of artistic expressions relating to the dog figure conserved until today, or the different findings of canids in different archaeological contexts in history, show us the great attention and interest for introducing this animal in the habits and different aspects in the life of these societies.

Given the importance acquired by this animal in ancient times, we consider that this line of research allows us to go further in the most transcendental issues of dogs and its links with humans. Therefore, this work's main aim is to be a starting point within a more detailed research, where it can be reviewed, in a general way, the historiography, uses and meanings of this animal in some areas of the Greek and Roman cultures.

## References

### *Ancient sources*

All Latin abbreviations used have been taken from Hornblower, S., Spawforth, A., Eidinow, E. (2012): Oxford Classical Dictionary 4th edition, Oxford. The Greek abbreviations have been taken from Liddell, G., Scott, R. (1940): A Greek-English Lexicon, Oxford.


Aelianus, *De Natura Animalium,* XI, 20; III, 2.

Aristotle, *History animalium,* 608–28, 31.

Arrian, *Cynegeticus*, 4, 2; 5, 9; 5, 11; 6, 2; 7, 7.

Cassiodorus, *Variae*, V, 42.

Cicero, *De Natura Deorum*, II, 158.

Columella, *De Re Rustica*, VII, 12, 13.

Homer, *Odyssey*, XVII, 300; XVII, 309.

Juvenal, *Satires*, VI, 653.

Martial, *Spectacula*, 14, 1.

Ovid, *Fasti*, IV, 763–766.

Oppianus Apamensis, *Cynegetica*, I, 14.

Petronius, *Satyrica*, 64, 6.

Pliny the Elder, *Historia Naturalis,* XXXIV, 38; VIII, 61.

Plutarch, *Moralia*, 991F; 969D ss. *De Sollertia Animalium*, 969 E, F.

Polyaenus, *Strategemata.*

Seneca, *Tragedies*, *Thyestes*, 497–505. *Dialogi*, *Of anger*, III, 42, 2.

SHA, *Historia Augusta*, Com, XV; Hdt, I, 15.

Strabo, *Geographica*, IV, 5, 2.

Symmachus, *Epistulae*, II, LXXVII.

Varro, *Rerum rusticarum*, II.

Vegetius, *De Re Militari*, IV, XXVI.

Xenophon, *Hiero*, 7, 3.


### *Modern sources*


AA. VV. 2005. *Attenti al cane! Storia e archeologia di un legame millenario.* Catalogo della mostra (6 luglio-31 maggio 2006). Comune di Milano: Milano.

AA. VV. 2010. *The Mastiff. A complete anthology of the Breed, 1850–1940.* London.

Altuna, J. 1994. El perro en los yacimientos arqueológicos del norte de la Península Ibérica. *Monografías. Homenaje al Dr. Joaquín González Echegaray* 17: 159–162.

Arnim, von H. 2002. *Stoici antichi. Tutti i frammenti.* Milano.

Barigazzi, A. 1992. Implicanze morali nella polemica plutarchea sulla psicologia degli animali, in Italo Gallo (ed.) *Plutarco e le scienze*: 297–315. Genova.

Bennett, D. and R.M. Timm 2016. The dogs of Roman Vindolanda, Part II: Time-stratigraphic occurrence, ethnographic comparison and biotype reconstruction. *Archaeofauna,* 25: 107–126.

Brewer, D., T. Clark and A. Phillips 2001. *Dogs in Antiquity. Anubis to Cerberus: the origins of the domestic dog,* Warminster. Aris & Phillips.

Bodson, L. (ed.) 1997. *L'animal de compagnie. Ses rôles et leurs motivations au regard de l'histoire*, Liège.

Bodson, L. 1980. Place et fonctions du chien dans le monde antique. In *Ethnozootechnie* 25: 13–21.

Brodrick, A.H. (ed.) 1972. *Animals in Archaeology*. London.

Bueno, A. 2011. Los perros en la conquista de América: historia e iconografía. In *Chronica Nova* 37: 177–204.

Burris, E.E. 1935. The place of the dog in superstition as revealed in Latin literature. In *Classical Philology* 30: 32–42.

Bussutil, J. 1969. The Maltese dog. In *Greece & Rome* 16: 205–208.

Camps i Rabadà, J. 2005. Lo que el hispano-romano Lucio J.M. Columela describió sobre perros, en su obra 'De re rustica'. Visto por un veterinario. In *Actas del XI Congreso Nacional de Historia de la Veterinaria* (Octubre, 2005, Murcia). Murcia: 319–326.

Carreras, F.F. 2013. Perros de guerra. In *Asociación Argentina de Historia de la Veterinaria*, Año XI, 74: 1–5.

Castignone, S. and G. Lanata (eds) 1994. *Filosofi e animali nel mondo antico*. Pisa.

Colini, A.M. and L. Cozza 1962. *Ludus Magnus*, Roma.

Dal Pra, M. 1950. *Lo scetticismo greco*. Milano.

Ferri, A. 2017. *Cane corso. The best*. Firenze.

Forster, E.S. 1940/41. Dogs in ancient warfare. In *Greece & Rome* 10: 114–117.

Franco, C. 2003. *Senza ritegno. Il cane e la donna nell'immaginario della Grecia antica*. Bologna.

Gasti, F. and E. Romano (eds) 2003. Buoni per pensare: Gli animali nel pensiero e nella letteratura dell'antichita. In *Atti della II Giornata Ghisleriana di Filologia classica* (Pavia, 18–19 aprile 2002), Como-Pavia.

Gautier, A. 1990. *La domestication. Et l'homme créa ses animaux*. Paris.

Guérin, C. 1994. *L'homme et la domestication des animaux*. Lyon.

Haraway, D.J. 2003. *The Companion Species Manifesto. Dogs, People, and Significant Othernes*. Chicago.

Hyland, A. 1990. *Equus. The Horse in the Roman World*, London.

Isnardi, M. (ed.) 1989. *Stoici antichi*, Torino.

Joubert, C.J. 1958. *Le chien dans le monde Antique*. Toulouse.

Mainoldi, C. 1984. L'image du loup et du chien dans la Grèce ancienne d'Homère à Platon. Paris.

Mancioli, D. 1987. *Giochi e Spettacoli*, Vita e costumi dei romani antichi. Museo della Civiltà Romana. Roma.

Marchesini, R. and S. Tonutti 2000. *Animali magici*, Milano.

Martin, T.R. and M. Berti 2012. Cani, lepri, barbari e Alessandro il Macedone. Nobiltà d'animo ed'azione nell opera di Arriano. In Costa, V. (ed.), *Tradizione e trasmissione degli storici greci frammentari.* Atti del Terzo Workshop Internazionale (Roma, 24–26 Febbraio 2011), vol II, Roma: 390–410.

Maspero, F. 1997. *Bestiario antico: gli animali-simbolo e il loro significato nell'imaginario dei popoli antichi.* Casale Monferrato. Pienme.

Méniel, P. 2006. Le chien en Gaule. In Curci, A. e Vitali, D. (eds), *Animali tra uomini e dei. Archeozoologia del mondo preromano.* Atti del Convegno Internazionale (8–9 novembre 2002), Roma: 45–52.

Merlen, H.A. 1971. *De canibus. Dog and Hound in Antiquity*. London.

Moreno, P. 1981. Modelli lisippei nell'arte decorativa di età repubblicana ed augustea. In *L'Art décoratif à Rome à la fin de la République et au debut du principat.* ÈcoleFrançaise de Rome, Roma: 173–227.

Oliver, A. 2014. El perro en el culto, la economía y el prestigio de los íberos. In *Quad. Preh. Arq. Cast.* 32: 43–61.

Ory, T. 1982. *Le chien, sa place et son image à Rome sous la République et le Haut Empire*. Paris.

Rea, R. 1987. L' Anfiteatro Flavio. Competizioni atletiche e spettacoli anfiteatrali: il punto di vista dell'intellettuale. In *Lo sport nel mondo antico. Ludi, munera, certamina a Roma*, Mostra (Roma 27–8 al 25–10 1987). Museo della Civiltà Romana. Roma.

Richter, G. 1930. *Animals in Greek Sculpture. A Survey*. London.

Robert, R. 1993. Rites de protection et défense. A propos des ossement d'un chien découverts au pied du rempart de Paestum. In *AnnOrNapFil* 25: 120–142.

Ryan, K. and J. Pam (eds) 1995. *The Symbolic Role of Animals in Archaeology*. Philadelphia.

Santese, G. 1994. Animali e razionalità in Plutarco. In Silvana Castignone e Giuliana Lanata (eds), *Filosofi e animali nel mondo antico*. Pisa: 139–170.

Toynbee, J. 1973. *Animals in Roman Life and Art*. London

Vauclair, J. 1992. *L'intelligence de l'animal*. Paris, Ed. du Seuil.

# 6.4 Dog in Philippine Life, Ritual and Creation Myths: In a Spirit of Hunting

Maria V. Stanyukovich

Peter the Great Museum of Anthropology and Ethnography (Kunstkamera),
Russian Academy of Sciences, 3 Universitetskaya Emb., St Petersburg, Russia,
mstan@kunstkamera.ru

*El primer hijo de Vhígan llamado Cabigát, fué á cazar*
*desde el cielo Judóc á la tierra con perros...*
*(Kabigat, the eldest son of [the god] Wigan, goes down*
*from the Hudok Sky layer to the Earth with his dogs...)*
Juan Villaverde 1911: 319

**Abstract**

The paper deals with the role of the dog in everyday life, past and present, as well as in Philippine folk beliefs, ritual and mythology. There is a particular emphasis of its importance as a hunting companion of old, and as one of the main agents in the creation of the present-day world, of human beings in general and of certain ethnic groups in particular. We start with a short overview of the history of the dog and its perception through the lens of language.

**Key words**: dog, Philippines, hunting, sacrificial pyramid, myths.

## 1 Dog as the primary animal companion of the Austronesians

East Asia is a critical region for dog domestication. The issue of the dissemination of dogs in the Austronesian world is a matter of interest in archaeology, palaeontology, genetics, anthropology and linguistics. The domestic dog, *Canis familiaris*, was introduced to the Philippines from the Asian mainland via Taiwan. The issue of dog introduction into the wider Pacific region is more complicated and implies several routes (Greig *et al.* 2016; Greig *et al.* 2018). Semi-wild dog species of the South, i.e., the recently discovered New Guinean singing dog *Canis hallstromi* and its close relative, the Australian dingo *Canis lupus dingo*, are believed to have travelled from the Southeast Asian mainland via the Sunda shelf to Sahul. In the small islands of Micronesia and Polynesia domestic dogs were eaten into extinction (Williams *et al.* 2018), and re-introduced only recently. The fact that domesticated dogs, pigs and chicken were the only animal companions of the Austronesians at the time of their expansion from Taiwan to the Philippines, Indonesia and beyond is not under discussion.

### 1.1 Ancient Austronesian dog-words and their modern reflexes

The list of the most ancient Proto-Austronesian (PAN) reconstructed roots contains quite a number of dog-words. Firstly, there is a root *asu* (dog - PAN *$asu_1$), reflexes of which are still widespread in modern Austronesian languages of the Insular Southeast Asia (but not in Oceania). Statistically **asu/wasu* is the most widespread of the present-day terms for 'dog' in Austronesian languages. In the Philippines alone, reflexes of that form are found in 22 languages (Reid 1971), which amounts to at least 50% of all the 'dog-terms'. The situation is pretty much the same in Taiwan, Sulawesi and Borneo (Blust 2002); the same applies to the protoform *tutu*, 'puppy'. The second group includes the terms that define 'dog-voices': various ways of howling, yelping and barking: to howl, of a dog - PAN **qauŋ*,[1] howling or yelping of a dog - PAN *$quaŋ_2$, to bark, of a dog - PAN *$hulhul_1$, to bark, of a dog - PAN **q<um>aŋqaŋ*, to bark, of a dog - PAN **qawqaw*, barking of a dog - PAN **qaŋqaŋ*. The importance of defining different kinds of 'dog-voices' was essential, as humans largely depended on dogs for safety from wild animals, human enemies and dangerous evil spirits (whose presence is visible only to dogs - a belief widespread all over the world). They were also essential for subsistence: hunting - especially wild boar and deer, the most precious game and main source of meat – was impossible without dogs.

The term **qayam* in Proto-Malayo-Polynesian (PMP) and Proto-Western Malayo-Polynesian (PWMP) is of special interest. It has a general meaning of a 'domestic

[1] q stands for glottal stop.

Figure 1. Ifugao dogs. Kiangan, Ifugao province, Northern Luzon, the Philippines, 2011 (Photo by M.V. Stanyukovich).

animal', meaning primarily the above mentioned triumvirate of the first-domesticates: dog, pig and chicken. To avoid going deeper into the issues of historical linguistics,[2] it can be stated that among the reflexes of that term in modern Austronesian languages we can find those used as umbrella terms, as well as ones applied to each of the three animal companions of the Austronesians. 'The History of Faunal Terms in Austronesian Languages' gives the following survey of umbrella terms: 'Gaddang *ayam* 'domesticated animal', Western Bukidnon Manobo *ayam* 'a pet', Tiruray *ayam* 'domesticated animal', Sarangani Manobo *ayam-ayam* 'pet, domesticated animal', Rungus Dusun *azam* 'livestock, domesticated animals', Kadazan *tazam* 'tame animals', Iban *ayam* 'pet', Dampelas *n-eaŋ* 'tame'' (Blust 2002: 91). We can add Tuwali Ifugao *ayum* 'to tame an animal or bird, e.g., a dog, pig, carabao, birds' (Hohulin and Hohulin 2014) (Figure 1).

As to non-umbrella terms, they designate a dog, a pig or a chicken in different languages, with a significant predominance of 'dog'. In other words, the dog was the domesticated animal per se for Austronesian speakers as long as hunting remained an essential activity. On the other hand, 'There is no doubt that the main role of the dog in most Polynesian societies was to provide a source of food < > the word for food became synonymous with the word for dog: poi in Hawaiian and '*ina'i* in Tahitian' (Williams *et al.* 2018), with a reference to linguistic data in Pollock (1986).

The image of the dog in mythology and folklore bears no connotations to wild canine species, like wolves, jackals, coyotes, foxes, as they are not represented in insular Southeast Asia and Oceania.[3]

## 2 Dog consumption and sacrifice

### *2.1 Dog consumption*

Difference in food habits has always been one of the major points in singling out the 'ethnic other' and, especially, of stigmatising non-Europeans. Nicolo Conti, one of the first Europeans who visited Java and Sumatra early in the 15th cent., wrote: 'These islands are populated by the most cruel and inhumane inhabitants. They eat mice, dogs, cats and other, even dirtier

[2] See Austronesian database https://www.trussel2.com/ACD/acd-s_q.htm#27688 and the works of R. Blust for details.

[3] There is evidence for the Philippine wild dog in the South of the Philippine archipelago: 'It has sharp claws, climbs trees, hunts cobras, and could be 36,000 years old. It is called tiger dog and aso ng gubat by locals [by the Bukidnon of Mindanao - MS]. It is also called bird catcher in Luzon and witch dog in the Visayas. The aso ng gubat in Bukidnon has a brindle coat—dark-brown with black stripes' (Limos 2020). Every statement cited above from a popular publication 'The Philippine Enquirer' needs to be verified. It's not clear whether it is really a wild dog, or a kind of *askal*, stray village or street dog (term derived from the Austronesian root *aso*, 'dog', and *kalye*, from Spanish *calle*, street, with lesser-known name *aspin*, *aso+Pinoy*, 'Philippine dog'). Or else, it could be a reference to a mythological tiger dog, an avatar of a tiger, beliefs in which in the neighbouring Borneo are so brilliantly analysed by Sellato (2019).

animals, and they exceed all the other mortals in their cruelty' (Bracciolini 1723, cite after Vozchikov 2019: 79). Such an attitude survived up to the present day, notwithstanding the fact that dog meat consumption was banned in Europe only recently and not everywhere (in Switzerland it is still regarded as a Christmas snack by 3% of the population). Archaeological evidence shows that among the Gauls, the Irish and some other ethnic groups the tradition goes back to ancient times, and the scale of dog meat consumption in France in modern times is really impressive (Mahler and Denis 1989; Milliet 1995).

*2.1.1 Sensitive points*

Treatment of the issues of dog sacrifice and dog meat consumption have been controversial. Sensitivity is due to colonial discourse on the one hand and present-day animal-rights organisations activities on the other. These are the main reasons why the practices are denied, despite abundant evidence that proves that both are deeply rooted and still very much alive. 'Dog-eaters' is a wide-spread pejorative epithet for 'Asian' and 'Eastern', as opposed to 'Western'. Prohibited officially (as it has been since 1998 in the Philippines) or not, dog meat remains part of the diet there (Podberscek 2009), just like in Vietnam, China, Thailand (Podberscek 2007), Indonesia (Parker 1991; Weichart 2004; Eijkelkamp 2015) and especially in South Korea – the only country that raised its voice against the prohibition as a 'globalization of morality' and openly defends its 'cultural rights' in that regard, calling for 'the acceptance of cultural diversity' (Lien 2004; Oh and Jackson 2011; Yoon 2016).

In the Philippines, the issue of national identity is deeply connected with the country's position as the only Christian nation and a showcase of Western civilisation and democracy in Asia. Moreover, in the painful period of change of coloniser, the representation of Filipinos as 'dog-eaters' during the Saint Louis World Fair in the USA has left an ever-bleeding scar on the national pride. It has also been the source of long-term stigmatisation of highlanders by the lowland population (Afable 1995). The event has been much-discussed by historians, anthropologists and the general public, enhanced by a documentary movie 'Bontok Eulogy' (1995) shot by Marlon Fuentes, an American film director of Filipino descent. He was the grandson of one of the Bontok warriors that were brought in 1904 to the USA for a show depicting singeing and consuming dogs. Even today, the anti-colonial discussion of the effect of the Saint Louis World Fair suffers from ascribing dog-eating exclusively to the highlanders of Northern Luzon.[4]

Another important cultural landmark was the publication of 'Dogeaters', 'the quintessential Filipino American novel' (Hagedorn 1990). As the author puts it in the interview given on the 30th anniversary of the event: 'And there was one book that I stumbled upon in the library called Little Brown Brother, by a historian named Leon Wolff, which is where I encountered the term. As a kid, I knew there were these cringy jokes in the Philippines about how people think we eat dogs, but I didn't know the root of all the shame around it. So in this book, there were these references to how the American soldiers would call the Filipinos 'dogeaters' and other things - like the fact that 'gook' came from the Philippine-American War and not the Vietnam War, as is commonly assumed' (Flora 2020).

Every now and then there are publications by the representatives of the Igorot (highlanders of the Cordillera) communities that find that attitude insulting:

> 'As an Igorot, I vehemently do not accept dog eating as my culture. I was not raised to eat dogs. Dog meat is not a part of my diet, nor has it ever been. I find it insulting that Igorots are branded as dog-eaters, not only in the Philippines but abroad. It is a shame, and because Igorots are Filipinos, dog-eating is a Philippine national shame' (Dawang 2003).

Publications dealing with the above-mentioned controversy in Indonesia and especially in Korea have started a debate on 'colonialist stigmatisation of dog eating', when Asian communities are forced to condemn dog-eating practices, shocking to Westerners, but to ignore, for instance, cow meat consumption, 'shocking to most East Indians' (Yoon 2016: 357, 358). An Indonesian researcher describes it,

> '... animal rights organizations in the East tend to focus on individual animal species, such as dogs. The key question is: why ban the consumption of dog meat while still allowing the consumption of meat from other animals?' (Resolute 2016: 150).

*2.1.2 Dog meat as food*

Popular practices of dog meat consumption in Indonesia contradict the rulings of Islam. Here is a sample of an ethnogenic myth about the origin of the Sama/Bajao,

[4] 'Examining the portrayal of Filipino natives in American newspaper articles that were published in 1904, a multitude of methods that were intended to attract readers as well as potentially influence their views on the U.S. occupation of the Philippines are revealed. Common words and phrases that appeared in these newspaper articles included 'savage', 'dog-eaters' and/or 'barbarian', which invoked a feeling of separation between the American readers and the Filipino natives and placed the latter in a negative light. The use of such words and phrases, along with the other methods that were deployed in these newspapers, creates the impression that all Filipinos are dog-eaters. The Filipinos' reputation as dog-eaters has persisted even today and it can be argued that it is a result of the American newspaper reports on the Filipino natives eating habits during the 1904 World's Fair' (Heinrich 2017: 38).

the so-called 'sea gypsies', seafaring groups that used to live in boats, engaged in trading and fishing. Once well-to-do merchants with a sophisticated culture and very rich folklore, these peaceful nomadic groups travelled widely among the Philippines, Indonesia, Malaysia and beyond. They were reduced to the present despised and miserable state due to a decline of possibilities for private international trade, intimidation by Tausug and other warlike Moslem groups and lack of their own authorised ancestral lands. Variants of the myth in question are found in many sources, including the fundamental works by Kiefer (1972), Sather (1984) etc. Quoting one of them, dealing with the loss of the Bajao mythical homeland due to serving dog meat:

> 'The wife of the Prophet, Siti Aisiah, is molested by a Bajau Laut fisherman when she visits him to purchase fish. For this wrongdoing God causes the Bajau Laut to suffer. The fisherman asks the Prophet for forgiveness and the Prophet advises him to prepare a feast. The Bajau does so but as he has insufficient meat, he butchers a dog and prepares its 'unclean' flesh. The Prophet and his followers arrive. The Prophet recites prayers over the food and the meat begins to bark. The Prophet promptly leaves and the original homeland of the Bajau Laut sinks into the sea. The few Bajau Laut who survive are saved by clinging to drifting debris. From this time on the Bajau Laut are excluded by God from the society of the faithful and are compelled as a punishment to live in boats, drifting like the debris to which their ancestors clung (Morrison 1993: 46, with a reference to Sather 1984: 13). The sources [5] state that the Tausug use a derogatory term *luwaan* (spit out or vomited out) to designate the Bajao – on the account of the myth mentioned above.

My own field experience in the Philippines since 1994 confirms that despite the prohibition, the attitude towards eating dog meat in the Philippines (members of animal rights organisations, UP and corresponding universities excluded), is generally neutral or positive, especially in the older generation, - although it is not something foreigners would easily happen to see because of the stigmatisation of that practice. In 'Ghosts of Manila', a book by James Hamilton-Paterson, a prominent British writer, a part-time resident of the Philippines, there is a matter-of-fact discussion about dogs served in the eateries jokingly called '*adobong payong*' (stewed umbrellas):

> ' 'Look at the meat on it,' - he said, slapping a bloody flank. 'That's pedigree dog meat, tons of it. Actually, we've been thinking it's probably too good for your customers seeing how they're used to the starving mongrels you normally serve. All ribs and skin. They must think they're eating stewed umbrellas.' This phrase, *adobong payong*, made Gringo lean helplessly against the taxi...' (Hamilton-Paterson 1994: 92).

However, the scale of dog meat consumption is rather low, and has always been so. It can not be compared to the levels, stated in other places, e.g., in a native Evenki settlement of Southern Siberia, near the Mongolian border, where some informants claim eating dog meat twice a week (Davydov and Simonova 2008: 219). We must take into consideration that meat is a staple food for the Evenki and other indigenous groups of Siberia, living in a cold climate and, historically, depending primarily on deer-herding and hunting, whereas the Philippine diet has always been mostly vegetarian (rice and other cereals, vegetables, and fish as the main source of protein), meat being a rare festive addition. Nevertheless, there are places in the Philippines, just as in Indonesia, where dog meat is on sale, even if illegally, and dishes can be found in the eateries, although you should know where to look for them and what euphemisms to use to order.

The exceptions to the rule are very few. Among them we can name the indigenous ethnic groups of inland Mindoro. Dog meat is a taboo food among the Hanunoo (Conklin 1975: 29) and the Buid, their neighbours. They consider the dog to be an 'inedible animal' on the grounds that dogs are 'too close' to men: 'The Buid do not avoid dog meat because it is thought to be ritually 'unclean,' but because dogs are in some sense companions to humans. When a dog dies, its body must be disposed of far from human habitation, because the smell of its decaying flesh is thought to attract evil spirits in the same way as does a human corpse' (Gibson 2015: 171). It is seen by lowlanders as an idiosyncratic trait of the named tribes: 'By contrast, dog meat is regarded as a great delicacy by the Christians, who serve it on social occasions such as christenings or as an accompaniment (*pulutan*) to the consumption of alcohol' (Gibson 2015: 171). Another example is to be found in Borneo: 'dogs among Dayak groups are traditionally not eaten at all<>. Only recently, with influence from Batak, Minahassa, Toraja, or Timorese newcomers, have younger Dayak started to eat dog meat. Black dogs are preferred. They are put in a sack and killed by beating them with sticks or clubs, in order to not shed blood, so the meat is not tough' (Bernard Sellato, e-mail letter to the author from 08.05.2021).

The situation is similar among the Philippine hunter-gatherers. Bion Griffin, one of the major authorities on Philippine Agta, states: 'Foragers and swiddeners typically do not consume dog flesh. Firstly, dogs are too valuable to eat. They are essential for hunting in the humid tropics. Secondly, dogs are essential to all tribal

[5] Cf. also (Roxas-Lim 2017: 53).

Figure 2. The Agta woman Taytayan carrying a puppy as she sets off to travel. Near the Nanadukan River in Isabela Province, the Philippines, 1981 (Photo by Bion P. Griffin).

Figure 3. Keladi, a Korowai girl, carrying her dog in a bag while travelling in a forest near Afiumriver. Kolof Braza, Asmat area, Papua province, Indonesia, 2015 (Photo by E. Beliakova).

people in Southeast Asia because they are so effective in warning of approaching strangers' (Griffin 1998: 35). However, it is the observation of a modern situation, when small hunter-gathering groups are driven by population pressure into the least favourable areas and surrounded by agriculturalists, from whom they get several items, essential for livelihood, including dogs, as it is with Agta Negritos: 'dog populations are not self-sustaining. Dogs are occasionally killed by pigs, and due to malnutrition, they are often sickly and die from disease; animals must therefore be purchased from farmers who can feed their dogs with grains' (Griffin 1998: 29). In the times preceding agriculture, hunter-gatherers were even more dependent on dogs and possibly shared food with them more generously.

Anthropologists have documented very special attitudes towards dogs, that included breast-feeding of puppies by local women (e.g., in New Guinea) and carrying them like babies (Figures 2–3). Traces of a similar attitude can be found in the language of the Cordillera people, where just two generations ago dogs were absolutely essential for hunting: e.g., '*am'o* bathe 1.1trans. to bathe someone or something, e.g., a child or dog' in Tuwali Ifugao (Hohulin and Hohulin 2014).

In highland agricultural communities dogs are fed with rice, but most of them are skinny and hungry, they walk freely everywhere hunting, looking for additional food (Figures 4–6).

To summarise, as far as the Philippines is concerned, archaeological evidence confirms that dog flesh was consumed all over the archipelago, both as everyday food and, more importantly, as a ceremonial one, the flesh of a sacrificed prey. To broaden the perspective, we can state that historically the attitude towards dogs primarily depended on the importance of hunting as a way of subsistence, and whether the dog was a hunting companion or not. In the latter case in the societies where sources of protein were scarce, dogs could be eaten into extinction, as happened in some areas of the Pacific.

In the modern world, we can use the popular phraseology: top dogs, i.e., the countries dominating the world due to their economic, political and military power, who dictate their standards to underdogs, i.e., poorer states dependent in every respect. That is a general part of the globalisation process, in which the attitude towards dog meat consumption is dictated to Asia by the West just like a lot of other standards, e.g., the Westernised ideals of beauty praising white skin, white teeth (as opposed to traditions of blackening and filing teeth) and 'round' European-shaped eyes. As such, the campaign against dog meat consumption should be regarded as a form of neo-colonialism.

Figure 4. A dog in rice fields of Asipulo, Ifugao province, Northern Luzon, the Philippines, 2007 (Photo by M.V. Stanyukovich).

Figure 5. Dog accompanies men who carry a pig for a funeral sacrifice. Kiangan, Ifugao province, Northern Luzon, the Philippines, 2011 (Photo by M.V. Stanyukovich).

Figure 6. Dog lying on drying rice. Kiangan, Ifugao province, Northern Luzon, the Philippines, 2012 (Photo by M.V. Stanyukovich).

### *2.1 Dog sacrifice*

*2.2.1 Government regulations*

To dismiss the current suggestion that dog sacrifice has never existed in the Philippines - or that it belonged to ancient times - let us turn to anthropological sources, archaeological findings and present-day government regulations. For this last, we can cite the Administrative Order No. 25 of the Philippine Department of Agriculture (2007) that contains rules and regulations regarding the use of animals during rituals (italics are mine - MVS):

> 'To promote positive integration of animal welfare and humane practices in the use of animals such as but not limited to dogs, chicken, pigs and carabaos [6], used as sacrifice or offering during various religious ritual of an established religion or sect or rituals required by a tribal or ethnic custom of indigenous cultural communities'. 2.1
>
> 'To control the indiscriminate slaughter of *non-food animals such as dogs* in the guise of spiritual, religious, tribal or ethnic custom of indigenous cultural communities *limiting the consumption of meat of said animals to those that have participated in the ceremony or ritual only*'. 2.2.1
>
> 'During the celebration of the religious ritual or ritual required by tribal or ethnic custom, the tribal leader or any person/s who shall perform the sacrifice of the animal/s, shall ensure that only humane means or methods shall be applied to sacrifice the animal used as an offering. *For dogs, cutting through the carotid arteries with one swift stroke with a sharp knife is the only accepted method.*' 6.2.1
>
> (Administrative Order 2007)

*2.2.2 Anthropological evidence*

Anthropological literature contains a wealth of material on occasions and particular details of dog sacrifices, most of them from the early 20th cent. Northern Luzon, but not limited to that time and geographic area. Here are a few examples from different ethnic groups. A. Jenks, who did his fieldwork among the Bontoc, states that 'Funerals and marriages are more often celebrated by a dog feast than are any other of their ceremonials' (Jenks 1905: 203). According to F.-C. Cole, the Tinguian sacrificed dogs at a whole range of ceremonies, primarily those aiming to cure illness: 'Since early morning a dog has been tied at the end of the house. It is now brought up to the bundle of leaves, and is knocked on the head with a club, its throat is cut, and some of its blood is applied with a head-axe to the

[6] Carabao is a water buffalo in the Philippines.

backs of the man and woman' (Cole 1922: 350); 'Arrived at the water's edge, the oldest relative will cut off the dog's head as a final payment for the life of the invalid. Since the act is carried out beside the river, the spirits will either witness the act, or see the blood as it floats away...' (Cole 1922: 354–355); 'On the third and last day, the medium leads a big dog to the edge of the village, and then kills it with a club. A piece of the animal's ear is cut off, is wrapped in a cloth, and is hung around the patient's neck as a protection against evil, and as a sign to all spirits that this ceremony has been held' (Cole 1922: 356).

Dog sacrifice in the Ifugao province is also not to be debated, as already mentioned (Stanyukovich 2018: 260). Just like a hundred years ago (Barton 1946), dogs are sacrificed to cure those whose illness is believed to be caused by sorcery. In such case one should ask *mumbaki* (a ritual specialist) to perform an *anap* rite to define the culprit, after which a ritual called *humyang* (in Yattuka) or *hagoho* (in Tuwali) is performed. As my informant RG from the South of the province puts it, 'They offer one dog and three to five chicken to return to the family who did such bad thing. Mostly pure black dog five months old and above. ... To return back a revenge because it [*kulam*, witchcraft – MS] caused a certain sickness in the family. Witch doctor cannot cure. So, they consult the mumbaki... so they perform the *anap* then if they found someone who did something ... the dog was used to fight the family who did something bad. The *mumbaki* will say 'I send or command you, black dog, to be the one to fight for us to the one who did something bad to a member of this family. Our ancestors and old mumbaki please help us. So, help us."

Here is a depiction of the Kankana-ey death ritual: 'While the coffin was still being made under the house <> the next offering was prepared: a dog was slaughtered and its liver inspected; its blood was kept for an unseasoned blood soup for the coffin-makers as payment in kind' (Kohnen and Kohnen 1986: 53).

The reference to blood is of special importance. Body liquids are universally regarded as the most potent powerful substances, beneficial or malevolent according to the way they are used (cf. Stanyukovich 2016). In his classical books on Indochina Georges Condominas repeatedly pointed out that blood plays the principal role in the connection between the human world and the world of spirits, being the best protection against evil. G. Sprenger states that even the members of 'societies without domestic animals, as, for example, among the Malaysian Orang Asli' practice blood sacrifice: they 'sacrifice their own blood to appease the thunder god (Endicott 1979: 156–159, Needham 1963, see also Valeri 1994)' (Sprenger 2016a: 35). The equation of blood and beer, although made on Amerindian material ('A jaguar's beer is blood in human perspective' (Sprenger 2016b: 76)) is also worth being mentioned, as pouring rice wine (beer) on animated as well as unanimated objects and smearing blood over them are essential for ritual procedures in Southeast Asian indigenous communities.

*2.2.3 Archaeological evidence*

Excavations in Santa Ana, part of present-day Metro Manila (Vitales 2018), as well as in the South (Palawan) and the North of the archipelago (Nagsabaran site in the Cagayan valley (Amano *et al.* 2013)) give us evidence of at least two uses of dogs. One is dog meat consumption, as it is clear from numerous cut marks on remains. The other shows dogs buried just like humans, untouched: 'The oldest dog remains identified were those of an adult individual that had been placed within its own 'grave cut' dug into the upper surfaces of the sterile clay below the basal layers of the shell midden <>. Though not directly related, this burial was overlain by an inhumation, indicating that the dog had been buried within the same general area as people' (Amano *et al.* 2013: 324). However, the interpretation in conclusion is insufficient: 'The deliberate burial of a dog demonstrates emotional ties between it, and one or more people' (Amano *et al.* 2013: 330). Although there could and surely were emotional ties with best friends and hunting companions, on whom life was vastly dependent, it was not the reason why dogs were buried close to humans. A dog is a burial companion of a human in many cultures from the Philippines and to the North, in Taiwan, China, Eastern Siberia and the Russian Far East (Kudinova 2014; Startsev 2017). A sacrificial animal can accompany a human to the underworld in several capacities. As I wrote elsewhere, negotiating with the gods in every traditional 'pagan' ritual includes a wide range of methods, not just pleading, worshiping and begging, but reasoning, convincing, distracting, warning, even deceiving and threatening (Stanyukovich 2011; Stanyukovich 2013).

*2.2.4 Dog's place in the sacrificial systems*

Every sacrificed item is primarily regarded as a messenger from humans to spirits, sent on a journey with a special assignment. That is the reason why the sacrificed animal is induced to cry out loudly[7] to summon the attention of those to whom it is sent. That is true not only of animals, but even of inanimate objects, of which betel nut is the most important. All over the area of Areca catechu growth, where betel chewing is customary (that is, all of the

[7] Hence point 6.2.1 cited above, with additional points in 6.2.1 and point 6.2.3 listing those 'inhumane means or methods' that make the victim give out loud cries (Administrative Order 2007).

Southeast Asia and beyond) betel is number one among sacrificed items. Here some of the functions of alcohol as an intermediary between humans and spirits, are delegated to betel. On one hand, betel is connected with the oral sphere (being the main part of the chewing quid) and eloquence; on the other, with the idea of fertility and body liquids, namely saliva, sperm and blood (for detailed discussion of betel in mythology and ritual see (Stanyukovich 2014)). In Philippine myths, sacrificed betel is often depicted as talking to the spirits, conveying messages; it can even hurt them by growing into a huge betel palm on the god's knee until the deity agrees to fulfil the demand of the sacrifice-maker (Cole 1915: 40). The role given to betel in myths is very similar to the mode of addressing the sacrificed animal, as recorded in R.F. Barton's classical work: "Be thou not dumb or stupid, pig, but speak up and know what to say. Tell the deities, 'I was not killed by a fall, I was not trapped or speared; I am a pig that was sacrified (sic! – MVS) by A ___ and a ___ [names of the husband and wife who are giving the feast]. Exhort the deities there in the Skyworld to increase the rice so that its grains become like the sands and swell up (in cooking) and be rough in the throats of the children and hired help and so that its bundles increase as if it were alive as it was before (the harvest)' (Barton 1946). There are no gender restrictions regarding those offerings without bloodshed, whereas sacrificing animals is a male job.

We can imagine the general scheme of a sacrifice as a pyramid. It's basic lowest level is 'food-offerings', that is, betel and rice wine/beer, to which several inanimate objects (textiles, ornaments of gold or precious stones) are added (they will be withdrawn and returned to the owners after the ritual). The next three storeys of the sacrifice pyramid are connected with bloodshed, in the following order: chickens, pigs and carabaos (water buffalo). Every succeeding storey is based on the previous ones, that is, it is just added, signifying the advancement of the offer due to the higher importance of ritual. In other words, blood sacrifice does not replace betel, rice wine and textile offerings, but is being added to it. During the most important rituals all four storeys are in place: gods and spirits are offered large quantities of betel, dozens of chickens, dozens of pigs and several carabaos. My Ifugao informants emphasise the importance of chicken as a main sacrificial animal; so do the indigenous groups of Mindanao. Hans Brandeis, a prominent researcher of Mindanao musical instruments, especially boat lutes, and ritual, has done profound fieldwork among the Tboli and other Mindanao IPs since early 1970s. According to him, pig is sacrificed when a lot of meat is required, and is regarded as a body, 'composed of many chickens' (personal communication, 2020).

Dogs have no place in that sacrificial pyramid. In the Philippines, in Borneo[8] (and seemingly in other Asian areas, - but that I can only suggest), the dog is sacrificed outside that usual pattern, primarily in rituals aimed to cure illness, and in death rituals[9]. Being closest to humans, a companion, guardian and helper, dogs seem to have a very special position as a sacrifice as well. It seems that while all the other animals are simply offerings and messengers to the gods and spirits, the dog is not to be wasted as an ordinary sacrificial victim suitable almost for any ritual. The dog is singled out; it retains its special place near humans, as well as its functions as a helper, friend and guardian, even once being sacrificed.

*2.2.5 Dog sacrifice in flood myth: a case of Mnong*

There is a motif, although not a Philippine one, related to origin and flood myths that is worth mentioning here. It belongs to the Mnong, Bahnaric people of Indochina, and deals with the choice of the sacrificial animal to stop the flood. It was recorded by Georges Condominas in Vietnam and published in his early French-language publication (Condominas 1957: 72–73). Here is its English rendition:

'Man learned to make effective sacrifices, say the M'nong, from two legendary heroes, Mot Dlong and Mot Dlaang, during the time of the great flood. When it seems as if the rain would never stop, Mot Dlong and Mot Dlaang sacrificed a dog, a crocodile and an iguana[10] to stop the rain; however, these sacrifices were ineffective and the rain continued to pour down. Mot Dlong and Mot Dlaang sat in the rain and pondered, wondering what had gone wrong. Then they had an inspiration: they decided to sacrifice buffaloes. They captured some buffaloes and had their servants prepare the animals for the sacrifice. A great sacrifice was held, and Mot Dlong and Mot Dlaang prayed for the rain to stop. The sacrifice was successful, and the rain stopped; since that time, buffaloes have been the preferred sacrificial animals among the M'nong' (Minority 1966: 499).

It is a typical explanatory myth with the logic 'from the opposite': in dreamtime[11] it was men, not women, who gave birth, animals could talk etc. It deals with the

[8] To cite from an e-mail letter of B. Sellato to the author, 'You write 'Dogs have no place in that sacrificial pyramid'; this is true in Borneo, too, and they have no place in sacrifice. But a good hunting dog may be killed to accompany its master in his grave. Also, a dog may be killed in the context of an oath: swearing on a dog's corpse (cf. a tiger's corpse)' (e-mail letter to the author from 08.05.2021).

[9] That is a general tendency. Cases of dog sacrifice are documented in other rituals (Jenks 1905: 203; Condominas 1957).

[10] Sic! There are no iguanas in Indochina, just lizards.

[11] 'Dreamtime' is a term accepted in comparative mythology studies. It dates back to Australian mythology where it defines the times of beginning, the chaotic era that preceded the present-time way of things.

reason why the buffalo, the main sacrificial animal among the Bahnaric peoples, should have been sacrificed for that major purpose. Crocodiles and iguanas, non-sacrificial animals, are both reptiles, connected with water, which perhaps explains their mention in a rain-flood myth. As to the dog, which is a sacrificial animal among the Bahnars (of which many examples are cited by Condominas), its presence in the flood (origin) myth, though not in the capacity of an offering, is quite widespread in Southeast Asian mythology, as we will see in the final part of the present paper.

## 3 The use of dogs' body parts for ornaments

This point is worth special study, that is beyond the limits of the present paper. It is important to mention, though, that dog teeth, claws, skin and fur were used for the ornamentation of traditional garments, including headgear and all kinds of necklaces, bracelets etc. (cf. e.g. (Vanoverbergh 1929: 192, 2013 etc.)). They were traditionally outnumbered by objects made of other living species, esp. crocodiles, pigs, shells and birds, but nevertheless they existed and are represented in museum collections (cf. e.g., https://philippines.fieldmuseum.org/heritage/catalogue/1255875; Fanged *et al.*. 2018).

## 4 Dogs and gods

### *4.1 Dog spirits and spirit dogs*

Ancient Philippine gods were hunters, just like ancient Filipinos. We are talking not only about the Agta, Alta and other 'Negrito' hunter-gatherers who were the first inhabitants of the islands, but of all the Austronesians, early tropical agriculturalists, who actively practiced hunting in the forests that surrounded their taro and, later, rice fields. Dogs were their primary companions and helpers, often regarded as gods themselves. They could be malevolent as well as benevolent. With the expansion of Hinduism, Buddhism, Islam and Christianity divine dogs were reduced to spirits of lower mythology, usually evil ones.

In the mythology of the Igorot (the highlanders of Northern Luzon) and the Lumad (the indigenous ethnic groups of Mindanao), that preserved their independence and their folk beliefs until the 20th cent., we find dog spirits and spirit dogs. E.g., in the Ifugao pantheon, one of the most gorgeously elaborate, there is a class of gods and spirits named Monduntug ('Mountaineers', class No 24, according to Barton): 'These deities live in the mountains downstream. They are conceived as carrying spears and as attended by spirit dogs. They are nocturnal in their roamings; unusual night noises are attributed to them and their dogs' (Barton 1946).

In Ifugao, a ritual clapper (musical instrument) is perceived as a dog of the rice gods and called so in ritual speech. For instance, at the *ingngilin* ritual, formerly performed in May-June by every field owner on harvest day, the invoked spirit of a son of the rice-granary idol (*bulul*) acts and speaks through a possessed medium. First, he rubs the fat of a sacrificed chicken on the ritual clapper and then ends his party with the words 'I enter this granary and I lay down my dog [the clapper] because ye are harvesting, ye earth people; *ya-bebe-mo Hiye-e-e-e-e-eh*!' (Barton 1946).

### *4.2 Celestial dog*

In Asia, solar and lunar eclipses are believed to be caused by an animal that devours those luminaries (motif a12a in the Berezkin and Duvakin catalogue). Among the Udege and other ethnic groups of the Russian Far East, it is a dog or a wolf (Startsev 2017: 35–36). Among the Austronesians we find it among the Tetum of East Timor and in Borneo. Bernard Sellato in his valuable work on tiger in Borneo beliefs and folklore, states that mythological tiger has very special connections with the moon (and thunder), and that a dog is one of the avatars of a tiger, the others being a bear and a dragon (Sellato 2019). The cases of Maguindanao dog-tiger (Ramos 1953) and the Maranao tiger-lion (Madale 1966) of Mindanao, Philippines, are connected with that mythological creature of Borneo. In Maguindanao tale a dog named *Arimaunga* chases the Moon to let the sun catch up with it (Ramos 1953: 9–11). The name of that celestial dog is doubtlessly derived from the word 'tiger'. B. Sellato gives two sets of corresponding words for 'tiger': '*harimaung, horomaung, remaung, halimaung, rima'ung*, derived from reconstructed PWMP4 **qari-maquŋas* 'wild feline' (ACD 2017); and *halimau, rimau, limau, horoma'u* (?), which may derive from a protoform **harimaw* and whose distribution is likely due to borrowing from Malay' (Sellato 2019: 7). In the Philippines, where no tigers existed in historical times, the name of that powerful animal in different forms (*sarimao* etc.) is attributed to mythological monsters that bear no connection to a tiger or dog (e.g., the *Sarimaw* monster in Bicol epics). In the Philippines it is generally a snake/dragon who is responsible for the eclipse. Among different ethnic groups that creature bears the names of *Bakunawa, Irago, Oryol, Naga, Ibingan.*

However, the Philippine celestial dog still exists in various forms, connected with thunder and lightning. That fits into motif i5 in the Berezkin and Duvakin catalogue, which encompasses the cases when thunder is represented as an animal (not necessarily dog).

Different regional traditions of the Tingguian have either thunder or a lightning in the shape of a dog. In the first case, thunder is perceived as 'the dog of Kadaklan, the greatest of all the spirits', 'by the barking of this dog, the god makes known his desires' (Cole 1916: 95). In the second, it is a dog-shaped lightning: 'Kīmat, lightning, came and demanded a drink, which was given. As he is usually considered as a dog, the writer inquired why he had appeared as a man' (Cole 1922: 342). We can add that the words meaning 'lightning' (Kimat, Kidul etc.) are popular dog names in the present-day Philippines.

The connection of mythological lightning/thunder with a dog can be traced much further to the North in continental Asia (see publications on the 'dogs of the god of thunder' in China (Kudinova 2014; Komissarov and Kudinova 2017)). On the other hand, to the South of the Philippines, in New Guinea, we find the overlapping of two functions of dogs in myths, those of thunder and of the keeper of the underworld: 'among the Papuans of Kiwai, thunder is the bark of two dogs announcing the arrival of new souls in the land of the dead (Landtman 1927: 311–312, cite after Berezkin 2005: 136).

### *4.3 Dog guardian of the Underworld*

Ilocano ethnic territory lies at the very North of the archipelago, therefore Ilocano folklore (just like that of the nearby Cordillera traditions) is largely free of Indian and Islamic influences, but severely affected by the early Spanish presence in the area. E.g., the father of the main epic hero, Lam-ang, is called Don Juan, and his fiancée is Doña Ines Kannoyan. In Ilocano mythology a large dog known under the Spanish name of Lobo ('wolf') guards the entrance to the underworld (Alacacin 1952, cited after Gaverza 2014: 61). Similar views, however, are typical of the Visayan area that lies in the central part of the country: a dog 'with one mammary gland and two sets of genitals' guards captured souls in a cave (Demetrio and Cordero-Fernando 1991, cited after Gaverza 2014: 135). The latter case brings simultaneously international and local reminiscences. On one hand, the motif of a spirit dog that guards the souls is a very widespread one, as shown in detail in 'The Black Dog at the river of tears. Some Amerindian Representations of the Passage of the land of the dead and their Eurasian roots' (Berezkin 2005). On the other hand, inside the Philippine archipelago the hermaphrodite spirit that comprises the power of both sexes and commands the souls reminds us of the Tagalog goddess Lakapati, Christianised as Santa Clara of Assisi. Still now, she supplies the childless with the souls of unborn babies in Obando church, built at the former *dambana* (worship place) of that androgenous deity (Stanyukovich 2020).

### *4.4 Lower mythology: shapeshifters and related spirits*

In Ifugao, among the mountaineers of Northern Luzon, there is a class of *gatui* — Harpy Deities: 'the *gatui* is conceived as normally in appearance a sort of hybrid between a dog and a bird but having often a human face and able to assume human form at will. The *gatui* preys in the same way as the Flying Monster (*taiyaban*), on souls and soul-stuff; it is often attended by a 'dog,' the *kilkilan*. These deities are strictly pathogenic' (Barton 1946).

Throughout the lowland Philippines, dogs are regarded as shapeshifters, analogous to the European werewolf, and other creatures of urban and rural legends, including the Suban-on *multo* (dog, pig or bird-shaped creatures with feet like humans', but reversed) and the legendary aswang (Lieban, 1967). The shapeshifters named above remind us once again of the triumvirate of first-domesticates, which we discussed in the linguistic section of the present paper. Although being widespread mostly in Christian lowland communities, the beliefs in those spirits, bearing a Spanish name (*multo* comes from Sp. '*muerto*', dead) are definitely deeply rooted in Austronesian mythology. Here is a corresponding shapeshifter from the Tingguian, one of those called '*igorrotes infieles*' by the Spanish:

'*Apdel* is the spirit who resides in the guardian stones (*pīnaing*) at the gate of the town. During a ceremony, or when the men are away for a fight, it becomes his special duty to protect the village from sickness and enemies. He has been known to appear as a red rooster or as a white dog' (Cole 1922: 296).

## 5 Dog in creation myths

### *5.1 In a spirit of hunting: Ifugao version*

The epigraph comes from one of the earliest records of the flood myth of the Ifugao, highlanders of Northern Luzon, set down by Juan Villaverde, a Spanish missionary. Here Kabigat, the eldest son of the god Wigan, goes on a hunting trip with his dogs and divine companions. Coming down from the Sky world (its *hudok* - layers - being a well-organised mountainous habitat), he finds the flat earth absolutely unsuitable for hunting: dogs' barking cannot be heard. He decides to improve the surface to make it liveable, which primarily implies huntable, i.e., to create mountains so that the sounds of barking echo from the rocks of the mountain slopes. That was achieved by closing off the rivers' outlets to the sea, combined with heavy rain that lasted for three days. After the flood, the land gained the proper shape, good for hunting with dogs. All the humans who previously inhabited the earth drowned and were replaced by direct descendants of gods from

Skyworld: Kabigat and his sister Bugan were sent down by their divine father Wigan. The descendants of that incestuous couple populated the mountains of Silipan (Villaverde 1911: 319).

That motif, presented in Villaverde's text and a number of others, that we will discuss here, seemingly fits into C12a, 'The dog and the world cataclysm' in the Mythological Catalogue (Berezkin, Duvakin). According to that largest online mythological database, the motif is found in Central and Southern America only. The Philippines is not listed under C12a, the description of which goes as follows: 'A dog warns people about approaching flood or world conflagration, instructs them what should be done'. The reason is because in the Philippine flood myths, the dog's role is different. Here the divine/ancestral character, responsible for creation or modification of the present-day landscape, is a hunter, accompanied by one or several dogs. Creation of mountains, the prevalent landscape of the Philippine archipelago, is attributed to the hunter's need to be able to hear the hunting dog's barking echoing from the mountain slopes. Inducing a flood is chosen by the hunting gods as a means to change the terrain from plain into mountains. Flood causes the death of all the inhabitants of the earth except for a brother and a sister, who are freezing on top of the mountain. The god then sends a dog and a deer (his helper in the hunt and his prospective prey) to bring fire to the siblings, in order to keep warm the ancestors to-be of present-day mankind.

### *5.2 In a spirit of hunting: two versions of a Bontoc story about Lumáwig*

'The sons of Lumáwig went hunting. In all the world there were no mountains, for the world was flat, and it was impossible to catch the wild pigs and the deer. Then said the elder brother: 'Let us flood the world so that mountains may rise up.' <>

Then the world was flooded. Then said the elder brother: 'Let us go and set a trap.' They used as a trap the head-basket at Mabúd-bodóbud. Then they raised the head-basket and there was much booty: wild pigs and deer and people—for all the people had perished.

There were alive only a brother and sister on Mt. Pókis. Then Lumáwig looked down on Pókis and saw that it was the only place not reached by the water, and that it was the abode of the solitary brother and sister.

Then Lumáwig descended and said: 'Oh, you are here!' And the man said: 'We are here, and here we freeze!'

Then Lumáwig sent his dog and his deer to Kalauwítan to get fire. They swam to Kalauwítan, the dog and the deer, and they got the fire. Lumáwig awaited them. He said: 'How long they are coming!' Then he went to Kalauwítan and said to his dog and the deer: 'Why do you delay in bringing the fire? Get ready!'

Then they went into the middle of the flood, and the fire which they had brought from Kalauwítan was put out! Then said Lumáwig: 'Why do you delay the taking? Again you must bring fire; let me watch you!'

Then they brought fire again, and he observed that that which the deer was carrying was extinguished, and he said: 'That which the dog has yonder will surely also be extinguished.'

Then Lumáwig swam and arrived and quickly took the fire which his dog had brought. He took it back to Pókis and he built a fire and warmed the brother and sister. Then said Lumáwig: 'You must marry, you brother and sister!'' (Seidenadel 1909; Beyer 1912).

A similar variant of the Igorot flood story was published by Mabel Cook Cole in 'Philippine Folk Tales' - adapted, but a reliable and valuable source (Cole 1916: 102–104). These texts contain one more dog-motif, that is, 'The dog and the fire', D4E1 in Berezkin and Duvakin Database, according to which D4E1 is found mostly in Africa and in the Asia-Pacific area, including the Philippines.

### *5.3 'Humans are dog's descendants'*

The last dog-motif among the origin myths that we discuss here is C12B, 'Humans are dog's descendants'. According to Berezkin and Duvakin database, it is to be found mostly in the Americas and in the Southern part of Asia, including Insular Southeast Asia, to which we can now add the Philippines.

The dog's involvement with a group of myths that depicts the vegetative origin of humans (i.e., humans originating from a certain plant) in the Philippine mythology results from a combination of two different motifs: 'Dog mates with a human' and 'First humans come out of bamboo'. Both are also represented separately, the latter being the most widespread version all over the archipelago. For example:

'During a great drought, Mampolompon could grow nothing on his clearing except one bamboo, and during a high wind this was broken. From this bamboo came a dog and a woman, who were the ancestors of the Moro' (Cole 1916: 99–101).

Corresponding origin myths with a dog as an ancestor of mankind, in general or of specific ethnic groups, are to be found in a number of Southeastern and Chinese traditions (Van 1993; Alimov 1994).

Figure 7. Guard/pet dog in a cage. Dumaguete, Negros Oriente, the Philippines, 2006 (Photo by M.V. Stanyukovich).

Figure 8. Korowai man named Bay resting with his dog during the midday heat. Kolof Braza, Asmat area, Papua province, Indonesia, 2015 (Photo by E. Beliakova).

### 6 Broadening the Austronesian context: human origin of a dog

The 'dog motif' in mythological narratives about Maui of Polynesia can be regarded as an inversion of the Philippine one that treats a dog as an ancestor of humans. Maui, the Polynesian trickster demigod, transformed his brother-in-law Irawaru, husband of Hina, into the first dog which was used to explain the human characteristics of dogs (Luomala 1958). Multiple similarities in the stories about Lumáwig, the major mythological hero of the North of the Philippines, and of Maui, the Polynesian trickster demigod, have been noted long ago (Luomala 1949, 1958). A comparative study of the Philippine and Polynesian traditions could be fruitful for defining 'dog motifs' variations in the Austronesian world.

### 7 Dogs as pets in the present-day Philippines

In fashionable parts of Manila one can see servants, walking dogs of the best breeds in the early morning or late evening. Even modest households, mostly urban, nowadays keep dogs as pets, often in the cages inside the living quarters (as traditional pets - monkeys, singing birds, pigeons and fighting roosters used to be kept). However, it is quite a new pattern. Sometimes a dog kept in a cage is regarded both as guard and pet

(Figure 7). Until recently dogs were not perceived as pets in the present-day usual meaning. They were servants, even tools, like a knife or a rope. Most dogs are severely undernourished by their owners, not to mention the *ascals* - street dogs, many of whom end up as the 'stewed umbrellas' mentioned above. Even those lucky enough to be well fed are infested with all kinds of parasites and suffer from skin deceases. The situation with cats is even worse. On the other hand, rabbits that were introduced after the II World War 'have come to be cared for as pets instead of being produced as meat sources' (Veneracion 2017). Filipinos react to rabbit consumption just like most Europeans react to dog-eating. To cite a Facebook post of a Russian, living permanently in the Philippines, 'When I mentioned that I had rabbit for dinner, my Filipino friends looked at me as if I told them that I have eaten my own son'.

**8 Conclusions**

Being one of the three domesticated animals that accompanied Austronesians on their conquering insular Southeast Asia and the Pacific, and perhaps the most important one, the dog has always been of immense value for the Filipinos and all the rest of the Austronesians. Historical linguistics gives us evidence that the dog was perceived as a domesticated animal per se, and traces of that attitude are still living in modern languages. Multiple functions of the dog include help in hunting (most essential), guarding, serving as a sacrificial animal with special properties, providing a source of protein, a source of material used for garments and ornaments, a pet. The aspect considered most important shifted according to the change of life circumstances.

Ancient Austronesians treasured dogs as only they could guarantee successful hunting. The reminiscences of that predominant dog function are still alive. In Ifugao the dog's name, just like that of the human child, was traditionally defined by *mumbaki*, the ritual specialist, in the course of a special ritual. According to the author's informants, 'hunting dogs are named with meanings according to *mumbaki*. During old times with the Tuwali [Ifugao] people call their dogs by the names: 1. *Alawin*- means dog that always follows his master wherever he goes hunting. 2. *Manabong* - dog that does not select any animal in the forest to hunt. 4. *Munduwong* - dog that never gives in until it catches animals in the forest. 5. *Paguyon* - has a strong sense of smell which it uses to follow the animal, wild chicken, squirrel, big lizard and others' (Informant MG, f., Kiangan, Ifugao 2021). Special attitudes towards dogs nowadays can be seen in those societies where hunting is still an important means of subsistence. It has faded but is still present in those ethnic groups where hunting with dogs was abandoned only recently (like in Ifugao). There, bonding between humans and their best friend and companion is the closest, it is stronger than those found in living patterns of the urban well-to-do Westernised families who keep expensive pedigree dogs as pets. All the rest of the Philippine dog population is less fortunate. Hunter-gatherers and early agriculturalists have the closest bonds with dogs, their helpers and companions (Figure 8). Once hunting functions are lost, dog tends to be regarded not as a personality, but rather as a useful tool for alarming and guarding. Most of the personal ties between the dog and the family are lost, but that loss is not to be absolutised. Peasants, farmers throughout the world demonstrate a purely utilitarian attitude towards dogs. That is the fashion. However, there might well be warm interpersonal relations between humans and dogs, concealed by customary rudeness of treatment in public, as sentimental love showed towards a dog is considered to be a childish behaviour.

Archaeological, ethnographic and linguistic data show that dog sacrifice and dog meat consumption have always been present among Austronesians (just as among most, probably all, other cultures in the world). However, it never reached the scale that was documented for American Indians, not to mention the Northern and Arctic hunting and herding groups, where meat has always constituted a much bigger share of nutrition than in tropical areas. In the Pacific, in some areas where hunting, the main dog function, was impossible for the lack of prey, dogs were eaten into extinction; elsewhere in Southeast Asia and beyond, dogs were sacrificed and eaten occasionally. Dog has never constituted a part of the 'regular' Southeast Asian sacrificial pyramid, that includes chicken, pig and buffalo. Dog sacrifices are traditionally a part of rituals aiming to cure a patient, and funeral rites. Such an attitude to dog can be traced through insular Southeast Asia to the continent, and as far as to the extreme North: most cultures of East Asia and the Far East consider dog meat as medicinal, dog fat for treating TB etc., and dog sacrifice as a means to ensure the health of an ill patient and the well-being of the deceased. Philippine and Southeast Asian materials, however, show more emphasis on the other reason why dog is sacrificed for the sick. Here the butchered dog is regarded as a 'secret weapon', sent to attack the family of a person who is believed to have caused the sickness by sorcery.

Mythology reflects the same pivotal points: dog, in its hunting capacity, is regarded to be one of those responsible for creating the present-day landscape, causing the flood, supplying the first humans with fire, and even to be itself the ancestor of all the humanity or of specific ethnic groups.

**Acknowledgements**

I am grateful to my Filipino long-time friends and informants, primarily those from Ifugao, for decades of happy collaboration; to Katerina Beliakova and P. Bion Griffin who generously allowed me to use their field photos and shared their field experiences with me; to Caroline Stone, Galina Sychenko, Sergey Klimenko, Hans Brandeis, Bernard Sellato and Thomas Gibson for fruitful discussions and valuable comments, and for Francesca Lugli for her kind patience. The work was funded by a grant No 20-012-00325 'Traditional material culture realia in the languages and folklore texts of the peoples of Southeast Asia' from the Russian Foundation for Basic Research.

**References**

Alimov, I.A. 1994. Drevnekitayskiye pamiatniki o sobake. (Древнекитайские памятники о собаке. (Ancient Chinese documento litterario about the dog)). *Kunstkamera. Etnograficheskiye tetradi.* (*Кунсткамера. Этнографические тетради (Kunstkamera. Ethnographic natebooks))* 5–6: 257–261.

Afable, P.O. 1995. The peoples of Eduardo Masferre's photographs. *Discovery, the Magazine of the Yale Peabody Museum of Natural History* 25(2): 10–20.

Amano, N., P.J. Piper, H. Hung, and P. Bellwood 2013. Introduced Domestic Animals in the Neolithic and Metal Age of the Philippines: Evidence from Nagsabaran, Northern Luzon. *Journal of Island and Coastal Archaeology* 8: 317–335.

Århem, K. and G. Sprenger (eds) 2016. *Animism in Southeast Asia.* Routledge Contemporary Southeast Asia Series 77.

Barton, R.F. 1946. The Religion of the Ifugaos. *American Anthropologist* 48, no. 4, pt. 2: 1–219.

Berezkin, Yu. 2005 'The Black Dog at the River of Tears': Some Amerindian Representations of the Passage to the Land of the Dead and Their Roots. *Forum for Anthropology and Culture* 2: 130–170.

Berezkin Yu.E. and E.N. Duvakin. World mythology and folklore: thematic classification and areal distribution of motifs. Analytical catalogue , viewed 1 May 2021, http://www.ruthenia.ru/folklore/berezkin/

Beyer, H.O. 1913. Origin Myths Among the Mountain Peoples of the Philippines. *Philippine Journal of Science*: 85–117.

Blust, R. 2002. The History of Faunal Terms in Austronesian Languages. *Oceanic Linguistics* 41(1): 89–139.

Bracciolini, P. 1723. *Historiae de varietate fortunae libri quatuor*. Paris.

Cole, F.-C. 1915. Traditions of the Tinguian: A study in Philippine Folklore. *Field Museum of Natural History publications 180, Anthropological series* 14(1): 1–226.

Cole, F.-C. 1922. *The Tinguian. Social, Religious, and Economic Life of a Philippine Tribe. Field Museum Anthropological Series* 14(2): 1–267.

Cole, M.C. 1916. *Philippine Folk Tales.* Chicago: A. C. McClurg and Company.

Condominas, J. 1957. *Nous avons mangé la forêt de la Pierre-Génie Gôo.* Paris: Mercure.

Conklin, H.C. 1975. *Hanunoo Agriculture. A Report on an Integral System of Shifting Cultivation in the Philippines.* Reprinted by Elliot's Books Northford, Connecticut, 1975. (First publication by Food and Agriculture organization of the United Nations. Rome, 1957).

Davydov, V.N. and V.V. Simonova 2008. Sobachye serdtse: antropologiya sobakoyedeniya v postsovetskoy evenkiyskoy derevne (Собачье сердце: антропология собакоедения в постсоветской эвенкийской деревне (Dog's heart: anthropology og dog-eating in postsoviet Evenki village)). *Vesti laboratorii drevnikh tekhnologiy (Вести лаборатории древних технологий) 1(6)*: 213–230.

Dawang, B.A. 2003. Dog-eating and my culture. *Animal People*, November 1. A Project of Animal People, Inc., 1992–2013, viewed 6 February 2021, <animalpeopleforum.org>.

Demetrio, F.R. and Cordero-Fernando, G. 1991. *The Soul Book*. Quezon City: GCF Books.

Eijkelkamp, W. 2015. Your food is my friend! An anthropological investigation in Yogyakarta's dog meat trade and the animal activism against this practice. Unpublished MA Dissertation, Leiden University.

Fanged, N., J.A. Carvajal, A. Salvador-Amores and D. Tolentino Jr. 2018. Catalogue of Objects in the Feasts of Merit Exhibition, in D. Tolentino Jr. (ed.) *Feasts of Merit: Wealth, Status, and Feasting in the Luzon Cordillera*: 119–198. Baguio City: Museo Kordilyera, University of the Philippines Baguio.

Flora, N. Jessica Hagedorn Looks Back on the Legacy of 'Dogeaters', viewed 1 February 2021, https://www.thenation.com/article/culture/jessica-hagedorn-dogeaters-anniversary-interview/

Gaverza, J.K.M. 2014. The myths of the Philippines. Unpublished BA Dissertation, University of the Philippines, Diliman, Quezon City , viewed 8 February 2021, https://www.academia.edu/36248979/THE_MYTHS_OF_THE_PHILIPPINES_2014

Gibson, T. 2015. *Sacrifice and sharing in the Philippine highlands: religion and society among the Buid of Mindoro*. Quezon City: Ateneo de Manila University Press. (Reprint. Originally published: London: Athlone Press, 1986).

Greig, K., R. Walter and E.A. Matisoo-Smith 2016. Dogs and people in Southeast Asia and the Pacific. *The Routledge Handbook of Bioarchaeology in Southeast Asia and the Pacific Islands*: 462–482. Routledge.

Greig, K., A. Gosling, C. J. Collins, J. Boocock, K. McDonald, D.J. Addison, M.S. Allen, B. David, M.

Gibbs, C. F. W. Higham, F. Liu, I. J. McNiven, S. C. O'Connor,. H. Tsang, R. Walter and E. Matisoo-Smith 2018. *Complex history of dog (Canis familiaris) origins and translocations in the Pacific revealed by ancient mitogenomes.* DOI: https://doi.org/10.1038/s41598-018-27363-8

Griffin, P.B. 1998 An ethnographic view of the pig in selected traditional Southeast Asian societies, in S.M. Nelson (ed.) *Ancestors for the Pigs: Pigs in Prehistory:* 27–37.

Hamilton-Paterson, J. 1994. *Ghosts of Manila.* London: Jonathan Cape.

Heinrich, S. 2017. The 'Savage' Filipino Natives and Their Dog-Eating Habits. *Western Illinois Historical Review* 8: 25–41.

Lien, M.A. 2004. Dogs, whales and kangaroos: Transnational activism and food taboos, in M.A. Lien and B. Nerlich (eds) *The politics of food*: 179–199. Oxford: Berg Publishers.

Hagedorn, J. 1990. *Dogeaters.* NY: Pantheon Books.

Hohulin, R.M. and E.L. Hohulin 2014. *Tuwali Ifugao Dictionary and Grammar Sketch.* Manila: Linguistic Society of the Philippines.

Jenks, A. 1905. *The Bontoc Igorot.* Manila: Bureau of Public Printing.

Kiefer, T.M. 1972. *The Taosug, Violence and Law in a Philippine Muslim Society.* NY, Chicago, Atlanta: Holt, Rinehart and Winston, Inc.

Kohnen, N. and P. Kohnen1986. *Igorot. Traditional Ways of Life and Healing among Philippine Mountain Tribes.* Köln, West Germany: SDK Systemdruck Köln GmbH.

Komissarov, S.A. and M.A. Kudinova 2017. Psy boga groma (eshche raz o znachenii kamennykh sobak Leizhou, Kitai) (Псы бога грома (еще раз о значении каменных собак Лейчжоу, Китай) (The hounds of the god of thunder (once again about the meaning of stone dogs' sculptures in Leizhou, China)). *Vestnik of Novosib. State Univ. Series: History, philology* 16(10): 25–31. Oriental Studies.

Kudinova, M.A. 2014. Ritual'no-mifologicheskaya semantika kamennykh izvayanii sobak v traditsionnoi kul'ture yuga Kitaya (na materialakh poluostrova Leizhou)(Ритуально-мифологическая семантика каменных изваяний собак в традиционной культуре юга Китая (на материалах полуострова Лэйчжоу) (Ritual and mythological semantics of stone sculptures of dogs in traditional culture of South China (based on a study of sculptures of Leizhou Peninsula)). *Vestnik of Novosib. State Univ. Series: History, philology* 13(4): 60–69. Oriental Studies.

Lieban, R.W. 1967. *Cebuano Sorcery.* Berkeley: University of California Press.

Limos, M.A. 2020. Philippine 'Witch Dog' Could Be 36,000 Years Old. It climbs trees and hunts cobras. *Esquire Philippines* 20.05.2020. Viewed 26 March 2021.

Luomala, K. 1949. *Maui-of-a-thousand-tricks: His Oceanic and European Biographers.* Honolulu: Bernice P. Bishop Museum.

Luomala, K. 1958. Polynesian Myths about Maui and the Dog. *Fabula* 2(1): 139–162. Berlin: Walter de Gruyter.

Mahler, X. and B. Denis. 1989. Le chien, animal de boucherie. *Ethnozootechnie* 43: 81–84.

Milliet, J. 1995. Manger du chien? C'est bon pour les sauvages! *L'Homme* 35(136): 75–94.

*Minority Groups in the Republic of Vietnam.* 1966. Cultural Information Analysis Center Headquarters, Department of the Army. American University (Washington, D.C.).

Morrison, J. 1993. Bajau gender: A study of the effects of socio-economic change on gender relations in a fishing community of Sabah, East Malaysia. Unpublished Ph.D. Dissertation. University of Hull.

Oh, M. and J. Jackson 2011. Animal rights vs. cultural rights: Exploring the dog meat debate in South Korea from a world polity perspective. *Journal of Intercultural Studies* 32(1): 31–56.

Parker, L. 1991. The dog-eaters of Bali. *Canberra Anthropology* 14(1): 1–23.

Podberscek, A. L. 2007. Dogs and cats as food in Asia, in M. Bekoff (ed.) *Encyclopedia of human animal relationships: A global exploration of our connections with animals:* 24–34. Westport, CT: Greenwood Press.

Podberscek, A.L. 2009. Good to pet and eat: The keeping and consuming of dogs and cats in South Korea. *Journal of Social Issues* 65(3): 615 632.

Pollock, N.J. 1986. Food classification in three pacific societies: Fiji, Hawaii and Tahiti. *Ethnology* 25(2): 107–117.

Ramos, M. 1953 *Tales of Long Ago in the Philippines.* Manila: Alip & Sons.

Resolute, P. 2016. Humanizing the Non-Human Animal: the Framing Analysis of Dogs' Rights Movement in Indonesia. *Masyarakat: Jurnal Sosiologi* 21(2): 149–172.

Reid, L.A. (ed.) 197I. *Philippine minor languages: Word-lists and phonologies.* Oceanic Linguistics Special Publication No. 8. Honolulu: University of Hawai'i Press.

Roxas-Lim, A. 2017. Marine-oriented Sama-Bajao people and their search for human rights. *Public Policy (Philippines)* 18: 49–66.

Sather, C.A. 1984. Sea and Shore People: Ethnicity and Ethnic Interaction in Southeastern Sabah, in Tan Chee Beng (ed.). *Contributions to Southeast Asian Ethnography* 3: 5–26.

Seidenadel, C.W. 1909. *The First Grammar of the Language Spoken by the Bontoc Igorot, with a Vocabulary and Texts.* Chicago: The Open Court Publishing Co.

Sellato, B. 2019. *The Other Tiger: History, Beliefs, and Rituals in Borneo.* Temasek Working Paper 1: 1–69.

Sprenger, G. 2016a. Dimensions of animism in Southeast Asia, in: K. Århem and G. Sprenger (eds). *Animism in Southeast Asia:* 31–51. London: Routledge.

Sprenger, G. 2016b. Graded personhood Human and non-human actors in the Southeast Asian uplands, in: K. Århem and G. Sprenger (eds). *Animism in Southeast Asia:* 73–90. London: Routledge.

Stanyukovich, M.V. 2011. Epos i pamyat' zhivikh i mertvykh (Эпос и память живых и мертвых (Epics and the memory of the living and the dead)). T.G. Ivanova, (ed.). *Klassicheskiy folklor segodnya: Materialy konferentsii, posvyaschchennoy 90-letiyu B.N. Putilova (Классический фольклор сегодня: Материалы конференции, посвященной 90-летию со дня рождения Б.Н. Путилова (Classic folklore today. Proceedings of the conference dedicated to the 90th anniversary of B.N. Putilov)]*: 472–491. St Petersburg: Dmitri Bulanin publishing house.

Stanyukovich, M.V. 2013. Epic as a Means to Control Memory and Emotions of Gods and Humans: Ritual Implications of Hudhud Among the Yattuka and Tuwali Ifugao, in: N. Revel (ed.) Songs of Memory in Islands of Southeast Asia: 167–197. Newcastle-upon-Tyne: Cambridge Scholars Publishing.

Stanyukovich, M.V. 2014. Betel, 'Lonely Heroes' and Magic Birth in the Philippines and Beyond: Comparative Mythology, Field Work and Folklore Corpora, in K. Antoni and D. Weiss (eds) *Sources of Mythology: Ancient and Contemporary Myths. Proceedings of the Seventh Annual International Conference on Comparative Mythology (15–17 May 2013, Tuebingen)*: 179–206. LIT, Zurich-Berlin.

Stanyukovich, M.V. 2016. Vystel v moloko (Выстрел в молоко (Disgust and Milk of Kindness) A Review of Valerie Curtis, Don't Look, Don't Touch: The Science behind Revulsion. Oxford: Oxford University Press, 2013, 165 pp. *Antropologicheskiy Forum (Антропологический Форум Forum for anthropology)* 29: 247–268.

Stanyukovich, M.V. 2018. The 'Swine turn': A review of Jon Henrik Ziegler Remme. Pigs and Persons in the Philippines. Human-Animal Entanglements in Ifugao Rituals. Lanham: Lexington, 2014. 161 p. *Forum for anthropology and culture* 14: 252–269.

Stanyukovich M.V. 2020. Plyaski plodorodiia. Tserkov' Presviatoy Devy rybolovnoy seti, tantsuyushchego Sv. Paskhaliya i Sv. Klary Assizskoy na Filippinakh i yeyo yazycheskoye naslediye. (Пляски плодородия. Церковь Пресвятой Девы рыболовной сети, Танцующего св. Пасхалия и св. Клары Ассизской на Филиппинах и ее языческое наследие (Fertility dance. The church of Our Lady of a Fishnet, of Dancing St. Pascual and of St. Clara of Assisi (the Philippines) and its ancient pagan heritage'. *Vestnik RGGU (Вестник РГГУ. Серия: Литературоведение. Языкознание. Культурология (RGGU Bulletin. Literary Theory. Linguistics. Cultural Studies Series))* 5: 112–139.

Startsev A.F. 2017. *Etnicheskiye predstavleniya tunguso-manchzhurov o prirode i obshchestve (Этнические представления тунгусо-манчжур о природе и обществе (Ethnic views of the Manchu-Tungus on the nature and society))*. Vladivostok, Dal'nauka.

Van, D. Ng. 1993. The Flood Myth and the Origin of Ethnic Groups in Southeast Asia. *The Journal of American Folklore* 106 (421): 304–337. American Folklore Society.

Weichart, G. 2004. Minahasa identity: A culinary practice. *Antropologi Indonesia 28(74)*: 55 74.

Vanoverbergh, M. 1929. *Dress and adornment in the Mountain Province of Luzon, Philippine Islands.* Washington: Catholic Anthropological Conference.

Veneracion, A.M. An Overview of the Rabbit Industry in the Philippines. Monthly agriculture. November 21, 2017, viewed 7 March 2021, https://www.agriculture.com.ph/2017/11/21/an-overview-of-the-rabbit-industry-in-the-philippines/?fbclid=IwAR3VYl6Tc8CNlLOXgyS2NKzZcF3YqGq0aVtMpsuvlo5FVKx9mZuLyELPShk.

Villaverde, J. 1911 (1894) Supersticiones de los Igorrotes Ifugaos. *El Correo Sino-Annamita* 38: 281–455.

Vitales, T.J. 2018. A howl from the grave: Osteological analysis of 12th- to 15th-century dogs from Santa Ana, Manila, Philippines. *International Journal of Osteology* 28(2): 170–178.

Vozchikov, D.V. 2019. Amok, popugai i volshebnoye derevo: Mir Nusantary v opisanii venetsianskogo kuptsa 15 v. (Амок, попугаи и волшебное дерево: Мир Нусантары в описании венецианского купца XV в. (Amok, Parrots and a magic tree: the world of Nusantara as described by a 15th century Venetian merchant). *Etnografia* 1(3): 73–98.

Williams, C.L., S.M. Mazzola G. Curone and G.Q. Pastorino 2018. What We Have Lost: Domestic Dogs of the Ancient South Pacific. *Annual Research and Review in Biology* 25(2): 1–11.

Yoon, H. 2016. Disappearing Bitches: Canine Affect and Postcolonial Bioethics Configurations. *Johns Hopkins University Press and the Society for Literature, Science, and the Arts* 24: 351–374.

# 6.5 Demonic Dogs of Mongolian Stag Stones and their Chinese Counterparts

Andrey V. Varenov

Novosibirsk State University, Pirogov st. 1, 630090, Novosibirsk, Russia. avvarenov@mail.ru

**Abstract**

The article analyses images of dog-like or tiger-like (leopard-like) animals, depicted on the surface of stag stones of Mongolian-Trans-Baikal style. According to D.G. Savinov they represent 'chthonic predators'. The victims of chthonic predators' attacks vary greatly (from a human to a fish) and are met in various contexts: stag-stone No. 15 in Ushkijn-Uver (Khövsgöl aimag, Mongolia); rock carving in Suyukou gorge of Helan Mountains (Ningxia-Hui Autonomous Region, PRC), a bronze mirror found in the No. 1612 burial of Guo state cemetery in Shangcunling (Henan province, PRC); a Neolithic painted pottery jar from the Dadiwan site (Gansu province, PRC). The author argues that all these scenes depict trials of a human soul in the Underworld.

**Keywords:** China, stag stones, rock carvings, chthonic predators, semantics.

## 1 Introduction: stag stones

Stag stones (or deer stones) are vertical steles, representing highly stylised sculptures of warriors.

On their surfaces earrings, necklaces and belts with weapons (swords or daggers, shields, knives, battle axes, bows with arrows or quivers) hanging from it are carved. The face is represented by two or three oblique strokes (lines). Realistic face features in stag stones are very rare. Many steles are all covered with deer images which is how they gained their name.[1]

According to the differences in how the deer are depicted, all stag stones are divided into three types (styles):

I. Mongolian-Trans-Baikal (deer are depicted with very long bent bodies, bent legs and long beak-shaped muzzles) (Figure 1, 1–2);
II. Sayano-Altai (deer are standing on straight stretched legs - on so called 'tip-toes position') (Figure 1, 3);
III. All-Eurasian (without figures of deer or other animals) (Figure 1, 4–7) (Novgorodova 1989: 185).

Stag stones are spread mainly in Central and Western Mongolia and in adjacent regions of Russia (Altai, Tuva, Trans-Baikal), Kazakhstan and China (Xinjang-Uyghur autonomous region) in the first half of the I millennium BC (Volkov 2002: 14). Stag stones of Mongolian-Trans-Baikal style are dated to the X-VII centuries BC (Savinov 1994: 110–113).

[1] The paper is prepared with the support of Russian Foundation for Basic Research (RFBR), project 18-09-00557: 'The Study of Rock Art Monuments in the Archaeology of China (Periods of Antiquity and Middle Ages)'.

## 2 Uushkijn-Uver stag stone site

One of the best known sites with stag stones of Mongolian-Trans-Baikal style is Uushkijn-Uver in Khövsgöl aimag (province) near Mörön city in Northern Mongolia. 15 beautiful stag-stones were studied there in 1970 by V.V. Volkov and E. A. Novgorodova (Figure 1, 1).

On the upper part of the stone No. 15 a scene, depicting two beasts of prey devouring a horse is engraved (Figure 2, 3). V.V. Volkov and E.A. Novgorodova (1975: 81) described them as feline predators, probably due to the spots, decorating their skins.

On the stag-stone No. 4 from the same site one can see a pack of five such beasts chasing a horse. There is no picture of the stag stone No. 4 in the article by V.V. Volkov and E.A. Novgorodova (1975).

In the book about the Mongolian stag stones by V.V. Volkov (2002) the pictures of predators from the stag stone No 4 are too small to see any details (Figure 3, 1). E.A. Novgorodova's pictures of predators from the stag stone No. 4 published in 1980 and in 1984 differ from each other and from V.V. Volkov's variant (Figure 3, 2, 3).

## 3 Species of hunting beasts

The photos of the stag stone No 4 made by the author in 2013 (Figure 4, 1) provide more details of this hunting scene (Figure 4, 2).

V.V. Volkov and E.A. Novgorodova called the beasts running after the horse 'spotty predators, most likely snow leopards'. They wrote: 'Obviously, it is a pack of snow leopards, represented in the moment of hunting a horse, running before them' (Volkov and Novgorodova

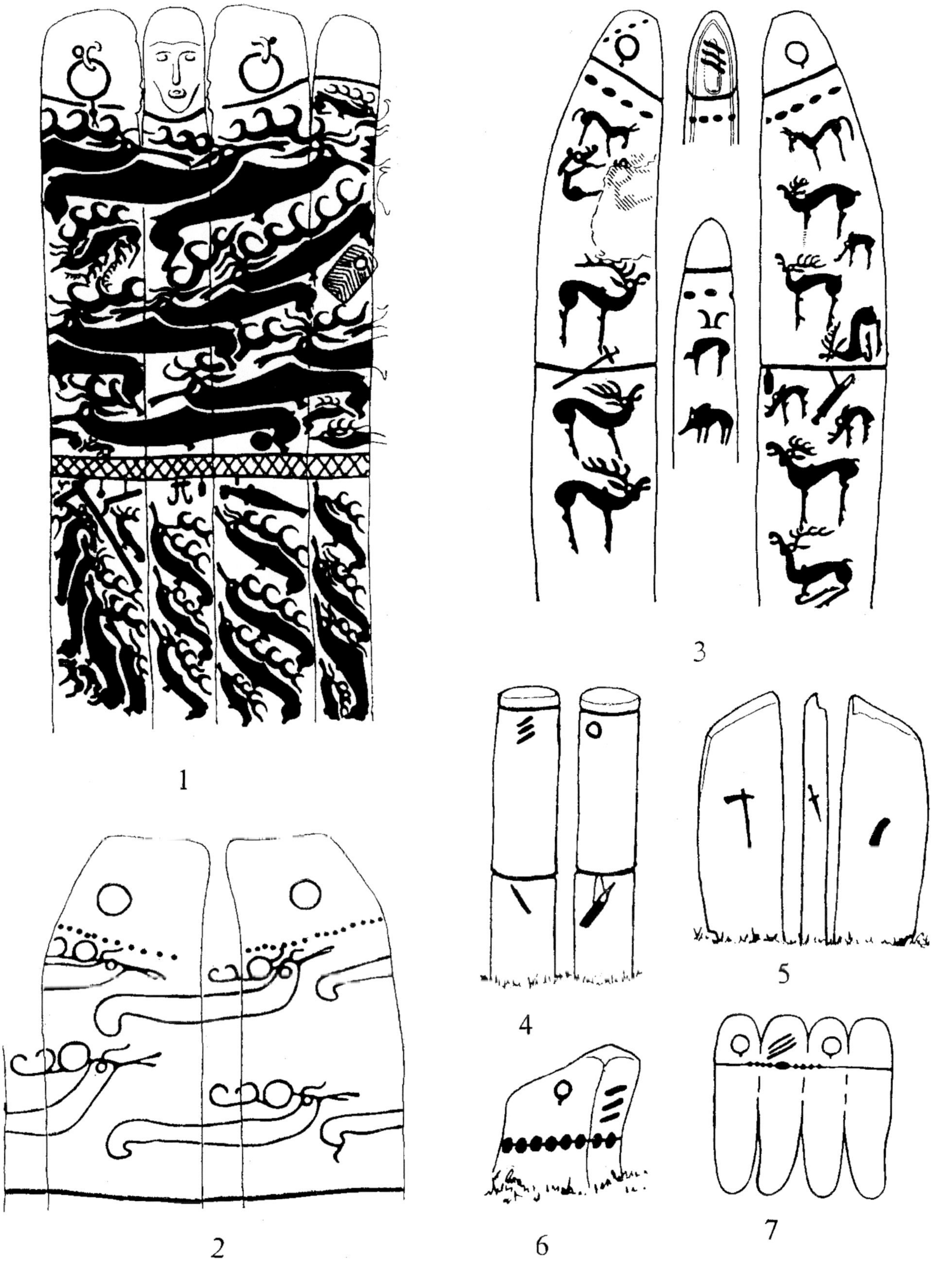


Figure 1. Stag-stones of three types: 1–2, stones of I (Mongolian-Transbaikal) type (1 - stag-stone No. 14 from Ushkijn-Uver); 3, stone of II (Sayano-Altai) type; 4–7, stones of III (All-Eurasian) type. Not to scale (After Gryaznov 1984: 77, 78).

Figure 2. Scenes with two chthonic predators: 1, 2, from Suyukou (Helan county); 3, from Ushkijn-Uver (Khövsgöl aimag, Mongolia); 4, from Shangcunling (Henan province); 5, from Dadiwan (Qingan county, Gansu province); 6, from Shizhaishan (Yunnan province). All figures are of different scales. After 1, 2, Gai Shanlin, Gai Zhihao 2002: 408; 3, Nowgorodowa 1980: 178; 4, Zhongguo tongqi quanji 2005: 4; 5, drawing after photo made by the author in Gansu provincial museum; 6, Pirazzoli Michele 1990: 81).

1975: 82). V.V. Volkov and E.A. Novgorodova continued to write about 'snow leopards' and 'feline predators' even much later, in their individual monographs (Novgorodova 1989: 206; Volkov 2002: 79, 83).

However, all real feline predators (except cheetahs) never chase their prey or (except lions) hunt in groups. They prefer individual ambushes. The chase of prey in packs is specific for dogs and dog-like creatures.

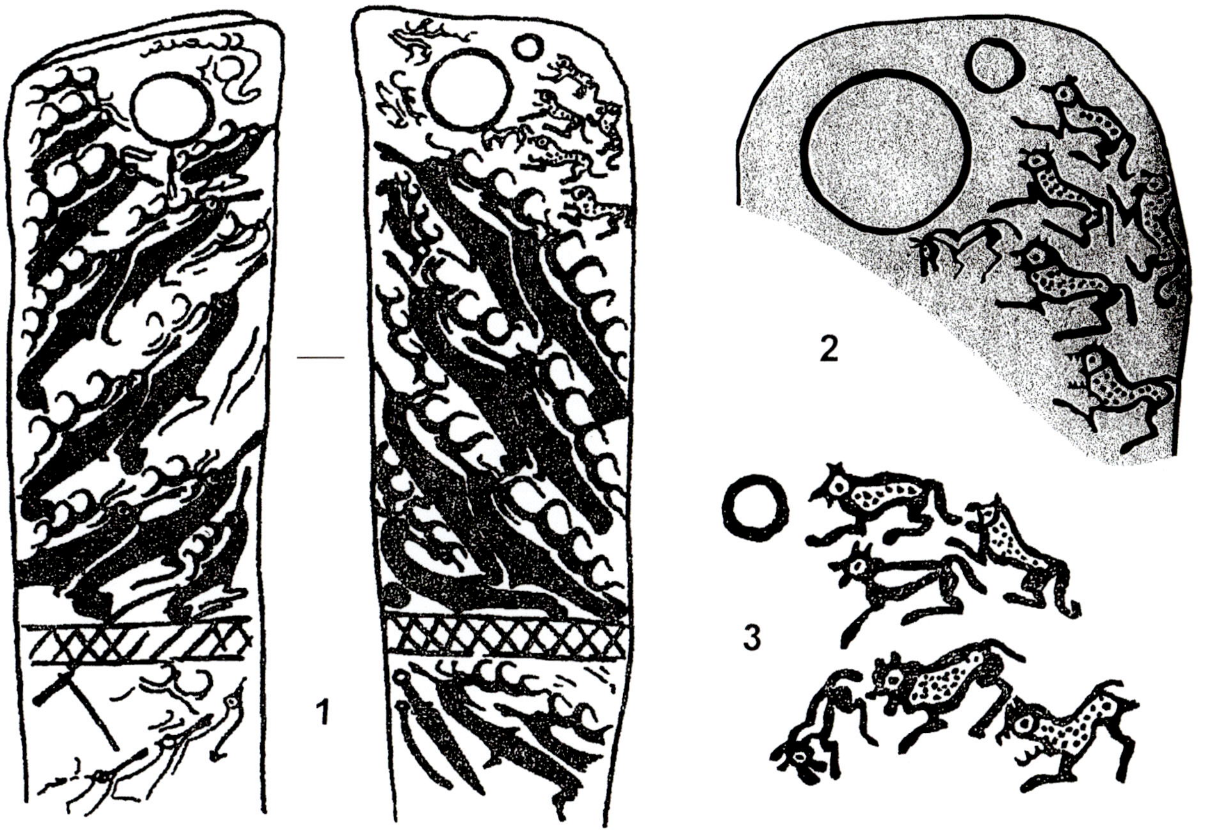


Figure 3. Stag-stone No. 4 from Ushkijn-Uver (Khövsgöl aimag, Mongolia): 1, general view; 2, 3, detail with predators chasing a horse (After 1, Volkov 2002: 189; 2, Nowgorodowa 1980: 134; 3, Novgorodova 1984: 99).

Figure 4. Stag-stone No. 4 from Ushkijn-Uver (Khövsgöl aimag, Mongolia): 1, general view; 2, detail with predators chasing a horse (Photo by A.V. Varenov 2013).

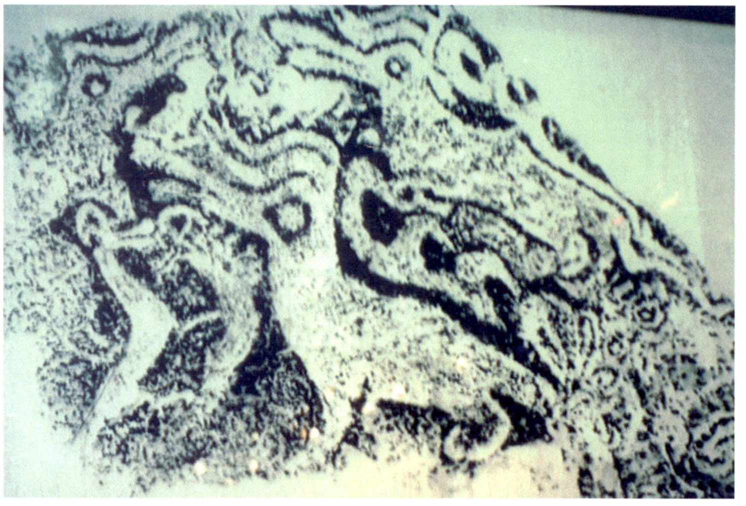

Figure 5. Scene with two chthonic predators: the rubbing from Suyukou gorge (Helan county), photo by A.V. Varenov taken in the Ningxia Provincial Museum.

Figure. 6. Scenes with two chthonic predators: 1, from Shangcunling (Henan province); 2, from Dadiwan (Qingan county, Gansu province); 3, from Shizhaishan (Yunnan province). All figures are of different scales. After 1, Zhongguo tongqi quanji 2005: 6; 2, from the Gansu Provincial Museum (Photo by A.V. Varenov 2012; 3 Zhongguo tongqi quanji 1993: 88).

According to D.G. Savinov (1994: 132) who analysed the pictures of predators on the surface of stag stones, 'especially noteworthy are the images of fantastic animals which combine the signs of different beasts of prey, first of all, feline predators and wolves (or dogs), being highly likely personages of chthonic origin'.

### 4 Suyukou gorge analogies

Compositionally much the same scene as on the stag-stone No. 15 from Uushkijn-Uver is depicted in Suyukou gorge of Helan Mountains in Ningxia-Hui Autonomous Region of the PRC (Gai Shanlin and Gai Zhihao 2002: 408).

It shows a human figure, being attacked by two dog-like beasts (Figure 2, 1). The species of beast is not very clear, but the whole scene is surrounded, like in Uushkijn-Uver, with figures of deer with beak-shaped muzzles (Figure 2, 2).

The author had no opportunity to visit Suyukou gorge and there are no photos of the scene with two dog-like beasts surrounded with figures of deer with beak-shaped muzzles in books and albums on Helan Mountains' rock art, published in China.

However, the photos of the rubbing of the rock-art scene with two dog-like beasts and a human figure amidst them, taken by the author at the Ningxia Province Historical Museum in 2000 and again in 2019 prove that it did really exist (Figure 5).

### 5 Dating by the Shangcunling mirror

Compositionally much the same scene is cast on the back surface of a bronze mirror, found in tomb M1612 of Shangcunling cemetery in Henan province of China (Figure 2, 4) (Shangcunling 1959: 27).

It depicts two beasts ready to devour (or fight for) a deer (Figure 6, *1*). These beasts 'look more like tigers or leopards', than dogs (Guo Moruo 1959: 14).

The mirror from Shangcunling gives a more exact date to Uushkijn-Uver and Suyukou compositions. Shangcunling was the cemetery of Guo state, which in 655 BC had been annexed by another state called Jin.

Not a single grave from Shangcunling could be later, than the middle of the VII century BC. The majority of its graves belong to the same period - second half of the IX - the first half of the VIII centuries BC (Komissarov 1985: 9–10).

### 6 Neolithic roots of stag stones

However, the roots of the tradition to depict two dog-like beasts, ready to tear apart their victim lie in China much earlier than the Zhou period.

The origin of the iconographical tradition to depict deer with very long bent bodies, specific for Mongolian-Trans-Baikal stag stones, can be traced back to *zun* pottery vessels of the Zhaobaogou Neolithic culture dating to the VI-V millennia BC, decorated with dragons with deer heads (Varenov 2016: 198–206).

The Majiayao culture anthropomorphic painted pottery vessels, namely *ping* bottles with necks representing bas-relief human faces, have bodies ornamented with three or four horizontal registers filled with oblique geometric patterns (Figure 7) that resemble the oblique positions of deer on the surface of deer-stones of Mongolian-Trans-Baikal style (Figure 8) (Varenov 2016: 207–208).

The picture on a Neolithic painted pottery jar, found at the Dadiwan site of the same Majiayao culture (dating

Figure 7. Majiayao culture anthropomorphic painted pottery vessel: The ping bottle with the neck representing bas-relief human face from the Dadiwan site (Photo by A.V. Varenov taken in the Gansu Provincial Museum 2019).

Figure 8. Stag-stone No 14 from Ushkijn-Uver, Khövsgöl aimag, Mongolia (Photo by A.V. Varenov 2013).

to the second half of IV - beginning of III millennia BC) in Gansu province shows two dogs, ready to start a fight over a fish, which is lying in between them (Figure 2, 5).

## 7 Semantics of Neolithic painted pottery vessels

It is generally accepted that ornamental belts on Neolithic pottery of China represented different levels of the Universe.

The belt with dogs and fish on the jar from Dadiwan is marked on the top with two straight lines, so it may represent the Underworld (Figure 6, 2) (Kudinova 2016: 36). According to V.V. Evsukov, who studied ancient Chinese mythology on the materials of Neolithic painted pottery, the image of a fish symbolised in this tradition a human soul (Evsyukov 1988: 84–85, 89–90). Therefore, the scene on the Dadiwan jar depictured its trials in the Underworld.

The ancient Chinese iconographical tradition to depict the trials of a fish-looking human soul survived until at least the IV-III centuries BC. A fish lying between two pangolins could be seen on the heel of a tubular socket of a battle axe of the Dian culture from grave 12 of the Shizhaishan cemetery (Figure 2, 6) (Pirazzoli 1990: 81). D.G. Savinov has also already noted the Shizhaishan and Ordos bronzes among eastern analogies to fantastic beasts from the Mongolian-Trans-Baikal stag stones (Figure 6, 3) (Savinov 1994: 132).

## 8 Conclusions. Different images of a human soul

Rock carvings in Suyukou and Uushkijn-Uver, as well as the Shangcunling mirror, could have the same meaning, as the picture on the Chinese Neolithic painted pottery. They showed the trials of a human soul in the Underworld.

However, a human soul in iconographical traditions of these cultures was presented differently. The creators of stag stones presented souls of their relatives like the horse images. For the owners of the Shangcunling mirror their souls were presented by another herbivore and at Helan Mountains the soul was presented by a small human figure.

## References


Evsyukov, V.V. 1988. *Mythology of China's Neolithic Age. After the Patterns of Yangshao Culture's Ceramics. Mifologiya kitaiskogo neolita. Po materialam rospisei na keramike kultury Yanshao.* Novosibirsk: Nauka. Евсюков, В.В. 1988. Мифология китайского неолита. По материалам росписей на керамике культуры Яншао.

Gai Shanlin, Gai Zhihao 2002. *Deciphering the Culture of Petroglyphs of Inner Mongolia.* Beijing: Beijing Tushuguan. *Nei Menggu yanhua de wenhua jiedu.* 盖山林，盖志浩。内蒙古岩画的文化解读。

Gryaznov, M.P. 1984. On the monumental art at the dawn of Scytho-Siberian cultures in steppe Asia. *Archaeological Bulletin of the State Hermitage Museum.* O monumentalnom iskusstve na zare skifo-sibirskih kultur v stepnoi Azii. *Arkheologicheskii sbornik Gosudarstvennogo Ermitazha* 25: 76–82. Грязнов,

М.П. 1984. О монументальном искусстве на заре скифо-сибирских культур в степной Азии. Археологический сборник Государственного Эрмитажа 25: 76–82.

Guo Moruo, 1959. Two-three Notes on the Bronzes, Found in Sanmenxia. Sanmenxia chutu tongqi ersan shi. 郭沫若。三门峡出土铜器二三事。*Wenwu* 1: 13–15.

Komissarov, S.A. 1985. Shangcunling - the Basic Site of the End of the Western Zhou Shangcunling - opornyi pamyatnik kontsa Zapadnogo Chzhou. *Dalnij Vostok i Tsentralnaya Asiya* (*Far East and Central Asia*): 3–12. Moscow: Nauka. Комиссаров С.А. Шанцуньлин - опорный памятник конца Западного Чжоу. Дальний Восток и Центральная Азия.

Kudinova, M.A. 2016. Image of a Dog in Neolithic Art of China. Obraz sobaki v neoliticheskom iskusstve Kitaya. *Vestnik of Novosibirsk State University. Series: History, Philology* 15(4), Oriental Studies: 33–41. Кудинова М.А. Образ собаки в неолитическом искусстве Китая. Вестник Новосибирского государственного университтета.

Nowgorodowa, E. 1980. *Alte Kunst der Mongolei*. Leipzig: E.A. Seemann Verlag.

Novgorodova, E.A. 1989. *Ancient Mongolia* (*Some Problems of Chronology and Ethno Cultural History*) *Drevnyaya Mongoliya* (*Nekotorye problemy hronologii i etnokulturnoi istorii*). Moscow: Nauka. Новгородова Э. А. Древняя Монголия (Некоторые проблемы хронологии и этнокультурной истории).

Pirazzoli, M. 1990. The Dian Culture: a Problem of Chronology. Dian wenhua de niandai wenti. 皮拉左里·米歇尔。滇文化的年代问题。*Kaogu* 1: 78–86.

Savinov, D.G. 1994. *Stag Stones in the Culture of Eurasian Nomads. Olennye kamni v kulture kochevnikov Evrasii.* Saint-Petersburg: Saint-Petersburg State University. Савинов Д.Г. 1994. Оленные камни в культуре кочевников Евразии.

*Shangcunling Guoguo mudi*. 1959. *The Cemetery of the State of Kuo at Shang Ts'un Ling*上村嶺虢國墓地。Peking: Science Press.

Varenov, A.V. 2016. On the Chinese Neolithic Roots of Stag Stones of Mongolian-Trans-Baikal Style and their Social and Semantic Interpretation. O kitaiskih neoliticheskih kornyah olennyh kamnei mongolo-zabaikalskogo stilya i ih sotsialno-semanticheskoi interpretatsii. *Stratum plus. Archaeology and Cultural Anthropology*. 2: 195–211. Варенов А. В. 2016. О китайских неолитических корнях оленных камней монголо-забайкальского стиля и их социально-семантической интерпретации. Археология и культурная антропология. *Stratum plus. Archaeology and Cultural Anthropology* 2: 195–211.

Volkov, V.V. 2002. *Stone Stelaes from Mongolia* ('*deer stones*'). *Olennye kamni Mongolii.* Moscow: Nauchnyi Mir. Волков В.В. Оленные камни Монголии.

Volkov, V.V. and E.A. Novgorodova 1975. Stag Stones of Uushkijn-Uver (Mongolia) Olennye kamni Ushkijn-Uvera (Mongoliya). *Prehistoric Archaeology of Siberia*: 78–84. Leningrad: Nauka. Волков В.В. и Э. А. Новгородова Оленные камни Ушкийн-Увэра (Монголия). Первобытная археология Сибири. *Pervobytnaya arheologiya Sibiri.*

*Zhongguo tongqi quanji*. 1993. *The Complete Collection of Chinese Bronzes.* 中国铜器全集。第14卷。滇昆明。Vol. 14. *Dian - Kunming*. Beijing: Wenwu publishers.

*Zhongguo tongqi quanji*. 2005. *The Complete Collection of Chinese Bronzes* 中国铜器全集。第16卷。銅鏡。Vol. 16. *Bronze Mirrors*. Beijing: Wenwu publishers.

# 6.6 A Few Days with Mongolian Dogs and their Herders

Graziano Capitini and Francesca Lugli

Associazione Italiana di Etnoarcheologia, Via Principe Umberto 41, 00185 Roma, Italy.
luglifrance@gmail.com grazcapitini@gmail.com
Corresponding author: Francesca Lugli, luglifrance@gmail.com

**Abstract**

The Italian Association for Ethnoarchaeology, with the sponsorship of the Italian Ministry of Ministry of Foreign Affairs and International Cooperation – Italy MFA began the project 'Camps of Mongolian nomads - an Ethnoarchaeological perspective' in 2002. Life at winter camps became the main goal of the mission in 2007 and since 2009 the missions were held during cold months in order to observe and document the camps during their use. The presence of dogs is always expected and assumed in the camps as they are considered crucial and indispensable. Families usually have 2–3 dogs no matter how many other animals they have. During the period of staying with nomads (more than three hundred days), the life of dogs was observed and it has been a special part of the research since 2011. In the article, a few days with nomads are described.

**Keywords**: dogs, nomads, pastoralism, methodology, Mongolia.

## 1 Introduction

Dogs can be noticed to always be present in all the ethnographic pictures and videos of Mongolian nomads' camps. To travel throughout the steppe means to meet dogs and avoid being bitten and many tourists and scholars have lived this difficult experience. The dogs' main task is to guard against wolves, other predators and thieves. They are indispensable and their importance was supposed to be one of the crucial points that allowed or at least helped the spread and the success of pastoralism in ancient times (Lugli 2016).[1] They are not 'shepherd dogs' and do not help the herders to control and to manage the livestock. However, they can go with the herders as companions during the day.

The Italian Association for Ethnoarchaeology with the sponsorship of the Ministry of Foreign Affairs and International Cooperation – Italy MFA began the project 'Camps of Mongolian nomads - an Ethnoarchaeological perspective' in 2002.[2] The project's purpose was the study and documentation of nomadic camps, with a particular focus on ethnographic and archaeological problems which can contribute to a better understanding of the success of steppe nomadism. Over the years, the research considered various crucial points of steppe pastoralism such as the tent (*ger*), camps, livestock, roaming, pasture, water and salt sources, food production and consumption, and climate change.

Since 2007 the research has been especially focused on winter life and the missions were prevalently conducted during the cold months in Bulgan, Arkanghai and Dundgovi regions (*aimag*).[3] The nomads can change summer and autumn camps every year, but in winter they always use the same camp where they stay put throughout the winter. To observe and to document nomads' camps and their lives is important in order to obtain useful models for the past, for the archaeology of pastoralism and also for the future. In fact, Mongolian pastoralism is quickly changing and is a threatened culture which is therefore urgent to study and document.

During the research of the cold months, the importance of dogs became apparent. In fact, 100% of interviewed nomads said that their lives would be impossible without dogs. Consequently, since 2011 dogs have become part of the research and are considered one of the crucial points of steppe nomadism. The life of dogs was essentially observed and documented during the harsh seasons. (Lugli 2014: 19–21; Lugli 2016: 125–139).

Dogs are part of the life in the camp and nomads' families usually have 2–3 dogs, no matter how many other animals they have. The presence of dogs is always expected and assumed. Anyone approaching a nomads' camp says '*nohoi horioo*'' (нохой хорио) 'Hold your dog!'.

[1] The article 'Mongolian Nomads and their Dogs' (Lugli 2016) concerns the relationships between nomads and dogs, the traditions, the life and the death of dogs and their indispensable role and importance for the success of steppe pastoralism. The article also concerns their presence in Mongolian camps in an ethnoarchaeological perspective and analyses the traces that they can leave in an archaeological point of view (Fiore 2016; Vidale 2016).

[2] The project was conducted with the collaboration of prof. Tserenkhand (Mongolian Academy of Sciences) (2002; 2005–2006) and with Prof. Dulam Sendenjav (National University of Ulaanbaatar; Mongolian University of Arts and Culture of Ulaanbaatar) since 2007.

[3] From 2002 to 2006 the missions were conducted in summer and autumn and in 2008 in late spring and various regions were visited in order to obtain a wide view and understanding of the different Mongolian ecosystems as well as its traditions.

Figure 1. Map of Mongolia with the localisation of Mogod district.

These words are also a greeting to a family and its dogs, a sort of idiomatic phrase which confirms the constant presence of dogs in the camps. Even if it has currently lost its original meaning and it can now also be used in situations where there are no dogs. The sentence implies 'is anybody in?' and it can certainly be considered proof of the constant presence and importance of dogs in the camps (Bamana 2014: 3).

The main task of Mongolian dogs is essentially to guard against wolves - which are a constant danger for nomads - and also against other predators and thieves.

Dogs bark to warn the herders that something dangerous or unusual is occurring. This is their main task. Mongolian dogs usually do not help the herders in managing the animals but can be a companion for herders during the day.[4]

The dog is considered part of the family. It has a name[5], it has its own place close to the tent and has a specific funeral rite when it dies (it is usually placed in a protected place, its tail is cut off and placed under its head, and fat, milk or butter is put into its mouth) and the herder often pronounces the same words which are used for the death of human beings.

It is also peculiar that Mongolians and dogs share a mythological kinship. In fact, a wolf (male) and a doe (female) are the mythological ancestors of Mongolians and a wolf is also the dog's ancestor (Bamana 2014: 8).

Dogs can reincarnate as humans and humans as dogs (Bianquis *et al.* 2013: 303–321; Bamana: 2014; Lugli and Capitini 2018: 63). This is another clue to the strong relationship which exists between Mongolian nomads and their dogs as well as their crucial position in Mongolian nomadism.

The observation, analysis and documentation of the life of dogs in nomads' camps, can substantially contribute to identifying the role that dogs have played in the nomadism of central Asia since the dawn of time.

In this article, we present some fragments of our experience with Mongolian dogs, a summary of the interviews and the observations we made about the dogs' 'everyday life' in the Mogod area where we have returned every year in winter since 2009.

We will give an example of the information that we were able to obtain in a very short period (1st–3rd December 2012) at the beginning of our 2012 mission when we arrived in the area of Mogod[6] - out of more than three

[4] Mongolian dogs are not currently 'herding dogs' or 'sheep dogs' and are not in the shepherd dogs breed list.

[5] Nomads' dogs are always males and always have a name. Females are essentially in the village and they do not always have a name. But females with a name have been documented. For example, once a puppy followed us in the village of Mogod in a very cold December. We were told that nobody wanted the puppy because nobody wants females. We asked many people and somebody told us that a woman, the owner of the *guanz* (canteen), loved dogs and that she also loved to have females. So, we went to her house asking her to take the puppy. She accepted and asked us what name we wanted to give the puppy. We said 'Bella' and she approved with enthusiasm. Every year when we came back to Mogod we went to meet Bella and her master. Her owner said that she thinks that Bella had a very happy life, she was free but was loved and had many splendid puppies.

[6] The mission in 2012 was from the middle of November to the middle of December.

hundred days we spent with nomads (Figure 1). Not because anything particular happened during those days, but as an example[7] of our routine and the validity of an ethnographic and ethnoarchaeological perspective in the study of dogs. Everyday with nomads is special and every two minutes it is possible to understand or observe something that proves or dismantles the models previously elaborated.[8]

## 2 Material and methods

We considered dogs to be as one of the crucial points for the study of the pastoralism of Central-Asian steppes. Therefore, we tried to analyse and document their importance, their role, the relationships that they have with their owners, their everyday life and the traditions that set their presence.

Our research was mainly fieldwork that took into account the ethnographic documentation from an anthropological and ethnoarchaeological point of view. So, the methodology combines the strategies that are typical of these three disciplines; graphic and photographic documentation and videos of the life in the camps were done as well as interviews.

The life of the dogs, therefore, was documented in this light. We tried to study the space they use in the camps, the traces that they can leave on the ground from an archaeological perspective and the relationships that they have with their masters.

Spending long periods with nomads allowed us to observe the relationship that nomads have with their dogs and vice versa in everyday life. We could easily make friends with Mongolian dogs because we have always had dogs and we know their behaviour. We also gave them some food to curry favour for their goodwill.[9] Nomads do not usually like strangers to give food to their dogs but in our case, they have always kindly allowed us to do so. They understood that to do our research dogs could not be aggressive towards us.

Once Bruno Marcolongo - who was director of a geoarchaeological mission in Ovorkhangai together with his wife Giovanna Fuggetta Marcolongo - asked me how we could study the dogs and why they loved us so much. I explained that we had the right attitude to be accepted and that we complemented this attitude with lavish eating gifts. Bruno decided that from that moment he too would keep food for the dogs, not because he was particularly interested in their friendship, but to avoid being bitten by them. He concluded that it was one of the best suggestions he had had for Mongolian missions.

A fundamental part of the research is represented by the interviews we spent a lot of time on, especially in the winter months.[10] In fact, during the cold months, nomads are less busy than during the rest of the year, they spend a lot of time inside the tent and are more willing and happy to talk about their lives. The interviews can take several hours. They are often done with the same person several times and are also repeated year after year. It is important to record events and any changes that occurred during the year.

The interviews include a first general part dedicated to information on the family and its members. The various points that we have identified as crucial are then investigated, namely:

- The camp.
  The choice of the place, its use over the years, the management of the spaces, the fixed structures (in the case of cold months camps), the relationship with the neighbouring camps;
- The *ger* (tent).
  The structure, internal use of the spaces by the various family members and guests;
- The cattle.
  The management in the various months, the production of milk and derivatives as well as the slaughter of the animals and the meat consumed, the relationship that the nomads have with the different animals;
- The seasonal roaming;
- Water sources and salt;
- Strategies to face macro and micro climate changes;
- Dogs.

Every year, interviews were mainly dedicated to a specific different point but dogs have always been an important theme of the conversation.

Since 2011, part of the research was focused on dogs and numerous interviews were done with particular attention to this topic.[11] A series of questions were then developed focusing attention mainly on that. Since 2013 the interview template has been used for the mission 'Siberian Nomads and their Dogs' which was promoted by the Italian Association of Ethnoarchaeology with the sponsorship of MAECI (see the article by Lugli and Sychenko in the volume).

[7] At the beginning of our research on dogs.
[8] In this article, we will only give a few hints at the issues involved in the other 'crucial' points we investigated. For reasons of space and contextualisation of the topic.
[9] Nomad's dogs sometimes accept dry food but it is better to give meat to be certain that they will appreciate it.
[10] More than 200 complete interviews were done during the missions (2002–2019).
[11] The results of the research on dogs were firstly presented at the conference 'Intangible Elements in Ethnoarchaeological Research' (Rome, 21–23 November 2012 -promoted by the Italian Association for Ethnoarchaeology) and published by Lugli (2014; 2016).

The 'crucial points' were always used to allow the informant to speak freely without feeling harnessed and constrained by rigid questions. It is also for this reason that, whenever possible, the interviews were carried out in several stages in order to create the right degree of confidence on the part of the interviewee.

## 3 A few days with our friends (1st-4th December 2012)

Before 2011, going to Mogod in the middle of winter was like an adventure. In fact, the road that today comfortably connects the capital to Ulaangom (Uvs province), from which you can easily get to Mogod, had not been built yet and it was necessary to go north to Bulgan, which is the capital of the Aimag of the same name, and from there go south over the mountains. It took approximately three days to go to Mogod, if everything went right and there were no hitches on the road such as bad weather conditions, blocked tracks or the car not getting stuck in the snow. From the capital to Mogod it currently takes about five to six hours including some stops along the way to refuel.

### *3.1 Camp 37/2011 of Narantsogt and Bolorman (N 48°03′158″; E 103°01′131″ 1736 metres above sea level) (1st–2nd December 2012)*

#### *3.1.1 Interview summary*

Usually, when we arrive in the Mogod area we initially stop at Camp 18, but that year Boyoo was out looking for horses and his wife was in hospital in Ulaanbaatar. So, we stopped at the camp of his brother Narantsogt and his wife Bolorman (Camp 37/2011 'Kharaat' N 48°03′158″; E 103°01′131″ - 1736 metres above sea level)[12] that we had already visited in 2011. The camp is not in the mountains but is in its proximity and is considered a good camp because it is well protected and there are not too many stones and rocks like in the mountains. Naratsogt and Bolorman got there on November 10th. When we arrived at around six pm, the dogs surrounded our car and our driver sounded the horn. The nomads came out of the tent and told us that we could relax because the dogs remembered us. Probably because we had fed them. They also said that we could stay for the night. In the warmth of the tent, we took the opportunity to chat, get up-to-date with the major events and changes that had occurred during the year and, of course, we got some information on the dogs.[13]

They have five dogs: Bankhar (7 years old, black), Baatar (Hero) (3 years old, black and white), Baavgai (Bear) (10 years old, yellowish), Arslan (Lyon) (8 years old, black with white chest). There is also Kurtan (Fast) (2 months old, black with the 'four eyes') who is a new dog of Boyoo and his wife. The puppy lives in the tent. The dogs were not related to each other and were all taken from the village.[14] Narantsogt and Bolorman said that the dogs are indispensable and that life in the steppe without them is impossible and unimaginable. When they choose a puppy, they take the first of the litter or the black one with a white chest and legs and 'four-eyes'. But colour is not considered crucial. Nobody pays to have a dog, you just take it. They smile when we ask them if a dog is a precious gift. No, it is not and traditionally the puppy is just taken. They usually take the puppy in their tent for two-three months if it is too cold in winter. If there are too many puppies in a litter, a few of them are usually killed by throwing them on the ground and then the bodies are taken away from home. There is not a funeral rite for the puppies. On the other hand, when a dog dies it is usually taken to a sheltered place, perhaps close to a small mound, its tail is cut off and placed under the head and milk is put in its mouth. Words can be said to wish that the dog reincarnates as a human in the next life.

The dog must stand guard and protect the animals, especially goats and sheep in the summer. In winter they do not have to do anything, only to guard at night, they can rest all day long. Dogs work harder in spring, summer and autumn to defend animals from wolves. During the warm months, they accompany the cattle. In winter, sheep and goats are usually close to the camp and they can even come back on their own.

There are many wolves and there is a special area where wolf hunting is practised. They show us the area on the map. Sometimes wolves come close to the camp and once ate a lamb belonging to Boyoo.

The puppy usually learns how to behave by observing adult dogs. But if there aren't any, it learns by itself. Narantsogt hunts marmots in the summer and wolves in the winter but never uses dogs for hunting. Dogs do not confront wolves because they know that it is dangerous. They only face them when they truly can't help it. Their task is essentially to bark to warn their master that there is a danger. Last year he killed four wolves. They are happy that there are wolves and they would not wish there were fewer of them even if last year two sheep were killed. Dogs hunt for pleasure but sometimes also for food, mainly squirrels and marmots.

[12] In 2012, camp 37 had approximately 700 sheep, 300 goats, 200 horses and 30 cows. In 2011 they were more or less the same numbers, that confirms that they had had a good year without any catastrophic events.

[13] We always make a sketch of the tent to document the arrangement of the furniture and supplies and the use of the space.

[14] The dogs in the camps are nearly always males. It is very rare to find a female.

Figure 2. Graziano Capitini with Bankhar, Baatar, Baavgai, Arslan and Kurtan (camp 37/2011, December 2012) (Photo by F. Lugli).

Some families also use dogs for hunting but they do not know if particular breeds are used for this.

Sometimes dogs do not like to stay at the camp if there are festivals such as *naddam* or celebrations for the nearby sacred mountain 'Avzaga Khairkhan' (April 5).[15] Naratsogt ad Bolorman think that festivals and celebrations are good opportunities to catch good food and 'social gatherings', but their dogs, on the other hand, do not go to parties.

*3.1.2 Observations*

Now and then we went out of the tent to observe what the dogs are doing. Two were on either side of the tent door and the other two were close by and stuck to the felt wall. During the night we got up a couple of times to see what they do. The first time they were where we had seen them, the second time two had moved behind the mounds of dung placed on the side of the tent and stored for use as fuel. When we went out they looked at us but remained curled up.

In the morning, before going to Bold's camp, we quickly documented the camp, made some observations and fraternised with the dogs (Figure 2). We confirmed what we had already observed before going to sleep and during the night, namely that two dogs slept on the sides of the door. The earth is still depressed by 8–10 centimetres where they slept. Baatar was still there and allowed us to observe his station without taking offence. We noticed where the other two dogs had slept close to the heaps of dung. An ellipsoidal patch of loose ground could still be seen. Nomads told us that dogs took turns to stay on the sides of the door. Although it was usually Baatar and Bankhar who stayed there. Right in front of the tent, 15 bones that were the leftovers of the dog's meal, could be seen at a macroscopic level. We took some samples and we marked their position on our plan. They were then analysed by Ivana Fiore.

During our work, the dogs and the children stayed with us. They were very curious and close-knit. It was clear that children and dogs are used to playing together. The dogs did not ask for food, but sometimes they wanted cuddles and attention. They observed our work. One of the children arrived with Baatar, the puppy that Boyoo and his wife had in their tent last year. They had told us that it was to make him healthy and strong (Figure 3). He was cuddled and was always treated with affection and no harshness. His name was also Baatar, which is a very common name for dogs. He was now a big and beautiful adult. He greeted us. He jumped and wagged his tail. Maybe he remembered us or trusted the example of the other four dogs that surrounded us.

We took hair samples (except of the former Baatar puppy) for Daria Sanna and her staff of the University of Sassari for the analysis of mitochondrial DNA (Sanna *et al.* 2018: 89–90; Sanna *et al.* in the volume).[16] After

[15] April 5th is said to be the day when the animals come out of their den and the birds arrive. The festival of the sacred mountain is only for men. The women wait at the foot of the mountain. The mountain is off limits to women throughout the year.

[16] We took samples in all the camps that we visited.

Figure 3. Boyoo and his wife with Baatar (camp 18/2011, December 2011) (Photo by F. Lugli).

greeting and thanking them, we gave presents to the nomads and some food to the dogs.

At one o'clock we left to go to the Bolds' camp where we arrived around twenty minutes later.

### *3.2 Bold and Suvdaa's camp, 'Khirliin ovojoo' (camp 06/2012 N 48°10'447"'; E 102°51'755" 1740 metres above sea level ) (2nd December 2012)*

We met Bold and his family in 2009 when we decided that life in winter camps should be the focus of our research and we had chosen to start in southwestern Bulgan.[17] Bold at the time was still working at the municipality of Mogod. He was the uncle of Alice, a student of the National University who was studying Italian and came with us.[18]

They were nomads but they had lived in the village where they worked. They decided to return to nomadism after their retirement. Suvdaa retired first and Bold a few years later. However, the house in Mogod was still used by the grandchildren who went to school. Their children started to be nomads 17 years ago. They always had animals and the life of the village is close to that of the nomads so the change was not a real change and it was not traumatic.

We always stay at their camp that we study yearly and we use it as a base place for going to the other camps of Mogod. Bold knows the area very well and he and his family have often helped us.

When we arrived at the camp, our interpreter Gansukh Solongo Tserem and our driver Tserenjamts Gerelee who had been with us in the missions since 2009, got out of the car and shouted the traditional greeting '*nohoi horioo*' (нохой хорио) that means 'hold the dogs'. In the case of Bold's camp, the sentence completely loses its literal value. In fact, not only the nomads know us, but also the dogs that run towards us when we return every year (Figure 4). However, during our absence, a newcomer could arrive without knowing that we are known and loved by his 'colleagues'. It is always better to pay attention.

The dogs were the first to notice our arrival. One dog, named Khvder, was the first to come to greet us, and began to bark joyfully and run towards us.[19] It is a real joy to return to Bold and his wife and be welcomed by the dogs. It makes us feel like part of the family. Over the years, friendship and affection have built up and when the lockdown broke out in Italy in March 2020, Bold went to the village to call our interpreter for news and to let us know that they were worried and thinking about us.

[17] In 2009 we also conducted our research in northeastern Arkhangai and in Dundgovi to have a comparison between different ecosystems: steppe, desert steppe and desert. In 2007 we made a first mission in winter in Arkhangai and Dundgovi where we could see the camps that we had already studied in 2002 and 2005 and to update our data as well as to visit new families.

[18] The camp is in the mountains and has been in use for 17 years but was pre-existing. It is in a splendid location. It was originally described by Lugli (2008; 2013; 2021).

[19] Khvder died at 18 years old in 2020.

Figure 4. Khvder runs towards Francesca Lugli when she arrives (camp 06/2012, December 2012) (Photo by G. Capitini).

Bold was in the village. Suvdaa, Bold's wife was there and their son Bayartogtokh who was not married. He is Batsaikhan's twin who stayed in camp 21 with his wife. In 2009 they were all in the same camp but after a year they split up and chose a different place close to Bold anyway. After greeting the dogs and giving them some goodies, we were immediately welcomed into the tent.

*3.2.1 Interview summary*

The first thing we do is to catch up on all the news since we last saw each other. We tell them about our life and they tell us about theirs. They too confirm that it was a good year and that the number of animals is more or less the same as last year. The wolves were also generous and did not kill any animals. They no longer have the worker they had last year, they prefer to be alone. They update us on the grandchildren, on the daughter in Ulaanbaatar and we ask them some ritual questions on the quantities of salt, grass and various other things on the management of the camp and livestock.

Obviously, we take the opportunity to also ask about dogs. We are told that Khvder, who lost a leg a few years ago in a trap, is doing quite well even though he is getting older. He is already ten years old and the other dogs are starting not to respect him as much as before. But he is Bold's favourite dog and it was Bold who saved him when he lost his paw, cared for him and loves him very much. When they move from one camp to another one, if it is too far, they take him by car, although each time it is a challenge because he does not like travelling by car. Khoilog is still a 'child' and so is Baatar, who is his son's favourite dog. The dogs are not related to each other, at least that they know of, and they were taken from the village. It is important not to choose the first puppy of the litter, but the biggest and the most beautiful. They usually choose the black one with four eyes, but this is not a fixed rule. When they get a puppy, they keep it in the tent for three months in winter and only one month in summer.

Bold's family also tells us about the traditional funeral ritual on the occasion of the death of a dog. They believe it is important to do this out of respect for the dog who has been a loyal lifelong friend. They don't choose a particular place, but it must be far from the camp. Of course, the dog is indispensable and a great friend. But they believe that man's best friend is not the dog, but the horse.

Life without dogs is impossible. At least two dogs are essential in a camp. Three is even better. Four are too many. But it depends on the camp and on the number of animals the family has. Dogs are part of the family and are good friends but man's best friend is certainly the horse. The dogs must watch the livestock at night. During the day they don't have to do anything. They eat and sleep and occasionally go for a little walk, perhaps with their owner when he goes to round up the animals. But they don't have to. Only if they feel like it.

Their dogs don't go hunting with their masters. But the dogs hunt on their own, usually, they hunt foxes, certainly never wolves which are too fierce. But only Khailog hunts. Khvder never hunted. It is not in his nature.

The sacred mountain 'Otsonk hangai' is about 4–5 km from the camp. Women can go there but not during the celebration that they organise in June which is only for men. Monks pray for rain. For a wish to be granted, it is necessary to go back there three times. All dogs love to go there when there is a celebration. They find food, meet friends and maybe even a 'girl'.

According to Bayartogtokh, dogs like to stay a couple of metres away from the tent in an easterly and southerly direction. We go out together to check.

*3.2.3 Observations in the camp*

It was a bit chilly outside, around -16°C, so we could not stay outside for a long time. We checked the macroscopic changes that had occurred in the camp. The most important was that there was only one tent and not two as in the previous year. Some improvements had been made to the animal shelters. We made photographic documentation and updated the plan.

The dogs were not there at that moment but it was possible to see the stations that Bayartogtokh had told us about. The two sub-circular depressions were of soft ground about twenty centimetres thick. At night, however, they stayed close to the door of the tent, at its sides and stuck to the tent wall to take advantage of the warmth. Two stratigraphic columns were taken from the stations. Massimo Vidale could then analyse them under the microscope (Vidale 2016).

The dogs ate every day in the area immediately in front of the tent but there were no bone fragments on a macroscopic level. The area was constantly cleaned and the rubbish was always piled-up in the same place. We went with Bayartogtokh, to look for the dogs' faeces. He knew very well where to find them because dogs usually like using the same places. He helped us to collect the samples that we positioned on the plan. Ivana Fiore analysed the fragments of bones (Fiore 2016).[20]

In agreement with Bold's family, we decide to stay at the camp of their son Batsaikhan and his wife which is nearby.

[20] The fact that dogs are usually creatures of habit can be told by any person who has dogs. The samples were taken in all the camps that we studied.

***3.3 Camp 'Urd khunkar' 21/2012 of Batsaikhan and Enkhjargal (N 48°09'153"; E 102°52'079" 1656 metres above sea level) (2nd, 3rd, 4th December 2012)***

In camp 21 lived Batsaikhan, Bayartogtokh's twin brother, his wife Enkhjargaland their daughter Enkhmend (5 years old). The other two children Tuvshinjargal (12 years old) and Enkhtuvshin (10 years old) had greeted us but had gone to the village because they had school the next day, so they had left the camp.

They moved here last year. Before that, they stayed in the same camp with Bold and his wife. It was a very windy and cold place but they said that there was good grass because the wind prevents the ice from forming.[21]

They had three dogs, Khurdan (Fast) (10 years old), Arslan (Lion) (10 years old) and Arslan (Lion) (2 years old), but at that moment none of them were about. They told us they were closeby, perhaps meeting some friends. We took photographs and made notes about the camp. We often needed to go inside to tent to warm up. Around seven o'clock we finally moved to the tent. At half-past seven, there was not any dog in sight yet. In our experience, it is highly unusual, but the nomads did not seem to be astonished or worried. They said that the dogs are free to come and go. They certainly would be back soon.

Our friends watched the movie 'ET' on TV, but the battery died just a few minutes before the end. Everyone laughed because we told them that always happened at Graziano's parents home. At half-past nine everybody fell asleep. The dogs did not come back. The nomads said that perhaps there was a female in heat. We woke up twice to check in the night. No dogs at all.

We all woke up around half-past eight the following day. We immediately opened the door to see if the dogs were there. Only one was there sleeping exhausted on an old blanket on the east side of the area in front of the tent.

We made graphic and photographic documentation of the camp and noted our observations. We documented the place where the dog had slept. He did not move but watched us.

We observed that on the west side of the door there was snow. Certainly, dogs did not stay there during the night because it was windy and not sheltered. Perhaps the east side was better.

The night before and in the morning we chatted with the nomads. They told us that the last year was good and that they did not lose animals as during the previous year. But we decided to postpone the interview because

[21] They had 400 sheep, 100 goats, 30 cows and 50 horses.

Figure 5. The old dog Khoilog near the tent (camp 040/2012, December 2012) (Photo by G. Capitini).

the weather was good and it was better to go to the camps (022-029) in the mountains to meet the families and to plan our work there in the next few days. We left at 10 o'clock.

### *3.4 Camp 040/2012* 'Ulaan Khalga' *of Erkhembayar and Narantuya (040/2012 N 48°12'451"; E 102°48'037") (3rd December 2012)*

We arrived at camp 022/2011 half an hour later. The mountains here are splendid and evocative and the camp is in a very protected place. It is considered to be a good winter camp. A new number was given from inside the tent that had been moved slightly to the east in 2012.[22] The old dog Khoilog (15 years old) recognised us and immediately came to greet us. Erkhembayar (33 years old) and Narantuya (30 years old) welcomed us and received us in the tent (Figure 5).

Here too we asked how the year had been and whether there had been any special events. They too agreed that it had been a good winter and they did not lose any animals. But in 2011 they got premiums as good herders and in 2012, they did not.

#### *3.4.1 Interview and some observations*

We ask about their dogs. They had three last year but one is in the village with its parents and one disappeared, perhaps it went with a family who had a bitch and was attacked by the other dogs. They will get another one in a few days, a male of course. The important thing is that it is male and strong. They will not consult a lama, that was an old tradition. They like Mongolian dog breeds, but they have been hybridised with other breeds and it is difficult to find a pure one. But they do not think that it is important to have a purebred dog.

We are told that Arslan, the light golden dog of his brother Enkhbayar and Altanzul (camp 23 N48°12'375"; E 102°47'984") died last year. We are very sad because he was a dear friend of ours.

Dogs are indispensable and life is impossible without them. Their task is essentially to guard against the wolves. They must bark to warn their masters. They think that the most dangerous period of the year is Autumn because wild animals go into their lairs and wolves remain without any wild prey. However, the dogs stay close to the camp, they never accompany the animals. The dog is 'only' a guard and he is not requested to fight wolves. They have never heard that dogs can confront a wolf. They say that they like dogs. They are nice and friendly. Erkhembayar does not hunt and his dogs do not hunt. They love wolves because when a man sees a wolf he becomes stronger. They do not use wolf organs to treat themselves. It was an old tradition.

When a dog dies, if he dies close to the camp they do the usual funeral ritual, if it dies far away they usually leave it there. They do not wish him anything, not even to be reincarnated as a human being. The dogs eat every day but not always at the same time. In winter they give

[22] The previous number was 022/2011.

them leftovers and bones and offal when they kill an animal. There is no specific place where they feed them, usually it is near the door of the tent. Dogs are strong. If they are sick they do not treat them and they certainly do not take them to the vet. If a dog accidentally cuts his leg in a trap, he will cure itself with saliva.

They say that their dog is almost always on the western side of the tent door.

We try to identify the location that he used last year. It was quite possible to see some traces and perhaps an excavation could confirm it.

### *3.5 Camp 24 'Bandiin tolgai' of Ankhaa and Dalgar (N 48°12'316"; E 102°47'901") – Interview (3rd December 2012)*

We also talked with Dalgar (40 years old) from nearby camp 24 whose husband Ankhaa (43 years old) had gone to Mogod and should return in the evening. We spoke outside, it was not very cold and it was nice to spend some minutes in the open air.

They used this camp for many years and then abandoned it for three years. They moved three-four kilometres away, near Ankhaa's brother and returned four years ago. Nobody used it when they were away. It is a good camp, very protected and the neighbours are all friends. Nearby there is a sacred mountain, *Otson Khangai* where in spring and June there is a celebration for the mountain. It is only for men but dogs are also admitted.

#### *3.5.1 Interview summary*

They also only have one dog, Arslan 'lion' (10 years old) and they also want to get another one but they don't know when yet. For the first two months if it is cold he will stay in the tent with them. They usually choose the biggest puppy, preferably a black one. The dog is indispensable because it guards. He does not fight wolves. He barks. She thinks that the most dangerous time for wolves is winter, and not autumn. According to her, there are not many wolves and they do not approach their animals. It is difficult to see a wolf. But one August a few years ago a wolf ate a sheep. Dalgar does not like wolves. Arslan is always with her, if she moves the dog follows her. She really loves the dog. The dog is certainly indispensable against wolves, but above all, he is a great friend. He must not be aggressive. He must be well balanced and know when there is a danger and he must do his work, which means that he can frighten people and animals if necessary.

Arslan is fed three times a day with soup, bones and various leftovers. They usually put an old pot near the tent but the dog takes the bones away to gnaw on them at his leisure. At the moment there is a second dog in the camp. He belongs to another family but he is a very good friend of their Khoilog and he spends a long time with him. Sometimes Arslan visits him too.

Arslan also usually sleeps close to the door of the tent.

As we talked, outside the tent, Arslan was standing next to his mistress and they made eye contact from time to time. She smiled at him and so did he. It was evident that they had a close relationship. Arslan followed every word and flicked his ears when he heard his name and realised that we were talking about him. When we walked into the camp to make our observations, he was very curious but he did not want to leave his mistress and preferred to stay close to her. And she certainly loved that the dog was so affectionate.

We gave something to Arslan who came over to us and then immediately went back beside Dalgar.

The dog was very shy. But we thought that we could become familiar with him in the next few days. We gave him a piece of dry meat, he appreciated it and ate it without moving which means that he trusted us. Before leaving the camp we documented the position where the dog stayed near the tent.

### *3.6 Camp 54 'Shinjit' of Banzragch and his daughter-in-law Narangerel, Ganzarig and Tsevelmaa*

We made a quick stop at camp 54 in the tent of Banzragch and Narangel. She had two sons, Ganzul (17 years old) and Ganjargal (11 years old). In the second tent, there were Ganzarig, who is the son of Banzragch, and his wife Tselmaa, they had three sons, Togsjargal (7 years old), Batkhun (5 years old) and Mungunkhun (1 year and three months). This camp had been used for at least twenty years; the previous camp is about 20 km away. Banzragch said that he had always been in this area. He came back here when his son died to help Narangerel. They arrived here in November.

It is a good camp because there is always grass and because it is very protected.[23]

They have always been nomads.

#### *3.6.1 Interview summary*

They had two dogs but they died. Without dogs, it is very dangerous and this year they will certainly take two new ones. They will get them in the village. They must be male and preferably black. For two months they keep them in their tent.

[23] They have 300 sheep, 200 goats, 11 cows and 50 horses.

The dogs guard against the wolves, they never follow the animals but always stay close to the tent. There are many wolves now because they are hunted less than previously and they have so many pups. Dogs do not fight the wolves and are never used for hunting. Now without dogs, it is a problem. They take turns waking up during the night to check on the animals because there are no dogs. At night they put up 'scarewolves', one in the camp and one on top of the mountain, but during the day they remove them because the animals bump into them. But without dogs it is not a very big problem because wolves usually kill the weakest and sickest animals. They all like dogs. They are great friends. They confirm the funeral ritual and say that sometimes they say a few words for the dog and sometimes not. They say that in winter the dogs are almost always on the right and or left side of the tent door. But they are free to stand where they like.

### *3.7 Camp 21/2011 (point 3.3) (3rd-4th December)*

We went back to camp 21 of Batsaikhan and Enkhjurgal at around six o'clock in the afternoon. The dogs were not there but they said they had come for a short rest during the day. They were certain that the dogs would come back soon. We spent the evening chatting with our kind hosts. We did the same thing the next day when we woke up.

#### *3.7.1 Interview summary*

When they take a puppy it must be male, preferably black and with four eyes. For forty-five days they keep it with them in their tent. The puppy and also the adult dogs eat leftovers and bones. Dogs are very important and life is impossible without them. They guard and protect the family. Their task is essentially to bark when a wolf or danger is coming close to the camp. They claim that they are always near the tent, even if they had been somewhere else since yesterday. But they think that if there is a danger the dogs will come back immediately. Dogs are never used for hunting.

They are fed twice a day and eat bones but also the soup if it is leftover and eat near the tent in the area in front of it. They always sleep close to the door of the tent. Dogs never have a shelter. Their dogs are good dogs. A dog is good when it guards well and barks and controls, especially at night. According to them, the most dangerous period for wolves is from spring to autumn. Winter is not so dangerous, because wolves gather and do not need to eat because they have already eaten. They like wolves and believe that if there are many, the animals will also increase. The elders say this. They explain that if there are so many animals there are so many wolves because they have so much to eat. It is all connected.

This year the wolves ate two animals but they always kill the weakest ones. Domestic animals are never attacked by dogs. Batsaikhan hunts wolves but does not really like it, only if he goes with his friends. Last year he did not kill any wolves.

If a dog gets sick, they don't treat it. Dogs are good friends and when they die they give them a funeral ritual.

Occasionally we went out to see if the dogs were in the camp. At the end of the morning, one came back to the camp with a considerable hip injury. He rubbed it against the snow several times. The nomads gave him some leftovers to feed him. The dog was exhausted, but ate voraciously and immediately after wiping his wound again in the snow, he curled up on the old blanket near the tent in pain. We tried to pet him but we understood that he wanted to rest. His 'friends' only came back the following day, they looked to be in a bad way.

## 4. Discussion

The presence of dogs is constant in Mongolian camps and various authors have highlighted their importance for Mongolian herders and their relationships (Fijn 2011; Bianquis *et al.* 2013; Altansan 2014; Bamana 2014; Lugli 2014; Terbish 2015; Lugli 2016; Fijn 2018; Lugli and Capitini Lugli 2018).

It is significant that the Mongolian mythical ancestors are thought to be a couple formed by a wolf and a doe. In fact, the wolf is also the dog's ancestor and that establishes a shared mythological kinship between Mongols and dogs. For this reason, Bamana (2014: 8) points out that during his field research, herders said that dogs and humans are made from the same bones. In our experience of many years, only a few people explicitly said that, but often we were told that humans and dogs are part of the same family, even if with different roles. Mythology was rarely mentioned but often during our conversations, it was implicit. The funeral ritual that is celebrated when a dog dies has also a relevant place in traditional Mongolian society (Fijn 2011; Bamana 2014; Lugli 2016; Lugli and Capitini 2018).

The importance of dogs is usually attributed to their main task which is to guard against dangers, especially wolves. This task was interpreted as a crucial point of steppe nomadism which contributed to the success of pastoralism in the region (Lugli 2014; 2016).

The short report of the few days we spent with nomads (1st-4th December 2012) and the summarised interviews are a quick cross-section of the role and status of dogs among the herders and above all of its repetitiveness and variability. A few considerations can be deduced.

Figure 6. Many families told us that winter is dangerous and for that reason, the 'scarewolves' are usually used in the cold months. (Camp 67/2012 *BorTalgai* (N 48.27679; E 103.17708) Khishig Öndör-Bulgan, December 2010 (Photo by G. Capitini).

### *4.1 Wolves*

Wolves are considered a great danger and they are always mentioned to be the main reason to have dogs. Dogs are requested to stay in the camps, to guard and bark if danger is moving close to the camp. That has always been confirmed by our informers over the years. But there are subtle contradictions. Sheep, goats, bovines and horses graze alone on the steppe and nobody protects them during the day. The dogs are primarily in the camps and the nomads visit their animals only a few times during the day to be certain that they are in the right area and do not stray too far. Actually, wolves are mainly nocturnal predators but they can also hunt during the day when hungry. Therefore, the animals can also be attacked during the day. Only Narantsogt and Bolorman (camp 37 - 1st-2nd December) said that their dogs go with the animals during spring, summer and autumn but when we went to Mogod in September 2017 their dogs were permanently in the camp or were closeby but not with the animals. In our experience, we never documented dogs going with the animals during the day. Dogs often accompany their masters and mistress, but it depends on their will.

There is not a general agreement on the more dangerous period of the year as concerns the wolves. For Narantsogt and Bolorman (camp 37) warm months are more dangerous ones, for Erkhembayar and Narantuya (camp 21) it is Autumn *'because the wild animals go into their lairs and the wolves remain without any wild prey'*, for Dalgar *'Winter, definitely not Autumn...... Winter is not so dangerous, because wolves gather and do not eat because they have already eaten'*. Many families told us that winter is dangerous and for that reason, the 'scarewolves' are usually used in the cold months and indeed we only documented them in the winter (Figure 6).

All the nomads we met in Mongolia said dogs are indispensable and that life without them is unimaginable. The reason is always the presence of wolves. For example, when we went to Delgerkhangai (Dundgovi) in 2018, not all the families that we had already met in 2005 and 2007 had dogs anymore. They said that wolves had almost been exterminated by Mongolian and Western hunters and by climate change, therefore dogs were not indispensable anymore. But wolves are not always described as dangerous and they are often said to eat only the weakest animals (Batsaikhan and Enkhjurgal camp 21). When we asked how many animals were killed by the wolves, they were always very few indeed.

Generally, nomads like wolves and according to Narantsogt and Bolorman *'they would not wish there were fewer of them even if last year two sheep were killed' (camp 37)* and Erkhembayar and Narantuya (camp 22) said that *'They love wolves because when a man sees a wolf he becomes stronger'.*

### 4.2 The presence of dogs in the camps

Another controversial point is the indispensable presence of dogs in the camps. All the informers said the same (and it has been confirmed everywhere over the years). But in camp 21 of Batsaikhan and Enkhjurgal, their dogs disappeared for three days and nobody was very worried about their absence. In camp 54 the dogs had died and had not been replaced yet. So, they had to wake up in the night to check on the animals in their camp.

### 4.3 Puppies, choice criteria and other

There are various nuances in the criteria to choose a puppy. The strength of a puppy is important but also its colour, black with a white chest are usually mentioned as well as having 'four eyes'. In addition, Narantsogt and Bolorman (camp 37) prefer the first of the litter. But Suvdaa (camp 04) said that she prefers the biggest and the most beautiful of the litter and that the colour is less important. They usually take the puppy in the tent with them during the first two-three months, especially in winter. We documented the presence of puppies in the tents at various times. They are usually treated with tenderness and their bad behaviour is considered with goodwill. The puppies learn to behave watching the adult dogs but they can also learn by themselves.

The puppies are always taken from the village where there are females. Usually, nomads prefer to know the owners but it is not indispensable. In our experience, dogs are usually allowed to roam freely and to mate whenever they want and with whomever they want and there are not many dog 'marriages' that are prearranged by nomads. Therefore, there is not a strict selective interference made by humans. The puppies usually live two-three months with nomads in their tent.

Nomads sometimes hunt but they usually do not have hunting dogs. Dogs can hunt but only for pleasure.

### 4.4 Dogs are not tied up

In camps 37, 06, 21, 40 and 24, nomads never tied up their dogs and that behaviour has been confirmed over the years during our stay with nomads in various seasons and regions (Bulgan, Arkhangai, Zavkan, Ovorkhangai, Uvs, Khovd, Khenti, Gobi Altai, Bayan Khongor and Dundgovi). Dogs can be tied up when nomads are in the camp and visitors arrive. This happens especially if the dog has an impetuous nature that does not necessarily mean being aggressive. Dogs are naturally curious and they often surround the newcomer. Obviously, they are dangerous if their masters are not in the camps but in this case, they are always released.

The traditional bankhar breed is currently very appreciated in the city and a few nomads have preferred to have a pure Mongol dog since around 2017. Sometimes they tightly tie up the poor puppy to force him to become mad and aggressive but it is not a traditional custom.[24] Lugli and Sychenko observed that also in Tuva in 2013 (see Lugli and Sychenko in the volume).

### 4.5 Other observations

The families of the camps 37, 04, 21, 40, 24 and 54 said that their dogs are friends even if the horse is usually considered human's best friend. Our informers declared that they love dogs and that they are part of the family. They like their loyalty but also their affection. They feed them two-three times a day. Dogs eat leftovers and bones.

Dogs are not nursed if they get sick but Bold took care of Khvder when he lost his paw in a trap.

We met many old dogs in the camps. For example, Khvder (camp 04) died at the age of 18, and Khoilog (camp 40) was 15 years old in 2012. Old dogs are documented to move by car from one camp to another if they have problems. This was the case of Khvder (camp 04). The funeral ritual is always reported by our informers who always said that a protected place far from the camp is usually preferred and that has always been confirmed over the years by other families. In 2016 we documented a dead frozen dog (his name was Paatsag) who had had the traditional funeral ritual. He was near the camp and in a place with no shelter or protection (Lugli and Capitini 2018).

Dogs often play with children who consider them to be friends. When children are requested to go to check on the animals dogs usually go with them. Dogs are their affectionate friends and trustworthy protectors.

During our missions, the words that nomads normally use with dogs were asked and documented. Dogs usually do not know many words but nomads speak to them

[24] That was documented also in Tuva (see Lugli and Sychenko in the volume).

clearly and easily to let their friends understand them. We did not observe a different tone of voice speaking to a dog as it is used for children, especially babies. That was observed also by N. Fijn (2018). It could be because even if we are friends with the nomads, we are not members of the family. We think that in the presence of unknown people, also in western countries where dogs are loved pets, dog owners do not use all the words and the different tone of voice they usually use with beloved four-legged animals.

### *4.6 Dog positions in the camp*

The spatial presence of the dogs in the camps has been a special topic of our research and it has always been observed and documented during our missions (see Lugli 2014 and Lugli 2016). Wherever we observed them, the dogs were usually found near the tent and in winter they sleep close to it at the sides of the door. Sometimes, an old blanket can be put on the ground for dogs by nomads (camp 21). They usually eat in the area in front of the tent that is constantly kept clean. The remains of bones and other food remains are moved to an area close to the camp that is usually used for that purpose every year.

## 5 Conclusions - Dogs as relatives and friends

The report of the few days (1st-4th December) with nomads that we spent in Mogod (Bulgan) at the beginning of our mission in 2012 is a significant observation of the variability and repetitiveness of the presence of dogs in Mongolian camps.

The dog is said to be indispensable and his task is to guard against danger especially wolves and that has always been confirmed by all the informers we met in various regions in Mongolia over the years. Wolves are feared and respected by nomads who sometimes hunt them. But the real danger of wolves is not constantly agreed upon by nomads.

Nomads consistently say that dogs are very good friends even if usually the horse is considered man's best friend.

Dogs share a mythical kinship with Mongols and they certainly are considered good friends and are part of the family. Mongolian dogs are traditionally independent. They must be obedient and do their work, especially during the night, but no more than that is requested. They can express their true nature and talents. For example, they hunt and go with their masters but only if they wish to. They are never forced to. They play with children because they are part of the same sphere and share with them a merry and fun time. Puppies usually spend their first two-three months with the new family in the tent. This contributes to creating a strong mutual bond between nomad dogs and vice versa. Nomads consider dogs as individuals and they like to tell special stories about them. Certainly, not all nomads love them and not all the dogs live a pleasant life. But nomads usually respect dogs as beings who are very close to humans. They believe that a dog can reincarnate into a human being and a human being can reincarnate as a dog. The relationship between nomads and dogs is almost equal: they are different beings that are part of the family, and they must be respectful and efficient but nothing more is requested. In return, dogs have food and usually affection. Therefore, dogs are not slaves to human beings. That was also observed by Lugli and Sychenko in Tuva in 2013 (see Lugli and Sychenko in the volume).

Dogs are a crucial component of Mongolian camps that is important to study and document before this changes as it is unfortunately already occurring. The understanding of the phenomena can illustrate the dynamic of the success of steppe pastoralism in ancient times.

## Acknowledgments

First of all, we thank all the nomads who always welcome us and gave us their friendship and Prof. Dulam and his wife Prof. Nandinbiling for their warm friendship and crucial help. We are grateful to the Italian Embassy in Mongolia and the Ministry of Foreign Affairs and International Cooperation – Italy MFA for supporting our mission. We thank our friend Alfredo Savino who generously gave us important suggestions and the translator Gansukh Solongo Tserem and the driver Tserenjamts Gerelee who nearly always came with us and were splendid travelling companions, without their *Sain saana* cooperative our research could not have been so fruitful and pleasant. Special thanks to our friend Bold and his family for their constant help and for letting us feel part of the family. Last but not least we thank Khvder, Baatar, Khoilog, Arslan and all the dogs that we met in Mongolia who always helped us with their friendship and affection.

## References

Altansan, A. 2014. Le pastoralisme aujourd'hui en Mongolie - Continuité et changement. Unpublished PhD dissertation, Faculté des lettres de l'Université de Fribourg (Suisse).

Bamana, G. 2014. Dogs and Herders: Mythical Kinship, Spiritual Analogy, and Sociality in Rural Mongolia. *Sino-Platonic Papers* 245: 1–16.

Bianquis, I., F. Aubin and S. Dulam 2013. Le chien et le bru, deux êtres luminaires en Mongolie, in K. Buffetrille, J.-L. Lambert, N. Luca and A. de Sale (eds) *D'une Anthropologie du chamanisme vers une anthropologie du croire: Hommage à l'œuvre de Roberte Hamayon*: 303–321. Paris: Centre d'Études Mongoles & Sibériennes -École Pratique des Haute Études.

Fijn, N. 2011. *Living with herds: human-animal co-existance in Mongolia*. Cambridge, New York: Cambridge University Press.

Fijn, N. 2018. Dogs Ears and Tails. Different Relational Ways of Being with Canines in Aborigenal Australia and Mongolia, in H.A. Swanson, M.E. Lien and G.B. Ween (eds) *Domestication gone wild: Politics and practices of multispecies relations*: 72–93. Durham: Duke University Press.

Fiore, I. 2016. Note 21 In Mongolian Nomads and their dogs (Lugli F.), in S. Biagetti and F. Lugli (eds) *The Intangible Elements of Culture in Ethnoarchaeological Research*:136–137. Springer.

Lugli, F. 2008. The nomadic camps of Mid-West Mongolia: The case of the spring site of Hulihin Shilin (Ih Tamir), in F. Lugli and A.A. Stoppiello (eds) *Proceeding of the 3rd Italian congress of Ethnoarchaeology (Rome, 17-19 March 2004)* (British Archaeological Reports International Series 1841): 135–145.

Lugli, F. 2013. Winter camps in Mongolia. In F. Lugli, A.A. Stoppiello and S. Biagetti (eds) *Proceedings of the 5th Italian congress of* ethnoarchaeology (Rome, 13–14 May 2010) (British Archaeological Reports International Series 2472): 209–216.

Lugli, F. 2014. Nomads and dogs: a crucial bond, in *Proceedings of Ural-Altai: through the centuries into the future (Gorno Altaisk 2nd-5th of July 2014)*: 19–21. Gorno Altaisk.

Lugli, F. 2016. Mongolian Nomads and their dogs, in S. Biagetti and F. Lugli (eds) *The Intangible Elements of Culture in Ethnoarchaeological Research*: 125–140. Springer.

Lugli, F. 2021. Ispol'zovaniye prostranstva na zimnikh stoybishchakh stepnykh skotovodov -etnoarkheologicheskaya tochka zreniya, XVIII Mezhdunarodnoy Zapadnosibirskoy arkheologo-etnograficheskoy konferentsii 'Zapadnaya Sibir' v transkul'turnom prostranstve Severnoy Yevrazii: itogi i perspektivy 50 let issledovaniy ZSAEK', sostoyavsheysya: 16–18.

Луйли, Ф. 2021. Использование пространства на зимних стойбищах степных скотоводов -тноархеологическая точка зрения, XVIII Международной Западносибирской археолого-этнографической конференции «Западная Сибирь в транскультурном пространстве Северной Евразии: итоги и перспективы 50 лет исследований ЗСАЭК», состоявшейся: 16–18.

Lugli, F. and G. Capitini 2018. Mongolian steppe nomads and their dogs (Bulgan, Arkhangai and Dundgovi) in I. Fiore and F. Lugli (eds) *Abstract Dogs, Past & Present - An Interdisciplinary perspective - 1st International Conference* (Annali dell'Università degli Studi di Ferrara Museologia Scientifica e Naturalistica 14): 63–64.

Sanna, D., G.L. Dedola, P. Cossu, T. Lai, F. Scarpa, A. Canu, M. Scandura, M. Apollonio, F. Lugli, P. Francalacci, P. Mereu and M. Casu 2018. Mitochondrial DNA variation among dogs of Mongolian, Tuvian and Altaic nomads in I. Fiore and F. Lugli (eds) *Abstract book Dogs, Past & Present - An Interdisciplinary perspective - 1st International Conference* (Annali dell'Università degli Studi di Ferrara Museologia Scientifica e Naturalistica 14): 89–90.

Terbish, B. 2015. The Mongolian Dogs as an Intimate 'Other'. In *Inner Asia* 17, 141–159.

Vidale M. 2016. Note 19 In Mongolian Nomads and their dogs (Lugli F.), in S. Biagetti and F. Lugli (eds) *The Intangible Elements of Culture in Ethnoarchaeological Research* : 132–134. Springer.

# 6.7 Dog and Wolf in the Non-Tale Prose of the Turkic Peoples of Siberia

Galina B. Sychenko

Associazione Italiana di Etnoarcheologia, Via Principe Umberto 41, 00185 Roma, Italy.
phurdok@gmail.com

**Abstract**

The article is devoted to a comparative analysis of the relationship between man, wolf and dog in the Turkic cultures of Siberia. As the main source of research, the author selected texts of non-tale prose – myths, legends, everyday stories - of the Siberian Turks (Altais, Tuvas, Khakasses, Shors, Yakuts and Dolgans), published in the academic series 'Monuments of Folklore of Peoples of Siberia and the Far East'. All mentions of these animals are highlighted in the corpus texts and all contexts of such mentions are analysed. Both similarities and differences, parallels and oppositions in relation to the wolf and the dog were found. On the whole, the texts of this group of folklore give a picture which is close to the ethnographic reality. They give an idea of the dog's special role, its distinguished position among domestic animals and the most ancient nature of its connection with man.

**Keywords**: dog, wolf, folklore, human-animal relationship, comparative studies.

## 1 Introduction

Starting from one of the first publications about dogs in Eurasia (Miller 1876) numerous folkloristic and ethnographic works have been published. It is difficult to calculate how many motives related to dogs have already been discovered and how many different aspects related to the dog and its role in human life have already been touched upon in research. In Turkic cultures of Siberia and in neighbouring areas, the important role of the dog, which reveals itself in protective and funeral rites, in mythology and folklore as well as in everyday practical life comes into light in many publications (Romanova 2000, Darzha 2009; Burnakov 2012; Ilimbetova and Ilimbetov 2012: 131–229; Muytueva 2015). Some comparative studies that through the image of a dog reveal ancient connections between Indo-Iranic and Turkic-Mongol, as well as other cultures have already been made (Cheremisin 1997; Cheremisin 2009; Beryozkin 2012). The same is true of the wolf - dog's wild relative (Kubarev and Cheremisin 1987; Golden 1997; Salmin 2011; Ilimbetova and Ilimbetov 2012: 38–130; Burnakov and Tsydenova 2015).[1] Nevertheless, there are some more aspects, which have not yet been studied enough. How dogs and wolves are seen in different folklore genres, how this vision relates to ethnographic realities and how deep historical memory is embedded in folklore texts - this angle could be one possible perspective for observing the issue.

If we regard folkloristic heritage as a collective memory keeping some important ideas from time immemorial, we may suppose that they reflect, always in an indirect way, not only ethnographic, but also historic reality, or, at least, human ideas about history. In other words, folkloristic studies have to deal with a conceptual, ideological level of the culture.

These ideas are represented mostly in narrative folklore texts of different genres.

In many traditional cultures narrative folklore is divided into two main groups of genres:

1. texts of imaginary character - epics, fairytales, fables, parables, anecdotes etc.;
2. texts of 'true' character - myths, legends, personal stories - all that which is defined as 'non-fairy tale prose', or just 'non-tale prose'.

This division is reflected in folk terminology. Table 1 shows the basic terms (without phonetic and dialectal variants) for 1) fairy tale (the first line of Table); 2) epic (the second line of the Table); and 3) non-fairy tale prose (the third line of the Table). As we can see, often the same term is used for fairy tale and epic (with the

[1] Just a few works that touch on mainly Turkic cultures are cited. Of particular note is the universal electronic Catalogue of folkloric and mythological motifs by Yuri Beryozkin: http://www.ruthenia.ru/folklore/berezkin which contains many folklore motifs about the personages of interest, not to mention the numerous websites devoted to the accumulation of a wide variety of material about the dog, for example http://www.bordercolliemuseum.org/BCCousins/Asia/Russia.html

Table 1. Ethnic terms for different genres of narrative folklore of the Turkic peoples of Siberia

| | Altai | Shor | Khakas | Tuva | Yakut | Dolgan |
|---|---|---|---|---|---|---|
| **Tale** | *chörchök* | *nybag* | *nymakh* | *tool* | *olongkho/ ostuoruya* | *hatyy olongko* |
| **Epics** | *kai chörchök* | *alyptyg nybag* | *alyptykh nymakh* | *maadyrlyg tool* | *olongkho* | *yryalaak olongko* |
| **Non-Tale Prose** | *kep-kuuchyn* | *purungu chook* | *kip-chookh* | *burun chugaa/ tøøgu-chugaa* | *kepseen* | *bylyrgy øster* |

Figure 1. Turkic-Speaking Peoples of Siberia.

addition of a clarifying definition), while a completely different term is used for non-tale prose (Table 1).

The translation and etymology of terms used for non-tale prose is very distinctive: narration about ancient times (*purungu chook, burun chugaa, bylyrgy øster*), historical narration (*tøøgu-chugaa*), and traditional narration (*kep-kuuchyn, kip-chookh, kepseen*). Texts in this category are understood to be truthful, telling what really happened – whether it was yesterday, a week or a century ago, or at the time of the original creation. They can be contrasted with texts of a fictional character, like a fairy tale or an epic. From this point of view there is no difference between a myth or legend and a personal story. They are distinguished only by their temporal distance: myths took place at the time of creation, legends - in the historical past, and personal stories took place in a recent time. This categorisation suggests that the texts of this group will reflect a more realistic view of the world.

In this article I will focus on them, limiting myself by examining non-tale prose of the Turkic peoples of Siberia. This choice is determined by my intention in the future to study the connection between folkloristic and ethnographic data collected in a frame of a 5-years project headed by Francesca Lugli[2] amongst this group of the Siberians. At this stage it is important to reveal basic ideas about two animals, having an important place in the thoughts as well in the life of the peoples in question.

## 2 Material and methodology

The main sources for this research are several text corpora of non-tale folklore published in the series 'Monuments of Folklore of the Peoples of Siberia and Far East' (Alekseev *et al.* 1995; Efremov 2000; Alekseev *et al.* 2010; Arbachakova 2010; Oynotkinova *et al.* 2011; Mindibekova and Sychenko 2016). These volumes embrace six ethnic groups: Altais, Dolgans, Khakasses, Shors, Tuvas and Yakuts who belong to the Turkic language family. The Altais, Khakasses, Shors and the Tuvas belong to the area of South Siberia, the Dolgans and the Yakuts – to North-East Siberia (Figure 1).

I extracted from texts of these volumes all the motifs and even simple mentions of dogs and wolves, and analysed this information. In addition, variants given for some of the texts and scholarly comments were taken into account.

In this section the materials obtained from the sources are represented and grouped into several sections.

Analysis is based on comparative methodology. Grouping the motifs already reveals ethnic and areal distinctions as well as common traits.

### *2.1 Appearance*

#### *2.1.1 External features of a dog*

The appearance of dogs is noted in many myths and legends. For instance, in the creation myths, the dog, appointed by the main deity (Ul'gen, Üch-Kurbustan, Khuday, or Burkhan) to watch over newly created humans, is naked, i.e., devoid of hair. This motive is found in Altai and Khakass folklore (Oynotkinova *et al.* 2011: 52–53, 72–73, 80–85, 86–87, 94–99; Mindibekova and Sychenko 2016: 76–77). Sometimes the epithet 'red' is also used, which refers to the skin of a naked dog. However, more often the colour epithets refer to the colour of the dog's coat. So, in Altai legends the girl-*almys* (dangerous malevolent spirit) turns into a red-headed puppy (Oynotkinova *et al.* 2011: 218–219); the *baatyr* (hero) Sartakpai had a greyhound with moon-sunny eyes (Oynotkinova *et al.* 2011: 234–235); bay-coloured dogs with four white marks on the forehead are mentioned in the legend of the famous hero Shunu (Oynotkinova *et al.* 2011: 320–321). In the Khakass' stories, the mythical hound is pale in colour; two brother-hunters cannot decide to whom the yellow hunting dog belongs (Mindibekova and Sychenko 2016: 166–167, 308–309).

The Tuvas in case of a solar eclipse consider that the only man in the clan should scream, the black dog with red tan marks should bark, and it is necessary to beat into a Chinese cauldron (Alekseev *et al.* 2010: 46–47). In the Shor tale, 'senior-junior blue dogs' - magical guards are mentioned (Arbachakova 2010: 276–277).

Yakut shaman Tuluurdakh follows the trail of a lost soul in the image of an iron yellowish dog (Alekseev *et al.* 1995: 262–263). In case of persecution by a harmful spirit, it is necessary to sacrifice a bluish-grey dog with red spots at the base of the ears and reddish tan marks on the knees (Alekseev *et al.* 1995: 232–233).

A Khakass myth mentions a paw structure feature with the presence of a thumb in a wolf and a dog (Mindibekova and Sychenko 2016: 126–127).

Nothing has been found about the appearance of the dog among the northernmost Turkic people, the Dolgans.

#### *2.1.2 External features of a wolf*

In the Tuvan myth, a harmful spirit *diiren* is mentioned that turns into an ash-grey she-wolf (Alekseev *et al.* 2010: 128–129).

Khakassian myths provide the greatest number of references to the appearance of wolves. For instance, one of them mentions a black-foreheaded shewerewolf (Mindibekova and Sychenko 2016: 168–169). In variants of the legend about a hero Ir-Tokhchyn, white and grey wolves; a mighty white-foreheaded she-wolf and her seven children; and nine wolves: eight white wolves and a ninth mighty black white-foreheaded she-wolf are mentioned (Mindibekova and Sychenko 2016: 290–291; 290–301; 302–305). A white wolf chases seven mares in the myth about Ursa Major (Mindibekova and Sychenko 2016: 325).

[2] The project took place from 2013 to 2017, see Lugli and Sychenko in the present volume. In reality, we were working among the Turkic peoples living in Southern Siberia, i.e., the Altai, the Shors and the Tuvas. To complete the picture, this article also includes materials on the Khakasses, as well as on the two peoples living in North-Eastern Siberia, the Dolgans and the Yakuts. The project was promoted by the Italian Association for Ethnoarchaeology with the sponsorship of the Ministry of Foreign Affairs and International Cooperation – Italy MFA.

### *2.2 Functions*

*2.2.1 Functions of a dog*

2.2.1.1 Man's companion

The dog is often presented as a human companion. This function is sometimes expressed explicitly, sometimes implicitly. For example, the toponymic legend speaks about *baatyr* Sartakpai who had as his companion a greyhound with moon-sun eyes. Traces of them are left on a stone in the valley of the Kara-Suu River in Altai (Oynotkinova *et al.* 2011: 234–235). The puppy of the Yakut fisherman from the Viliui River is his only companion, and it is hard to part with him (Alekseev *et al.* 1995: 214–217).

In the Altai and Tuvinian cosmogonic myth of the origin of the constellation *Üch-Myigak* 'Three Female *Marals*' (Orion) three dogs or two dogs and one puppy - are its indispensable element together with a hunter and three female *marals* (*Cervus elaphus sibiricus*) (Alekseev *et al.* 2010: 42–43; Oynotkinova *et al.* 2011: 106–107). Some variants speak about a dog, a golden eagle (*Aquila chrysaetos*) and an arrow, or a horse, arrow and a dog, instead of three dogs (Alekseev *et al.* 2010: 293; Oynotkinova *et al.* 2011: 419). In Khakass' version about *Chetigen* 'Seven Khaans' (Ursa Major) the protagonists are three brother-hunters, two dogs and two female *marals* (Mindibekova and Sychenko 2016: 84–85; 290–301; 302–305).

2.2.1.2 Protecting a human

This important function of a dog is expressed in numerous ways and contexts. Probably the most characteristic are variants of a myth about the origin of humankind. This famous myth tells how a supreme deity, who occurs in texts under different names (Ül'gen, Kuday, Burkhan, Üch-Kurbustan), created the world and then created a man, or men from clay, who had no soul. The deity distances himself in search of a *tyn* 'soul' and leaves the dog to watch over the man / men. At this point, malicious deity Erlik - the antagonist of the main deity and sometimes his brother - appears. His role is to corrupt everything the Creator has created (land, flora, fauna and men), so that the perfect world becomes imperfect - as we know it. As for the men, Erlik only gets close to them by seducing the dog guarding them, which, as mentioned above, was created naked, without hair. Erlik promises him fur and never-ending food. The dog agrees and allows Erlik to get close to the men. The harmful deity animates them by blowing the soul through the anus. The men come to life, but become mortal, whereas by the Creator's design they were supposed to be immortal. In another version of the myth, Erlik spits on the man and when the main deity returns, he finds his creation covered in saliva and slime. The dog repents, asks for forgiveness, and the upper deity agrees. He turns the man inside out - so that now all the slime is inside his body. The dog gets his fur and endless food - human excrement. From now on, it is always on guard for the man and serves him. These myths are distributed among the Altais in numerous versions (Oynotkinova *et al.* 2011: 72–75; 82–87; 94–99; 116–119; 407–411), and also among the Khakasses (Mindibekova and Sychenko 2016: 76–77; 322).[3]

Some Altai and Khakass myths tell the story of how a dog once saved people from starvation. People who were living well and were well-fed ceased to value their affluence. Women started wiping their children's bottoms with baked cakes. The deity Kuday became very angry and caused a famine: the barley ceased to grow. Then the dog began to howl, asking the god for food. The deity took pity and threw a few grains to the dog. The barley began to grow again. But while the grains used to cover the whole plant, they now only grow on the top of the plant. People began to eat it and appreciate it more. They were saved from starvation by a dog, but the dog himself was only given the husks (Oynotkinova *et al.* 2011: 86–89,412–413; Mindibekova and Sychenko 2016: 76–79; 323).

2.2.1.3 Hunting, herding and fishing

Hunting and cattle, sheep and horse breeding are the main economic activities of the Turkic peoples of Siberia. The Dolgans have also developed reindeer breeding, and fishing is of commercial importance to the Yakuts as well as cattle and horse breeding. In myths and legends, economic contexts in which the dog appears are represented in varying degrees. Contexts connected with hunting occur rather regularly. This is found in the cosmological myths of the celestial hunters (see above).

A dog helps a hunter sitting in an ambush to track down an evil spirit and then helps him fight the *almys* (Oynotkinova *et al.* 2011: 216–217). In another story, a hunting dog barks at a squirrel near a shamanic larch tree (Oynotkinova *et al.* 2011: 294–295). A woman-*almys* who turns into a red-haired puppy lives with a hunter who eventually kills her (Oynotkinova *et al.* 2011: 218–219).

In a Khakass legend, two hunting brothers, unable to decide which one of them owns their only dog, kill it and throw it into a lake (Mindibekova and Sychenko 2016: 166–167, 350–351).

There is also a particular legend in Khakass folklore, which tells of the mythical dog *Khubai Khus* (literally, 'Pale Bird'). It hatched from the egg of a *turpan* (*Melanitta*

[3] It also occurs in the Chuvash folklore (Salmin 2011).

*fusca*) and was distinguished by its extraordinary hunting qualities. Having slayed all the beasts, it was taken by the spirit of the lake, as it was impossible for such an outstanding hound to remain on the earth. Some variants add to this plot some pastoral motives: a rich man exchanges a hunter's dog for fifteen cows; before *Khubai Khus* dies, he fills the courtyard of the previous owner with all kinds of cattle (Mindibekova and Sychenko 2016: 308–309, 386).[4] Perhaps that is why some legends not related to this plot use a paired word expression *aday-khus* (literally 'dog-bird'), which collectively means 'livestock' (Mindibekova and Sychenko 2016: 216, 292, 369).[5]

In the voluminous text of the Khakass myth of the council of all beasts and birds, which decided who of them gets the intellect, there is a multi-part episode in which a cow, horses, a camel, a sheep and geese and ducks take turns complaining about difficulties and dangers from wolves, asking to give them to the *khyrna khulakh / goloukhiy* (khak. / rus., literally, 'bare-eared'), i.e., a man. He would protect them and use their milk and other goods and skills. The one reason is that it was he who has got the intellect, and the other reason is he has a dog, which is regarded as an antagonist of a wolf. The wolf himself says he's afraid of the dog. In addition to the shepherding context, there is also a hunting motif in this rich text: the hare says that he has the hardest life because he is hunted by all predators, and also by the dog, man's friend (Mindibekova and Sychenko 2016: 122–127).

The dog appears in a pure shepherding context in a Shor tale. We consider this case here because it is the only story involving a dog in the Shor volume, and its genre features (tale - non-tale) are blurred. Having learnt to understand the language of animals, a shepherd hears two mythical dogs - which are descended from the sky to help him to herd the sheep - talking to a wolf. The wolf is trying to beg for a sheep, the older dog takes pity on him and wants to give him a sheep, the younger dog shows loyalty to his master and refuses to give the predator a sheep (Arbachakova 2010: 276–281).

In a Yakut legend a female spirit visits the fisherman and demands to have a puppy in exchange for fishing success (Alekseev *et al.* 1995: 214–217).

Dolgan myths often mention neighbouring peoples: Nganasans, Nenets, Evenks, as well as unknown mythical peoples. All of them are known as hunters of wild deer, fox and other animals of a *tundra*, however a dog seldom appears in a hunting context. Thus, one of the texts says that one Nganasanian left to hunt wild deer, his dogs remained to guard the house (Efremov 2000: 370–371). Another story tells of two Dolgans who went with their dogs to set traps for Arctic foxes. However, the dogs are not involved in the hunting process as such (Efremov 2000: 330–333). They guard the tent outside and warn of the arrival of strangers, as in other texts.

2.2.1.4 Transporting

Of all the Turkic peoples of Siberia, only the Dolgans use dogs as a sledding animal. This motif occurs regularly in various texts. Dog sledges are mainly used for hunting, other business, visits, and to escape from the persecution of enemies (Efremov 2000: 332–335; 338–343). The dog as a sledding animal is mentioned more often than deer, while the horse is rarely mentioned. Various spirits and mythical creatures also ride on dog sleds; the spirit of the Earth uses the fox, a wild substitute for the dog, for this purpose (Efremov 2000: 328–335).

The distance in the shooting competition is measured in dog sleds (Efremov 2000: 372–373).

2.2.1.5 Healing and ritual use of a dog

We combine the functions of healing and ritual use because they are closely intertwined and often it is impossible to separate them, as is typical for many traditional cultures.

The ritual use of the dog is documented in the texts of some of the peoples under consideration. Thus, in a Dolgan legend, a man saw two women who later turned out to be *matushki ospy* (Rus.) or smallpox spirits (lit., 'Mothers'). He spontaneously began to sing and gave them three dogs in his possession as an offering. These women went away on the road that opened up and the people of the *nasleg* (settlement) were saved from the epidemic (Efremov 2000: 330–331).

The dog is sacrificed to the spirit-master of the *locus* in the legend of the Yakut people. A female spirit visits the fisherman and asks for a puppy in exchange for fishing luck. The puppy is his only companion and it is hard to part with him, but nothing can be done, as the spirit threatens to rob him not only of his luck, but also of his life. The fisherman is compelled, after the third time, to agree and bury the puppy in the snow (Alekseev *et al.* 1995: 214–217).

In another legend of the Yakuts, a man called Chaadai has his dog and his older brother has a bull which were stolen and they both starve to death. The younger brother becomes a malevolent spirit and when he starts chasing someone, a shamanic ritual *kamlaniye* is performed and a bluish-grey dog, with red spots at

[4] Similar Kazakh motives see in Shaygozova and Sultanova (2012).

[5] V.K. Darzha derived the expression in the Tuvinian language *yt-kush* 'dog-bird' from the particular type of hunting which involved the tandem of a bird of prey and a hound dog *taygan* (Darzha 2009: 499–500).

the base of the ears and reddish patches on the knees, is dedicated to him. During the ritual, Chaadai is instructing people to take good care of the dog. The dog now should be worshiping: kept in honour and respect, fed plentifully and deliciously, and not forced to work (Alekseev *et al.* 1995: 214–217).

Traces of the ritual of an offering of a dog to the spirit of the lake can be found in the Khakass legends about two hunting brothers, as well as about the wonderful dog *Khubay Khus.*

In Tuvas' text, a solar eclipse comes from the Sun being trapped and eaten by a *mangys* (the dragon-like mythical creature) Ara-Khoo. For it to come out, the only man in the family has to shout loudly, a black dog with red underpants has to bark and somebody should beat a Chinese cast-iron cauldron (Alekseev *et al.* 2010: 46–47).

One Dolgan text also mentions the anti-ritual context of using a dog: instead of the usual funeral ritual, the remains of an enemy are scattered to be eaten by dogs and birds, which is considered a very bad ending to human life (Efremov 2000: 360–361). In a more neutral manner, the motif of the hero Sartakpai's body being eaten after his death by dogs is presented in an Altai text (Oynotkinova *et al.* 2011: 472).

2.2.1.6 Toponymic

We have already given a rather detailed account of the legend of two brothers who killed their puppy, here we will only mention the fact that the lake, where they have thrown the puppy, is called *Saraa Aday Köl* 'Yellow Dog's Lake' (Mindibekova and Sychenko 2016: 166–167).

We conclude this section by referring to an Altai text that tells of twelve animals who escaped from the Flood and decide where they will live and what they will do (Oynotkinova *et al.* 2011: 119–127). The dog becomes man's first assistant and his greatest friend. Of himself he says: 'What you drink, I will drink. If an enemy comes, having noticed him, I will announce you. As many teeth as I have, I will fight the enemy, I will help you. I will guard your door unarmed. I'll take your wounded animal and catch him for you. With your child, whom you raise, warming him, I will sleep' (Oynotkinova *et al.* 2011: 125).

*2.2.2 Functions of a wolf*

2.2.2.1 Competition with humans for food resources

The wolf's functions, unlike that of a dog, are not numerous. The wolf is the most vicious of predators and is feared by all animals. In relation to man, whom the wolf, in turn fears, and whose main function is to compete for livestock, as stated, for example, in the Khakass' myth of the council of all beasts and birds, discussed above. This text also refers to the distribution of intellect and other qualities between living beings. Intellect was given to man, and most of the animals came under his protection. Man's power is enhanced by the fact that, in addition to intelligence, he has a faithful helper - a dog, as the wolf says with annoyance (Mindibekova and Sychenko 2016: 122–127).

However, the wolf cannot destroy livestock uncontrollably. The text says that the wolf's howl signifies his request to the deity Kudai to allow him to take some livestock.

In the Yakut legend the inhabitants of Bötün *nasleg* were once saved from starvation by finding a pit with food supplies made by a wolf. In return for being saved from starvation they began to worship him as a deity (Alekseev *et al.* 1995: 196–197).

2.2.2.2 Human ancestor (totemic-like function)

The totemic function of the wolf in Turkic societies has been discussed many times in the scientific literature (Golden 1997; Ilimbetova and Ilimbetov 2012: 38–130). However, in the collections of mythological texts of the South Siberian Turks we find a very limited number of myths and legends talking about it. Thus, only one text was found in the Altai corpus, saying that the ancestor of all Kypchaks was reared by a serpent, and became yellow like a serpent himself. He could not get along with any girl, killing them and their children, and eventually found a companion she-wolf, by whom he had 15 sons. From them came the many groups of Kypchaks (Oynotkinova *et al.* 2011: 260–265).

A totemic motif is also found in the Khakass toponymic legend (see below), but in an inverted version. Having assumed her real human form, the she-wolf tells the hero Saizan that she is destined to marry him. The hero, enraged at the wolf for destroying his cattle, kills her with an arrow from a bow. The girl foretells his death (Mindibekova and Sychenko 2016: 168–169; 347–348).

In one of the Khakass ethnic groups, the Sagays, the clan *Chitti Pür* 'Seven Wolves' exists (Mindibekova and Sychenko 2016: 156–157; 347–348). The origin of this name is not very clear, but we can assume that in this case, the wolf is regarded as the ancestor of a particular clan group as well.

We have not yet found any references to the wolf as an ancestor of man among the northern group of Turks.

2.2.2.2 Toponymic function

Previously, speaking of the appearance of wolves, we mentioned Khakassian toponymic legends in which one, two or several wolves appear. Thus, the black-foreheaded she-wolf, rescuing the hero Saizan from pursuit, gives the mountains Tumzukh, Irze, and Khyspanakh their names (Mindibekova and Sychenko 2016: 168–171).

Legends about a hero Ir-Tokhchyn mention names of such places as Kyzyl-gash, Tuim, mountains Sunduk, Khara-khaya, rivers Segirtym, Tuim, Karysh, Bir'a, Uibat, Kamyshta, Syr, Askiz, Baza, Es', Tëya, Tashtyp, Abakan, Kan-Tegir. All these names are given by the hero during his pursuit of the wolves (Mindibekova and Sychenko 2016: 288–291; 290–301; 302–307).

### 2.3 Magical / spiritual contexts

There are still some references to the dog and the wolf that do not belong to any of the categories listed above. We have highlighted a separate magical context where much of what has already been discussed can be attributed.

#### *2.3.1 Magical / spiritual contexts where dogs are mentioned*

On several occasions we have already mentioned motifs of the dog's 'werewolfism', his particular sensitivity to the presence of spirits and his close connection with their world. In one story, the author said that when he used to graze cattle in the Kaya-Bazhy area, he could hear the clatter of horses' hooves, the clatter of harnesses and people's voices every night, but no one was in sight. The dog kept barking as it sensed the presence of the spirits of the area (Oynotkinova *et al.* 2011: 196–197).

Another story about a shaman ancestor says that when he was young, he fell ill with the 'shaman's disease' and disappeared because he was taken by spirits to be trained. He was later found in the bush by a dog (Oynotkinova *et al.* 2011: 286–279).

One Altai text says that in ancient times an invasion of seven voracious *almyses* took place in Altai. They were caught and then in order to destroy them they were tied to four poles; their scalps were removed and dog scalps were put on instead. The meat was picked off the bones with horse whips, etc. Unable to endure such tortures, the *almyses* left the Altai. In this text a dog's scalp is used as a powerful magical remedy against malevolent mythical creatures (Oynotkinova *et al.* 2011: 226–229).

Several legends about the Altai hero Shunu-*baatyr* say that in order to reduce his strength before putting him in a prison pit, enemies cut out his shoulder blades and poked out his eyes and replaced them with blades and the eyes of a dog (in some versions, a horse) and when the moment came, put them back in. By this action, Shunu's excessive damaging power was reduced and he became a 'normal' *baatyr* Oynotkinova *et al.* 2011: 320–321, 328–329, 502).

In the same legend bay-haired dogs with four white marks on the forehead that bark west, together with a child looking at a red sunset and running around the *ail* ('yurt') in a counter clockwise direction, a red-haired cow with its horns clenched and mooing to the west, and a three-humped camel represent a very powerful evil sorcery cast by the hero's adversary (Oynotkinova *et al.* 2011: 320–321).

Shunu, in another text, for his belligerence, receives curses from the Burkhan: 'Let your dogs that smell evil, outside howling, run away unknown!'. This is very grave, because a dog should not run away from his master. Constant howling of dogs, and running away from home are a bad omen (Oynotkinova *et al.* 2011: 300–301, 499).

#### *2.3.2 Magical / spiritual contexts where wolves are mentioned*

The wolf or she-wolf appears quite consistently in the context of werewolfism, as mentioned above. Without explicitly referring to the wolf as the spirit-master of the area, myths and legends provide enough evidence to establish such a connection.

Thus, almost all toponymic legends of the Khakasses present the image of the wolf / she-wolf with some prominent external features as a master / mistress of this or that territory.

A wolf's howl is considered a bad omen. In one story, the hero Boor, hearing a wolf howling, says: 'You have eaten all my horses, now you want to eat my shit?' (that is, to take a life from the hero). However, it turns out that this howl was emitted by an enemy army. Apparently, it was a means of psychological influence or had a magical meaning (Oynotkinova *et al.* 2011: 352–353, 508).

### 3 Discussion

Our materials show that the texts, which belong to the category of non-tale prose, contain quite a lot of references to various animals, including wolves and dogs. However, among the texts we have analysed there are no texts specifically devoted to these animals, as probably no scholars have collected the texts from the point of view of a targeted study of the mythology of these animals. Related motifs appear in myths about the creation of the earth and man, cosmic objects, totemic and toponymic myths, stories of mythological creatures and legendary heroes, etc. Their presence in these important varieties of texts indicates the important role of these animals.

In different ethnic collections, however, these animals are mentioned with varying frequency. The greatest number of mentions (up to 20 and more) we find in the Altai and Khakasses. Less - 6–7 mentions - among the Yakuts and Dolgans. Only 5 mentions are found among the Tuvas and none among the Shors, which can be explained by the fact that the Shor volume does not include all the texts of the non-tale prose plots.

Ethnic preferences will still be explored in the future, drawing on other sources. What is more significant at this stage seems to be the comparison of views on the dog and the wolf.

### *3.1 External features of a dog and a wolf*

When comparing the external description of a dog and a wolf, the main difference is that in general the appearance of the dog is described more frequently and with more variety. With dogs, not only the overall colouration is noted, but also the colour of individual body parts such as the head, paws and ears. For example, the presence of 'four eyes', i.e., light spots above the eyes proper, is important. Dogs with such markings are still distinguished as carriers of special abilities.

A wolf or a she-wolf with distinctive external features is mentioned in the Khakass legends of Ir-Tokhchin. Of particular importance here is a broad powerful forehead of the predator - most probably, a symbol of its power and leading position.

However, while the wolf's colour palette is in the black-and-white range, with the dog it includes reddish, yellowish and bluish hues. In our opinion, this testifies that the Siberian Turks had ancient local breeds of dogs which are quite far from the external resemblance to the wolf. This can also be confirmed through ethnographic and genetic material (Darzha 2009: 501–504).

Some details, such as the presence / absence of a rudimental thumb on the front paws, are found in the ethnographic material as well (see Lugli and Sychenko in this volume).

### *3.2 Relationship of a human with a dog*

By analysing the material cited, we can see that the human-dog relationship is multi-faceted and varied.

A dog accompanies a human in different situations and economic contexts (hunting, herding, fishing, transporting; the specific circumstances depend on the type of economy of an ethnic group). It cannot be overlooked that the hunting context clearly prevails. This motif is ever-present in myths about the origins of constellations (Orion, Pleiades, the Ursa Major). However, analysing the texts, we can conclude that the presence of a dog in the hunting context does not only have an applied function. When hunting, a man is not in his inhabited world, but in the world of wildlife, inhabited by both wild animals and spirit-masters of natural objects. It is the dog, as man's closest companion, that is able to alert to their presence and generally make his stay in this environment more comfortable and safer. The constant presence of a dog reinforces man's position in relation to the rest of the animal world.

It seems that the protection of humans is one of the most important tasks prescribed to a dog by the supreme deities since the creation of the world and man. This protection manifests itself in warning of the arrival of strangers, enemies, predators, the presence of spirits and so on. A dog's attitude towards its responsibilities is characterised by loyalty and, in some situations, altruism (up to and including sacrificing itself for its master's interests). No other animal demonstrates such selflessness towards humans.

The dog, on the other hand, sometimes fails in its duty, and gives in to persuasion. Perhaps this behaviour explains some of the ambivalence towards the dog. On the one hand, the dog is regarded as a valuable object possessed by man, but which the spirits also seek to possess. Sometimes spirits threaten to force a man to part with his dog / puppy. In the texts of the Siberian Turks, we find quite significant traces of dog sacrifice to local spirits, although it is not mentioned directly and in real ethnographic practice such cases have not been recorded.

On the other hand, although a dog is a valued object, an inseparable companion and man's best friend, it sometimes seems to be treated in a dismissive and humiliating way. The dog is placed outside the house, its food - at least in the myth - are human excrements. The word 'dog' is used as a swearword in the Khakass' legend of the hero Öchen-pig: so he is called by his wife, who mourns the dead husband and their son (Mindibekova and Sychenko 2016: 228–239).

Such ambivalence seems to underline the dog's special position among domestic animals. No matter what, the bond between man and dog only grows stronger.

### *3.3 Relationship of a human with a wolf*

While the dog is mentioned by all the Turkic peoples of Siberia, the wolf is not mentioned by the Dolgans. It is necessary to note that the grey wolf (*Canis lupus lupus*) in these latitudes is not found, here it is the other variety - the polar wolf (*Canis lupus tundrarum*) of light, almost white colours. However, instead of the wolf, the fox or the polar fox (*Vulpes lagopus* or *Alopex lagopus*) acts as the dog's wild substitute.

Table 2. Common and distinguishing contexts of references to the wolf and the dog in texts of non-tale prose of the Siberian Turks

| | Creation of the Universe and a man | Constellations (Orion, Pleiads, Ursa Major) | Totem | Toponymics | Rituals | Magic context | Healing | Economic |
|---|---|---|---|---|---|---|---|---|
| **Dog** | + | + | | + | + | + | + | + |
| **Wolf** | | | + | + | + | + | | |

The relationship between man and wolf is characterised in two ways. On the one hand, the wolf harms domestic animals, making him an undesirable rival for nomadic herders. In the myths, complaints of domestic animals against the wolf are often heard, and its anger is emphasised. It is even called the most ferocious of all animals.

On the other hand, the relationship is imbued with mutual respect, and the wolf is never spoken of disparagingly. Despite the wolf's savagery and anger, man has no fear of it. On the contrary, the wolf itself is afraid of a man with a clever mind, a weapon and a dog.

The wolf also performs important spiritual functions. He is often the embodiment of the spirit of an area, which occurs, for example, in toponymic legends. Equally significant is the she-wolf's totemic role for some clans.

### *3.4 Parallelism between a dog and a wolf*

In the cultures in question, the wolf-dog kinship that existed in earlier times is well understood, but it is rarely expressed explicitly. In a Shor tale the wolf tries to play on the dog's kinship feelings by begging for food. Such humiliating behaviour is explained by the poetics of the tale genre, as was stated earlier.

More often this connection implicitly manifests itself in a certain parallelism of these figures. Thus, both animals often appear in different magical contexts with the dog being the stronger magical tool. Perhaps the reason for this is precisely the dog's closeness to the animal and the human world at the same time.

Both animals are connected to the spirit world. But while the dog most often feels, senses, perceives their presence, the wolf / she-wolf / pack of wolves are often spirits of the *locus* themselves.

Khakass folklore has a variant of the myth of the creation of the Ursa Major, in which the White Wolf acts instead of a hunter with dogs chasing female *marals.* The Wolf chases seven horses, which by the will of the deity Khudai become a constellation.

Finally, both the dog and the wolf are linked to the origin of some toponyms, with the wolf being the spirit of the area, literally covering it with parts of his body and giving it appropriate names.

### *3.5 Opposition between a dog and a wolf*

A dog has been guarding and protecting man since creation, he primarily serves man, whereas a wolf never does. If people happen to be saved by the wolf's supplies, it is only by accident, but the dog shares its part with man with altruistic willingness.

The wolf, in turn, is the totem of some Altai (Kypchaks) and Khakass (Sagays) people. Table 2 summarises the discussion on the role of the dog and the wolf.

## 4 Conclusions

The analysis of non-tale prose texts makes it possible to draw some significant conclusions about the nature of the relationship between man and dog, as well as the wolf, in Turkic cultures. Thus, the dog was created before man, and at the time of man's creation was already assigned by the deity to serve him. Already at this point their connection becomes unbreakable. This reflects the idea of the primordial human - dog bond.

Such connection determines the dog's extreme closeness to man, distinguishing its position from that of all other domestic animals. Utilitarianism is combined with an attitude towards the dog as an inseparable companion to man, accompanying him in virtually all contexts of life.

A close relationship is combined with ambivalence, in which a very positive attitude towards the dog is sometimes accompanied by a pejorative context, disdain, etc. This may be due to a psychology of priority, dominance, and power, derived from man's confidence in his friend. Both in life and in mythological texts the dog - after the initial betrayal - never betrays the human again, serves him loyally and displays genuine altruism towards the human and his family.

The wolf is a human competitor for resources, nevertheless, he always evokes respect and reverence. The she-wolf is the ancestor of some tribes, but is always a character alien to humans, although the latter is the only living creature that does not experience fear of the predator.

Man's position in relation to the wolf is greatly reinforced by the presence of a dog around him. This is particularly evident in the context of pastoralists, who are characterised by a constant struggle to keep their herds and livestock safe.

The dog is perceived as a distant relative of the wolf, having gone over to the side of man, living in a kind of symbiosis with him, and in some ways having become humanised. However, its connection with the wild world continues to persist, conditioning the dual nature of this animal. It combines the human, the domestic with the natural and the wild, allowing the dog to feel comfortable both within inhabited territory and moving through uninhabited forest-steppe spaces.

This explains the numerous parallels between the dog and the wolf, the most important of which, in the author's opinion, lie in the field of magical powers of these animals, their sensitivity to the spirit world.

The material demonstrates that different economic contexts show differences in the relationship between man, dog and wolf. In the pastoral context, the antagonism of the two animals prevails; in the hunting context, the parallelism between them reveals itself much more often.

Consequently, it may well be that it was in pastoral societies that the dog was finally separated from the world of wildlife and became an integral component of the nomadic economic and cultural complex. Such a conclusion is also confirmed by modern ethnographic research (see the article by Lugli and Sychenko in this volume).

## References


Alekseev, N.A., N.V. Emel'anov and V.T. Petrov (eds) 1995. *Predaniya, legendy i mify sakha (yakutov)*. Novosibirsk: Nauka (Алексеев, Н.А., Н.В. Емельянов и В.Т. Петров (сост.). *Предания, легенды и мифы саха (якутов)*. Новосибирск: Наука) (Памятники фольклора народов Сибири и Дальнего Востока - ПФНСДВ 9).

Alekseev, N.A., D.S. Kuular, Z.B. Samdan and Zh.M. Yusha (eds) 2010. *Mify, legendy, predaniya tuvintsev*. Novosibirsk: Nauka (Алексеев, Н.А., Д.С. Куулар, З.Б. Самдан и Ж.М. Юша (сост.). *Мифы, легенды, предания тувинцев*. Новосибирск: Наука) (ПФНСДВ 28).

Arbachakova, L.N. (ed.) 2010. *Fol'klor shortsev*. Novosibirsk: Nauka (Арбачакова Л.Н. (сост.) *Фольклор шорцев*. Новосибирск: Наука) (ПФНСДВ 29).

Beryozkin, Yu.E. 2012. Sibirsko-yuzhnoaziatskiye fol'klorniye paralleli i mifologiya evraziyskoy stepi. *Arkheologiya, etnografiya iantropologiya Evrazii* 4: 144–155 (Берёзкин Ю.Е. Сибирско-южноазиатские фольклорные параллели и мифология евразийской степи. *Археология, этнография и антропология Евразии* 4: 144–155).

Burnakov, V.A. 2012. Traditsionnye predstavleniya khakasov o sobake (konets XIX – seredina XX veka). *Arkheologiya, etnografiya i antropologiya Evrazii* 2: 114–123 (Бурнаков В.А. Традиционные представления хакасов о собаке (конец XIX - середина XX века. *Археология, этнография и антропология Евразии* 2: 114–123).

Burnakov, V.A. and D.Ts. Tsydenova 2015. Obraz volka v religiozno-mifologicheskih predstavlenijah khakasov (konec XIX – XX vek). *Vestnik Novosibirskogo gosudarstvennogo universiteta. Seriya Istoriya i filologiya* 14–3: 121–132 (Бурнаков В.А. и Д.Ц. Цыденова. Образ волка в религиозно-мифологических представлениях хакасов (конец XIX - XX век). *Вестник Новосибирского государственного университета. Серия История и филология* 14–3: 121–132).

Cheremisin, D.V. 1997. K irano-t'urkskim sv'az'am v oblasti mifologii. Bogin'a Umay i mificheskaya ptitsa. *Narody Sibiri. Istoriya i kul'tura*: 31–43. Novosibirsk: IAET (Черемисин Д.В. К ирано-тюркским связям в области мифологии. Богиня Умай и мифическая птица. *Народы Сибири. История и культура*: 31–43. Новосибирск: ИАЭТ).

Cheremisin, D.V. 2009. O semantike ornitomorfnyh personazhey zverinogo stil'a v ritual'noy atributike pazyrykskih kurganov. *Arkheologiya, etnografiya i antropologiya Evrazii* 1: 85–94 (Черемисин Д.В. О семантике орнитоморфных персонажей звериного стиля в ритуальной атрибутике пазырыкских курганов. *Археология, этнография и антропология Евразии* 1: 85–94).

Darzha, V.K. 2009. *Traditsionnye muzhskie zan'atiya tuvintsev*. Kyzyl: Tuvinskoe knizhnoe izdatel'stvo (Даржа В.К. *Традиционные мужские занятия тувинцев*. Кызыл: Тувинское книжное издательство).

Efremov, P.E. (ed.) 2000. *Fol'klor dolgan*. Novosibirsk: Nauka (Ефремов П.Е (сост.). *Фольклор долган*. Новосибирск: Наука) (ПФНСДВ 19).

Golden, Peter B. 1997. Wolves, Dogs and Qipčaq Religion. *Acta Orientalia Akademiae Scientiarum Hungaricae* L (1–3): 87–97.

Ilimbetova, A.F. and F.F. Ilimbetov 2012. *Kul't zhivotnyh v miforitual'noy traditsii bashkir*. Ufa: Gilem (Илимбетова

А.Ф. и Ф.Ф. Илимбетов. *Культ животных в мифоритуальной традиции башкир*. Уфа: Гилем).

Kashtanova, S.V. and Yu. A. Zakharov 2009. Tuvinskaya ovcharka - aborigennaya pastush'ya sobaka Tuvy. *Novye issledovaniya Tuvy* 4: 225–244 (Каштанова С.В. и Ю.А. Захаров. Тувинская овчарка - аборигенная пастушья собака Тувы. *Новые исследования Тувы* 4: 225–244).

Kubarev, V.D. and D.V. Cheremisin 1987. Volk v iskusstve i verovaniyah kochevnikov Central'noy Azii. *Traditsionnye verovaniya i byt narodov Sibiri*: 98–117. Novosibirsk: Nauka (Кубарев В.Д. и Д.В. Черемисин. Волк в искусстве и верованиях кочевников Центральной Азии. *Традиционные верования и быт народов Сибири*: 98–117. Новосибирск: Наука).

Miller, V.F. 1876. *Znachenie sobaki v mifologicheskih verovaniyah*. Moskva: Sinodal'naya tipografiya (Миллер В.Ф. *Значение собаки в мифологических верованиях*. Москва: Синодальная типография).

Mindibekova, V.V. and G.B. Sychenko (eds) 2016. *Neskazochnaya proza khakasov*. Novosibirsk: Nauka (Миндибекова В.В. и Г.Б. Сыченко (сост.) *Несказочная проза хакасов*. Новосибирск: Наука) (ПФНСДВ 34).

Muytueva, I.N. 2015. Sobaka v ustnom narodnom tvorchestve altaytsev. *Natsional'naya assotsiatsiya uchenyh* VIII (15) *Filologicheskie nauki*: 9–13 (Муйтуева И.Н. Собака в устном народном творчестве алтайцев. *Национальная ассоциация ученых* VIII (15) *Филологические науки*: 9–13).

Oynotkinova, N.R., I.B. Shinzhin, K.V. Yadanova and E.E. Yamaeva (eds) 2011. *Neskazochnaya proza altaytsev*. Novosibirsk: Nauka (Ойноткинова Н.Р., И.Б. Шинжин, К.В. Яданова и Е.Э. Ямаева (сост.) *Несказочная проза алтайцев*. Новосибирск: Наука)(ПФНСДВ 30).

Romanova, S.V. 2000. Sobaka v obr'adah i predstavleniyah t'urkoyazychnyh narodov Sibiri. *Kul'turnoe nasledie narodov Sibiri i Severa. Materialy Chetvertyh Sibirskih chteniy*: 210–215. StPetersburg: MAE (Романова С.В. Собака в обрядах и представлениях тюркоязычных народов Сибири. *Культурное наследие народов Сибири и Севера. Материалы Четвертых Сибирских чтений*: 210–215. Санкт-Петербург: МАЭ).

Salmin, A.K. 2011. Sobaka v traditsionnyh predstavleniyah chuvashey. *Arkheologiya, etnografiya i antropologiya Evrazii* 1: 124–128 (Салмин А.К. Собака в традиционных представлениях чувашей. *Археология, этнография и антропология Евразии*1: 124–128).

Shaygozova, Zh.N. and M.E. Sultanova 2012. Kul't sobaki v kazakhskoy traditsionnoy kul'ture. *Intangible culture heritage* 2: 69–79 (Шайгозова Ж.Н. и М.Е. Султанова. Культ собаки в казахской традиционной культуре. *Нематериальное культурное наследие* 2: 69–79).